Computational Methods in (with Maple

Ralph E. White and Venkat R. Subramanian

Computational Methods in Chemical Engineering with Maple

 Springer

Prof. Dr. Ralph E. White
University of South Carolina
Dept. Chemical Engineering
Columbia SC 29208
3 C 15 Swearingen Eng. Bldg.
USA
E-mail: white@cec.sc.edu

Dr. Venkat R. Subramanian
Associate Professor
Department of Energy Environmental &
Chemical Engineering
Washington University in Saint Louis
One Brookings Drive, Box 1180
Saint Louis, MO 63130
USA
E-mail: vsubramanian@seas.wustl.edu

ISBN 978-3-662-51887-8   ISBN 978-3-642-04311-6 (eBook)

DOI 10.1007/978-3-642-04311-6

© 2010 Springer-Verlag Berlin Heidelberg
Softcover reprint of the hardcover 1st edition 2010

This work is subject to copyright. All rights are reserved, whether the whole or part of the material is concerned, specifically the rights of translation, reprinting, reuse of illustrations, recitation, broadcasting, reproduction on microfilm or in any other way, and storage in data banks. Duplication of this publication or parts thereof is permitted only under the provisions of the German Copyright Law of September 9, 1965, in its current version, and permission for use must always be obtained from Springer. Violations are liable to prosecution under the German Copyright Law.

The use of general descriptive names, registered names, trademarks, etc. in this publication does not imply, even in the absence of a specific statement, that such names are exempt from the relevant protective laws and regulations and therefore free for general use.

Typesetting: Scientific Publishing Services Pvt. Ltd., Chennai, India

Cover Design: WMX Design, Heidelberg, Germany

Printed in acid-free paper

9 8 7 6 5 4 3 2 1

springer.com

Preface

This book presents Maple solutions to a wide range of problems relevant to chemical engineers and others. Many of these solutions use Maple's symbolic capability to help bridge the gap between analytical and numerical solutions. The readers are strongly encouraged to refer to the references included in the book for a better understanding of the physics involved, and for the mathematical analysis.

This book was written for a senior undergraduate or a first year graduate student course in chemical engineering. Most of the examples in this book were done in Maple 10. However, the codes should run in the most recent version of Maple. We strongly encourage the readers to use the classic worksheet (*.mws) option in Maple as we believe it is more user-friendly and robust.

In chapter one you will find an introduction to Maple which includes simple basics as a convenience for the reader such as plotting, solving linear and nonlinear equations, Laplace transformations, matrix operations, 'do loop,' and 'while loop.' Chapter two presents linear ordinary differential equations in section 1 to include homogeneous and nonhomogeneous ODEs, solving systems of ODEs using the matrix exponential and Laplace transform method. In section two of chapter two, nonlinear ordinary differential equations are presented and include simultaneous series reactions, solving nonlinear ODEs with Maple's 'dsolve' command, stop conditions, differential algebraic equations, and steady state solutions. Chapter three addresses boundary value problems. Section one of chapter three discusses the matrix exponential method in solving linear and nonlinear boundary value problems, semi-infinite domains, the matrizant method, and has examples of heat transfer in a fin, cylindrical and spherical catalyst pellet. Chapter three's section two discusses nonlinear boundary value problems and includes series solutions for diffusion of a second order reaction, multiple steady states, finite difference solutions for nonlinear boundary value problems, shooting technique for nonlinear boundary problem, and eigenvalue problems, and includes examples of nonlinear heat transfer, multiple steady states in a catalyst pellet, Blasius equation in an infinite domain, diffusion with a second order reaction, the Graetz problem using the finite difference method and the shooting technique. In chapter four you will find solution techniques for partial differential equations in semi-infinite domains in semi-infinite domains, Laplace transform, similarity solution techniques for Parabolic and elliptical PDEs as well as nonlinear partial differential equations. Some examples found in chapter four are for heat conduction in a rectangular slab, heat conduction with transient boundary conditions, heat conduction with radiation at the surface and plane flow past a flat plate, the Blasius equation. Chapter five presents the method of lines for parabolic partial differential equations and has two sections. Section one discusses the semianalytical method for parabolic partial differential equations and section two discusses the numerical method of lines for parabolic partial differential equations. Section one has some examples which include a semianalytical method for heat conduction in a rectangular slab, nonhomogeneous, partial differential equations, the Graetz problem, composite domains, and the calculation of an exponential matrix. Section two includes examples for diffusion with second order reaction, variable diffusivity, nonlinear radiation at the surface, stiff nonlinear partial differential equations, exothermal reaction in a sphere, etc. Chapter six contains semianalytical and numerical methods of lines for elliptical partial differential equations and includes several examples. Some of the examples are heat transfer in a rectangle, the Graetz problem with a fixed wall temperature, nonlinear radiation boundary condition, numerical solution for heat transfer for nonlinear elliptic partial differential equations. In chapter seven, you find a discussion of partial differential equations in finite domains. Some of the examples include separation of variables for heat conduction in a rectangle, heat conduction with an insulator boundary condition, separation of variables for heat conduction in a rectangle with an initial profile, diffusion with a reaction, and numerical separation of variable for diffusion in a cylinder. Chapter nine discusses parameter estimation and includes the least squares method, confidence intervals, nonlinear least squares, a one parameter model and a two parameter model. Chapter ten contains miscellaneous topics on numerical methods some of the examples include a finite difference solution for boundary values problems, and elliptical partial differential equations, etc.

Acknowledgements

The authors would like to thank Sandra Knotts and Long Cai for their editorial assistance.

This book is dedicated to my wife, Marjorie, and our four children (Robert, Priscilla, Lillian, and Samuel). Ralph E. White

This book is dedicated to my wife Gomathi, daughter Anupama and parents Subramanian and Suriaprabha. Venkat R. Subramanian

Contents

1 Introduction..1
  1.1 Introduction to Maple 1
    1.1.1 Getting Started with Maple...1
    1.1.2 Plotting with Maple 3
    1.1.3 Solving Linear and Nonlinear Equations 5
    1.1.4 Matrix Operations..6
    1.1.5 Differential Equations..11
    1.1.6 Laplace Transformations 16
    1.1.7 Do Loop..18
    1.1.8 While Loop 19
    1.1.9 Write Data Out Example..19
    1.1.10 Reading in Data from a Text File...............................23
    1.1.11 Summary...24
    1.1.12 Problems...24
  References 27

2 Initial Value Problems..29
  2.1 Linear Ordinary Differential Equations 29
    2.1.1 Introduction..29
    2.1.2 Homogeneous Linear ODEs...29
    2.1.3 First Order Irreversible Series Reactions...................31
      Example 2.1. Irreversible Series Reactions
        (see equations (2.8))..32
    2.1.4 First Order Reversible Series Reactions 37
      Example 2.2. Reversible Series Reactions
        (see equations (2.10))..38
    2.1.5 Nonhomogeneous Linear ODEs 47
      Example 2.3. Heating of Fluid in a Series of Tanks 49
      Example 2.4. Time Varying Input to a CSTR with a Series
        Reaction 56
    2.1.6 Higher Order Linear Ordinary Differential Equations...............63

			Example 2.5 A Second Order ODE 65
	2.1.7	Solving Systems of ODEs Using the Laplace Transform Method..72
			Example 2.6. Laplace Solution of Example 2.1 Equations........73
			Example 2.7. Laplace Solution for Second Order System with Dirac forcing Function..76
	2.1.8	Solving Linear ODEs Using Maple's 'dsolve' Command.......80
			Example 2.8. Solving Linear ODEs Using Maple 80
			Example 2.9. Heat Transfer in a Series of Tanks, 'dsolve'.........81
	2.1.9	Summary..83
	2.1.10	Problems 84
2.2	Nonlinear Ordinary Differential Equations...87
	2.2.1	Introduction..87
			Example 2.2.1. Simultaneous Series Reactions 88
	2.2.2	Solving Nonlinear ODEs Using Maple's 'dsolve' Command..94
	2.2.3	Series Solutions for Nonlinear ODEs 98
			Example 2.2.2. Fermentation Kinetics......................................99
			Example 2.2.3 101
	2.2.4	Stop Conditions 103
			Example 2.2.4. Stop Conditions 103
	2.2.5	Stiff ODEs 107
			Example 2.2.5. Stiff Ordinary Differential Equations 107
	2.2.6	Differential Algebraic Equations 112
			Example 2.2.6. Differential Algebraic Equations 112
	2.2.7	Multiple Steady States 116
			Example 2.2.7. Multiple Steady States 117
	2.2.8	Steady State Solutions 124
			Example 2.2.8. Steady State Solutions 124
			Example 2.2.9. Phase Plane Analysis 139
	2.2.9	Summary..148
	2.2.10	Problems 149
	Appendix A: Matrix Exponential Method 155
	Appendix B: Matrix Exponential by the Laplace Transform Method..161
References 167

3 Boundary Value Problems..169
3.1 Linear Boundary Value Problems...169
3.1.1 Introduction..169
3.1.2 Exponential Matrix Method for Linear Boundary Value Problems 169
Example 3.1 171
Example 3.2 175
3.1.3 Exponential Matrix Method for Linear BVPs with Semi-infinite Domains...180

		Example 3.3 181
	3.1.4	Use of Matrizant in Solving Boundary Value Problems..........184
		Example 3.4 185
		Example 3.5 187
		Example 3.6 189
	3.1.5	Symbolic Finite Difference Solutions for Linear Boundary Value Problems..195
		Example 3.7 196
		Example 3.8. Cylindrical Catalyst Pellet 203
	3.1.6	Solving Linear Boundary Value Problems Using Maple's 'dsolve' Command 208
		Example 3.9. Heat Transfer in a Fin 208
		Example 3.10. Cylindrical Catalyst Pellet 209
		Example 3.11. Spherical Catalyst Pellet 210
	3.1.7	Summary..212
	3.1.8	Exercise Problems..213
3.2	Nonlinear Boundary Value Problems 217
	3.2.1	Introduction..217
	3.2.2	Series Solutions for Nonlinear Boundary Value Problems......218
		Example 3.2.1. Series Solutions for Diffusion with a Second Order Reaction 218
		Example 3.2.2. Series Solutions for Non-isothermal Catalyst Pellet – Multiple Steady States 223
	3.2.3	Finite Difference Solutions for Nonlinear Boundary Value Problems 229
		Example 3.2.3. Diffusion with a Second Order Reaction 229
	3.2.4	Shooting Technique for Boundary Value Problem 233
		Example 3.2.4. Nonlinear Heat Transfer 233
		Example 3.2.5. Multiple Steady States in a Catalyst Pellet 238
	3.2.5	Numerical Solution for Boundary Value Problems Using Maple's 'dsolve' Command 244
		Example 3.2.6. Diffusion with Second Order Reaction...........245
		Example 3.2.7. Heat Transfer with Nonlinear Radiation Boundary Conditions 247
		Example 3.2.8. Diffusion of a Substrate in an Enzyme Catalyzed Reaction – BVPs with Removable Singularity..250
		Example 3.2.9. Multiple Steady States in a Catalyst Pellet 253
		Example 3.2.10. Blasius Equation – Infinite Domains 256
	3.2.6	Numerical Solution for Coupled BVPs Using Maple's 'dsolve' Command..259
		Example 3.2.11. Axial Conduction and Diffusion in a Tubular Reactor 259
	3.2.7	Solving Boundary Value Problems and Initial Value Problems 262
		Example 3.2.12. Diffusion with a Second Order Reaction 262

	3.2.8	Multiple Steady States 266
		Example 3.2.13. Multiple Steady States in a Catalyst Pellet - η vs. Φ 266
	3.2.9	Eigenvalue Problems 272
		Example 3.2.14. Graetz Problem–Finite Difference Solution 272
		Example 3.2.15. Graetz Problem–Shooting Technique 278
	3.2.10	Summary 286
	3.2.11	Exercise Problems 288
References		293

4 Partial Differential Equations in Semi-infinite Domains 295
- 4.1 Partial Differential Equations (PDEs) in Semi-infinite Domains 295
- 4.2 Laplace Transform Technique for Parabolic PDEs 295
  - Example 4.1. Heat Conduction in a Rectangular Slab 296
  - Example 4.2. Heat Conduction with Transient Boundary Conditions 301
  - Example 4.3. Heat Conduction with Flux Boundary Conditions 305
  - Example 4.4. Heat Conduction with an Initial Profile 308
  - Example 4.5. Heat Conduction with a Source Term 311
- 4.3 Laplace Transform Technique for Parabolic PDEs – Advanced Problems 314
  - Example 4.6. Heat Conduction with Radiation at the Surface 314
  - Example 4.7. Unsteady State Diffusion with a First-Order Reaction 318
- 4.4 Similarity Solution Technique for Parabolic PDEs 324
  - Example 4.8. Heat Conduction in a Rectangular Slab 325
  - Example 4.9. Laminar Flow in a CVD Reactor 328
- 4.5 Similarity Solution Technique for Elliptic Partial Differential Equations 333
  - Example 4.10. Steady State Heat Conduction in a Plate 333
  - Example 4.11. Current Distribution in an Electrochemical Cell 336
- 4.6 Similarity Solution Technique for Nonlinear Partial Differential Equations 339
  - Example 4.12. Variable Diffusivity 340
  - Example 4.13. Plane Flow Past a Flat Plate – Blassius Equation 342
- 4.7 Summary 348
- 4.8 Exercise Problems 348

References 352

5 Method of Lines for Parabolic Partial Differential Equations 353
- 5.1 Semianalytical Method for Parabolic Partial Differential Equations (PDEs) 353
  - 5.1.1 Introduction 353
  - 5.1.2 Semianalytical Method for Homogeneous PDEs 353
    - Example 5.1. Heat Conduction in a Rectangular Slab 356

	5.1.3	Semianalytical Method for Nonhomogeneous PDEs	365
		Example 5.2	366
		Example 5.3	374
		Example 5.4	382
		Example 5.5	390
		Example 5.6. Semianalytical Method for the Graetz Problem	401
		Example 5.7. Semianalytical Method for PDEs with Known Initial Profiles	414
	5.1.4	Semianalytical Method for PDEs in Composite Domains	425
		Example 5.8	425
	5.1.5	Expediting the Calculation of Exponential Matrix	437
		Example 5.9	438
		Example 5.10	442
		Example 5.11	448
	5.1.6	Summary	451
	5.1.7	Exercise Problems	452
5.2	Numerical Method of Lines for Parabolic Partial Differential Equations (PDEs)		456
	5.2.1	Introduction	456
	5.2.2	Numerical Method of Lines for Parabolic PDEs with Linear	456
		Example 5.2.1. Diffusion with Second Order Reaction	458
		Example 5.2.2. Variable Diffusivity	464
	5.2.3	Numerical Method of Lines for Parabolic PDEs with Nonlinear Boundary	469
		Example 5.2.3. Nonlinear Radiation at the Surface	470
	5.2.4	Numerical Method of Lines for Stiff Nonlinear PDEs	474
		Example 5.2.4. Exothermal Reaction in a Sphere	474
	5.2.5	Numerical Method of Lines for Nonlinear Coupled PDEs	480
		Example 5.2.5. Two Coupled PDEs	480
	5.2.6	Numerical Method of Lines for Moving Boundary Problems	491
		Example 5.2.6. The Shrinking Core Model for Catalyst Regeneration	491
	5.2.7	Summary	501
	5.2.8	Exercise Problems	502
References			505

6 Method of Lines for Elliptic Partial Differential Equations 507
6.1 Semianalytical and Numerical Method of Lines for Elliptic PDEs 507
6.1.1 Introduction 507
6.1.2 Semianalytical Method for Elliptic PDEs in Rectangular Coordinates 507
Example 6.1. Heat Transfer in a Rectangle 508
Example 6.2 520

6.1.3 Semianalytical Method for Elliptic PDEs in Cylindrical
Coordinates – Graetz Problem 536
Example 6.3. Graetz Problem with a Fixed Wall
Temperature 536
6.1.4 Semianalytical Method for Elliptic PDEs with Nonlinear
Boundary Conditions 547
Example 6.4. Nonlinear Radiation Boundary Condition 547
6.1.5 Semianalytical Method for Elliptic PDEs with Irregular
Shapes 556
Example 6.5. Potential Distribution in a Hull Cell 556
6.1.6 Numerical Method of Lines for Elliptic PDEs in
Rectangular Coordinates 564
Example 6.6. Numerical Solution for Heat Transfer in a
Rectangle 565
Example 6.7. Numerical Solution for Heat Transfer for
Nonlinear Elliptic PDEs 573
6.1.7 Summary 581
References 585

7 Partial Differential Equations in Finite Domains 587
7.1 Separation of Variables Method for Partial Differential Equations
(PDEs) in Finite Domains 587
7.1.1 Introduction 587
7.1.2 Separation of Variables for Parabolic PDEs with
Homogeneous Boundary Conditions 587
Example 7.1. Heat Conduction in a Rectangle 587
Example 7.2. Heat Conduction with an Insulator Boundary
Condition 599
Example 7.3. Mass Transfer in a Spherical Pellet 604
7.1.3 Separation of Variables for Parabolic PDEs with an Initial
Profile 609
Example 7.4. Heat Conduction in a rectangle with an Initial
Profile 609
Example 7.5. Heat Conduction in a Slab with a Linear Initial
Profile 613
7.1.4 Separation of Variables for Parabolic PDEs with Eigenvalues
Governed by Transcendental Equations 618
Example 7.6. Heat Conduction in a Slab with Radiation
Boundary Conditions 618
7.1.5 Separation of Variables for Parabolic PDEs with
Nonhomogeneous Boundary Conditions 623
Example 7.7. Heat Conduction in a slab with
Nonhomogeneous Boundary Conditions 623
Example 7.8. Diffusion with Reaction 629
7.1.6 Separation of Variables for Parabolic PDEs with Two
Flux Boundary Conditions 635

Contents   XIII

		Example 7.9. Diffusion in a Slab with Nonhomogeneous Flux Boundary Conditions 635
	7.1.7	Numerical Separation of Variables for Parabolic PDEs 643
		Example 7.10. Heat Transfer in a Rectangle 643
	7.1.8	Separation of Variables for Elliptic PDEs 649
		Example 7.11. Heat Transfer in a Rectangle 649
		Example 7.12. Diffusion in a Cylinder 655
		Example 7.13. Heat Transfer with Nonhomogeneous Boundary Conditions 660
		Example 7.14. Heat Transfer with a Nonhomogeneous Governing Equation 667
	7.1.9	Summary 672
	7.1.10	Exercise Problems 672
References 678

8 Laplace Transform Technique for Partial Differential Equations 679

8.1	Laplace Transform Technique for Partial Differential Equations (PDEs) in Finite Domains 679
	8.1.1 Introduction 679
	8.1.2 Laplace Transform Technique for Hyperbolic PDEs 679
	Example 8.1. Wave Propagation in a Rectangle 679
	Example 8.2. Wave Propagation in a Rectangle 682
	8.1.3 Laplace Transform Technique for Parabolic Partial Differential Equations – Simple Solutions 685
	Example 8.3. Heat Transfer in a Rectangle 685
	Example 8.4. Transient Heat Transfer in a Rectangle 688
	8.1.4 Laplace Transform Technique for Parabolic Partial Differential Equations – Short Time Solution 690
	Example 8.5. Heat Transfer in a Rectangle 691
	Example 8.6. Mass Transfer in a Spherical Pellet 696
	8.1.5 Laplace Transform Technique for Parabolic Partial Differential Equations – Long Time Solution 701
	Example 8.7. Heat Conduction with an Insulator Boundary Condition 703
	Example 8.8. Diffusion with Reaction 709
	Example 8.9. Heat Conduction with Time Dependent Boundary Conditions 714
	8.1.6 Laplace Transform Technique for Parabolic Partial Differential Equations – Heaviside Expansion Theorem for Multiple Roots 719
	Example 8.10. Heat Transfer in a Rectangle 720
	Example 8.11. Diffusion in a Slab with Nonhomogeneous Flux Boundary Conditions during Charging of a Battery 725
	Example 8.12. Distribution of Overpotential in a Porous Electrode 729

		Example 8.13. Heat Conduction in a Slab with Radiation Boundary Conditions 736
	8.1.7	Laplace Transform Technique for Parabolic Partial Differential Equations in Cylindrical Coordinates...................742
		Example 8.14. Heat Conduction in a Cylinder 742
	8.1.8	Laplace Transform Technique for Parabolic Partial Differential Equations for Time Dependent Boundary Conditions – Use of Convolution Theorem 747
		Example 8.15. Heat Conduction in a Rectangle with a Time Dependent Boundary Condition 748
	8.1.9	Summary..755
	8.1.10	Exercise Problems...755
References 760

9 Parameter Estimation...761
9.1 Introduction 761
9.2 Least Squares Method..762
    9.2.1 Summation Form or Classical Form 769
    9.2.2 Confidence Intervals: Classical Approach 775
    9.2.3 Prediction of New Observations 776
    9.2.4 A One Parameter through the Origin Model 777
9.3 Nonlinear Least Squares 778
    Example 9.1. Parameter Estimation 783
9.4 Hessian Matrix Approach 789
9.5 Confidence Intervals..795
9.6 Sensitivity Coefficient Equations 797
9.7 One Parameter Model 807
9.8 Two Parameter Model 812
9.9 Exercise Problems 819
References 819

10 Miscellaneous Topics 821
10.1 Miscellaneous Topics on Numerical Methods.............................821
    10.1.1 Introduction..821
    10.1.2 Iterative Finite Difference Solution for Boundary Value Problems...821
        Example 10.1. Diffusion with a Second Order Reaction.....821
        Example 10.2. Nonisothermal Reaction in a Catalyst Pellet – Multiple Steady States 825
    10.1.3 Finite Difference Solution for Elliptic PDEs 827
        Example 10.3. Heat Transfer in a Rectangle......................827
        Example 10.4. Heat Transfer in a Cylinder........................832
    10.1.4 Iterative Finite Difference Solution for Elliptic PDEs.........833
        Example 10.5. Heat Transfer in a Rectangle – Nonlinear Elliptic PDE 833

	10.1.5	Numerical Method of Lines for First Order Hyperbolic PDEs	838
		Example 10.6. Wave Propagation in a Rectangle with Consistent Initial/Boundary Conditions.	839
		Example 10.7. Wave Propagation in a Rectangle with inconsistent Initial/Boundary Conditions	844
	10.1.6	Numerical Method of Lines for Second Order Hyperbolic PDEs	848
		Example 10.8. Wave Equation with Consistent Initial/Boundary Conditions	848
		Example 10.9. Wave Equation with Inconsistent Initial/Boundary Conditions	852
	10.1.7	Summary	855
	10.1.8	Exercise Problems	855
References			856

Subject Index 857

Chapter 1
Introduction

1.1 Introduction to Maple

1.1.1 Getting Started with Maple

Some Maple basics are presented in this chapter as a convenience for the reader. Two Maple books[1, 2] that have proven to be useful are given as references 1 and 2 at the end of this chapter. Maple can be started either from the shortcut on the desktop or from Start → Programs → Maple 12. This opens a new Maple worksheet in the Maple environment. You should usually type 'restart' as the first command in your Maple worksheets.

> restart;

This restart command clears all the stored variables and restarts the worksheet every time it is executed.

Numerical values can be assigned to variables in Maple by using the characters ':=' after x, for example. That is, to assign the value 2 to the variable x, the colon and equal sign ':=' characters are used together. You can use the # sign to add comments

> x:=2; # an assignment statement.
$$x := 2$$

Note that ':=' is the assignment operator which assigns an expression or number (2) to a variable named x. If the colon is not used, the value is not assigned. For example, 2 is not assigned to y by using '=' only. For example, type

> y=2;
$$y = 2$$

Now type both x and y to see their values.

> x;
$$2$$

> y;
$$y$$

This shows that ':=' assigned the value 2 to x whereas '=' did not assign 2 to y.

One can use Maple to do numerical and symbolic calculations. A few examples are shown next.

> x^2;
$$4$$

> x^2.;
$$4.$$

> sqrt(x);
$$\sqrt{2}$$

> x^0.5;
$$1.414213562$$

> abs(x);
$$2$$

> -x;
$$-2$$

> x+y;
$$-2+y$$

> abs(-2);
$$2$$

The imaginary number $\sqrt{-1}$ is designated as I in Maple:

> (-1)^(1/2);
$$I$$

The Maple command 'evalf' provides numeric evaluation and the 'eval' command yields a symbolic evaluation:

> evalf(sqrt(2));
$$1.414213562$$

> eval(sqrt(2));
$$\sqrt{2}$$

Symbolic variables can also be assigned to names as follows:

> z:=y;
$$z := y$$

> z;
$$y$$

1.1 Introduction to Maple

Differentiation can be done by using the 'diff' command:

> diff(y,y);

$$1$$

> diff(y^2,y);

$$2y$$

Integration can be done by using the 'int' command:

> int(y,y);

$$\frac{y^2}{2}$$

Maple can also do definite integration:

> int(y,y=0..1);

$$\frac{1}{2}$$

1.1.2 Plotting with Maple

Plots can be made in Maple using the 'plot' command:

> plot(y,y=0..1);

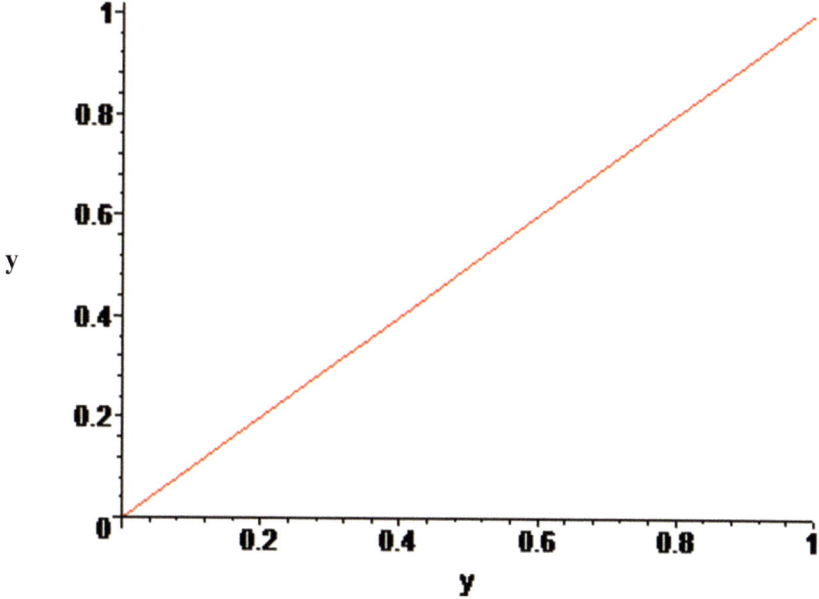

Fig. 1.1 Maple plot of y = y

```
> plot(y^2,y=0..1);
```

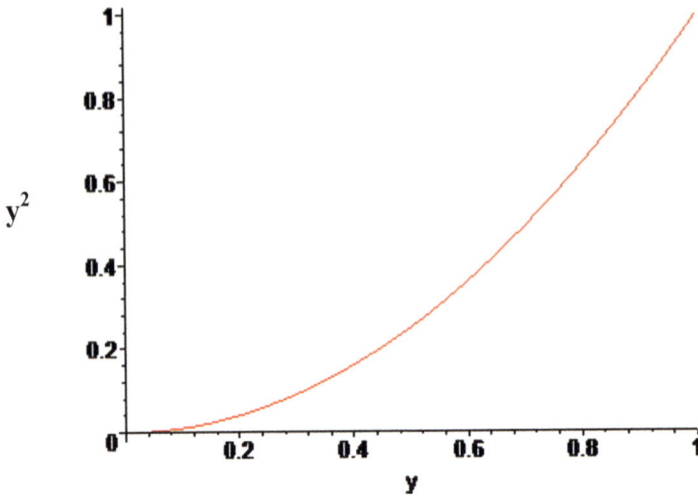

Fig. 1.2 Maple plot of $y^2 = y$

To plot both curves on the same graph in a box use the following command.

```
> plot([y,y^2],y=0..1,axes=boxed);
```

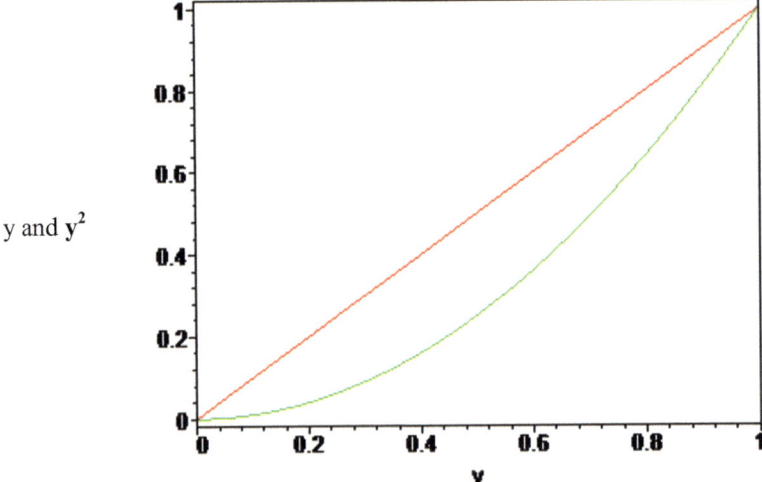

Fig. 1.3 Maple plot of y and y^2 vs y

1.1 Introduction to Maple

1.1.3 Solving Linear and Nonlinear Equations

One can solve equations in Maple using the 'solve' and 'fsolve' commands. The 'solve' command is used to solve linear equations in symbolic form and the 'fsolve' command is used to solve linear and nonlinear equations numerically. For example,

> restart:

> eq:=x+2;

$$eq := x + 2$$

> solve(eq);

$$-2$$

Maple can solve equations in symbolic form also:

> eq:=x-a;

$$eq := x - a$$

> solve(eq);

$$\{a = x, x = x\}$$

This solution says that either x = x or a = x. To solve specifically for x

> solve(eq,x);

$$a$$

Note that a has not been assigned to x which can by seen by typing x:

> x;

$$x$$

One can assign the value of a to x by solving the above equation for x:

> eq:=x-a;

$$eq := x - a$$

> x:=solve(eq,x);

$$x := a$$

One can use the 'fsolve' command in Maple to solve equations numerically:

> eq1:=y+1;

$$eq1 := y + 1$$

> fsolve(eq1,y);

$$-1.$$

Note that 'fsolve' returns a floating point number with a decimal point.

Two or more nonlinear equations can be solved by using 'fsolve'. For example, consider finding the solutions (x and y) for the following two equations.

> restart:

> eq1:=x+tan(y)=1;
$$eq1 := x + \tan(y) = 1$$

> eq2:=y^2+tan(x)=1;
$$eq2 := y^2 + \tan(x) = 1$$

> fsolve({eq1,eq2},{x,y});
$$\{x = -3.858064894, y = 1.367788596\}$$

One can find other solutions to these equations by restricting the ranges of x and y:

> fsolve(f#,{x=1..3,y=1..3});
$$\{x = 1.760535729, y = 2.491382707\}$$

1.1.4 Matrix Operations

Maple has a package for solving linear algebra problems which can be called by using the 'with(linalg)' command.

> restart:

> with(linalg):

Warning, the protected names norm and trace have been redefined and unprotected.

Maple is capable of doing a variety of matrix operations. For example, let A and B be 2 x 2 matrices which can be entered as follows:

> A:=matrix(2,2,[1,2,3,4]);
$$A := \begin{bmatrix} 1 & 2 \\ 3 & 4 \end{bmatrix}$$

> B:=matrix(2,2,[1,1,3,2]);
$$B := \begin{bmatrix} 1 & 1 \\ 3 & 2 \end{bmatrix}$$

Use the 'evalm' command to perform matrix operations. For example, matrix addition and subtraction can be done:

> evalm(A+B);
$$\begin{bmatrix} 2 & 3 \\ 6 & 6 \end{bmatrix}$$

1.1 Introduction to Maple

> evalm(A-B);

$$\begin{bmatrix} 0 & 1 \\ 0 & 2 \end{bmatrix}$$

Multiplication of matrices requires using evalm and '&*':

> evalm(A&*B);

$$\begin{bmatrix} 7 & 5 \\ 15 & 11 \end{bmatrix}$$

The determinant of a matrix can be found by using 'det':

> det(A);

$$-2$$

and

> det(B);

$$-1$$

Matrices can be inverted by using the 'inverse command':

> inverse(A);

$$\begin{bmatrix} -2 & 1 \\ \frac{3}{2} & \frac{-1}{2} \end{bmatrix}$$

> inverse(inverse(A));

$$\begin{bmatrix} 1 & 2 \\ 3 & 4 \end{bmatrix}$$

The transpose of a matrix can be obtained also

> transpose(A);

$$\begin{bmatrix} 1 & 3 \\ 2 & 4 \end{bmatrix}$$

A particular element of a matrix can be printed easily:

> A[1,1];

$$1$$

A matrix can be raised to a power by using the 'evalm' command:

> evalm(A^2);

$$\begin{bmatrix} 7 & 10 \\ 15 & 22 \end{bmatrix}$$

The characteristic polynomial, eigenvalues, and eigenvectors of a matrix can be obtained as follows:

> charpoly(A,lambda);

$$\lambda^2 - 5\lambda - 2$$

> eigenvalues(A);

$$\frac{5}{2} + \frac{\sqrt{33}}{2}, \frac{5}{2} - \frac{\sqrt{33}}{2}$$

> eigenvectors(A);

$$\left[\frac{5}{2} + \frac{\sqrt{33}}{2}, 1, \left\{\left[1, \frac{3}{4} + \frac{\sqrt{33}}{4}\right]\right\}\right], \left[\frac{5}{2} - \frac{\sqrt{33}}{2}, 1, \left\{\left[1, \frac{3}{4} - \frac{\sqrt{33}}{4}\right]\right\}\right]$$

or

> eigenvects(A);

$$\left[\frac{5}{2} + \frac{\sqrt{33}}{2}, 1, \left\{\left[1, \frac{3}{4} + \frac{\sqrt{33}}{4}\right]\right\}\right], \left[\frac{5}{2} - \frac{\sqrt{33}}{2}, 1, \left\{\left[1, \frac{3}{4} - \frac{\sqrt{33}}{4}\right]\right\}\right]$$

Matrices can be raised to various powers and added. For example, let

> eq:=A+A^2+A^3;

$$eq := A + A^2 + A^3$$

> evalm(eq);

$$\begin{bmatrix} 45 & 66 \\ 99 & 144 \end{bmatrix}$$

Maple's 'Id' command can be used to generate an identity matrix:

> Id:=band([1],2);

$$Id := \begin{bmatrix} 1 & 0 \\ 0 & 1 \end{bmatrix}$$

1.1 Introduction to Maple

Elements of a matrix can be in symbolic form and a variety of matrix operations can be performed:

> A:=matrix(2,2,[a,b,c,d]);

$$A := \begin{bmatrix} a & b \\ c & d \end{bmatrix}$$

> transpose(A);

$$\begin{bmatrix} a & c \\ b & d \end{bmatrix}$$

> inverse(A);

$$\begin{bmatrix} \dfrac{d}{ad-bc} & -\dfrac{b}{ad-bc} \\ -\dfrac{c}{ad-bc} & \dfrac{a}{ad-bc} \end{bmatrix}$$

> evalm(A&*B);

$$\begin{bmatrix} a+3b & a+2b \\ c+3d & c+2d \end{bmatrix}$$

A matrix can be multiplied with a scalar:

> evalm(2*A);

$$\begin{bmatrix} 2a & 2b \\ 2c & 2d \end{bmatrix}$$

Eigenvalues can be obtained:

> eigenvalues(A);

$$\frac{a}{2} + \frac{d}{2} + \frac{\sqrt{a^2 - 2ad + d^2 + 4bc}}{2}, \; \frac{a}{2} + \frac{d}{2} - \frac{\sqrt{a^2 - 2ad + d^2 + 4bc}}{2}$$

Eigenvectors can be obtained:

> eigenvects(A);

$$\left[\frac{a}{2} + \frac{d}{2} + \frac{\sqrt{a^2 - 2ad + d^2 + 4bc}}{2},\, 1,\, \left\{ \left[-\frac{-\frac{a}{2} + \frac{d}{2} - \frac{\sqrt{a^2 - 2ad + d^2 + 4bc}}{2}}{c},\, 1 \right] \right\} \right],$$

$$\left[\frac{a}{2} + \frac{d}{2} - \frac{\sqrt{a^2 - 2ad + d^2 + 4bc}}{2},\, 1,\, \left\{ \left[-\frac{-\frac{a}{2} + \frac{d}{2} + \frac{\sqrt{a^2 - 2ad + d^2 + 4bc}}{2}}{c},\, 1 \right] \right\} \right]$$

The exponential matrix of a matrix can be obtained as follows:

> exponential(B,t);

$$\left[\frac{1}{2}e^{\left(\frac{t(3+\sqrt{13})}{2}\right)}+\frac{1}{26}\sqrt{13}\,e^{\left(-\frac{t(-3+\sqrt{13})}{2}\right)}-\frac{1}{26}\sqrt{13}\,e^{\left(\frac{t(3+\sqrt{13})}{2}\right)}+\frac{1}{2}e^{\left(-\frac{t(-3+\sqrt{13})}{2}\right)},\right.$$

$$\left.-\frac{1}{13}\sqrt{13}\,e^{\left(-\frac{t(-3+\sqrt{13})}{2}\right)}+\frac{1}{13}\sqrt{13}\,e^{\left(\frac{t(3+\sqrt{13})}{2}\right)}\right]$$

$$\left[-\frac{3}{13}\sqrt{13}\,e^{\left(-\frac{t(-3+\sqrt{13})}{2}\right)}+\frac{3}{13}\sqrt{13}\,e^{\left(\frac{t(3+\sqrt{13})}{2}\right)},\right.$$

$$\left.\frac{1}{2}e^{\left(\frac{t(3+\sqrt{13})}{2}\right)}-\frac{1}{26}\sqrt{13}\,e^{\left(-\frac{t(-3+\sqrt{13})}{2}\right)}+\frac{1}{26}\sqrt{13}\,e^{\left(\frac{t(3+\sqrt{13})}{2}\right)}+\frac{1}{2}e^{\left(-\frac{t(-3+\sqrt{13})}{2}\right)}\right]$$

The 'map' command can be used to differentiate and integrate each element in a matrix:

> A:=matrix(2,2,[x,a*x,1/x,c]);

$$A := \begin{bmatrix} x & ax \\ \dfrac{1}{x} & c \end{bmatrix}$$

> map(diff,A,x);

$$\begin{bmatrix} 1 & a \\ -\dfrac{1}{x^2} & 0 \end{bmatrix}$$

> map(int,A,x);

$$\begin{bmatrix} \dfrac{x^2}{2} & \dfrac{ax^2}{2} \\ \ln(x) & cx \end{bmatrix}$$

> map(int,A,x=0..1);

$$\begin{bmatrix} \dfrac{1}{2} & \dfrac{a}{2} \\ \infty & c \end{bmatrix}$$

1.1.5 Differential Equations

Maple's 'dsolve' command can be used to obtain analytical and series solutions for differential equations. Differential equations are discussed in more detail in chapters 2 and 3. In this section, some Maple commands are introduced to solve relatively simple differential equations.

> restart:

You have to use y(x) if you are trying to solve y as a function of x (y is the dependent variable and x is the independent variable)

> eq:=diff(y(x),x)=x;

$$eq := \frac{d}{dx} y(x) = x$$

> dsolve(eq,y(x));

$$y(x) = \frac{x^2}{2} + \_C1$$

Note that the constant _C1 is returned as part of the solution. If you specify the initial condition, Maple can be used to obtain the complete solution:

> dsolve({eq,y(0)=1},y(x));

$$y(x) = \frac{x^2}{2} + 1$$

Second order equations can also be solved with 'dsolve':

> eq:=y(x)+diff(y(x),x$2)=x^3;

$$eq := y(x) + \left(\frac{d^2}{dx^2} y(x)\right) = x^3$$

> dsolve(eq,y(x));

$$y(x) = \sin(x) \_C2 + \cos(x) \_C1 + x(-6 + x^2)$$

Note that there are two constants, _C2 and _C1, in this case. The D(y)(x) command can be used to set the derivate of y as an initial condition at x=0 $\left(\frac{dy}{dx} = 0, \text{e.g.}\right)$, and the other initial condition $(y(0) = 1)$ can be set easily also:

> dsolve({eq,y(0)=1,D(y)(0)=0},y(x));

$$y(x) = 6 \sin(x) + \cos(x) + x(-6 + x^2)$$

Next, store the right hand side (rhs) in ya and then plot ya:

> ya:=rhs(dsolve({eq,y(0)=1,D(y)(0)=0},y(x)));

$$ya := 6\sin(x) + \cos(x) + x(-6 + x^2)$$

> plot(ya,x=0..1);

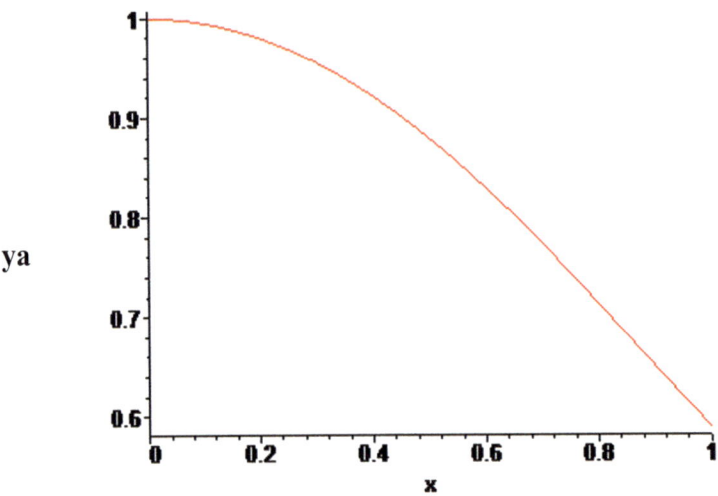

Fig. 1.4 Maple plot of ya vs x

Maple's 'dsolve' can be used to solve nonlinear equations. For example, consider the following equation:

> eq:=diff(y(x),x$2)=y(x)^2;

$$eq := \frac{d^2}{dx^2} y(x) = y(x)^2$$

Solve this equation by using 'dsolve':

> dsolve(eq,y(x));

$$\int^{y(x)} \frac{3}{\sqrt{6\_a^3 - 3\_C1}} d\_a - x - \_C2 = 0, \int^{y(x)} -\frac{3}{\sqrt{6\_a^3 - 3\_C1}} d\_a - x - \_C2 = 0$$

1.1 Introduction to Maple

Maple gives the solution as an integral. Instead one can get a series solution by specifying 'type = series' in 'dsolve' as follows:

> dsolve(eq,y(x),type =series);

$$y(x) = y(0) + D(y)(0) x + \frac{1}{2} y(0)^2 x^2 + \frac{1}{3} y(0) D(y)(0) x^3 + \left(\frac{1}{12} y(0)^3 + \frac{1}{12} D(y)(0)^2\right) x^4 + \frac{1}{12} y(0)^2 D(y)(0) x^5 + O(x^6)$$

Consider another nonlinear differential equation.

> eq:=diff(y(x),x)=-tan(x)+exp(-y(x));

$$eq := \frac{d}{dx} y(x) = -\tan(x) + e^{(-y(x))}$$

> dsolve(eq,y(x));

$$y(x) = -\ln\left(\frac{1}{\cos(x)(\_C1 + \ln(\sec(x) + \tan(x)))}\right)$$

Use the type = series option to obtain a series solution.

> ya:=rhs(dsolve({eq,y(0)=1},y(x),type=series));

$$ya := 1 + e^{(-1)} x + \left(-\frac{1}{2}(e^{(-1)})^2 - \frac{1}{2}\right) x^2 + \frac{1}{3}\left((e^{(-1)})^2 + \frac{1}{2}\right) e^{(-1)} x^3 + \left(-\frac{1}{12} - \frac{1}{4}(e^{(-1)})^4 - \frac{1}{6}(e^{(-1)})^2\right) x^4 + \left(\frac{1}{24} e^{(-1)} + \frac{1}{5}(e^{(-1)})^5 + \frac{1}{6}(e^{(-1)})^3\right) x^5 + O(x^6)$$

> ya:=evalf(ya);

$$ya := 1. + 0.3678794412 x - 0.5676676416 x^2 + 0.07790892966 x^3 - 0.1104681236 x^4 + 0.02497374418 x^5 + O(x^6)$$

One can remove the order term $\left(O(x^6)\right)$ in the series by using the 'convert' command:

> ya:=convert(ya,polynom);

$$ya := 1. + 0.3678794412 x - 0.5676676416 x^2 + 0.07790892966 x^3 - 0.1104681236 x^4 + 0.02497374418 x^5$$

> plot(ya,x=0..1);

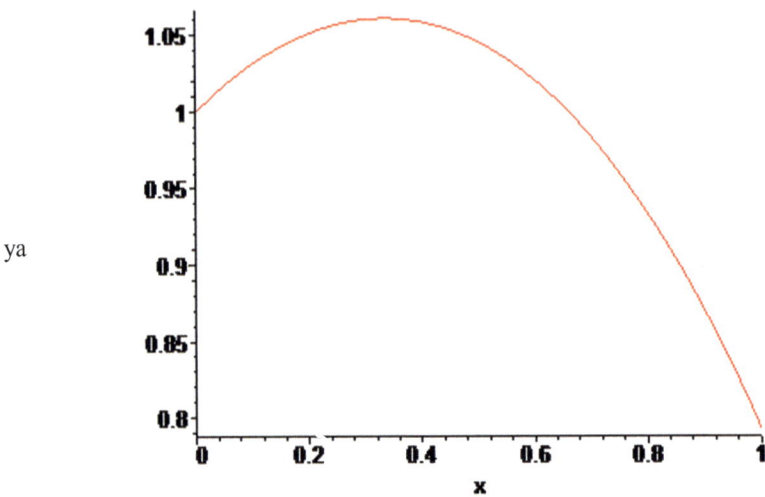

Fig. 1.5 Maple plot of ya vs x

One can also use 'dsolve' to solve boundary value problems. Consider heat transfer in a fin:[3]

> eq:=diff(y(x),x$2)=H^2*y(x);

$$eq := \frac{d^2}{dx^2} y(x) = H^2 y(x)$$

where H is a parameter. The governing equation can be solved without specifying the boundary conditions as:

> dsolve(eq,y(x));

$$y(x) = \_C1\, e^{(Hx)} + \_C2\, e^{(-Hx)}$$

Suppose the boundary conditions are at x=0, y=1, and at x=1, $\frac{dy}{dx} = 0$. If one of the boundary conditions is specified, Maple gives a solution with one constant.

> ya:=rhs(dsolve({eq,y(0)=1},y(x)));

$$ya := (-\_C2 + 1)\, e^{(Hx)} + \_C2\, e^{(-Hx)}$$

1.1 Introduction to Maple

The constant _C2 can be obtained by using the boundary condition at x = 1:

> diff(ya,x);

$$(-\_C2+1)H e^{(Hx)} - \_C2\, H e^{(-Hx)}$$

> bc:=subs(x=1,diff(ya,x));

$$bc := (-\_C2+1)H e^{H} - \_C2\, H e^{(-H)}$$

> _C2:=solve(bc,_C2);

$$\_C2 := \frac{e^{H}}{e^{H}+e^{(-H)}}$$

The complete solution is obtained by using Maple's 'simplify' command as follows:

> ya:=simplify(ya);

$$ya := \frac{e^{(-H+Hx)} + e^{(-H(-1+x))}}{e^{H}+e^{(-H)}}$$

Plot the solution ya with H=1:

> plot(subs(H=1,ya),x=0..1);

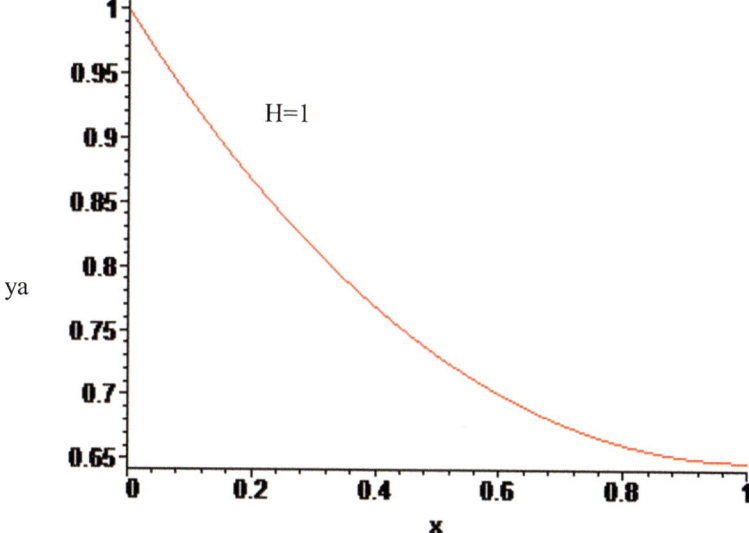

Fig. 1.6 Maple plot of ya vs x for H=1

Next, plot the solution ya with H=3 and use points instead of a line.

> plot(subs(H=3,ya),x=0..1,style=point);

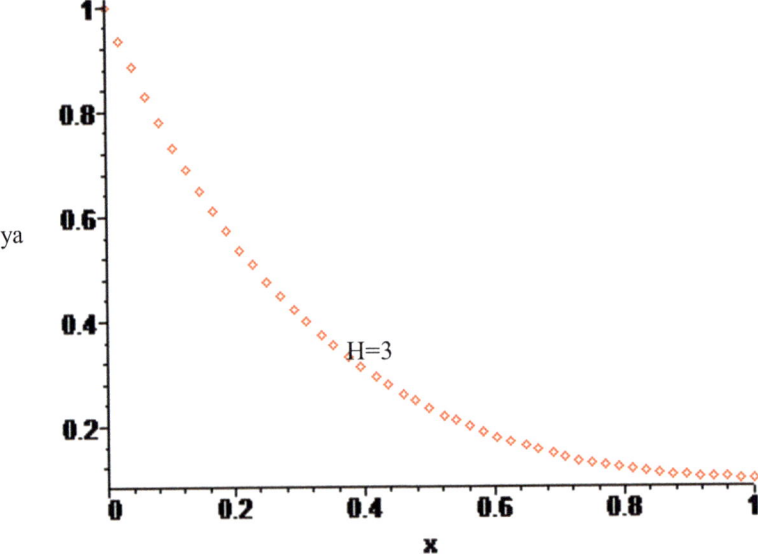

Fig. 1.7 Maple point plot of ya vs x for H=3

1.1.6 Laplace Transformations

Maple can be used to obtain Laplace transforms and inverse Laplace transforms of functions symbolically. For this purpose, the package 'with(inttrans)' is used:

> restart:

> with(inttrans):

Suppose we want to find the Laplace transform of t, we use

> f(t):=t;

$$f(t) := t$$

> laplace(f(t),t,s);

$$\frac{1}{s^2}$$

Laplace transforms for different functions can be obtained easily:

> laplace(f(t)*t,t,s);

$$\frac{2}{s^3}$$

1.1 Introduction to Maple

> f(t):=exp(-t);

$$f(t) := e^{(-t)}$$

> laplace(f(t),t,s);

$$\frac{1}{1+s}$$

Both Laplace and inverse Laplace transforms can be obtained with Maple.

> f(t):=sin(t);

$$f(t) := \sin(t)$$

> f(s):=laplace(f(t),t,s);

$$f(s) := \frac{1}{s^2+1}$$

> invlaplace(f(s),s,t);

$$\sin(t)$$

Inverse Laplace transforms for different functions can be also obtained:

> f(s):=1/sqrt(s);

$$f(s) := \frac{1}{\sqrt{s}}$$

> invlaplace(f(s),s,t);

$$\frac{1}{\sqrt{\pi t}}$$

> f(s):=1/(s)^(3/2);

$$f(s) := \frac{1}{s^{(3/2)}}$$

> invlaplace(f(s),s,t);

$$\frac{2\sqrt{t}}{\sqrt{\pi}}$$

> f(s):=exp(-sqrt(s));

$$f(s) := e^{(-\sqrt{s})}$$

> invlaplace(f(s),s,t);

$$\frac{1}{2} \frac{e^{\left(-\frac{1}{4t}\right)}}{\sqrt{\pi}\, t^{(3/2)}}$$

Unfortunately, Maple cannot find the inverse Laplace transform for complicated functions:

> f(s):=1/sinh(sqrt(s));

$$f(s) := \frac{1}{\sinh(\sqrt{s})}$$

> invlaplace(f(s),s,t);

$$\text{invlaplace}\left(\frac{1}{\sinh(\sqrt{s})}, s, t\right)$$

This does not mean that the inverse Laplace transform does not exist; instead, one has to use advanced techniques for finding the desired inverse Laplace transform (see chapter 8 for details).

1.1.7 Do Loop

It is possible to carry out a sequence of steps using a 'do loop.' The syntax is

> **for** *variables* **in** *expresion* **do** *statement sequence* **end do**;

statement sequence

We can prepare a set of differential equations by using a 'do loop'. Use the shift enter keys to add the extra line after the "do" as shown in the following worksheet.

> restart:

> N:=3;

$$N := 3$$

> for i from 1 to N do #Use the shift enter keys to add each new line in a do loop.

diff(y[i](t),t)=(y[i+1](t),-y[i-1](t));

od;

$$\frac{d}{dt} y_1(t) = (y_2(t), -y_0(t))$$

$$\frac{d}{dt} y_2(t) = (y_3(t), -y_1(t))$$

$$\frac{d}{dt} y_3(t) = (y_4(t), -y_2(t))$$

1.1.8 While Loop

It is possible to carry out a sequence of statements or commands until a prescribed condition is satisfied. The 'while' command can be used to do this. The general statement is of the form:

> **while** *conditions* **do** *statement sequence* **end do**;

For example, one could use the following worksheet to determine the square of a number.

> *restart* :

> $N := 1$;

$$N := 1$$

> **while** $N < 3$ **do** $A := N^2$; $N := N + 1$; **end do**;

$$A := 1$$
$$N := 2$$
$$A := 4$$
$$N := 3$$

>

1.1.9 Write Data Out Example

Data can be generated and written out to a text file (i.e., a . txt file). For example, we can use Maple to solve the second order ordinary differential equation

$$\frac{d^2 u}{dx^2} = u \tag{1.1}$$

with the following boundary conditions:

$$u(0) = 0.21 \tag{1.2}$$

and

$$\left.\frac{du}{dx}\right|_{x=1} = 0 \tag{1.3}$$

The result is

$$u(x) = 0.21 \frac{\cosh(x-1)}{\cosh(1)} \tag{1.4}$$

Values for this analytical solution at various values of x can be generated and exported to a text file as shown in the worksheet below.

> *restart:with(plots):with(linalg):*

Specify the values for x at which the analytical solution will be calculated and later exported. Make sure you have the whole range (i.e., 0 to 1) of x included.

> $xdata := matrix\left(11, 1, \left[seq\left(\dfrac{k}{10.0}, k = 0..10\right)\right]\right);$

$$xdata := \begin{bmatrix} 0. \\ 0.1000000000 \\ 0.2000000000 \\ 0.3000000000 \\ 0.4000000000 \\ 0.5000000000 \\ 0.6000000000 \\ 0.7000000000 \\ 0.8000000000 \\ 0.9000000000 \\ 1.000000000 \end{bmatrix}$$

Input the Governing Equation and Boundary Conditions:

> $eq := diff(u(x), x\$2) = u(x);$

$$eq := \dfrac{d^2}{dx^2} u(x) = u(x)$$

> $bcs := u(0) = 0.21, (D(u))(1) = 0;$

$$bcs := u(0) = 0.21, D(u)(1) = 0$$

Solve the differential equation and rearrange the results to the desired form:

> $u := rhs(dsolve(\{eq, bcs\}));$

$$u := \dfrac{21}{100} \dfrac{e\, e^{-x}}{e^{-1} + e} + \dfrac{21}{100} \dfrac{e^{-1} e^{x}}{e^{-1} + e}$$

> $u := convert(u, trig);$

$$u := \dfrac{21}{200} \dfrac{(\cosh(1) + \sinh(1))(\cosh(x) - \sinh(x))}{\cosh(1)}$$
$$+ \dfrac{21}{200} \dfrac{(\cosh(1) - \sinh(1))(\cosh(x) + \sinh(x))}{\cosh(1)}$$

1.1 Introduction to Maple

> *simplify(%);*

$$\frac{21}{100} \frac{\cosh(1)\cosh(x) - \sinh(1)\sinh(x)}{\cosh(1)}$$

> *u := combine(%);*

$$u := \frac{21}{100} \frac{\cosh(-1+x)}{\cosh(1)}$$

Make sure the solution satisfies the boundary conditions:

> *evalf(subs(x = 0, u));*

$$0.2100000000$$

> *evalf(subs(x = 1, (D(u))(1)));*

$$0.$$

Plot the results:

> *plot(u, x = 0..1);*

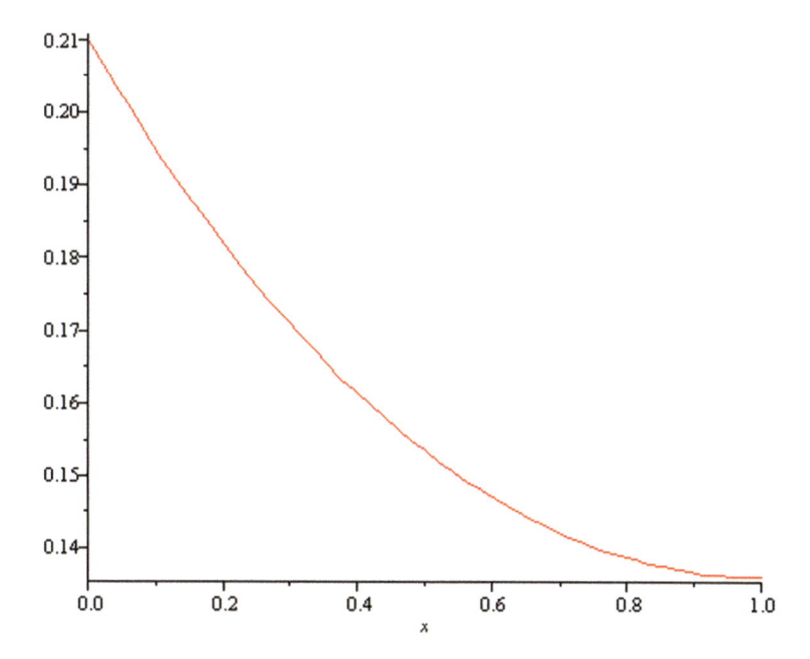

Fig. 1.8

Save the analytical solution at the x values specified at the beginning of this worksheet as uana[i].

> for i to rowdim(xdata) do `:=`
(uana[i], evalf(subs(x = xdata[i, 1], u)))
end do; 1

$$uana_1 := 0.2100000000$$
$$uana_2 := 0.1950307288$$
$$uana_3 := 0.1820133908$$
$$uana_4 := 0.1708177041$$
$$uana_5 := 0.1613316181$$
$$uana_6 := 0.1534601934$$
$$uana_7 := 0.1471246498$$
$$uana_8 := 0.1422615792$$
$$uana_9 := 0.1388223103$$
$$uana_{10} := 0.1367724217$$
$$uana_{11} := 0.1360913975$$

Export the Analytical Solution from Maple into the Text File:

> $uanalytical := evalm(matrix(rowdim(xdata), 2, [seq([xdata[kd, 1], uana[kd]], kd = 1..rowdim(xdata))]));$

$$uanalytical := \begin{bmatrix} 0. & 0.2100000000 \\ 0.1000000000 & 0.1950307288 \\ 0.2000000000 & 0.1820133908 \\ 0.3000000000 & 0.1708177041 \\ 0.4000000000 & 0.1613316181 \\ 0.5000000000 & 0.1534601934 \\ 0.6000000000 & 0.1471246498 \\ 0.7000000000 & 0.1422615792 \\ 0.8000000000 & 0.1388223103 \\ 0.9000000000 & 0.1367724217 \\ 1.000000000 & 0.1360913975 \end{bmatrix}$$

Note: To export to a text file, the formats compatible with Maple are Arrays, Matrices, etc. For additional help please type ? writedata

1.1 Introduction to Maple   23

> $fd := fopen$ ("maple_output.txt", $WRITE, TEXT$):
  $writedata\,(fd, uanalytical\,)$;
  $close\,(fd)$:

Now, the output from the analytical solution is stored into a file called "maple_output.txt" in the folder where you saved this original Maple file.

1.1.10 Reading in Data from a Text File

Data can be read into Maple from a text file as shown below.
This worksheet is entitled ReadDataInExp.mws.

> restart:

Read in data from a text file named "maple_output.txt" located on the D drive under the folder named "ECHE700Sp09".

> fd := fopen("D:\\ECHE700Sp09\\maple_output.txt",READ);

data:=readdata(fd,2);

fclose(fd);

$$fd := 0$$

$data := [\,[0., 0.21\,], [0.1, 0.1950307288\,], [0.2, 0.1820133908\,], [0.3, 0.1708177041\,], [0.4, 0.1613316181\,], [0.5, 0.1534601934\,], [0.6, 0.1471246498\,], [0.7, 0.1422615792\,], [0.8, 0.1388223103\,], [0.9, 0.1367724217\,], [1., 0.1360913975\,]\,]$

Print the data.

> data:=evalm(data);

$$data := \begin{bmatrix} 0. & 0.21 \\ 0.1 & 0.1950307288 \\ 0.2 & 0.1820133908 \\ 0.3 & 0.1708177041 \\ 0.4 & 0.1613316181 \\ 0.5 & 0.1534601934 \\ 0.6 & 0.1471246498 \\ 0.7 & 0.1422615792 \\ 0.8 & 0.1388223103 \\ 0.9 & 0.1367724217 \\ 1. & 0.1360913975 \end{bmatrix}$$

1.1.11 Summary

In this chapter, some useful Maple commands were introduced. In section 1.1.1, basic Maple commands for assignment, evaluation, differentiation and integration were introduced. In section 1.1.2, commands for plotting were introduced. In section 1.1.3, linear and nonlinear equations were solved using Maple. Linear equations were solved symbolically (exactly) and nonlinear equations were solved numerically. In section 1.1.4, Maple's matrix operations such as addition, subtraction, finding the inverse, eigenvalues, etc. were introduced. In section 1.1.5, simple linear differential equations were solved using Maple's 'dsolve' command to obtain a closed form analytical solution. In addition, series solutions were obtained for certain nonlinear differential equations. In section 1.1.6, Laplace and inverse Laplace transforms for simple functions were obtained using Maple. In section 1.1.7, using a 'do loop' to carry out a sequence of steps using Maple was explained. In section 1.1.8, using a 'while loop' to carry out a sequence of statements or commands was explained using Maple. In section 1.1.9, steps for writing out data from Maple into a text file was discussed. In section 1.1.10, reading data into Maple from a text file was explained.

1.1.12 Problems

Create a different Maple worksheet for each of the following problems. Start each worksheet with the restart command.

1. Assign x = 4 and obtain the following results using Maple:

2. (1) x^2 (2) $1 + y/x$ (3) $\sqrt{x}$ (4) $\dfrac{1+y}{x}$

   (5) $1.2^x + x^2 - x^{1.2}$

3. Assign x = 2 and y = 3 and obtain results for the following using Maple:

   (1) sin(x)   (b) arcsin(x) (i.e., $\sin^{-1}(x)$) (3) log(x)   (4) log(y/x)

   (5) exp(x)   (6) exp(x) + exp(y) – exp(xy)   (7) log(y-x) + log(x-y)

4. Use Maple to find the derivatives of the following functions:

   (1) $x^2 - x \sin(x)$ (2) $\left(\dfrac{1}{1+x+x^2}\right) \log(x)$

5. Plot $\exp(-x^2)$ from x = 0 to 5. Use Maple to find the definite integral

   $$\int_0^L \exp(-x^2) dx$$

   for L = 0.1, 0.5, 1, and 2.

1.1 Introduction to Maple

6. Use Maple to plot the following functions for x varying from 0 to 1:
   (1) exp(x) (2) exp(-x) (3) $1 - x + x^2$ (4) x (1-x) (5) $x^2 - \log(x)$

7. Use Maple to plot the following functions for x varying from 0 to 1:
   (1) $\sin(\pi x)$ (2) $\cos(\pi x)$ (3) arcsin(x) (4) $\sin\left(\dfrac{\pi}{2} x\right) \exp(-x)$

8. Use Maple to solve the following equations symbolically (use the 'solve' command).
   (1) $ax^2 + bx + c = 0$
   (2) $x^3 - 1 = 0$
   (3) $x^4 - x^2 = 0$
   (4) x + y = 3; x – y = 2
   (5) x + y = a; x – y = a – b
   (6) x + y + z = a; 2x + 3y + 4z = a + b + c; x – y - z = b
   (7) x + y + z = 6; xyz = 6; xy + yz + zx = 11

9. Use Maple to solve the following equations numerically by using 'fsolve'. Find all the possible roots.
   (1) $x^3 - \tan(y) = xy$; $y^3 - \tan(x) = 1$
   (2) $x^2 + y^2 = 1$; $x^2 - y^2 = ¼$
   (3) $x^3 - 2x - \dfrac{1}{x} = 0$

10. Define the following matrices $A = \begin{bmatrix} 1 & 2 & 3 \\ 2 & 3 & 4 \\ 3 & 4 & 6 \end{bmatrix}$; $B = \begin{bmatrix} 1 & 2 & 0 \\ 0 & 1 & 2 \\ 0 & 0 & 1 \end{bmatrix}$ in Maple.

    (1) Find A+B, A-B, AB and BA.
    (2) Find the determinant of A, B and AB.
    (3) Find A^{-1}, B^{-1}, A/B and B/A.
    (4) Find the eigenvalues and eigenvectors of A, B, AB and BA.
    (5) Find A^3, A + B + AB-BA.

11. Consider the matrix $A = \begin{bmatrix} \alpha & 1 & 0 \\ 0 & 1 & 1 \\ 0 & -1 & \alpha \end{bmatrix}$.

   (1) Substitute $\alpha = 4$ in A and find the determinant, characteristic polynomial, eigenvalues, eigenvectors and inverse of A using Maple.
   (2) Substitute $\alpha = 2$ in A and find the determinant, characteristic polynomial, eigenvalues, eigenvectors and inverse of A using Maple.
   (3) Substitute $\alpha = 1$ in A and find the determinant, characteristic polynomial, eigenvalues, eigenvectors and inverse of A using Maple.
   (4) Substitute $\alpha = 3$ in A and find the determinant, characteristic polynomial, eigenvalues, eigenvectors and inverse of A using Maple.

12. Consider the differential equation $\dfrac{dy}{dx} = -yx$; $y(0) = 1$. Use Maple to solve this differential equation by using the 'dsolve' command to obtain a closed-form solution. Obtain a series solution for the same. Plot the profiles.

13. Consider diffusion with a first order reaction in a rectangular catalyst pellet.[4] The governing equation in dimensionless form is $\dfrac{d^2y}{dx^2} = \Phi^2 y$; $\dfrac{dy}{dx}(0) = 0$; $y(1) = 1$. Solve this differential equation using the 'dsolve' command to obtain a closed-form solution. Plot the profile.

14. Use Maple to find the Laplace transforms of the following functions.

    (1) $\sinh(at) + \cosh(at)$

    (2) $\exp(at) \sinh(at)$

    (3) $\dfrac{1}{1+t}$

15. Use Maple to find the inverse Laplace transforms of the following functions.

    (1) $\dfrac{1}{s(s+1)} - \dfrac{1}{1+s^2}$

    (2) $\dfrac{1}{1+s\,\exp(s)}$

    (3) $\dfrac{1}{1+\sqrt{s}}$

References

1. Garvan, F.: The Maple Book, p. 479. Chapman & Hall/CRC, Boca Raton (2002)
2. Abel, M.L., Braselton, J.P.: Differential Equations with Maple V, 3rd edn. Academic Press, London (2001)
3. Davis, M.E.: Numerical Methods and Modeling for Chemical Engineers. John Wiley & Sons, Chichester (1984)
4. Rice, R.G., Do, D.D.: Applied Mathematics and Modeling for Chemical Engineers. John Wiley & Sons, Inc., Chichester (1995)

Chapter 2
Initial Value Problems

Engineers develop mathematical models to describe processes of interest to them. For example, the process of converting a reactant A to a product B in a batch chemical reactor can be described by a first order, ordinary differential equation with a known initial condition. This type of model is often referred to as an initial value problem (IVP), because the initial conditions of the dependent variables must be known to determine how the dependent variables change with time. In this chapter, we will describe how one can obtain analytical and numerical solutions for linear IVPs and numerical solutions for nonlinear IVPs.

2.1 Linear Ordinary Differential Equations

2.1.1 Introduction

First order series/parallel chemical reactions and process control models are usually represented by a linear system of coupled ordinary differential equations (ODEs). Single first order equations can be integrated by classical methods (Rice and Do, 1995). However, solving more than two coupled ODEs by hand is difficult and often involves tedious algebra. In this chapter, we describe how one can arrive at the analytical solution for linear first order ODEs using Maple, the matrix exponential, and Laplace transformations.

2.1.2 Homogeneous Linear ODEs

Consider two linear ordinary differential equations:

$$\frac{dy_1}{dt} = a_1 y_1 + a_2 y_2$$
$$\frac{dy_2}{dt} = a_3 y_1 + a_4 y_2 \tag{2.1}$$

with the following initial conditions

$$y_1(0) = y_{10} \text{ and } y_2(0) = y_{20} \tag{2.2}$$

where y_1 and y_2 are the dependent variables; a_1, a_2, a_3, and a_4 are constants. This system of linear differential equations (equation(2.1)) can be written in matrix form as

$$\frac{dY}{dt} = AY \qquad (2.3)$$

where the dependent variables are expressed as a 2 x 1 matrix:

$$Y = \begin{bmatrix} y_1 \\ y_2 \end{bmatrix} \qquad (2.4)$$

The 2 x 2 coefficient matrix **A** is written in this case as

$$A = \begin{bmatrix} a_1 & a_2 \\ a_3 & a_4 \end{bmatrix} \qquad (2.5)$$

and Y_0 is the vector (2 x 1 matrix) of initial conditions:

$$Y_0 = \begin{bmatrix} y_{10} \\ y_{20} \end{bmatrix} \qquad (2.6)$$

The solution of equation (2.3) is given by:[2, 3]

$$Y = \exp(At) Y_0 \qquad (2.7)$$

where $\exp(At)$ is a n x n matrix and is called the matrix exponential or the exponential matrix of A (see Appendix A for a detailed derivation of equation (2.7)). It is of interest to note that Professor Amundson introduced the exponential matrix to chemical engineers in 1966 on page 166 of his book.[2] At that time it was very difficult to obtain symbolically $\exp(At)$. We believe that the need to find the exponential matrix $(\exp(At))$ has discouraged chemical engineers from using the exponential matrix to solve coupled systems of ODEs. Maple provides a command to find the exponential matrix as a function of t, the independent variable. Maple can also find the exponential matrix as a function of parameters such as a_1, a_2, a_3, and a_4 which we will illustrate by solving some classical problems in chemical engineering. In general, for a linear system of n simultaneous coupled first order differential equations, **Y** is a n x 1 matrix, **A** is a n x n matrix and, again, $\exp(At)$ is a n x n matrix.

2.1.3 First Order Irreversible Series Reactions

Consider the first order series reactions[4] $\left(A \xrightarrow{k_1} B \xrightarrow{k_2} C\right)$. The governing equations for this reaction scheme can be written as

$$\frac{dC_A}{dt} = -k_1 C_A$$
$$\frac{dC_B}{dt} = k_1 C_A - k_2 C_B \tag{2.8}$$

where k_1 and k_2 are rate constants and the initial conditions are $C_A(0) = 1$ mol/l, $C_B(0) = 0$, and $C_C(0) = 0$. The concentration of species C ($C_C(t)$) at any time is given by the material balance:

$$C_C(t) = 1 \text{ mol/l} - C_A(t) - C_C(t) \tag{2.9}$$

A Maple procedure for solving equations (2.8) subject to the initial conditions is presented below. This procedure is useful for developing the correct coefficient matrix **A** to avoid "by hand" errors.

Procedure

1. Start the Maple program with a 'restart' command to clear all variables.
2. Call Maple's linear algebra package by using the 'with(linalg)' command.
3. Call Maple's plotting package by using the 'with(plots)' command.
4. Use Maple to enter the governing equations (equation (2.8)).
5. Name the variables vars.
6. Name the right hand side of the equations eqs.
7. Next, use Maple to generate the coefficient matrix (**A**) using Maple's 'genmatrix' command.
8. Store the initial conditions for the dependent variables in the vector **Y0**. The first row of **Y0** corresponds to the initial condition for the first dependent variable (c_a). The second row of **Y0** corresponds to the initial condition for the second variable (c_b).
9. Find the matrix exponential of **A** (i.e., exp(**A**t)) as a function of the parameters (rate constants) and the independent variable (t) using the Maple command 'exponential(A,t)'. Call this matrix mat.
10. Next, find the solution (sol) by multiplying mat by **Y0** using Maple's 'evalm' command.
11. The first row of the sol vector corresponds to $c_{a(t)}$, and the second row corresponds to $c_{b(t)}$.
12. Note that the analytical solutions for c_a and c_b are obtained as functions of the parameters (rate constants) and the independent variable, t. Lower case letters are used for these variables with k_1 and k_2 as parameters.
13. For a given set of rate constants, plot the concentration profiles.

This procedure is illustrated in Example 2.1.

Example 2.1. Irreversible Series Reactions (see equations (2.8))

> restart:

> with(linalg):

> with(plots):

Enter the governing equations as follows:

> eq[1]:=diff(C[A](t),t)=-k1*C[A](t);

$$eq_1 := \frac{d}{dt} C_A(t) = -k1 \, C_A(t)$$

> eq[2]:=diff(C[B](t),t)=k1*C[A](t)-k2*C[B](t);

Store the variables in 'vars':

$$eq_2 := \frac{d}{dt} C_B(t) = k1 \, C_A(t) - k2 \, C_B(t)$$

> vars:=[C[A](t),C[B](t)];

$$vars := [C_A(t), C_B(t)]$$

Next, store the right hand sides of equations 1 and 2 in 'eqs.'

> eqs:=[rhs(eq[1]),rhs(eq[2])];

$$eqs := [-k1 \, C_A(t), k1 \, C_A(t) - k2 \, C_B(t)]$$

Now use the 'genmatrix' Maple command to produce the coefficient matrix **A**.

> A:=genmatrix(eqs,vars);

$$A := \begin{bmatrix} -k1 & 0 \\ k1 & -k2 \end{bmatrix}$$

Specify the initial conditions.

> Y0:=matrix(2,1,[1,0]);

$$Y0 := \begin{bmatrix} 1 \\ 0 \end{bmatrix}$$

Use Maple's 'exponential(A,t)' command to produce the exponential (**At**) matrix.

> mat:=exponential(A,t);

$$mat := \begin{bmatrix} e^{-k1\,t} & 0 \\ -\dfrac{k1\left(-e^{-k2\,t} + e^{-k1\,t}\right)}{-k2 + k1} & e^{-k2\,t} \end{bmatrix}$$

2.1 Linear Ordinary Differential Equations

Obtain the solution vector using Maple's 'evalm' command and matrix multiplication (&*).

> sol:=evalm(mat&*Y0);

$$sol := \begin{bmatrix} e^{-k1\,t} \\ -\dfrac{k1\left(-e^{-k2\,t}+e^{-k1\,t}\right)}{-k2+k1} \end{bmatrix}$$

The first row of 'sol' is the solution for the concentration of species A (C_A) and the second row is for the concentration of species B (C_B).

> ca:=sol[1,1];

$$ca := e^{-k1\,t}$$

> cb:=sol[2,1];

$$cb := -\dfrac{k1\left(-e^{-k2\,t}+e^{-k1\,t}\right)}{-k2+k1}$$

The concentration of species C is obtained from the material balance equation.

> cc:=1-ca-cb;

$$cc := 1 - e^{-k1\,t} + \dfrac{k1\left(-e^{-k2\,t}+e^{-k1\,t}\right)}{-k2+k1}$$

Note that the concentrations of the species are obtained as functions of the rate constants (k_1 and k_2) and the independent variable t. To illustrate the results, enter values for the rate constants (pars).

> pars:={k1=1.,k2=0.,k3=2.,k4=3.};

$$pars := \{k2 = 1, k1 = 2\}$$

Next, substitute these values for the rate constants into the concentrations store as C_A, B_B, and C_C.

> Ca:=subs(pars,ca);

$$Ca := e^{-2t}$$

> Cb:=subs(pars,cb);

$$Cb := 2\,e^{-t} - 2\,e^{-2t}$$

> Cc:=subs(pars,cc);

$$Cc := 1 + e^{-2t} - 2\,e^{-t}$$

Plot the concentration profiles for these values of the rate constants as shown in Figure 2.1.

>p1:=plot(Ca,t=0..10,thickness=3,color=blue,labels=["t(min)","Concentration(mols/l)"],

labeldirections=[horizontal, vertical],legend='C[A]'):

>p2:=plot(Cb,t=0..10,linestyle=1,thickness=3,labels=["t(min)","Concentration(mols/l)"], labeldirections=[horizontal, vertical],legend='C[B]'):

>3:=plot(Cc,t=0..10,linestyle=1,thickness=3,color=green,labels=["t(min)","Concentration (mols/l)"],labeldirections=[horizontal, vertical],legend='C[C]'):

> p4:=textplot([6,.9,k1=2.0*Unit(min^(-1))]):

> p5:=textplot([6,.7,k2=1.0*Unit(min^(-1))]):

> display([p1,p2,p3,p4,p5],title="Figure 2.1 Concentrations of A, B, and C as functions of time.");

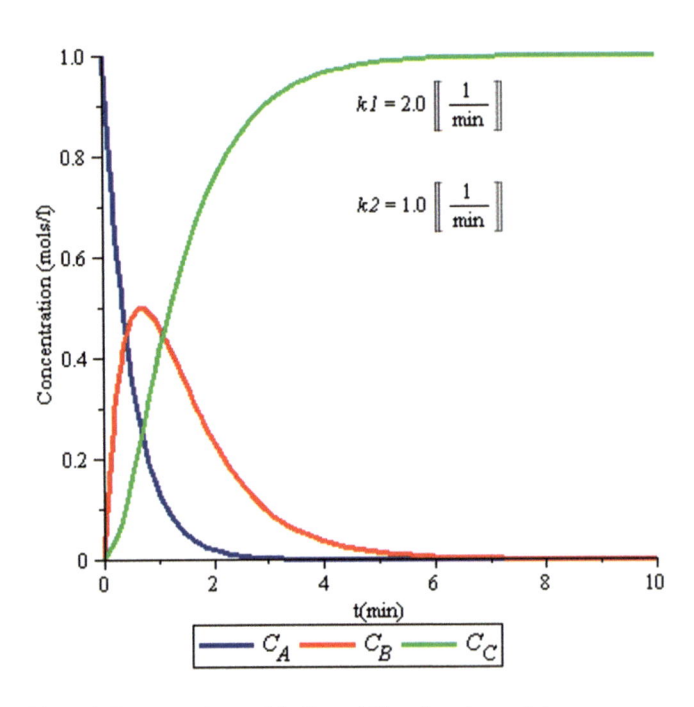

Fig. 2.1 Concentrations of A, B, and C as functions of time

Suppose $k_1 = k_2 = 1 \min^{-1}$. We can obtain an expression for Ca with these values for the parameters:

> pars:={k1=1,k2=2};

$$pars := \{k1 = 1, k2 = 2\}$$

> Ca:=subs(pars,ca);

$$Ca := e^{-t}$$

2.1 Linear Ordinary Differential Equations

We get division by zero because we have k_1 and k_2 in the denominator of the expression for Cb. To handle this problem, we can apply Maple's 'limit' command when both the rate constants approach 1 (i.e., as $k_1 \rightarrow 1$ and as $k_2 \rightarrow 1$). First, the 'limit' command is applied with respect to the first parameter, $k_1 \rightarrow 1$.

> **Cb:=subs(pars,cb);**

$$Cb := -e^{-2t} + e^{-t}$$

Next, the 'limit' command is applied with respect to the second parameter, for $k_2 \rightarrow 1$ also:

> Cc:=subs(pars,cc);

$$Cc := 1 + e^{-2t} - 2e^{-t}$$

> Cb:=limit(cb,k1=1);

$$Cb := -\frac{e^{-k2\,t} - e^{-t}}{k2 - 1}$$

> Cb:=limit(Cb,k2=1);

$$Cb := \frac{t}{e^t}$$

Similarly, limits can be applied for Cc also:

> Cc:=limit(Cc,k1=1);

$$Cc := 1 + e^{-2t} - 2e^{-t}$$

> Cc:=limit(Cc,k2=1);

$$Cc := 1 + e^{-2t} - 2e^{-t}$$

We can use these results to plot the concentration profiles for the case when $k_1 = k_2$ as shown in Figure 2.2.

>p1:=plot(Ca,t=0..10,thickness=3,color=blue,labels=["t(min)","Concentration(mol/l)"],labeldirections=[horizontal, vertical],legend='C[B]'):

>p2:=plot(Cb,t=0..10,linestyle=1,thickness=3,color=red,labels=["t(min)","Concentration(mol/l)"],labeldirections=[horizontal, vertical],legend='C[B]'):

>p3:=plot(Cc,t=0..10,linestyle=1,thickness=3,color=green,labels=["t(min)","Concentration (mol/l)"],labeldirections=[horizontal, vertical],legend='C[C]'):

> p4:=textplot([6,.8,(k1=1.0*Unit(min^(-1)))]):

> p5:=textplot([6,.6,(k2=1.0*Unit(min^(-1)))]):

> display([p1,p2,p3,p4,p5],title="Figure 2.2 Concentrations of A, B, and C as functions of time.");

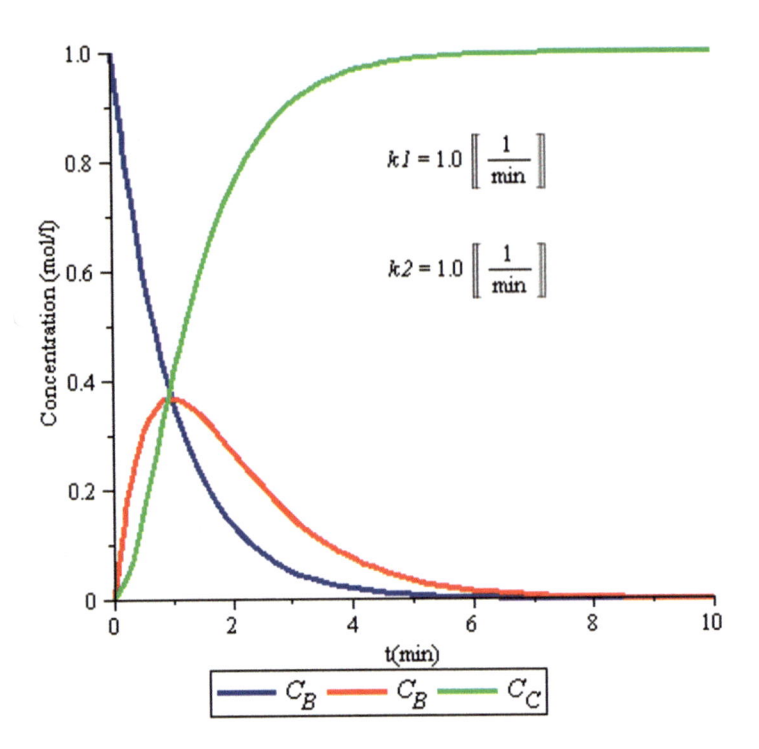

Fig. 2.2 Concentrations of A, B, and C as functions of time

We can solve for the time at which B is at its maximum by differentiating Cb with respect to t:

> Eqmax:=diff(cb,t);

$$Eqmax := -\frac{k1\left(k2\ e^{-k2\ t} - k1\ e^{-k1\ t}\right)}{-k2 + k1}$$

Next, solve 'eqmax' to find the time at which Cb is at a maximum in terms of the rate constants k_1 and k_2:

> tmax:=solve(Eqmax,t);

$$tmax := -\frac{\ln\left(\dfrac{k2}{k1}\right)}{-k2 + k1}$$

Substitute this value for time into the concentration equation for B(cb(t)) to find the maximum value of Cb as a function of the rate constants k_1 and k_2.

2.1 Linear Ordinary Differential Equations

> cbmax:=subs(t=tmax,cb);

$$cbmax := -\frac{k1\left(-e^{\frac{k2\ln\left(\frac{k2}{k1}\right)}{-k2+k1}} + e^{\frac{k1\ln\left(\frac{k2}{k1}\right)}{-k2+k1}}\right)}{-k2+k1}$$

The equation for 'cbmax' can be simplified further by using Maple's 'simplify' command.

> cbmax:=simplify(cbmax);

$$cbmax := \frac{k1\left(\left(\frac{k2}{k1}\right)^{\frac{k2}{-k2+k1}} - \left(\frac{k2}{k1}\right)^{\frac{k1}{-k2+k1}}\right)}{-k2+k1}$$

A maximum for the concentration of B(cbmax) for the special case of $k_1 = k_2$ can be found as:

>limit(cbmax,k2=k1);

$$e^{-1}$$

The time needed for the concentration of B to reach a maximum, for the case when $k_1 = k_2$ can be found as

> limit(tmax,k2=k1);

$$\frac{1}{k1}$$

Note that when $k_1 = k_2$, 'cbmax' is independent of the rate constants, the time taken to reach this maximum value e^{-1} is inversely proportional to the rate constant k_1.

2.1.4 First Order Reversible Series Reactions

In Example 2.1, Maple was used to solve two simultaneous first order ODEs. The same methodology can be used to solve more than two simultaneous ODEs. For example, the material balance equations for the time dependent concentration of each species (A, B, and C) in an isothermal batch reactor with reversible series reactions ($A \underset{k_2}{\overset{k_1}{\rightleftarrows}} B \underset{k_4}{\overset{k_3}{\rightleftarrows}} C$) can be written as follows:[5]

$$\frac{dC_a}{dt} = -k_1 C_a + k_2 C_b$$

$$\frac{dC_b}{dt} = k_1 C_a - k_2 C_b - k_3 C_b + k_4 C_c \qquad (2.10)$$

$$\frac{dC_c}{dt} = k_3 C_b - k_4 C_c$$

In this case, the initial conditions are $C_a(0) = 1$ mol/l; $C_b(0) = 0$ and $C_c(0) = 0$. One might ask "What are the values of parameters ($k_1 \ldots k_4$), if any, that would produce a maximum in concentration of species B?" This question can be answered by using Maple to obtain a solution to the equations in (2.10) given initial conditions for the concentrations.

Constantinides and Mostoufi[6] solved this system of ODEs (eq. (2.10)) for a given set of values for the parameters ($k_1 = 1$ min^{-1}, $k_2 = 0$ min^{-1}, $k_3 = 2$ min^{-1}, $k_4 = 0$ min^{-1}). They did this by two different methods. They found the exponential matrix using MATLAB's command 'expm' and used MATLAB to predict the concentration profiles from $t = 0$ to 5 minutes. The disadvantage of this approach is that one has to solve the problem for every set of parameters ($k_1 \ldots k_4$). In addition, special care is needed for rate constant values that yield a coefficient matrix that has eigenvectors that are repeated or are very small. We illustrate below how Maple can be used to find the exponential matrix and solve this problem symbolically and efficiently. In addition, we demonstrate how the symbolic solution can be used to analyze the effect of the parameters in determining the maximum concentration of the intermediate species, B. It should be noted that, similar symbolic solutions could be obtained using MATLAB or other symbolic software also by applying the same methodology presented here.

This reaction scheme (equation (2.10) is simulated below by following the procedure described for the previous example (see Example 2.1).

Example 2.2. Reversible Series Reactions (see equations (2.10))

> restart:

> with(linalg):

> with(plots):

Specify the equations.

> eq[1]:=diff(Ca(t),t)=-k1*Ca(t)+k2*Cb(t);

$$eq_1 := \frac{d}{dt} Ca(t) = -k1\ Ca(t) + k2\ Cb(t)$$

> eq[2]:=diff(Cb(t),t)=k1*Ca(t)-k2*Cb(t)-k3*Cb(t)+k4*Cc(t);

$$eq_2 := \frac{d}{dt} Cb(t) = k1\ Ca(t) - k2\ Cb(t) - k3\ Cb(t) + k4\ Cc(t)$$

> eq[3]:=diff(Cc(t),t)=k3*Cb(t)-k4*Cc(t);

$$eq_3 := \frac{d}{dt} Cc(t) = k3\ Cb(t) - k4\ Cc(t)$$

Specify the variables.

> vars:=[Ca(t),Cb(t),Cc(t)];

$$vars := [Ca(t), Cb(t), Cc(t)]$$

2.1 Linear Ordinary Differential Equations

Specify the right hand side of the equations.

> eqs:=[seq(rhs(eq[i]),i=1..3)];

$$eqs := [-k1\ Ca(t) + k2\ Cb(t), k1\ Ca(t) - k2\ Cb(t) - k3\ Cb(t) + k4\ Cc(t), k3\ Cb(t) - k4\ Cc(t)]$$

Generate the coefficient matrix A.

> A:=genmatrix(eqs,vars,A);

$$A := \begin{bmatrix} -k1 & k2 & 0 \\ k1 & -k2-k3 & k4 \\ 0 & k3 & -k4 \end{bmatrix}$$

Generate the exponential matrix of A.

> mat:=exponential(A,t):

Set the initial conditions vector:

> Y0:=matrix(3,1,[1,0,0]);

$$Y0 := \begin{bmatrix} 1 \\ 0 \\ 0 \end{bmatrix}$$

Obtain the solution.

> sol:=evalm(mat&*Y0):

Pull out the components of the solution.

> ca:=sol[1,1]:

> cb:=sol[2,1]:

> cc:=sol[3,1]:

The solution vector 'sol' and its elements (sol [1,1], e.g.) are too long to present here. They are functions of the rate constants (k1, k2, k3, and k4) and the independent variable t. One can obtain the solution from the CD with this book by changing the colons to semicolons at the end of the command line in the worksheet. As in Example 2.1, the solution obtained can be plotted for a particular set of values for the rate constants. Let's set these to be the same as those used on page 282 of Constantinides and Mostoufi. [6]

> pars:={k1=1.,K2=0.,k3=2.,k4=3.};

$$pars := \{k1 = 1., k2 = 0., k3 = 2., k4 = 3.\}$$

Obtain the concentrations as functions of time by substituting these values for the rate constants into the components of the solution vector:

> Ca:=subs(pars,ca);

$$Ca := 1.000000000 e^{-1.000000000\, t}$$

> Cb:=subs(pars,cb);

$$Cb := -0.5000000000\, e^{-1.000000000\, t} - 0.1000000000\, e^{-5.000000000\, t} + 0.6000000000$$

> Cc:=subs(pars,cc);

$$Cc := 0.1000000000\, e^{-5.000000000\, t} + 0.4000000000 - 0.5000000000\, e^{-1.000000000\, t}$$

Plot the results.

>p[1]:=plot(Ca,t=0..10,thickness=3,linestyle=1,color=blue,labels=["t(min)"," Concentration (mol/1)"],labeldirections=[horizontal, vertical],legend='C[A]'):

>p[2]:=plot(Cb,t=0..10,thickness=3,linestyle=1,labels=["t(min)","Concentrations(mol/1)"], labeldirections=[horizontal, vertical],legend='C[B]'):

>p[3]:=plot(Cc,t=0..10,thickness=3,linestyle=1,color=green,labels=["t(min)", "Concentration (mol/1)"],labeldirections=[horizontal,vertical],legend='C[C]'):

> p[4]:=textplot([4,.9,k1=1.0*Unit(min^(-1))]),textplot([6,.9,k2=0]):

> p[5]:=textplot([4,.7,k3=2.0*Unit(min^(-1))]),textplot([7,.7,k4=3.0*Unit(min^(-1))]):

> display(f, title="Figure 2.3 Concentrations of A, B, and C as functions of time.", axes=boxed);

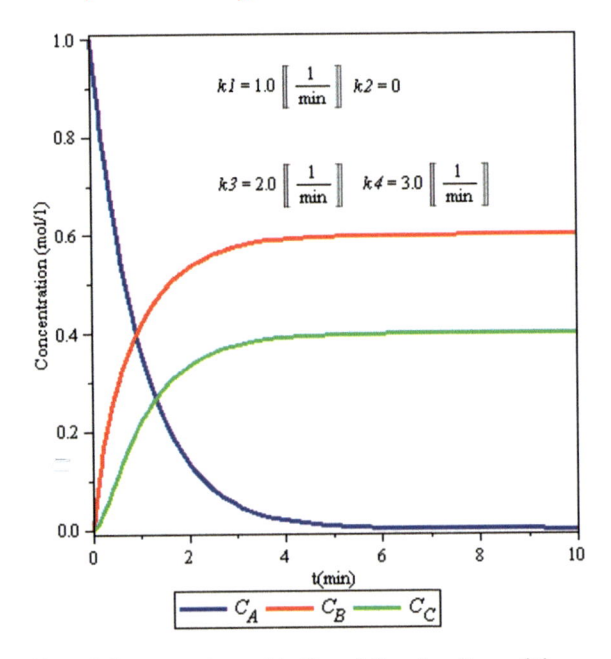

Fig. 2.3 Concentrations of A, B, and C as functions of time

2.1 Linear Ordinary Differential Equations

This figure matches Figure Exp. 5.2 on page 282 of Constantinides and Mostoufi, 1999.[6] For a different set of values for the parameters, the solution can be obtained by just substituting the numerical values of the rate constants into the solution vector components. We did not observe a maximum for Cb for this particular set of values of the rate constants. However, we can use different values of the rate constants to illustrate that a maximum can exist in the concentration of species B:

> pars:={k1=10.,k2=0.5,k3=2.,k4=3.};

$$pars := \{k3 = 2., k4 = 3., k1 = 10., k2 = 0.5\}$$

> Ca:=subs(pars,ca);

$$Ca := 0.9084683542\, e^{-10.67617498\, t} + 0.06240543150\, e^{-4.823825022\, t} + 0.02912621380$$

> Cb:=subs(pars,cb);

$$Cb := 0.6460428672\, e^{-4.823825022\, t} - 1.228567139\, e^{-10.67617498\, t} + 0.5825242716$$

> Cc:=subs(pars,cc);

$$Cc := 0.3200987842\, e^{-10.67617498\, t} - 0.7084482988\, e^{-4.823825022\, t} + 0.3883495144$$

>p[1]:=plot(Ca,t=0..5,thickness=3,color=blue,linestyle=1,labels=["t(min)","Concentration (mol/1)"],labeldirections=[horizontal, vertical],legend='C[A]'):

>p[2]:=plot(Cb,t=0..5,thickness=3,linestyle=1,labels=["t(min)","Concentration (mol/1)"],labeldirections=[horizontal, vertical],legend='C[B]'):

>p[3]:=plot(Cc,t=0..5,thickness=3,color=green,linestyle=1,labels=["t(min)","Concentration (mol/1)"],labeldirections=[horizontal, vertical],legend='C[C]'):

> p[4]:=textplot([2,.9,k1=10.0*Unit(min^(-1))]),textplot([3.5,.9,k2=0.5*Unit(min^(-1))]):

> p[5]:=textplot([2,.7,k3=2.0*Unit(min^(-1))]),textplot([3.5,.7,k4=3.0*Unit(min^(-1))]):

> display({seq(p[i],i=1..5)},title="Figure 2.4 Concentrations of A, B, and C as functions of time.",axes=boxed);

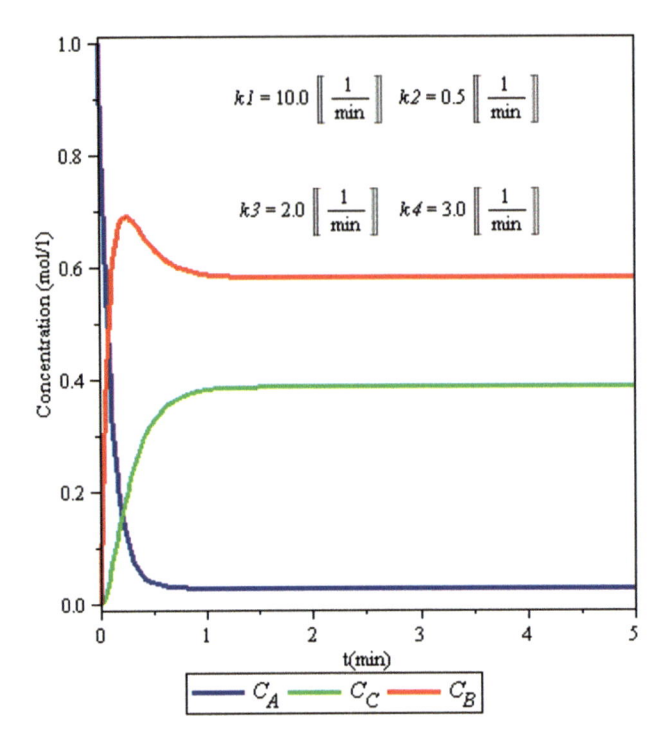

Fig. 2.4 Concentrations of A, B, and C as functions of time

We can find an equation in terms of the rate constants that will provide a means of finding the time at which the maximum in Cb occurs. We do this by using Maple's 'simplify' command and by finding ∂cb/∂t by using Maple's 'diff' command.

> eq:=simplify(diff(cb,t)):

Next, we use Maple's 'solve' command to set the equation (eq) equal to zero and solve for 'tmax':

> tmax:=solve(eq,t);

$$tmax := -\left(\ln\left(\left(k1 + k3 - \sqrt{k1^2 + 2k1k2 - 2k1k3 - 2k1k4 + k2^2 + 2k3k2 - 2k2k4 + k3^2 + 2k4k3 + k4^2} + k2 - k4\right) \middle/ \left(k1 + k3 + \sqrt{k1^2 + 2k1k2 - 2k1k3 - 2k1k4 + k2^2 + 2k3k2 - 2k2k4 + k3^2 + 2k4k3 + k4^2} + k2 - k4\right)\right)\right) \middle/ \sqrt{k1^2 + 2k1k2 - 2k1k3 - 2k1k4 + k2^2 + 2k3k2 - 2k2k4 + k3^2 + 2k4k3 + k4^2}$$

The concentration of B at 'tmax' can be obtained by substitution of the 'tmax' expression into the solution for the concentration of B as a function of time.

2.1 Linear Ordinary Differential Equations

> cbmax:=simplify(subs(t=tmax,cb)):

The solution 'cbmax' is not printed to conserve space. We can find if a maximum exists or not by substituting numerical values for the rate constants. If Cb has a maximum, 'tmax' should be real and positive. A value for 'tmax' can be obtained by substituting the same rate constants into the derived equations.

> simplify(subs(k1=1.,k2=0.,k3=2.,k4=3.0,tmax));

$$0.7853981635 I$$

> simplify(subs(k1=2.,k2=0.,k3=2.,k4=3.0,tmax));

$$0.2310490602 + 1.047197551 I$$

> simplify(subs(k1=10.,k2=1.,k3=1.,k4=1.0,tmax));

$$0.263546690:$$

An interesting case is when $k_1 = k_3$ and $k_2 = k_4$.

> pars:={k1=2,k2=1/2,k3=2,k4=1/2};

$$pars := \left\{ k1 = 2, k3 = 2, k2 = \frac{1}{2}, k4 = \frac{1}{2} \right\}$$

> A1:=evalm(subs(pars,evalm(A)));

$$A1 := \begin{bmatrix} -2 & \frac{1}{2} & 0 \\ 2 & -\frac{5}{2} & \frac{1}{2} \\ 0 & 2 & -\frac{1}{2} \end{bmatrix}$$

> eigenvalues(A1);

$$0, -\frac{3}{2}, -\frac{7}{2}$$

One of the eigenvalues is zero in this case.

> Ca:=evalf(subs(pars,ca));

$$Ca := 0.04761904762 + 0.6666666668\, e^{-1.500000000\, t} + 0.2857142858\, e^{-3.500000000\, t}$$

> Cb:=evalf(subs(pars,cb));

$$Cb := 0.6666666668\, e^{-1.500000000\, t} - 0.8571428573\, e^{-3.500000000\, t} + 0.1904761905$$

> Cc:=evalf(subs(pars,cc))

$$Cc := 0.5714285716\, e^{-3.500000000\, t} - 1.333333334\, e^{-1.500000000\, t} + 0.7619047620$$

```
>p[1]:=plot(Ca,t=0..5,thickness=3,color=blue,linestyle=1,labels=["t(min)","C
oncentration (mols/1)"],labeldirections=[horizontal, vertical],legend='C[A]'):
>p[2]:=plot(Cb,t=0..5,thickness=3,linestyle=1,labels=["t(min)","Concentratio
n (mols/1)"],labeldirections=[horizontal, vertical],legend='C[B]'):
>p[3]:=plot(Cc,t=0..5,thickness=3,color=green,linestyle=1,labels=["t(min)","C
oncentration (mols/1)"],labeldirections=[horizontal, vertical],legend='C[C]'):
>p[4]:=textplot([1.5,.9,k1=1.0*Unit(min^(-1))]);
```

$$p_4 := PLOT(\ldots)$$

```
>p[5]:=textplot([3.5,.9,k2=0.5*Unit(min^(-1))]);
```

$$p_5 := PLOT(\ldots)$$

```
>p[6]:=textplot([3.5,.6,k3=1.0*Unit(min^(-1))]);
```

$$p_6 := PLOT(\ldots)$$

```
>p[7]:=textplot([3.5,.4,k4=0.5*Unit(min^(-1))]);
```

$$p_7 := PLOT(\ldots)$$

```
>display({seq(p[i],i=1..7)},title="Figure 2.5 Concentrations of A, B, and C
as functions of time.",axes=boxed);
```

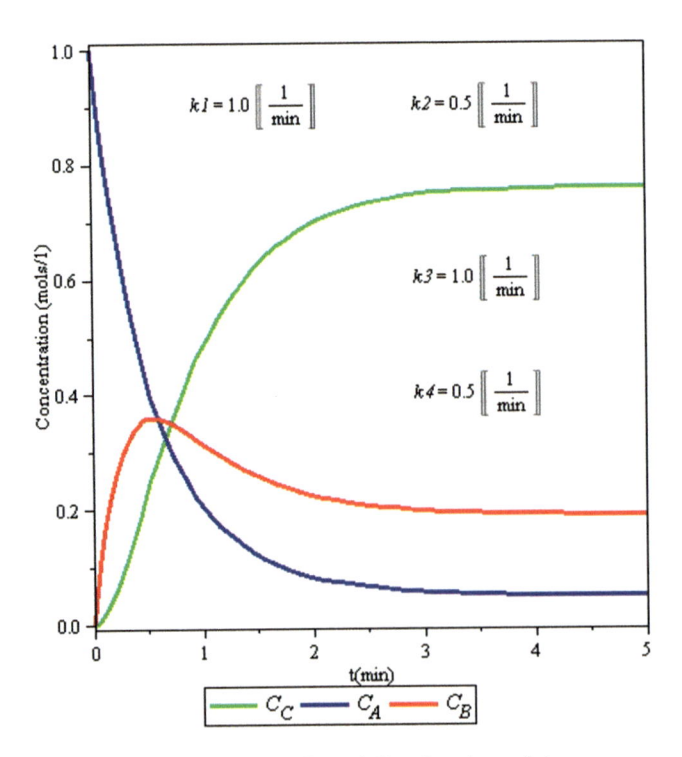

Fig. 2.5 Concentrations of A, B, and C as functions of time

2.1 Linear Ordinary Differential Equations

Another interesting case is when $k_2=k_4=0$.

> pars:={k1=1,k2=0,k3=1,k4=0};

$$pars := \{k1 = 1, k2 = 0, k4 = 0, k3 = 1\}$$

> A1:=evalm(subs(pars,evalm(A)));

$$A1 := \begin{bmatrix} -1 & 0 & 0 \\ 1 & -1 & 0 \\ 0 & 1 & 0 \end{bmatrix}$$

> eigenvalues(A1);

$$-1, -1, 0$$

One of the eigenvalues is zero and the other two eigenvalues are repeated in this case.

> Ca:=evalf(subs(pars,ca));

Error, numeric exception: division by zero

We get division by zero again as we did in Example 2.1. The solution can be obtained by using Maple's 'limit' command.

> Ca:=limit(ca,k1=1):

> Ca:=limit(Ca,k2=0):

> Ca:=limit(Ca,k3=1):

> Ca:=limit(Ca,k4=0);

$$Ca := e^{-t}$$

Alternatively, the four lines can be successively applied in the same statement as

> Cb:=limit(limit(limit(limit(cb,k1=1),k2=0),k3=1),k4=0);

$$Cb := \frac{t}{e^t}$$

> Cc:=limit(limit(limit(limit(cc,k1=1),k2=0),k3=1),k4=0);

$$Cc := \frac{e^t - t - 1}{e^t}$$

>p[1]:=plot(Ca,t=0..5,thickness=3,linestyle=1,color=blue,labels=["t(min)","Concentration(mols/1)"],labeldirections=[horizontal, vertical],legend='C[A]'):

> p[2]:=plot(Cb,t=0..5,thickness=3,linestyle=1,labels=["t(min)","Concentration (mols/1)"],

labeldirections=[horizontal, vertical],legend='C[B]'):

>[3]:=plot(Cc,t=0..5,thickness=3,color=green,linestyle=1,labels=["t(min)","Concentration (mols/1)"],labeldirections=[horizontal, vertical],legend='C[C]'):

> p[4]:=textplot([1.5,.9,k1=1.0*Unit(min^(-1))]);

$$p_4 := PLOT(...)$$

> p[5]:=textplot([1.5,.7,k2=0*Unit(min^(-1))]);

$$p_5 := PLOT(...)$$

> p[6]:=textplot([4,.6,k3=1.0*Unit(min^(-1))]);

$$p_6 := PLOT(...)$$

> p[7]:=textplot([4,.4,k4=0*Unit(min^(-1))]);

$$p_7 := PLOT(...)$$

> display({seq(p[i],i=1..7)},title="Figure 2.6 Concentrations of A, B, and C as functions of time.",axes=boxed);

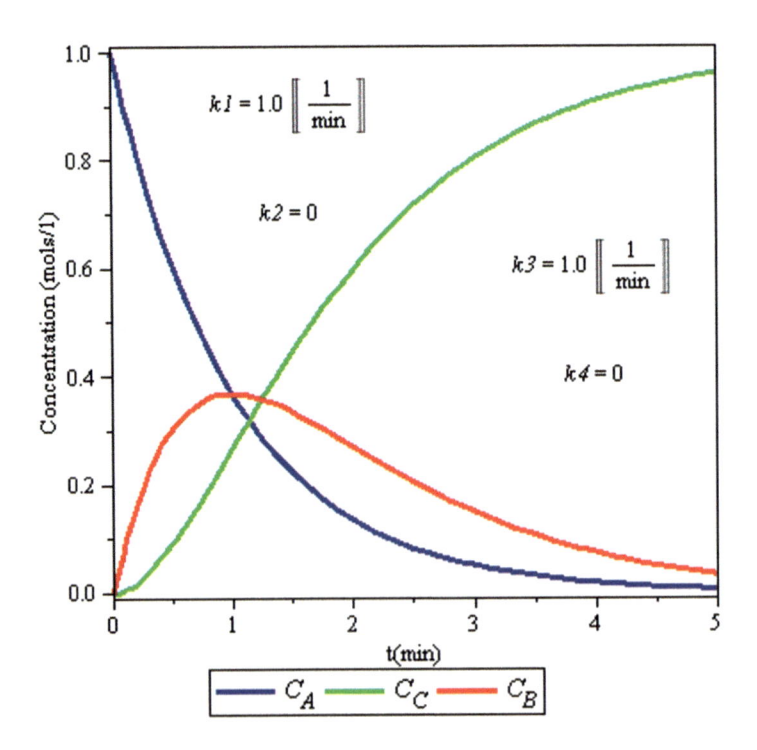

Fig. 2.6 Concentrations of A, B, and C as functions of time

2.1 Linear Ordinary Differential Equations

As can be seen by comparison, Figure 2.6 is the same as Figure 2.2 as expected.

2.1.5 Nonhomogeneous Linear ODEs

A system of n nonhomogeneous first order linear ODEs can be written in matrix forms as follows:

$$\frac{d\mathbf{Y}}{dt} = \mathbf{AY} + \mathbf{b}(t) \tag{2.11}$$

where $\mathbf{b}(t)$ is an $n \times 1$ forcing function matrix, which makes the equation nonhomogeneous. The solution to this matrix differential equation is given by[2], [3], [6], [7]

$$\mathbf{Y} = \exp(\mathbf{A}t)\mathbf{Y}_0 + \int_0^t \exp[-\mathbf{A}(\tau - t)]\,\mathbf{b}(\tau)\,d\tau \tag{2.12}$$

(see Appendix A for a detailed derivation of equation (2.12)).

When $\mathbf{b}$ is a constant the vector equation (2.12) can be simplified by removing $\mathbf{b}$ from under the integral:

$$\mathbf{Y} = \exp(\mathbf{A}t)\mathbf{Y}_0 + \left[\int_0^t \exp[-\mathbf{A}(\tau - t)]\,d\tau\right]\mathbf{b} \tag{2.13}$$

Equation (2.13) can be integrated to obtain

$$\mathbf{Y} = \exp(\mathbf{A}t)\mathbf{Y}_0 + \left[-\exp[-\mathbf{A}(\tau - t)]\right]_0^t \mathbf{A}^{-1}\mathbf{b} \tag{2.14}$$

Equation (2.14) can be expanded to read:

$$\mathbf{Y} = \exp(\mathbf{A}t)(\mathbf{Y}_0 + \mathbf{A}^{-1}\mathbf{b}) - \mathbf{A}^{-1}\mathbf{b} \tag{2.15}$$

Equations (2.12) and (2.15) simplify to equation (2.7) (the homogeneous equations solution) when the forcing function $\mathbf{b}$ vector is the zero vector. The procedure for solving nonhomogeneous linear ODEs is presented next.

Calculation Procedure for Nonhomogeneous, Linear ODEs

1. Start Maple with a 'restart' command to clear all variables.
2. Call 'with(linalg)' and 'with(plots)' commands.

3. Enter the governing equations and store them in eq[1], eq[2], etc.
4. Store the variables are stored as an array in vars.
5. Store he right hand sides of eq[1], eq[2], etc. in eqs.
6. Use Maple's 'genmatrix' command to generate the **A** and **b** matrices from eqs and vars.
7. Note that Maple generates **A** as needed and the **B** vector that satisfies **Ax** = **B**, so one has to add a minus sign to find the **b** vector that is needed for equation (2.11) (*i.e.*, **b**=−**B**).
8. When **b** is a constant vector, equation (2.15) can be used to obtain the solution.
9. When **b**(t) is a function of time, equation (2.12) is used to obtain the solution.
10. Store the initial conditions for the dependent variables in the vector **Y0**.
11. Use Maple to obtain the exponential matrix (exp(**A**t)) as a function of the parameters and the independent variable (t), store as mat.
12. Find the solution vector (sol) by multiplying mat by **Y0** and adding the nonhomogeneous solution according to equation (2.12) or (2.15) depending on whether or not **b** is a constant vector.
13. The first row of sol corresponds to the first dependent variable; the second row corresponds to the second dependent variable, etc.
14. Note that analytical solutions are obtained as functions of the parameters and the independent variable (t).
15. For a given set of parameters, one can plot the profiles by using Maple's 'plot' command.

This procedure is illustrated in Examples 2.3 and 2.4.

Heated Tanks in a Series

The exponential matrix method can be used to determine the temperatures in a series of three heated tanks used to preheat a multi-component oil solution. The energy balance equations for the tanks are as follows:[8]

$$\frac{dT_1}{dt} = \frac{W}{M}(T_0 - T_1) + \frac{UA}{MC_p}(T_{steam} - T_1)$$

$$\frac{dT_2}{dt} = \frac{W}{M}(T_1 - T_2) + \frac{UA}{MC_p}(T_{steam} - T_2) \quad (2.16)$$

$$\frac{dT_3}{dt} = \frac{W}{M}(T_2 - T_3) + \frac{UA}{MC_p}(T_{steam} - T_3)$$

where T_1, T_2 and T_3 are the temperatures in °C in Tanks 1, 2, and 3; T_{steam} is the temperature of the saturated steam (250°C) used to heat the tanks and T_0 (20°C) is

2.1 Linear Ordinary Differential Equations

the temperature of the oil fed to the first tank; W is the mass flow rate; M is the mass of the fluid in the tank; C_p is the specific heat capacity of the oil; U is the overall heat transfer coefficient; and A is the heat transfer area in each tank. We introduce two parameters $\alpha = \dfrac{W}{M}$ and $\beta = \dfrac{UA}{MC_p}$ and values for T_{steam} and T_0 into equations (2.16) to obtain the following equations.

$$\frac{dT_1}{dt} = \alpha(20 - T_1) + \beta(250 - T_1)$$

$$\frac{dT_2}{dt} = \alpha(T_1 - T_2) + \beta(250 - T_2) \qquad (2.17)$$

$$\frac{dT_3}{dt} = \alpha(T_2 - T_3) + \beta(250 - T_3)$$

All the tanks are at an initial temperature of 20°C. Find the time it will take for the third tank to reach 99% of its steady state value. The values of the constants are W = 100 kg/min, M = 1000 kg, C_p = 2kJ/kg°C, and UA = 10kJ/min°C. Determine how this time varies with the parameters α and β. Equations (equation (2.17)) can be solved using Maple and the procedure described above for nonhomogeneous simultaneous linear ODEs follows.

Example 2.3. Heating of Fluid in a Series of Tanks

> restart:

> with(linalg):with(plots):

Enter the number of differential equations.

> N:=3;

$$N := 3$$

Since all the equations are in the same form, a 'for' loop can be used to generate the equations.

> for i to N do eq[i]:=diff(T[i](t),t)=eval(alpha*(T[i-1](t)-T[i](t))+beta*(T[steam]-T[i](t)));od;

$$eq_1 := \frac{d}{dt} T_1(t) = \alpha \left(T_0(t) - T_1(t) \right) + \beta \left(T_{steam} - T_1(t) \right)$$

$$eq_2 := \frac{d}{dt} T_2(t) = \alpha \left(T_1(t) - T_2(t) \right) + \beta \left(T_{steam} - T_2(t) \right)$$

$$eq_3 := \frac{d}{dt} T_3(t) = \alpha \left(T_2(t) - T_3(t) \right) + \beta \left(T_{steam} - T_3(t) \right)$$

Next, enter the temperature of the inlet steam to the first tank T_0 and the temperature of the steam 'Tsteam' as parameters (pars).

> pars:={T[0](t)=20,T[steam]=250};

$$pars := \{ T_0(t) = 20, T_{steam} = 250 \}$$

Specify the dependent variables.

> vars:=[seq(T[i](t),i=1..N)];

$$vars := \left[T_1(t), T_2(t), T_3(t) \right]$$

Specify the right hand sides of the governing equations.

> eqs:=[seq(rhs(eq[i]),i=1..N)];

$$eqs := \left[\alpha \left(T_0(t) - T_1(t) \right) + \beta \left(T_{steam} - T_1(t) \right), \alpha \left(T_1(t) - T_2(t) \right) \right. \\ \left. + \beta \left(T_{steam} - T_2(t) \right), \alpha \left(T_2(t) - T_3(t) \right) + \beta \left(T_{steam} - T_3(t) \right) \right]$$

Update these right hand sides by substituting the parameters:

> eqs:=subs(pars,eqs);

$$eqs := \left[\alpha \left(20 - T_1(t) \right) + \beta \left(250 - T_1(t) \right), \alpha \left(T_1(t) - T_2(t) \right) \right. \\ \left. + \beta \left(250 - T_2(t) \right), \alpha \left(T_2(t) - T_3(t) \right) + \beta \left(250 - T_3(t) \right) \right]$$

Generate the coefficient matrix **A** and the vector **B** (recall that **b** = −**B**):

> A:=genmatrix(eqs,vars,'B');

$$A := \begin{bmatrix} -\alpha - \beta & 0 & 0 \\ \alpha & -\alpha - \beta & 0 \\ 0 & \alpha & -\alpha - \beta \end{bmatrix}$$

> evalm(B);

$$\begin{bmatrix} -20\alpha - 250\beta & -250\beta & -250\beta \end{bmatrix}$$

The forcing function vector **b** is the negative of the **B** vector.

2.1 Linear Ordinary Differential Equations

> b:=matrix(N,1);for i to N do b[i,1]:=-B[i];od:evalm(b);

$$b := array(1..3, 1..1, [\ ])$$

$$\begin{bmatrix} 20\alpha + 250\beta \\ 250\beta \\ 250\beta \end{bmatrix}$$

Next, find the exponential matrix for the coefficient matrix **A**:

> mat:=exponential(A,t);

$$mat := \begin{bmatrix} e^{-(\alpha+\beta)t} & 0 & 0 \\ \alpha t e^{-(\alpha+\beta)t} & e^{-(\alpha+\beta)t} & 0 \\ \frac{1}{2}\alpha^2 t^2 e^{-(\alpha+\beta)t} & \alpha t e^{-(\alpha+\beta)t} & e^{-(\alpha+\beta)t} \end{bmatrix}$$

Specify the initial tank temperatures.

> Y0:=matrix(3,1,[20,20,20]);

$$Y0 := \begin{bmatrix} 20 \\ 20 \\ 20 \end{bmatrix}$$

Since the **b** vector is the constant vector, find the intermediate vector s1.

> s1:=evalm(Y0+inverse(A)&*b);

$$s1 := \begin{bmatrix} 20 - \dfrac{20\alpha + 250\beta}{\alpha + \beta} \\ 20 - \dfrac{\alpha(20\alpha + 250\beta)}{(\alpha+\beta)^2} - \dfrac{250\beta}{\alpha+\beta} \\ 20 - \dfrac{\alpha^2(20\alpha + 250\beta)}{(\alpha+\beta)^3} - \dfrac{250\alpha\beta}{(\alpha+\beta)^2} - \dfrac{250\beta}{\alpha+\beta} \end{bmatrix}$$

Next, find the solution vector:

```
> sol:=evalm(mat&*s1-inverse(A)&*b);
```

$$sol := \left[\left[e^{-(\alpha+\beta)t}\left(20 - \frac{20\alpha + 250\beta}{\alpha+\beta}\right) + \frac{20\alpha + 250\beta}{\alpha+\beta}\right],\right.$$
$$\left[\alpha t e^{-(\alpha+\beta)t}\left(20 - \frac{20\alpha + 250\beta}{\alpha+\beta}\right) + e^{-(\alpha+\beta)t}\left(20\right.\right.$$
$$\left.- \frac{\alpha(20\alpha + 250\beta)}{(\alpha+\beta)^2} - \frac{250\beta}{\alpha+\beta}\right) + \frac{\alpha(20\alpha + 250\beta)}{(\alpha+\beta)^2}$$
$$\left.+ \frac{250\beta}{\alpha+\beta}\right],$$
$$\left[\frac{1}{2}\alpha^2 t^2 e^{-(\alpha+\beta)t}\left(20 - \frac{20\alpha + 250\beta}{\alpha+\beta}\right)\right.$$
$$+ \alpha t e^{-(\alpha+\beta)t}\left(20 - \frac{\alpha(20\alpha + 250\beta)}{(\alpha+\beta)^2} - \frac{250\beta}{\alpha+\beta}\right)$$
$$+ e^{-(\alpha+\beta)t}\left(20 - \frac{\alpha^2(20\alpha + 250\beta)}{(\alpha+\beta)^3} - \frac{250\alpha\beta}{(\alpha+\beta)^2}\right.$$
$$\left.\left.- \frac{250\beta}{\alpha+\beta}\right) + \frac{\alpha^2(20\alpha + 250\beta)}{(\alpha+\beta)^3} + \frac{250\alpha\beta}{(\alpha+\beta)^2} + \frac{250\beta}{\alpha+\beta}\right]\right]$$

The elements of the solution vector 'sol' are the desired temperatures of the three tanks. Store these in T1, T2, and T3:

```
> T1:=sol[1,1];
```

$$T1 := e^{-(\alpha+\beta)t}\left(20 - \frac{20\alpha + 250\beta}{\alpha+\beta}\right) + \frac{20\alpha + 250\beta}{\alpha+\beta}$$

```
> T2:=sol[2,1];
```

$$T2 := \alpha t e^{-(\alpha+\beta)t}\left(20 - \frac{20\alpha + 250\beta}{\alpha+\beta}\right) + e^{-(\alpha+\beta)t}\left(20\right.$$
$$\left.- \frac{\alpha(20\alpha + 250\beta)}{(\alpha+\beta)^2} - \frac{250\beta}{\alpha+\beta}\right) + \frac{\alpha(20\alpha + 250\beta)}{(\alpha+\beta)^2}$$
$$+ \frac{250\beta}{\alpha+\beta}$$

2.1 Linear Ordinary Differential Equations

> T3:=sol[3,1];

$$T3 := \frac{1}{2}\alpha^2 t^2 e^{-(\alpha+\beta)t}\left(20 - \frac{20\alpha + 250\beta}{\alpha+\beta}\right)$$
$$+ \alpha t e^{-(\alpha+\beta)t}\left(20 - \frac{\alpha(20\alpha+250\beta)}{(\alpha+\beta)^2} - \frac{250\beta}{\alpha+\beta}\right)$$
$$+ e^{-(\alpha+\beta)t}\left(20 - \frac{\alpha^2(20\alpha+250\beta)}{(\alpha+\beta)^3} - \frac{250\alpha\beta}{(\alpha+\beta)^2}\right.$$
$$\left. - \frac{250\beta}{\alpha+\beta}\right) + \frac{\alpha^2(20\alpha+250\beta)}{(\alpha+\beta)^3} + \frac{250\alpha\beta}{(\alpha+\beta)^2} + \frac{250\beta}{\alpha+\beta}$$

Next, determine values for α and β and then substitute them into the expression for T1, T2, and T3.

> pars:={alpha=(100./1000.),beta=10./(1000*2)};

$$pars := \{\beta = 0.005000000000, \alpha = 0.1000000000\}$$

> TT1:=subs(pars,T1);

$$TT1 := -10.95238095 e^{-0.1050000000\, t} + 30.9523809$$

> TT2:=subs(pars,T2);

$$TT2 := -1.095238095 t\, e^{-0.1050000000\, t} - 21.38321995 e^{-0.1050000000\, t} + 41.38321995$$

> TT3:=subs(pars,T3);

$$TT3 := -0.05476190475 t^2\, e^{-0.1050000000\, t}$$
$$- 2.138321995 t\, e^{-0.1050000000\, t} - 31.31735233 e^{-0.1050000000\, t} + 51.31735233$$

Finally, plot the temperature profiles:

> p[1]:=plot(TT1,t=0..60,thickness=2,linestyle=1,color=blue,legend="T1"):

> p[2]:=plot(TT2,t=0..60,thickness=2,linestyle=1,legend="T2"):

> p[3]:=plot(TT3,t=0..60,thickness=2,linestyle=1,color=green,legend="T3"):

>display({seq(p[i],i=1..3)},title="Figure 2.7 Temperatures for three tanks.",axes=boxed,

labels=["t(min)","Temperature(degreesC)"],labeldirections=[HORIZONTAL, VERTICAL]);

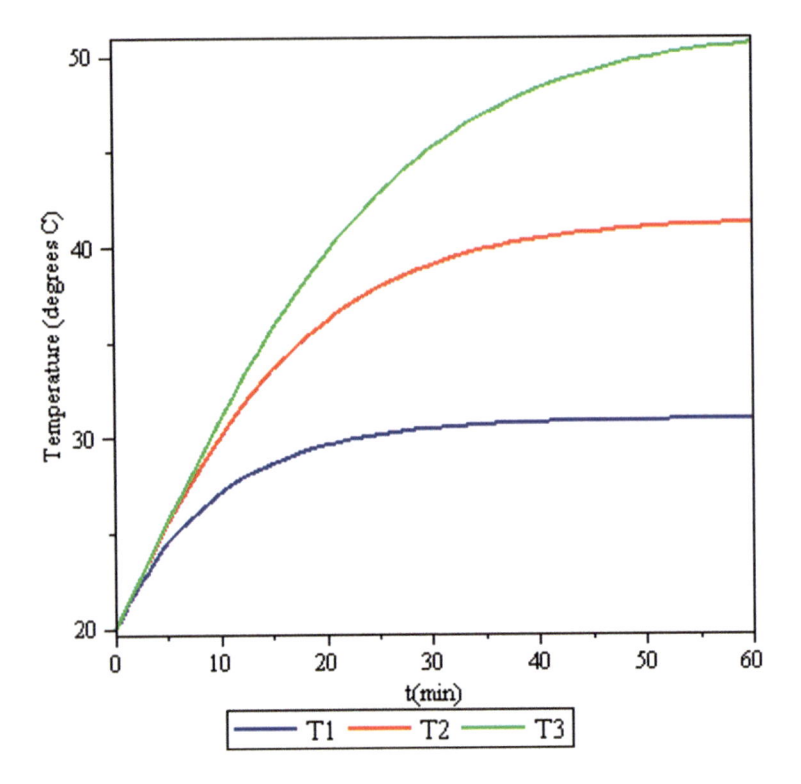

Fig. 2.7 Temperatures for three tanks

The time taken for the third tank to reach 99% of the steady state value can be obtained from the TT3 equation by first finding its value for infinity by using Maple's 'limit' command:

> T3steadystate:=limit(TT3,t=infinity);

$$T3steadystate := 51.3173523.$$

Next, the time taken to reach 99% of this steady state value is determined by using 'fsolve' to find the time T when TT3 is 99% of it steady state value. First, define eq3:

> eq3:=0.99*T3steadystate=TT3;

$$eq3 := 50.80417881 = -0.05476190475 t^2 \, e^{-0.1050000000\, t}$$
$$- 2.138321995 t \, e^{-0.1050000000\, t} - 31.31735233 e^{-0.1050000000\, t}$$
$$+ 51.31735233$$

Next, solve eq3 for t and call that Timereqd.

> Timereqd:=fsolve(eq3,t);

$$Timereqd := 63.0143955.$$

2.1 Linear Ordinary Differential Equations

Alternatively, one can find the steady state value of the temperature analytically by applying the 'limit' command on T3 instead of TT3.

> T3steadystate:=limit(T3,t=infinity);

$$T3steadystate := \lim_{t \to \infty} \left(\frac{1}{2} \alpha^2 t^2 e^{-(\alpha+\beta)t} \left(20 - \frac{20\alpha + 250\beta}{\alpha + \beta} \right) \right.$$
$$+ \alpha t e^{-(\alpha+\beta)t} \left(20 - \frac{\alpha(20\alpha + 250\beta)}{(\alpha+\beta)^2} - \frac{250\beta}{\alpha+\beta} \right)$$
$$+ e^{-(\alpha+\beta)t} \left(20 - \frac{\alpha^2(20\alpha + 250\beta)}{(\alpha+\beta)^3} - \frac{250\alpha\beta}{(\alpha+\beta)^2} \right.$$
$$\left. - \frac{250\beta}{\alpha+\beta} \right) + \frac{\alpha^2(20\alpha + 250\beta)}{(\alpha+\beta)^3} + \frac{250\alpha\beta}{(\alpha+\beta)^2} + \frac{250\beta}{\alpha+\beta} \right)$$

Maple is unable to find the limit, as it does not know the sign of alpha and beta. We can specify that alpha and beta are greater than zero by using Maple's 'assume' command:

> assume(alpha>0,beta>0);

> T3steadystate:=eval(T3steadystate);

$$T3steadystate := \frac{750\,\alpha\mathtt{\sim}^2\,\beta\mathtt{\sim} + 750\,\alpha\mathtt{\sim}\,\beta\mathtt{\sim}^2 + 250\,\beta\mathtt{\sim}^3 + 20\,\alpha\mathtt{\sim}^3}{\alpha\mathtt{\sim}^3 + 3\,\alpha\mathtt{\sim}^2\,\beta\mathtt{\sim} + 3\,\alpha\mathtt{\sim}\,\beta\mathtt{\sim}^2 + \beta\mathtt{\sim}^3}$$

Note that the trailing tildes (tildes symbol) indicate that the α and β must be greater than zero.

Time Varying Input to a CSTR with a Series Reaction

Consider a continuously stirred tank reactor (CSTRs) sustaining the series reaction ($A \xrightarrow{k_1} B \xrightarrow{k_2} C$).[1], [9] The material balance equations are as follows:

$$\frac{dC_A}{dt} = \frac{1}{\tau}(C_{A,in} - C_A) - k_1 C_A$$

$$\frac{dC_B}{dt} = -\frac{1}{\tau} C_B + k_1 C_A - k_2 C_B \qquad (2.18)$$

$$\frac{dC_C}{dt} = -\frac{1}{\tau} C_C + k_2 C_B$$

with the initial conditions $C_A(0) = C_B(0) = C_C(0) = 0$ mol/l. Let the input to the tank be $C_{A,in} = 1 + \sin(2t)$ in mol/l−min). The values of the parameters are $\tau = 1$ min; $k_1 = 1$ min^{-1} and $k_2 = 1/4$ min^{-1}. The concentrations as a function of time in this CSTR reactor can be obtained by following the procedure described earlier for non-homogeneous linear ODEs.

Example 2.4. Time Varying Input to a CSTR with a Series Reaction

> restart:

> with(linalg):with(plots):with(Units):

Specify the number of ordinary differential equations (ODEs).

> N:=3;

$$N := 3$$

Use Maple commands to generate the three governing equations.

> eq[1]:=diff(C[A](t),t)=1/tau*(C[Ain]-C[A](t))-k[1]*C[A](t);

$$eq_1 := \frac{d}{dt} C_A(t) = \frac{C_{Ain} - C_A(t)}{\tau} - k_1 C_A(t)$$

> eq[2]:=diff(C[B](t),t)=-1/tau*(C[B](t))+k[1]*C[A](t)-k[2]*C[B](t);

$$eq_2 := \frac{d}{dt} C_B(t) = -\frac{C_B(t)}{\tau} + k_1 C_A(t) - k_2 C_B(t)$$

> eq[3]:=diff(C[C](t),t)=-1/tau*(C[C](t))+k[2]*C[B](t);

$$eq_3 := \frac{d}{dt} C_C(t) = -\frac{C_C(t)}{\tau} + k_2 C_B(t)$$

Specify the input to the reactor as pars here.

> pars:={C[Ain]=1+sin(2*t)};

$$pars := \{C_{Ain} = 1 + \sin(2t)\}$$

Name the dependent variables.

> vars:=[C[A](t),C[B](t),C[C](t)];

$$vars := [C_A(t), C_B(t), C_C(t)]$$

Specify the right hand sides of the governing equations.

2.1 Linear Ordinary Differential Equations

> eqs:=[seq(rhs(eq[i]),i=1..3)];

$$eqs := \left[\frac{C_{Ain} - C_A(t)}{\tau} - k_1 C_A(t), -\frac{C_B(t)}{\tau} + k_1 C_A(t) - k_2 C_B(t), \right.$$
$$\left. -\frac{C_C(t)}{\tau} + k_2 C_B(t) \right]$$

Form the set of right hand sides of the equations with the time dependent input to the tank.

> eqs:=subs(pars,eqs);

$$eqs := \left[\frac{1 + \sin(2t) - C_A(t)}{\tau} - k_1 C_A(t), -\frac{C_B(t)}{\tau} + k_1 C_A(t) \right.$$
$$\left. - k_2 C_B(t), -\frac{C_C(t)}{\tau} + k_2 C_B(t) \right]$$

Use 'eqs' and 'vars' to find the coefficient matrix **A** and the forcing function vector **b**. In this case, we find **b1** and then **b** since **b** = −**b**. The **A** matrix and **b1** vector are generated as described in the procedure for nonlinear homogeneous ODEs:

> A:=genmatrix(eqs,vars,'b1');

$$A := \begin{bmatrix} -\frac{1}{\tau} - k_1 & 0 & 0 \\ k_1 & -\frac{1}{\tau} - k_2 & 0 \\ 0 & k_2 & -\frac{1}{\tau} \end{bmatrix}$$

Use **b1** to find **b**.

> b:=matrix(N,1);for i to N do b[i,1]:=-b1[i];od:evalm(b);

$$b := array(1..3, 1..1, [\,])$$

$$\begin{bmatrix} \frac{1 + \sin(2t)}{\tau} \\ 0 \\ 0 \end{bmatrix}$$

In this case the forcing function vector **b** depends on time, which means we will need to use

$$Y = \exp(At)Y_0 + \int_0^t \exp[-A(\tau-t)]\, b(\tau)\, d\tau$$

to find the solution vector. Next, find the matrix exponential of **A**.

> mat:=exponential(A,t);

$$mat := \left[\left[e^{-\frac{(1+k_1\tau)t}{\tau}}, 0, 0\right],\right.$$

$$\left[-\frac{k_1\left(-e^{-\frac{(1+k_2\tau)t}{\tau}} + e^{-\frac{(1+k_1\tau)t}{\tau}}\right)}{-k_2 + k_1}, e^{-\frac{(1+k_2\tau)t}{\tau}}, 0\right],$$

$$\left[\frac{-k_1 e^{-\frac{(1+k_2\tau)t}{\tau}} + k_2 e^{-\frac{(1+k_1\tau)t}{\tau}} - e^{-\frac{t}{\tau}}k_2 + e^{-\frac{t}{\tau}}k_1}{-k_2 + k_1},\right.$$

$$\left.\left.-e^{-\frac{(1+k_2\tau)t}{\tau}} + e^{-\frac{t}{\tau}}, e^{-\frac{t}{\tau}}\right]\right]$$

Specify the initial conditions.

> Y0:=matrix(3,1,[0,0,0]);

$$Y0 := \begin{bmatrix} 0 \\ 0 \\ 0 \end{bmatrix}$$

Since τ is a parameter in the system of governing equations, we use t1 as the dummy integration variable in

$$Y = \exp(At)Y_0 + \int_0^t \exp[-A(\tau-t)]\, b(\tau)\, d\tau$$

We will need to generate a **b** vector that depends on time. Call this vector **b2**.

> b2:=subs(t=t1,evalm(b));

$$b2 := \begin{bmatrix} \dfrac{1 + \sin(2\, t1)}{\tau} \\ 0 \\ 0 \end{bmatrix}$$

2.1 Linear Ordinary Differential Equations

The matrix exponential under the integral sign the above equation is obtained next and named mat2.

> mat2:=subs(t=t-t1,evalm(mat)):

> mat3:=evalm(mat2&*b2):

> mat4:=map(int,mat3,t1=0..t):

> sol:=evalm(mat&*Y0+mat4):

The solution is not printed for brevity. Store the concentration expressions Ca, Cb, and Cc.

> ca:=sol[1,1];

> cb:=sol[2,1]:

> cc:=sol[3,1]:

$$
\begin{aligned}
ca := & \Bigg(\Bigg(-1 - 2k_1\tau - k_1^2\tau^2 - 4\tau^2 + 2\tau + 2k_1\tau^2 + e^{\frac{(1+k_1\tau)t}{\tau}} \\
& + 2e^{\frac{(1+k_1\tau)t}{\tau}} k_1\tau + e^{\frac{(1+k_1\tau)t}{\tau}} k_1^2\tau^2 + 4e^{\frac{(1+k_1\tau)t}{\tau}} \tau^2 \\
& - 4e^{\frac{(1+k_1\tau)t}{\tau}} \tau \cos(t)^2 + 2e^{\frac{(1+k_1\tau)t}{\tau}} \tau \\
& - 4e^{\frac{(1+k_1\tau)t}{\tau}} k_1\tau^2 \cos(t)^2 + 2e^{\frac{(1+k_1\tau)t}{\tau}} k_1\tau^2 \\
& + 2e^{\frac{(1+k_1\tau)t}{\tau}} \sin(t)\cos(t) \\
& + 4e^{\frac{(1+k_1\tau)t}{\tau}} k_1\tau\sin(t)\cos(t) + 2e^{\frac{(1+k_1\tau)t}{\tau}} \\
& k_1^2\tau^2\sin(t)\cos(t)\Bigg)e^{-\frac{(1+k_1\tau)t}{\tau}}\Bigg) \Big/ \Big(1 + 3k_1\tau + 3k_1^2\tau^2 \\
& + 4\tau^2 + k_1^3\tau^3 + 4\tau^3 k_1\Big)
\end{aligned}
$$

> cb:=sol[2,1]:

> cc:=sol[3,1]:

Again cb and cc are not printed for brevity. Next, numerical values for the rate constants (k1 and k2) and the time constant τ are specified and substituted in the expressions for Ca, Cb, and Cc:

> pars:={tau=1,k[1]=1,k[2]=1/4};

$$pars := \left\{\tau = 1, k_1 = 1, k_2 = \frac{1}{4}\right\}$$

> Ca:=simplify(subs(pars,ca));

$$Ca := -\frac{1}{4}\left(1 - 3e^{2t} + 2e^{2t}\cos(t)^2 - 2e^{2t}\sin(t)\cos(t)\right)e^{-2t}$$

> Cb:=simplify(subs(pars,cb));

$$Cb := -\frac{1}{1335}e^{-2t}\left(390e^{2t}\cos(t)^2 + 90e^{2t}\sin(t)\cos(t) - 445 + 784e^{\frac{3}{4}t} - 729e^{2t}\right)$$

> Cc:=simplify(subs(pars,cc));

$$Cc := -\frac{1}{5340}e^{-2t}\left(42e^{2t}\cos(t)^2 + 174e^{2t}\sin(t)\cos(t) + 445 - 3136e^{\frac{3}{4}t} + 3204e^t - 555e^{2t}\right)$$

Next, plot the concentration profiles. Specify the time you want to use to plot the results.

> tf:=20;

$$tf := 20$$

2.1 Linear Ordinary Differential Equations

> p1:=plot(Ca,t=0..tf,thickness=3,color=blue,linestyle=1,labels=["t(min)","Ca(mol/1)"],labeldirections=[HORIZONTAL,VERTICAL],title="Figure 2.8 Concentraiton of species A as a function of time.",axes=boxed);

$$p1 := PLOT(...)$$

> pt1:=textplot([10,0.1,{tau=1.0*Unit(min),k[1]=1.0*Unit(min^(-1)),k[2]=0.25*Unit(min^(-1))}]);

$$pt1 := PLOT(...)$$

> display([p1,pt1]);

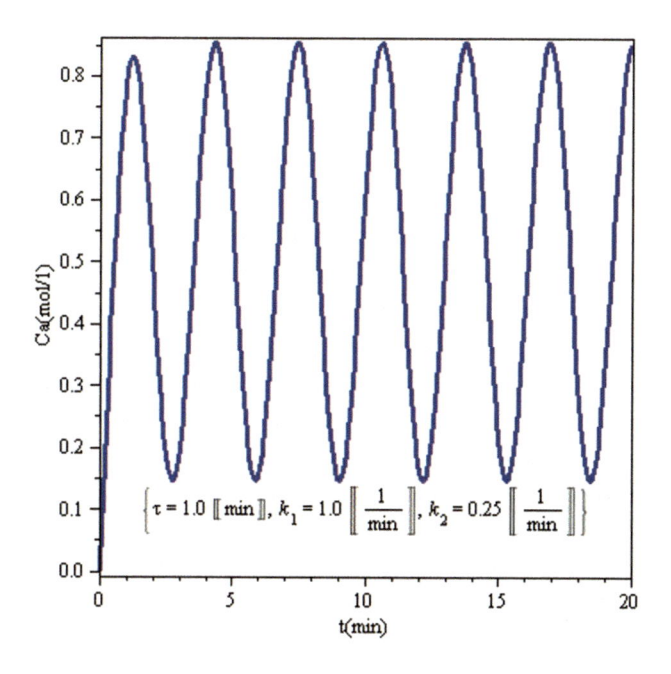

Fig. 2.8 Concentration of species A as a function of time

> p2:=plot(Cb,t=0..tf,thickness=3,color=red,linestyle=1,labels=["t(min)","Cb(mol/1)"],
labeldirections=[HORIZONTAL,VERTICAL],title="Figure 2.9 Concentration of species B as a function of time.",axes=boxed);

$$p2 := PLOT(...)$$

> pt1:=textplot([10,0.1,[tau=1.0*Unit(min),k[1]=1.0*Unit(min^(-1)),k[2]=0.25*Unit(min^(-1))]]);

$$pt1 := PLOT(...)$$

> display([p2,pt1]);

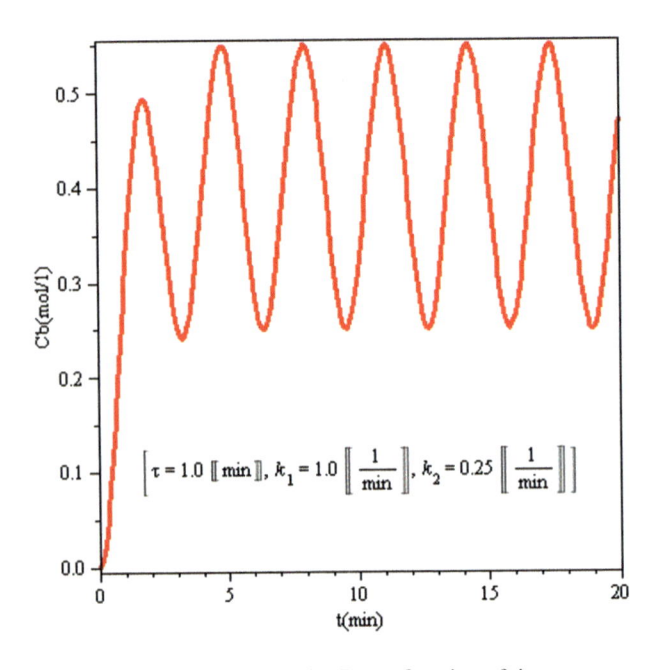

Fig. 2.9 Concentration of species B as a function of time

2.1 Linear Ordinary Differential Equations

```
> p3:=plot(Cc,t=0..tf,thickness=3,color=green,linestyle=1,labels=["t(min)","
Cc(mol/1)"],
labeldirections=[HORIZONTAL,VERTICAL],title="Figure 2.10
Concentration of species C as a function of time.",axes=boxed);
```

$$p3 := PLOT(...)$$

```
> pt3:=textplot([10,0.02,[tau=1.0*Unit(min),k[1]=1.0*Unit(min^(-1)),k[2]=
0.25*Unit(min^(-1))]]);
```

$$pt3 := PLOT(...)$$

```
> display([p3,pt3]);
```

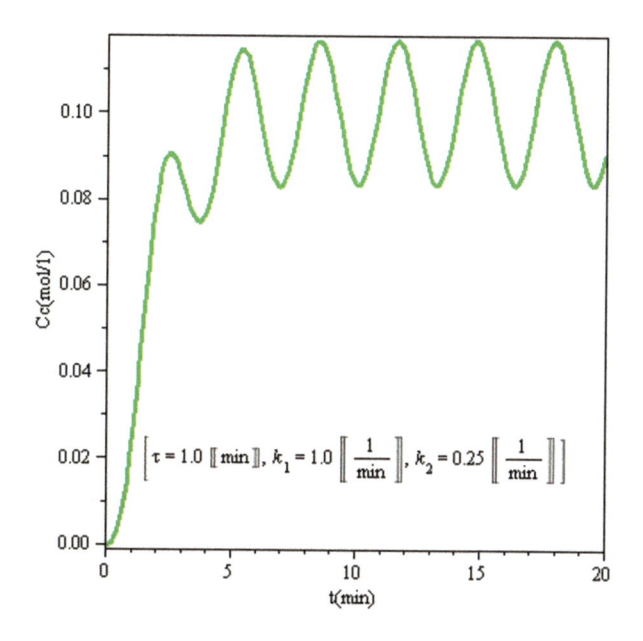

Fig. 2.10 Concentration of species C as a function of time

We observe that all the concentrations start from zero and after about six minutes oscillate at.

2.1.6 Higher Order Linear Ordinary Differential Equations

Higher order linear ODEs can also be solved by changing them into a system of first order ODEs and using the exponential matrix approach discussed earlier. The most general form of a linear ODE of nth order is[1]

$$\frac{d^n y}{dt^n} + a_{n-1}\frac{d^{n-1} y}{dt^{n-1}} + \ldots a_1 \frac{dy}{dt} + a_0 y = f(t) \qquad (2.19)$$

Introducing the variable transformations,

$$Y_1 = y, \; Y_2 = \frac{dY_1}{dt} = \frac{dy}{dt},\ldots, Y_n = \frac{dY_{n-1}}{dt} = \frac{d^{n-1} y}{dt^{n-1}} \qquad (2.20)$$

yields the following n first order ODEs

$$\frac{dY_1}{dt} = Y_2$$

$$\frac{dY_2}{dt} = Y_3$$

$$\ldots\ldots \qquad (2.21)$$

$$\frac{dY_{n-1}}{dt} = Y_n$$

and

$$\frac{dY_n}{dt} = -a_0 Y_1 - a_1 Y_2 \ldots\ldots - a_{n-2} Y_{n-1} - a_{n-1} Y_n + f(t)$$

differential equations where the dependent variables are

$$Y = [Y_1, Y_2, Y_3, \ldots Y_{n-1}, Y_n]^T$$
$$= \left[y, \frac{dy}{dt}, \frac{d^2 y}{dt^2}, \ldots \frac{d^{n-2} y}{dt^{n-2}}, \frac{d^{n-1} y}{dt^{n-1}}\right]^T \qquad (2.22)$$

the coefficient matrix is

$$A = \begin{bmatrix} 0 & 1 & 0 & 0 & \ldots & 0 \\ 0 & 0 & 1 & 0 & \ldots & 0 \\ \cdot & \cdot & \cdot & \cdot & \ldots & \cdot \\ \cdot & \cdot & \cdot & \cdot & \ldots & \cdot \\ 0 & 0 & 0 & \ldots & 0 & 1 \\ -a_0 & -a_1 & -a_2 & -a_3 & \ldots & -a_{n-1} \end{bmatrix} \qquad (2.23)$$

2.1 Linear Ordinary Differential Equations

and the forcing function matrix is

$$b = [0, 0, 0, ...0, 0, f(t)]^T \qquad (2.24)$$

Again, the solution is obtained by finding the exponential matrix and the non-homogeneous part (see equation 2.12). The procedure used to solve higher order linear ODEs can be summarized as follows:

1. Start the Maple program with a 'restart' command to clear all variables.
2. Call 'with(linalg)' and 'with(plots)' commands.
3. Enter the governing equations.
4. Enter the coefficient matrix (**A**) based on equation (2.23).
5. Enter the forcing function matrix (**b**) based on equation (2.24).
6. Store the initial conditions for the dependent variables in the vector Y0.
7. The first row of **Y0** corresponds to the initial condition for y, the second row corresponds to the initial condition for the first derivative of y, etc.
8. Find the matrix exponential (exp(At)) as a function of the parameters and the independent variable (t) using the 'exponential(A,t)' command in Maple.
9. Store the matrix exponential in mat. Next, find the solution (sol) by multiplying mat with **Y0** and adding the non-homogenous solution according to equation (2.14) or (2.15) depending on the b vector.
10. The first row of sol corresponds to the dependent variable y; the second row corresponds to the first derivative of y, etc. The nth row of sol corresponds to $(n-1)^{th}$ derivative of y.

Note that this procedure yields analytical solutions as functions of the parameters and the independent variable (t).

Second Order Ordinary Differential Equation (ODE)

Consider a second order system subject to a k step input,[10]

$$\frac{d^2y}{dt^2} + \frac{2\varsigma}{\tau}\frac{dy}{dt} + \frac{1}{\tau^2}y = \frac{k}{\tau^2} \qquad (2.25)$$

with the initial conditions $y(0) = 0$ and $\frac{dy}{dt}(0) = 0$, and τ, δ, and k are constant parameters. Equation (2.25) can be solved by following the procedure described for higher order linear ODEs.

Example 2.5 A Second Order ODE

> restart:

> with(linalg):with(plots):

Enter the order of the differential equation.

> N:=2;

$$N := 2$$

> eq:=diff(y(t),t$2)+2*zeta/tau*diff(y(t),t)+1/tau^2*y(t)=k/tau^2;

$$eq := \frac{d^2}{dt^2} y(t) + \frac{2\zeta \left(\frac{d}{dt} y(t)\right)}{\tau} + \frac{y(t)}{\tau^2} = \frac{k}{\tau^2}$$

Enter the corresponding A matrix and b vector.
> A:=matrix(N,N,[0,1,-1/tau^2,-2*zeta/tau]);

$$A := \begin{bmatrix} 0 & 1 \\ -\frac{1}{\tau^2} & -\frac{2\zeta}{\tau} \end{bmatrix}$$

Enter the forcing function (i.e., the b matrix).
> b:=matrix(N,1,[0,k/tau^2]);

$$b := \begin{bmatrix} 0 \\ \frac{k}{\tau^2} \end{bmatrix}$$

Find the matrix exponential of the coefficient matrix.
> mat:=exponential(A,t);

$$mat := \left[\left[-\frac{1}{2} \frac{1}{\sqrt{\zeta^2-1}} \left(-e^{-\frac{\left(\zeta-\sqrt{\zeta^2-1}\right)t}{\tau}} \sqrt{\zeta^2-1}\right.\right.\right.$$

$$+ \zeta e^{-\frac{\left(\zeta+\sqrt{\zeta^2-1}\right)t}{\tau}} - \zeta e^{-\frac{\left(\zeta-\sqrt{\zeta^2-1}\right)t}{\tau}}$$

$$\left. -e^{-\frac{\left(\zeta+\sqrt{\zeta^2-1}\right)t}{\tau}} \sqrt{\zeta^2-1}\right),$$

$$\left.-\frac{1}{2} \frac{\tau\left(e^{-\frac{\left(\zeta+\sqrt{\zeta^2-1}\right)t}{\tau}} - e^{-\frac{\left(\zeta-\sqrt{\zeta^2-1}\right)t}{\tau}}\right)}{\sqrt{\zeta^2-1}}\right],$$

2.1 Linear Ordinary Differential Equations

$$\left[\frac{1}{2} \frac{e^{-\frac{\left(\zeta+\sqrt{\zeta^2-1}\right)t}{\tau}} - e^{-\frac{\left(\zeta-\sqrt{\zeta^2-1}\right)t}{\tau}}}{\tau\sqrt{\zeta^2-1}}, \right.$$

$$\frac{1}{2}\frac{1}{\sqrt{\zeta^2-1}}\left(e^{-\frac{\left(\zeta-\sqrt{\zeta^2-1}\right)t}{\tau}}\sqrt{\zeta^2-1}\right.$$

$$+\zeta e^{-\frac{\left(\zeta+\sqrt{\zeta^2-1}\right)t}{\tau}} - \zeta e^{-\frac{\left(\zeta-\sqrt{\zeta^2-1}\right)t}{\tau}}$$

$$\left.\left.+ e^{-\frac{\left(\zeta+\sqrt{\zeta^2-1}\right)t}{\tau}}\sqrt{\zeta^2-1}\right)\right]$$

Specify the initial conditions.

> Y0:=matrix(N,1,[0,0]);

$$Y0 := \begin{bmatrix} 0 \\ 0 \end{bmatrix}$$

In this case, b is a constant vector, but we can use an equation that will work even when the b vector is a function of time. Recall that τ is a parameter (equation (2.12)). We will use b2 with the dummy integration variable t1 (even though it is not needed) to illustrate the procedure.

> b2:=subs(t=t1,evalm(b));

$$b2 := \begin{bmatrix} 0 \\ \dfrac{k}{\tau^2} \end{bmatrix}$$

The next step is to generate the matrix exponential under the integral sign (see equation (2.12)) and name this matrix mat2.

> mat2:=subs(t=t-t1,evalm(mat)):

Next, multiply the vector b2 by mat2 to obtain the vector to be integrated. Call this vector mat3.

> mat3:=evalm(mat2&*b2):

The next step is to use Maple's 'map' command to integrate the elements in mat3. Call the result mat4.

> mat4:=map(int,mat3,t1=0..t):

Obtain the solution vector.

> sol:=evalm(mat&*Y0+mat4):

> sol:=map(simplify,sol);

$$sol := \left[\left[-\frac{1}{2}\frac{1}{\sqrt{\zeta^2-1}}\left(k\left(-\zeta+\sqrt{\zeta^2-1}+\zeta e^{\frac{2\sqrt{\zeta^2-1}\,t}{\tau}}\right.\right.\right.\right.$$
$$\left.\left.+e^{\frac{2\sqrt{\zeta^2-1}\,t}{\tau}}\sqrt{\zeta^2-1}-2e^{\frac{(\zeta+\sqrt{\zeta^2-1})t}{\tau}}\sqrt{\zeta^2-1}\right)\right.$$
$$\left.\left.e^{-\frac{(\zeta+\sqrt{\zeta^2-1})t}{\tau}}\right]\right],$$

$$\left[\frac{1}{2}\frac{k\left(-1+e^{\frac{2\sqrt{\zeta^2-1}\,t}{\tau}}\right)e^{-\frac{(\zeta+\sqrt{\zeta^2-1})t}{\tau}}}{\tau\sqrt{\zeta^2-1}}\right]$$

The first row of sol corresponds to the dependent variable y.

> y:=sol[1,1];

$$y := -\frac{1}{2}\frac{1}{\sqrt{\zeta^2-1}}\left(k\left(-\zeta+\sqrt{\zeta^2-1}+\zeta e^{\frac{2\sqrt{\zeta^2-1}\,t}{\tau}}\right.\right.$$
$$\left.+e^{\frac{2\sqrt{\zeta^2-1}\,t}{\tau}}\sqrt{\zeta^2-1}-2e^{\frac{(\zeta+\sqrt{\zeta^2-1})t}{\tau}}\sqrt{\zeta^2-1}\right)$$
$$\left.e^{-\frac{(\zeta+\sqrt{\zeta^2-1})t}{\tau}}\right)$$

2.1 Linear Ordinary Differential Equations

Next, y is made dimensionless by dividing by k.
> theta:=y/k;

$$\theta := -\frac{1}{2} \frac{1}{\sqrt{\zeta^2-1}} \left(\left(-\zeta + \sqrt{\zeta^2-1} + \zeta e^{\frac{2\sqrt{\zeta^2-1}\,t}{\tau}}\right.\right.$$
$$\left.+ e^{\frac{2\sqrt{\zeta^2-1}\,t}{\tau}} \sqrt{\zeta^2-1} - 2 e^{\frac{\left(\zeta+\sqrt{\zeta^2-1}\right)t}{\tau}} \sqrt{\zeta^2-1}\right)$$
$$\left. e^{-\frac{\left(\zeta+\sqrt{\zeta^2-1}\right)t}{\tau}}\right)$$

Next, the dimensionless time T=t/τ is introduced.
> theta:=subs(t=tau*T,theta);

$$\theta := -\frac{1}{2} \frac{1}{\sqrt{\zeta^2-1}} \left(\left(-\zeta + \sqrt{\zeta^2-1} + \zeta e^{2\sqrt{\zeta^2-1}\,T}\right.\right.$$
$$\left.+ e^{2\sqrt{\zeta^2-1}\,T} \sqrt{\zeta^2-1} - 2 e^{\left(\zeta+\sqrt{\zeta^2-1}\right)T} \sqrt{\zeta^2-1}\right)$$
$$\left. e^{-\left(\zeta+\sqrt{\zeta^2-1}\right)T}\right)$$

Now one can plot θ for values of ζ greater than 1:
> pars:=[1.5,2,2.5,3];

$$pars := [1.5, 2, 2.5, 3]$$

> for i from 1 to 4 do
 p[i]:=plot(subs(zeta=pars[i],theta),T=0..15):
od:
> pt[1]:=textplot([3.2,evalf(subs({T=4.0,zeta=pars[1]},theta)),'zeta=pars[1]']):
pt[2]:=textplot([5.4,evalf(subs({T=5.0,zeta=pars[2]},theta)),pars[2]]):
pt[3]:=textplot([6.0,evalf(subs({T=5.5,zeta=pars[3]},theta)),pars[3]]):
pt[4]:=textplot([6.5,evalf(subs({T=6.0,zeta=pars[4]},theta)),pars[4]]):
>display([seq(p[i],i=1..4),seq(pt[i],i=1..4)],thickness=3,axes=boxed,labels= ["T(t/tau)","theta (y/k)"],title="Figure 2.11 Dimensionless variables theta (y/k) as a function of dimensionless time T(t/tau) for zeta>1.");

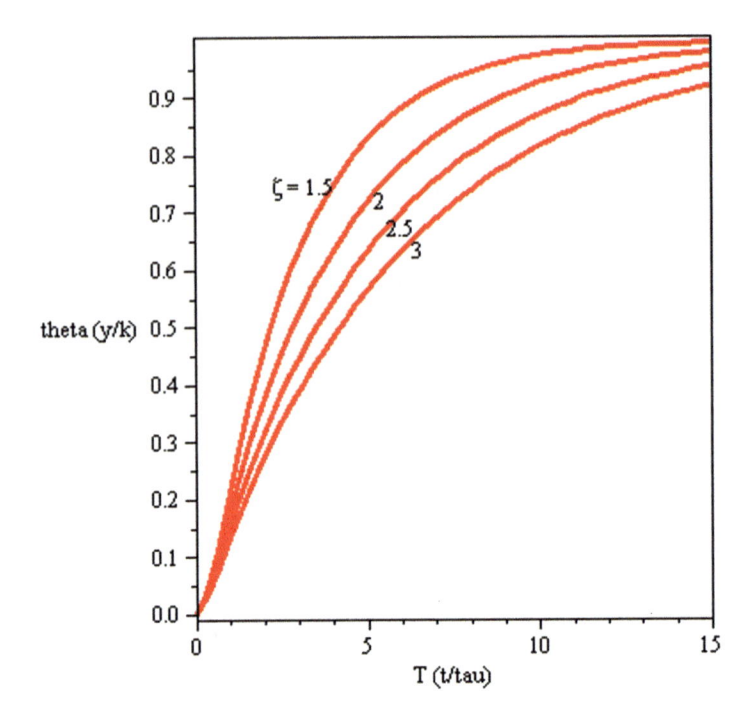

Fig. 2.11 Dimensionless variables theta (y/k) as a function of dimensionless time T(t/tau) for zeta > 1

Figure 2.11 shows that the system is over damped for the values of $\zeta > 1$. Next, specify values for $\zeta < 1$.

> spars:=[0.2,0.3,0.4,0.5,0.6,0.75];

$$spars := [0.2, 0.3, 0.4, 0.5, 0.6, 0.75]$$

> for i from 1 to 6 do p2[i]:=plot(evalf(Re(subs(zeta=spars[i],theta))),T=0..15):od:

> pt[1]:=textplot([1.3,Re(subs({T=2.8,zeta=spars[1]},theta)),'zeta=spars[1]']):

pt[2]:=textplot([3.5,Re(subs({T=2.8,zeta=spars[2]},theta)),spars[2]]):

pt[3]:=textplot([3.5,Re(subs({T=2.8,zeta=spars[3]},theta)),spars[3]]):

pt[4]:=textplot([3.5,Re(subs({T=2.8,zeta=spars[4]},theta)),spars[4]]):

pt[5]:=textplot([3.5,Re(subs({T=2.8,zeta=spars[5]},theta)),spars[5]]):

pt[6]:=textplot([3.5,Re(subs({T=2.8,zeta=spars[6]},theta)),spars[6]]):

>display([seq(p2[i],i=1..6),seq(pt[i],i=1..6)],thickness=3,axes=boxed,labels= ["T(t/tau)","theta (y/k)"],title="Figure 2.12 Dimensionless variable theta (y/k) as a function of dimensionless time T(t/tau) for zeta<1.");

2.1 Linear Ordinary Differential Equations 71

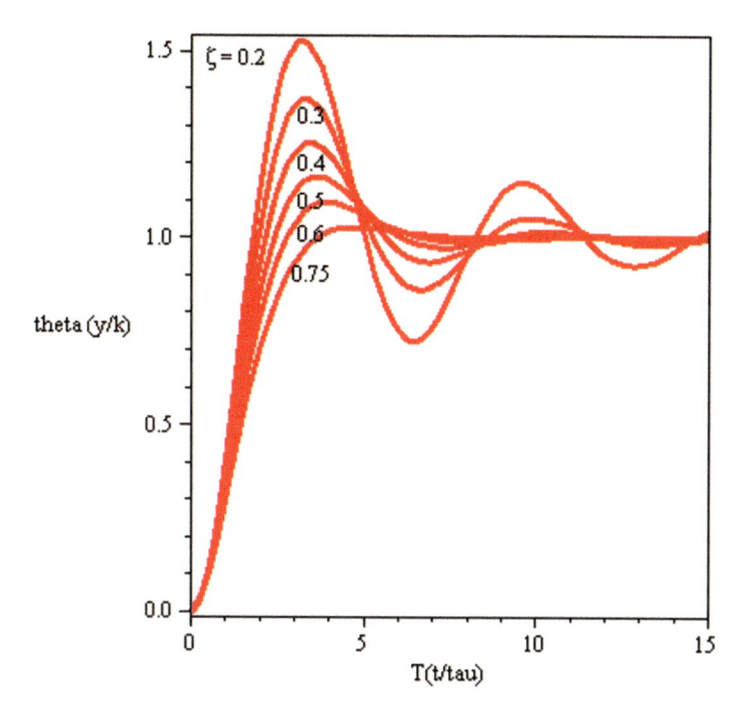

Fig. 2.12 Dimensionless variable theta (y/k) as a function of dimensionless time T(t/tau) for zeta < 1

We observe that the system is under damped for the values of $\zeta<1$. Now let us consider the special case of $\zeta=1$.

> subs(zeta=1,theta);

Error, numeric exception: division by zero

We get division by zero and hence we apply the limit T=1.

> theta1:=limit(theta,zeta=1);

$$\theta 1 := \frac{-1 - T + e^T}{e^T}$$

> pt[1]:=textplot([10,0.8,(zeta=1.0)]):

>pt[2]:=plot(theta1,T=0..15,thickness=3,axes=boxed,labels=["T (t/tau)","theta (y/k)"],title="Figure 2.13 Dimensionless variables theta (y/k) as a function of dimensionless time T(t/tau) for zeta=1.0."):

display(pt[1],pt[2]);

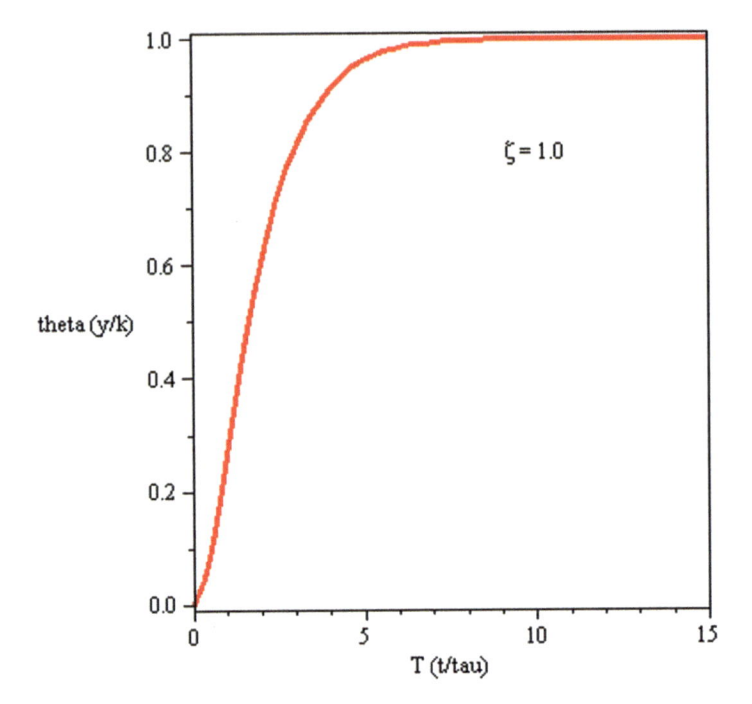

Fig. 2.13 Dimensionless variables theta (y/k) as a function of dimensionless time T(t/tau) for zeta = 1.0

Figure 2.13 shows that the system is critically damped for $\zeta=1$.

2.1.7 Solving Systems of ODEs Using the Laplace Transform Method

ODEs can be solved by applying Laplace transform technique. Consider a set of linear ODEs:

$$\frac{dy_i}{dt} = \sum_{j=1}^{n} a_{i,j} y_j + b_i(t) \qquad i = 1..n \qquad (2.26)$$

This set of linear ODEs can be converted to the Laplace domain by using Maple. The resulting set of linear algebraic equations can be written in matrix form as follows:

$$P\overline{Y} = b \qquad (2.27)$$

where $\overline{Y}$, **P** and **b** are functions of s, the Laplace variable. $\overline{Y}$, is the solution in the Laplace domain and can be found by inverting **P**:

$$\overline{Y} = P^{-1}b \qquad (2.28)$$

2.1 Linear Ordinary Differential Equations

Then the solution in the time domain Y(t) can be obtained by inverting the solution obtained in the Laplace domain $\overline{Y}$ (s). The procedure to solve linear ODEs using the Laplace transform technique and Maple is as follows:

1. Start the Maple program with a 'restart' command to clear all variables.
2. Call 'with(linalg)' and 'with(plots)' commands.
3. Call Maple's Laplace transformation package by using the 'with(inttrans)' command.
4. Enter the governing equations and number of equations (n).
5. Enter the initial conditions.
6. Apply the Laplace transformation to the equations using the command laplace("equations",t,s) which converts the equations from time domain to the Laplace (s) domain.
7. Store all the Laplace domain in eqs.
8. Use dummy variables in the Laplace domain (ua, ub, etc) for brevity of notation.
9. Store the dummy variable in vars.
10. Substitute these dummy variables in eqs.
11. Generate the **P** and **b** matrices.
12. Find the solution in Laplace domain by inverting the **P** matrix.
13. Convert the solution in the Laplace domain to the time domain using invlaplace("equations,"s,t) which converts the equations from the s domain to the time domain.

The first row of sol corresponds to the first dependent variable; the second row corresponds to the second dependent variable, etc. Note that analytical solutions are obtained as functions of the parameters and the independent variable (t).

Laplace Solution of Example 2.1 Equations

The IVP solved in example 2.1 is solved below using the Laplace transform technique following the procedure described above.

Example 2.6. Laplace Solution of Example 2.1 Equations

> restart:

> with(linalg):with(plots):

Warning, the protected names norm and trace have been redefined and unprotected
Warning, the name changecoords has been redefined

> with(inttrans):

Warning, the name hilbert has been redefined

> eq[1]:=diff(ca(t),t)=-k1*ca(t);

$$eq_1 := \frac{d}{dt}\,ca(t) = -k1\,ca(t)$$

> eq[2]:=diff(cb(t),t)=k1*ca(t)-k2*cb(t);

$$eq_2 := \frac{d}{dt}\, cb(t) = k1\ ca(t) - k2\ cb(t)$$

> N:=2;

$$N := 2$$

> ca(0):=1;cb(0):=0;

$$ca(0) := 1$$
$$cb(0) := 0$$

> eq[1]:=laplace(eq[1],t,s);

$$eq_1 := s\ laplace(ca(t), t, s) - 1 = -k1\ laplace(ca(t), t, s)$$

> eq[2]:=laplace(eq[2],t,s);

$$eq_2 := s\ laplace(cb(t), t, s) = k1\ laplace(ca(t), t, s)$$
$$- k2\ laplace(cb(t), t, s)$$

> eqs:=[eq[1],eq[2]];

$$eqs := [s\ laplace(ca(t), t, s) - 1 = -k1\ laplace(ca(t), t, s),$$
$$s\ laplace(cb(t), t, s) = k1\ laplace(ca(t), t, s)$$
$$- k2\ laplace(cb(t), t, s)]$$

Dummy variables are introduced for convenience.

> dummyvar:={laplace(ca(t),t,s)=ua,laplace(cb(t),t,s)=ub};

$$dummyvar := \{laplace(ca(t), t, s) = ua, laplace(cb(t), t, s) = ub\}$$

> eqs:=subs(dummyvar,eqs);

$$eqs := [s\ ua - 1 = -k1\ ua,\ s\ ub = k1\ ua - k2\ ub]$$

> vars:=[ua,ub];

$$vars := [ua, ub]$$

> P:=genmatrix(eqs,vars,'B');

$$P := \begin{bmatrix} s + k1 & 0 \\ -k1 & s + k2 \end{bmatrix}$$

2.1 Linear Ordinary Differential Equations

```
> b:=matrix(N,1):for i to N do b[i,1]:=B[i]:od:evalm(b);
```

$$\begin{bmatrix} 1 \\ 0 \end{bmatrix}$$

```
> sol:=evalm(inverse(P)&*b);
```

$$sol := \begin{bmatrix} \dfrac{1}{s + k1} \\ \dfrac{k1}{(s + k1)(s + k2)} \end{bmatrix}$$

The vector 'sol' is the solution in the Laplace domain. This is converted to the time domain by using Maple's 'map' command with the 'invlaplace' command.

```
> solt:=map(invlaplace,sol,s,t);
```

$$solt := \begin{bmatrix} e^{-k1\,t} \\ \dfrac{k1\left(-e^{-k1\,t} + e^{-k2\,t}\right)}{k1 - k2} \end{bmatrix}$$

```
> ca:=solt[1,1];
```

$$ca := e^{-k1\,t}$$

```
> cb:=solt[2,1];
```

$$cb := \dfrac{k1\left(-e^{-k1\,t} + e^{-k2\,t}\right)}{k1 - k2}$$

```
> cc:=1-ca-cb;
```

$$cc := 1 - e^{-k1\,t} - \dfrac{k1\left(-e^{-k1\,t} + e^{-k2\,t}\right)}{k1 - k2}$$

These results are consistent with those presented in Example 2.1.

Laplace Solution for Second Order System with Dirac forcing Function

Solve the IVP discussed in Example 2.5 with a Dirac(t) function as the forcing function.

$$\frac{d^2 y}{dt^2} + \frac{2\varsigma}{\tau}\frac{dy}{dt} + \frac{1}{\tau^2}y = \frac{\text{Dirac}(t)}{\tau^2} \tag{2.29}$$

This IVP can be solved easily using the Laplace transform technique. The procedure presented above for the Laplace transformation technique can be used for solving this example. When solving a second order differential equation, the initial conditions for both y and $\dfrac{dy}{dt}$ must be provided.

Example 2.7. Laplace Solution for Second Order System with Dirac forcing Function

> restart:

> with(linalg):with(plots):

> with(inttrans):

> eq:=diff(y(t),t$2)+2*zeta/tau*diff(y(t),t)+1/tau^2*y(t)=Dirac(t)/tau^2;

$$eq := \frac{d^2}{dt^2} y(t) + \frac{2\zeta \left(\frac{d}{dt} y(t)\right)}{\tau} + \frac{y(t)}{\tau^2} = \frac{Dirac(t)}{\tau^2}$$

Even though the equation is second order there is only one equation to be solved.

> N:=1;

$$N := 1$$

The initial conditions are entered here.

> y(0):=1;D(y)(0):=0;

$$y(0) := 1$$

$$D(y)(0) := 0$$

> eq:=laplace(eq,t,s);

$$eq := s^2\, laplace(y(t), t, s) - s + \frac{2\zeta s\, laplace(y(t), t, s)}{\tau} - \frac{2\zeta}{\tau} + \frac{laplace(y(t), t, s)}{\tau^2} = \frac{1}{\tau^2}$$

> eqs:=[eq];

$$eqs := \left[s^2\, laplace(y(t), t, s) - s + \frac{2\zeta s\, laplace(y(t), t, s)}{\tau} - \frac{2\zeta}{\tau} + \frac{laplace(y(t), t, s)}{\tau^2} = \frac{1}{\tau^2} \right]$$

> dummyvar:={laplace(y(t),t,s)=u};

$$dummyvar := \{laplace(y(t), t, s) = u\}$$

> eqs:=subs(dummyvar,eqs);

$$eqs := \left[s^2 u - s + \frac{2\zeta s\, u}{\tau} - \frac{2\zeta}{\tau} + \frac{u}{\tau^2} = \frac{1}{\tau^2} \right]$$

2.1 Linear Ordinary Differential Equations

> vars:=[u];

$$vars := [u]$$

> PP:=genmatrix(eqs,vars,'b1');

$$PP := \left[s^2 + \frac{2\zeta s}{\tau} + \frac{1}{\tau^2} \right]$$

> b:=matrix(N,1):for i to N do b[i,1]:=b1[i]:od:evalm(b);

$$\left[s + \frac{2\zeta}{\tau} + \frac{1}{\tau^2} \right]$$

> sol:=evalm(inverse(PP)&*b);

$$sol := \left[\frac{\tau^2 \left(s + \frac{2\zeta}{\tau} + \frac{1}{\tau^2} \right)}{s^2 \tau^2 + 2\zeta s \tau + 1} \right]$$

> solt:=map(invlaplace,sol,s,t);

$$solt := \left[e^{-\frac{t\zeta}{\tau}} \left(\cosh\left(\frac{t\sqrt{\tau^6 (\zeta^2 - 1)}}{\tau^4} \right) \right.\right.$$
$$\left.\left. + \frac{\sinh\left(\frac{t\sqrt{\tau^6 (\zeta^2 - 1)}}{\tau^4} \right) \sqrt{\tau^6 (\zeta^2 - 1)} \, (\zeta \tau + 1)}{(\zeta^2 - 1) \tau^4} \right) \right]$$

> y:=solt[1,1];

$$y := e^{-\frac{t\zeta}{\tau}} \left(\cosh\left(\frac{t\sqrt{\tau^6 (\zeta^2 - 1)}}{\tau^4} \right) \right.$$
$$\left. + \frac{\sinh\left(\frac{t\sqrt{\tau^6 (\zeta^2 - 1)}}{\tau^4} \right) \sqrt{\tau^6 (\zeta^2 - 1)} \, (\zeta \tau + 1)}{(\zeta^2 - 1) \tau^4} \right)$$

78　　　　　　　　　　　　　　　　　　　　　　　2 Initial Value Problems

The value for τ is entered here for plotting purposes.

> tau:=1;

$$\tau := 1$$

> y:=eval(simplify(y));

$$y := \frac{1}{\sqrt{\zeta^2 - 1}} \left(e^{-t\zeta} \left(\cosh\left(t\sqrt{\zeta^2 - 1}\right) \sqrt{\zeta^2 - 1} \right.\right.$$
$$\left.\left. + \sinh\left(t\sqrt{\zeta^2 - 1}\right) \zeta + \sinh\left(t\sqrt{\zeta^2 - 1}\right) \right) \right)$$

> pars:=[1.5,2,2.5,3];

$$pars := [1.5, 2, 2.5, 3]$$

> for i from 1 to 4 do p[i]:=plot(subs(zeta=pars[i],y),t=0..15):od:
> display(seq(p[i],i=1..4),thickness=3,axes=boxed,labels=["t","y"],title="Figure 2.14");

Fig. 2.14

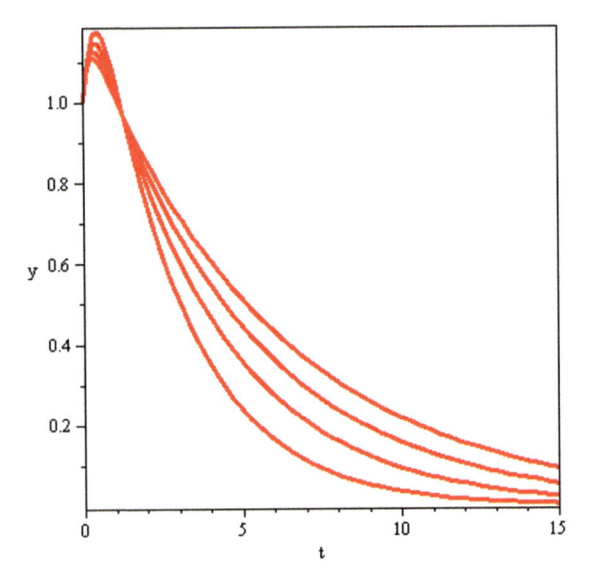

> pars:=[0.2,0.3,0.4,0.5,0.6,0.75];

$$pars := [0.2, 0.3, 0.4, 0.5, 0.6, 0.75]$$

2.1 Linear Ordinary Differential Equations

> for i from 1 to 6 do p2[i]:=plot(evalf(Re(subs(zeta=pars[i],y))),t=0..15):od:

> display(seq(p2[i],i=1..6),thickness=3,axes=boxed,labels=["t","y"],title=
"Figure 2.15");

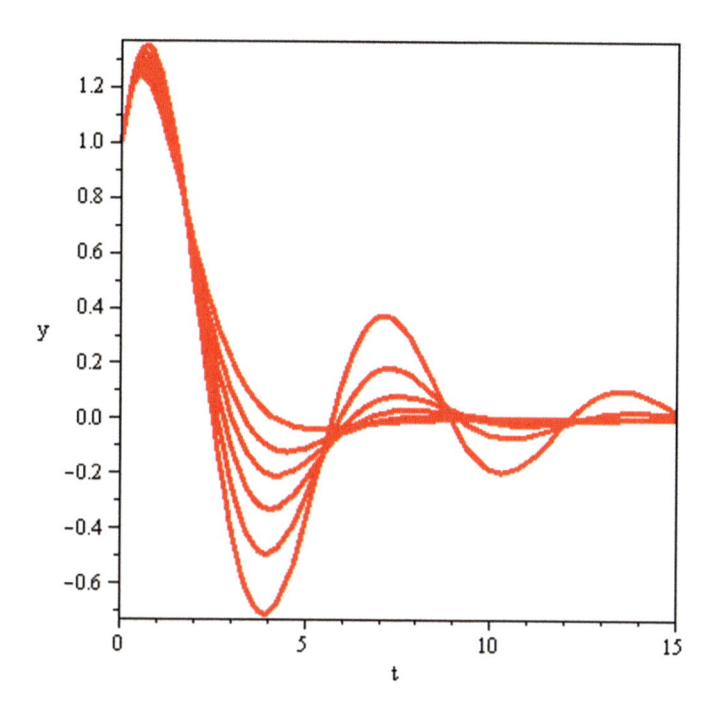

Fig. 2.15

> subs(zeta=1,y);

Error, numeric exception: division by zero

> y1:=limit(y,zeta=1);

$$y1 := \frac{2t+1}{e^t}$$

> plot(y1,t=0..15,thickness=3,axes=boxed,labels=["t","y"],title="Figure 2.16");

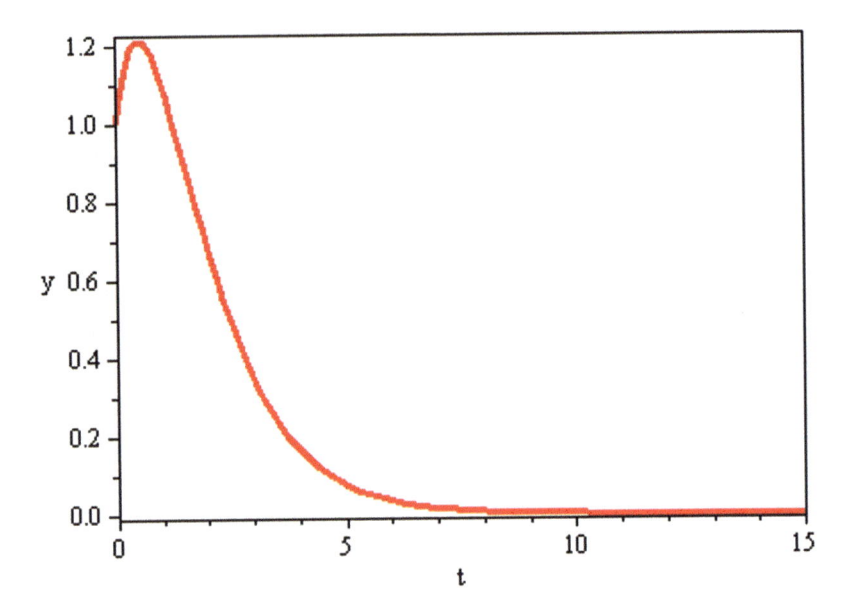

[Figure 2.16: plot of y vs t, curve rising to ~1.2 near t=1 then decaying to 0 by t≈10]

Fig. 2.16

We observe that the system oscillates for $\zeta<1$ and does not oscillate for $\zeta\geq 1$.

2.1.8 Solving Linear ODEs Using Maple's 'dsolve' Command

In the previous sections we solved linear ODEs using exponential matrix (section 2.1.2 – 2.1.4) and the Laplace transform technique (section 2.1.5). Alternatively, Maple's dsolve command can be used to solve linear ODEs. However, the analytical solution obtained from the dsolve command may not be in a simplified form.

The reaction scheme described in example 2.1 is solved below using the 'dsolve' command.

Example 2.8. Solving Linear ODEs Using Maple

The reaction scheme described in example 2.1 is solved below using the 'dsolve' command.

> restart:

> with(plots):

Warning, the name changecoords has been redefined

> eq[1]:=diff(ca(t),t)=-k1*ca(t);

$$eq_1 := \frac{d}{dt} ca(t) = -k1\ ca(t)$$

2.1 Linear Ordinary Differential Equations

> eq[2]:=diff(cb(t),t)=k1*ca(t)-k2*cb(t);

$$eq_2 := \frac{d}{dt} cb(t) = k1 \ ca(t) - k2 \ cb(t)$$

> vars:=(ca(t),cb(t));

$$vars := ca(t), cb(t)$$

> eqs:=(eq[1],eq[2]);

$$eqs := \frac{d}{dt} ca(t) = -k1 \ ca(t), \frac{d}{dt} cb(t) = k1 \ ca(t) - k2 \ cb(t)$$

> ICs:=(ca(0)=1,cb(0)=0);

$$ICs := ca(0) = 1, cb(0) = 0$$

> sol:=dsolve({eqs,ICs},{vars});

$$sol := \left\{ cb(t) = \frac{k1 \ e^{-k2 \ t}}{k1 - k2} - \frac{k1 \ e^{-k1 \ t}}{k1 - k2}, ca(t) = e^{-k1 \ t} \right\}$$

> assign(sol):

> ca(t);

$$e^{-k1 \ t}$$

> cb(t);

$$\frac{k1 \ e^{-k2 \ t}}{k1 - k2} - \frac{k1 \ e^{-k1 \ t}}{k1 - k2}$$

Higher order linear ODEs can also be solved using the dsolve command. It should be noted that Maple solves equations in symbolic form. Therefore, even if the constants are numerical, the output is in symbolic form. Sometimes, this output can be messy. It should be noted that when more than two equations are solved the 'dsolve' command may not be able to give an elegant solution. For illustration, the heat transfer problem solved in example 2.3 is solved below using Maple's 'dsolve' command.

Example 2.9. Heat Transfer in a Series of Tanks, 'dsolve'

> restart:

> with(linalg):with(plots):

Warning, the protected names norm and trace have been redefined and unprotected
Warning, the name changecoords has been redefined

> N:=3;

$$N := 3$$

82   2 Initial Value Problems

> for i to N do eq[i]:=diff(T[i](t),t)=eval(alpha*(T[i-1](t)-T[i](t))+
beta*(T[steam]-T[i](t)));od;

$$eq_1 := \frac{d}{dt} T_1(t) = \alpha \left(T_0(t) - T_1(t) \right) + \beta \left(T_{steam} - T_1(t) \right)$$

$$eq_2 := \frac{d}{dt} T_2(t) = \alpha \left(T_1(t) - T_2(t) \right) + \beta \left(T_{steam} - T_2(t) \right)$$

$$eq_3 := \frac{d}{dt} T_3(t) = \alpha \left(T_2(t) - T_3(t) \right) + \beta \left(T_{steam} - T_3(t) \right)$$

> pars:={T[0](t)=20,T[steam]=250};

$$pars := \{ T_0(t) = 20, T_{steam} = 250 \}$$

> vars:=[seq(T[i](t),i=1..N)];

$$vars := [T_1(t), T_2(t), T_3(t)]$$

> for i to 3 do eq[i]:=subs(pars,eq[i]);od;

$$eq_1 := \frac{d}{dt} T_1(t) = \alpha \left(20 - T_1(t) \right) + \beta \left(250 - T_1(t) \right)$$

$$eq_2 := \frac{d}{dt} T_2(t) = \alpha \left(T_1(t) - T_2(t) \right) + \beta \left(250 - T_2(t) \right)$$

$$eq_3 := \frac{d}{dt} T_3(t) = \alpha \left(T_2(t) - T_3(t) \right) + \beta \left(250 - T_3(t) \right)$$

>dsolve({eq[1],eq[2],eq[3],T[1](0)=20,T[2](0)=20,T[3](0)=20});

$$\left\{ T_3(t) = \frac{1}{2} \frac{1}{(\beta + \alpha)^3} \left(\left(-\frac{230 \alpha^5 \beta t^2}{\beta + \alpha} + \left(-\frac{690 \beta^2 t^2}{\beta + \alpha} \right. \right. \right. \right.$$

$$\left. - \frac{460 t \beta (2\alpha + \beta)}{\beta^2 + 2\alpha\beta + \alpha^2} \right) \alpha^4 + \left(-\frac{690 \beta^3 t^2}{\beta + \alpha} \right.$$

$$\left. \left. - \frac{460 \beta (3\alpha^2 + 3\alpha\beta + \beta^2)}{\alpha^3 + 3\alpha^2\beta + 3\beta^2\alpha + \beta^3} - \frac{1380 \beta^2 (2\alpha + \beta) t}{\beta^2 + 2\alpha\beta + \alpha^2} \right) \alpha^3 \right.$$

2.1 Linear Ordinary Differential Equations

$$+\left(-\frac{1380\beta^2\left(3\alpha^2+3\alpha\beta+\beta^2\right)}{\alpha^3+3\alpha^2\beta+3\beta^2\alpha+\beta^3}-\frac{230\beta^4 t^2}{\beta+\alpha}\right.$$

$$\left.-\frac{1380\beta^3\left(2\alpha+\beta\right)t}{\beta^2+2\alpha\beta+\alpha^2}\right)\alpha^2+\left(\vphantom{\frac{1}{1}}\right.$$

$$\left.-\frac{1380\beta^3\left(3\alpha^2+3\alpha\beta+\beta^2\right)}{\alpha^3+3\alpha^2\beta+3\beta^2\alpha+\beta^3}-\frac{460\beta^4\left(2\alpha+\beta\right)t}{\beta^2+2\alpha\beta+\alpha^2}\right)\alpha$$

$$\left.-\frac{460\beta^4\left(3\alpha^2+3\alpha\beta+\beta^2\right)}{\alpha^3+3\alpha^2\beta+3\beta^2\alpha+\beta^3}\right)e^{-(\beta+\alpha)t}\Bigg]$$

$$+\frac{1}{2}\frac{40\alpha^3+1500\alpha^2\beta+1500\beta^2\alpha+500\beta^3}{(\beta+\alpha)^3},\ T_2(t)$$

$$=\frac{1}{(\beta+\alpha)^2}\Bigg(\Bigg(-\frac{230\alpha^3\beta t}{\beta+\alpha}+\Bigg(-\frac{460\beta^2 t}{\beta+\alpha}\Bigg.$$

$$\left.-\frac{230\beta\left(2\alpha+\beta\right)}{\beta^2+2\alpha\beta+\alpha^2}\right)\alpha^2+\left(-\frac{230\beta^3 t}{\beta+\alpha}\right.$$

$$\left.-\frac{460\beta^2\left(2\alpha+\beta\right)}{\beta^2+2\alpha\beta+\alpha^2}\right)\alpha-\frac{230\beta^3\left(2\alpha+\beta\right)}{\beta^2+2\alpha\beta+\alpha^2}\Bigg)e^{-(\beta+\alpha)t}\Bigg]$$

$$+\frac{20\alpha^2+500\alpha\beta+250\beta^2}{(\beta+\alpha)^2},\ T_1(t)=-\frac{230 e^{-(\beta+\alpha)t}\beta}{\beta+\alpha}$$

$$+\frac{10\left(2\alpha+25\beta\right)}{\beta+\alpha}\Bigg\}$$

We observe that solutions obtained for T2(t) and T3(t) using the 'dsolve' command are long and messy compared to the solution obtained using the exponential matrix approach (Example 2.3). When more than three differential equations are to be solved it is recommended that the exponential matrix method be used. As an exercise, readers can verify that the solution obtained using the 'dsolve' command is equivalent to the solution obtained in example 2.3.

2.1.9 Summary

In this chapter analytical solutions were derived for linear ODEs using three methods: the matrix exponential, Laplace transform, and dsolve. In section 2.1.2, the given linear coupled system of homogenous ODEs was converted to matrix form. The analytical solution for this matrix differential equation was then found using the matrix exponential. Maple provides the exponential matrix as a function of the independent variable and the parameters in the governing equations. This approach yields an elegant solution for a given system of linear coupled homogeneous ODEs. This methodology was then extended to non-homogenous coupled linear ODEs in section 2.1.3. This approach yields analytical solutions for linear coupled ODEs with time-dependent forcing functions.

Higher order ODEs (of order n) were converted to a system of n coupled linear first order ODEs in section 2.1.4. This system was then solved using the exponential matrix developed earlier. This approach yields analytical solutions for linear ODEs of any order. In section 2.1.5, the given system of coupled linear ODEs was converted to Laplace domain. The resulting linear system of algebraic equations was then solved for the solution in the Laplace domain. The solution obtained in the Laplace domain was then converted to the time domain.

Maple's 'dsolve' command was used to solve linear ODEs in section 2.1.6. In our opinion, exponential matrix method is the best method to arrive at an elegant analytical solution. The Laplace transform technique illustrated in section 2.1.5 could be used for integro-differential equations. Maple's 'dsolve' command has to be used if the exponential matrix method fails.

2.1.10 Problems

1. Consider two tanks in a series (Pushpavanam, 1998), [11] in which the height of each tank is governed by

2.
$$\frac{dh_1}{dt} = \frac{F}{A_1} - \frac{h_1 - h_2}{R_1 A_1}$$
$$\frac{dh_2}{dt} = \frac{h_1 - h_2}{R_1 A_2} - \frac{h_2}{R_2 A_2}$$

where, h is the height of liquid in the tank, R_1 and R_2 are the resistances of the values, A is the cross sectional area, and F is the flow rate. The subscripts 1 and 2 are the subscripts for tank 1 and tank 2 respectively. Obtain an analytical solution for the height of liquid in each tank as a function of the parameters. Plot your profiles for the set of parameters F = 1, A_1 = 1, R_1 = ½, A_2 = 2 and R_2 = ¼.

3. Consider the series reaction scheme modeled in example 2.2 (equation 2.10), eliminate the concentration of species C, C_C in the first two equations using material balance. This will yield a system of two non-homogeneous ODEs. Solve this system to get an analytical solution for the individual species as a function of rate constants and time.

4. A steel ball initially at a uniform temperature of 100°C is dropped into an insulated vessel containing water at 20°C (Pushpavanam, 1998)[11]. Determine the steady state temperature of water and steel ball. The energy

2.1 Linear Ordinary Differential Equations

balance gives the governing equations for the temperature of steel ball (T_b) and the temperature of water (T_w) as:

$$m_b C_b \frac{dT_b}{dt} = -UA(T_b - T_w)$$

$$m_w C_w \frac{dT_w}{dt} = UA(T_b - T_w)$$

where m is the mass, C is the specific heat capacity and UA is the heat transfer rate. And b and w are the subscripts for the steel ball and water respectively. Find transient and steady state analytical solutions for the temperature of water and steel ball as a function of the parameters. What do you observe? Plot your profiles for the set of parameters $m_b = 1.25$ kg, $m_w = 5$ kg, $C_b = 3360$ J/g/°C, $C_w = 4200$ J/kg/°C and UA = 4200 J/s/°C.

5. Consider the series-parallel reaction scheme in a batch reactor

$$A \xrightarrow{k_1} B \xrightarrow{k_3} C$$
$$\downarrow k_2$$
$$C$$

Assuming that all the reactions are first order, write down the governing equations for this reaction. Assuming 1, 0, and 0 mol/cm³ as the initial conditions for A, B, and C develop analytical expressions for the concentration of A, B and C. Do you observe a maximum for B? Plot the concentration profiles for the parameters $k_1 = 2$ min⁻¹; $k_2 = 1$ min⁻¹; and $k_3 = 2$ min⁻¹.

6. Consider the series reaction scheme in a constant batch reactor (Bequette, 1998)[12]

$$A \xrightarrow{k_1} B \xrightarrow{k_2} C \xrightarrow{k_3} D$$

Assuming that all the reactions are first order, write down the governing equations for the concentration of A, B and C. Assuming 1, 0, and 0 mol/cm³ as the initial conditions for A, B, and C develop analytical expressions for the concentration of A, B and C. Plot the concentration profiles for the parameters $k_1 = 2$ min⁻¹; $k_2 = 1$ min⁻¹; and $k_3 = 2$ min⁻¹. Do you observe a maximum for B? Explain how you would obtain the concentration of D.

7. Consider the second order system solved in example 2.5

$$\frac{d^2y}{dt^2} + \frac{2\varsigma}{\tau}\frac{dy}{dt} + \frac{1}{\tau^2}y = \frac{k}{\tau^2}u(t)$$

$$y(0) = 0; \quad \frac{dy}{dt}(0) = 0$$

subject to a sinusoidal input u(t) = sin(ωt). Obtain an analytical solution for y(t) and analyze this problem for different values of ς for τ = 1 and ω = 1. Solve this problem using the exponential matrix and by using the Laplace transform technique.

8. Consider the second order system in problem 6, redo the problem if the input is $u(t) = \sin(\omega t)e^{-t}$.

9. Consider the CSTR problem discussed in the chapter (example 2.4). Solve the governing equations if the reaction scheme is given by

$$A \xrightarrow{k_1} B \underset{k_3}{\overset{k_2}{\rightleftharpoons}} C$$

Plot your concentration profiles for $k_2 = 1/8$ and $k_3 = 1/2$. All other parameters are same as that of example 2.4. What do you observe?

10. Consider heating of a fluid stream by steam coils in a series of tanks (see example 2.3). Write down the differential equations describing the evolution of temperature in a system of four such tanks in series. Find the evolution of temperature with time in each tank using the exponential matrix method. Plot your temperature profiles for the parameter values [α, β] = [0.1, 0.005] and [0.1, 0.01]. Find the time taken by the last tank to reach 99% of its steady state value. How does this time compare with the time for a 3-tank system and a 2-tank system when the total weight of all the tanks remains the same? All other parameters are same as that of example 2.3.

11. Consider the first order series reaction taking place in a plug flow reactor. Optimize the length of the reactor to maximize the concentration of B in the outlet stream. Take initial conditions from problem 5.

12. Consider the pharmacokinetics problem (Bequette, 1998)[12]

$$\frac{dc_1}{dt} = -k_1 c_1 + k_2 c_2 + u$$

$$\frac{dc_2}{dt} = k_3 c_1 - k_4 c_2$$

$$c_1(0) = c_2(0) = 0$$

where c_1 and c_2 are the concentration of the drug in compartment 1 and 2 respectively, u is the injection rate, 5.2 ppm/min. Obtain an analytical solution for this problem as a function of the rate constants. Plot your profiles for the following set of parameters $[k_1, k_2, k_3, k_4] = [0.26, 0.1, 0.1, 0.094]$ min^{-1}.

13. Consider a typical gas absorption column (Bequette, 1998; Varma and Morbidelli, 1997)[3, 12] solute balance gives the governing equations

2.2 Nonlinear Ordinary Differential Equations

$$\frac{dx_1}{dt} = -\frac{L+Va}{M}x_1 + \frac{Va}{M}x_2 + \frac{L}{M}x_f \quad \text{stage 1}$$

$$\frac{dx_i}{dt} = \frac{L}{M}x_{i-1} - \frac{L+Va}{M}x_i + \frac{Va}{M}x_{i+1} \quad \text{stage } i = 2..N-1$$

$$\frac{dx_N}{dt} = \frac{L}{M}x_{N-1} - \frac{L+Va}{M}x_N + \frac{V}{M}y_f \quad \text{stage } N$$

where N is the number of stages in the absorption column, L is the liquid feed flow rate, V is the vapor feed flow rate, M is the mass of liquid molar hold up per stage, a is the equilibrium constant, x_f and y_f are the feed mole fraction of the liquid stream and the vapor stream respectively. Values of the parameters are L = 4/3 kgmol inert oil/min, V = 5/3 kgmol air/min, M = 20/3 kgmol, x_f = 0, y_f = 0.1 and a = 0.5. At time t = 0, the mole fraction x in all the stages is zero. Plot transient profiles for the mole fraction (x) of all the stages for N = 3. What is the steady state mole fraction of the liquid stream (x_N) leaving the absorption column? What is the minimum number of stages N to be used to make sure that the steady state mole fraction of the liquid stream leaving the absorption column (x_N) is greater than 0.12? Also plot x vs. the stage number (from 0 to N) for different values of time (0, 10, 20, .. till steadystate).

14. Redo problem 12 for N = 9 and x_f = 0.1. What do you observe?
15. Consider the irreversible series reaction A→B taking place in N CSTRs in series. The governing equation for the concentration of A in a particular tank is given by:

$$\frac{dca_i}{dt} = \frac{1}{\tau}(ca_{i-1} - ca_i) - k_1 ca_i$$

where τ is the time constant and k_1 is the rate constant. The feed concentration to the first tank (ca_0) is 1 and the parameters are $k_1 = \tau = 1$ and N = 3. Solve this problem to obtain a transient analytical solution for the concentration of A in each tank. Increase N to 10 and obtain the transient plots. In addition plot the concentration as a function of the tank number for different times.

2.2 Nonlinear Ordinary Differential Equations

2.2.1 Introduction

Chemical reactions with reaction orders other than one and many practical systems are typically represented by nonlinear ordinary differential equations

(ODEs). In this section analytical, series and numerical solutions are developed for nonlinear IVPs using Maple.

Example 2.2.1. Simultaneous Series Reactions

Consider a second order reaction

$$2A \xrightarrow{k_1} B \xrightarrow{k_2} C \qquad (2.30)$$

governed by the nonlinear ODEs:

$$\frac{dc_a}{dt} = -2k_1 c_a^2$$
$$\frac{dc_b}{dt} = -k_1 c_b^2 - k_2 c_b \qquad (2.31)$$

with the initial conditions $c_a(0) = 1$; $c_b(0) = 0$ and $c_c(0) = 0$; and, k_1 and k_2 are the rate constants. The concentration of species $C(c_c)$ at any time is given by the material balance

$$c_a + c_b + c_c = c_a(0) = 1 \qquad (2.32)$$

The equation above is solved below in Maple:

> restart:

> with(plots):

The governing equations are entered here:

> eq[1]:=diff(ca(t),t)=-k1*ca(t)^2;

$$\frac{d}{dt} ca(t) = -k1 \ ca(t)^2$$

> eq[2]:=diff(cb(t),t)=k1*ca(t)^2-k2*cb(t);

$$\frac{d}{dt} cb(t) = k1 \ ca(t)^2 - k2 \ cb(t)$$

The variables are entered here:

> vars:=(ca(t),cb(t));

$$ca(t), cb(t)$$

2.2 Nonlinear Ordinary Differential Equations

The equations are stored in eqs.

```
> eqs:=(eq[1],eq[2]);
```

$$\frac{d}{dt}\,ca(t) = -k1\,ca(t)^2,\ \frac{d}{dt}\,cb(t) = k1\,ca(t)^2 - k2\,cb(t)$$

The initial conditions are stored in ICs:

```
> ICs:=(ca(0)=1,cb(0)=0);
```

$$ca(0) = 1,\ cb(0) = 0$$

```
> sol:=dsolve({eqs,ICs},{vars});
```

$$\left\{ cb(t) = \left(\frac{k2\left(-\dfrac{e^{k2\,t}}{k2\,t + \dfrac{k2}{k1}} - e^{-\dfrac{k2}{k1}}\,\mathrm{Ei}\!\left(1,\,-k2\,t - \dfrac{k2}{k1}\right) \right)}{k1} \right.\right.$$

$$\left.\left. + \frac{k1\,e^{\dfrac{k2}{k1}} + \mathrm{Ei}\!\left(1,\,-\dfrac{k2}{k1}\right)k2}{k1\,e^{\dfrac{k2}{k1}}} \right) e^{-k2\,t},\ ca(t) = \frac{1}{k1\,t + 1} \right\}$$

```
> assign(sol):
> ca(t);
```

$$\frac{1}{k1\,t + 1}$$

```
> cb(t);
```

$$\left(\frac{k2\left(-\dfrac{e^{k2\,t}}{k2\,t + \dfrac{k2}{k1}} - e^{-\dfrac{k2}{k1}}\,\mathrm{Ei}\!\left(1,\,-k2\,t - \dfrac{k2}{k1}\right) \right)}{k1} \right.$$

$$\left. + \frac{k1\,e^{\dfrac{k2}{k1}} + \mathrm{Ei}\!\left(1,\,-\dfrac{k2}{k1}\right)k2}{k1\,e^{\dfrac{k2}{k1}}} \right) e^{-k2\,t}$$

The 'help' command in Maple is invoked to describe Ei. The following description for Ei is given in Maple's help file.

> ?Ei

Ei - The Exponential Integral
Calling Sequence
   Ei(z)
   Ei(a, z)
Parameters
   z - algebraic expression
   a - algebraic expression
Description
- The exponential integrals, Ei(a, z), are defined for $Re(z) > 0$ by
> Ei(a, z) = convert(Ei(a, z), Int) assuming Re(z) > 0;
This classical definition is extended by analytic continuation to the entire complex plane using
> Ei(a, z) = z^(a-1)*GAMMA(1-a, z);
with the exception of the point 0 in the case of Ei(1, z).
- For all of these functions, 0 is a branch point and the negative real axis is the branch cut. The values on the branch cut are assigned such that the functions are continuous in the direction of increasing argument (equivalently, from above).
- The classical definition for the 1-argument exponential integral is a Cauchy Principal Value integral, defined for real arguments x, as the following
> convert(Ei(x),Int) assuming x::real;
> value(%);
for $x < 0$, Ei(x) = -Ei(1, -x). This classical definition is extended to the entire complex plane using
$$Ei(z) = -Ei(1, -z) + (\ln(z) - \ln(1/z))/2 - \ln(-z)$$

Note that this extension has its branch cut on the negative real axis, but unlike for the 2-argument Ei functions this extension is not continuous onto the branch cut from either above or below. That is, this extension provides an analytic continuation of Ei(z) from the positive real axis, but not in any direction from the negative real axis. If you want a continuation from the negative real axis, use -Ei(1, -z) in place of Ei(z).

Reference:
Abramowitz, M. and Stegun, I. Handbook of Mathematical Functions. New York: Dover Publications Inc., 1965.

An additional reference for the function of Ei is F. B. Hildebrand, Advanced Calculus for Applications, 2d Ed, 1976, Prentice-Hall, page 50.[13]

2.2 Nonlinear Ordinary Differential Equations

The concentration of species C is found using the material balance.

> cc(t):=1-ca(t)-cb(t);

$$1 - \frac{1}{k1\ t + 1}$$
$$- \left(\frac{k2 \left(-\dfrac{e^{k2\ t}}{k2\ t + \dfrac{k2}{k1}} - e^{-\dfrac{k2}{k1}} \operatorname{Ei}\left(1, -k2\ t - \dfrac{k2}{k1}\right) \right)}{k1} \right.$$
$$\left. + \frac{k1\ e^{\dfrac{k2}{k1}} + \operatorname{Ei}\left(1, -\dfrac{k2}{k1}\right) k2}{k1\ e^{\dfrac{k2}{k1}}} \right) e^{-k2\ t}$$

Plots can be made for different values of rate constants.

> pars:={k1=1,k2=1};

$$\{k2 = 1, k1 = 1\}$$

> Ca:=subs(pars,ca(t));

$$\frac{1}{t+1}$$

> Cb:=subs(pars,cb(t));

$$\left(-\frac{e^t}{t+1} - e^{-1} \operatorname{Ei}(1, -t - 1) + \frac{e + \operatorname{Ei}(1, -1)}{e} \right) e^{-t}$$

> Cc:=subs(pars,cc(t));

$$1 - \frac{1}{t+1} - \left(-\frac{e^t}{t+1} - e^{-1} \operatorname{Ei}(1, -t - 1) + \frac{e + \operatorname{Ei}(1, -1)}{e} \right) e^{-t}$$

> p1:=plot(eval(Ca),t=0..10,thickness=3,color=green):

> p2:=plot(eval(Re(Cb)),t=0..10,linestyle=1,thickness=3,axes=boxed):

> p3:=plot(eval(Re(Cc)),t=0..10,linestyle=2,thickness=3,color=magenta):

To get rid of the residual errors while calculating the Ei functions only the real part is plotted.

> display({p1,p2,p3},labels=[t,C],title=" Figure 2.17");

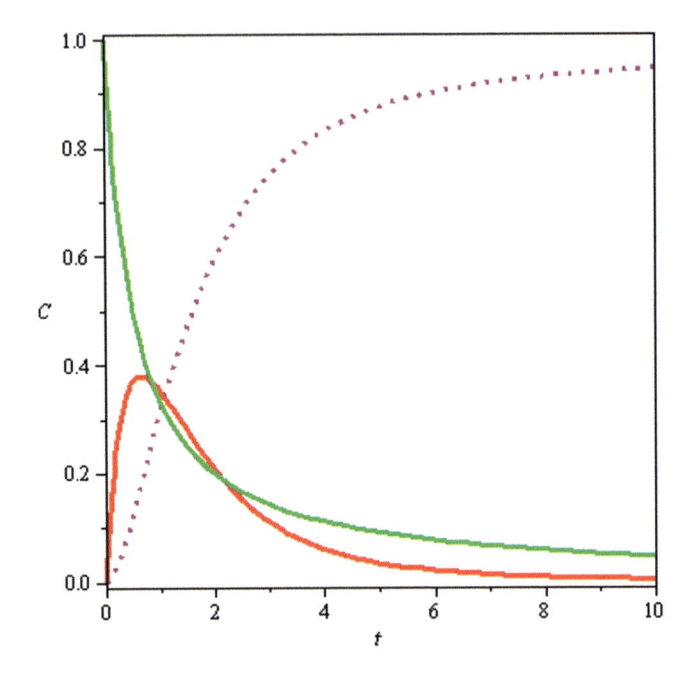

Fig. 2.17

> pars:={k1=2,k2=1};

$$\{k2 = 1, k1 = 2\}$$

> Ca:=subs(pars,ca(t));

$$\frac{1}{2t+1}$$

> Cb:=subs(pars,cb(t));

$$\left(-\frac{1}{2}\frac{e^t}{t+\frac{1}{2}} - \frac{1}{2}e^{-\frac{1}{2}}\operatorname{Ei}\left(1, -t-\frac{1}{2}\right)\right.$$
$$\left. + \frac{1}{2}\frac{2e^{\frac{1}{2}} + \operatorname{Ei}\left(1, -\frac{1}{2}\right)}{e^{\frac{1}{2}}}\right)e^{-t}$$

2.2 Nonlinear Ordinary Differential Equations

> Cc:=subs(pars,cc(t));

$$1 - \frac{1}{2t+1} - \left(-\frac{1}{2}\frac{e^t}{t+\frac{1}{2}} - \frac{1}{2}e^{-\frac{1}{2}}\text{Ei}\left(1, -t-\frac{1}{2}\right) \right.$$
$$\left. + \frac{1}{2}\frac{2e^{\frac{1}{2}} + \text{Ei}\left(1, -\frac{1}{2}\right)}{e^{\frac{1}{2}}} \right) e^{-t}$$

> p1:=plot(Ca,t=0..10,thickness=3,color=green):
> p2:=plot(Re(Cb),t=0..10,linestyle=1,thickness=3,axes=boxed):
> p3:=plot(Re(Cc),t=0..10,linestyle=2,thickness=3,color=magenta):
> display({p1,p2,p3},labels=[t,C],title="Figure 2.18");

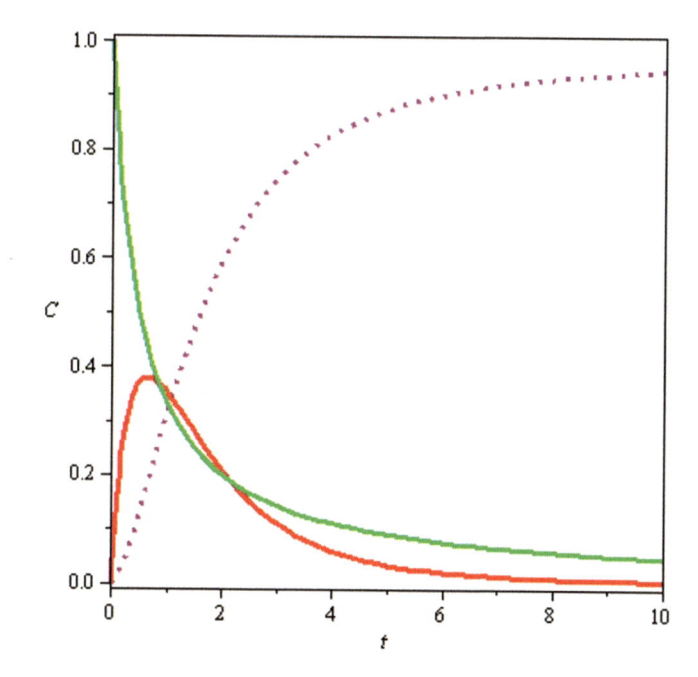

Fig. 2.18

We observe that a maximum exists for the concentration of species B. Sometimes, Maple gives implicit solutions, i.e., independent variable (t), as a function of the dependent variable (y).

2.2.2 Solving Nonlinear ODEs Using Maple's 'dsolve' Command

Consider an nth order reaction

$$A \xrightarrow{k} Products \quad (2.33)$$

governed by the ODE:

$$\frac{dc}{dt} = -kc^n \quad (2.34)$$

with the initial conditions c(0)=1; k is the rate constant, and n is the order of the reaction. Series solutions are obtained in Maple below:

> restart:

> with(plots):

Enter the governing equation:

> eq:=diff(c(t),t)=-k*c(t)^n;

$$eq := \frac{d}{dt} c(t) = -k \, c(t)^n$$

The analytical solution is found as:

> ca:=rhs(dsolve({eq,c(0)=1},c(t)));

$$ca := \frac{1}{(-k\,t + k\,t\,n + 1)^{\frac{1}{n-1}}}$$

Next, the series solution can be obtained as:

> sol:=dsolve({eq,c(0)=1},{c(t)},type=series);

$$sol := c(t) = 1 - k\,t + \frac{1}{2} k^2 n\, t^2 + \left(-\frac{1}{3} k^3 n^2 + \frac{1}{6} k^3 n\right) t^3$$
$$+ \frac{1}{24} k^4 n \left(6 n^2 - 7 n + 2\right) t^4 + \left(-\frac{1}{5} k^5 n^4 + \frac{23}{60} k^5 n^3\right.$$
$$\left. - \frac{29}{120} k^5 n^2 + \frac{1}{20} k^5 n\right) t^5 + O(t^6)$$

By default, Maple returns series solutions accurate to the order of t^6. The order can be increased as:

> Order:=8;

$$Order := 8$$

2.2 Nonlinear Ordinary Differential Equations

```
> sol:=dsolve({eq,c(0)=1},{c(t)},type=series);
```

$$sol := c(t) = 1 - k\,t + \frac{1}{2}k^2\,n\,t^2 - \frac{1}{6}k^3\,n\,(2\,n - 1)\,t^3$$
$$+ \left(\frac{1}{4}k^4\,n^3 - \frac{7}{24}k^4\,n^2 + \frac{1}{12}k^4\,n\right)t^4 - \frac{1}{120}k^5\,n\,(24\,n^3$$
$$- 46\,n^2 + 29\,n - 6)\,t^5 + \left(\frac{1}{6}k^6\,n^5 - \frac{163}{360}k^6\,n^4\right.$$
$$+ \frac{329}{720}k^6\,n^3 - \frac{73}{360}k^6\,n^2 + \frac{1}{30}k^6\,n\right)t^6 + \left(-\frac{1}{7}n^6\,k^7\right.$$
$$+ \frac{71}{140}n^5\,k^7 - \frac{901}{1260}n^4\,k^7 + \frac{2521}{5040}n^3\,k^7 - \frac{437}{2520}n^2\,k^7$$
$$+ \frac{1}{42}k^7\,n\right)t^7 + O(t^8)$$

The order obtained is converted to polynomial form for plotting purposes.

```
> assign(sol):
> c(t):=convert(c(t),polynom);
```

$$c(t) := 1 - k\,t + \frac{1}{2}k^2\,n\,t^2 - \frac{1}{6}k^3\,n\,(2\,n - 1)\,t^3 + \left(\frac{1}{4}k^4\,n^3\right.$$
$$- \frac{7}{24}k^4\,n^2 + \frac{1}{12}k^4\,n\right)t^4 - \frac{1}{120}k^5\,n\,(24\,n^3 - 46\,n^2 + 29\,n$$
$$- 6)\,t^5 + \left(\frac{1}{6}k^6\,n^5 - \frac{163}{360}k^6\,n^4 + \frac{329}{720}k^6\,n^3 - \frac{73}{360}k^6\,n^2\right.$$
$$+ \frac{1}{30}k^6\,n\right)t^6 + \left(-\frac{1}{7}n^6\,k^7 + \frac{71}{140}n^5\,k^7 - \frac{901}{1260}n^4\,k^7\right.$$
$$+ \frac{2521}{5040}n^3\,k^7 - \frac{437}{2520}n^2\,k^7 + \frac{1}{42}k^7\,n\right)t^7$$

Next, the series solution obtained is plotted for different values of parameters and compared with the analytical solution.

```
> pars:={k=1,n=1};
```

$$pars := \{n = 1, k = 1\}$$

```
> C:=subs(pars,c(t));
```

$$C := 1 - t + \frac{1}{2}t^2 - \frac{1}{6}t^3 + \frac{1}{24}t^4 - \frac{1}{120}t^5 + \frac{1}{720}t^6 - \frac{1}{5040}t^7$$

```
> Ca:=subs(pars,ca);
```

Error, numeric exception: division by zero

Since division by zero occurs, the limit is obtained.

> Ca:=limit(ca,n=1);

$$Ca := e^{-kt}$$

> Ca:=subs(k=1,Ca);

$$Ca := e^{-t}$$

> tf:=3;

$$tf := 3$$

> plot([C,Ca],t=0..tf,thickness=3,axes=boxed,title="Figure 2.19",labels=[t,"C"]);

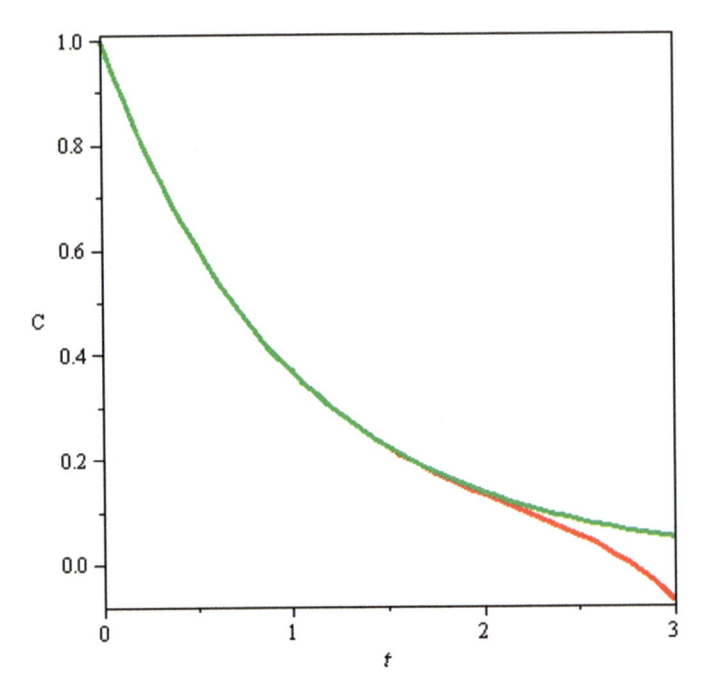

Fig. 2.19

We observe that the series solution diverges for values of t greater than 2. Next, plots are made for different values of the parameters:

> pars:={k=1,n=1/3};

$$pars := \left\{ n = \frac{1}{3}, k = 1 \right\}$$

2.2 Nonlinear Ordinary Differential Equations

> C:=subs(pars,c(t));

$$C := 1 - t + \frac{1}{6}t^2 + \frac{1}{54}t^3 + \frac{1}{216}t^4 + \frac{1}{648}t^5 + \frac{7}{11664}t^6 + \frac{1}{3888}t^7$$

> Ca:=subs(pars,ca);

$$Ca := \left(-\frac{2}{3}t + 1\right)^{3/2}$$

> tf:=1.5;

$$tf := 1.5$$

> plot([C,Ca],t=0..tf,thickness=3,axes=boxed,title="Figure 2.20",labels=[t,"C"]);

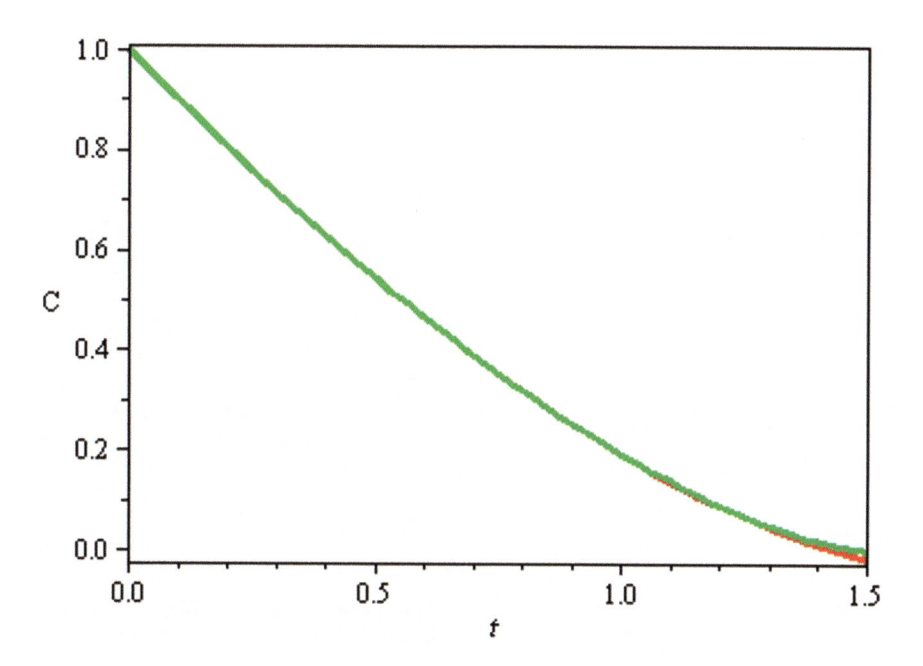

Fig. 2.20

For these values, both the exact analytical solution and the series solution match exactly until t=1.4. Next, a second order reaction is considered.

> pars:={k=1,n=2};

$$pars := \{n = 2, k = 1\}$$

> C:=subs(pars,c(t));

$$C := 1 - t + t^2 - t^3 + t^4 - t^5 + t^6 - t^7$$

> Ca:=subs(pars,ca);

$$Ca := \frac{1}{t+1}$$

> tf:=1;

$$tf := 1$$

> plot([C,Ca],t=0..tf,thickness=3,axes=boxed,title="Figure 2.21",labels=[t,"C"]);

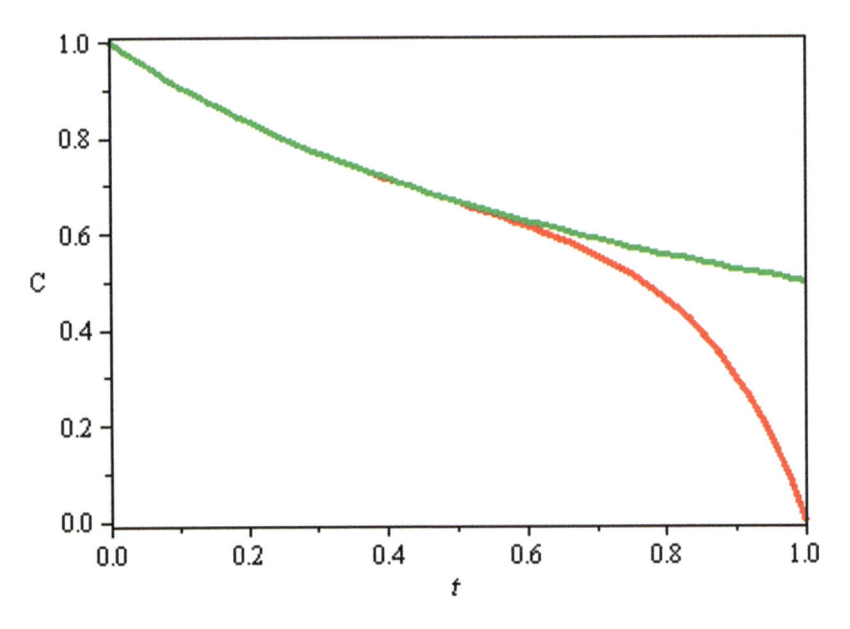

Fig. 2.21

For this case, the series solution starts to diverge after t is greater than 0.4. Hence, one has to be careful while using series solutions. The divergence of the series solution obtained depends upon the problem and values of the parameters. Nevertheless, Maple can give series solutions to the order t^{100} also.

2.2.3 Series Solutions for Nonlinear ODEs

Series solutions for nonlinear ODEs can be obtained using Maple's 'dsolve' command. The syntax is:

dsolve({"differential equations, initial conditions"},{"dependent variable"}, type=series).

2.2 Nonlinear Ordinary Differential Equations

The series solution obtained may be convergent or divergent depending on the problem.

Example 2.2.2. Fermentation Kinetics

> restart:

> with(plots):

Enter the governing equations.

> eq[1]:=diff(y[1](t),t)=k[1]*y[1](t)*(1-y[1](t)/k[2]);

$$eq_1 := \frac{d}{dt} y_1(t) = k_1 y_1(t) \left(1 - \frac{y_1(t)}{k_2}\right)$$

> eq[2]:=diff(y[2](t),t)=k[3]*y[1](t)-k[4]*y[2](t);

$$eq_2 := \frac{d}{dt} y_2(t) = k_3 y_1(t) - k_4 y_2(t)$$

Enter the dependent variables:

> vars:=(y[1](t),y[2](t));

$$vars := y_1(t), y_2(t)$$

Enter the values for the parameters:

> pars:={k[1]=0.04,k[2]=3.92,k[3]=.018,k[4]=0.022};

$$pars := \{k_1 = 0.04, k_2 = 3.92, k_3 = 0.018, k_4 = 0.022\}$$

> eqs:=(subs(pars,eq[1]),subs(pars,eq[2]));

$$eqs := \frac{d}{dt} y_1(t) = 0.04 y_1(t) \left(1 - 0.2551020408 y_1(t)\right), \frac{d}{dt} y_2(t)$$
$$= 0.018 y_1(t) - 0.022 y_2(t)$$

Enter the initial conditions:

> ICs:=(y[1](0)=0.29,y[2](0)=0);

$$ICs := y_1(0) = 0.29, y_2(0) = 0$$

> sol:=dsolve({eqs,ICs},{vars},type=numeric);

$$sol := \mathbf{proc}(x\_rkf45) \ ... \ \mathbf{end \ proc}$$

Next, the plots are made.

> odeplot(sol,[t,y[1](t)],0..400,title="Figure 2.22",axes=boxed,thickness=3);

Fig. 2.22

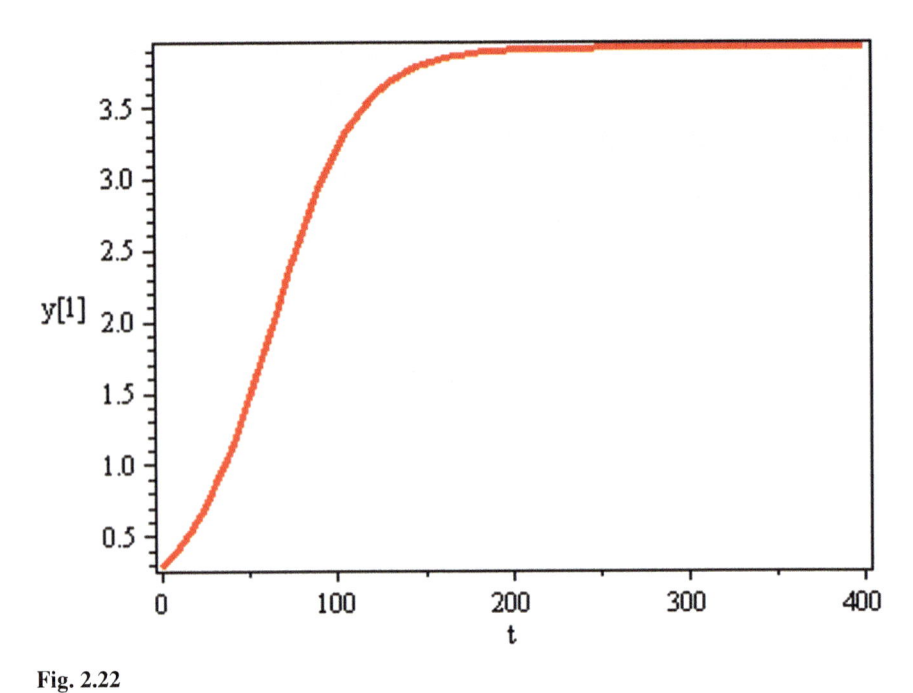

> odeplot(sol,[t,y[2](t)],0..400,title="Figure 2.23",axes=boxed,thickness=3);

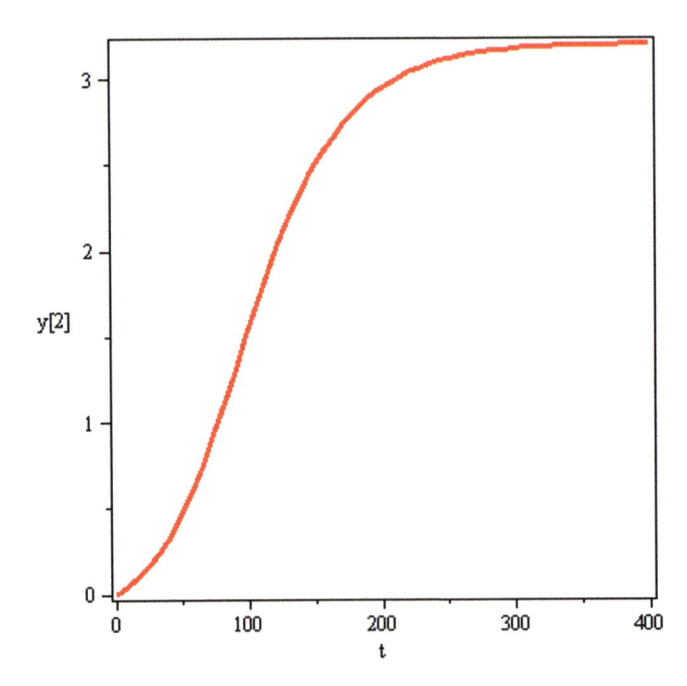

Fig. 2.23

2.2 Nonlinear Ordinary Differential Equations

Next, the solution at a particular time can be obtained as:

> sol(0);
$$[t = 0., y_1(t) = 0.29000000000000, y_2(t) = 0.]$$

> sol(100.);
$$[t = 100., y_1(t) = 3.18890696007849872, y_2(t) = 1.54409378497148175]$$

> sol(200);
$$[t = 200., y_1(t) = 3.90360922790476028, y_2(t) = 2.94610336316537502]$$

Similarly higher order ODEs can be solved using Maple's 'dsolve' command as shown in the next example.

Example 2.2.3

> restart:

> with(plots):

Enter the governing equation:

> eq:=diff(y(t),t$2)+2*zeta/tau*diff(y(t),t)+y(t)=1/tau^2;

$$eq := \frac{d^2}{dt^2} y(t) + \frac{2\zeta\left(\frac{d}{dt} y(t)\right)}{\tau} + y(t) = \frac{1}{\tau^2}$$

Enter the parameters:

> pars:={tau=1,zeta=1/2};

$$pars := \left\{\tau = 1, \zeta = \frac{1}{2}\right\}$$

> eq:=subs(pars,eq);

$$eq := \frac{d^2}{dt^2} y(t) + \frac{d}{dt} y(t) + y(t) = 1$$

> ICs:=(y(0)=0,D(y)(0)=0);

$$ICs := y(0) = 0, D(y)(0) = 0$$

> sol:=dsolve({eq,ICs},{y(t)},type=numeric,method=gear);

$$sol := \mathbf{proc}(x\_gear) \; ... \; \mathbf{end\ proc}$$

Note that Gear's method is used for this example. The following methods are available in Maple:

rkf45, rosenbrock, dverk78, 1sode, gear, taylorseries, or classical.

Next, the dependent variable is plotted:

> odeplot(sol,[t,y(t)],0..50,axes=boxed,title="Figure 2.24",thickness=3, labels=[t,"y(t)"]);

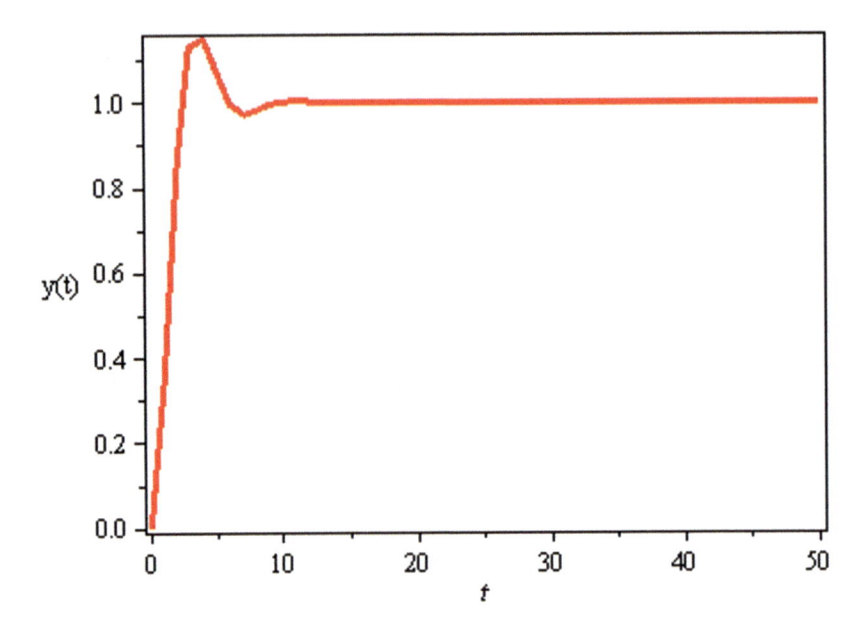

Fig. 2.24

Next, the derivative is plotted:
> odeplot(sol,[t,diff(y(t),t)],0..20,title="Figure 2.25",axes=boxed,thickness=3);

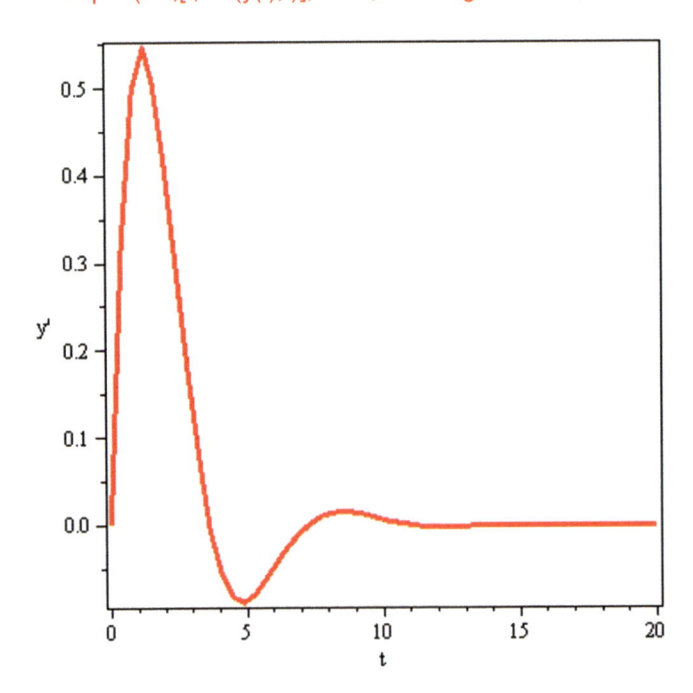

Fig. 2.25

2.2 Nonlinear Ordinary Differential Equations

2.2.4 Stop Conditions

Maple can be asked to stop the numerical calculation based on a criterion on the dependent variable. The syntax is:

dsolve({"differential equations, initial conditions"},{"dependent variables"}, type=numeric, stop_cond=["function to be satisfied"]). This is best illustrated by the next example.

Example 2.2.4. Stop Conditions

> restart:

> with(plots):

Enter the governing equations:

> eq[1]:=diff(y[1](t),t)=-10*y[1](t)^2+y[2](t);

$$eq_1 := \frac{d}{dt} y_1(t) = -10 y_1(t)^2 + y_2(t)$$

> eq[2]:=diff(y[2](t),t)=10*y[1](t)^2-2*y[2](t);

$$eq_2 := \frac{d}{dt} y_2(t) = 10 y_1(t)^2 - 2 y_2(t)$$

Enter the variables:

> vars:=(y[1](t),y[2](t));

$$vars := y_1(t), y_2(t)$$

> eqs:=(eq[1],eq[2]);

$$eqs := \frac{d}{dt} y_1(t) = -10 y_1(t)^2 + y_2(t), \frac{d}{dt} y_2(t) = 10 y_1(t)^2 - 2 y_2(t)$$

> ICs:=(y[1](0)=1.,y[2](0)=0);

$$ICs := y_1(0) = 1., y_2(0) = 0$$

The governing equations are numerically solved as:

> sol:=dsolve({eqs,ICs},{vars},type=numeric);

$$sol := \mathbf{proc}(x\_rkf45) \ \ldots \ \mathbf{end\ proc}$$

The concentration profiles are plotted:

> odeplot(sol,[t,y[1](t)],0..2,title="Figure 2.26",axes=boxed,thickness=3);

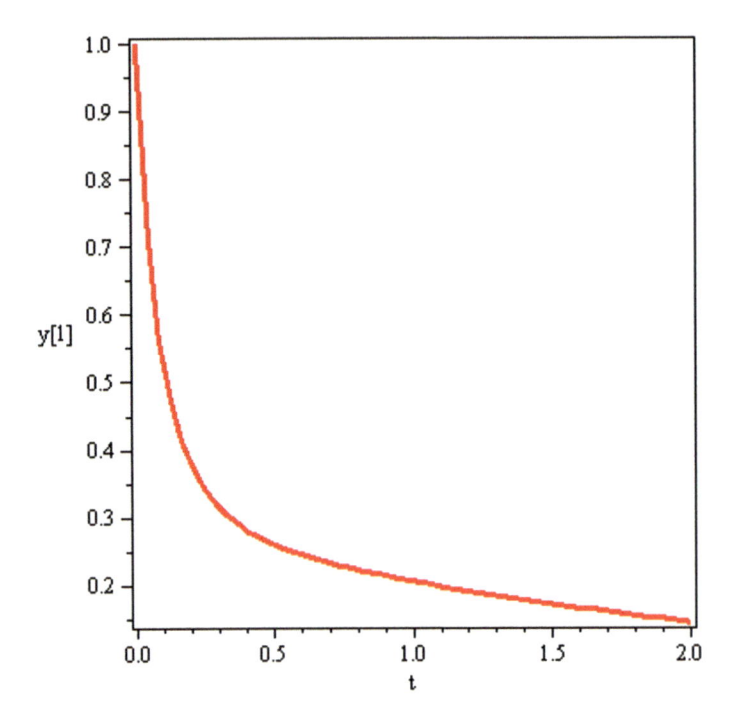

Fig. 2.26

> odeplot(sol,[t,y[2](t)],0..2,title="Figure 2.27",axes=boxed,thickness=3);

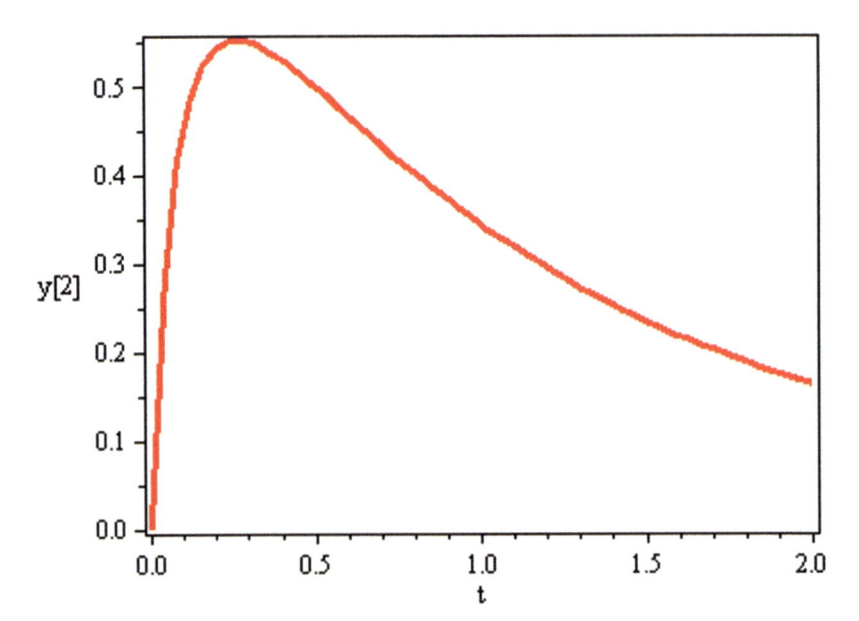

Fig. 2.27

2.2 Nonlinear Ordinary Differential Equations

The objective is to find the time at which the maximum occurs. When y_2 attains the maximum value, $\dfrac{dy_2}{dt}$ becomes zero and hence the right hand side of the equation becomes zero.

> sol:=dsolve({eqs,ICs},{vars},type=numeric,stop_cond=[-2*y[2](t)+ 10*y[1](t)^2]);

$$sol := \mathbf{proc}(x\_rkf45) \; ... \; \mathbf{end\ proc}$$

If we try to evealute the solution at t=1, we get:

> ssol:=sol(1);

Warning, cannot evaluate the solution further right of .26421692, stop condition #1 violated

$$ssol := \big[t = 0.264216920632211638,\ y_1(t) = 0.332432541535803594,$$
$$y_2(t) = 0.552556973359769055 \big]$$

The numerical calculation stops when the residual is satisfied. The time at which the maximum occures is given by:

> ssol[1];

$$t = 0.26421692063221163$$

The maximum value of y_2 is given by:

> ssol[3];

$$y_2(t) = 0.55255697335976905$$

When we plot the solution, the profiles stop when the maximum value of y_2 is reached:

> odeplot(sol,[t,y[1](t)],0..2,title="Figure 2.28",axes=boxed,thickness=3);

Warning, cannot evaluate the solution further right of .26421692, stop condition #1 violated

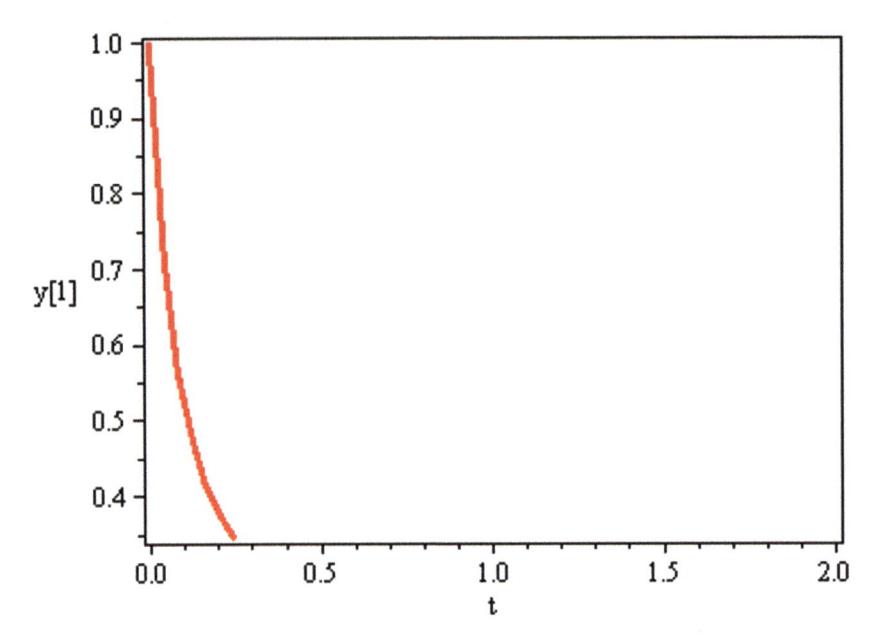

Fig. 2.28

> odeplot(sol,[t,y[2](t)],0..2,title="Figure 2.29",axes=boxed,thickness=3);

Warning, cannot evaluate the solution further right of .26421692, stop condition #1 violated

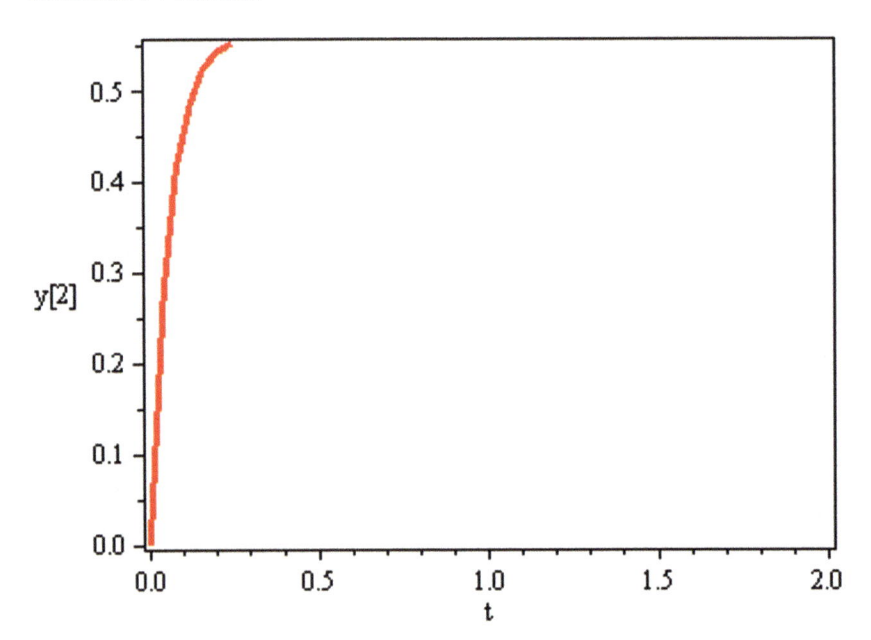

Fig. 2.29

2.2 Nonlinear Ordinary Differential Equations

2.2.5 Stiff ODEs

The standard Euler methods and Runge-Kutta methods do not converge for stiff ODE's. A still system can be defined as one in which the stability of the numerical methods used becomes an issue. Maple has an inbuilt stiff solver.

Example 2.2.5. Stiff Ordinary Differential Equations

> restart:

> with(plots):

The governing equations are entered here:

> eq[1]:=diff(B(t),t)=k*B(t)*S(t)/(K+S(t));

$$eq_1 := \frac{d}{dt} B(t) = \frac{k\, B(t)\, S(t)}{K + S(t)}$$

> eq[2]:=diff(S(t),t)=-0.75*k*B(t)*S(t)/(K+S(t));;

$$eq_2 := \frac{d}{dt} S(t) = -\frac{0.75\, k\, B(t)\, S(t)}{K + S(t)}$$

The dependent variables are entered here:

> vars:=(B(t),S(t));

$$vars := B(t), S(t)$$

The parameters are entered here:

> pars:={k=0.3,K=1e-6};

$$pars := \{k = 0.3, K = 0.000001\}$$

> eqs:=(subs(pars,eq[1]),subs(pars,eq[2]));

$$eqs := \frac{d}{dt} B(t) = \frac{0.3\, B(t)\, S(t)}{0.000001 + S(t)},\ \frac{d}{dt} S(t) = -\frac{0.225\, B(t)\, S(t)}{0.000001 + S(t)}$$

The initial conditions are entered here:

> ICs:=(B(0)=0.05,S(0)=5);

$$ICs := B(0) = 0.05, S(0) = 5$$

Next, the numerical solution is found and plotted until t=20:

> sol:=dsolve({eqs,ICs},{vars},type=numeric);

$$sol := \textbf{proc}(x\_rkf45\,)\ \ldots\ \textbf{end proc}$$

> tf:=20;

$$tf := 20$$

> odeplot(sol,[t,B(t)],0..tf,title="Figure 2.30",color=blue,axes=boxed,thickness=3);

108   2 Initial Value Problems

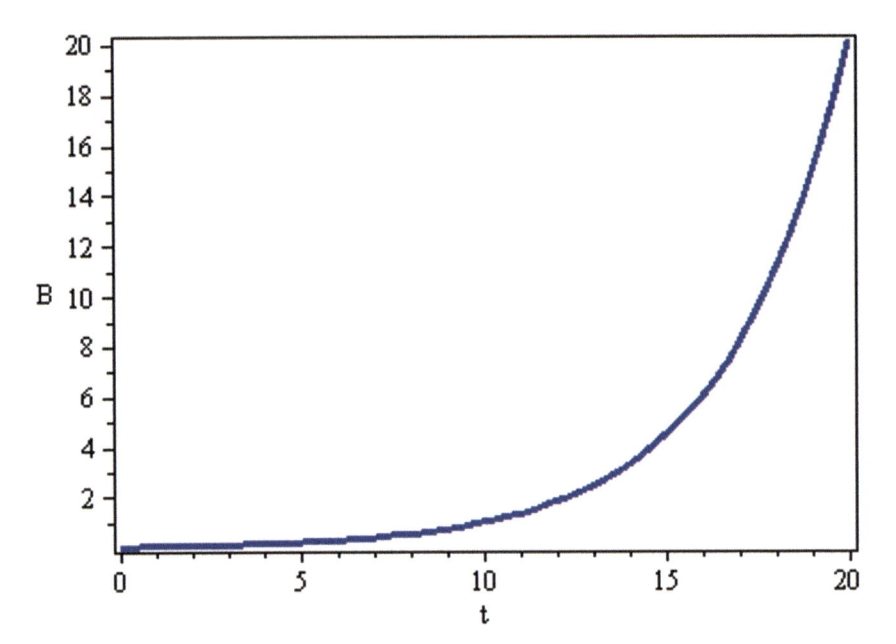

Fig. 2.30

> odeplot(sol,[t,S(t)],0..tf,title="Figure 2.31",color=blue,axes=
boxed,thickness=3);

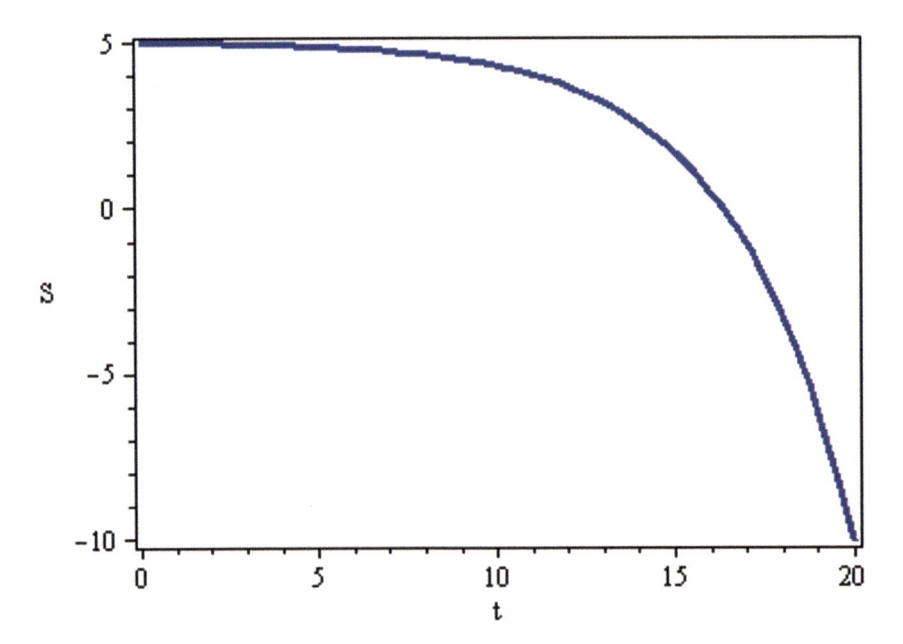

Fig. 2.31

2.2 Nonlinear Ordinary Differential Equations

We observe that Maple predicts negative concentration. This is because the default absolute error in 'dsolve' numeric is only 1d-6, which can be decreased to predict more accurate solutions:

> sol:=dsolve({eqs,ICs},{vars},type=numeric,abserr=1e-10);

$$sol := \mathbf{proc}(x\_rkf45) \ ... \ \mathbf{end\ proc}$$

> odeplot(sol,[t,B(t)],0..tf,title="Figure 2.32",axes=boxed,thickness=3);

Warning, cannot evaluate the solution further right of 16.734694, maxfun limit exceeded (see ?dsolve,maxfun for details)

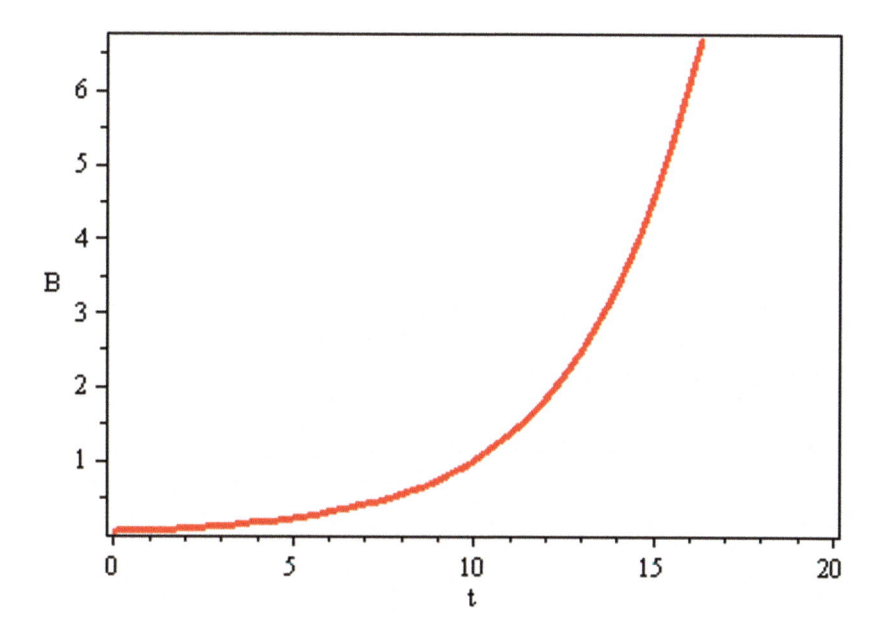

Fig. 2.32

> odeplot(sol,[t,S(t)],0..tf,title="Figure 2.33",axes=boxed,thickness=3);

Warning, cannot evaluate the solution further right of 16.734694, maxfun limit exceeded (see ?dsolve,maxfun for details)

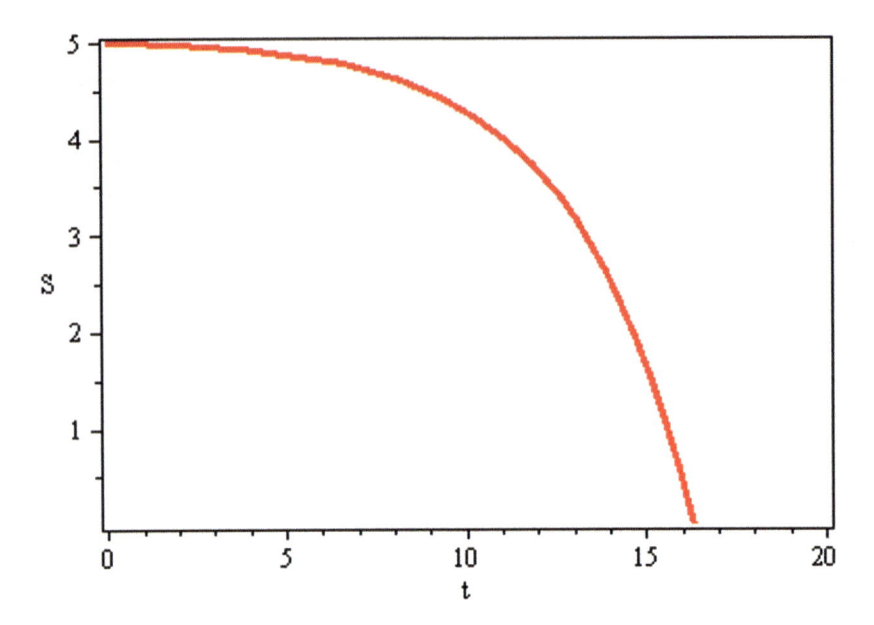

Fig. 2.33

Maple's default Runge-Kutta method cannot predict the profiles after T=16.3. In addition, the program takes too long to run. This is a still problem and can be conveniently solved by using Maple's still solver.

> sol:=dsolve({eqs,ICs},{vars},type=numeric,stiff=true,abserr=1e-10);

$$sol := \mathbf{proc}(x\_rosenbrock) \ldots \mathbf{end\ proc}$$

> odeplot(sol,[t,B(t)],0..tf,title="Figure 2.34", color=green,axes= boxed,thickness=3);

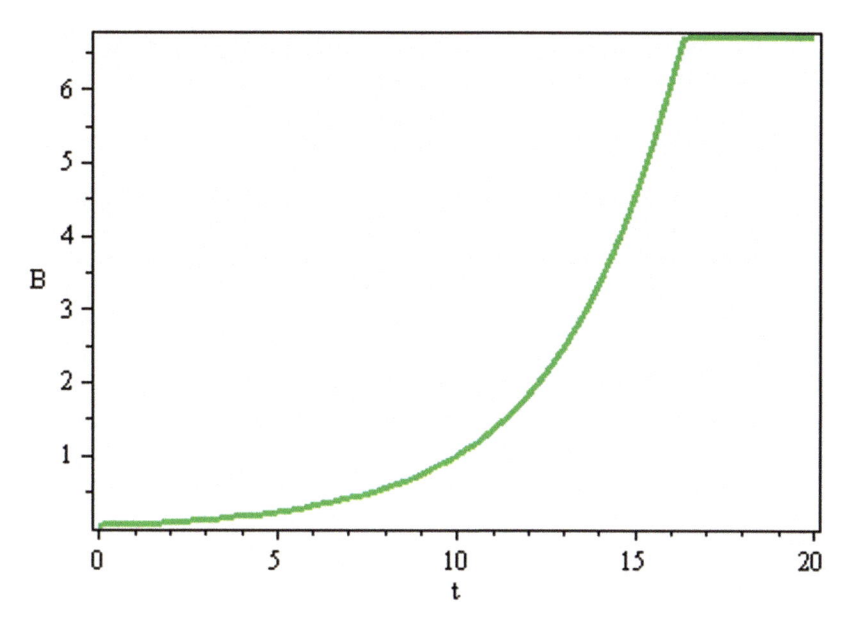

2.2 Nonlinear Ordinary Differential Equations

Fig. 2.34

> odeplot(sol,[t,S(t)],0..tf,title="Figure 2.35",color=green,axes=boxed,thickness=3);

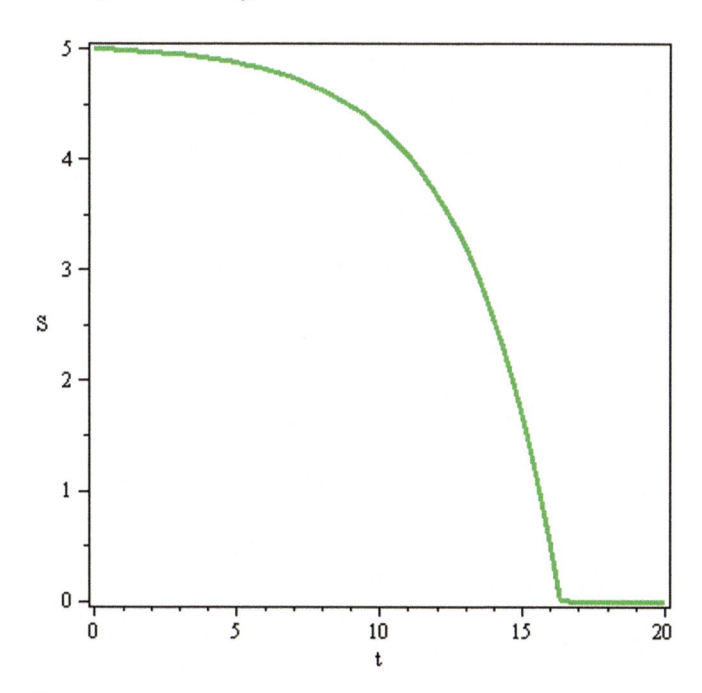

Fig. 2.35

2.2.6 Differential Algebraic Equations

Often times modeling of chemical systems involves solving differential equations coupled with algebraic equations. This system is usually referred as differential algebraic systems (DAEs). DAEs can be solved by converting the algebraic equations to differential equations. The initial condition for the converted differential equation is found using the algebraic equation. This is best illustrated using the next example.

Example 2.2.6. Differential Algebraic Equations

There are two dependent variables L and T. Here x_2 is the independent variable as x_1 can be replaced using the relation $x_1 + x_2 = 1$. The algebraic equation $k_1x_1 + k_2x_2 = 1$ governs the temperature. The differential equation is obtained using the differentiating equation $k_1x_1 + k_2x_2 = 1$. The initial condition is found by substituting values for constants and x_2 into equation $k_1x_1 + k_2x_2 = 1$.

> restart:

> with(plots):

The differential equation is entered here:

> eq[1]:=diff(L(x2),x2)=L(x2)/x2/(k[2]-1);

$$eq_1 := \frac{d}{dx2} L(x2) = \frac{L(x2)}{x2\,(k_2 - 1)}$$

The algebraic equation is entered here:

> eq[2]:=k[1]*x1+k[2]*x2-1;

$$eq_2 := k_1\, x1 + k_2\, x2 - 1$$

Relations for the equilibrium rations and the mole fraction are entered:

> k[1]:=P1/P;k[2]:=P2/P;x1:=1-x2;

$$k_1 := \frac{P1}{P}$$

$$k_2 := \frac{P2}{P}$$

$$x1 := 1 - x2$$

The Antoine equation is entered here.

> P1:=10^(A[1]+B[1]/(T(x2)+C[1]));P2:=10^(A[2]+B[2]/(T(x2)+C[2]));

$$P1 := 10^{A_1 + \frac{B_1}{T(x2) + C_1}}$$

2.2 Nonlinear Ordinary Differential Equations

$$P2 := 10^{A_2 + \frac{B_2}{T(x2) + C_2}}$$

Numerical values for the constants are entered:

> P:=760*1.2;A[1]:=6.90565;B[1]:=-1211.033;C[1]:=220.79;A[2]:=6.95464;B[2]:=-1344.8;C[2]:=219.482;

$$P := 912.0$$
$$A_1 := 6.90565$$
$$B_1 := -1211.033$$
$$C_1 := 220.79$$
$$A_2 := 6.95464$$
$$B_2 := -1344.8$$
$$C_2 := 219.482$$

The algebraic equation simplifies as:

> eq[2];

$$0.0010964912281 \, 10^{6.90565 - \frac{1211.033}{T(x2) + 220.79}} (1 - x2)$$
$$+ 0.0010964912281 \, 10^{6.95464 - \frac{1344.8}{T(x2) + 219.482}} x2 - 1$$

The initial condition for the temperature is found as:

> Ti:=fsolve(subs(T(x2)=Ti,x2=0.4,eq[2]),Ti);

$$Ti := 95.58508724$$

> eq[1];

$$\frac{d}{dx2} L(x2) = \frac{L(x2)}{x2 \left(0.0010964912281 \, 10^{6.95464 - \frac{1344.8}{T(x2) + 219.482}} - 1 \right)}$$

The differential equation for T is obtained by differentiating the algebraic equation:

> eq[3]:=diff(eq[2],x2);

$$eq_3 :=$$

$$\frac{1}{(T(x2)+220.79)^2}\left(1.32788706110^{6.90565-\frac{1211.033}{T(x2)+220.79}}\right.$$

$$\left.\left(\frac{1}{dx2}d\,T(x2)\right)\ln(10)\,(1-x2)\right)$$

$$-0.0010964912281\,0^{6.90565-\frac{1211.033}{T(x2)+220.79}}$$

$$+\frac{1}{(T(x2)+219.482)^2}\left(1.474561403\right.$$

$$10^{6.95464-\frac{1344.8}{T(x2)+219.482}}\left(\frac{d}{dx2}T(x2)\right)\ln(10)\,x2\right)$$

$$+\,0.0010964912281\,0^{6.95464-\frac{1344.8}{T(x2)+219.482}}$$

The dependent variables are entered here:

> vars:=(L(x2),T(x2));

$$vars := L(x2),\,T(x2)$$

The governing equations are entered here:

> eqs:=(eq[1],eq[3]);

$$eqs := \frac{d}{dx2}L(x2)$$

$$=\frac{L(x2)}{x2\left(0.0010964912281\,0^{6.95464-\frac{1344.8}{T(x2)+219.482}}-1\right)},$$

$$\frac{1}{(T(x2)+220.79)^2}\left(1.32788706110^{6.90565-\frac{1211.033}{T(x2)+220.79}}\right.$$

$$\left.\left(\frac{1}{dx2}d\,T(x2)\right)\ln(10)\,(1-x2)\right)$$

$$-0.0010964912281\,0^{6.90565-\frac{1211.033}{T(x2)+220.79}}$$

$$+\frac{1}{(T(x2)+219.482)^2}\left(1.474561403\right.$$

$$10^{6.95464-\frac{1344.8}{T(x2)+219.482}}\left(\frac{d}{dx2}T(x2)\right)\ln(10)\,x2\right)$$

$$+\,0.0010964912281\,0^{6.95464-\frac{1344.8}{T(x2)+219.482}}$$

2.2 Nonlinear Ordinary Differential Equations

The initial conditions are entered here.

> ICs:=(L(0.4)=100,T(0.4)=Ti);

$$ICs := L(0.4) = 100, \ T(0.4) = 95.5850872$$

The numerical solution is obtained as:

> sol:=dsolve({eqs,ICs},{vars},type=numeric);

$$sol := \mathbf{proc}(x\_rkf45) \ ... \ \mathbf{end \ proc}$$

The solution obtained is plotted here:

> odeplot(sol,[x2,L(x2)],0.4..0.9,title="Figure 2.36",axes=boxed,thickness=3);

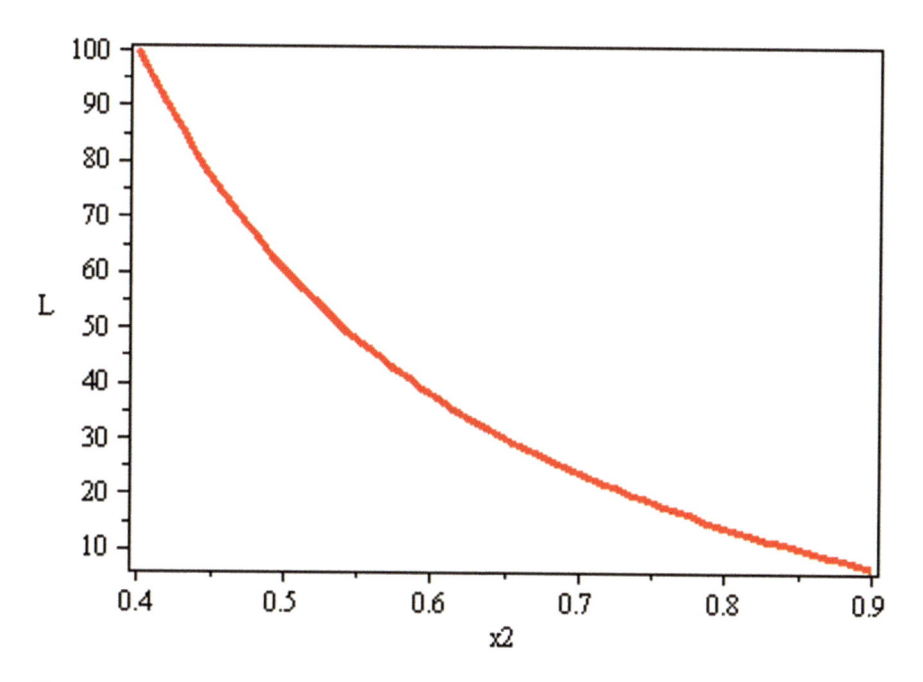

Fig. 2.36

```
> odeplot(sol,[x2,T(x2)],0.4..0.9,title="Figure 2.37",axes=boxed,thickness=3);
```

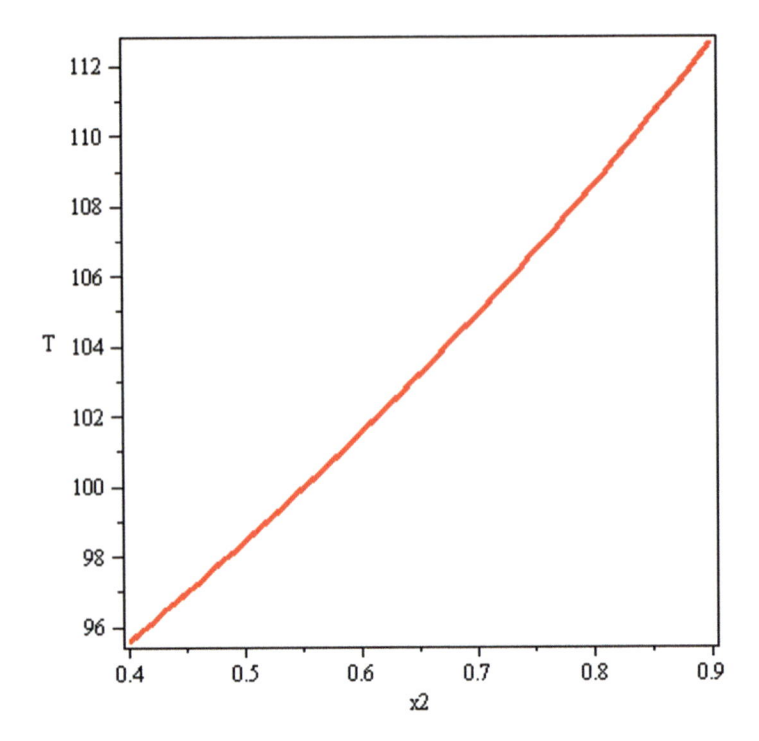

Fig. 2.37

The moles of liquid remaining are found as:

```
> sol(0.9);
```

$$[x2 = 0.9, L(x2) = 6.83621864266518564, T(x2) = 112.647603031846856]$$

```
> sol(0.9)[2];
```

$$L(x2) = 6.8362186426651856$$

At the end of distillation, 6.83 moles of liquid remain.

2.2.7 Multiple Steady States

A certain class of initial value problems exhibits multiple steady states. Depending on the initial condition, the problems converge to different steady state values.

2.2 Nonlinear Ordinary Differential Equations

Multiplicity of steady states is a separate science by itself and different systems have been analyzed for multiple states in the literature (Aris, 1999).[14] The theory of bifurcation analysis is commonly used in the literature and is beyond the scope of this book. In this book, we restrict ourselves to finding and plotting the multiple states numerically.

Example 2.2.7. Multiple Steady States

> restart:

> with(plots):

The governing equation is entered here:

> eq:=diff(theta(t),t)=P*(1-theta(t))-theta(t)*exp(-alpha*theta(t));

$$eq := \frac{d}{dt}\theta(t) = P(1-\theta(t)) - \theta(t)\,e^{-\alpha\,\theta(t)}$$

The steady states are found by equating the right hand side to zero:

> Eq:=subs(theta(t)=theta,rhs(eq));

$$Eq := P(1-\theta) - \theta\,e^{-\alpha\,\theta}$$

Even though P is the parameter and theta is the dependent variable, the steady state equation cannot be solved using theta as a function of P. However, P can be solved as a function of theta as follows:

> Ps:=solve(Eq,P);

$$Ps := -\frac{\theta\,e^{-\alpha\,\theta}}{-1+\theta}$$

Next, P can be plotted as a function of theta (steady state solution):

> p1:=plot(subs(alpha=6,Ps),theta=0..1,view=[0..1,0..0.1],labels=[theta,P], thickness=3,title="Figure 2.38",axes=boxed):

> p2:=plot(0.05,theta=0..1):

> p3:=textplot([0.2,.055,(0.05)]):

> p4:=textplot([0.58,.055,(0.05)]):

> p5:=textplot([0.9,.055,(0.05)]):

> display({p1,p2,p3,p4,p5});

118   2 Initial Value Problems

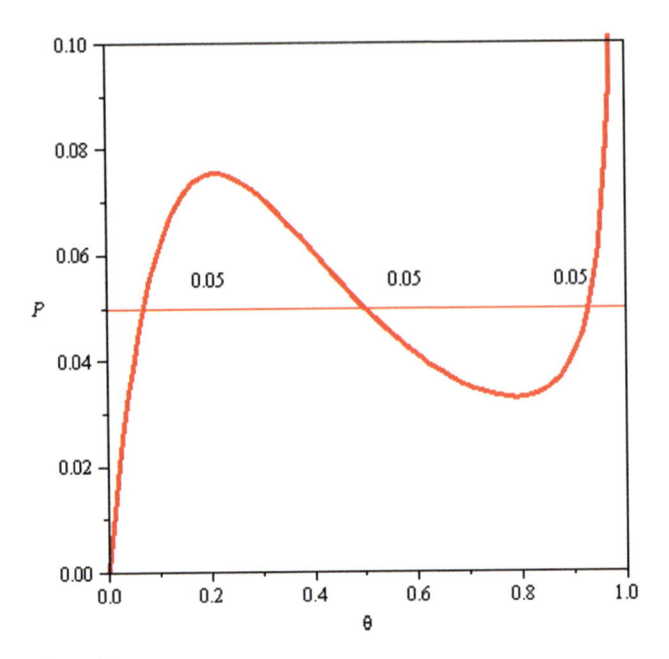

Fig. 2.38

In the above plot we observe that the line P=0.05 cuts the curve at three different points. Theta vs P can be made by using the 'implicitplot' command in Maple.

> implicitplot(P-subs(alpha=6,Ps),P=0..0.1,theta=0..0.97,thickness=3, color=green,title="Figure 2.39",axes=boxed);

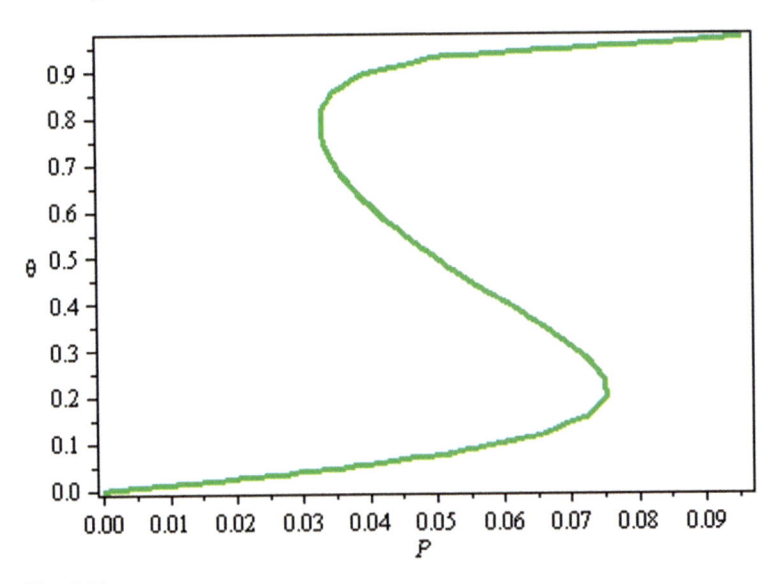

Fig. 2.39

2.2 Nonlinear Ordinary Differential Equations

Hence, there are three different steady states. These three different states can be obtained by equating the pressure to 0.5.

> Eqtheta:=subs(alpha=6,Ps)=0.05;

$$Eqtheta := -\frac{\theta e^{-6\theta}}{-1+\theta} = 0.05$$

There are three different roots for the above equation. They can be obtained by providing different initial guesses:

> st1:=fsolve(Eqtheta,theta=0.);

$$st1 := 0.0711855682$$

> st2:=fsolve(Eqtheta,theta=0.5);

$$st2 := 0.497866110$$

> st3:=fsolve(Eqtheta,theta=0.8);

$$st3 := 0.929740579$$

Next, the transient equation is solved for different initial conditions:

> eqtheta:=subs(alpha=6,P=0.05,eq);

$$eqtheta := \frac{d}{dt}\theta(t) = 0.05 - 0.05\,\theta(t) - \theta(t)\,e^{-6\theta(t)}$$

> sol:=dsolve({eqtheta,theta(0)=0},theta(t),type=numeric);

$$sol := \mathbf{proc}(x\_rkf45\,) \ \ldots \ \mathbf{end\ proc}$$

> odeplot(sol,[t,theta(t)],0.0..20,axes=boxed,color=blue,title="Figure 2.40", thickness=3);

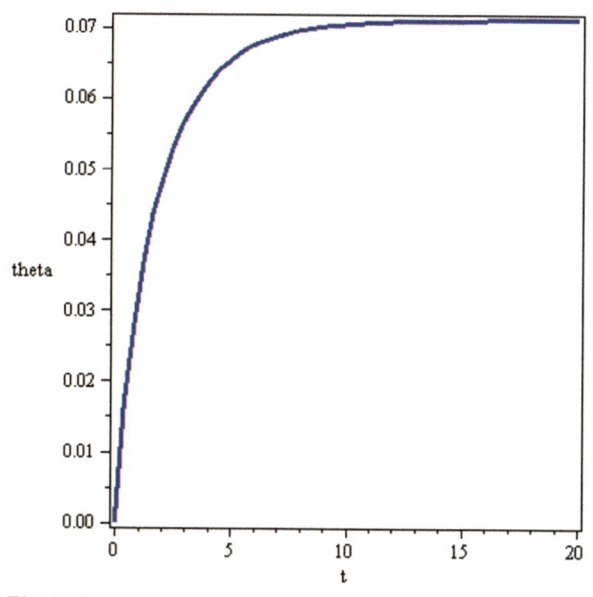

Fig. 2.40

> sol:=dsolve({eqtheta,theta(0)=1},theta(t),type=numeric);

$$sol := \mathbf{proc}(x\_rkf45) \; ... \; \mathbf{end\; proc}$$

> odeplot(sol,[t,theta(t)],0.0..500,axes=boxed,color=magenta,title="Figure 2.41",thickness=3);

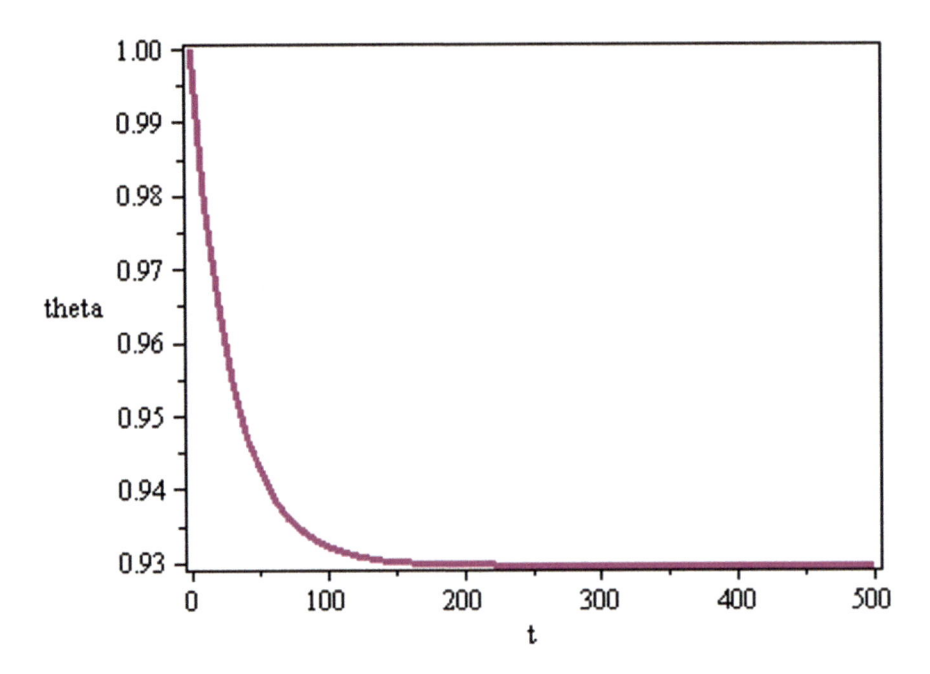

Fig. 2.41

> sol:=dsolve({eqtheta,theta(0)=0.5},theta(t),type=numeric);

$$sol := \mathbf{proc}(x\_rkf45) \; ... \; \mathbf{end\; proc}$$

> odeplot(sol,[t,theta(t)],0.0..500,axes=boxed,title="Figure 2.42",thickness=3);

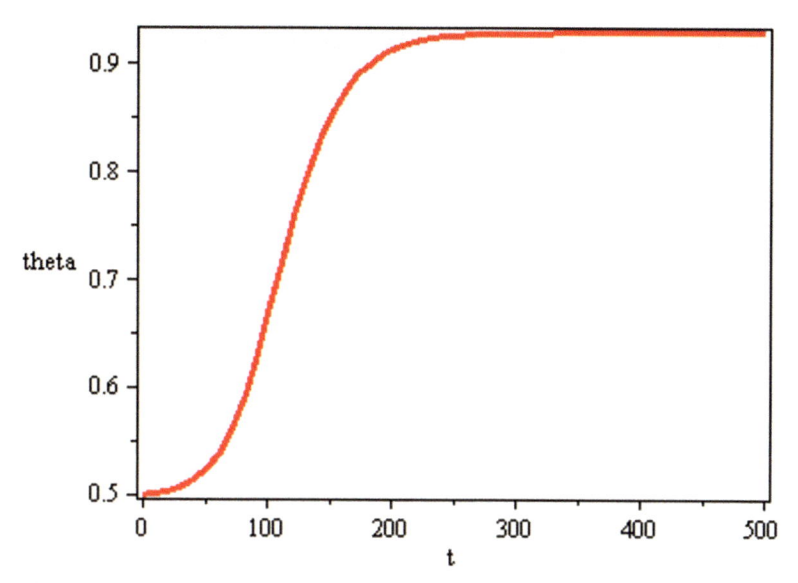

2.2 Nonlinear Ordinary Differential Equations

Fig. 2.42

> sol:=dsolve({eqtheta,theta(0)=0.4},theta(t),type=numeric);

$$sol := \mathbf{proc}(x\_rkf45) \ ... \ \mathbf{end \ proc}$$

> odeplot(sol,[t,theta(t)],0.0..50,axes=boxed,color=green,title="Figure 2.43",thickness=3);

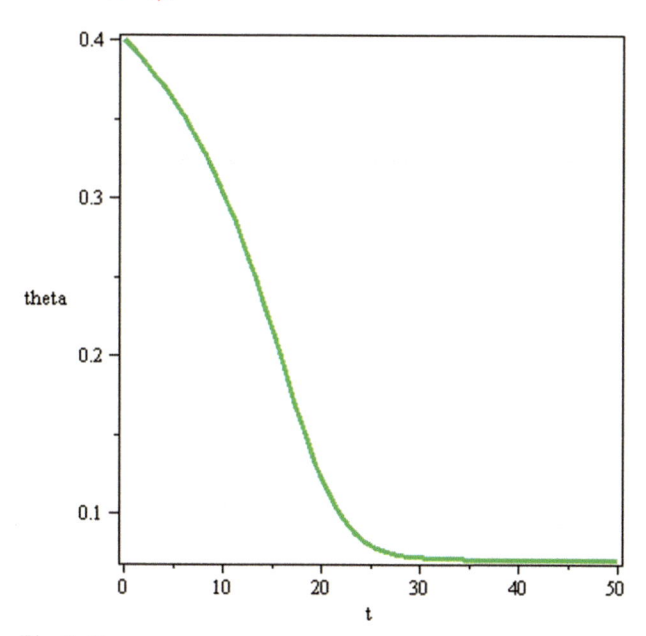

Fig. 2.43

We obtained three different steady states. The stability of these states can be verified by assigning these values as the initial conditions. If we start with a stable steady state solution as the initial condition, the process remains at the stable steady state solution. If we start with an unstable steady state solution, the process moves to one of the steady state solutions.

> sol:=dsolve({eqtheta,theta(0)=st1},theta(t),type=numeric);

$$sol := \mathbf{proc}(x\_rkf45\ )\ ...\ \mathbf{end\ proc}$$

> odeplot(sol,[t,theta(t)],0..1000,axes=boxed,color=blue,title="Figure 2.44", thickness=3,

view=[0..1000,0..1]);

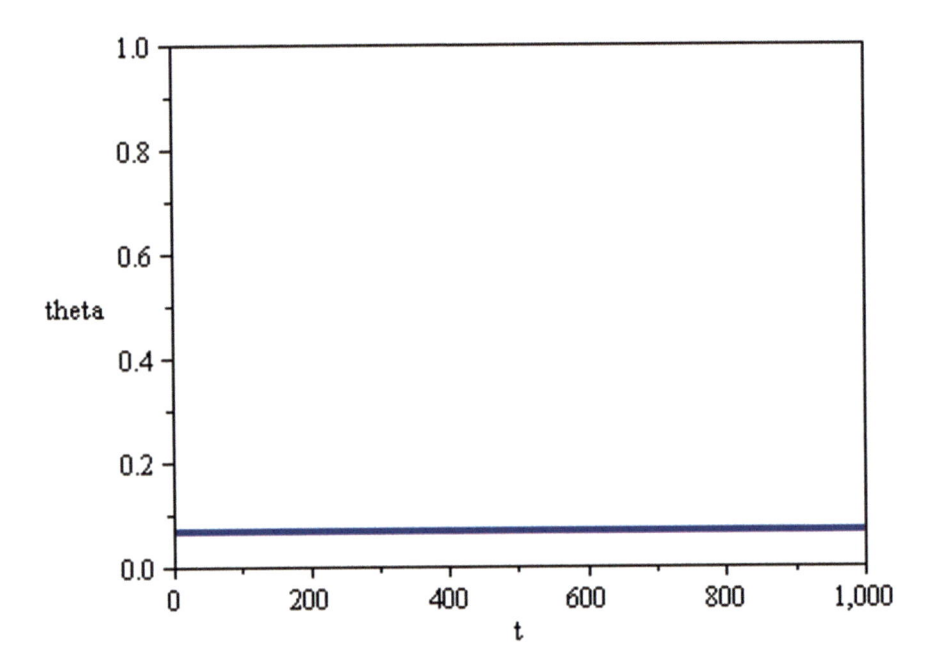

Fig. 2.44

> sol:=dsolve({eqtheta,theta(0)=st3},theta(t),type=numeric);

$$sol := \mathbf{proc}(x\_rkf45\ )\ ...\ \mathbf{end\ proc}$$

> odeplot(sol,[t,theta(t)],0.0..1000,axes=boxed,color=magenta,title= "Figure 2.45",thickness=3,view=[0..1000,0..1]);

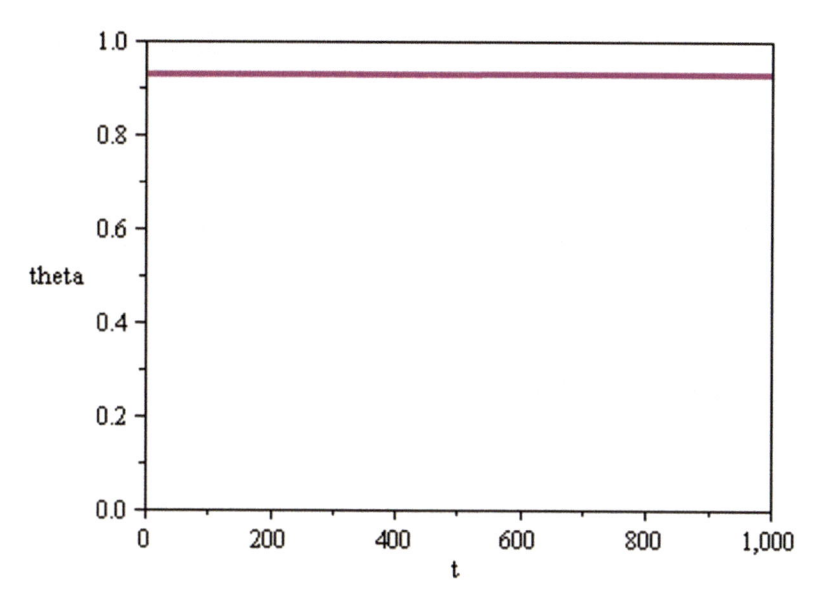

2.2 Nonlinear Ordinary Differential Equations

Fig. 2.45

> sol:=dsolve({eqtheta,theta(0)=st2},theta(t),type=numeric);

$$sol := \mathbf{proc}(x\_rkf45) \ \ldots \ \mathbf{end \ proc}$$

> odeplot(sol,[t,theta(t)],0.0..1000,axes=boxed,title="Figure 2.46",thickness=3);

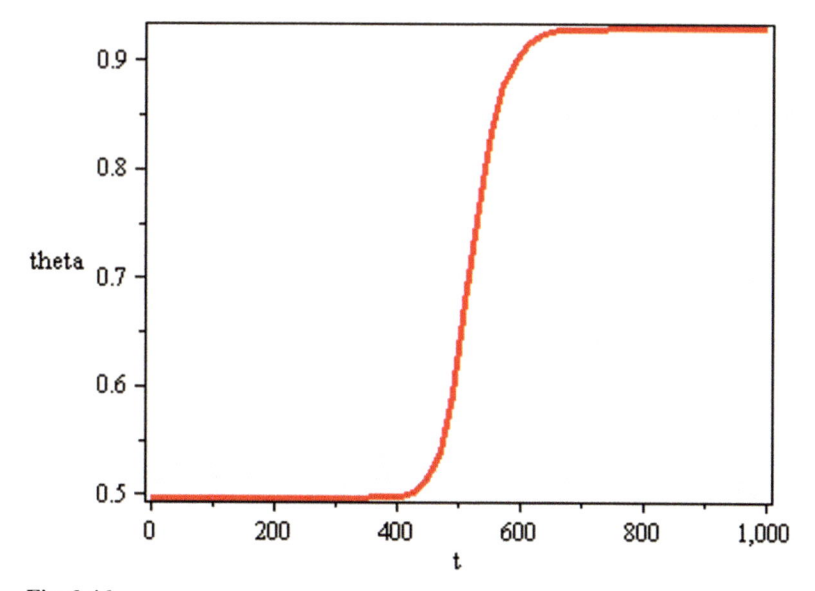

Fig. 2.46

We observe that both st1=0.071 and st3=0.93 are stable steady states. However, st2=0.498 is an unstable steady state. The dependent variable stays at st2 only until t=400 and then the process approaches the stable steady state st3.

2.2.8 Steady State Solutions

When there are two dependent variables (as in example 2.2.8), the independent variable (t) can be eliminated. One of the dependent variable can be solved as function of another dependent variable. This analysis is possible only if the independent variable, t is not present explicitly in the governing equations.

Example 2.2.8. Steady State Solutions

> with(plots):

Enter the governing equations for concentration and temperature:

> eq[1]:=diff(C(t),t)=F/V*(Cf-C(t))-k*exp(-E/R/T(t))*C(t);

$$eq_1 := \frac{d}{dt} C(t) = \frac{F\,(Cf - C(t))}{V} - k\,e^{-\frac{E}{R\,T(t)}}\,C(t)$$

> eq[2]:=diff(T(t),t)=F/V*(Tf-T(t))+(-H/rho/cp)*k*exp(-E/R/T(t))*C(t)-UA/V/rho/cp*(T(t)-Tj);

$$eq_2 := \frac{d}{dt} T(t) = \frac{F\,(Tf - T(t))}{V} - \frac{H\,k\,e^{-\frac{E}{R\,T(t)}}\,C(t)}{\rho\,cp}$$

$$-\frac{UA\,(T(t) - Tj)}{V\,\rho\,cp}$$

The steady state equations are stored in Eq[1] and Eq[2]:

> vars:={C(t)=Cs,T(t)=Ts};

$$vars := \left\{ C(t) = \frac{F\,Cf}{F + k\,e^{-\frac{E}{R\,Ts}}\,V},\ T(t) = Ts \right\}$$

> Eq[1]:=subs(vars,rhs(eq[1]));

$$Eq_1 := \frac{F\left(Cf - \frac{F\,Cf}{F + k\,e^{-\frac{E}{R\,Ts}}\,V}\right)}{V} - \frac{k\,e^{-\frac{E}{R\,Ts}}\,F\,Cf}{F + k\,e^{-\frac{E}{R\,Ts}}\,V}$$

2.2 Nonlinear Ordinary Differential Equations

> Eq[2]:=subs(vars,rhs(eq[2]));

$$Eq_2 := \frac{F\,(Tf - Ts)}{V} - \frac{H\,k\,e^{-\frac{E}{R\,Ts}}\,F\,Cf}{\rho\,cp\,\left(F + k\,e^{-\frac{E}{R\,Ts}}\,V\right)} - \frac{UA\,(Ts - Tj)}{V\,\rho\,cp}$$

The steady state concentration can be solved as a function of the steady state temperature Ts as:

> Cs:=solve(Eq[1],Cs);

Warning, solving for expressions other than names or functions is not recommended.

$$Cs := \frac{F\,Cf}{F + k\,e^{-\frac{E}{R\,Ts}}\,V}$$

The equation for steady state temperature simplifies as:

> Eq[2];

$$\frac{F\,(Tf - Ts)}{V} - \frac{H\,k\,e^{-\frac{E}{R\,Ts}}\,F\,Cf}{\rho\,cp\,\left(F + k\,e^{-\frac{E}{R\,Ts}}\,V\right)} - \frac{UA\,(Ts - Tj)}{V\,\rho\,cp}$$

Values for the parameters are substituted:

> pars:={F=V,k=9703*3600,H=-5960,E=11843,rho=500/cp,Cf=10,R=1.987, UA=V*150};

$$pars := \left\{F = V, k = 34930800, E = 11843, H = -5960, \rho = \frac{500}{cp}, Cf\right.$$
$$\left. = 10, R = 1.987, UA = 150\,V\right\}$$

The equation for Ts becomes:

> steq:=subs(pars,Eq[2]);

$$steq := Tf - \frac{13}{10}\,Ts + \frac{4163751360\,e^{-\frac{5960.241570}{Ts}}}{V + 34930800\,e^{-\frac{5960.241570}{Ts}}\,V} + \frac{3}{10}\,Tj$$

V is taken as 1:

> steq:=subs(V=1,steq);

$$steq := Tf - \frac{13}{10}\,Ts + \frac{4163751360\,e^{-\frac{5960.241570}{Ts}}}{1 + 34930800\,e^{-\frac{5960.241570}{Ts}}} + \frac{3}{10}\,Tj$$

The values for Tf and Tj are substituted to solve for Ts:

> neq:=subs(Tf=298,Tj=298,steq);

$$neq := \frac{1937}{5} - \frac{13}{10} Ts + \frac{4163751360 e^{-\frac{5960.241570}{Ts}}}{1 + 34930800 e^{-\frac{5960.241570}{Ts}}}$$

The nonlinear equation is plotted as a function of Ts, and we observe that the function crosses the x axis at 3 distinct points:

> plot(neq,Ts=298..400,thickness=3,title="Figure 2.47",labels=[Ts,"neq"]);

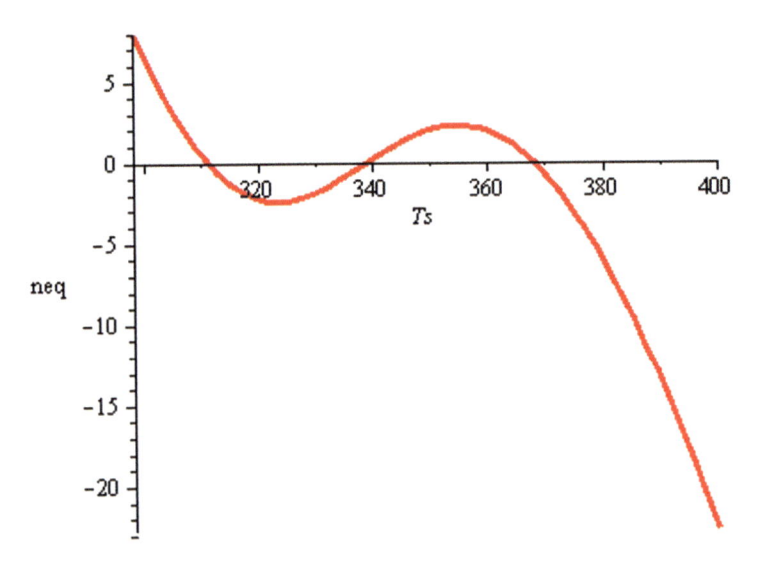

Fig. 2.47

The steady state values are solved as:
> st1:=fsolve(neq,Ts=300);

$$st1 := 311.1709984$$

> st2:=fsolve(neq,Ts=350);

$$st2 := 339.097123$$

> st3:=fsolve(neq,Ts=400);

$$st3 := 368.062852$$

The corresponding steady state values for the concentration are stored as:
> cst1:=evalf(subs(pars,V=1,Ts=st1,Cs));

$$cst1 := 8.56356560$$

> cst2:=evalf(subs(pars,V=1,Ts=st2,Cs));

$$cst2 := 5.51793114$$

2.2 Nonlinear Ordinary Differential Equations

```
> cst3:=evalf(subs(pars,V=1,Ts=st3,Cs));
```
$$cst3 := 2.35891711\!.$$

```
> steq;
```
$$Tf - \frac{13}{10}Ts + \frac{4163751360 e^{-\frac{5960.241570}{Ts}}}{1 + 34930800 e^{-\frac{5960.241570}{Ts}}} + \frac{3}{10}Tj$$

The feed temperature Tf can be solved as a function of Ts shown below:

```
> Tfs:=solve(steq,Tf);
```

$$Tfs := -\frac{1}{1. + 3.493080010^7\, e^{-\frac{5960.241570}{Ts}}}\bigg(0.1000000000\big(-13.\,Ts$$

$$- 4.541004010^8\, Ts\, e^{-\frac{5960.241570}{Ts}}$$

$$+ 4.16375136010^{10}\, e^{-\frac{5960.241570}{Ts}} + 3.\,Tj$$

$$+ 1.047924010^8\, Tj\, e^{-\frac{5960.241570}{Ts}}\bigg)\bigg)$$

Tf can be plotted as a function of Ts as shown below:

```
> plot(subs(Tj=298,Tfs),Ts=298..400,thickness=3,color=green,title=
"Figure 2.48",axes=boxed,labels=[Ts,Tf]);
```

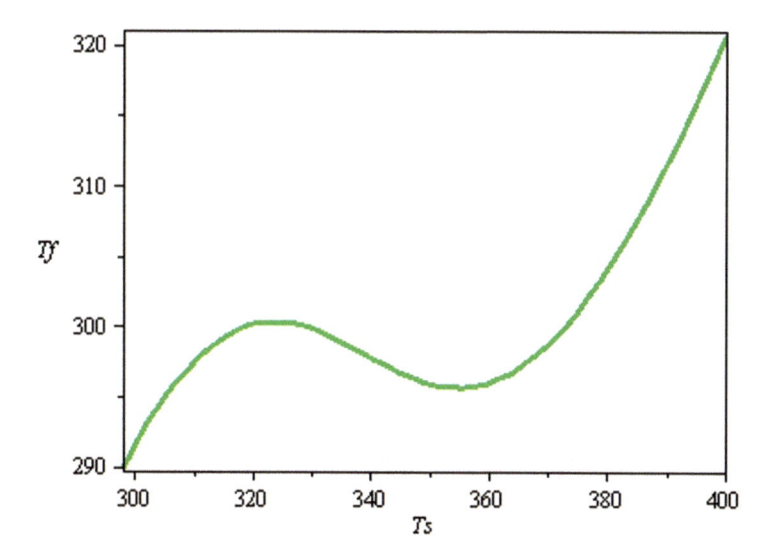

Fig. 2.48

128   2 Initial Value Problems

```
> implicitplot(Tf-subs(Tj=298,Tfs),Tf=290..306,Ts=290..400,thickness=3,
color=blue,title="Figure 2.49",axes=boxed);
```

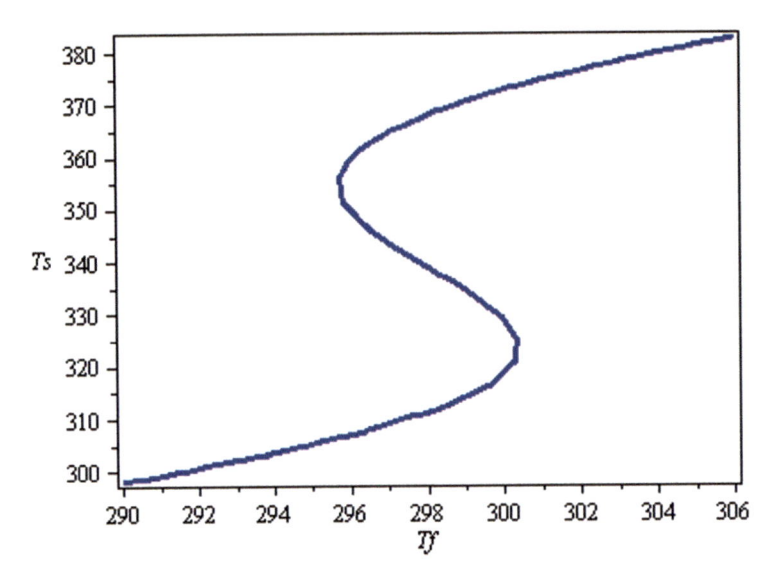

Fig. 2.49

Similarly Ts can be plotted as a function of Tj for the constant value of Tf.

```
> Tjs:=solve(steq,Tj);
```

$$Tjs := -\frac{1}{1. + 3.493080010^7 \, e^{-\frac{5960.241570}{Ts}}} \left(0.3333333333 \left(10. \, Tf \right.\right.$$
$$+ 3.493080010^8 \, Tf \, e^{-\frac{5960.241570}{Ts}} - 13. \, Ts$$
$$- 4.541004010^8 \, Ts \, e^{-\frac{5960.241570}{Ts}}$$
$$\left.\left. + 4.16375136010^{10} \, e^{-\frac{5960.241570}{Ts}} \right)\right)$$

```
> implicitplot(Tj-subs(Tf=298,Tjs),Tj=275..320,Ts=290..400,thickness=3,
color=magenta,title="Figure 2.50",axes=boxed);
```

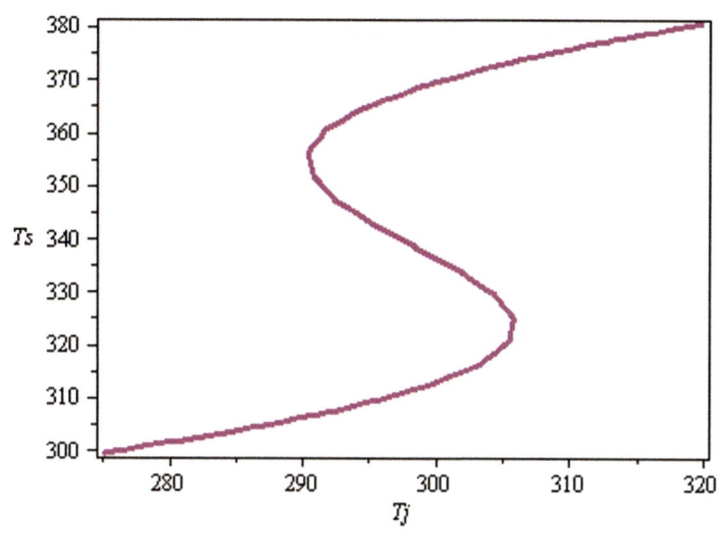

2.2 Nonlinear Ordinary Differential Equations

Fig. 2.50

From the above two plots we observe that there multiple steady states exist. For a particular value of Tj or Tf there can be three distinct values for Ts. Next, the dynamic equations are solved for different initial conditions.

```
> for i to 2 do eq[i]:=subs(Tf=298,Tj=298,subs(pars,eq[i]));od;
```

$$eq_1 := \frac{d}{dt} C(t) = 10 - C(t) - 34930800 e^{-\frac{5960.241570}{T(t)}} C(t)$$

$$eq_2 := \frac{d}{dt} T(t) = \frac{1937}{5} - \frac{13}{10} T(t) + 416375136 e^{-\frac{5960.241570}{T(t)}} C(t)$$

```
> vars:=(C(t),T(t));
```

$$vars := C(t), T(t)$$

```
> eqs:=(eq[1],eq[2]);
```

$$eqs := \frac{d}{dt} C(t) = 10 - C(t) - 34930800 e^{-\frac{5960.241570}{T(t)}} C(t), \frac{d}{dt} T(t)$$

$$= \frac{1937}{5} - \frac{13}{10} T(t) + 416375136 e^{-\frac{5960.241570}{T(t)}} C(t)$$

```
> Ics:=(C(0)=0,T(0)=298);
```

$$Ics := C(0) = 0, T(0) = 298$$

```
> sol:=dsolve({eqs,Ics},{vars},type=numeric);
```

$$sol := \mathbf{proc}(x\_rkf45) \; ... \; \mathbf{end \; proc}$$

> odeplot(sol,[t,C(t)],0..10,thickness=3,color=red,title="Figure. 2.51", axes=boxed);

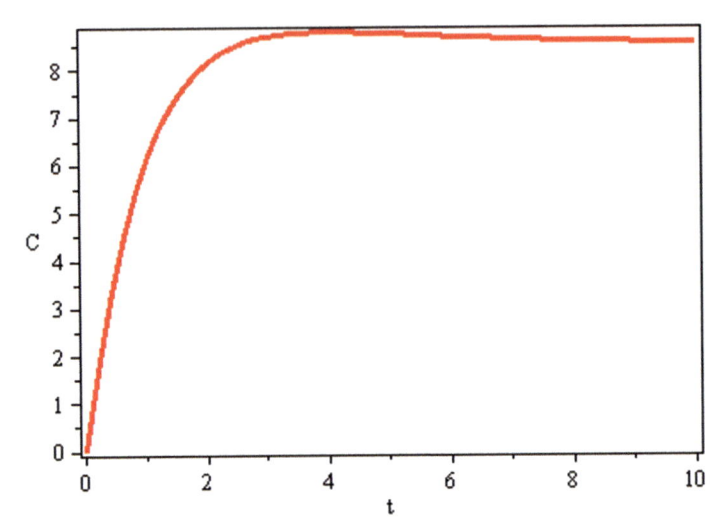

Fig. 2.51

> odeplot(sol,[t,T(t)],0..20,thickness=3,color=green,title="Figure 2.52", axes=boxed);

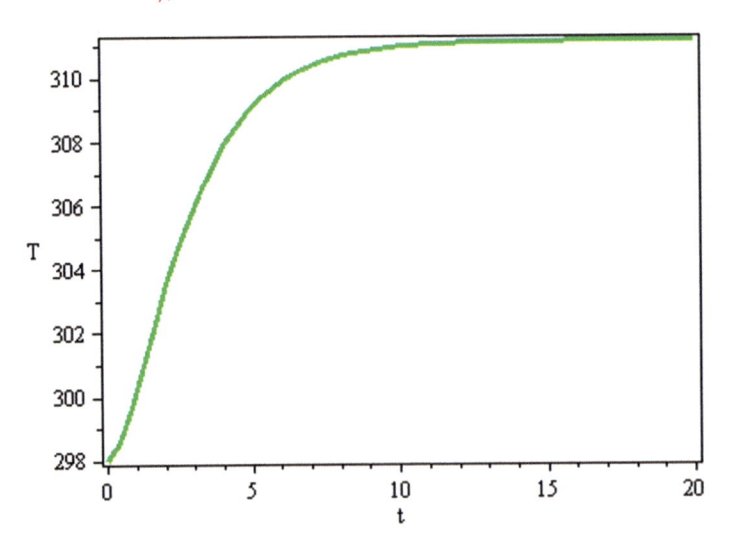

Fig. 2.52

For this case, the process reaches the lower steady state condition st1.

> Ics:=(C(0)=5,T(0)=330);

$$Ics := C(0) = 5, T(0) = 330$$

2.2 Nonlinear Ordinary Differential Equations

> sol:=dsolve({eqs,lcs},{vars},type=numeric);

$$sol := \mathbf{proc}(x\_rkf45\ )\ ...\ \mathbf{end\ proc}$$

> odeplot(sol,[t,C(t)],0..10,thickness=3,color=blue,title="Figure 2.53", axes=boxed);

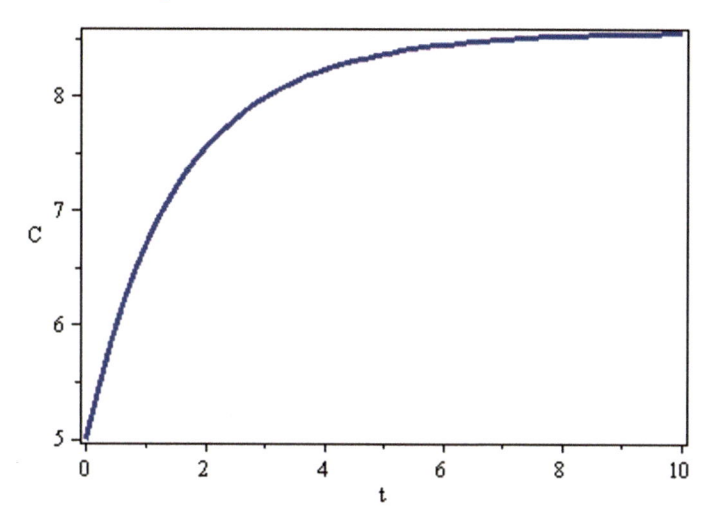

Fig. 2.53

> odeplot(sol,[t,T(t)],0..20,thickness=3,color=magenta,title="Figure 2.54", axes=boxed);

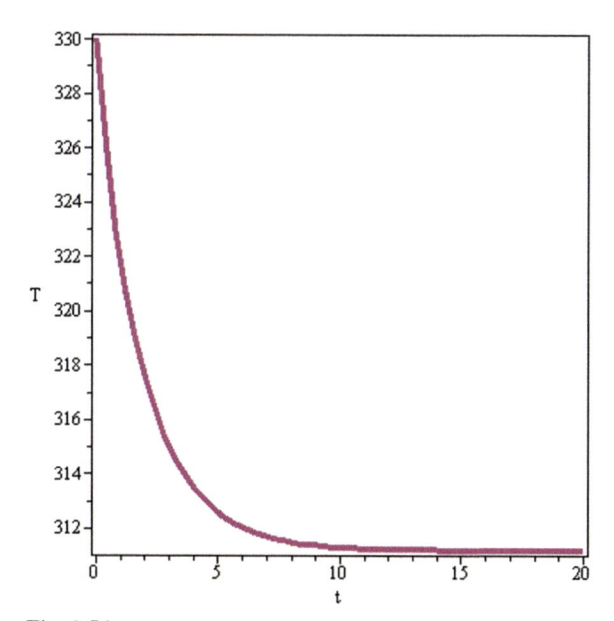

Fig. 2.54

132   2 Initial Value Problems

For this case, also, the process reaches the lower steady state condition st1.
> Ics:=(C(0)=9,T(0)=340);
$$Ics := C(0) = 9, T(0) = 340$$
> sol:=dsolve({eqs,Ics},{vars},type=numeric);
$$sol := \mathbf{proc}(x\_rkf45) \ ... \ \mathbf{end\ proc}$$
> odeplot(sol,[t,C(t)],0..10,thickness=3,color=red,title="Figure 2.55", axes=boxed);

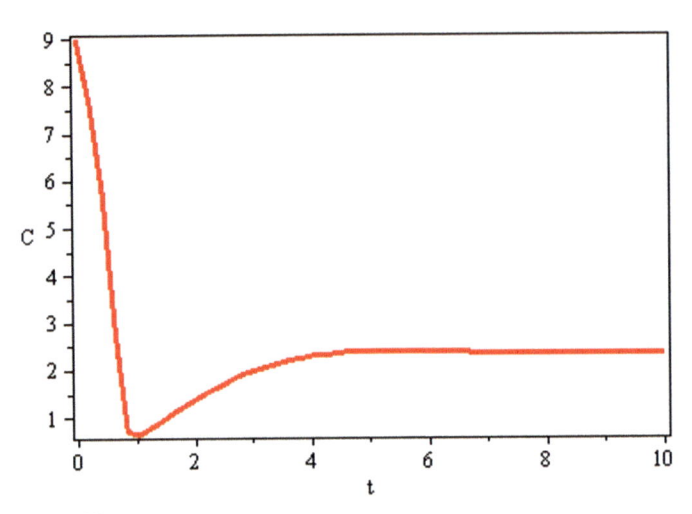

Fig. 2.55

> odeplot(sol,[t,T(t)],0..20,thickness=3,color=green,title="Figure 2.56", axes=boxed);

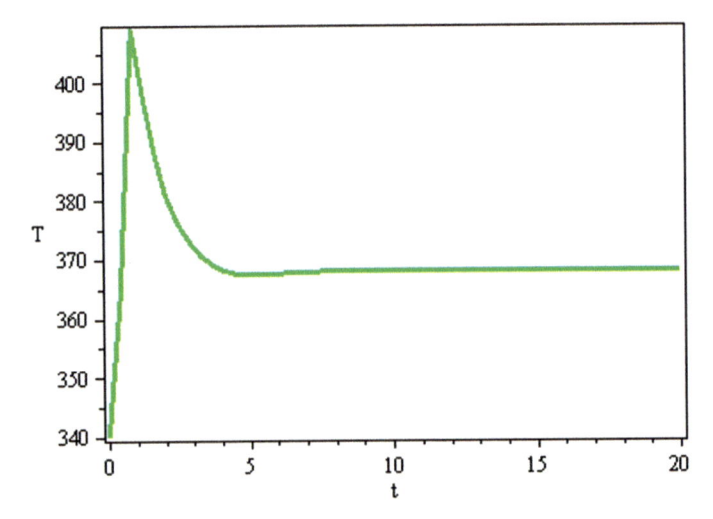

Fig. 2.56

2.2 Nonlinear Ordinary Differential Equations

For this case, the process reaches the lower steady state condition st3. Next, the differential equations are solved taking the steady state values as the initial conditions:

> Ics:=(C(0)=cst1,T(0)=st1);

$$Ics := C(0) = 8.563565608, T(0) = 311.170998$$

> sol:=dsolve({eqs,Ics},{vars},type=numeric);

$$sol := \mathbf{proc}(x\_rkf45) \ ... \ \mathbf{end \ proc}$$

> odeplot(sol,[t,C(t)],0..50,thickness=3,color=magenta,title="Figure 2.57", axes=boxed,view=[0..50,0..10]);

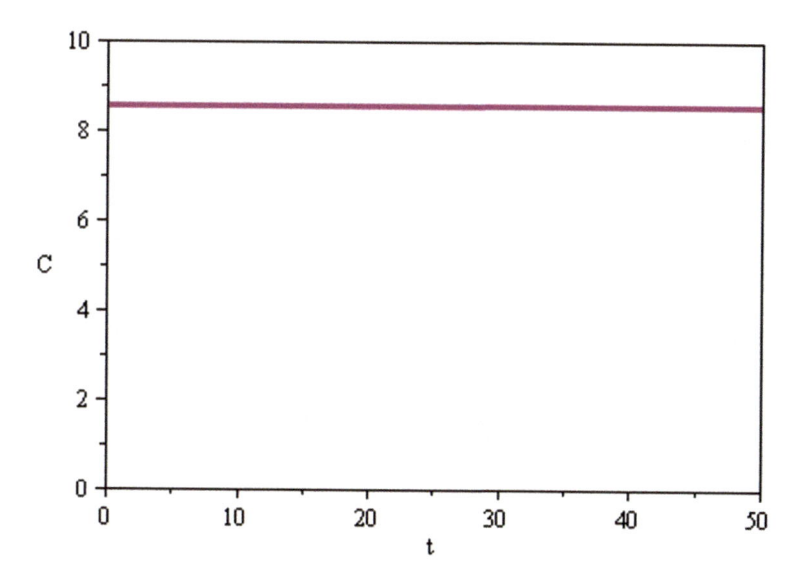

Fig. 2.57

> odeplot(sol,[t,T(t)],0..50,thickness=3,color=red,title="Figure 2.58", axes=boxed,view=[0..50,298..400]);

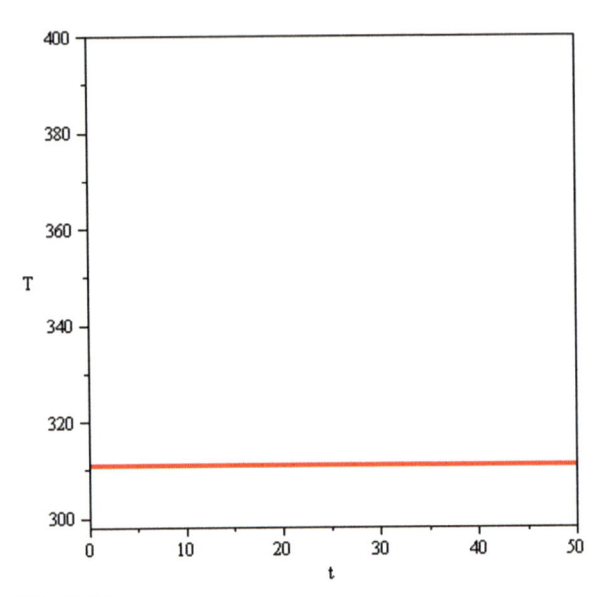

Fig. 2.58

Hence, the lower steady state condition st1 is stable.

> lcs:=(C(0)=cst2,T(0)=st2);

$$lcs := C(0) = 5.517931147, T(0) = 339.097123\text{'}$$

> sol:=dsolve({eqs,lcs},{vars},type=numeric);

$$sol := \mathbf{proc}(x\_rkf45) \ ... \ \mathbf{end\ proc}$$

> odeplot(sol,[t,C(t)],0..50,thickness=3,color=green,title="Figure 2.59", axes=boxed,view=[0..50,0..10]);

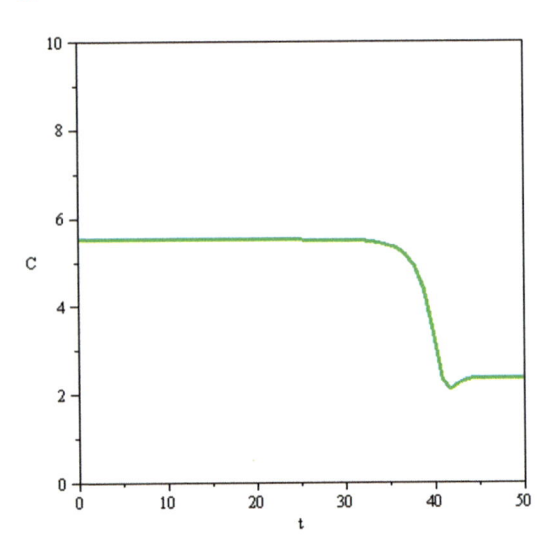

Fig. 2.59

2.2 Nonlinear Ordinary Differential Equations

Hence, the middle steady state condition st2 is unstable.

> odeplot(sol,[t,T(t)],0..50,thickness=3,color=blue,title="Figure 2.60", axes=boxed,view=[0..50,298..400]);

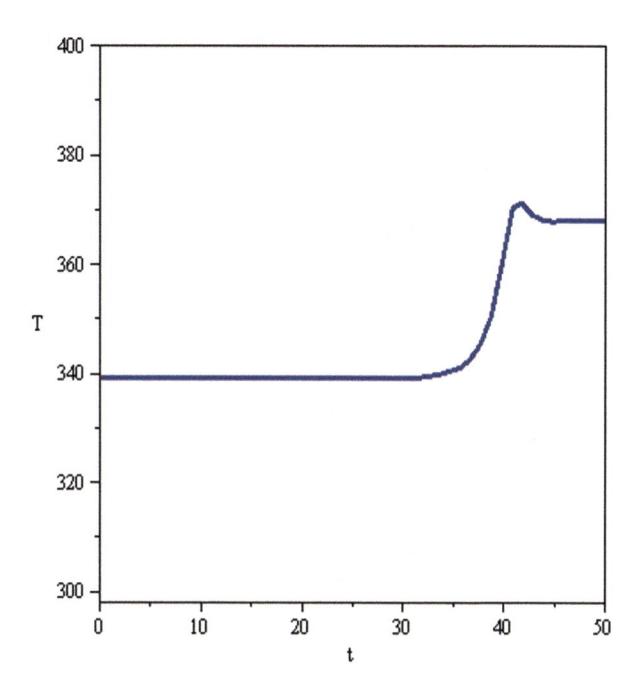

Fig. 2.60

> Ics:=(C(0)=cst3,T(0)=st3);

$$Ics := C(0) = 2.358917112, T(0) = 368.062852$$

> sol:=dsolve({eqs,Ics},{vars},type=numeric);

$$sol := \mathbf{proc}(x\_rkf45) \ ... \ \mathbf{end\ proc}$$

> odeplot(sol,[t,C(t)],0..50,thickness=3,color=magenta,title="Figure 2.61", axes=boxed,view=[0..50,0..10]);

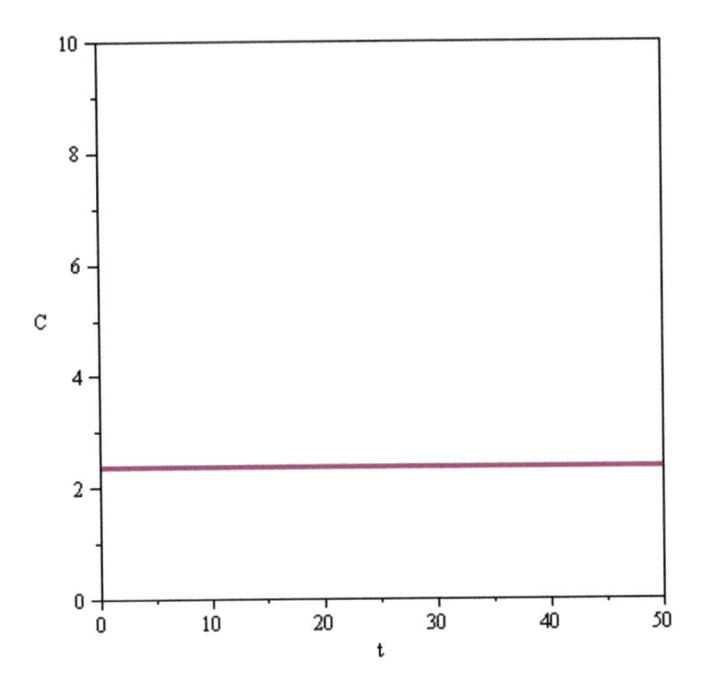

Fig. 2.61

> odeplot(sol,[t,T(t)],0..50,thickness=3,title="Figure 2.62",axes=boxed,
view=[0..50,298..400]);

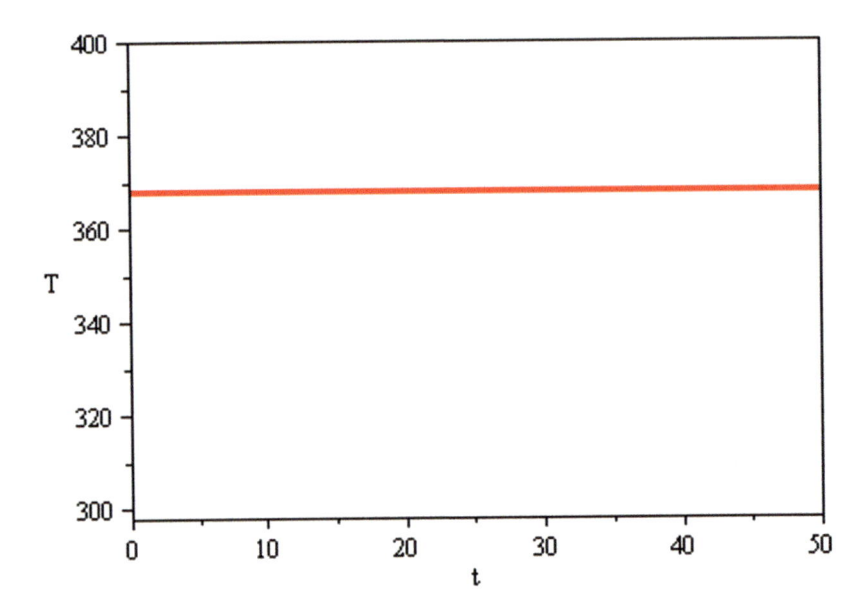

Fig. 2.62

2.2 Nonlinear Ordinary Differential Equations

Hence, the higher steady state condition st3 is stable. Next, initial conditions that are very close to the second steady states are taken:

> Ics:=(C(0)=cst2*1.001,T(0)=st2*1.001);

$$Ics := C(0) = 5.523449078, T(0) = 339.436220\text{\textcent}$$

> sol:=dsolve({eqs,Ics},{vars},type=numeric);

$$sol := \mathbf{proc}(x\_rkf45) \ ... \ \mathbf{end\ proc}$$

> odeplot(sol,[t,C(t)],0..20,thickness=3,color=green,title="Figure 2.63", axes=boxed,view=[0..20,0..10]);

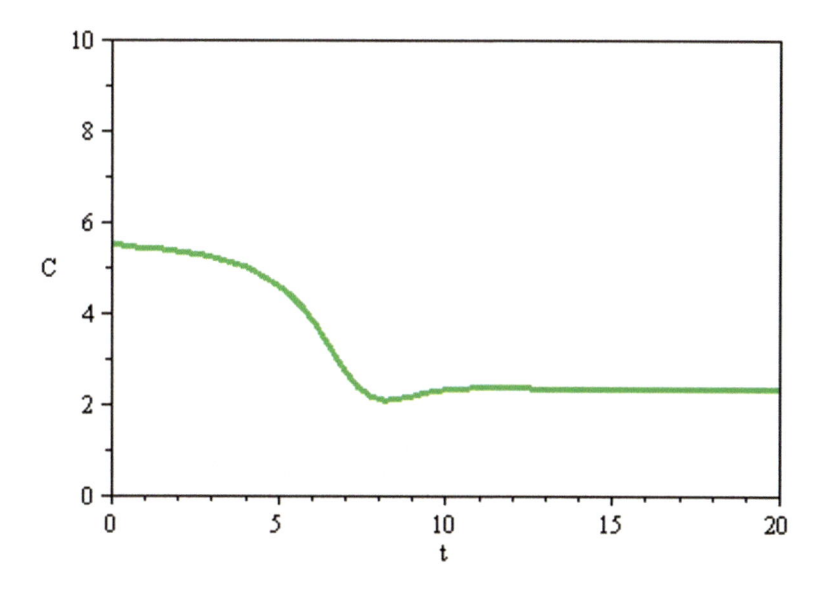

Fig. 2.63

> odeplot(sol,[t,T(t)],0..20,thickness=3,color=blue,title="Figure 2.64", axes=boxed,view=[0..20,298..400]);

Fig. 2.64

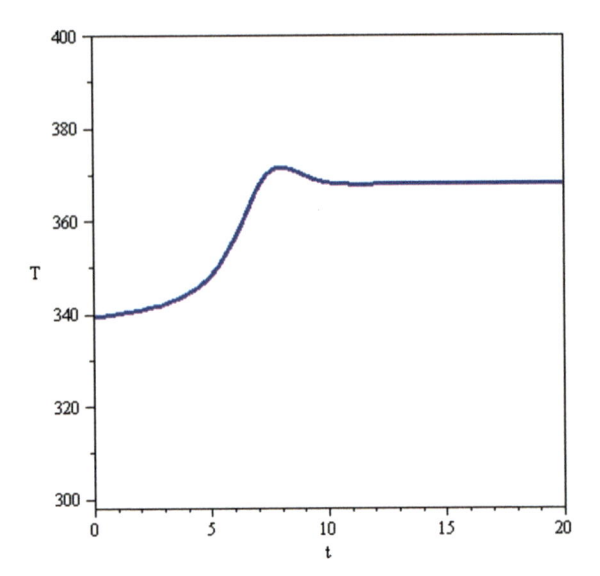

> Ics:=(C(0)=cst2*0.999,T(0)=st2*0.999);

$$Ics := C(0) = 5.512413216, T(0) = 338.758026$$

> sol:=dsolve({eqs,Ics},{vars},type=numeric);

$$sol := \mathbf{proc}(x\_rkf45) \ ... \ \mathbf{end \ proc}$$

> odeplot(sol,[t,C(t)],0..20,thickness=3,color=magenta,title="Figure 2.65", axes=boxed,view=[0..20,0..10]);

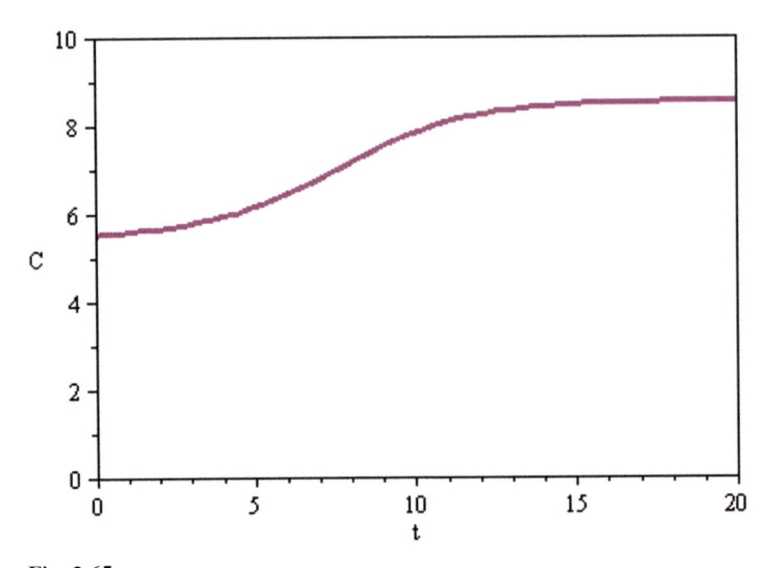

Fig. 2.65

2.2 Nonlinear Ordinary Differential Equations

```
> odeplot(sol,[t,T(t)],0..20,thickness=3,color=red,title="Figure 2.66",
axes=boxed,view=[0..20,298..400]);
```

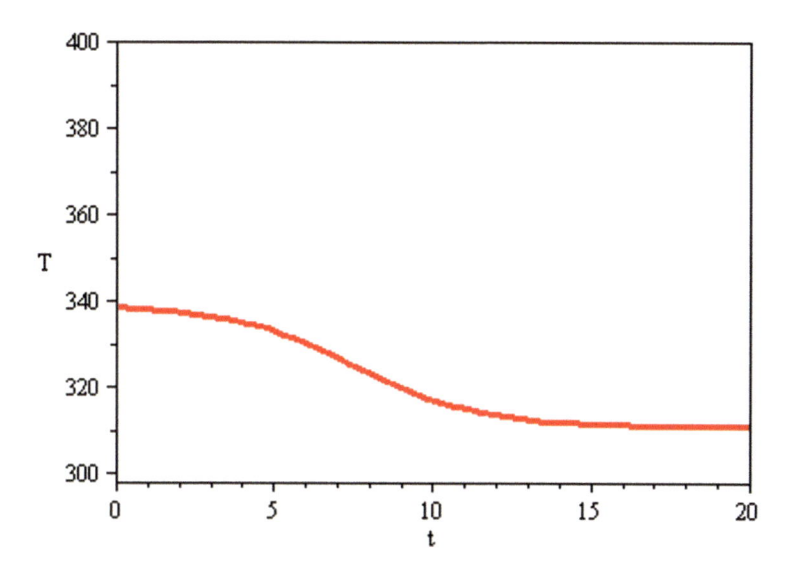

Fig. 2.66

For the values closer to the middle steady state st2, the process reaches st1 or st3 at the end. Hence, we conclude that st2 is unstable.

Example 2.2.9. Phase Plane Analysis

Example 2.2.8 is reviewed here using the phase plane analysis. For this purpose the independent variable is eliminated and temperature T is solved as a function of C, the concentration.

```
> restart:
```

```
> with(plots):
```

```
> eq[1]:=diff(C(t),t)=F/V*(Cf-C(t))-k*exp(-E/R/T(t))*C(t);
```

$$eq_1 := \frac{d}{dt} C(t) = \frac{F\,(Cf - C(t))}{V} - k\,e^{-\frac{E}{R\,T(t)}}\,C(t)$$

> eq[2]:=diff(T(t),t)=F/V*(Tf-T(t))+(-H/rho/cp)*k*exp(-E/R/T(t))*C(t)-UA/V/rho/cp*(T(t)-Tj);

$$eq_2 := \frac{d}{dt} T(t) = \frac{F (Tf - T(t))}{V} - \frac{H k e^{-\frac{E}{R T(t)}} C(t)}{\rho\, cp}$$
$$- \frac{UA (T(t) - Tj)}{V \rho\, cp}$$

The governing equation for T(C) is derived here:

> eq:=diff(T(C),C)=subs(C(t)=C,T(t)=T(C),rhs(eq[2])/rhs(eq[1]));

$$eq := \frac{d}{dC} T(C)$$
$$= \frac{\frac{F (Tf - T(C))}{V} - \frac{H k e^{-\frac{E}{R T(C)}} C}{\rho\, cp} - \frac{UA (T(C) - Tj)}{V \rho\, cp}}{\frac{F (Cf - C)}{V} - k e^{-\frac{E}{R T(C)}} C}$$

> pars:={F=V,k=9703*3600,H=-5960,E=11843,rho=500/cp,Cf=10,R=1.987,UA=V*150};

$$pars := \left\{ F = V, k = 34930800, E = 11843, H = -5960, \rho = \frac{500}{cp}, Cf \right.$$
$$\left. = 10, R = 1.987, UA = 150\, V \right\}$$

Values for parameters are entered here:

> Eq:=subs(pars,eq);

$$Eq := \frac{d}{dC} T(C)$$
$$= \frac{Tf - \frac{13}{10} T(C) + 416375136 e^{-\frac{5960.241570}{T(C)}} C + \frac{3}{10} Tj}{10 - C - 34930800 e^{-\frac{5960.241570}{T(C)}} C}$$

2.2 Nonlinear Ordinary Differential Equations

Both the jacket temperature and the feed stream temperature are taken to be 298 K.

> Deq:=subs(Tj=298,Tf=298,Eq);

$$Deq := \frac{d}{dC}T(C)$$
$$= \frac{\frac{1937}{5} - \frac{13}{10}T(C) + 416375136 e^{-\frac{5960.241570}{T(C)}} C}{10 - C - 34930800 e^{-\frac{5960.241570}{T(C)}} C}$$

The differential equation is solved below and the numerical simulation is stopped when the denominator becomes zero. The simulation is performed for different initial conditions:

> sol:=dsolve({Deq,T(0)=298},T(C),type=numeric,stop_cond=
[10-C-34930800*exp(-5960.241570*1/T(C))*C]);

$$sol := \mathbf{proc}(x\_rkf45) \ ... \ \mathbf{end\ proc}$$

> odeplot(sol,[C,T(C)],0..10,thickness=3,color=red,title="Figure 2.67", axes=boxed);

Warning, cannot evaluate the solution further right of 8.7992960, stop condition #1 violated

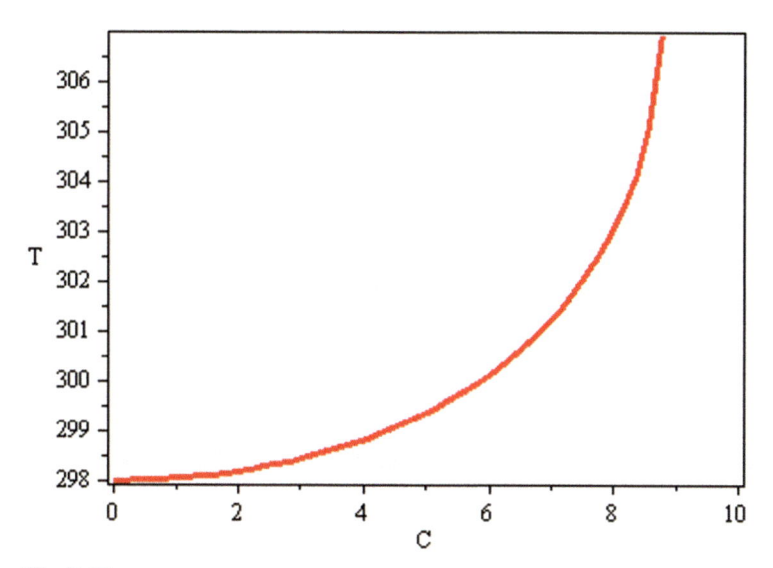

Fig. 2.67

> sol:=dsolve({Deq,T(5)=350},T(C),type=numeric,stop_cond=
[10-C-34930800*exp(-5960.241570*1/T(C))*C]);

$$sol := \mathbf{proc}(x\_rkf45)\ ...\ \mathbf{end\ proc}$$

> odeplot(sol,[C,T(C)],0..10,thickness=3,color=green,title="Figure 2.68",
axes=boxed);

Warning, cannot evaluate the solution further left of 1.7682405, stop condition #1 violated

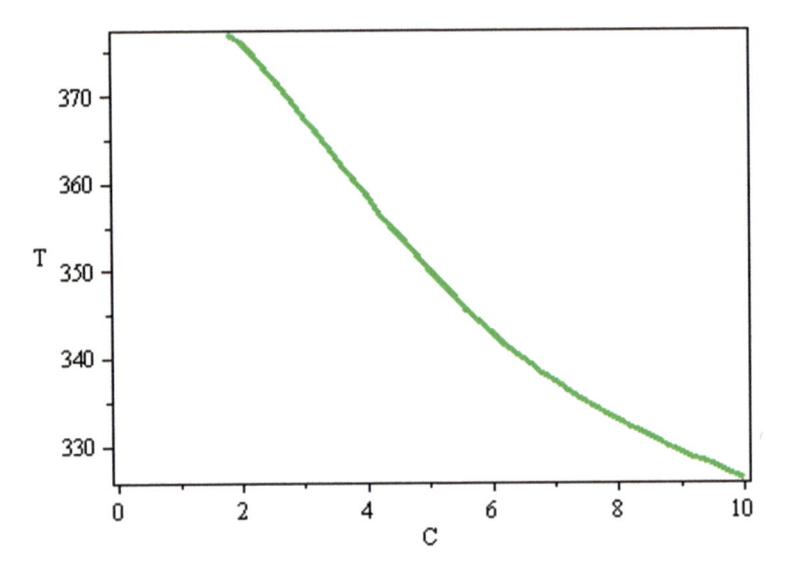

Fig. 2.68

> sol:=dsolve({Deq,T(9)=400},T(C),type=numeric,stop_cond=
[10-C-34930800*exp(-5960.241570*1/T(C))*C]);

$$sol := \mathbf{proc}(x\_rkf45)\ ...\ \mathbf{end\ proc}$$

> odeplot(sol,[C,T(C)],0..10,thickness=3,color=blue,title="Figure 2.69",
axes=boxed);

Warning, cannot evaluate the solution further left of .46053734e-1, stop condition #1 violated

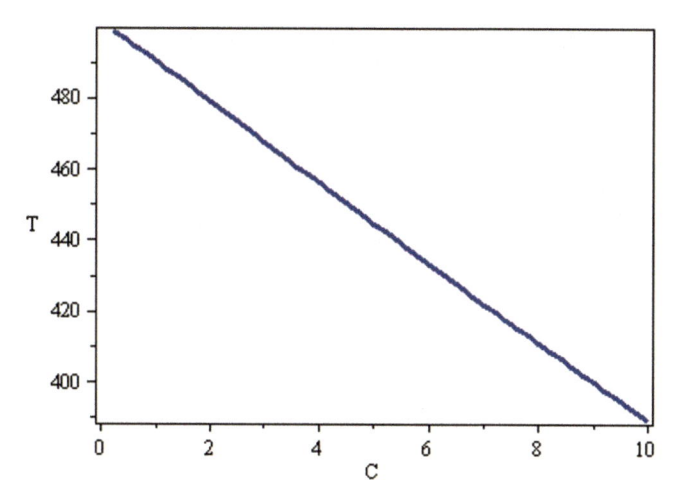

2.2 Nonlinear Ordinary Differential Equations

Fig. 2.69

> sol:=dsolve({Deq,T(5.5179)=339.0971},T(C),type=numeric,stop_cond=
[10-C-34930800*exp(-5960.241570*1/T(C))*C]);

$$sol := \mathbf{proc}(x\_rkf45) \ ... \ \mathbf{end\ proc}$$

> odeplot(sol,[C,T(C)],0..10,thickness=3,color=magenta,title="Figure 2.70",
axes=boxed);

Warning, cannot evaluate the solution further right of 8.5634663, stop condition #1 violated

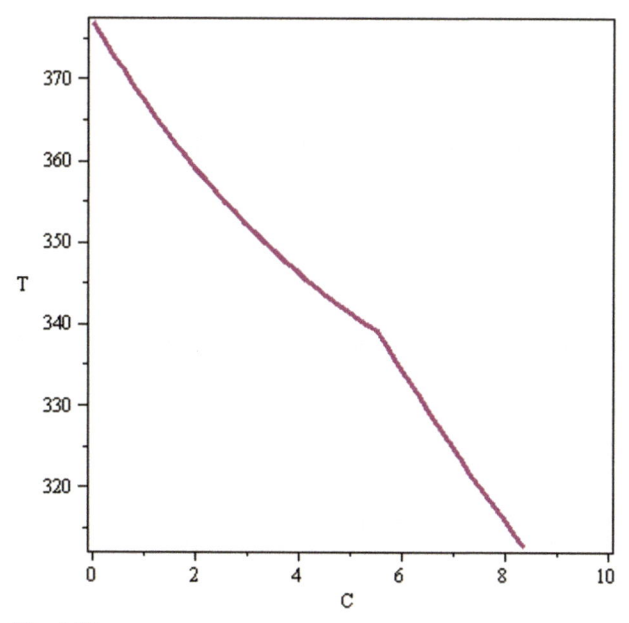

Fig. 2.70

> sol:=dsolve({Deq,T(5.5234)=339.44},T(C),type=numeric,stop_cond=
[10-C-34930800*exp(-5960.241570*1/T(C))*C]);

$$sol := \mathbf{proc}(x\_rkf45\ )\ ...\ \mathbf{end\ proc}$$

> odeplot(sol,[C,T(C)],0..10,thickness=3,color=brown,title="Figure 2.71",
axes=boxed);

Warning, cannot evaluate the solution further left of .23729690e-1, stop condition #1 violated

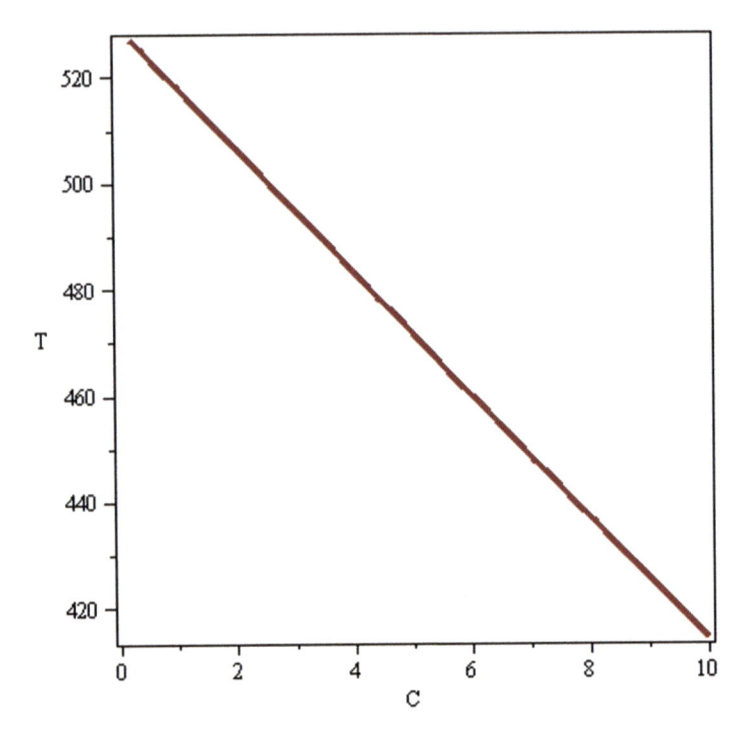

Fig. 2.71

Temperature can increase or decrease with the concentration C, depending on the initial conditions. The slope of the curve and the maximum for T depend on the initial conditions. A sequence of runs is performed for three different initial concentrations of 0.5, 5, and 9.5. The initial condition for T is varied in the range of 300 K - 420 K.

> MM:=20;

$$MM := 20$$

2.2 Nonlinear Ordinary Differential Equations 145

```
> TT:=seq(300+i/MM*120,i=0..MM);
```

$TT := 300, 306, 312, 318, 324, 330, 336, 342, 348, 354, 360, 366, 372, 378,$
$384, 390, 396, 402, 408, 414, 420$

```
> for i from 0 to MM do sol:=dsolve({Deq,T(0.5)=TT[i+1]},T(C),
type=numeric,stop_cond=[10-C-34930800*exp(-5960.241570*1/T(C))*C]);
p[i]:=odeplot(sol,[C,T(C)],0..10,thickness=3,color=red,title="Figure 2.72",
axes=boxed);od:
```

```
> for i from 0 to MM do sol:=dsolve({Deq,T(0.5)=TT[i+1]},T(C),
type=numeric,stop_cond=[10-C-34930800*exp(-5960.241570*1/T(C))*C]);
p1[i]:=odeplot(sol,[C,T(C)],0..10,thickness=3,color=green,title="Figure 2.72",
axes=boxed);od:
```

```
> for i from 0 to MM do sol:=dsolve({Deq,T(9.5)=TT[i+1]},T(C),
type=numeric,stop_cond=[10-C-34930800*exp(-5960.241570*1/T(C))*C]);
p2[i]:=odeplot(sol,[C,T(C)],0..10,thickness=3,color=blue,title="Figure 2.72",
axes=boxed);od:
```

```
> display({seq(p[i],i=0..MM),seq(p1[i],i=0..MM),seq(p2[i],i=0..MM)},view=[0..10,
```

```
> 295...420]);
```

Warning, cannot evaluate the solution further right of 8.7899223, stop condition #1 violated
Warning, cannot evaluate the solution further right of 8.7599625, stop condition #1 violated
Warning, cannot evaluate the solution further right of 8.7278642, stop condition #1 violated
Warning, cannot evaluate the solution further right of 8.6943952, stop condition #1 violated
Warning, cannot evaluate the solution further right of 8.6576248, stop condition #1 violated
Warning, cannot evaluate the solution further right of 8.6197738, stop condition #1 violated
Warning, cannot evaluate the solution further right of 8.5875974, stop condition #1 violated
Warning, cannot evaluate the solution further right of 8.5656287, stop condition #1 violated
Warning, cannot evaluate the solution further right of 8.5634928, stop condition #1 violated
Warning, cannot evaluate the solution further right of 8.5634834, stop condition #1 violated
Warning, cannot evaluate the solution further right of 8.5635784, stop condition #1 violated
Warning, cannot evaluate the solution further right of 8.5635741, stop condition #1 violated

Warning, cannot evaluate the solution further right of 5.3754375, stop condition #1 violated
Warning, cannot evaluate the solution further right of 3.6509855, stop condition #1 violated
Warning, cannot evaluate the solution further right of 2.8295708, stop condition #1 violated
Warning, cannot evaluate the solution further right of 2.4573345, stop condition #1 violated
Warning, cannot evaluate the solution further right of 2.4078413, stop condition #1 violated
Warning, cannot evaluate the solution further right of 2.4048821, stop condition #1 violated
Warning, cannot evaluate the solution further right of 2.4070544, stop condition #1 violated
Warning, cannot evaluate the solution further left of .49764692, stop condition #1 violated
Warning, cannot evaluate the solution further left of .43228936, stop condition #1 violated
Warning, cannot evaluate the solution further right of 8.7899223, stop condition #1 violated
Warning, cannot evaluate the solution further right of 8.7599625, stop condition #1 violated
Warning, cannot evaluate the solution further right of 8.7278642, stop condition #1 violated
Warning, cannot evaluate the solution further right of 8.6943952, stop condition #1 violated
Warning, cannot evaluate the solution further right of 8.6576248, stop condition #1 violated
Warning, cannot evaluate the solution further right of 8.6197738, stop condition #1 violated
Warning, cannot evaluate the solution further right of 8.5875974, stop condition #1 violated
Warning, cannot evaluate the solution further right of 8.5656287, stop condition #1 violated
Warning, cannot evaluate the solution further right of 8.5634928, stop condition #1 violated
Warning, cannot evaluate the solution further right of 8.5634834, stop condition #1 violated
Warning, cannot evaluate the solution further right of 8.5635784, stop condition #1 violated
Warning, cannot evaluate the solution further right of 8.5635741, stop condition #1 violated
Warning, cannot evaluate the solution further right of 5.3754375, stop condition #1 violated
Warning, cannot evaluate the solution further right of 3.6509855, stop condition #1 violated

2.2 Nonlinear Ordinary Differential Equations 147

Warning, cannot evaluate the solution further right of 2.8295708, stop condition #1 violated
Warning, cannot evaluate the solution further right of 2.4573345, stop condition #1 violated
Warning, cannot evaluate the solution further right of 2.4078413, stop condition #1 violated
Warning, cannot evaluate the solution further right of 2.4048821, stop condition #1 violated
Warning, cannot evaluate the solution further right of 2.4070544, stop condition #1 violated
Warning, cannot evaluate the solution further left of .49764692, stop condition #1 violated
Warning, cannot evaluate the solution further left of .43228936, stop condition #1 violated
Warning, cannot evaluate the solution further right of 9.7410116, stop condition #1 violated
Warning, cannot evaluate the solution further left of 8.5635577, stop condition #1 violated
Warning, cannot evaluate the solution further left of 8.5638089, stop condition #1 violated
Warning, cannot evaluate the solution further left of 8.4876502, stop condition #1 violated
Warning, cannot evaluate the solution further left of 8.0713076, stop condition #1 violated
Warning, cannot evaluate the solution further left of 6.9575718, stop condition #1 violated
Warning, cannot evaluate the solution further left of 1.3218078, stop condition #1 violated
Warning, cannot evaluate the solution further left of .70000766, stop condition #1 violated
Warning, cannot evaluate the solution further left of .43175821, stop condition #1 violated
Warning, cannot evaluate the solution further left of .29193900, stop condition #1 violated
Warning, cannot evaluate the solution further left of .20969420, stop condition #1 violated
Warning, cannot evaluate the solution further left of .15713260, stop condition #1 violated
Warning, cannot evaluate the solution further left of .12152136, stop condition #1 violated
Warning, cannot evaluate the solution further left of .96245238e-1, stop condition #1 violated
Warning, cannot evaluate the solution further left of .77677669e-1, stop condition #1 violated
Warning, cannot evaluate the solution further left of .63659244e-1, stop condition #1 violated

Warning, cannot evaluate the solution further left of .52827584e-1, stop condition #1 violated
Warning, cannot evaluate the solution further left of .44303373e-1, stop condition #1 violated
Warning, cannot evaluate the solution further left of .37491839e-1, stop condition #1 violated
Warning, cannot evaluate the solution further left of .31979299e-1, stop condition #1 violated
Warning, cannot evaluate the solution further left of .27462576e-1, stop condition #1 violated
Warning, cannot evaluate the solution further left of .23729690e-1, stop condition #1 violated

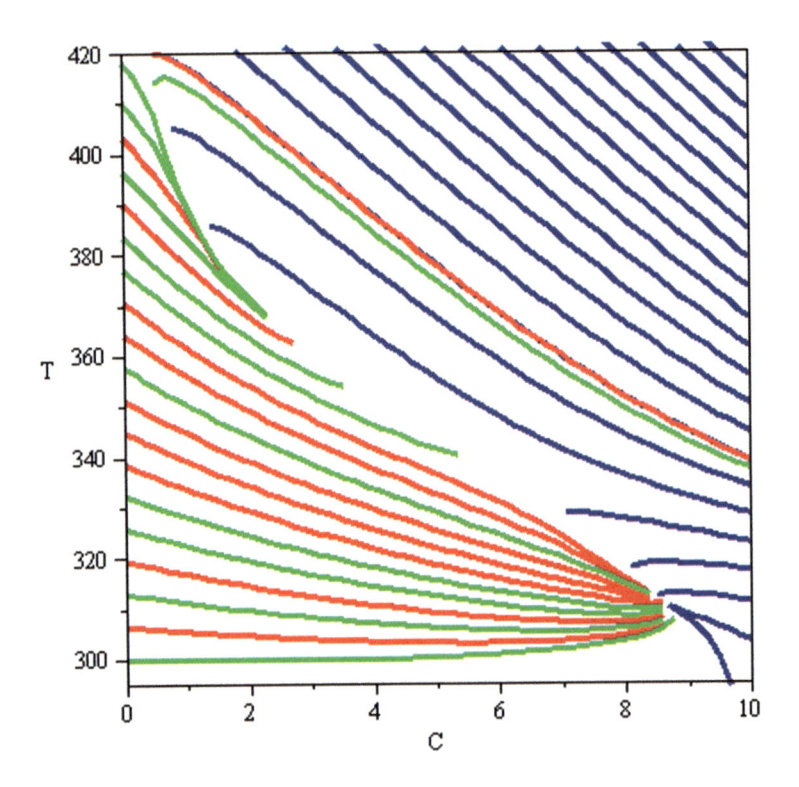

Fig. 2.72

2.2.9 Summary

In this chapter, nonlinear IVPs were solved numerically. In section 2.2.2 a nonlinear IVP was solved analytically using Maple's 'dsolve' command. This approach is limited to very few nonlinear ODEs. In section 2.2.3, series solutions were obtained using Maple's 'dsolve' command. This approach is valid for all

2.2 Nonlinear Ordinary Differential Equations

nonlinear ODEs. However, the solution obtained may not always converge. One has to use these solutions cautiously and check the convergence of the series solution obtained.

In section 2.2.4 nonlinear IVPS were solved numerically using Maple's 'dsolve' command. Maple provides different options for the numerical solution, Runge-Kutta, Gear, backsode, etc. Maple's default Runge-Kutta method accurate to the order $(\Delta t)^4$ is good enough for a variety of problems. Maple' 'dsolve' command is very convenient to use and the solution obtained can be plotted easily. There are different ways to control the error. The default absolute error set in Maple is 1e-6. For complex problems, this can be reduced by setting abserr = 1e-10, etc. Similarly, the relative error can be set as relerr = 1e-6.

In section 2.2.4, stop condition was used to predict the maximum yield in a chemical reaction. In section 2.2.5, a stiff problem was solved using Maple's default numerical solver. We concluded that the conventional numerical methods fail for this stiff problem. Maple's stiff solver was found to be superior for this stiff problem. Generally, one has to use a stiff solver only if the conventional methods fail, as stiff solvers take more time to solve ordinary IVPs than the conventional solvers.

In section 2.2.6, DAEs were solved by converting the algebraic equations to differential equations. This methodology should be valid for any system of DAEs as long as one can find the initial conditions for the variables governed by the algebraic differential equations.

In section 2.2.7, multiple steady states in a heterogeneous chemical reaction (one dependent variable) and a jacketed stirred tank reactor (two dependent variables) were analyzed. Both stable and unstable steady states were obtained. The transient behavior of the system was found to depend on the initial conditions. The methodology and Maple programs presented in this chapter should be valid for any system of IVPs with multiple steady states. In section 2.2.8, phase plane behavior of a jacketed stirred tank reactor was analyzed. The program provided should be of use for analyzing phase plane behavior of different chemical systems. A total of ten different examples were presented in this chapter.

2.2.10 Problems

1. The dissolution rate of a salt is governed by

$$\frac{dy}{dt} = ky(c - c_{sat})$$

$$y(0) = y_0 = 10$$

where, y is the amount of salt dissolved (kg), k is a proportionality constant, V is the volume of water used for dissolving the salt (90 liters), c is the concentration of the salt (kg/liter) at any instant given by $c = \dfrac{y_0 - y}{V}$ and csat is the concentration of the saturated salt, 1/3 kg/liter. Solve this problem to get a closed form solution for the amount of salt dissolved (see example 2.2.1). If it takes one hour to dissolve 50% of the initial salt, what is the rate constant k? Once k is obtained plot the transient profile for the dissolved salt. How much of the salt can be dissolved in one hour if the volume of the water used is doubled?

2. Consider the series reaction scheme modeled in example 2.2.1. Redo this problem if the first reaction is first order and the second reaction is second order (Rice and Do, 1995).[1] Obtain a closed form solution and plot the concentration profiles for typical values of rate constants.

3. Redo example 2.2.1 if both the reactions are second order. Obtain a closed form solution if possible and plot the concentration profiles for typical values of rate constants.

4. Consider a CSTR experiencing slow catalyst decay (Rice and Do, 1995).[1] The governing equation for the reactant concentration (x) and catalyst activity (y) are given by:

$$\frac{dx}{dt} = 1 - x - xy$$

$$\frac{dy}{dt} = -\varepsilon xy$$

$$x(0) = 0 \text{ and } y(0) = 1$$

where ε is a proportionality constant (0.5). Solve this problem numerically and plot the profiles (see example 2.2.2).

5. Consider parallel deactivation in a well-stirred reactor (Rice and Do, 1995).[1] The governing equation for the reactant concentration (x) and catalyst activity (y) are given by:

$$\frac{dx}{dt} = \alpha(1 - x) - x^2 y$$

$$\frac{dy}{dt} = -\varepsilon xy$$

$$x(0) = 0 \text{ and } y(0) = 1$$

where ε is the ratio of deactivation rate constant to kinetic rate constant (0.1) and $\alpha = F/(kVC0)$. F is the flow rate, k is the rate constant, V is the volume and C0 is the feed concentration. Solve this problem numerically and plot the concentration profiles for $\alpha = 1$.

6. Consider catalytic cracking of a gas oil (A) to gasoline (B). Gas oil (A) cracks catalytically to gasoline (B) according to second order kinetics. In a parallel reaction A also forms Coke (C) according to second order kinetics. Gasoline (B) once formed also cracks according to a first order reaction to form Coke (C). The governing equations for A, B and C is given by (Rice and Do, 1995):[1]

2.2 Nonlinear Ordinary Differential Equations

$$\frac{dC_A}{dt} = -k_1 C_A^2 - k_2 C_A^2$$

$$\frac{dC_B}{dt} = k_1 C_A^2 - k_3 C_B$$

$$C_A + C_B + C_C = 1$$

$$C_A(0) = 1 \text{ and } C_B(0) = C_C(0) = 0$$

Solve this problem to obtain a closed form solution for the concentration of gasoline as a function of rate constants and time. When will you stop the experiment to obtain maximum concentration of gasoline (B)?

7. Also, solve this problem numerically for typical values of rate constants ($k1 = 1$, $k2 = 0.1$ and $k3 = 0.5$). Plot the concentration profiles using both analytical and numerical solution.

8. Consider the Van der Pol equation (Rice and Do, 1995)[1]

$$\frac{d^2 y}{dt^2} - \mu(1 - y^2)\frac{dy}{dt} + y = 0$$

$$y(0) = 1; \quad \frac{dy}{dt}(0) = 0$$

Solve this problem numerically for $\mu = 0.1$, 1, and 10. Plot the transient profiles.

9. Consider an adiabatic tubular reactor (Davis, 1984)[15] with the following data: length $L = 2$ m, radius $Rp = 0.1$ m, inlet reactant concentration $c0 = 30$ moles/m3, inlet temperature $T0 = 700$K, enthalpy $\Delta H = -10000$ J/mole, specific heat capacity $Cp = 1000$ J/kg/K, activation energy $Ea = 100$ J/mole, $\rho = 1200$ kg/m3, velocity $u0 = 3$ m/s, and rate constant $k0 = 5$ s-1. Dimensionless concentration (y) and dimensionless temperature (θ) are governed by material and energy balances as:

$$\frac{dy}{dz} = -\text{Da } y \exp\left[\delta\left(1 - \frac{1}{\theta}\right)\right]$$

$$\frac{d\theta}{dz} = \beta \text{Da } y \exp\left[\delta\left(1 - \frac{1}{\theta}\right)\right]$$

$$\theta(0) = 1 \text{ and } y(0) = 1$$

where

$$\text{Da} = \frac{Lk_0}{u_0}; \quad \beta = \frac{c_0(-\Delta H)}{\rho C p T_0}; \quad \delta = \frac{E_a}{R_g T_0}$$

where Rg is the gas constant, 8.314 J/mole/K. Solve this system of equations numerically and plot dimensionless concentration and temperature profiles across the length of the reactor.

10. Consider the reaction scheme (Davis, 1984)[15]

$$\frac{dC_A}{dt} = -k_1 C_A + k_2 C_B C_C$$

$$\frac{dC_B}{dt} = k_1 C_A - k_2 C_B C_C - k_3 C_B^2$$

$$\frac{dC_C}{dt} = k_3 C_B^2$$

$$C_A(0) = 1, \ C_B(0) = C_C(0) = 0$$

$$k_1 = 0.08, \ k_2 = 2 \times 10^4, \text{ and } k_3 = 6 \times 10^7$$

Solve this stiff system of equations and plot the concentration profiles.

11. Material and energy balance in a reactor gives the following dimensionless equations (Finlayson, 1980):[16]

$$\frac{dc}{dt} = -10.5(c-1) - \phi^2 c \exp\left(\gamma \left[1 - \frac{1}{T}\right]\right)$$

$$\frac{dT}{dt} = -10.5(T-1) + \beta \phi^2 c \exp\left(\gamma \left[1 - \frac{1}{T}\right]\right)$$

$c(0) = 0.73$ and $T(0) = 1$

$\phi^2 = 1.21; \ \beta = 0.15; \ \gamma = 30$

Solve this system of equations numerically and plot the concentration and temperature profiles.

12. Consider the dynamics of a catalytic fluidized bed in which an irreversible gas phase reaction takes place (Aiken and Lapidus, 1974;[17] Cutlip and Shacham, 1999).[9] Partial pressure of reactant in fluid (P), temperature of reactant in fluid (T), partial pressure of reactant at the catalyst surface (Pp) and partial pressure of reactant at the catalyst surface (Tp) are governed by the following equations:

2.2 Nonlinear Ordinary Differential Equations

$$\frac{dP}{dt} = 0.1 + 320P_p - 321P$$

$$\frac{dT}{dt} = 1752 - 269T + 267T_p$$

$$\frac{dP_p}{dt} = 1880\left(P - P_p(1+K)\right)$$

$$\frac{dT_p}{dt} = 1.3\left(T - T_p\right) + 10400KP_p$$

$$K = 0.0006 \exp\left(20.7 - \frac{15000}{T_p}\right)$$

This problem exhibits multiple steady states. Obtain all the steady states by equating the transient term to zero in all the equations. For mathematical convenience, express steady state P, T, and Pp in terms of steady state Tp using the first three equations. Use the steady state equation for Tp (after eliminating all other dependent variables) to obtain the multiple steady states.

13. Solve the dynamic problem using the initial conditions $P(0) = 0.1$, $T(0) = 600$, $Pp(0) = 0$ and $Tp(0) = 761$ and plot the dynamic profiles for $t = 0..15$. Can you change the initial conditions to obtain a different steady state? (see examples 2.2.6 and 2.2.7)

14. Consider radiation to a thin copper plate in a furnace (Cutlip and Shacham, 1999),[9] the temperature (T) of the thin plate is governed by:

$$\frac{dT}{dt} = \frac{\sigma A\left(T_F^4 - T^4\right)}{V\rho C_p}$$

$$T(0) = 293K$$

where σ is the Stefan-Boltzman constant (= 5.678×10^{-8} W/(m2K4)), A = 0.5 m2, V = 3.75×10^{-4} m3, ρ = 8950 kg/m3, Cp = 383 J/kg/K and TF = 1273K. Solve this problem numerically and plot the temperature profile.

15. Design of a chemical flow reactor involves the following nonlinear differential equation (Hanna and Sandall, 1995)[18]

$$\frac{dC}{d\phi} = \frac{C}{1+\beta C}$$

$$C(0) = 1$$

Solve this problem analytically and plot C vs. ϕ for $\beta = 0.1$ and 1.

16. The following nonlinear equation governs the concentration in an isothermal batch reactor (Hanna and Sandall, 1995):[18]

$$\frac{dC}{dt} = -\left(\frac{C}{\sqrt{(1+C)}} - 0.3C^2\right)$$

$$C(0) = 0.6$$

where t is the time in hours. Solve this problem analytically if possible. If analytical solution is not possible, obtain a series solution. Plot the transient concentration using 'dsolve numeric' command and calculate the time taken for the concentration to become 0.1. How does the series solution compare with the exact solution (numerical) for this problem?

17. Consider a CSTR with a second order kinetics (Bequette, 1998),[12] the governing equation is:

$$\frac{dC}{dt} = \frac{1}{\tau}(C_{in} - C) - kC^2$$

$$C(0) = 0$$

Obtain an analytical solution for this problem if possible. Plot the concentration profiles if Cin = 1, τ = 1 and k = 1. Can you obtain a closed form solution for C if Cin = sin(t)? If an analytical solution is not possible, solve the equation numerically to obtain the concentration profile.

18. Consider a surge tank with an outlet flow rate that depends on the square root of the height of liquid in the tank (Bequette, 1998).[12] When there is no inlet flow, the governing equation is:

$$\frac{dh}{dt} = -a\sqrt{h}$$

$$h(0) = 4$$

Solve this equation to obtain the height of liquid as a function of time. Plot the transient behavior for a = 0.8.

19. Consider two interacting tanks in series (Bequette, 1998).[12] The governing equations for height in the tanks are:

$$\frac{dh_1}{dt} = \frac{F}{A_1} - \frac{\beta_1}{A_1}\sqrt{h_1 - h_2}$$

$$\frac{dh_2}{dt} = \frac{\beta_1}{A_1}\sqrt{h_1 - h_2} - \frac{\beta_2}{A_2}\sqrt{h_2}$$

$$h_1(0) = 12; \; h_2(0) = 7$$

$$F = 5 \frac{ft^3}{min} \quad \beta_1 = 2.5 \frac{ft^{2.5}}{min}; \quad \beta_2 = \frac{5}{\sqrt{6}} \frac{ft^{2.5}}{min}; \quad A_1 = 5 \; ft^2; \; A_2 = 10 \; ft^2$$

Plot the transient profiles for t = 0..100 minutes.

20. Consider the predator-prey problem (Bequette, 1998):[12]

$$\frac{dy_1}{dt} = \alpha(1-y_2)y_1$$

$$\frac{dy_2}{dt} = -\beta(1-y_1)y_2$$

$$y_1(0) = 1.5; \ y_2(0) = 0.75$$

$$\alpha = \beta = 1$$

where t is in days. Plot the dynamic profiles for t = 0..100 days. Do phase plane analysis for this problem and plot y1 vs. y2 (see example 2.2.8).

21. A series parallel reaction takes place in a CSTR (Bequette, 1998).[12] The governing equations are:

$$\frac{dC_A}{dt} = \frac{1}{\tau}(C_{A,\,in} - C_A) - k_1 C_A - k_3 C_A^2$$

$$\frac{dC_B}{dt} = \frac{1}{\tau}(-C_B) + k_1 C_A - k_3 C_B$$

$$C_A(0) = C_B(0) = 0$$

$$k_1 = \frac{5}{6} \text{ min}^{-1}; \ k_2 = \frac{5}{3} \text{ min}^{-1}; \ k_3 = \frac{1}{6} \frac{\text{liter}}{\text{mol min}}; \ C_{A,\,in} = 10 \frac{\text{mol}}{\text{liter}}$$

Find the time constant, τ if the steady state concentration of A is 3 mol/liter (Hint: Use the steady state version of the first equation to find the steady state concentration of A). Once τ is obtained, plot the transient profile.

References

1. Rice, R.G., Do, D.D.: Applied Mathematics and Modeling for Chemical Engineers. John Wiley & Sons, Inc., Chichester (1995)
2. Amundson, N.R.: Mathematical Methods in Chemical Engineering: Matrices and Their Applications. Prentice Hall, Inc., Englewood Cliffs (1966)
3. Fogler, H.S.: Elements of Chemical Reaction Engineering, 3rd edn. Prentice-Hall, Englewood Cliffs (1998)
4. Abell, M.L., Braselton, J.P.: Differential Equations with Maple V, 3rd edn. Academic Press, London (2001)
5. Constantinides, A., Mostoufi, N.: Numerical Methods for Chemical Engineers with MATLAB Applications. Prentice-Hall PTR, Englewood Cliffs (1999)
6. Taylor, R., Krishna, R.: Multicomponent Mass Transfer, 579 pages. John Wiley & Sons, Inc., New York (1993)
7. Varma, A., Morbidelli, M.: Mathematical Methods in Chemical Engineering. Oxford University Press, Inc., Oxford (1997)

8. Subramanian, V.R., White, R.E.: Solving Differential Equations with Maple. Chemical Engineering Education, 328–336 (Fall 2000)
9. Cutlip, M.B., Shacham, M.: Problem Solving in Chemical Engineering with Numerical Methods. Prentice Hall PTR, Englewood Cliffs (1999)
10. Jain, M.C.: Vector Spaces and Matrices in Physics. CRC Press, Boca Raton (2001)
11. Pushpavanam, S.: Mathematical Methods in Chemical Engineering. Prentice-Hall of India Private Limited, Englewood Cliffs (2001)
12. Bequette, B.W.: Process Dynamics: Modeling, Analysis, and Simulation. Prentice-Hall PTR, Englewood Cliffs (1998)
13. Hildebrand, F.B.: Advanced Calculus for Applications, 2nd edn. Prentice-Hall, Englewood Cliffs (1976)
14. Aris, R.: Mathematical Modeling: A Chemical Engineer's Perspective. Academic Press, London (1999)
15. Davis, M.E.: Numerical Methods and Modeling for Chemical Engineers. John Wiley & Sons, Chichester (1984)
16. Finlayson, B.A.: Nonlinear Analysis in Chemical Engineering. McGraw-Hill, New York (1980)
17. Aiken, R.C., Lapidus, L.: Effective Numerical Integration Method for Typical Stiff Systems. AIChE Journal 20(2), 368–375 (1974)
18. Hanna, O.T., Sandall, O.C.: Computational Methods in Chemical Engineering. Prentice Hall, Inc., Englewood Cliffs (1995)

Appendix A: Matrix Exponential Method

This appendix presents two methods of obtaining an analytical solution to a system of first order ordinary differential equations. Both methods (power series and the Laplace transform) yield a solution in terms of the matrix exponential. That is, we seek a solution to

$$\frac{dY}{dt} = AY + b(t) \tag{A.1}$$

where $\frac{dY}{dt}$ is a n x 1 vector and is the derivative of Y which is a n x 1 vector of dependent variables, A is a n x n matrix of constants, and b is n x 1 vector which may depend on time. Equation (A.1) can be solved by assuming a power series solution of the form

$$Y(t) = e_0 + e_1 t + e_2 t^2 + \cdots \tag{A.2}$$

where e_j represents the jth n x 1 vector which is to be determined. Substitution of t = 0 into equation (A.2) yields

$$Y(0) = e_0 \tag{A.3}$$

Substitution of equation (A.2) into equation (A.1) with $b = 0$ yields

Appendix A: Matrix Exponential Method 157

$$\mathbf{e}_1 + 2\mathbf{e}_2 t + \cdots = \mathbf{A}(\mathbf{e}_0 + \mathbf{e}_1 t + \mathbf{e}_2 t^2 + \cdots) \tag{A.4}$$

By equating coefficients of like powers of t in equation (A.4) we obtain

$$\mathbf{e}_1 = \mathbf{A}\mathbf{e}_0$$

$$\mathbf{e}_2 = \frac{1}{2}\mathbf{A}\mathbf{e}_1 = \frac{1}{2}\mathbf{A}^2\mathbf{e}_0$$

$$\mathbf{e}_3 = \frac{1}{3}\mathbf{A}\mathbf{e}_2 = \frac{1}{(3)(2)}\mathbf{A}^3\mathbf{e}_0 \tag{A.5}$$

$$\vdots$$

$$\mathbf{e}_k = \frac{1}{k!}\mathbf{A}^k\mathbf{e}_0$$

Substituting equations (A.3) and (A.5) into equation (A.2) yields

$$\mathbf{Y}(t) = (\mathbf{I} + \mathbf{A}t + \frac{1}{2!}\mathbf{A}^2 t^2 + \cdots + \frac{1}{k!}\mathbf{A}^k t^k + \cdots)\mathbf{Y}(0) \tag{A.6}$$

Since

$$\exp(\mathbf{A}t) = \mathbf{I} + \mathbf{A}t + \frac{1}{2!}\mathbf{A}^2 t^2 + \cdots \tag{A.7}$$

equation (A.6) can be written as

$$\mathbf{Y}(t) = \exp(\mathbf{A}t)\mathbf{Y}(0) \tag{A.8}$$

To handle the case where $\mathbf{b}$ is not equal to $\mathbf{0}$, rewrite equation (A.1) as follows:

$$\mathbf{Y}' - \mathbf{A}\mathbf{Y} = \mathbf{b} \tag{A.9}$$

Next, premultiply both sides of equation (A.9) by $\exp(-\mathbf{A}t)$ to obtain

$$e^{-\mathbf{A}t}(\mathbf{Y}' - \mathbf{A}\mathbf{Y}) = e^{-\mathbf{A}t}\mathbf{b} \tag{A.10}$$

Now, rewrite equation (A.10) with the left hand side shown as a complete differential:

$$\frac{d}{dt}\left(e^{-\mathbf{A}t}\mathbf{Y}(t)\right) = e^{-\mathbf{A}t}\mathbf{b} \tag{A.11}$$

Integration of equation (A.11) with τ as a dummy integration variable yields

Evaluation of the left hand side of equation (A.12) yields

$$e^{-At}\mathbf{Y}(\tau)\Big|_{\tau=t_0}^{\tau=t} = \int_{\tau=t_0}^{\tau=t} e^{-A\tau}\mathbf{b}(\tau)d\tau \qquad (A.12)$$

$$e^{-At}\mathbf{Y}(t) - e^{-At_0}\mathbf{Y}(t_0) = \int_{\tau=t_0}^{\tau=t} e^{-A\tau}\mathbf{b}(\tau)d\tau \qquad (A.13)$$

Premultiplying both sides of equation (A.13) by e^{At} followed by rearrangement yields:

$$\mathbf{Y} = e^{At}e^{-At_0}\mathbf{Y}(t_0) + e^{At}\int_{\tau=t_0}^{\tau=t} e^{-A\tau}\mathbf{b}(\tau)d\tau \qquad (A.14)$$

or

$$\mathbf{Y} = e^{A(t-t_0)}\mathbf{Y}(t_0) + e^{At}\int_{\tau=t_0}^{\tau=t} e^{-A\tau}\mathbf{b}(\tau)d\tau \qquad (A.15)$$

Equation (A.15) can also be written as follows:

$$\mathbf{Y} = e^{A(t-t_0)}\mathbf{Y}(t_0) + \int_{\tau=t_0}^{\tau=t} e^{-A(\tau-t)}\mathbf{b}(\tau)d\tau \qquad (A.16)$$

Note that for one dependent variable equation (A.1) becomes

$$\frac{dy}{dt} = ay + b(t) \qquad (A.17)$$

and equation (A.16) becomes

$$y = e^{a(t-t_0)}y(t_0) + \int_{\tau=t_0}^{\tau=t} e^{-a(\tau-t)}b(\tau)d\tau \qquad (A.18)$$

Equation (A.18) was obtained by using the so-called integrating factor (IF) where for equation (A.17)

$$\text{IF} = \exp(-at) \qquad (A.19)$$

Note that the integrating factor $\exp(-At)$ was used to solve equation (A.9).

To illustrate the use of equation (A.15), let

Appendix A: Matrix Exponential Method

$$\mathbf{A} = \begin{pmatrix} 0 & 1 \\ -2 & -3 \end{pmatrix} \tag{A.20}$$

and

$$\mathbf{b} = \begin{pmatrix} 0 \\ 1 \end{pmatrix} \tag{A.21}$$

Also, let $t_0 = 0$ and

$$\mathbf{Y}(0) = \begin{pmatrix} 0 \\ 0 \end{pmatrix} \tag{A.22}$$

Thus, equation (A.15) becomes

$$\mathbf{Y}(t) = e^{\mathbf{A}t} \int_0^t e^{-\mathbf{A}\tau} \mathbf{b} \, d\tau \tag{A.23}$$

Equation (A.7) yields the matrix exponential of $\mathbf{A}t$

$$e^{\mathbf{A}t} = \begin{pmatrix} 2e^{-t} - e^{-2t} & e^{-t} - e^{-2t} \\ -2e^{-t} + 2e^{-2t} & -e^{-t} + 2e^{-2t} \end{pmatrix} \tag{A.24}$$

and the matrix exponential of $-\mathbf{A}\tau$ as

$$e^{-\mathbf{A}\tau} = \begin{pmatrix} 2e^{\tau} - e^{2\tau} & e^{\tau} - e^{2\tau} \\ -2e^{\tau} + 2e^{2\tau} & -e^{\tau} + 2e^{2\tau} \end{pmatrix} \tag{A.25}$$

Thus, equation (A.23) becomes

$$\mathbf{Y}(t) = \int_0^t \begin{pmatrix} 2e^{-(t-\tau)} - e^{-2(t-\tau)} & e^{-(t-\tau)} - e^{-2(t-\tau)} \\ -2e^{-(t-\tau)} + 2e^{-2(t-\tau)} & -e^{-(t-\tau)} + 2e^{-2(t-\tau)} \end{pmatrix} \begin{pmatrix} 0 \\ 1 \end{pmatrix} d\tau \tag{A.26}$$

which yields

$$\mathbf{Y}(t) = \int_0^t \begin{pmatrix} e^{-(t-\tau)} - e^{-2(t-\tau)} \\ -e^{-(t-\tau)} + 2e^{-2(t-\tau)} \end{pmatrix} d\tau \tag{A.27}$$

Next, integrate each element in the vector in equation (A.27); the first element is

$$\int_0^t \left(e^{-(t-\tau)} - e^{-2(t-\tau)} \right) d\tau \tag{A.28}$$

or

$$e^{-t}\int_0^t e^\tau d\tau - e^{-2t}\int_0^t e^{2\tau}d\tau \qquad (A.29)$$

or

$$e^{-t}\left(e^\tau\right)\Big|_0^t - e^{-2t}\frac{e^{2\tau}}{2}\Big|_0^t \qquad (A.30)$$

Equation (A.30) becomes

$$e^{-t}e^t - e^{-t}e^0 - \frac{e^{-2t}}{2}(e^{2t} - e^0) \qquad (A.31)$$

or

$$1 - e^{-t} - \frac{1}{2} + \frac{e^{-2t}}{2} \qquad (A.32)$$

which can be written as

$$\frac{1}{2} - e^{-t} + \frac{e^{-2t}}{2} \qquad (A.33)$$

The second element in equation (A.27) can be written as

$$-e^{-t}\int_0^t e^\tau d\tau + 2e^{-2t}\int_0^t e^{+2\tau}d\tau \qquad (A.34)$$

or

$$-e^{-t}(e^\tau\big|_0^t) + 2e^{-2t}\left(\frac{1}{2}e^{2\tau}\right)_0^t \qquad (A.35)$$

or

$$-e^{-t}\left(e^t - 1\right) + 2e^{-2t}\left(\frac{1}{2}e^{2t} - \frac{1}{2}\right) \qquad (A.36)$$

Equation (A.36) becomes

$$-1 + e^{-t} + \frac{2}{2}e^{-2t+2t} - e^{-2t} \qquad (A.37)$$

or

Appendix A: Matrix Exponential Method

$$-1 + e^{-t} + 1 - e^{-2t} \tag{A.38}$$

which simplifies to

$$e^{-t} - e^{-2t} \tag{A.39}$$

Thus, equation (A.27) becomes

$$\mathbf{Y}(t) = \begin{pmatrix} y_1(t) \\ y_2(t) \end{pmatrix} = \begin{pmatrix} \dfrac{1}{2} - e^{-t} + \dfrac{1}{2} e^{-2t} \\ e^{-t} - e^{-2t} \end{pmatrix} \tag{A.40}$$

It is easy to solve this problem with $\mathbf{Y}(0) = \mathbf{0}$ by using Maple. First write the solution (see equation (A.15)):

$$\mathbf{Y} = e^{\mathbf{A}t} \int_0^t e^{-\mathbf{A}\tau} \mathbf{b} \, d\tau \tag{A.41}$$

and solve by using Maple:

> restart:with(linalg):

Warning, the protected names norm and trace have been redefined and unprotected

> A:=matrix(2,2,[0,1,-2,-3]);

$$A := \begin{bmatrix} 0 & 1 \\ -2 & -3 \end{bmatrix}$$

> mat:=exponential(A*t);

$$mat := \begin{bmatrix} -e^{(-2t)} + 2e^{(-t)} & e^{(-t)} - e^{(-2t)} \\ -2e^{(-t)} + 2e^{(-2t)} & 2e^{(-2t)} - e^{(-t)} \end{bmatrix}$$

> mattau:=exponential(A*(-tau));

$$mattau := \begin{bmatrix} 2e^{\tau} - e^{(2\tau)} & -e^{(2\tau)} + e^{\tau} \\ 2e^{(2\tau)} - 2e^{\tau} & -e^{\tau} + 2e^{(2\tau)} \end{bmatrix}$$

> b:=matrix(2,1,[0,1]);

$$b := \begin{bmatrix} 0 \\ 1 \end{bmatrix}$$

> mattaub:=evalm(mattau&*b);

$$mattaub := \begin{bmatrix} -e^{(2\tau)} + e^{\tau} \\ -e^{\tau} + 2e^{(2\tau)} \end{bmatrix}$$

> mati:=map(int,evalm(mattaub),tau=0..t);

$$mati := \begin{bmatrix} -\frac{1}{2} - \frac{1}{2}e^{(2t)} + e^{t} \\ -e^{t} + e^{(2t)} \end{bmatrix}$$

> sol:=evalm(mat&*mati);

$$sol := \begin{bmatrix} (-e^{(-2t)} + 2e^{(-t)})\left(-\frac{1}{2} - \frac{1}{2}e^{(2t)} + e^{t}\right) + (e^{(-t)} - e^{(-2t)})(-e^{t} + e^{(2t)}) \\ (-2e^{(-t)} + 2e^{(-2t)})\left(-\frac{1}{2} - \frac{1}{2}e^{(2t)} + e^{t}\right) + (2e^{(-2t)} - e^{(-t)})(-e^{t} + e^{(2t)}) \end{bmatrix}$$

> simplify(sol);

$$\begin{bmatrix} \frac{1}{2}e^{(-2t)} + \frac{1}{2}e^{(-t)} \\ e^{(-t)} - e^{(-2t)} \end{bmatrix}$$

Comparison of equation (A.40) to the solution above (sol) shows that Maple yields the same result for Y that we obtained by hand.

Appendix B: Matrix Exponential by the Laplace Transform Method

The matrix exponential can also be obtained by using the Laplace transform technique. Taking the Laplace transform of the governing equations now written in lower case x (see Ogata page 725).[1]

$$\frac{dY}{dt} = AY \quad (B.1)$$

yields

$$sY(s) - y(t=0) = AY(s) \quad (B.2)$$

or

$$(sI - A)Y(s) = y(0) \quad (B.3)$$

Thus,

Appendix B: Matrix Exponential by the Laplace Transform Method

$$Y(s) = (sI - A)^{-1} y(0) \tag{B.4}$$

and

$$y(t) = \mathcal{L}^{-1}\left[(sI - A)^{-1}\right] y(0) \tag{B.5}$$

Recall that (see Example 3 (with $a = 0$) on page 809 of Kreyszig)[2]

$$\frac{1}{c - bz} = \frac{1}{c} + \frac{b}{c^2} z + \frac{b^2}{c^3} z^2 + \cdots \tag{B.6}$$

Equation (B.6) can be used as a scaler example of how to expand the inverse of a matrix. That is, by analogy with $z = 1$ we can write

$$(sI - A)^{-1} = \frac{I}{s} + \frac{A}{s^2} + \frac{A^2}{s^3} + \cdots \tag{B.7}$$

This can be shown to be correct by premultiplying both sides of equation (B.7) by $sI - A$ to get

$$(sI - A)(sI - A)^{-1} = (sI - A)\left(\frac{I}{s} + \frac{A}{s^2} + \frac{A^2}{s^3} + \cdots\right) \tag{B.8}$$

or

$$I = sI\frac{I}{s} + sI\frac{A}{s^2} + sI\frac{A^2}{s^3} + \cdots - A\frac{I}{s} - \frac{A^2}{s^2} - \cdots = I^2 = I \tag{B.9}$$

Thus,

$$\mathcal{L}^{-1}\{(sI - A)^{-1}\} = I + At + \frac{A^2 t^2}{2!} + A^3 \frac{t^3}{3!} + \cdots \tag{B.10}$$

Comparison of the right hand side of equation (B.10) to the series definition of $\exp(At)$ reveals that

$$\mathcal{L}^{-1}\{(sI - A)^{-1}\} = \exp(At) \tag{B.11}$$

For our example

$$A = \begin{pmatrix} 0 & 1 \\ -1 & -2 \end{pmatrix} \tag{B.12}$$

So that

$$(s\mathbf{I} - \mathbf{A}) = \begin{pmatrix} s & 0 \\ 0 & s \end{pmatrix} - \begin{pmatrix} 0 & 1 \\ -1 & -2 \end{pmatrix} \tag{B.13}$$

or

$$(s\mathbf{I} - \mathbf{A}) = \begin{pmatrix} s & -1 \\ 1 & s+2 \end{pmatrix} \tag{B.14}$$

The inverse of equation (B.14) can be obtained by using the adjoint and determinant of $(s\mathbf{I} - \mathbf{A})$:

$$(s\mathbf{I} - \mathbf{A})^{-1} = \frac{adj\,(s\mathbf{I} - \mathbf{A})}{Det(s\mathbf{I} - \mathbf{A})} \tag{B.15}$$

or

$$(s\mathbf{I} - \mathbf{A})^{-1} = \frac{\begin{pmatrix} s+2 & 1 \\ -1 & s \end{pmatrix}}{(s)(s+2)+1} \tag{46a}$$

or

$$(s\mathbf{I} - \mathbf{A})^{-1} = \begin{pmatrix} \dfrac{s+2}{s^2+2s+1} & \dfrac{1}{s^2+2s+1} \\ \dfrac{-1}{s^2+2s+1} & \dfrac{s}{s^2+2s+1} \end{pmatrix} \tag{46b}$$

The inverse Laplace transform of the matrix elements in equation (46b) can be found from Laplace transform tables or by using the Heaveside expansion theorem. For example, consider the entry in the a_{12} position of equation (46b). We find from Table 8.1 of Varma and Morbidelli[3] that

$$\mathcal{L}^{-1}\left\{\frac{1}{(s+1)^2}\right\} = te^{-t} \tag{B.16}$$

This result can also be obtained from equation 8.3.20 in Varma and Morbidelli:

$$f(t) = \sum_{j=1}^{m} \frac{B_j}{(j-1)!} t^{j-1} e^{bt} + r(t) \tag{B.17}$$

Where B_j is given by equation 8.3.16 of Varma and Morbidelli[3] and $r(t)$ is zero for our case. For this case, $m = 2$ ($s = -1$ is a root of $(s+1)^2$ and is repeated twice, thus $b = -1$). Expanding equation (B.17) yields

Appendix B: Matrix Exponential by the Laplace Transform Method

$$f(t) = \frac{B_1}{0!} e^{-t} + \frac{B_2}{1!} te^{-t} \tag{B.18}$$

where

$$B_1 = \frac{1}{1!}\left(\frac{dW}{ds}\right)_{s=-1} \tag{B.19}$$

$$B_2 = \frac{1}{0!}(W)_{s=-1} \tag{B.20}$$

with

$$W = \frac{(s+1)^2}{(s+1)^2} = 1 \tag{B.21}$$

Thus

$$B_1 = 0 \tag{B.22}$$

since $dW/ds = 0$ and

$$B_2 = 1 \tag{B.23}$$

Thus, equation (B.18) becomes

$$f(t) = (0)e^{-t} + te^{-t} = te^{-t} \tag{B.24}$$

Now determine the inverse Laplace transform of the element in the a_{11} position in equation (46b):

$$\mathcal{L}^{-1}\left\{\frac{s+2}{s^2+2s+1}\right\} = ? \tag{B.25}$$

Equation (B.17) applies in this case also where now (see equation 8.3.12 of Varma and Morbidelli)[3]

$$W = \frac{(s+1)^2}{(s+1)^2}(s+2) = (s+2) \tag{B.26}$$

Thus, in this case

$$B_1 = \left.\frac{dW}{ds}\right|_{s=-1} = \left.\frac{d(s+2)}{ds}\right|_{s=-1} = 1 \tag{B.27}$$

or

$$B_1 = 1 \tag{B.28}$$

and

$$B_2 = (s+2)\big|_{s=-1} = 1 \tag{B.29}$$

So that

$$f(t) = e^{-t} + te^{-t} \tag{B.30}$$

Thus, after determining the inverse Laplace transform of the elements in the a_{21} and a_{22} positions in equation (46b) we find that

$$\mathcal{L}^{-1}\{(s\mathbf{I}-\mathbf{A})^{-1}\} = \begin{pmatrix} e^{-t}+te^{-t} & te^{-t} \\ -te^{-t} & e^{-t}-te^{-t} \end{pmatrix} \tag{B.31}$$

which is the same expression we obtained before (see equation (B.31)) for $\exp(\mathbf{A}t)$.

The solution to equation (B.1) with $\mathbf{A}$ given by using equation (B.12) is

$$\mathbf{y} = \exp(\mathbf{A}t)\,\mathbf{y}(0) \tag{B.32}$$

For example, if

$$\mathbf{y}(0) = \begin{bmatrix} 1 & 0 \end{bmatrix}^T \tag{B.33}$$

then

$$\mathbf{y} = \begin{pmatrix} y_1 \\ y_2 \end{pmatrix} = \begin{pmatrix} e^{-t}+te^{-t} \\ -te^{-t} \end{pmatrix} \tag{B.34}$$

As an additional observation, note that the equation used to determine the characteristic polynomial for determining the eigenvalues of a coefficient matrix

$$\mathrm{Det}(\lambda\mathbf{I}-\mathbf{A}) = 0 \tag{B.35}$$

is the same equation as

$$\mathrm{Det}(s\mathbf{I}-\mathbf{A}) = 0 \tag{B.36}$$

with λ replaced by s which is the denominator in equation (B.15). This tells us that the eigenvalues of a matrix $\mathbf{A}$ have the same values as the node or pole values (i.e., the s values). For example, in the characteristic polynomial for the eigenvalues for the coefficient $\mathbf{A}$ given in equation (B.2) is

$$\mathrm{Det}(\lambda\mathbf{I}-\mathbf{A}) = 0 \tag{B.37}$$

which becomes

$$\mathrm{Det}\left(\lambda\begin{pmatrix}1 & 0\\ 0 & 1\end{pmatrix}-\begin{pmatrix}1 & 0\\ -1 & -2\end{pmatrix}\right)=0 \quad (\text{B}.38)$$

$$\mathrm{Det}\begin{pmatrix}\lambda-1 & 0\\ +1 & \lambda+2\end{pmatrix}=0 \quad (\text{B}.39)$$

or

$$\lambda(\lambda+2)+1=0 \quad (\text{B}.40)$$

or

$$\lambda^2+2\lambda+1=0 \quad (\text{B}.41)$$

or

$$(\lambda+1)^2=0 \quad (\text{B}.42)$$

and finally we have

$$\lambda_1=-1 \text{ and } \lambda_2=-1 \quad (\text{B}.43)$$

Clearly, one can replace λ with s in equation (B.37) to find that we have repeated poles or nodes:

$$s_1=-1 \text{ and } s_2=-1 \quad (\text{B}.44)$$

These poles are the roots of the denominator of all the elements in equation (46b). It is important to note that both the eigenvalue characteristic polynomial and the characteristic polynomial of the denominator in a Laplace transform function are both based on the coefficient matrix A. The two procedures (matrix exponential and Laplace transform) yield the same time dependent solutions to the original system of differential equations, as expected. The matrix exponential approach requires finding the eigenvalues and eigenvectors, and the Laplace transform technique requires finding the inverse of the Laplace transform. This handout demonstrates that the methods are related.

References

1. Ogata, K.: Modern Control Engineering. Prentice-Hall, Englewood Cliffs (1997)
2. Kreyszig, E.: Advanced Engineering Mathematics. John Wiley & Sons, Inc., Chichester (1993)
3. Varma, A., Morbidelli, M.: Mathematical Methods in Chemical Engineering. Oxford University Press, Inc., Oxford (1997)

Chapter 3
Boundary Value Problems

Mathematical modeling of mass or heat transfer in solids involves Fick's law of mass transfer or Fourier's law of heat conduction. Engineers are interested in the steady state distribution of heat or concentration across the slab or the material in which the experiment is performed. This steady state process involves solving second order ordinary differential equations subject to boundary conditions at two ends. Whenever the problem requires the specification of boundary conditions at two points, it is often called a two point boundary value problem. Both linear and nonlinear boundary value problems will be discussed in this chapter. We will present analytical solutions for linear boundary value problems and numerical solutions for nonlinear boundary value problems.

3.1 Linear Boundary Value Problems

3.1.1 Introduction

Fick's law of diffusion and Fourier's law of conduction are usually represented by second order ordinary differential equations (ODEs). In this chapter, we describe how one can obtain analytical solutions for linear boundary value problems using Maple and the matrix exponential.

3.1.2 Exponential Matrix Method for Linear Boundary Value Problems

Consider a general boundary value problem

$$\frac{d^2y}{dx^2} + a_1 \frac{dy}{dx} + a_2 y = f(x) \tag{3.1}$$

with the boundary conditions

$$y(0) = 1 \tag{3.2}$$

and

$$\frac{dy}{dx}(1) = 0 \tag{3.3}$$

Equation (3.1) can be converted into two first order differential equations (see section 2.1.4) and can be cast into matrix form as follows:

$$\frac{d\mathbf{Y}}{dx} = \mathbf{A}\mathbf{Y} + \mathbf{b}(x) \tag{3.4}$$

where the dependent variable vector is

$$\mathbf{Y} = \begin{bmatrix} y \\ \dfrac{dy}{dx} \end{bmatrix} \tag{3.5}$$

the coefficient matrix is

$$\mathbf{A} = \begin{bmatrix} 0 & 1 \\ -a_2 & -a_1 \end{bmatrix} \tag{3.6}$$

and the forcing function matrix is

$$\mathbf{b} = \begin{bmatrix} 0 \\ f(x) \end{bmatrix} \tag{3.7}$$

Equation (3.4) is a nonhomogeneous vector differential equation (see section 2.1.3). The solution for equation (3.4) is given by (see Appendix A)[1] [2] [3] [4] [5]

$$\mathbf{Y} = \exp(\mathbf{A}x)\mathbf{Y}_0 + \int_0^x \exp[-\mathbf{A}(\zeta - x)] \, \mathbf{b}(\zeta) \, d\zeta \tag{3.8a}$$

or when **b** is a constant vector

$$\mathbf{Y} = \exp(\mathbf{A}x)(\mathbf{Y}_0 + \mathbf{A}^{-1}\mathbf{b}) - \mathbf{A}^{-1}\mathbf{b} \tag{3.8b}$$

or when **b** is zero

$$\mathbf{Y} = \exp(\mathbf{A}x)\mathbf{Y}_0 \tag{3.8c}$$

where ζ is a dummy variable of integration. The procedure involved for solving boundary value problems using the matrix exponential (exp(**A**x)) is as follows:

3.1 Linear Boundary Value Problems

1. Start the Maple program with a 'restart' command to clear all variables.
2. Call 'with(linalg)' and 'with(lplots)' commands.
3. Enter the governing equation (equation (3.1)).
4. Enter the coefficient matrix (**A**) based on equation (3.6).
5. Enter the forcing function matrix (**b**) based on equation (3.7).
6. Store the initial conditions for the dependent variables in the vector Y0.

    6a. The first row of Y0 corresponds to the initial condition for y, the second row corresponds to the initial condition for the derivative, $\frac{dy}{dx}$.

    6b. Usually either y or $\frac{dy}{dx}$ at x = 0 is known. The unknown initial condition at x = 0 is taken as an unknown constant c_1.
7. The matrix exponential (exp(**A**x)) is found as a function of the parameters and the independent variable (x) using the 'exponential(A,x)' command in Maple.
8. The matrix exponential found is stored in mat. Now the solution (sol) is found by multiplying mat times Y0 and adding the non-homogenous solution according to equation (3.8) depending on the **b** matrix.
9. The first row of sol corresponds to the dependent variable y; the second row corresponds to the first derivative of y, etc.
10. Next, the boundary condition at x =1 is used to solve for the unknown constant c_1.
11. Then c_1 is substituted into the solution.

Once the analytical solution is obtained, plots can be made for a particular value of the parameters.

Example 3.1

Consider the conduction of heat in a rectangular cooling fin. The governing differential equation[6] in dimensionless form is

$$\frac{d^2\theta}{dx^2} = H^2\theta$$

$$\theta(0) = 1 \text{ and } \frac{d\theta}{dx}(1) = 0$$

(3.9)

This boundary value problem is solved below by following the procedure described earlier.

> restart:
> with(linalg):with(plots):
> N:=2;

$$N := 2$$

> eq:=diff(theta(x),x$2)-H^2*theta(x);

$$eq := \frac{d^2}{dx^2}\theta(x) - H^2\,\theta(x)$$

> A:=matrix(N,N,[0,1,H^2,0]);

$$A := \begin{bmatrix} 0 & 1 \\ H^2 & 0 \end{bmatrix}$$

> mat:=exponential(A,x);

$$mat := \begin{bmatrix} \frac{1}{2}e^{Hx} + \frac{1}{2}e^{-Hx} & \frac{1}{2}\frac{-e^{-Hx}+e^{Hx}}{H} \\ -\frac{1}{2}He^{-Hx} + \frac{1}{2}He^{Hx} & \frac{1}{2}e^{Hx} + \frac{1}{2}e^{-Hx} \end{bmatrix}$$

The exponentials are converted to trigonometric form for convenience.

> mat:=map(convert,mat,trig);

$$mat := \begin{bmatrix} \cosh(Hx) & \frac{\sinh(Hx)}{H} \\ -\frac{1}{2}H(\cosh(Hx)-\sinh(Hx)) + \frac{1}{2}H(\cosh(Hx)+\sinh(Hx)) & \cosh(Hx) \end{bmatrix}$$

> Y0:=matrix(N,1,[1,c[1]]);

$$Y0 := \begin{bmatrix} 1 \\ c_1 \end{bmatrix}$$

> sol:=evalm(mat&*Y0);

$$sol := \begin{bmatrix} \cosh(Hx) + \frac{\sinh(Hx)\,c_1}{H} \\ -\frac{1}{2}H(\cosh(Hx)-\sinh(Hx)) + \frac{1}{2}H(\cosh(Hx)+\sinh(Hx)) + \cosh(Hx)\,c_1 \end{bmatrix}$$

The solution is obtained as a function of x, H, and the unknown initial condition c_1.

The first row of sol corresponds to θ and the second row corresponds to $\dfrac{\theta}{dx}$.

3.1 Linear Boundary Value Problems

> theta:=sol[1,1];

$$\theta := \cosh(Hx) + \frac{\sinh(Hx)\, c_1}{H}$$

> dthetadx:=sol[2,1];

$$dthetadx := -\frac{1}{2} H\,(\cosh(Hx) - \sinh(Hx)) + \frac{1}{2} H\,(\cosh(Hx) + \sinh(Hx)) + \cosh(Hx)\, c_1$$

The boundary condition at $x = 1$ is applied to solve for c_1.

> bc2:=subs(x=1,dthetadx)=0;

$$bc2 := -\frac{1}{2} H\,(\cosh(H) - \sinh(H)) + \frac{1}{2} H\,(\cosh(H) + \sinh(H)) + \cosh(H)\, c_1 = 0$$

> c[1]:=solve(bc2,c[1]);

$$c_1 := -\frac{H\sinh(H)}{\cosh(H)}$$

The value of the constant c_1 is substituted into the expression for θ.

> theta:=eval(theta);

$$\theta := \cosh(Hx) - \frac{\sinh(Hx)\,\sinh(H)}{\cosh(H)}$$

The expression for θ can be simplified further by using Maple's 'combine' command:

> theta:=combine(theta);

$$\theta := \frac{\cosh(Hx - H)}{\cosh(H)}$$

You can make plots for different values of the heat transfer coefficient, H, by using a 'pars' array and a 'do loop.'

> pars:=[1,2,5,10];

$$pars := [1, 2, 5, 10]$$

> clr:=[red,green,gold,blue];

> for i from 1 to 4 do p[i]:=plot(subs(H=pars[i],theta),x=0..1,color=clr[i]):od:
> display(seq(p[i],i=1..4),thickness=4,title="Figure 3.1",axes=boxed,labels=["X","theta"],labeldirections=[HORIZONTAL,VERTICAL]);

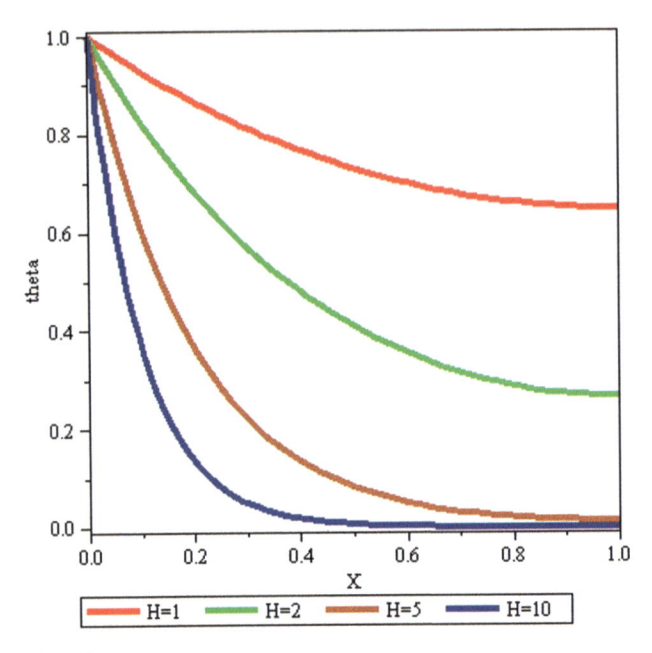

Fig. 3.1

A three dimensional plot for θ can be made for different values of H as x varies between zero and one.

```
> plot3d(theta,x=0..1,H=0..10,axes=boxed,title="Figure 3.2",
orientation=[45,45]);
```

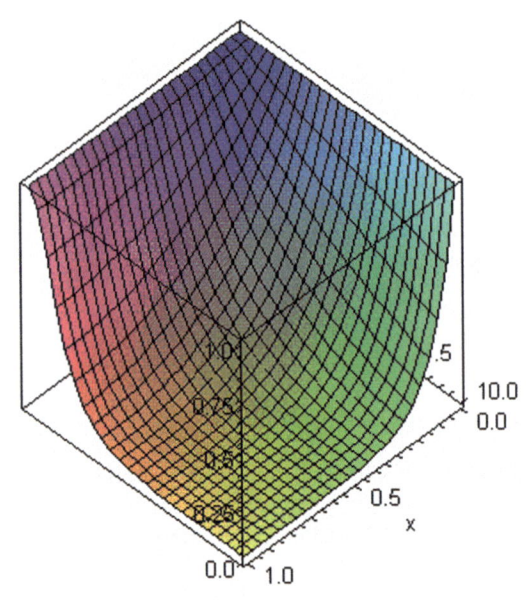

Fig. 3.2

3.1 Linear Boundary Value Problems

The above picture shows that as H increases the temperature distribution becomes nonuniform.

Example 3.2

Diffusion with a convection and simultaneous first order reaction in a rectangular plate can be simulated using the program described above by using minor modifications. Consider the composition profile in a packed tube reactor undergoing isothermal linear kinetics with axial diffusion. The governing equation is

$$\frac{d^2C}{dx^2} - Pe\frac{dC}{dx} - HaC = 0$$

$$\frac{dC}{dx}\bigg|_{x=0} + Pe(1-C(0)) = 0 \tag{3.10}$$

$$\frac{dC}{dx}(1) = 0$$

Solving this boundary value problem by hand is difficult and the solution for this problem is very long. This boundary value problem is solved below easily by following the procedure described earlier.

```
> restart:
> with(linalg):with(plots):
> N:=2;
```

$$N := 2$$

```
> eq:=diff(C(x),x$2)-Pe*diff(C(x),x)-Ha*C(x);
```

$$eq := \frac{d^2}{dx^2}C(x) - Pe\left(\frac{d}{dx}C(x)\right) - Ha\,C(x)$$

```
> A:=matrix(N,N,[0,1,Ha,Pe]);
> mat:=exponential(A,x);
```

$$mat := \Bigg[\Bigg| \frac{1}{2} \frac{1}{\sqrt{Pe^2+4Ha}} \left(e^{\frac{1}{2}(Pe+\sqrt{Pe^2+4Ha})x} \sqrt{Pe^2+4Ha} + Pe\,e^{\frac{1}{2}(Pe-\sqrt{Pe^2+4Ha})x} \right.$$

$$-Pe\,e^{\frac{1}{2}(Pe+\sqrt{Pe^2+4Ha})x} + e^{\frac{1}{2}(Pe-\sqrt{Pe^2+4Ha})x} \sqrt{Pe^2+4Ha} \Bigg),$$

$$-\frac{e^{\frac{1}{2}(Pe-\sqrt{Pe^2+4Ha})x} - e^{\frac{1}{2}(Pe+\sqrt{Pe^2+4Ha})x}}{\sqrt{Pe^2+4Ha}},$$

$$\left[-\frac{Ha\left(e^{\frac{1}{2}\left(Pe - \sqrt{Pe^2 + 4Ha} \right)x} - e^{\frac{1}{2}\left(Pe + \sqrt{Pe^2 + 4Ha} \right)x} \right)}{\sqrt{Pe^2 + 4Ha}}, -\frac{1}{2}\frac{1}{\sqrt{Pe^2 + 4Ha}} \left(-e^{\frac{1}{2}\left(Pe + \sqrt{Pe^2 + 4Ha} \right)x} \sqrt{Pe^2 + 4Ha} + Pe\, e^{\frac{1}{2}\left(Pe - \sqrt{Pe^2 + 4Ha} \right)x} - Pe\, e^{\frac{1}{2}\left(Pe + \sqrt{Pe^2 + 4Ha} \right)x} \right. \right.$$
$$\left. \left. -e^{\frac{1}{2}\left(Pe - \sqrt{Pe^2 + 4Ha} \right)x} \sqrt{Pe^2 + 4Ha} \right) \right]$$

Note that for this boundary value problem both C and $\frac{dC}{dx}$ at $x=0$ are not known. However, using the boundary condition at $x=0$, $\frac{dC}{dx}$ can be written in terms of the unknown constant c_1, which is the concentration at $x=0$.

> Y0:=matrix(N,1,[c[1],-Pe*(1-c[1])]);

$$Y0 := \begin{bmatrix} c_1 \\ -Pe\,(1 - c_1) \end{bmatrix}$$

> sol:=evalm(mat&*Y0):
> C:=sol[1,1];

$$C := \frac{1}{2} \frac{1}{\sqrt{Pe^2 + 4Ha}} \left(\left(e^{\frac{1}{2}\left(Pe + \sqrt{Pe^2 + 4Ha} \right)x} \sqrt{Pe^2 + 4Ha} + Pe\, e^{\frac{1}{2}\left(Pe - \sqrt{Pe^2 + 4Ha} \right)x} \right. \right.$$
$$\left. - Pe\, e^{\frac{1}{2}\left(Pe + \sqrt{Pe^2 + 4Ha} \right)x} + e^{\frac{1}{2}\left(Pe - \sqrt{Pe^2 + 4Ha} \right)x} \sqrt{Pe^2 + 4Ha} \right) c_1$$
$$\left. + \frac{\left(e^{\frac{1}{2}\left(Pe - \sqrt{Pe^2 + 4Ha} \right)x} - e^{\frac{1}{2}\left(Pe + \sqrt{Pe^2 + 4Ha} \right)x} \right) Pe\,(1 - c_1)}{\sqrt{Pe^2 + 4Ha}} \right)$$

> dCdx:=sol[2,1];

$$dCdx := -\frac{Ha\left(e^{\frac{1}{2}\left(Pe - \sqrt{Pe^2 + 4Ha} \right)x} - e^{\frac{1}{2}\left(Pe + \sqrt{Pe^2 + 4Ha} \right)x} \right) c_1}{\sqrt{Pe^2 + 4Ha}} + \frac{1}{2}\frac{1}{\sqrt{Pe^2 + 4Ha}} \left(\left(-e^{\frac{1}{2}\left(Pe + \sqrt{Pe^2 + 4Ha} \right)x} \sqrt{Pe^2 + 4Ha} + Pe\, e^{\frac{1}{2}\left(Pe - \sqrt{Pe^2 + 4Ha} \right)x} - Pe\, e^{\frac{1}{2}\left(Pe + \sqrt{Pe^2 + 4Ha} \right)x} \right. \right.$$
$$\left. \left. -e^{\frac{1}{2}\left(Pe - \sqrt{Pe^2 + 4Ha} \right)x} \sqrt{Pe^2 + 4Ha} \right) Pe\,(1 - c_1) \right)$$

3.1 Linear Boundary Value Problems

To find c_1, we can use the known boundary condition at $x = 0$.

> bc2:=subs(x=1,dCdx)=0;

$$bc2 := -\frac{Ha\left(e^{\frac{1}{2}Pe - \frac{1}{2}\sqrt{Pe^2 + 4Ha}} - e^{\frac{1}{2}Pe + \frac{1}{2}\sqrt{Pe^2 + 4Ha}}\right)c_1}{\sqrt{Pe^2 + 4Ha}} + \frac{1}{2}\frac{1}{\sqrt{Pe^2 + 4Ha}}\left(\left(-e^{\frac{1}{2}Pe + \frac{1}{2}\sqrt{Pe^2 + 4Ha}}\sqrt{Pe^2 + 4Ha} + Pe\,e^{\frac{1}{2}Pe - \frac{1}{2}\sqrt{Pe^2 + 4Ha}} - Pe\,e^{\frac{1}{2}Pe + \frac{1}{2}\sqrt{Pe^2 + 4Ha}}\right.\right.$$
$$\left.\left. -e^{\frac{1}{2}Pe - \frac{1}{2}\sqrt{Pe^2 + 4Ha}}\sqrt{Pe^2 + 4Ha}\right)Pe\,(1 - c_1)\right) = 0$$

The equation for c_1 can be solved easily by using Maple's 'solve' command.

> c[1]:=solve(bc2,c[1]);

$$c_1 := \left(Pe\left(-e^{\frac{1}{2}Pe + \frac{1}{2}\sqrt{Pe^2 + 4Ha}}\sqrt{Pe^2 + 4Ha} + Pe\,e^{\frac{1}{2}Pe - \frac{1}{2}\sqrt{Pe^2 + 4Ha}} - Pe\,e^{\frac{1}{2}Pe + \frac{1}{2}\sqrt{Pe^2 + 4Ha}}\right.\right.$$
$$\left. -e^{\frac{1}{2}Pe - \frac{1}{2}\sqrt{Pe^2 + 4Ha}}\sqrt{Pe^2 + 4Ha}\right)\Big/\left(2Ha\,e^{\frac{1}{2}Pe - \frac{1}{2}\sqrt{Pe^2 + 4Ha}} - 2Ha\,e^{\frac{1}{2}Pe + \frac{1}{2}\sqrt{Pe^2 + 4Ha}}\right.$$
$$\left. -Pe\,e^{\frac{1}{2}Pe + \frac{1}{2}\sqrt{Pe^2 + 4Ha}}\sqrt{Pe^2 + 4Ha} + Pe^2\,e^{\frac{1}{2}Pe - \frac{1}{2}\sqrt{Pe^2 + 4Ha}} - Pe^2\,e^{\frac{1}{2}Pe + \frac{1}{2}\sqrt{Pe^2 + 4Ha}}\right.$$
$$\left.\left. -Pe\,e^{\frac{1}{2}Pe - \frac{1}{2}\sqrt{Pe^2 + 4Ha}}\sqrt{Pe^2 + 4Ha}\right)\right)$$

Now that we have c_1, the complete solution can be determined for C by using Maple's 'eval' command:

> C:=eval(C);

$$C := \frac{1}{2}\left(\left(e^{\frac{1}{2}\left(Pe + \sqrt{Pe^2 + 4Ha}\right)x}\sqrt{Pe^2 + 4Ha} + Pe\,e^{\frac{1}{2}\left(Pe - \sqrt{Pe^2 + 4Ha}\right)x} - Pe\,e^{\frac{1}{2}\left(Pe + \sqrt{Pe^2 + 4Ha}\right)x}\right.\right.$$
$$\left. + e^{\frac{1}{2}\left(Pe - \sqrt{Pe^2 + 4Ha}\right)x}\sqrt{Pe^2 + 4Ha}\right)Pe\left(-e^{\frac{1}{2}Pe + \frac{1}{2}\sqrt{Pe^2 + 4Ha}}\sqrt{Pe^2 + 4Ha}\right.$$
$$\left.\left. + Pe\,e^{\frac{1}{2}Pe - \frac{1}{2}\sqrt{Pe^2 + 4Ha}} - Pe\,e^{\frac{1}{2}Pe + \frac{1}{2}\sqrt{Pe^2 + 4Ha}} - e^{\frac{1}{2}Pe - \frac{1}{2}\sqrt{Pe^2 + 4Ha}}\sqrt{Pe^2 + 4Ha}\right)\right)\Big/$$
$$\left(\sqrt{Pe^2 + 4Ha}\left(2Ha\,e^{\frac{1}{2}Pe - \frac{1}{2}\sqrt{Pe^2 + 4Ha}} - 2Ha\,e^{\frac{1}{2}Pe + \frac{1}{2}\sqrt{Pe^2 + 4Ha}}\right.\right.$$
$$\left.\left. -Pe\,e^{\frac{1}{2}Pe + \frac{1}{2}\sqrt{Pe^2 + 4Ha}}\sqrt{Pe^2 + 4Ha} + Pe^2\,e^{\frac{1}{2}Pe - \frac{1}{2}\sqrt{Pe^2 + 4Ha}} - Pe^2\,e^{\frac{1}{2}Pe + \frac{1}{2}\sqrt{Pe^2 + 4Ha}}\right.\right.$$

$$-Pe\, e^{\frac{1}{2}Pe-\frac{1}{2}\sqrt{Pe^2+4Ha}}\sqrt{Pe^2+4Ha}\bigg)\bigg) + \frac{1}{\sqrt{Pe^2+4Ha}}\bigg(\bigg(e^{\frac{1}{2}\left(Pe-\sqrt{Pe^2+4Ha}\right)x}$$

$$-e^{\frac{1}{2}\left(Pe+\sqrt{Pe^2+4Ha}\right)x}\bigg)Pe\bigg(1-\bigg(Pe\bigg(-e^{\frac{1}{2}Pe+\frac{1}{2}\sqrt{Pe^2+4Ha}}\sqrt{Pe^2+4Ha}$$

$$+Pe\, e^{\frac{1}{2}Pe-\frac{1}{2}\sqrt{Pe^2+4Ha}} - Pe\, e^{\frac{1}{2}Pe+\frac{1}{2}\sqrt{Pe^2+4Ha}} - e^{\frac{1}{2}Pe-\frac{1}{2}\sqrt{Pe^2+4Ha}}\sqrt{Pe^2+4Ha}\bigg)\bigg)\bigg/$$

$$\bigg(2Ha\, e^{\frac{1}{2}Pe-\frac{1}{2}\sqrt{Pe^2+4Ha}} - 2Ha\, e^{\frac{1}{2}Pe+\frac{1}{2}\sqrt{Pe^2+4Ha}} - Pe\, e^{\frac{1}{2}Pe+\frac{1}{2}\sqrt{Pe^2+4Ha}}\sqrt{Pe^2+4Ha}$$

$$+Pe^2\, e^{\frac{1}{2}Pe-\frac{1}{2}\sqrt{Pe^2+4Ha}} - Pe^2\, e^{\frac{1}{2}Pe+\frac{1}{2}\sqrt{Pe^2+4Ha}} - Pe\, e^{\frac{1}{2}Pe-\frac{1}{2}\sqrt{Pe^2+4Ha}}\sqrt{Pe^2+4Ha}\bigg)\bigg)\bigg)$$

Next, one can plot the concentration profile by substituting values for the parameters Ha and Pe:

> pars:={Ha=1,Pe=10};

$$pars := \{Ha = 1, Pe = 50\}$$

> plot(subs(pars,C),x=0..1,thickness=4,axes=boxed,labels=[x,"C"], title="Figure 3.3");

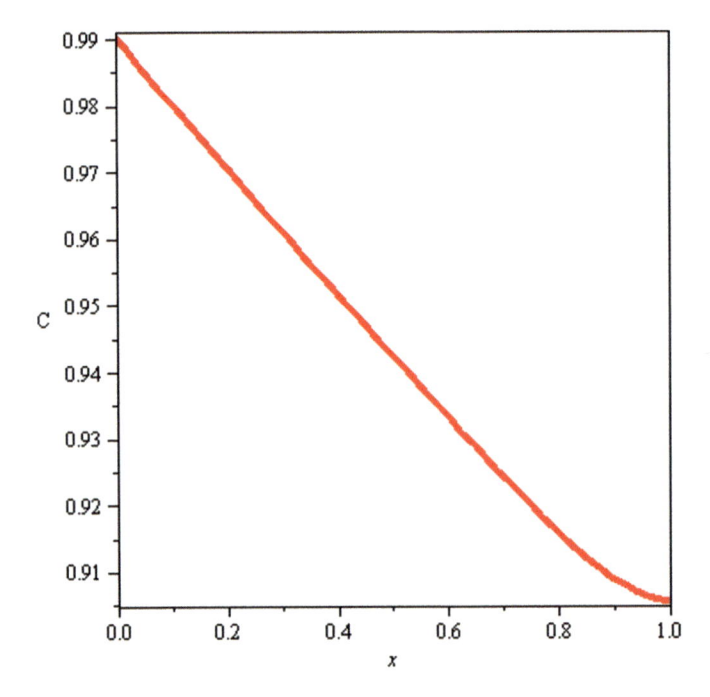

Fig. 3.3

3.1 Linear Boundary Value Problems

New plots can be made for different sets of parameters such as:

```
> pars:={Ha=1,Pe=50};
> plot(subs(pars,C),x=0..1,thickness=4,axes=boxed,labels=[x,"C"],
color=green,title="Figure 3.4");
```

$$pars := \{Ha = 1, Pe = 50\}$$

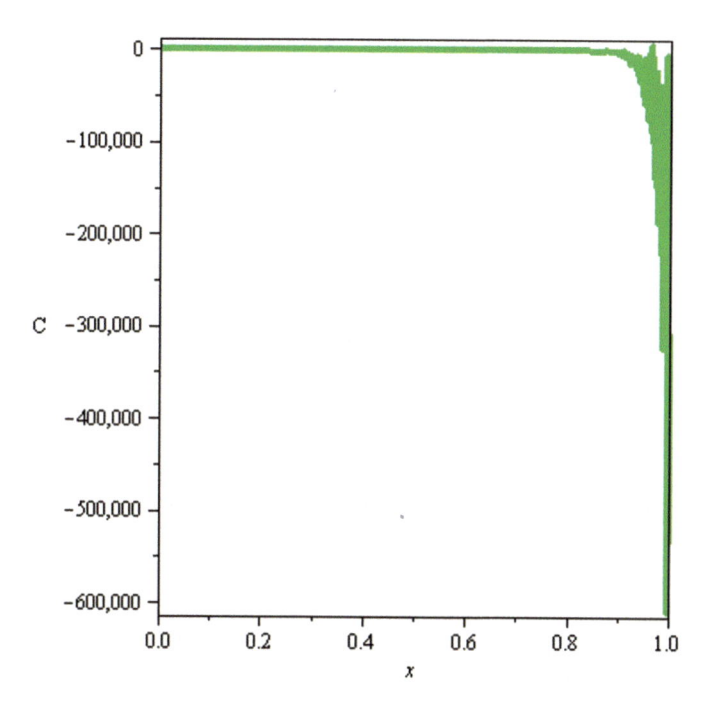

Fig. 3.4

When we get weird plots, we can solve this problem by increasing the number of digits as:

```
> Digits:=30;
```

$$Digits := 30$$

```
> plot(subs(pars,C),x=0..1,thickness=4,axes=boxed,labels=[x,"C"],color=gold,
title="Figure 3.5");
```

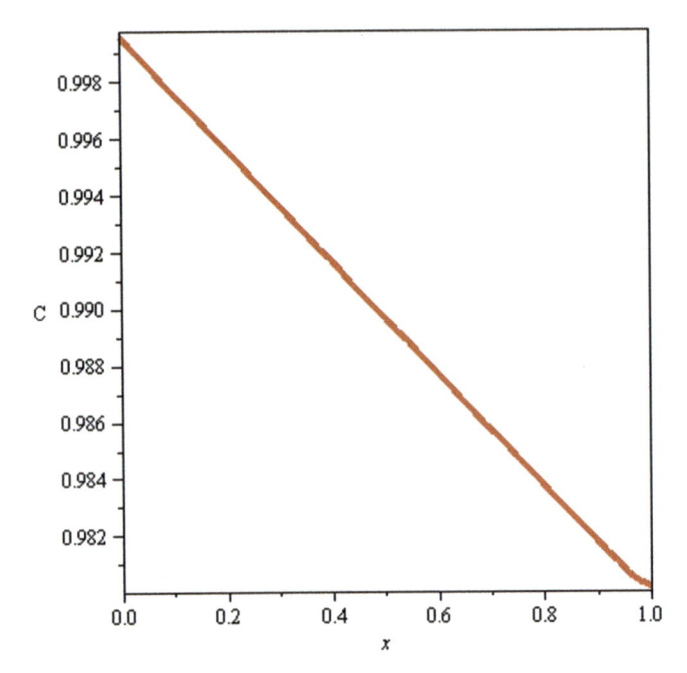

Fig. 3.5

3.1.3 Exponential Matrix Method for Linear BVPs with Semi-infinite Domains

The methodology developed in section 3.1.2 can be used for semi-infinite boundary conditions, also. The procedure for solving boundary value problems in semi-infinite domain is as follows:

1. Start the Maple program with a 'restart' command to clear all variables.
2. Call 'with(linalg)' and 'with(lplots)' commands.
3. Enter the governing equation.
4. Enter the coefficient matrix (**A**) based on equation (3.6).
5. Enter the forcing function matrix (**b**) based on equation (3.8).
6. Store the initial conditions for the dependent variables in the matrix Y0.
7. The first row of Y0 corresponds to the initial condition for y, the second row corresponds to the initial condition for the derivative, $\frac{dy}{dx}$.
8. Usually either y or $\frac{dy}{dx}$ at x = 0 is known. The unknown initial condition at x = 0 is taken as an unknown constant c_1.

3.1 Linear Boundary Value Problems

9. The matrix exponential (exp(**A**x)) is found as a function of the parameters and the independent variable (x) using the 'exponential(A,x)' command in Maple.
10. The matrix exponential found is stored in mat. Now the solution (sol) is found by multiplying mat with Y0 and adding the non-homogenous solution according to equation (3.8) or (3.9) depending on the **b** matrix.
11. The first row of sol corresponds to the dependent variable y; the second row corresponds to the first derivative of y, etc.
12. Next, the unknown constant c_1 is found by using the fact that the solution obtained is finite when x tends to infinity.
13. Then, c_1 is substituted into the solution.

Once the analytical solution is obtained, plots can be made for particular values of the parameters.

Example 3.3

Consider diffusion with a first order reaction in a semi-infinite plane:

$$D1\frac{d^2C}{dx^2} - kC$$

$$C(0) = 1 \text{ and } C(\infty) \text{ is defined}$$

(3.11)

where C is the dimensionless concentration, D_1, is the diffusion coefficient and k is the rate constant. This equation is solved below using the procedure described above.

```
> restart:
> with(linalg):with(plots):
> N:=2;
```

$$N := 2$$

```
> eq:=diff(C(x),x$2)-k/D1*C(x);
```

$$eq := \frac{d^2}{dx^2} C(x) - \frac{k\, C(x)}{D1}$$

```
> A:=matrix(N,N,[0,1,k/D1,0]);
```

$$A := \begin{bmatrix} 0 & 1 \\ \dfrac{k}{D1} & 0 \end{bmatrix}$$

> mat:=exponential(A,x);

$$mat := \begin{bmatrix} \frac{1}{2}e^{\frac{\sqrt{D1\,k}\,x}{D1}} + \frac{1}{2}e^{-\frac{\sqrt{D1\,k}\,x}{D1}} & \frac{1}{2}\frac{D1\left(-e^{-\frac{\sqrt{D1\,k}\,x}{D1}} + e^{\frac{\sqrt{D1\,k}\,x}{D1}}\right)}{\sqrt{D1\,k}} \\ \frac{1}{2}\frac{k\left(-e^{-\frac{\sqrt{D1\,k}\,x}{D1}} + e^{\frac{\sqrt{D1\,k}\,x}{D1}}\right)}{\sqrt{D1\,k}} & \frac{1}{2}e^{\frac{\sqrt{D1\,k}\,x}{D1}} + \frac{1}{2}e^{-\frac{\sqrt{D1\,k}\,x}{D1}} \end{bmatrix}$$

> Y0:=matrix(N,1,[1,c[1]]);

$$Y0 := \begin{bmatrix} 1 \\ c_1 \end{bmatrix}$$

> sol:=evalm(mat&*Y0);

$$sol := \begin{bmatrix} \frac{1}{2}e^{\frac{\sqrt{D1\,k}\,x}{D1}} + \frac{1}{2}e^{-\frac{\sqrt{D1\,k}\,x}{D1}} + \frac{1}{2}\frac{D1\left(-e^{-\frac{\sqrt{D1\,k}\,x}{D1}} + e^{\frac{\sqrt{D1\,k}\,x}{D1}}\right)c_1}{\sqrt{D1\,k}} \\ \frac{1}{2}\frac{k\left(-e^{-\frac{\sqrt{D1\,k}\,x}{D1}} + e^{\frac{\sqrt{D1\,k}\,x}{D1}}\right)}{\sqrt{D1\,k}} + \left(\frac{1}{2}e^{\frac{\sqrt{D1\,k}\,x}{D1}} + \frac{1}{2}e^{-\frac{\sqrt{D1\,k}\,x}{D1}}\right)c_1 \end{bmatrix}$$

> C:=sol[1,1];

$$C := \frac{1}{2}e^{\frac{\sqrt{D1\,k}\,x}{D1}} + \frac{1}{2}e^{-\frac{\sqrt{D1\,k}\,x}{D1}} + \frac{1}{2}\frac{D1\left(-e^{-\frac{\sqrt{D1\,k}\,x}{D1}} + e^{\frac{\sqrt{D1\,k}\,x}{D1}}\right)c_1}{\sqrt{D1\,k}}$$

This can be rewritten as

> C:=collect(C,exp(1/D1*(D1*k)^(1/2)*x));

$$C := \left(\frac{1}{2} + \frac{1}{2}\frac{D1\,c_1}{\sqrt{D1\,k}}\right)e^{\frac{\sqrt{D1\,k}\,x}{D1}} + \frac{1}{2}e^{-\frac{\sqrt{D1\,k}\,x}{D1}} - \frac{1}{2}\frac{D1\,e^{-\frac{\sqrt{D1\,k}\,x}{D1}}\,c_1}{\sqrt{D1\,k}}$$

> C:=collect(C,exp(-1/D1*(D1*k)^(1/2)*x));

$$C := \left(\frac{1}{2} - \frac{1}{2}\frac{D1\,c_1}{\sqrt{D1\,k}}\right)e^{-\frac{\sqrt{D1\,k}\,x}{D1}} + \left(\frac{1}{2} + \frac{1}{2}\frac{D1\,c_1}{\sqrt{D1\,k}}\right)e^{\frac{\sqrt{D1\,k}\,x}{D1}}$$

Since C is finite as x tends to infinity, the second parenthesis must go to zero because exp(x) goes to infinity as x goes to infinity. Consequently, the equation for c_1 can be found by setting the coefficient of the second term equal to zero:

3.1 Linear Boundary Value Problems

```
> eqbc:=coeff(C,exp(1/D1*(D1*k)^(1/2)*x));
```

$$eqbc := \frac{1}{2} + \frac{1}{2}\frac{D1\ c_1}{\sqrt{D1\ k}}$$

```
> c[1]:=solve(eqbc,c[1]);
```

$$c_1 := -\frac{\sqrt{D1\ k}}{D1}$$

Thus, the desired solution for C is simply:

```
> C:=eval(C);
```

$$C := e^{-\frac{\sqrt{D1\ k}\ x}{D1}}$$

Plots can be made by substituting values for the parameters D_1 and k:

```
> pars:={D1=1e-5,k=1};
```

$$pars := \{k = 1, D1 = 0.00001\}$$

```
> plot(subs(pars,C),x=0..1e-2,labels=[x,"C"],thickness=4,axes=boxed,
title="Figure 3.6");
```

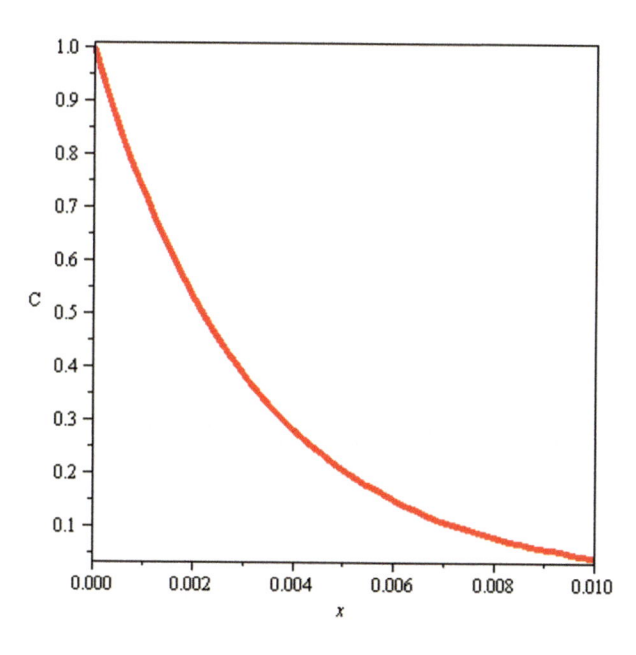

Fig. 3.6

New plots can be made by substituting different values for the parameters.

> pars:={D1=1e-7,k=1};

$$pars := \{k = 1, DI = 1.\ 10^{-7}\}$$

> plot(subs(pars,C),x=0..1e-2,labels=[x,"C"],thickness=4,axes=boxed, color=black,title="Figure 3.7");

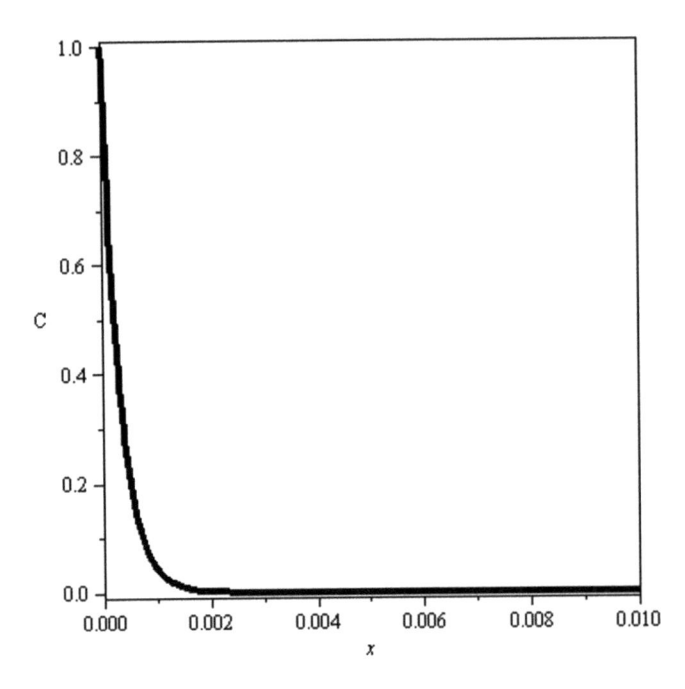

Fig. 3.7

We observe that as the diffusion coefficient decreases mass transfer limitations increase the length of the diffusion layer (distance required for C to drop to approximately 0) which decreases as expected.

3.1.4 Use of Matrizant in Solving Boundary Value Problems

Consider the matrix differential equation

$$\frac{dY}{dt} = A(t)Y \tag{3.12}$$

with a known initial condition of Y_0. Note that in equation (3.12) the coefficient matrix depends on t. The solution for this matrix equation is given by[1] [7] [4]

$$Y = \Omega(A)Y_0 \tag{3.13}$$

3.1 Linear Boundary Value Problems

where the matrizant $\Omega(\mathbf{A})$ is given by

$$\Omega(\mathbf{A}) = [\mathbf{I}] + \int_0^t [\mathbf{A}(t_1) dt_1] + \int_0^t \left[\mathbf{A}(t_1) \left[\int_0^{t_1} [\mathbf{A}(t_2) dt_2] \right] dt_1 \right] + \ldots \quad (3.14)$$

where **I** is the identity matrix. This matrizant $\Omega(\mathbf{A})$ reduces to the exponential matrix exp(**A**t) when **A** is a constant matrix. This method is also referred as Picard's method or the successive substitution method.

The procedure for solving linear initial value problems using the matrizant is the same as that in section 2.1.2 except that instead of finding the exponential matrix, the matrizant is found.

Example 3.4

To illustrate the process of using the matrizant, consider the initial value problem

$$\frac{dc}{dt} = -tc$$

$$c(0) = 1 \quad (3.15)$$

This equation is solved in Maple by finding the matrizant below.

```
> restart:
> with(linalg):with(plots):
```
Warning, the protected names norm and trace have been redefined and unprotected

Warning, the name changecoords has been redefined

```
> N:=1;
```

$$N := 1$$

Enter the number of terms used in calculating the matrizant. (Usually six terms are sufficient).

```
> nvars:=6;
```

$$nvars := 6$$

```
> Eq:=diff(c(t),t)=-t*c(t);
```

$$Eq := \frac{d}{dt} c(t) = -t\, c(t)$$

```
> A:=matrix(1,1,[-t]);
```

$$A := \begin{bmatrix} -t \end{bmatrix}$$

186   3 Boundary Value Problems

> Y0:=matrix(1,1,[1]);

$$Y0 := \begin{bmatrix} 1 \end{bmatrix}$$

> id:=Matrix(N,N,shape=identity);

$$id := \begin{bmatrix} 1 \end{bmatrix}$$

Define the two dummy variables X_1 and X_2.

> X1:=matrix(N,N);X2:=matrix(N,N);

$$X1 := array(1..1, 1..1, [\ ])$$

$$X2 := array(1..1, 1..1, [\ ])$$

A dummy variable t_1 is used in the integration. For matrix integration, Maple's 'map' command should be used.

> X1:=map(int,subs(t=t1,evalm(A)),t1=0..t);

$$X1 := \begin{bmatrix} -\frac{1}{2} t^2 \end{bmatrix}$$

> mat := evalm(id + X1) ;

$$mat := \begin{bmatrix} -\frac{1}{2} t^2 + 1 \end{bmatrix}$$

We now have the first two terms of the matrizant. The next step is to find the next five terms. A 'do loop' can be written to find the matrizant:

> for i from 2 to nvars do S:=evalm(subs(t=t1,evalm(A))&*subs(t=t1, evalm(X1))):X2:=map(int,S,t1=0..t):mat := evalm(mat +X2) : X1:=evalm(X2):od : evalm(mat) ;

$$\begin{bmatrix} -\frac{1}{2} t^2 + 1 + \frac{1}{8} t^4 - \frac{1}{48} t^6 + \frac{1}{384} t^8 - \frac{1}{3840} t^{10} + \frac{1}{46080} t^{12} \end{bmatrix}$$

> sol:=evalm(mat&*Y0);

$$sol := \begin{bmatrix} -\frac{1}{2} t^2 + 1 + \frac{1}{8} t^4 - \frac{1}{48} t^6 + \frac{1}{384} t^8 - \frac{1}{3840} t^{10} + \frac{1}{46080} t^{12} \end{bmatrix}$$

> C:=sol[1,1];

$$C := -\frac{1}{2} t^2 + 1 + \frac{1}{8} t^4 - \frac{1}{48} t^6 + \frac{1}{384} t^8 - \frac{1}{3840} t^{10} + \frac{1}{46080} t^{12}$$

Thus, the process yields a series solution in t for C. This solution can be compared to the series solution given by Maple's 'dsolve' command:

3.1 Linear Boundary Value Problems

> 'dsolve'({Eq,c(0)=1},c(t),type=series);

$$c(t) = 1 - \frac{1}{2}t^2 + \frac{1}{8}t^4 + O(t^6)$$

By default, Maple gives a series solution accurate to the order of t^6. The order of the series solution can be increased by using Maple's order specification as:

> Order:=14;

$$Order := 14$$

> 'dsolve'({Eq,c(0)=1},c(t),type=series);

$$c(t) = 1 - \frac{1}{2}t^2 + \frac{1}{8}t^4 - \frac{1}{48}t^6 + \frac{1}{384}t^8 - \frac{1}{3840}t^{10} + \frac{1}{46080}t^{12} + O(t^{14})$$

We observe that the series solution obtained using the matrizant method matches exactly with the series solution given by Maple's 'dsolve' command.

Example 3.5

The method described for example 3.4 can be used to solve boundary value problems. Consider the boundary value problem given by the Airy differential equation:

$$\frac{d^2y}{dx^2} = xy$$

(3.16)

$$y(0) = 1 \text{ and } \frac{dy}{dx}(1) = 0$$

The procedure for solving this boundary value problem is the same as that of section 3.1.2 but instead of the exponential matrix, the matrizant must be used.

> restart:
> with(linalg):with(plots):
Warning, the protected names norm and trace have been redefined and unprotected

Warning, the name changecoords has been redefined

> N:=2;

$$N := 2$$

> nvars:=6;

$$nvars := 6$$

188   3 Boundary Value Problems

> Eq:=diff(y(x),x$2)=x*y(x);

$$Eq := \frac{d^2}{dx^2} y(x) = x\, y(x)$$

> A:=matrix(2,2,[0,1,x,0]);

$$A := \begin{bmatrix} 0 & 1 \\ x & 0 \end{bmatrix}$$

> Y0:=matrix(N,1,[1,c[1]]);

$$Y0 := \begin{bmatrix} 1 \\ c_1 \end{bmatrix}$$

> id:=Matrix(N,N,shape=identity);

$$id := \begin{bmatrix} 1 & 0 \\ 0 & 1 \end{bmatrix}$$

> X1:=matrix(N,N);X2:=matrix(N,N);

$$X1 := array(1..2, 1..2, [\,])$$

$$X2 := array(1..2, 1..2, [\,])$$

> X1:=map(int,subs(x=x1,evalm(A)),x1=0..x);

$$X1 := \begin{bmatrix} 0 & x \\ \frac{1}{2} x^2 & 0 \end{bmatrix}$$

> mat := evalm(id + X1) ;

$$X1 := \begin{bmatrix} 0 & x \\ \frac{1}{2} x^2 & 0 \end{bmatrix}$$

> for i from 2 to nvars do
S:=evalm(subs(x=x1,evalm(A))&*subs(x=x1,evalm(X1))):X2:=
map(int,S,x1=0..x):mat := evalm(mat +X2) :
X1:=evalm(X2):od :
> evalm(mat) ;

$$\left[\left[1 + \frac{1}{6} x^3 + \frac{1}{180} x^6 + \frac{1}{12960} x^9, x + \frac{1}{12} x^4 + \frac{1}{504} x^7\right], \left[\frac{1}{2} x^2 + \frac{1}{30} x^5 + \frac{1}{1440} x^8, 1 + \frac{1}{3} x^3 + \frac{1}{72} x^6 + \frac{1}{4536} x^9\right]\right]$$

3.1 Linear Boundary Value Problems

```
> sol:=evalm(mat&*Y0);
```

$$sol := \begin{bmatrix} 1 + \frac{1}{6}x^3 + \frac{1}{180}x^6 + \frac{1}{12960}x^9 + \left(x + \frac{1}{12}x^4 + \frac{1}{504}x^7\right)c_1 \\ \frac{1}{2}x^2 + \frac{1}{30}x^5 + \frac{1}{1440}x^8 + \left(1 + \frac{1}{3}x^3 + \frac{1}{72}x^6 + \frac{1}{4536}x^9\right)c_1 \end{bmatrix}$$

```
> y:=sol[1,1];
```

$$y := 1 + \frac{1}{6}x^3 + \frac{1}{180}x^6 + \frac{1}{12960}x^9 + \left(x + \frac{1}{12}x^4 + \frac{1}{504}x^7\right)c_1$$

```
> dydx:=sol[2,1];
```

$$dydx := \frac{1}{2}x^2 + \frac{1}{30}x^5 + \frac{1}{1440}x^8 + \left(1 + \frac{1}{3}x^3 + \frac{1}{72}x^6 + \frac{1}{4536}x^9\right)c_1$$

Now the unknown constant c_1 is solved by using the boundary condition at $x = 1$.

```
> bc2:=subs(x=1,dydx)=0;
```

$$bc2 := \frac{769}{1440} + \frac{764}{567}c_1 = 0$$

```
> c[1]:=solve(bc2,c[1]);
```

$$c_1 := -\frac{48447}{122240}$$

```
> y:=eval(y);
```

$$y := 1 + \frac{1}{6}x^3 + \frac{1}{180}x^6 + \frac{1}{12960}x^9 - \frac{48447}{122240}x - \frac{16149}{488960}x^4 - \frac{769}{977920}x^7$$

Readers can verify this series solution with the series solution obtained by using Maple's 'dsolve' command.

Example 3.6

Next, the classical problem of diffusion with reaction in a cylindrical catalyst pellet is considered:[8] [4]

$$\frac{1}{x}\frac{d}{dx}\left(x\frac{dc}{dx}\right) - \phi^2 c = 0$$

(3.17)

$$\frac{dc}{dx}(0) = 0 \text{ and } c(1) = 1$$

where c is the dimensionless concentration and ϕ is the Thiele modulus. When this boundary value problem is cast into the matrix form (equation (3.6)), the matrizant involves integration of $\frac{1}{x}$ from 0 to 1. This problem can be treated by using a logarithm variable transform as shown by Subramanian, Haran, and White[4] or this can be handled by integrating from x0 to x and applying the limit command for x0. We introduce the following variables for convenience:

$$c = y_1$$
$$x\frac{dc}{dx} = y_2 \qquad (3.18)$$

This transformation converts equation (3.17) to the following system of first order ODEs:

$$\frac{dy_1}{dx} = \frac{dc}{dx} = \frac{1}{x}\left(x\frac{dc}{dx}\right) = \frac{1}{x}y_2$$
$$\frac{dy_2}{dx} = \frac{d}{dx}\left(x\frac{dc}{dx}\right) = \phi^2 xc = \phi^2 xy_1 \qquad (3.19)$$

Next, equation (3.19) can be converted to the matrix form as

$$\frac{d\mathbf{Y}}{dx} = \mathbf{A}(x)\mathbf{Y} \qquad (3.20)$$

where the dependent variables are

$$\mathbf{Y} = \begin{bmatrix} y_1 \\ y_2 \end{bmatrix} = \begin{bmatrix} c \\ x\dfrac{dc}{dx} \end{bmatrix} \qquad (3.21)$$

and the coefficient matrix **A** is given by

$$\mathbf{A} = \begin{bmatrix} 0 & \dfrac{1}{x} \\ \phi^2 x & 0 \end{bmatrix} \qquad (3.22)$$

The initial condition vector is

$$\mathbf{Y}_0 = \begin{bmatrix} c_1 \\ 0 \end{bmatrix} \qquad (3.23)$$

Equation (3.19) can be solved below by finding the matrizant in the manner described above.

3.1 Linear Boundary Value Problems

```
> restart:
> with(linalg):with(plots):
> N:=2;
```

For brevity, only four terms are used for calculating the matrizant in this example.

```
> nvars:=4;
> Eq:=1/x*diff(x*diff(c(x),x),x)=phi^2*c(x);
```

$$Eq := \frac{\frac{d}{dx} c(x) + x \left(\frac{d^2}{dx^2} c(x) \right)}{x} = \phi^2 c(x)$$

Enter the **A** matrix (equation (3.22)).

```
> A:=matrix(2,2,[0,1/x,phi^2*x,0]);
```

$$A := \begin{bmatrix} 0 & \frac{1}{x} \\ \phi^2 x & 0 \end{bmatrix}$$

```
> Y0:=matrix(2,1,[c[1],0]);
```

$$Y0 := \begin{bmatrix} c_1 \\ 0 \end{bmatrix}$$

```
> id:=Matrix(N,N,shape=identity);
```

$$id := \begin{bmatrix} 1 & 0 \\ 0 & 1 \end{bmatrix}$$

```
> X1:=matrix(N,N);X2:=matrix(N,N);
```

$$X1 := array(1..2, 1..2, [\ ])$$
$$X2 := array(1..2, 1..2, [\ ])$$

```
> X1:=map(int,subs(x=x1,evalm(A)),x1=0..x);
```

$$X1 := \begin{bmatrix} 0 & \infty \\ \frac{1}{2} \phi^2 x^2 & 0 \end{bmatrix}$$

To avoid the singularity in X_1, integrate from x_0 to x and later find the limit as x_0 goes to zero.

> X1:=map(int,subs(x=x1,evalm(A)),x1=x0..x)assuming x>0,x0>=0,x>=x0;

$$X1 := \begin{bmatrix} 0 & -\ln(x0) + \ln(x) \\ \dfrac{1}{2}\phi^2(x^2 - x0^2) & 0 \end{bmatrix}$$

> mat := evalm(id + X1) ;

$$mat := \begin{bmatrix} 1 & -\ln(x0) + \ln(x) \\ \dfrac{1}{2}\phi^2(x^2 - x0^2) & 1 \end{bmatrix}$$

> for i from 2 to nvars do
S:=evalm(subs(x=x1,evalm(A))&*subs(x=x1,evalm(X1))):X2:=
map(int,S,x1=x0..x):mat := evalm(mat +X2) :
X1:=evalm(X2):od :
> evalm(mat)assuming x>0,x0>=0,x>=x0;

$$\Bigg[\Big[1 - \tfrac{1}{4}\phi^2 x0^2 + \tfrac{1}{2}\phi^2 \ln(x0) x0^2 + \tfrac{1}{4}\phi^2 x^2 - \tfrac{1}{2}\phi^2 x0^2 \ln(x) + \tfrac{1}{16}\phi^4 x0^4 \ln(x0) - \tfrac{5}{64}\phi^4 x0^4$$
$$- \tfrac{1}{16}\phi^4 x0^4 \ln(x) + \tfrac{1}{8}\phi^4 \ln(x0) x0^2 x^2 + \tfrac{1}{64}\phi^4 x^4 - \tfrac{1}{8}\phi^4 x0^2 x^2 \ln(x) + \tfrac{1}{16}\phi^4 x0^2 x^2, -\ln(x0) + \ln(x)$$
$$- \tfrac{1}{4}\phi^2 \ln(x0) x0^2 + \tfrac{1}{4}\phi^2 x0^2 + \tfrac{1}{4}\phi^2 x0^2 \ln(x) - \tfrac{1}{4}\phi^2 \ln(x0) x^2 + \tfrac{1}{4}\phi^2 x^2 \ln(x) - \tfrac{1}{4}\phi^2 x^2\Big]$$
$$\Big[\tfrac{1}{2}\phi^2(x^2 - x0^2) - \tfrac{1}{16}\phi^4 x0^4 + \tfrac{1}{4}\phi^4 \ln(x0) x0^2 x^2 + \tfrac{1}{16}\phi^4 x^4 - \tfrac{1}{4}\phi^4 x0^2 x^2 \ln(x), 1 + \tfrac{1}{4}\phi^2 x0^2$$
$$- \tfrac{1}{2}\phi^2 \ln(x0) x^2 + \tfrac{1}{2}\phi^2 x^2 \ln(x) - \tfrac{1}{4}\phi^2 x^2 + \tfrac{1}{64}\phi^4 x0^4 - \tfrac{1}{8}\phi^4 \ln(x0) x0^2 x^2 + \tfrac{1}{16}\phi^4 x0^2 x^2$$
$$+ \tfrac{1}{8}\phi^4 x0^2 x^2 \ln(x) - \tfrac{1}{16}\phi^4 \ln(x0) x^4 + \tfrac{1}{16}\phi^4 x^4 \ln(x) - \tfrac{5}{64}\phi^4 x^4\Big]\Bigg]$$

> sol:=evalm(mat&*Y0);

$$sol := \begin{bmatrix} \left(1 + \int_{x0}^{x} \tfrac{1}{2}\dfrac{\phi^2(x1^2 - x0^2)}{x1} dx1 + \int_{x0}^{x} \dfrac{\int_{x0}^{x1}\phi^2 x1 \left(\int_{x0}^{x1}\tfrac{1}{2}\dfrac{\phi^2(x1^2 - x0^2)}{x1}dx1\right)dx1}{x1} dx1\right) c_1 \\ \left(\tfrac{1}{2}\phi^2(x^2 - x0^2) + \int_{x0}^{x}\phi^2 x1 \left(\int_{x0}^{x1}\tfrac{1}{2}\dfrac{\phi^2(x1^2 - x0^2)}{x1}dx1\right)dx1\right) c_1 \end{bmatrix}$$

> C:=sol[1,1];

$$C := \left(1 + \int_{x0}^{x} \tfrac{1}{2}\dfrac{\phi^2(x1^2 - x0^2)}{x1} dx1 + \int_{x0}^{x} \dfrac{\int_{x0}^{x1}\phi^2 x1 \left(\int_{x0}^{x1}\tfrac{1}{2}\dfrac{\phi^2(x1^2 - x0^2)}{x1}dx1\right)dx1}{x1} dx1\right) c_1$$

3.1 Linear Boundary Value Problems

> dCdx:=1/x*sol[2,1];

$$dCdx := \frac{\frac{1}{2}\phi^2(x^2-x0^2) + \int_{x0}^{x} \phi^2 x1 \left(\int_{x0}^{x1} \frac{\frac{1}{2}\phi^2(x1^2-x0^2)}{x1} dx1 \right) dx1 \, c_1}{x}$$

To find c_1 apply the boundary condition at $x = 1$:

> bc2:=eval(subs(x=1,C))=1 assuming x>0,x0>=0,x>=x0;

Warning, unable to determine if 0 is between x0 and x1; try to use assumptions or set _EnvAllSolutions to true

$$bc2 := \left(1 - \frac{1}{4}\phi^2 x0^2 + \frac{1}{2}\phi^2 \ln(x0) x0^2 + \frac{1}{4}\phi^2 + \frac{1}{16}\phi^4 x0^4 \ln(x0) - \frac{5}{64}\phi^4 x0^4 + \frac{1}{8}\phi^4 \ln(x0) x0^2 + \frac{1}{64}\phi^4 \right.$$
$$\left. + \frac{1}{16}\phi^4 x0^2 \right) c_1 = 1$$

> c[1]:=solve(bc2,c[1]);

$$c_1 := \frac{64}{64 - 16\phi^2 x0^2 + 32\phi^2 \ln(x0) x0^2 + 16\phi^2 + 4\phi^4 x0^4 \ln(x0) - 5\phi^4 x0^4 + 8\phi^4 \ln(x0) x0^2 + \phi^4 + 4\phi^4 x0^2}$$

> C:=eval(C);

Warning, unable to determine if 0 is between x0 and x; try to use assumptions or set _EnvAllSolutions to true

Warning, unable to determine if 0 is between x0 and x1; try to use assumptions or set _EnvAllSolutions to true

$$C := \frac{64 \left(1 + \int_{x0}^{x} \frac{\frac{1}{2}\phi^2(x1^2-x0^2)}{x1} dx1 + \int_{x0}^{x} \frac{\phi^2 x1 \left(\int_{x0}^{x1} \frac{\frac{1}{2}\phi^2(x1^2-x0^2)}{x1} dx1 \right)}{x1} dx1 \right)}{64 - 16\phi^2 x0^2 + 32\phi^2 \ln(x0) x0^2 + 16\phi^2 + 4\phi^4 x0^4 \ln(x0) - 5\phi^4 x0^4 + 8\phi^4 \ln(x0) x0^2 + \phi^4 + 4\phi^4 x0^2}$$

Now apply the limit command for x0.

> Warning, premature end of input, use <Shift> + <Enter> to avoid this message.

> C:=limit(C,x0=0);

$$C := \frac{64 + 16\phi^2 x^2 + \phi^4 x^4}{64 + 16\phi^2 + \phi^4}$$

Divide both numerator and denominator by 64. (Note when different values of 'nvars' are used, this number has to be changed accordingly.)

> n1:=numer(C)/64;

$$n1 := 1 + \frac{1}{4} \phi^2 x^2 + \frac{1}{64} \phi^4 x^4$$

> d1:=denom(C)/64;

$$d1 := 1 + \frac{1}{4} \phi^2 + \frac{1}{64} \phi^4$$

> C:=n1/d1;

$$C := \frac{1 + \frac{1}{4} \phi^2 x^2 + \frac{1}{64} \phi^4 x^4}{1 + \frac{1}{4} \phi^2 + \frac{1}{64} \phi^4}$$

One can verify that both the numerator and the denominator of C are modified Bessel functions of the order zero by using Maple.

> series(BesselI(0,phi*x),x);

$$1 + \frac{1}{4} \phi^2 x^2 + \frac{1}{64} \phi^4 x^4 + O(x^6)$$

> series(BesselI(0,phi),phi);

$$1 + \frac{1}{4} \phi^2 + \frac{1}{64} \phi^4 + O(\phi^6)$$

Next, plots can be obtained by substituting the parameters for the Thiele modulus ϕ.

> pars:=[0.1,1,2,10];

$$pars := [0.1, 1, 2, 10]$$

> clr:=[red,green,blue,brown];

$$clr := [red, green, blue, brown]$$

> for i to 4 do p[i]:=plot(subs(phi=pars[i],C),x=0..1,color=clr[i]):od:
> pt[1]:=textplot([0.1,evalf(subs({x=0.1,phi=pars[1]},C)),'phi=pars[1]'],
align=below):
pt[2]:=textplot([0.4,evalf(subs({x=.4,phi=pars[2]},C)),'phi=pars[2]'],
align=below):
pt[3]:=textplot([0.5,evalf(subs({x=.5,phi=pars[3]},C)),'phi=pars[3]'],
align=below):
pt[4]:=textplot([0.8,evalf(subs({x=0.8,phi=pars[4]},C)),'phi=pars[4]'],
align=below):
> display({seq(p[i],i=1..4),seq(pt[i],i=1..4)},axes=boxed,thickness=3,
title="Figure 3.8",labels=[x,"C"]);

3.1 Linear Boundary Value Problems

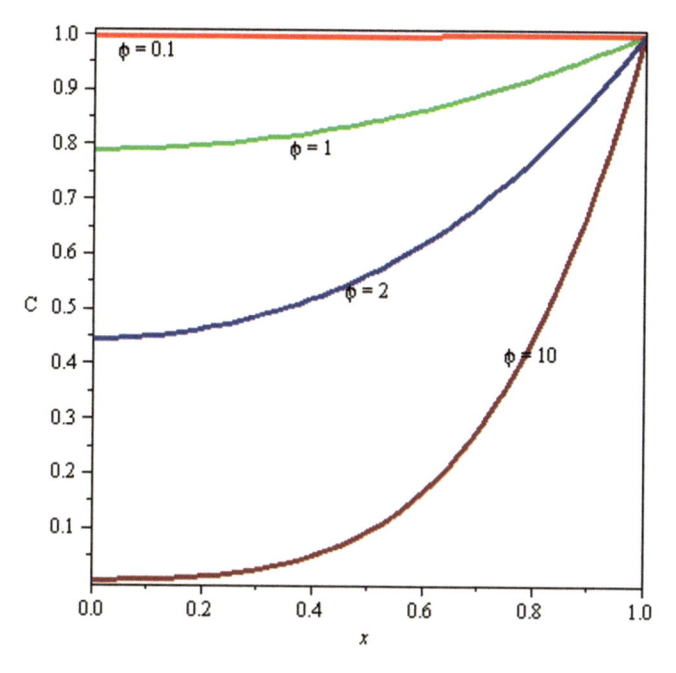

Fig. 3.8

For higher values of ϕ, more terms (nvars) in the matrizant series solution are needed for higher accuracy.

3.1.5 Symbolic Finite Difference Solutions for Linear Boundary Value Problems

Consider a second order differential equation (equation (3.1)). This equation can be converted to finite difference form (accurate to the order h^2) as follows:

$$\frac{y_{i+1} - 2y_i + y_{i-1}}{h^2} + a_1 \frac{y_{i+1} - y_{i-1}}{2h} + a_2 y_i = f(x = ih), \quad i = 1..N \quad (3.24)$$

where i is the index of the node points, N is the number of interior node points, and h is the node spacing defined by

$$h = \frac{L}{N+1} \quad (3.25)$$

where L is the length of the domain. Thus, x = 0 corresponds to the node point i = 0 and x = L corresponds to the node point i = N+1. The variable y_i corresponds to the dependent variable at node point i. Equation (3.24) is a system of N linear algebraic equations for N dependent variables (y_i, i = 1..N). The boundary values y_0 and y_{N+1} are eliminated using the boundary conditions. Equation (3.24) can be cast into matrix form as[5]

$$\mathbf{AX} = \mathbf{B} \qquad (3.26)$$

The solution to equation (3.26) can be obtained by inverting the **A** matrix ($\mathbf{X} = \mathbf{A}^{-1}\mathbf{B}$). The procedure for solving linear boundary value problems using finite difference is as follows:

1. Start the Maple program with a 'restart' command to clear all variables.
2. Call 'with(linalg)' and 'with(lplots)' commands.
3. Enter the governing equation.
4. Enter the number of interior node points, N.
5. Enter the length of the domain, L.
6. Convert the governing equations and boundary conditions to finite difference form.
7. Eliminate the boundary values (y_0 and y_{N+1}) using the boundary conditions.
8. Store the finite difference equations in eqs.
9. Store the dependent variables, y_i, i = 1..N in vars.
10. Generate **A** matrix and **B** vector using Maple's 'genmatrix' command.
11. Find the solution by inverting the **A** matrix.
12. Note that Maple can invert **A** when it is a function of parameters in the system (heat transfer coefficient, rate constants, etc.).

Once the symbolic finite difference solution is obtained, plots can be made for particular values of the parameters.

Example 3.7

Consider the convective diffusion problem (Finlayson, 1980)[11]

$$\frac{d^2c}{dx^2} - Pe\frac{dc}{dx} = 0$$
$$c(0) = 1 \qquad (3.27)$$
$$c(1) = 0$$

An analytical solution can be obtained using the exponential matrix method described in section 3.1.2:

$$c = \frac{e^{Pe} - e^{Pex}}{e^{Pe} - 1} \qquad (3.28)$$

This particular problem was chosen as the finite difference solution for this equation and shows oscillations for high Peclet numbers when the central difference expression is used for the first derivative. This equation is solved below using the procedure described above.

```
> restart:
> with(linalg):with(plots):
> N:=4;
```

$$N := 4$$

3.1 Linear Boundary Value Problems

```
> L:=1;
```
$$L := 1$$

```
> eq:=diff(y(x),x$2)-Pe*diff(y(x),x);
```
$$eq := \frac{d^2}{dx^2} y(x) - Pe \left(\frac{d}{dx} y(x) \right)$$

```
> bc1:=y(x)-1;
```

```
> bc2:=y(x);
```

Central difference expressions for the second and first derivatives are

```
> d2ydx2:=(y[m+1]-2*y[m]+y[m-1])/h^2;
```
$$d2ydx2 := \frac{y_{m+1} - 2 y_m + y_{m-1}}{h^2}$$

```
> dydx:=(y[m+1]-y[m-1])/2/h;
```
$$dydx := \frac{1}{2} \frac{y_{m+1} - y_{m-1}}{h}$$

The governing equation in finite difference form is:

```
> Eq[m]:=subs(diff(y(x),x$2)=d2ydx2,diff(y(x),x)=dydx,y(x)=y[m],x=m*h,eq);
```
$$Eq_m := \frac{y_{m+1} - 2 y_m + y_{m-1}}{h^2} - \frac{1}{2} \frac{Pe \left(y_{m+1} - y_{m-1} \right)}{h}$$

A 'for loop' can be written for the interior node points as

```
> for i to N do Eq[i]:=subs(m=i,Eq[m]);od;
```
$$Eq_1 := \frac{y_2 - 2 y_1 + y_0}{h^2} - \frac{1}{2} \frac{Pe \left(y_2 - y_0 \right)}{h}$$

$$Eq_2 := \frac{y_3 - 2 y_2 + y_1}{h^2} - \frac{1}{2} \frac{Pe \left(y_3 - y_1 \right)}{h}$$

$$Eq_3 := \frac{y_4 - 2 y_3 + y_2}{h^2} - \frac{1}{2} \frac{Pe \left(y_4 - y_2 \right)}{h}$$

$$Eq_4 := \frac{y_5 - 2 y_4 + y_3}{h^2} - \frac{1}{2} \frac{Pe \left(y_5 - y_3 \right)}{h}$$

```
> Eq[0]:=y[0]=1;
```
$$Eq_0 := y_0 = 1$$

> Eq[N+1]:=y[N+1]=0;
$$Eq_5 := y_5 = 0$$

> y[0]:=solve(Eq[0],y[0]);
$$y_0 := 1$$

> y[N+1]:=solve(Eq[N+1],y[N+1]);
$$y_5 := 0$$

> h:=L/(N+1);
$$h := \frac{1}{5}$$

> for i to N do Eq[i]:=eval(Eq[i]);od;
$$Eq_1 := 25 y_2 - 50 y_1 + 25 - \frac{5}{2} Pe (y_2 - 1)$$
$$Eq_2 := 25 y_3 - 50 y_2 + 25 y_1 - \frac{5}{2} Pe (y_3 - y_1)$$
$$Eq_3 := 25 y_4 - 50 y_3 + 25 y_2 - \frac{5}{2} Pe (y_4 - y_2)$$
$$Eq_4 := -50 y_4 + 25 y_3 + \frac{5}{2} Pe\, y_3$$

> eqs:=[seq(Eq[i],i=1..N)];
$$eqs := \left[25 y_2 - 50 y_1 + 25 - \frac{5}{2} Pe (y_2-1),\, 25 y_3 - 50 y_2 + 25 y_1 - \frac{5}{2} Pe (y_3 - y_1),\, 25 y_4 - 50 y_3 + 25 y_2 \right.$$
$$\left. - \frac{5}{2} Pe (y_4 - y_2),\, -50 y_4 + 25 y_3 + \frac{5}{2} Pe\, y_3 \right]$$

> vars:=[seq(y[i],i=1..N)];
$$vars := [y_1, y_2, y_3, y_4]$$

> A:=genmatrix(eqs,vars,'B1');
$$A := \begin{bmatrix} -50 & 25 - \frac{5}{2} Pe & 0 & 0 \\ 25 + \frac{5}{2} Pe & -50 & 25 - \frac{5}{2} Pe & 0 \\ 0 & 25 + \frac{5}{2} Pe & -50 & 25 - \frac{5}{2} Pe \\ 0 & 0 & 25 + \frac{5}{2} Pe & -50 \end{bmatrix}$$

3.1 Linear Boundary Value Problems

> evalm(B1);

$$\left[-25 - \frac{5}{2} Pe \quad 0 \quad 0 \quad 0 \right]$$

Maple generates a row vector, which can be converted to a column vector as:

> B:=matrix(N,1):for i to N do B[i,1]:=B1[i]:od:evalm(B);

$$\begin{bmatrix} -25 - \frac{5}{2} Pe \\ 0 \\ 0 \\ 0 \end{bmatrix}$$

The solution is obtained as:

> X:=evalm(inverse(A)&*B);

$$X := \begin{bmatrix} -\dfrac{16 \left(100 + Pe^2\right) \left(-25 - \frac{5}{2} Pe\right)}{50000 + 1000 Pe^2 + Pe^4} \\ -\dfrac{2}{5} \dfrac{(10 + Pe)\left(300 + Pe^2\right)\left(-25 - \frac{5}{2} Pe\right)}{50000 + 1000 Pe^2 + Pe^4} \\ -\dfrac{8 (10 + Pe)^2 \left(-25 - \frac{5}{2} Pe\right)}{50000 + 1000 Pe^2 + Pe^4} \\ -\dfrac{2}{5} \dfrac{(10 + Pe)^3 \left(-25 - \frac{5}{2} Pe\right)}{50000 + 1000 Pe^2 + Pe^4} \end{bmatrix}$$

> for i to N do y[i]:=X[i,1];od;

$$y_1 := -\dfrac{16 \left(100 + Pe^2\right)\left(-25 - \frac{5}{2} Pe\right)}{50000 + 1000 Pe^2 + Pe^4}$$

$$y_2 := -\dfrac{2}{5} \dfrac{(10 + Pe)\left(300 + Pe^2\right)\left(-25 - \frac{5}{2} Pe\right)}{50000 + 1000 Pe^2 + Pe^4}$$

$$y_3 := -\frac{8(10+Pe)^2\left(-25-\frac{5}{2}Pe\right)}{50000+1000Pe^2+Pe^4}$$

$$y_4 := -\frac{2}{5}\frac{(10+Pe)^3\left(-25-\frac{5}{2}Pe\right)}{50000+1000Pe^2+Pe^4}$$

> y[0]:=eval(y[0]);y[N+1]:=eval(y[N+1]);

$$y_0 := 1$$

$$y_5 := 0$$

Next, the result obtained is compared with the exact analytical solution:
> ya:=(exp(Pe)-exp(Pe*x))/(exp(Pe)-1);

$$ya := \frac{e^{Pe} - e^{Pe\,x}}{e^{Pe} - 1}$$

> p1:=plot([seq([i*h,subs(Pe=1,y[i])],i=0..N+1)],thickness=4,color=blue, axes=boxed):
> p2:=plot(subs(Pe=1,ya),x=0..1,thickness=8,color=brown,axes=boxed, linestyle=2):
> display({p1,p2},title="Figure 3.9",labels=[x,"y"]);

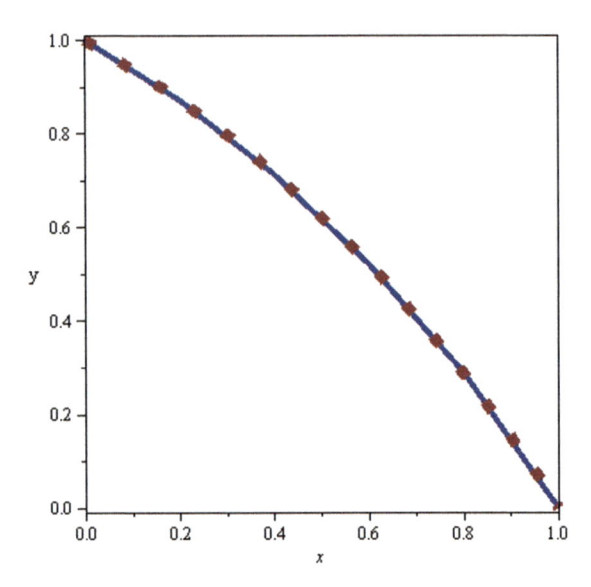

Fig. 3.9

3.1 Linear Boundary Value Problems

We observe that both the finite difference solution and the analytical solution match exactly when the Peclet number is 1. New plots can be obtained for different values of the Peclet number as follows:

```
> p1:=plot([seq([i*h,subs(Pe=50,y[i])],i=0..N+1)],color=blue,thickness=4,
axes=boxed):
> p2:=plot(subs(Pe=50,ya),x=0..1,thickness=5,color=brown,axes=boxed,
linestyle=2):
> display({p1,p2},title="Figure 3.10",labels=[x,"y"]);
```

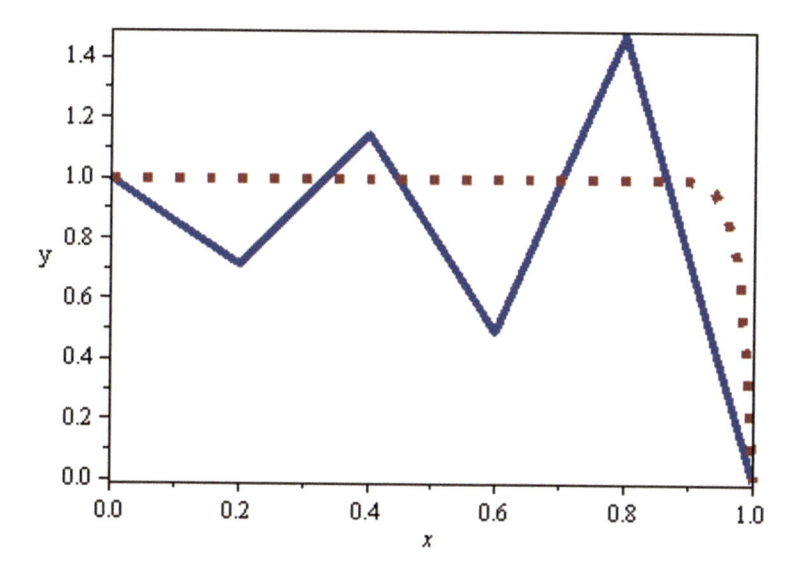

Fig. 3.10

This shows that for $Pe = 50$, 4 interior node points are not enough and we observe oscillations.[11] [12] This happens usually when central difference approximations are used for the convective term $\left(\dfrac{dc}{dx}\right)$. Use a forward difference approximation for the first derivative to solve this problem. Only dydx in the Maple program needs to be changed:

```
> dydx:=(y[m]-y[m-1])/h;
```

$$dydx := \frac{y_m - y_{m-1}}{h}$$

The results obtained using this approximation is given below:

Fig. 3.11

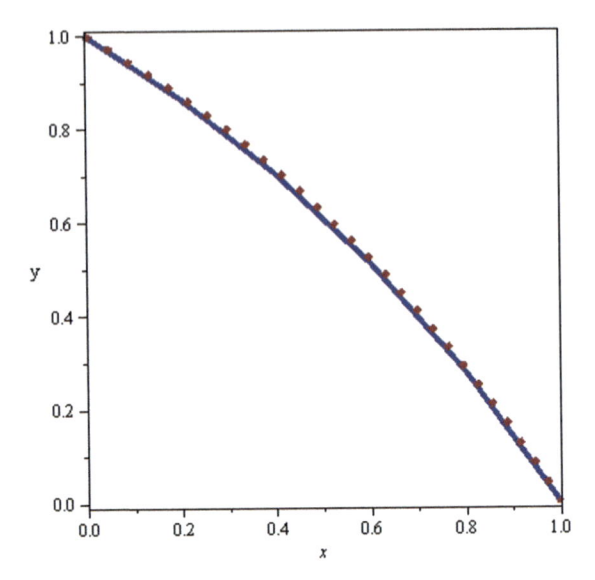

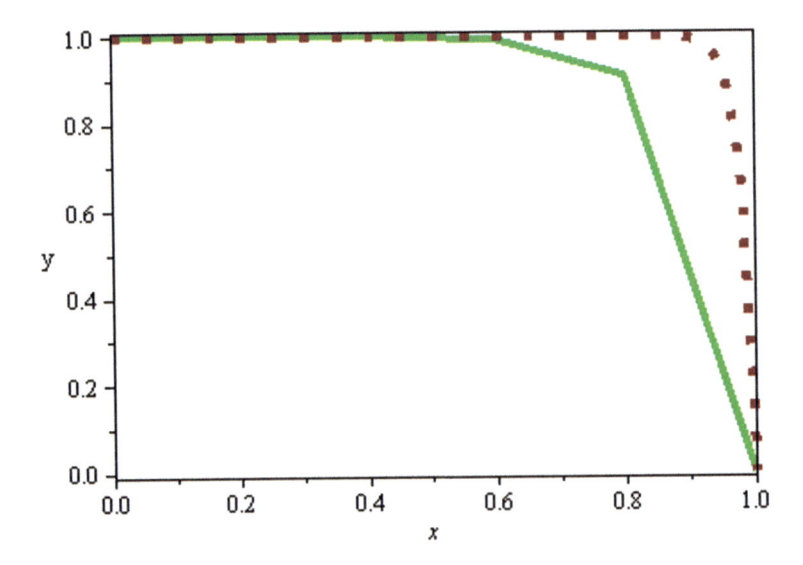

Fig. 3.12

We observe that when the forward difference accurate to the order h2 is used, even when the Peclet number is 1, there is a slight discrepancy between the finite difference solution and the analytical solution. However, when the Peclet number is high ($Pe = 50$) the forward difference scheme does not five an unrealistic

3.1 Linear Boundary Value Problems

oscillation like the central difference scheme. Note that for three digit accuracy with the analytical solution, $N=40$ and 110 interior node points are required for central difference and forward difference approximations, respectively.

>

Example 3.8. Cylindrical Catalyst Pellet

In the example discussed above, both the boundary conditions are of the Neumann type. However, many problems involve derivative boundary conditions. These problems can be handled by using the three point forward and backward differences at $x = 0$ and $x = 1$, respectively. This is illustrated by solving the cylindrical pellet problem solved in example 3.6 with a different boundary condition at the surface ($x = 1$):

$$\frac{dc}{dx}(1) = 1 - c(1) \tag{3.29}$$

The Maple program developed for the previous example can be modified to handle the derivative boundary conditions:

```
> restart:
> with(linalg):with(plots):
> N:=4;
```

> L:=1;

$$L := 1$$

> eq:=diff(y(x),x$2)+1/x*diff(y(x),x)-phi^2*y(x);

$$eq := \frac{d^2}{dx^2} y(x) + \frac{\frac{d}{dx} y(x)}{x} - \phi^2 y(x)$$

> bc1:=diff(y(x),x);

$$bc1 := \frac{d}{dx} y(x)$$

> bc2:=diff(y(x),x)-1+y(x);

$$bc2 := \frac{d}{dx} y(x) - 1 + y(x)$$

The central difference expression for the second and first derivatives are
> d2ydx2:=(y[m+1]-2*y[m]+y[m-1])/h^2;

$$d2ydx2 := \frac{y_{m+1} - 2y_m + y_{m-1}}{h^2}$$

> dydx:=(y[m+1]-y[m-1])/2/h;

$$dydx := \frac{1}{2} \frac{y_{m+1} - y_{m-1}}{h}$$

The three point forward and backward difference expressions for the derivative are:

> dydxf:=(-y[2]+4*y[1]-3*y[0])/(2*h);

$$dydxf := \frac{1}{2} \frac{-y_2 + 4y_1 - 3y_0}{h}$$

> dydxb:=(y[N-1]-4*y[N]+3*y[N+1])/(2*h);

$$dydxb := \frac{1}{2} \frac{y_3 - 4y_4 + 3y_5}{h}$$

The governing equation in finite difference form is:

> Eq[m]:=subs(diff(y(x),x$2)=d2ydx2,diff(y(x),x)=dydx,y(x)=y[m],x=m*h,eq);

$$Eq_m := \frac{y_{m+1} - 2y_m + y_{m-1}}{h^2} + \frac{1}{2} \frac{y_{m+1} - y_{m-1}}{mh^2} - \phi^2 y_m$$

The boundary conditions in finite difference form are:

> Eq[0]:=subs(diff(y(x),x)=dydxf,y(x)=y[0],bc1);

$$Eq_0 := \frac{1}{2} \frac{-y_2 + 4y_1 - 3y_0}{h}$$

> Eq[N+1]:=subs(diff(y(x),x)=dydxb,y(x)=y[N+1],bc2);

$$Eq_5 := \frac{1}{2} \frac{y_3 - 4y_4 + 3y_5}{h} - 1 + y_5$$

A 'for loop' can be written for the interior node points as

> for i to N do Eq[i]:=subs(m=i,Eq[m]);od;

$$Eq_1 := \frac{y_2 - 2y_1 + y_0}{h^2} + \frac{1}{2} \frac{y_2 - y_0}{h^2} - \phi^2 y_1$$

$$Eq_2 := \frac{y_3 - 2y_2 + y_1}{h^2} + \frac{1}{4} \frac{y_3 - y_1}{h^2} - \phi^2 y_2$$

$$Eq_3 := \frac{y_4 - 2y_3 + y_2}{h^2} + \frac{1}{6} \frac{y_4 - y_2}{h^2} - \phi^2 y_3$$

$$Eq_4 := \frac{y_5 - 2y_4 + y_3}{h^2} + \frac{1}{8} \frac{y_5 - y_3}{h^2} - \phi^2 y_4$$

> y[0]:=solve(Eq[0],y[0]);

$$y_0 := -\frac{1}{3} y_2 + \frac{4}{3} y_1$$

3.1 Linear Boundary Value Problems

> y[N+1]:=solve(Eq[N+1],y[N+1]);

$$y_5 := \frac{-y_3 + 4y_4 + 2h}{3 + 2h}$$

> h:=L/(N+1);

$$h := \frac{1}{5}$$

> for i to N do Eq[i]:=eval(Eq[i]);od;

$$Eq_1 := \frac{100}{3} y_2 - \frac{100}{3} y_1 - \phi^2 y_1$$

$$Eq_2 := \frac{125}{4} y_3 - 50 y_2 + \frac{75}{4} y_1 - \phi^2 y_2$$

$$Eq_3 := \frac{175}{6} y_4 - 50 y_3 + \frac{125}{6} y_2 - \phi^2 y_3$$

$$Eq_4 := \frac{925}{68} y_3 - \frac{575}{34} y_4 + \frac{225}{68} - \phi^2 y_4$$

> eqs:=[seq(Eq[i],i=1..N)];

$$eqs := \left[\frac{100}{3} y_2 - \frac{100}{3} y_1 - \phi^2 y_1, \frac{125}{4} y_3 - 50 y_2 + \frac{75}{4} y_1 - \phi^2 y_2, \frac{175}{6} y_4 - 50 y_3 + \frac{125}{6} y_2 - \phi^2 y_3, \frac{925}{68} y_3 - \frac{575}{34} y_4 + \frac{225}{68} - \phi^2 y_4 \right]$$

> vars:=[seq(y[i],i=1..N)];

$$vars := [y_1, y_2, y_3, y_4]$$

> A:=genmatrix(eqs,vars,'B1');

$$A := \begin{bmatrix} -\frac{100}{3} - \phi^2 & \frac{100}{3} & 0 & 0 \\ \frac{75}{4} & -50 - \phi^2 & \frac{125}{4} & 0 \\ 0 & \frac{125}{6} & -50 - \phi^2 & \frac{175}{6} \\ 0 & 0 & \frac{925}{68} & -\frac{575}{34} - \phi^2 \end{bmatrix}$$

> evalm(B1);

$$\begin{bmatrix} 0 & 0 & 0 & -\frac{225}{68} \end{bmatrix}$$

Maple generates a row vector, which can be converted to a column vector as:

```
> B:=matrix(N,1):for i to N do B[i,1]:=B1[i]:od:evalm(B);
```

$$\begin{bmatrix} 0 \\ 0 \\ 0 \\ -\dfrac{225}{68} \end{bmatrix}$$

The solution is obtained as

```
> X:=evalm(inverse(A)&*B);
```

$$X := \begin{bmatrix} \dfrac{82031250}{82031250 + 60703125\,\phi^2 + 5235000\,\phi^4 + 122600\,\phi^6 + 816\,\phi^8} \\[6pt] \dfrac{1640625}{2}\,\dfrac{100 + 3\,\phi^2}{82031250 + 60703125\,\phi^2 + 5235000\,\phi^4 + 122600\,\phi^6 + 816\,\phi^8} \\[6pt] \dfrac{26250\,(3125 + 250\,\phi^2 + 3\,\phi^4)}{82031250 + 60703125\,\phi^2 + 5235000\,\phi^4 + 122600\,\phi^6 + 816\,\phi^8} \\[6pt] \dfrac{75}{2}\,\dfrac{2187500 + 328125\,\phi^2 + 9600\,\phi^4 + 72\,\phi^6}{82031250 + 60703125\,\phi^2 + 5235000\,\phi^4 + 122600\,\phi^6 + 816\,\phi^8} \end{bmatrix}$$

```
> for i to N do y[i]:=X[i,1];od;
```

$$y_1 := \dfrac{82031250}{82031250 + 60703125\,\phi^2 + 5235000\,\phi^4 + 122600\,\phi^6 + 816\,\phi^8}$$

$$y_2 := \dfrac{1640625}{2}\,\dfrac{100 + 3\,\phi^2}{82031250 + 60703125\,\phi^2 + 5235000\,\phi^4 + 122600\,\phi^6 + 816\,\phi^8}$$

$$y_3 := \dfrac{26250\,(3125 + 250\,\phi^2 + 3\,\phi^4)}{82031250 + 60703125\,\phi^2 + 5235000\,\phi^4 + 122600\,\phi^6 + 816\,\phi^8}$$

$$y_4 := \dfrac{75}{2}\,\dfrac{2187500 + 328125\,\phi^2 + 9600\,\phi^4 + 72\,\phi^6}{82031250 + 60703125\,\phi^2 + 5235000\,\phi^4 + 122600\,\phi^6 + 816\,\phi^8}$$

```
> y[0]:=eval(y[0]);y[N+1]:=eval(y[N+1]);
```

$$y_0 := -\dfrac{546875}{2}\,\dfrac{100 + 3\,\phi^2}{82031250 + 60703125\,\phi^2 + 5235000\,\phi^4 + 122600\,\phi^6 + 816\,\phi^8}$$
$$+ \dfrac{109375000}{82031250 + 60703125\,\phi^2 + 5235000\,\phi^4 + 122600\,\phi^6 + 816\,\phi^8}$$

3.1 Linear Boundary Value Problems

$$y_5 := -\frac{131250}{17} \frac{3125 + 250\phi^2 + 3\phi^4}{82031250 + 60703125\phi^2 + 5235000\phi^4 + 122600\phi^6 + 816\phi^8}$$
$$+ \frac{750}{17} \frac{2187500 + 328125\phi^2 + 9600\phi^4 + 72\phi^6}{82031250 + 60703125\phi^2 + 5235000\phi^4 + 122600\phi^6 + 816\phi^8} + \frac{2}{17}$$

Now the result obtained is plotted for different values of the Thiele modulus Φ:

> pars:=[0.1,0.5,1,2,5];

$$pars := [0.1, 0.5, 1, 2, 5]$$

> clr:=[black,red,green,gold,blue];

$$clr := [black, red, green, gold, blue]$$

> for j from 1 to 5 do
p[j]:=plot([seq([i*h,subs(phi=pars[j],y[i])],i=0..N+1)],
thickness=3,color=clr[j]):od:

> pt:=textplot([seq([3*h,evalf(subs(phi=pars[j],y[3])-0.02),phi=pars[j]],j=1..5)]):
> display({seq(p[i],i=1..5),pt},title="Figure 3.13",axes=boxed,labels=[x,y]);

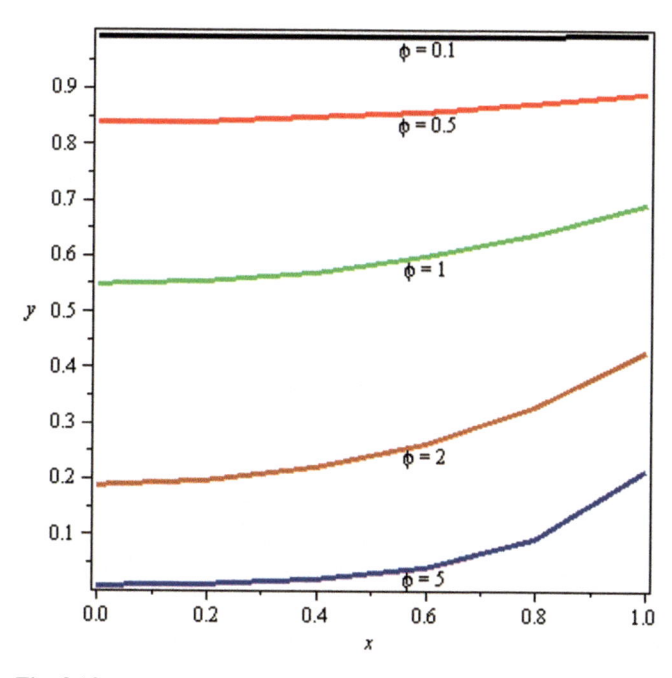

Fig. 3.13

Now the result obtained is plotted for different values of Thiele modulus Φ. The results obtained for $N=10$ interior node points are given below.

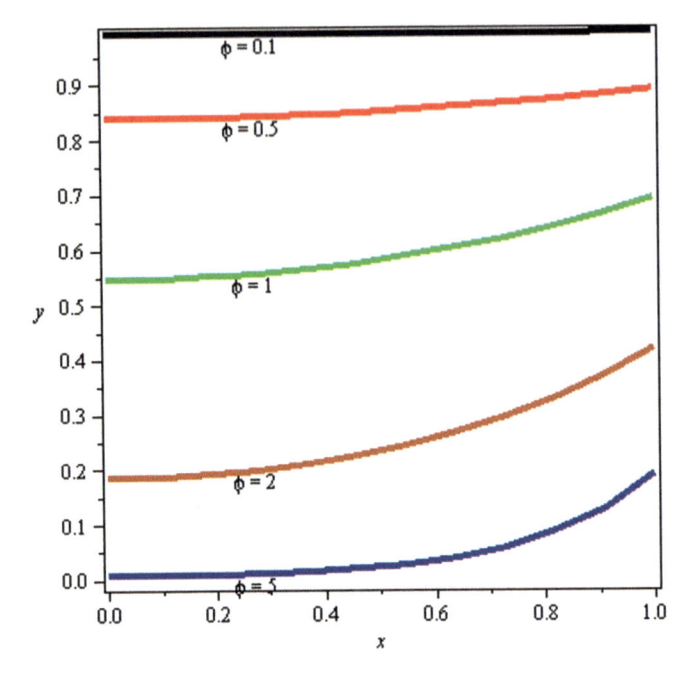

Fig. 3.14

Note that when the problem is stiff (Pe>1000, or Φ > 20), N > 20 node points might be needed for accurate solutions. Consequently, inverting the A matrix symbolically involves lot of computational effort as the order of the matrix increases with N. It is recommended that you specify the values for the parameters and convert the entries of the A matrix to decimal points using the following Maple command before the matrix inversion:

> A:=map(evalf,A);

3.1.6 Solving Linear Boundary Value Problems Using Maple's 'dsolve' Command

Maple's 'dsolve' command can be used to solve linear boundary value problems. One of the advantages of using Maple's 'dsolve' command is Maple can give Bessel and other special function solutions to linear boundary value problems. However, the analytical solution obtained from the 'dsolve' command may not be in simplified or elegant form. The syntax for using the 'dsolve' command is follows.

'dsolve'({"differential equations, boundary conditions"},{"dependent variable"})

Example 3.9. Heat Transfer in a Fin

The heat transfer problem solved in example 3.1 can be solved using Maple's 'dsolve' command as follows:

3.1 Linear Boundary Value Problems

```
> restart:
> with(plots):
> eq:=diff(y(x),x$2)-H^2*y(x);
```

$$eq := \frac{d^2}{dx^2} y(x) - H^2 y(x)$$

```
> BCs:=y(0)=1,D(y)(1)=0;
```

$$BCs := y(0) = 1, D(y)(1) = 0$$

```
> sol:='dsolve'({eq,BCs},y(x));
```

$$sol := y(x) = \frac{e^{-H} e^{Hx}}{e^H + e^{-H}} + \frac{e^H e^{-Hx}}{e^H + e^{-H}}$$

The solution obtained can be stored in ya as:

```
> ya:=rhs(sol);
```

$$ya := \frac{e^{-H} e^{Hx}}{e^H + e^{-H}} + \frac{e^H e^{-Hx}}{e^H + e^{-H}}$$

The solution can be converted to trigonometric form and simplified further as

```
> ya:=convert(ya,trig);
```

$$ya := \frac{1}{2} \frac{(\cosh(H) - \sinh(H))(\cosh(Hx) + \sinh(Hx))}{\cosh(H)}$$
$$+ \frac{1}{2} \frac{(\cosh(H) + \sinh(H))(\cosh(Hx) - \sinh(Hx))}{\cosh(H)}$$

```
> ya:=combine(ya);
```

$$ya := \frac{\cosh(-H + Hx)}{\cosh(H)}$$

We observe that 'dsolve' gives a long solution compared to the matrix exponential method (example 3.1). As an exercise readers can verify that the solution obtained by the matrix exponential method and the solution obtained here using Maple's 'dsolve' command are equivalent.

Example 3.10. Cylindrical Catalyst Pellet

The catalyst pellet problem solved in example 3.6 can be solved using Maple's 'dsolve' command as follows:

```
> restart:
> with(plots):
> eq:=diff(c(x),x$2)+1/x*diff(c(x),x)-phi^2*c(x);
```

$$eq := \frac{d^2}{dx^2} c(x) + \frac{\frac{d}{dx} c(x)}{x} - \phi^2 c(x)$$

> BCs:=D(c)(0)=0,c(1)=1;

$$BCs := D(c)(0) = 0, c(1) = 1$$

> sol:='dsolve'({eq,BCs},c(x));

$$sol := c(x) = \frac{\text{BesselI}(0, \phi x)}{\text{BesselI}(0, \phi)}$$

The solution obtained can be assigned as

> Ca:=rhs(sol);

$$Ca := \frac{\text{BesselI}(0, \phi x)}{\text{BesselI}(0, \phi)}$$

Example 3.11. Spherical Catalyst Pellet

Concentration distribution inside a spherical catalyst pellet is governed by the following equation:

$$\frac{d^2 c}{dx^2} + \frac{2}{x} \frac{dc}{dx} - \Phi^2 c = 0 \tag{3.30}$$

$$\frac{dc}{dx}(0) = 0 \text{ and } c(1) = 1$$

This equation is solved below using Maple's 'dsolve' command:

> restart:
> with(plots):
> eq:=diff(c(x),x$2)+2/x*diff(c(x),x)-phi^2*c(x);

$$eq := \frac{d^2}{dx^2} c(x) + \frac{2\left(\frac{d}{dx} c(x)\right)}{x} - \phi^2 c(x)$$

> BCs:=D(c)(0)=0,c(1)=1;

$$BCs := D(c)(0) = 0, c(1) = 1$$

3.1 Linear Boundary Value Problems

> sol:=dsolve({eq,BCs},c(x));

$$sol := c(x) = \frac{\sinh(\phi x)}{\sinh(\phi) x}$$

Maple is not able to solve this problem directly. We can solve this problem without specifying the boundary conditions:

> sol:=dsolve({eq},c(x));

$$sol := \left\{ c(x) = \frac{\_C1 \ \sinh(\phi x)}{x} + \frac{\_C2 \ \cosh(\phi x)}{x} \right\}$$

The solution obtained can be assigned as:

> Ca:=rhs(sol[1]);

$$Ca := \frac{\_C1 \ \sinh(\phi x)}{x} + \frac{\_C2 \ \cosh(\phi x)}{x}$$

Now if ya has to be finite at x = 0, _C2 should be zero.

> _C2:=0;

$$\_C2 := 0$$

> Ca:=eval(Ca);

$$Ca := \frac{\_C1 \ \sinh(\phi x)}{x}$$

Next, the boundary condition at x = 1 is used to solve for _C1.

> bc2:=subs(x=1,Ca)-1;

$$bc2 := -1 + \_C1 \ \sinh(\phi)$$

> _C1:=solve(bc2,_C1);

$$\_C1 := \frac{1}{\sinh(\phi)}$$

> Ca:=eval(Ca);

$$Ca := \frac{\sinh(\phi x)}{\sinh(\phi) x}$$

A three dimensional plot can be made as:

> plot3d(Ca,x=0..1,phi=0..10,axes=boxed,orientation=[120,60],
title="Figure 3.15",labels=[x,phi,"Ca"]);

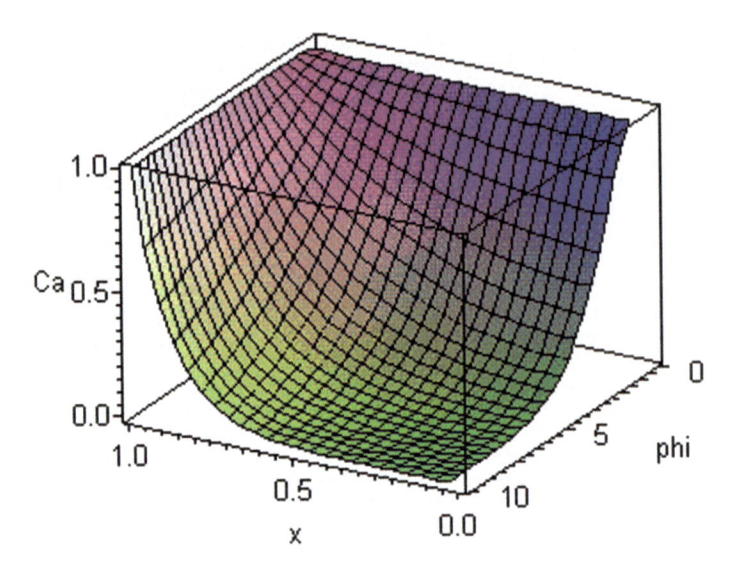

Fig. 3.15

3.1.7 Summary

In this chapter, analytical solutions were obtained for linear boundary value problems. In section 3.1.2, the given linear boundary value problem is converted to matrix form. The analytical solution for this matrix differential equation was found by using the matrix exponential. Maple provides the exponential matrix as a function of the independent variable and the parameters in the governing equations. The unknown initial condition at x = 0 was taken as an unknown constant and found later using the boundary condition at x = 1. This approach yields an elegant solution for linear boundary value problems. This methodology is valid only if the coefficient matrix is constant. This methodology was then extended for linear boundary value problems with semi-infinite domain in section 3.1.3. This is a powerful technique for solving linear boundary value problems in semi-infinite domains.

In section 3.1.4, an analytical series solution using the matrizant was developed for the case where the coefficient matrix is a function of the independent variable. This methodology provides series solutions for Boundary value problems without resorting to any conventional series solution technique. In section 3.1.5, finite difference solutions were obtained for linear Boundary value problems as a function of parameters in the system. The solution obtained is equivalent to the analytical solution because the parameters are explicitly seen in the solution. One has to be careful when solving convective diffusion equations, since the central difference scheme for the first derivative produces numerical oscillations.

In section 3.1.6, linear Boundary value problems were solved using Maple's 'dsolve' command. The solution obtained may not be in the simplified form. Maple gives the Bessel function and other special function solutions for linear

3.1 Linear Boundary Value Problems

boundary value problems. In our opinion, the exponential matrix method is the best method for linear boundary value problems. Maple's 'dsolve' command should be used only if the coefficient matrix is a function of the independent variable. While using the 'dsolve' command, it is better to find the constants separately instead of specifying the boundary conditions in the 'dsolve' command. Eleven examples were presented in this chapter.

3.1.8 Exercise Problems

1. Consider diffusion with a first order isothermal reaction in a rectangular pellet.[11] [8]. The governing equation and boundary conditions for concentration in dimensionless form are:

$$\frac{d^2c}{dx^2} = \Phi^2 c$$

$$\frac{dc}{dx}(0) = 0 \text{ and } c(1) = 1$$

where Φ is the Thiele modulus. Solve this equation using exponential matrix method and plot the profiles for $\Phi = 0.1, 1, 2$ and 10.

2. Consider the diffusion reaction problem (problem 1) with a mass transfer resistance at the surface.[6] [11] The governing equation and boundary conditions for dimensionless concentration are:

$$\frac{d^2c}{dx^2} = \Phi^2 c$$

$$\frac{dc}{dx}(0) = 0 \text{ and } \frac{dc}{dx}(1) = Bi[1-c(1)]$$

where Bi is the Biot number. Solve this equation using matrix exponential matrix method and plot the profiles for $\Phi = 0.1, 1, 2$ and 10 for two different values for the Biot number (Bi = 1, 100). Compare the results obtained with problem 1.

3. Consider gas absorption with chemical reaction in an agitated tank.[13] The governing equation and boundary conditions for dimensionless concentration are given by:

$$\frac{d^2c}{dx^2} = \Phi^2 c$$

$$c(0) = 1 \text{ and } c(1) = 0$$

where Φ is the Thiele modulus. Solve this equation using exponential matrix method and plot the profiles for $\Phi = 0$ and 1.

4. Consider steady state plug flow in a tubular reactor.[14] The governing equation and boundary conditions for dimensionless concentration are:

$$\frac{d^2c}{dx^2} - Pe\frac{dc}{dx} = (PeDa)c$$

$$\frac{dc}{dx}(0) = Pe(c(0)-1) \text{ and } \frac{dc}{dx}(1) = 0$$

where Pe is the Peclet number and Da is the Damkohler number. Solve this equation using exponential matrix method. Plot the concentration profiles for Pe = 5 and Da = 1. In addition plot the exit concentration (c(1)) as a function of Da for different values of Pe.

5. Redo problem 4 if the boundary condition at x = 0 is c(0) = 1. Plot the concentration profiles for Pe = 5 and Da = 1. How does the exit concentration compare with problem 4 for low and high values of Peclet number (for Da = 1)?

6. Consider a linear electrochemical reaction inside a porous electrode.[15] [16] The dimensionless solid phase potential (Φ_1) and electrolyte potential (Φ_2) are governed by the macroscopic porous electrode theory:

$$\frac{d^2\Phi_1}{dx^2} = \frac{\beta}{1+\beta}v^2(\Phi_1 - \Phi_2)$$

$$\frac{d^2\Phi_2}{dx^2} = -\frac{1}{1+\beta}v^2(\Phi_1 - \Phi_2)$$

where v is the dimensionless current density and β is ratio of electrolyte conductivity to solid phase conductivity. The boundary conditions are:

$$\frac{d\Phi_1}{dx}(0) = 0; \quad \frac{d\Phi_1}{dx}(1) = -\delta\beta$$

$$\frac{d\Phi_2}{dx}(0) = -\delta; \quad \frac{d\Phi_2}{dx}(1) = 0$$

where δ is the dimensionless current density. The voltage difference across the porous electrode is given by V = $\Phi_1(1)$ - $\Phi_2(0)$. Solve this coupled system of equations using matrix exponential method and prove that two unknown initial condition constants at x = 0 cannot be solved using the above set of boundary conditions. Alternatively, use the following set of boundary conditions:

$$\frac{d\Phi_1}{dx}(0) = 0; \quad \frac{d\Phi_1}{dx}(1) = -\delta\beta$$

$$\Phi_2(0) = 0; \frac{d\Phi_2}{dx}(0) = -\delta; \quad \frac{d\Phi_2}{dx}(1) = 0$$

3.1 Linear Boundary Value Problems

7. Solve the governing equations using the modified boundary conditions and obtain an analytical solution using the matrix exponential method. Show that one of the two boundary conditions at x = 1 becomes redundant (i.e., automatically satisfied). Plot the potential profiles inside the electrode for $v = \delta = 1$ and $\beta = 0.1$.

8. Subtract the two equations in problem 6 to obtain a single equation for $\eta = \Phi_1 - \Phi_2$. Obtain the boundary conditions for η and arrive at an analytical solution for η using the matrix exponential method.

9. Consider steady state diffusion in a long a cylindrical annulus.[17] The governing equation and boundary conditions are:

$$\frac{d^2c}{dx^2} + \frac{1}{x}\frac{dc}{dx} = 0$$

$$c(1) = 0 \text{ and } c(2) = 1$$

10. To solve this equation using Maple's 'dsolve' command, convert this equation to two first order equations for c and $x\frac{dc}{dx}$. Integrate this system of equations by finding the matrizant (see examples 3.4, 3.5 and 3.6. Note: integrate the equations from 1 to x instead of 0 to x). Show that you get the same solution as Maple's 'dsolve' command.

11. Solve the same problem by writing down two first order equations for c and $\frac{dc}{dx}$. Obtain the matrizant solution for 2, 4, and 6 terms. Show by plotting the profiles that the solution obtained approaches with the analytical solution as the number of terms in the matrizant increases. This problem shows that we get more efficient solution for problems in cylindrical coordinates by solving for c and $x\frac{dc}{dx}$ instead of solving for c and $\frac{dc}{dx}$.

12. Consider heat transfer in a thin metallic circular fin.[18] The dimensionless temperature is governed by:

$$\frac{d^2y}{dx^2} + \frac{1}{x}\frac{dy}{dx} = H^2 y$$

$$y(1) = 1 \text{ and } y(\lambda) = 0$$

where H is the dimensionless heat transfer coefficient and λ is the ratio of outer radius to inner radius. Obtain an analytical solution for this problem using Maple's 'dsolve' command.

13. Obtain the series solutions for problem 9 by using the matrizant method and writing down two first order equations for c and $\frac{dc}{dx}$.

14. Consider diffusion with reaction in a cylinder pellet.[18] The governing equation for dimensionless concentration is:

$$\frac{d^2c}{dx^2} + \frac{1}{x}\frac{dc}{dx} = \phi^2 y$$

$$\frac{dc}{dx}(0) = \text{and } \frac{dc}{dx}(1) = \text{Bi}(1-c(1))$$

where ϕ is the Thiele modulus and Bi is the Biot number. Obtain an analytical solution for this problem using Maple's 'dsolve' command.

15. Redo problem 12 and find series solutions using matrizant method.

16. Consider diffusion with reaction in a spherical pellet.[18] The governing equation for dimensionless concentration is:

$$\frac{d^2c}{dx^2} + \frac{2}{x}\frac{dc}{dx} = \phi^2 y$$

$$\frac{dc}{dx}(0) = 0 \text{ and } \frac{dc}{dx}(1) = \text{Bi}(1-c(1))$$

where ϕ is the Thiele modulus. Obtain an analytical solution for this problem using Maple's 'dsolve' command.

17. Consider the diffusion-reaction problem in a spherical particle discussed in example 3. Obtain the series solutions for this problem using the matrizant method. Write down first order equations for c and $x^2\frac{dc}{dx}$. Plot dimensionless concentration profiles for different values of Thiele modulus ϕ.

18. Redo problem 14 by writing down first order equations for c and $\frac{dc}{dx}$. How does the series solution obtained compare with that of problem 14? Which method is more efficient?

19. Consider limiting current conditions in a rotating disk electrode.[15] The governing equation and boundary conditions for concentration are:

$$D\frac{d^2c}{dz^2} - v\frac{dc}{dz} = 0$$

$$c(0) = 0 \text{ and } c(\infty) = c_{bulk}$$

where the velocity v is given by:

$$v = -0.51023 \, v^{-\frac{1}{2}} \, \Omega^{\frac{3}{2}} \, z^2$$

Define $\alpha = \dfrac{0.51023\, v^{-\frac{1}{2}} \Omega^{\frac{3}{2}}}{D}$ and $u = \dfrac{c}{c_{bulk}}$ to obtain:

$$\frac{d^2u}{dz^2} + \alpha z^2 \frac{du}{dz} = 0$$

$$u(0) = 0 \text{ and } u(\infty) = 1$$

Replace ∞ by $z = L$. Solve this problem by finding the matrizant. Obtain your profiles for $\alpha = 0.1$ and $L = 5$. (Note that you might have to include more than 30 terms in the matrizant depending on the values of α and L).

20. Solve problem 16 by using Maple's 'dsolve' command (without specifying the boundary conditions). Plot the two fundamental solutions obtained (f1, f2) from $z = 0$ to $z = 1$. One of the two functions will be undefined at $z = 0$ and the other function will be well-defined at $z = 0$ (one can verify this by plotting the fundamental solution). Hence, the analytical solution is an arbitrary constant multiplied by the well-defined fundamental solution. Find this constant using the boundary condition at $z = L$. Plot the solution obtained for $\alpha = 0.1$ and $L = 5$.

21. Consider a multi-component diffusion-reaction problem.[19] [4] [20] The governing equations for molar fractions of gas and liquid reactants inside a gas-fed porous electrode of a fuel cell are:

$$\frac{d^2 C1}{dx^2} = k_1 C1$$

$$\frac{d^2 C2}{dx^2} = k_2 C1$$

$$C1(0) = 0.21; D(C1)(1) = 0$$

$$D(C2)(0) = 0; C2(1) = 0.127$$

Obtain an analytical solution for this boundary value problem using matrix exponential method and plot the mole fraction profiles for $k_1 = 1$, $k_2 = 0.1$.

3.2 Nonlinear Boundary Value Problems

3.2.1 Introduction

Heat, mass or momentum transfer in solids is typically represented by boundary value problems (boundary value problems). Variable diffusivity or thermal conductivity, nonlinear source terms or nonlinear boundary conditions make the boundary value

218   3 Boundary Value Problems

problems nonlinear. In this chapter, series and numerical solutions are presented for nonlinear Boundary value problems using Maple.

3.2.2 Series Solutions for Nonlinear Boundary Value Problems

Series solutions for nonlinear boundary value problems can be obtained using Maple's 'dsolve' command. The syntax is

'dsolve'({"differential equations, initial conditions"},{"dependent variable"}, type=series);

Note that, the initial condition at x = 0 is provided for both the dependent variable and its derivative. The unknown initial conditions are taken as constants. These constants are then found using the boundary condition at x = 1. The series solution obtained may be convergent or divergent depending on the problem.

Example 3.2.1. Series Solutions for Diffusion with a Second Order Reaction

Consider diffusion with a second order reaction in a rectangular pellet.[18] The dimensionless concentration is governed by:

$$\frac{d^2 c}{dx^2} = \Phi^2 c^2 \qquad (3.31)$$

with the boundary conditions:

$$\frac{dc}{dx}(0) = 0 \qquad (3.32)$$

and

$$c(1) = 1 \qquad (3.33)$$

where Φ is the Thiele modulus. This boundary problem is solved in Maple below for different values of Φ. The unknown initial condition $c(0)$ is taken as an unknown constant c_1.

> restart:
> with(plots):

Enter the governing equation:

> eq:=diff(c(x),x$2)=Phi^2*c(x)^2;

$$eq := \frac{d^2}{dx^2} c(x) = \Phi^2 c(x)^2$$

3.2 Nonlinear Boundary Value Problems

Enter the boundary condition at x = 1:

> bc:=c(x)-1;

$$bc := c(x) - 1$$

The series solutions are obtained assuming that $c(0) = c_1$:

> sol:='dsolve'({eq,c(0)=c1,D(c)(0)=0},{c(x)},type=series);

$$sol := c(x) = c1 + \frac{1}{2}\Phi^2 c1^2 x^2 + \frac{1}{12}\Phi^4 c1^3 x^4 + O(x^6)$$

The order of the series is increased to get more accurate solutions:

> Order:=8;

$$Order := 8$$

> sol:='dsolve'({eq,c(0)=c1,D(c)(0)=0},{c(x)},type=series);

$$sol := c(x) = c1 + \frac{1}{2}\Phi^2 c1^2 x^2 + \frac{1}{12}\Phi^4 c1^3 x^4 + \frac{1}{72}\Phi^6 c1^4 x^6 + O(x^8)$$

The solution obtained is converted to polynomial form and stored in ca:

> ca:=convert(rhs(sol),polynom);

$$ca := c1 + \frac{1}{2}\Phi^2 c1^2 x^2 + \frac{1}{12}\Phi^4 c1^3 x^4 + \frac{1}{72}\Phi^6 c1^4 x^6$$

We observe that the series solution obtained is a function of Φ and c_1. Now the boundary condition is evaluated using the solution obtained and substituting x = 1:

> eqc:=subs(c(x)=ca,bc);

$$eqc := c1 + \frac{1}{2}\Phi^2 c1^2 x^2 + \frac{1}{12}\Phi^4 c1^3 x^4 + \frac{1}{72}\Phi^6 c1^4 x^6 - 1$$

> eqc:=subs(x=1,eqc);

$$eqc := c1 + \frac{1}{2}\Phi^2 c1^2 + \frac{1}{12}\Phi^4 c1^3 + \frac{1}{72}\Phi^6 c1^4 - 1$$

This is a nonlinear equation and cannot be solved explicitly:

> solve(eqc,c1);

$$\text{RootOf}(72\_Z + 36\Phi^2\_Z^2 + 6\Phi^4\_Z^3 + \Phi^6\_Z^4 - 72)$$

However, the solution can be obtained for a particular value of Φ:

> solve(subs(Phi=0.1,eqc));

0.9950412410, -167.2158854 + 493.4466901 I, −266.5632705, -167.2158854 − 493.4466901 I

The 'solve' command takes too long to solve and gives complex roots. Hence, the 'fsolve' command is used to solve for the real values.

> fsolve(subs(Phi=0.1,eqc),c1=1);

−266.5632705, 0.9950412410

Next, the constant c_1 is solved for different values of Φ:

> cc[1]:=fsolve(subs(Phi=0.1,eqc),c1=1)[2];

$$cc_1 := 0.9950412410$$

> cc[2]:=fsolve(subs(Phi=1,eqc),c1=1)[2];

$$cc_2 := 0.7124733325$$

> cc[3]:=fsolve(subs(Phi=2,eqc),c1=1)[2];

$$cc_3 := 0.4466946595$$

> cc[4]:=fsolve(subs(Phi=5,eqc),c1=1)[2];

$$cc_4 := 0.1722258996$$

Next, plots are made by substituting the values for c_1 and Φ:

p1:=plot(subs(c1=cc[1],Phi=0.1,ca),x=0..1,thickness=3,color=red, axes=boxed):
> p2:=plot(subs(c1=cc[2],Phi=1,ca),x=0..1,thickness=3,color=green, axes=boxed):
> p3:=plot(subs(c1=cc[3],Phi=2,ca),x=0..1,thickness=3,color=gold, axes=boxed):
> p4:=plot(subs(c1=cc[4],Phi=5,ca),x=0..1,thickness=3,color=blue, axes=boxed):
> pt:=textplot([[0.5,subs(x=0.5,c1=cc[1],Phi=0.1,ca)-0.05,[c1=cc[1],Phi=0.1]],[0.5,subs(x=0.5,c1=cc[2],Phi=1,ca)-0.08,[c1=cc[2],Phi=1]],[0.5,subs(x=0.5,c1=cc[3],Phi=2,ca)-0.1,[c1=cc[3],Phi=2]],[0.5,subs(x=0.5,c1=cc[4],Phi=5,ca)-0.1,[c1=cc[4],Phi=5]]]):
> display({p1,p2,p3,p4,pt},labels=[x,c],title="Figure 3.16");

3.2 Nonlinear Boundary Value Problems

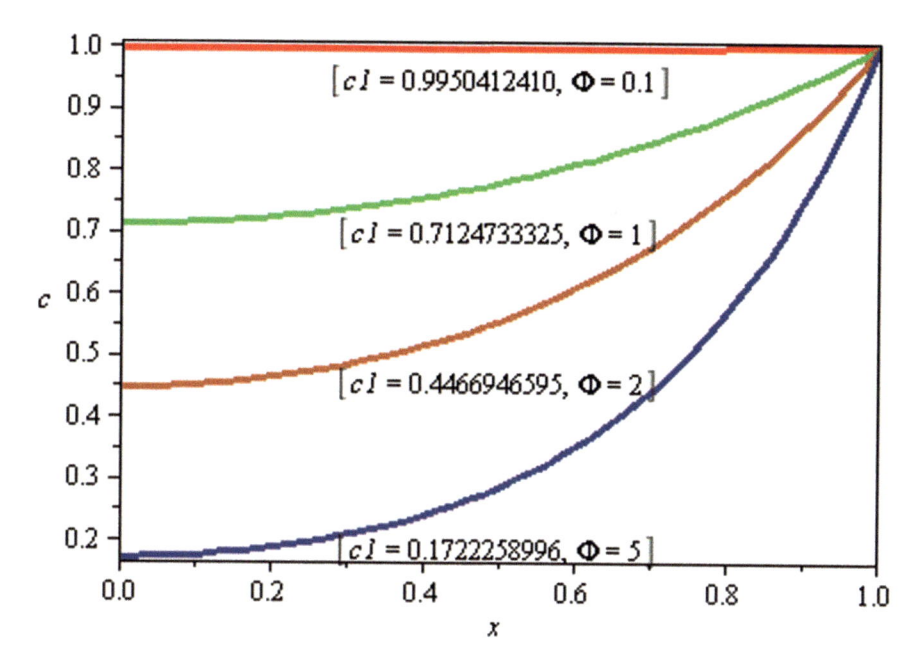

Fig. 3.16

We observe that as Φ increases, the concentration profiles become steeper. We have obtained the series solution for the given nonlinear boundary value problem. There is no guarantee that the solution has converged. Accuracy of the solution can be analyzed by increasing the number of terms in the series solution and checking the values for c_1:

> Order:=10;

$$Order := 10$$

> sol:='dsolve'({eq,c(0)=c1,D(c)(0)=0},{c(x)},type=series);

$$sol := c(x) = c1 + \frac{1}{2}\Phi^2 c1^2 x^2 + \frac{1}{12}\Phi^4 c1^3 x^4 + \frac{1}{72}\Phi^6 c1^4 x^6 + \frac{1}{504}\Phi^8 c1^5 x^8 + O(x^{10})$$

> ca:=convert(rhs(sol),polynom);

$$sol := c(x) = c1 + \frac{1}{2}\Phi^2 c1^2 x^2 + \frac{1}{12}\Phi^4 c1^3 x^4 + \frac{1}{72}\Phi^6 c1^4 x^6 + \frac{1}{504}\Phi^8 c1^5 x^8 + O(x^{10})$$

> eqc:=subs(c(x)=ca,bc);

$$eqc := c1 + \frac{1}{2}\Phi^2 c1^2 x^2 + \frac{1}{12}\Phi^4 c1^3 x^4 + \frac{1}{72}\Phi^6 c1^4 x^6 + \frac{1}{504}\Phi^8 c1^5 x^8 - 1$$

> eqc:=subs(x=1,eqc);

$$eqc := c1 + \frac{1}{2}\Phi^2 c1^2 + \frac{1}{12}\Phi^4 c1^3 + \frac{1}{72}\Phi^6 c1^4 + \frac{1}{504}\Phi^8 c1^5 - 1$$

> fsolve(subs(Phi=0.1,eqc),c1=1);

$$-458.7096161, -327.2897446, 0.9950412410$$

We observe that there are three different solutions of which only the last one makes sense:

> cc2[1]:=fsolve(subs(Phi=0.1,eqc),c1=1)[3];

$$cc2_1 := 0.9950412410$$

> fsolve(subs(Phi=1,eqc),c1=1);

$$0.7122776915$$

For higher values of Φ, 'fsolve' produces only the correct solution:

> cc2[2]:=fsolve(subs(Phi=1,eqc),c1=1);

$$cc2_2 := 0.7122776915$$

> cc2[3]:=fsolve(subs(Phi=2,eqc),c1=1);

$$cc2_3 := 0.4444311359$$

> cc2[4]:=fsolve(subs(Phi=5,eqc),c1=1);

$$cc2_4 := 0.1653141670$$

The constants c_1 obtained using eight terms and ten terms in the series are compared:

> for i to 4 do print(cc[i],cc2[i]);od;

$$0.9950412410, 0.9950412410$$
$$0.7124733325, 0.7122776915$$
$$0.4466946595, 0.4444311359$$
$$0.1722258996, 0.1653141670$$

For $\Phi = 0.1$ and 1, we observe that the solution has converged to the third digit. For $\Phi = 2$, the solution has converged to the first two digits. For $\Phi = 5$, we observe that the solution has not converged. Hence, more terms are required in the series for higher values of Φ. The series solution obtained using Maple may or may not converge. The convergence of the solution depends on the parameters (as illustrated in this example), and the nonlinearity of the problem. In the next example the non-isothermal reaction in a rectangular pellet is analyzed.

>

3.2 Nonlinear Boundary Value Problems

Example 3.2.2. Series Solutions for Non-isothermal Catalyst Pellet – Multiple Steady States

The dimensionless concentration in a non-isothermal catalyst pellet[8] is governed by:

$$\frac{d^2c}{dx^2} = \Phi^2 c \exp\left(\frac{\gamma\beta(1-c)}{1+\beta(1-c)}\right) \tag{3.34}$$

with the boundary conditions:

$$\frac{dc}{dx}(0) = 0 \tag{3.35}$$

and

$$c(1) = 1 \tag{3.36}$$

This boundary value problem has multiple solutions for $\Phi = 0.2$, $\beta = 0.8$ and $\gamma = 20$. The series solutions are obtained for this problem in Maple below:

> restart:
> with(plots):

The governing equation is entered here:

> eq:=diff(c(x),x$2)=Phi^2*c(x)*exp(gamma*beta*(1-c(x))/(1+beta*(1-c(x))));

$$eq := \frac{d^2}{dx^2} c(x) = \Phi^2 c(x) e^{\frac{\gamma\beta(1-c(x))}{1+\beta(1-c(x))}}$$

The values for the parameters are entered here:

> eq:=subs(gamma=20,beta=0.8,eq);

$$eq := \frac{d^2}{dx^2} c(x) = \Phi^2 c(x) e^{\frac{16.0(1-c(x))}{1.8-0.8c(x)}}$$

The boundary condition at $x = 1$ is entered here:

> bc:=c(x)-1;

$$bc := c(x) - 1$$

The order of the series solution is specified and the governing equation is solved as a function of c_1 the concentration at the center, and Φ:

> Order:=8;

$$Order := 8$$

224   3 Boundary Value Problems

> sol:='dsolve'({eq,c(0)=c1,D(c)(0)=0},{c(x)},type=series);

$$sol := c(x) = c1 + \frac{1}{2}\Phi^2 c1\, e^{\frac{80(-1+c1)}{-9+4c1}} x^2 + \frac{1}{24}\frac{\Phi^4 \left(e^{\frac{80(-1+c1)}{-9+4c1}}\right)^2 c1\,(81 - 472\,c1 + 16\,c1^2)}{(-9+4c1)^2} x^4$$

$$+ \frac{1}{720}\frac{\Phi^6 c1 \left(e^{\frac{80(-1+c1)}{-9+4c1}}\right)^3 (-270864\,c1 + 791776\,c1^2 - 15104\,c1^3 + 6561 + 256\,c1^4)}{(-9+4c1)^4} x^6 + O(x^8)$$

The series solution obtained is stored in ca:

> ca:=convert(rhs(sol),polynom);

$$ca := c1 + \frac{1}{2}\Phi^2 c1\, e^{\frac{80(-1+c1)}{-9+4c1}} x^2 + \frac{1}{24}\frac{\Phi^4 \left(e^{\frac{80(-1+c1)}{-9+4c1}}\right)^2 c1\,(81 - 472\,c1 + 16\,c1^2) x^4}{(-9+4c1)^2}$$

$$+ \frac{1}{720}\frac{\Phi^6 c1 \left(e^{\frac{80(-1+c1)}{-9+4c1}}\right)^3 (-270864\,c1 + 791776\,c1^2 - 15104\,c1^3 + 6561 + 256\,c1^4) x^6}{(-9+4c1)^4}$$

The boundary condition at $x = 1$ is evaluated using the series solution obtained:

> eqc:=subs(c(x)=ca,bc);

$$eqc := c1 + \frac{1}{2}\Phi^2 c1\, e^{\frac{80(-1+c1)}{-9+4c1}} x^2 + \frac{1}{24}\frac{\Phi^4 \left(e^{\frac{80(-1+c1)}{-9+4c1}}\right)^2 c1\,(81 - 472\,c1 + 16\,c1^2) x^4}{(-9+4c1)^2}$$

$$+ \frac{1}{720}\frac{\Phi^6 c1 \left(e^{\frac{80(-1+c1)}{-9+4c1}}\right)^3 (-270864\,c1 + 791776\,c1^2 - 15104\,c1^3 + 6561 + 256\,c1^4) x^6}{(-9+4c1)^4} - 1$$

> eqc:=subs(x=1,eqc);

$$eqc := c1 + \frac{1}{2}\Phi^2 c1\, e^{\frac{80(-1+c1)}{-9+4c1}} + \frac{1}{24}\frac{\Phi^4 \left(e^{\frac{80(-1+c1)}{-9+4c1}}\right)^2 c1\,(81 - 472\,c1 + 16\,c1^2)}{(-9+4c1)^2}$$

$$+ \frac{1}{720}\frac{\Phi^6 c1 \left(e^{\frac{80(-1+c1)}{-9+4c1}}\right)^3 (-270864\,c1 + 791776\,c1^2 - 15104\,c1^3 + 6561 + 256\,c1^4)}{(-9+4c1)^4} - 1$$

The equation for c_1 is plotted as a function of c_1. We observe that there are four solutions:

> plot(subs(Phi=0.2,eqc),c1=0..1,view=[0..1,-1..1],title="Figure 3.17",
thickness=4);

3.2 Nonlinear Boundary Value Problems

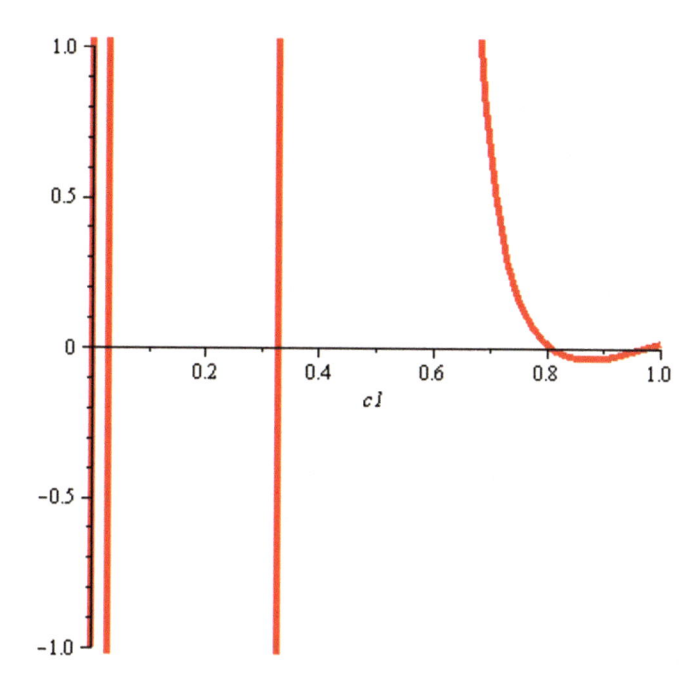

Fig. 3.17

The values for the constant c_1 are obtained using the 'fsolve' command.

> cc[1]:=fsolve(subs(Phi=0.2,eqc),c1=1);

$$cc_1 := 0.9716851561$$

> cc[2]:=fsolve(subs(Phi=0.2,eqc),c1=0.8);

$$cc_2 := 0.8079753866$$

> cc[3]:=fsolve(subs(Phi=0.2,eqc),c1=0.0);

$$cc_3 := 0.00002667544163$$

> cc[4]:=fsolve(subs(Phi=0.2,eqc),c1=0.03);

$$cc_4 := 0.02914597330$$

We obtain four different solutions for c_1. These values for c_1 and Φ are substituted in the series solution obtained and plots are made:

p1:=plot(subs(c1=cc[1],Phi=0.2,ca),x=0..1,thickness=4,color=black, axes=boxed):
> p2:=plot(subs(c1=cc[2],Phi=0.2,ca),x=0..1,thickness=4,color=blue, axes=boxed):
> p3:=plot(subs(c1=cc[3],Phi=0.2,ca),x=0..1,thickness=4,color=brown, axes=boxed):
> p4:=plot(subs(c1=cc[4],Phi=0.2,ca),x=0..1,thickness=4,color=red, axes=boxed):
>
pt:=textplot([[0.3,evalf(subs(x=0.3,c1=cc[1],Phi=0.2,ca)+0.04),[c1=cc[1],Phi=0.2]],[0.3,evalf(subs(x=0.3,c1=cc[2],Phi=0.2,ca)+0.04),[c1=cc[2],Phi=0.2]], [0.3,evalf(subs(x=0.3,c1=cc[3],Phi=0.2,ca)+0.04),[c1=cc[3],Phi=0.2]]]):
> display({p1,p2,p3,pt},title="Figure 3.18",labels=[x,c], view=[0..1,-0.01..1.05]);

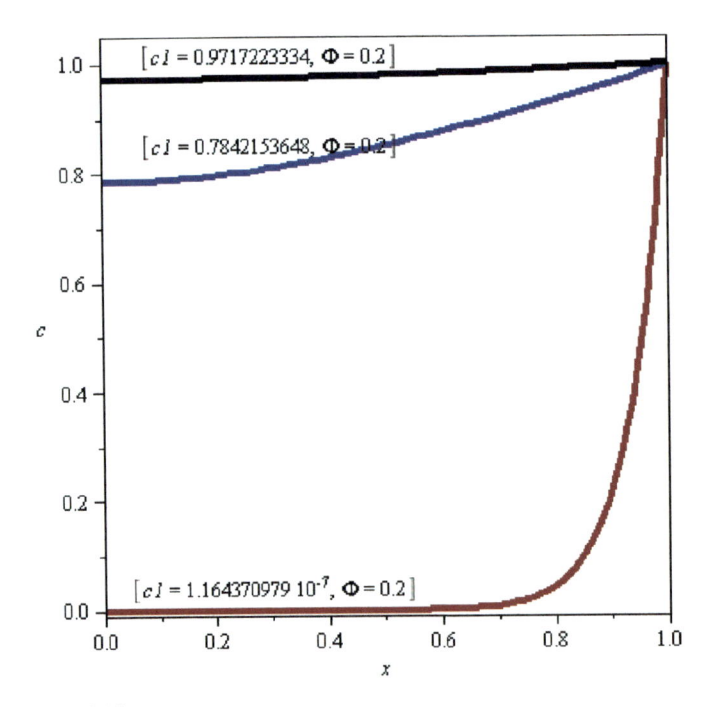

Fig. 3.18

3.2 Nonlinear Boundary Value Problems 227

> display(p4,labels=[x,c],title="Figure 3.19");

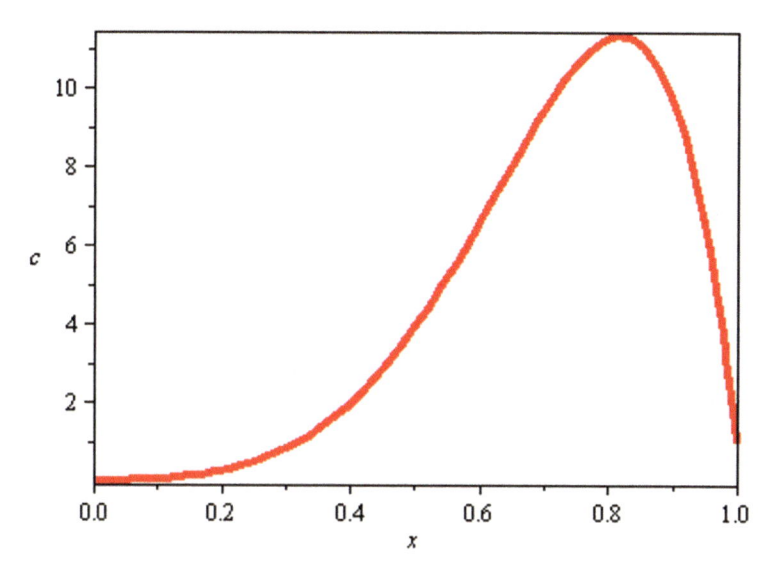

Fig. 3.19

We observe that the first three values of c_1 make sense and we discard the fourth value. There is no guarantee that the solution obtained is the converged one. The accuracy and convergence of the series solution are analyzed below by increasing the number of terms in the series.

> Order:=16;

$$Order := 16$$

> sol:='dsolve'({eq,c(0)=c1,D(c)(0)=0},{c(x)},type=series):
> ca:=convert(rhs(sol),polynom):
> eqc:=subs(c(x)=ca,bc):
> eqc:=subs(x=1,eqc):
> cc2[1]:=fsolve(subs(Phi=0.2,eqc),c1=1);

$$cc2_1 := 0.9717223260$$

> cc2[2]:=fsolve(subs(Phi=0.2,eqc),c1=0.8);

$$cc2_2 := 0.7894024799$$

> cc2[3]:=fsolve(subs(Phi=0.2,eqc),c1=0.0);

$$cc2_3 := 2.503698214 \, 10^{-7}$$

> for i to 3 do print(cc[i],cc2[i]);od;

$$0.9716851561, 0.9717223260$$

$$0.8079753866, 0.7894024799$$

$$0.00002667544163, 2.503698214\ 10^{-7}$$

We observe that the first value for c_1 has converged and the next two values have not converged. Next, the order is increased to 20.

> Order:=20;

$$Order := 20$$

```
> sol:='dsolve'({eq,c(0)=c1,D(c)(0)=0},{c(x)},type=series):
> ca:=convert(rhs(sol),polynom):
> eqc:=subs(c(x)=ca,bc):
> eqc:=subs(x=1,eqc):
> cc3[1]:=fsolve(subs(Phi=0.2,eqc),c1=1);
```

$$cc3_1 := 0.9717223334$$

> cc3[2]:=fsolve(subs(Phi=0.2,eqc),c1=0.8);

$$cc3_2 := 0.7842153648$$

> cc3[3]:=fsolve(subs(Phi=0.2,eqc),c1=0.0);

$$cc3_3 := 1.164370979\ 10^{-7}$$

> for i to 3 do print(cc[i],cc2[i],cc3[i]);od;

$$0.9716851561, 0.9717223260, 0.9717223334$$

$$0.8079753866, 0.7894024799, 0.7842153648$$

$$0.00002667544163, 2.503698214\ 10^{-7}, 1.164370979\ 10^{-7}$$

We observe that the first value for c_1 has converged (three digit accuracy). The second value of c_1 has converged to two digit accuracy. The third value has not converged yet. But, the order of magnitude for the third value of c_1 has converged. Next, plots are made for three different values of c_1.

```
p1:=plot(subs(c1=cc3[1],Phi=0.2,ca),x=0..1,thickness=4,color=black,
axes=boxed):
> p2:=plot(subs(c1=cc3[2],Phi=0.2,ca),x=0..1,thickness=4,color=blue,
axes=boxed):
> p3:=plot(subs(c1=cc3[3],Phi=0.2,ca),x=0..1,thickness=4,color=brown,
axes=boxed):
```

3.2 Nonlinear Boundary Value Problems

```
>
pt:=textplot([[0.3,evalf(subs(x=0.3,c1=cc3[1],Phi=0.2,ca)+0.04),[c1=cc3[1],
Phi=0.2]],[0.3,evalf(subs(x=0.3,c1=cc3[2],Phi=0.2,ca)+0.04),[c1=cc3[2],
Phi=0.2]],[0.3,evalf(subs(x=0.3,c1=cc3[3],Phi=0.2,ca)+0.04),[c1=cc3[3],
Phi=0.2]]]):
> display({p1,p2,p3,pt},labels=[x,c],title="Figure 3.20",view=[0..1,-0.01..1.05]);
```

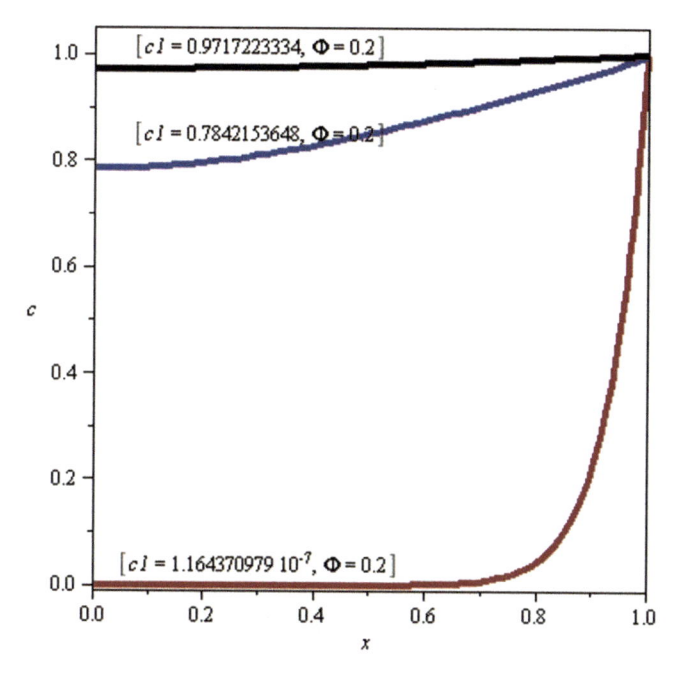

Fig. 3.20

Multiple steady states in a rectangular catalyst pellet were analyzed in this example. This problem will be revisited later in this chapter.

3.2.3 Finite Difference Solutions for Nonlinear Boundary Value Problems

The theory of finite difference solution for Boundary value problems was developed in section 3.1.5. When finite difference approximations are used, the given nonlinear boundary value problem is converted to a system of nonlinear algebraic equations. This resulting system is solved in this section using Maple's 'fsolve' command.

Example 3.2.3. Diffusion with a Second Order Reaction

Example 3.2.1 is solved using finite differences in Maple below. The program developed for example 3.8 is modified to solve this example. (Note that y is used as the dependent variable instead of c).

> restart:
> with(plots):

The number of node points is entered here:
> N:=4;
$$N := 4$$

The length of the domain is entered here:
> L:=1;
$$L := 1$$

The governing equation is entered below:
> eq:=diff(y(x),x$2)-Phi^2*y(x)^2;
$$eq := \frac{d^2}{dx^2} y(x) - \Phi^2 y(x)^2$$

The boundary conditions are entered here:
> bc1:=diff(y(x),x);
$$bc1 := \frac{d}{dx} y(x)$$

> bc2:=y(x)-1;
$$bc2 := y(x) - 1$$

Next, a general program is written to convert the governing equation and the boundary conditions to finite difference form. The central difference expression for the second and first derivatives are:
> d2ydx2:=(y[m+1]-2*y[m]+y[m-1])/h^2;
$$d2ydx2 := \frac{y_{m+1} - 2 y_m + y_{m-1}}{h^2}$$

> dydx:=(y[m+1]-y[m-1])/2/h;
$$dydx := \frac{1}{2} \frac{y_{m+1} - y_{m-1}}{h}$$

Three point forward and backward difference expressions for the derivative are:
> dydxf:=(-y[2]+4*y[1]-3*y[0])/(2*h);
$$dydxf := \frac{1}{2} \frac{-y_2 + 4 y_1 - 3 y_0}{h}$$

3.2 Nonlinear Boundary Value Problems

> dydxb:=(y[N-1]-4*y[N]+3*y[N+1])/(2*h);

$$dydxb := \frac{1}{2}\frac{y_3 - 4y_4 + 3y_5}{h}$$

The governing equation in finite difference form is:

> Eq[m]:=subs(diff(y(x),x$2)=d2ydx2,diff(y(x),x)=dydx,y(x)=y[m],x=m*h,eq);

$$Eq_m := \frac{y_{m+1} - 2y_m + y_{m-1}}{h^2} - \Phi^2 y_m^2$$

The boundary conditions in finite difference form are:

> Eq[0]:=subs(diff(y(x),x)=dydxf,y(x)=y[0],bc1);

$$Eq_0 := \frac{1}{2}\frac{-y_2 + 4y_1 - 3y_0}{h}$$

> Eq[N+1]:=subs(diff(y(x),x)=dydxb,y(x)=y[N+1],bc2);

$$Eq_5 := y_5 - 1$$

A 'for loop' can be written for the interior node points as

> for i to N do Eq[i]:=subs(m=i,Eq[m]);od;

$$Eq_1 := \frac{y_2 - 2y_1 + y_0}{h^2} - \Phi^2 y_1^2$$

$$Eq_2 := \frac{y_3 - 2y_2 + y_1}{h^2} - \Phi^2 y_2^2$$

$$Eq_3 := \frac{y_4 - 2y_3 + y_2}{h^2} - \Phi^2 y_3^2$$

$$Eq_4 := \frac{y_5 - 2y_4 + y_3}{h^2} - \Phi^2 y_4^2$$

The node spacing is given by:

> h:=L/(N+1);

$$h := \frac{1}{5}$$

The value for Φ is sustained in the governing equations. The governing equations are stored in eqs:

> eqs:=seq(eval(subs(Phi=1,Eq[i])),i=0..N+1);

$$eqs := -\frac{5}{2}y_2 + 10y_1 - \frac{15}{2}y_0, 25y_2 - 50y_1 + 25y_0 - y_1^2, 25y_3 - 50y_2 + 25y_1 - y_2^2, 25y_4 - 50y_3 + 25y_2 - y_3^2, 25y_5 - 50y_4 + 25y_3 - y_4^2, y_5 - 1$$

The variables are stored in vars:

> vars:=seq(y[i],i=0..N+1);

$$vars := y_0, y_1, y_2, y_3, y_4, y_5$$

The 'fsolve' command sometimes gives negative values when guess values for the dependent variables are not provided. To avoid this, an initial guess of 1 is provided:

> fsolve({eqs},{vars});

$$\{y_5 = 1.000000000, y_2 = -2.960037328, y_0 = -4.145136024, y_1 = -3.848861350, y_3 = -1.720740468, y_4 = -.3630056965\}$$

> vars:=seq(y[i]=1,i=0..N+1);

$$vars := y_0 = 1, y_1 = 1, y_2 = 1, y_3 = 1, y_4 = 1, y_5 = 1$$

> sol:=fsolve({eqs},{vars});

$$sol := \{y_5 = 1., y_2 = 0.7539844610, y_0 = 0.7122065043, y_1 = 0.7226509935, y_3 = 0.8080576312, y_4 = 0.8882490868\}$$

The solution obtained is assigned and plotted:

> assign(sol):
> plot([seq([i*h,y[i]],i=0..N+1)],thickness=4,axes=boxed,title="Figure 3.21", labels=[x,y]);

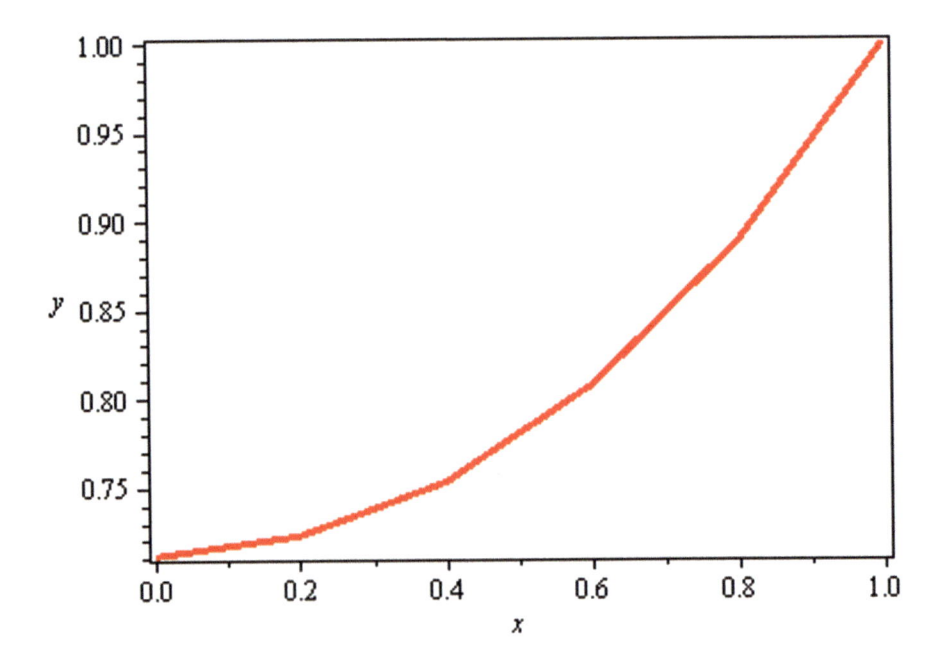

Fig. 3.21

3.2 Nonlinear Boundary Value Problems

> y[0];

The accuracy of the solution obtained can be checked by following the concentration at the center y[0] and increasing the number of node points:

> y[0];

$$0.7122065043$$

The value for y[0] obtained with N = 10 interior node points is:

> y[0];

$$.7123656510$$

Hence, we conclude that the solution obtained has converged. The finite difference solution is an easy technique to apply. However, the number of node points required might increase drastically for stiff boundary value problems.

3.2.4 Shooting Technique for Boundary Value Problem

The shooting technique involves converting the given boundary value problem to a system of initial value problems. The unknown initial conditions are guessed. These unknown conditions are then updated using the known boundary condition at $x = 1$. In this technique, the unknown initial condition at $x = 0$ is estimated using an optimization procedure. This is best illustrated using the next example.

Example 3.2.4. Nonlinear Heat Transfer

Consider a modification of the heat transfer discussed in example 3.1.1 with a nonlinear source term. The governing equation is

$$\frac{d^2y}{dx^2} = (1+0.1y)y \tag{3.37}$$

with the boundary conditions:

$$\frac{dy}{dx}(0) = 0 \tag{3.38}$$

and

$$y(1) = 1 \tag{3.39}$$

The initial condition, y(0) is taken as

$$y(0) = \alpha \tag{3.40}$$

An initial value for α is guessed and the governing equation (3.37) is solved as an initial value problem. Then, a new value of α is obtained using the following relation.[6] [21]

$$\alpha_{new} = \alpha_{old} + \left[\frac{\text{Boundary Condition at } x = 1_{expected} - \text{Boundary Condition at } x = 1_{predicted,old}}{d(\text{Boundary Condition at } x = 1)} \right]_{predicted,old}$$

$$= \alpha_{old} + \left[\frac{y_x - y(\alpha_{old})_{x=1}}{\frac{dy(\alpha_{old})}{d\alpha} \bigg|_{x=1}} \right]$$

$$= \alpha_{old} + \left[\frac{1 - y(\alpha_{old})_{x=1}}{\frac{dy(\alpha_{old})}{d\alpha} \bigg|_{x=1}} \right] \tag{3.41}$$

In equation (3.41), the Jacobian $\dfrac{dy(\alpha_{old})}{d\alpha}$ is often calculated numerically. However, the Jacobian can be predicted exactly by differentiating the governing equation (3.37) with respect to alpha as

$$\frac{d^2}{dx^2}\left(\frac{dy}{d\alpha}\right) = (1+0.1\frac{dy}{d\alpha})y + (1+0.1y)\frac{dy}{d\alpha} \tag{3.42}$$

Next, the Jacobian is treated as a variable y_2.

$$\frac{d^2 y2}{dx^2} = (1+0.1y2)y + (1+0.1y)y2 \tag{3.43}$$

The initial conditions for y_2 are obtained by differentiating equations (3.40) and (3.38) with respect to α:

$$y2(0) = 1 \tag{3.44}$$

and

$$\frac{dy2}{dx}(0) = 0 \tag{3.45}$$

This boundary value problem is solved in Maple below:

> restart:
> with(plots):

Enter the governing equation:

> eq:=diff(y(x),x$2)-(1+0.1*y(x))*y(x);

$$eq := \frac{d^2}{dx^2} y(x) - (1 + 0.1\, y(x))\, y(x)$$

3.2 Nonlinear Boundary Value Problems

The sensitivity equation is developed by treating y as a function of x and α:

> eqalpha:=subs(y(x)=Y(x,alpha),eq);

$$eqalpha := \frac{\partial^2}{\partial x^2} Y(x, \alpha) - (1 + 0.1\, Y(x, \alpha))\, Y(x, \alpha)$$

The governing equation for the Jacobian $\left(\frac{dy}{d\alpha}\right)$ is obtained by differentiating the governing equation with respect to α:

> eqalpha:=diff(eqalpha,alpha);

$$eqalpha := \frac{\partial^3}{\partial x^2 \partial \alpha} Y(x, \alpha) - 0.1 \left(\frac{\partial}{\partial \alpha} Y(x, \alpha) \right) Y(x, \alpha) - (1 + 0.1\, Y(x, \alpha)) \left(\frac{\partial}{\partial \alpha} Y(x, \alpha) \right)$$

A new variable, $y_2(x)$ is used to present the Jacobian $\left(\frac{dy}{d\alpha}\right)$:

> eqalpha:=subs(diff(Y(x,alpha),alpha)=y2(x),eqalpha);

$$eqalpha := \frac{d^2}{dx^2} y2(x) - 0.1\, y2(x)\, Y(x, \alpha) - (1 + 0.1\, Y(x, \alpha))\, y2(x)$$

> eqalpha:=subs(Y(x,alpha)=y(x),eqalpha);

$$eqalpha := \frac{d^2}{dx^2} y2(x) - 0.1\, y2(x)\, y(x) - (1 + 0.1\, y(x))\, y2(x)$$

The variables are stored in vars:

> vars:=(y(x),y2(x));

$$vars := y(x), y2(x)$$

The original governing equation and the sensitivity equation are stored in eqs:

> eqs:=(eq,eqalpha);

$$eqs := \frac{d^2}{dx^2} y(x) - (1 + 0.1\, y(x))\, y(x),\ \frac{d^2}{dx^2} y2(x) - 0.1\, y2(x)\, y(x) - (1 + 0.1\, y(x))\, y2(x)$$

The initial value for α is given here:

> alpha0:=0.5;

$$\alpha 0 := 0.5$$

The initial conditions are stored in ICs:

> ICs:=(y(0)=alpha0,D(y)(0)=0,y2(0)=1,D(y2)(0)=0);

$$ICs := y(0) = 0.5,\ D(y)(0) = 0,\ y2(0) = 1,\ D(y2)(0) = 0$$

Next the numerical solution is obtained and stored in sol:

> sol:=dsolve({eqs,ICs},{vars},type=numeric);

$$sol := \mathbf{proc}(x\_rkf45) \ \ldots \ \mathbf{end\ proc}$$

The solution is evaluated at $x = 1$:

> sol(1);

$$\left[x = 1., y(x) = 0.787720725926567834, \frac{d}{dx} y(x) = 0.628296298165053944, y2(x) = 1.60809619176739282, \right.$$
$$\left. \frac{d}{dx} y2(x) = 1.33935818442695509 \right]$$

The predicted value of y is stored in ypred:

> ypred:=rhs(sol(1)[2]);

$$ypred := 0.787720725926567834$$

The predicted value for the Jacobian is stored in $y_2$pred:

> y2pred:=rhs(sol(1)[4]);

$$y2pred := 1.60809619176739282$$

The new value for α is obtained as:

> alpha1:=alpha0+(1-ypred)/y2pred;

$$\alpha 1 := 0.6320065772$$

The error is calculated based on the value of α:

> err:=alpha1-alpha0;

$$err := 0.1320065772$$

The new value of α is then assigned to $\alpha 0$ for the next iteration.

> alpha0:=alpha1;

$$\alpha 0 := 0.6320065772$$

A program is written to update the values of α until the error is > 1e - 6. One can set stricter tolerance limits for higher accuracy.

> k:=1;

$$k := 1$$

> while err>1e-6 do
> ICs:=(y(0)=alpha0,D(y)(0)=0,y2(0)=1,D(y2)(0)=0);
> sol:=dsolve({eqs,ICs},{vars},type=numeric);
> ypred:=rhs(sol(1)[2]);
> y2pred:=rhs(sol(1)[4]);

3.2 Nonlinear Boundary Value Problems

```
> alpha1:=alpha0+(1-ypred)/y2pred;
> err:=abs(alpha1-alpha0);
> alpha0:=alpha1;k:=k+1;
> end;
```

$$ICs := y(0) = 0.6320065772, D(y)(0) = 0, y2(0) = 1, D(y2)(0) = 0$$

$$sol := \mathbf{proc}(x\_rkf45) \ \ldots \ \mathbf{end \ proc}$$

$$ypred := 1.00115116384101999$$

$$y2pred := 1.62555796254959861$$

$$\alpha 1 := 0.6312984117$$

$$err := 0.0007081655$$

$$\alpha 0 := 0.6312984117$$

$$k := 2$$

$$ICs := y(0) = 0.6312984117, D(y)(0) = 0, y2(0) = 1, D(y2)(0) = 0$$

$$sol := \mathbf{proc}(x\_rkf45) \ \ldots \ \mathbf{end \ proc}$$

$$ypred := 1.00000003296316220$$

$$y2pred := 1.62546395111771891$$

$$\alpha 1 := 0.6312983914$$

$$err := 2.03 \ 10^{-8}$$

$$\alpha 0 := 0.6312983914$$

$$k := 3$$

```
> k;
```

$$3$$

Three iterations were required for this problem. The error in α is found to be:

```
> err;
```

$$2.03 \ 10^{-8}$$

The solution obtained is then plotted:

```
> odeplot(sol,[x,y(x)],0..1,axes=boxed,color=brown,title="Figure 3.22",
thickness=4);
```

Fig. 3.22

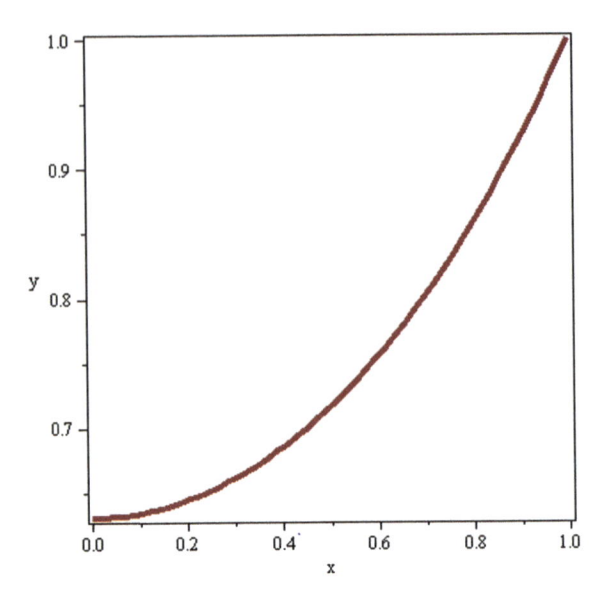

For this boundary value problem it takes only three iterations for the solution to converge. Depending on the problem, and the initial guess provided, the program might take any number of iterations to converge. In addition, for updating α (equation (3.41)) the Jacobian in the denominator might approach zero for certain problems. Sometimes, it is necessary to scale the update of α using the following relation

$$\alpha_{new} = \alpha_{old} + \rho \left[\frac{1 - y(\alpha_{old})_{x=1}}{\frac{dy(\alpha_{old})}{d\alpha}\bigg|_{x=1}} \right] \quad (3.46)$$

where ρ is the number between 0 and 1. The lower the value of ρ the higher is the number of iterations required for convergence. The value of ρ depends on the problem. This is illustrated in the next example.

Example 3.2.5. Multiple Steady States in a Catalyst Pellet

The catalyst pellet problem solved in example 3.2.2 is solved here using the shooting technique. The Maple program is given below:

```
> restart:
> with(plots):
```

The governing equation is entered here (after substituting the parameter values):

```
> eq:=diff(y(x),x$2)-0.04*y(x)*exp(16*(1-y(x))/(1+0.8*(1-y(x))));
```

3.2 Nonlinear Boundary Value Problems

$$eq := \frac{d^2}{dx^2} y(x) - 0.04 y(x) e^{\frac{16(1-y(x))}{1.8 - 0.8 y(x)}}$$

```
> eqalpha:=subs(y(x)=Y(x,alpha),eq):
> eqalpha:=diff(eqalpha,alpha):
> eqalpha:=subs(diff(Y(x,alpha),alpha)=y2(x),eqalpha):
```

The sensitivity equation is:

```
> eqalpha:=subs(Y(x,alpha)=y(x),eqalpha);
```

$$eqalpha := \frac{d^2}{dx^2} y2(x) - 0.04 y2(x) e^{\frac{16(1-y(x))}{1.8 - 0.8 y(x)}} - 0.04 y(x) \left(-\frac{16 y2(x)}{1.8 - 0.8 y(x)} \right.$$

$$\left. + \frac{12.8 (1 - y(x)) y2(x)}{(1.8 - 0.8 y(x))^2} \right) e^{\frac{16(1-y(x))}{1.8 - 0.8 y(x)}}$$

The variables are stored in vars:

```
> vars:=(y(x),y2(x));
```

$$vars := y(x), y2(x)$$

The governing equations are stored in eqs:

```
> eqs:=(eq,eqalpha);
```

$$eqs := \frac{d^2}{dx^2} y(x) - 0.04 y(x) e^{\frac{16(1-y(x))}{1.8 - 0.8 y(x)}}, \frac{d^2}{dx^2} y2(x) - 0.04 y2(x) e^{\frac{16(1-y(x))}{1.8 - 0.8 y(x)}} - 0.04 y(x) \left(-\frac{16 y2(x)}{1.8 - 0.8 y(x)} \right.$$

$$\left. + \frac{12.8 (1 - y(x)) y2(x)}{(1.8 - 0.8 y(x))^2} \right) e^{\frac{16(1-y(x))}{1.8 - 0.8 y(x)}}$$

The boundary value problem has multiple solutions. The solution obtained depends on the initial guess provided for α. An initial guess of 0.9 is given:

```
> alpha0:=0.9;
```

$$\alpha 0 := 0.9$$

```
> ICs:=(y(0)=alpha0,D(y)(0)=0,y2(0)=1,D(y2)(0)=0);
```

$$ICs := y(0) = 0.9, D(y)(0) = 0, y2(0) = 1, D(y2)(0) = 0$$

```
> sol:=dsolve({eqs,ICs},{vars},type=numeric,abserr=1e-10);
```

$$sol := \mathbf{proc}(x\_rkf45) \ldots \mathbf{end\ proc}$$

```
> sol(1);
```

$$\left[x = 1., y(x) = 0.968526696660516496, \frac{d}{dx} y(x) = 0.119738543547597192, y2(x) = 0.230740415090328915, \right.$$

$$\left. \frac{d}{dx} y2(x) = -1.20071140540140586 \right]$$

> ypred:=rhs(sol(1)[2]);

$$ypred := 0.9685266966660516496$$

> y2pred:=rhs(sol(1)[4]);

$$y2pred := 0.230740415090328915$$

The new value of α is obtained as:

> alpha1:=alpha0+(1-ypred)/y2pred;

$$\alpha 1 := 1.036401346$$

For this example, the error is calculated based on the boundary condition at x = 1.

> err:=abs(1-ypred);

$$err := 0.0314733033$$

> alpha0:=alpha1;

$$\alpha 0 := 1.036401346$$

> k:=1;

$$k := 1$$

The iteration is performed until the error becomes less than the tolerance limit 1e - 10.

> tol:=1e-10;

$$tol := 1.10^{-10}$$

```
> while err> tol do
> ICs:=(y(0)=alpha0,D(y)(0)=0,y2(0)=1,D(y2)(0)=0);
> sol:=dsolve({eqs,ICs},{vars},type=numeric);
> ypred:=rhs(sol(1)[2]);
> y2pred:=rhs(sol(1)[4]);
> alpha1:=alpha0+(1-ypred)/y2pred;
> err:=abs(1-ypred);
> alpha0:=alpha1;k:=k+1;
> end:
> k;
```

$$6$$

The problem has converged after six iterations. The concentration at the center of the particle $(x = 0)$ (x = 0) is given by:

> alpha1;

$$0.9717223372$$

3.2 Nonlinear Boundary Value Problems

The error obtained is:

> err;

$$0.$$

Next, the solution obtained is plotted and stored in p1.

> p1:=odeplot(sol,[x,y(x)],0..1,axes=boxed,thickness=3,color=blue):

The same steps are performed for a different initial guess of 0.5. The solution obtained is stored in p2.

> alpha0:=0.5;

$$\alpha 0 := 0.5$$

> ICs:=(y(0)=alpha0,D(y)(0)=0,y2(0)=1,D(y2)(0)=0);

$$ICs := y(0) = 0.5, D(y)(0) = 0, y2(0) = 1, D(y2)(0) = 0$$

> sol:=dsolve({eqs,ICs},{vars},type=numeric,abserr=1e-10);

$$sol := \mathbf{proc}(x\_rkf45) \ldots \mathbf{end\ proc}$$

> sol(1);

$$\left[x = 1., y(x) = 1.58640607813000578, \frac{d}{dx}y(x) = 1.25197150063613450, y2(x) = -3.60417190050730118, \frac{d}{dx}y2(x) = -4.84304330565780372 \right]$$

> ypred:=rhs(sol(1)[2]);

$$ypred := 1.58640607813000578$$

> y2pred:=rhs(sol(1)[4]);

$$y2pred := -3.60417190050730118$$

> alpha1:=alpha0+(1-ypred)/y2pred;

$$\alpha 1 := 0.6627020281$$

> err:=abs(1-ypred);

$$err := 0.586406078$$

> alpha0:=alpha1;

$$\alpha 0 := 0.6627020281$$

> k:=1;

$$k := 1$$

```
> while err> tol do
> ICs:=(y(0)=alpha0,D(y)(0)=0,y2(0)=1,D(y2)(0)=0);
> sol:=dsolve({eqs,ICs},{vars},type=numeric);
> ypred:=rhs(sol(1)[2]);
> y2pred:=rhs(sol(1)[4]);
> alpha1:=alpha0+(1-ypred)/y2pred;
> err:=abs(1-ypred);
> alpha0:=alpha1;k:=k+1;
> end:
> k;
```

$$8$$

The problem has converged after eight iterations. The concentration at the center of the particle (x = 0) is given by:

```
> alpha1;
```

$$0.7733977427$$

```
> err;
```

$$0.$$

```
> p2:=odeplot(sol,[x,y(x)],0..1,axes=boxed,thickness=3,color=green):
```

Next, an initial guess of 1e - 4 is used. For this case the updated α becomes a negative. Hence, a scaling factor of $\rho=0.2$ is used:

```
> alpha0:=1e-4;
```

$$\alpha 0 := 0.0001$$

```
> ICs:=(y(0)=alpha0,D(y)(0)=0,y2(0)=1,D(y2)(0)=0);
```

$$ICs := y(0) = 0.0001, D(y)(0) = 0, y2(0) = 1, D(y2)(0) = 0$$

```
> sol:=dsolve({eqs,ICs},{vars},type=numeric,abserr=1e-10);
```

$$sol := \mathbf{proc}(x\_rkf45) \ \dots \ \mathbf{end\ proc}$$

```
> sol(1);
```

$$\left[x = 1., y(x) = 2.05172720318686918, \frac{d}{dx} y(x) = 3.91761302634986297, y2(x) = 2299.96672316761942, \frac{d}{dx} y2(x) = -0.00636497432250418376 \right]$$

```
> ypred:=rhs(sol(1)[2]);
```

$$ypred := 2.05172720318686918$$

```
> y2pred:=rhs(sol(1)[4]);
```

$$y2pred := 2299.96672316761942$$

3.2 Nonlinear Boundary Value Problems 243

> alpha1:=alpha0+(1-ypred)/y2pred;

$$\alpha 1 := -0.0003572793130$$

> rho:=0.2;

$$\rho := 0.2$$

> alpha1:=alpha0+rho*(1-ypred)/y2pred;

$$\alpha 1 := 0.00000854413740$$

> err:=abs(1-ypred);

$$err := 1.051727203$$

> alpha0:=alpha1;

$$\alpha 0 := 0.00000854413740$$

> k:=1;

$$k := 1$$

> while err> tol do
> ICs:=(y(0)=alpha0,D(y)(0)=0,y2(0)=1,D(y2)(0)=0);
> sol:=dsolve({eqs,ICs},{vars},type=numeric);
> ypred:=rhs(sol(1)[2]);
> y2pred:=rhs(sol(1)[4]);
> alpha1:=alpha0+rho*(1-ypred)/y2pred;
> err:=abs(1-ypred);
> alpha0:=alpha1;k:=k+1;
> end:

The problem has converged after 93 iterations. The concentration at the center particle (x = 0) is given by:

> k;

$$93$$

> alpha1;

$$0.000001033486796$$

> err;

$$0.$$

> p3:=odeplot(sol,[x,y(x)],0..1,title="Figure 3.23",axes=boxed,thickness=4, color=brown):
> pt:=textplot([[0.2,0.95,"Steady State 1"],[0.2,0.75,"Steady State 2"], [0.2,0.02,"Steady State 3"]]);

$$pt := PLOT(\ldots)$$

> display({p1,p2,p3,pt});

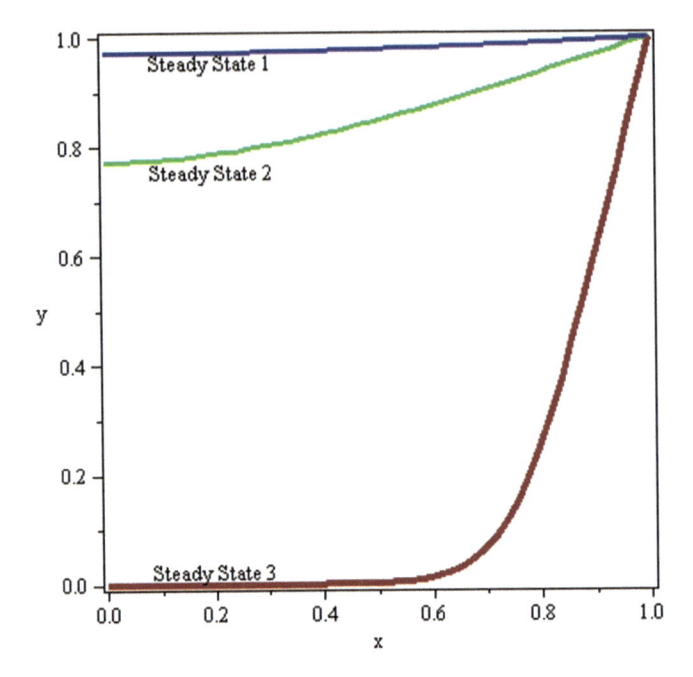

Fig. 3.23

Hence, we observe that the shooting technique can predict three multiple states in a catalyst pellet. The number of iterations required to obtain a converged solution depends on the initial guess and the scaling factor ρ.

3.2.5 *Numerical Solution for Boundary Value Problems Using Maple's 'dsolve' Command*

The series solution technique in section 3.2.2 may not produce converging series solutions for all of the boundary value problems. The finite difference technique discussed earlier in section 3.2.3 might be computationally intensive as the number of node points increase and an approximate initial guess has to be provided. The shooting technique described in section 3.2.4 is very robust, but involves more computational effort. In addition, one has to solve an additional number of differential equation and the solution might take large number of iterations to converge. Conveniently, boundary value problems can be solved numerically using Maple's 'dsolve' command. By default Maple uses the finite difference technique coupled with the Richardson interpolation technique. The syntax is:

'dsolve'({"differential equations, boundary conditions"},{"dependent variables"}, type=numeric).

3.2 Nonlinear Boundary Value Problems

The differential equation entered can be of any order. For a differential equation of order N, N boundary conditions have to be specified. The numerical solution can be stored in a variable and can be used later for plotting purposes as shown in the following examples.

Example 3.2.6. Diffusion with Second Order Reaction

Example 3.2.1 is solved here using Maple's 'dsolve' command. The boundary condition at the surface, $x = 1$ is taken as:

$$\frac{dc}{dx}(1) = 100(1 - c(1)) \tag{3.47}$$

Example 3.2.1 is solved below in Maple below with the modified boundary condition:

> restart:
> with(plots):

Enter the governing equation:

> Eq:=diff(c(x),x$2)=Phi^2*c(x)^2;

$$Eq := \frac{d^2}{dx^2} c(x) = \Phi^2 c(x)^2$$

The value of the parameter is substituted here:

> eq:=subs(Phi=1,Eq);

$$eq := \frac{d^2}{dx^2} c(x) = c(x)^2$$

The boundary conditions are entered here:

> BCs:=(D(c)(0),D(c)(1)=100*(1-c(1)));

$$BCs := D(c)(0), D(c)(1) = 100 - 100\, c(1)$$

The numerical solution is obtained here:

> sol:=dsolve({eq,BCs},{c(x)},numeric);

$$sol := \mathbf{proc}(x\_bvp) \ \ldots \ \mathbf{end\ proc}$$

The concentration profile obtained is plotted here:

>odeplot(sol,[x,c(x)],0..1,thickness=4,title="Figure 3.24",axes=boxed, color=gold);

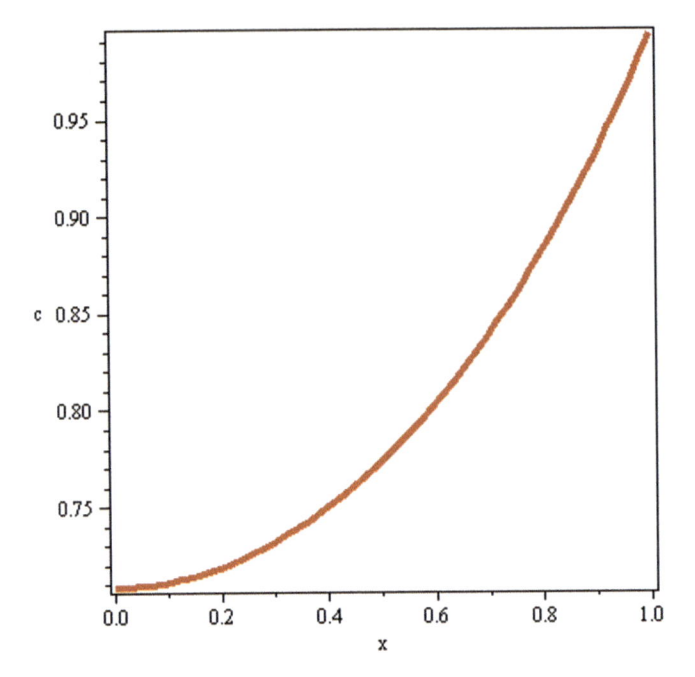

Fig. 3.24

Next, the problem is solved for a higher value of Φ:

> eq:=subs(Phi=10,Eq);

$$eq := \frac{d^2}{dx^2} c(x) = 100\, c(x)^2$$

> BCs:=(D(c)(0),D(c)(1)=100*(1-c(1)));

$$BCs := \mathrm{D}(c)(0),\ \mathrm{D}(c)(1) = 100 - 100\, c(1)$$

> sol:=dsolve({eq,BCs},{c(x)},numeric);

$$sol := \mathbf{proc}(x\_bvp)\ \ldots\ \mathbf{end\ proc}$$

>odeplot(sol,[x,c(x)],0..1,thickness=4,title="Figure 3.25",axes=boxed, color=brown);

3.2 Nonlinear Boundary Value Problems

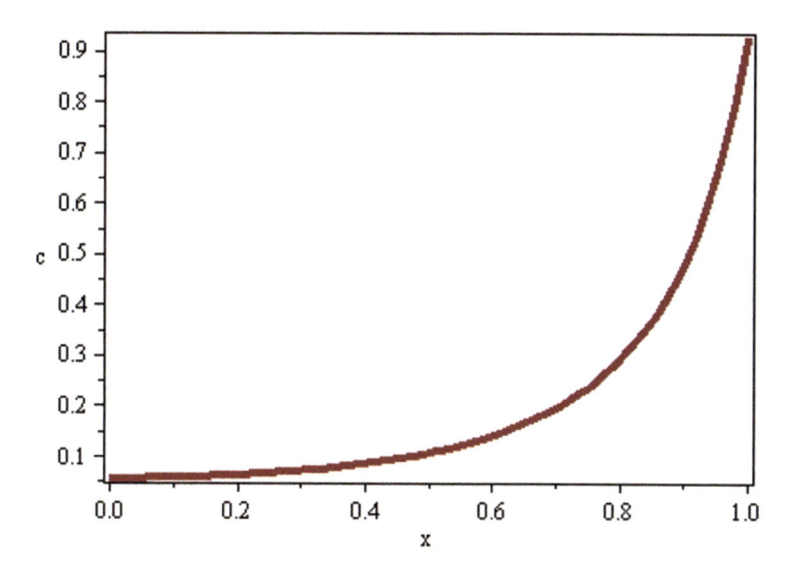

Fig. 3.25

We observe that as Φ increases, the profile becomes steeper and the time taken to solve the problem also increases.

Example 3.2.7. Heat Transfer with Nonlinear Radiation Boundary Conditions

Example 3.2.4 is solved here using Maple's 'dsolve' command. The boundary condition at the surface, $x = 1$ is taken as:

$$\frac{dy}{dx}(1) = 100\left(1 - y(1)^4\right) \tag{3.48}$$

This example is solved in Maple below:

```
> restart:
> with(plots):
> eq:=diff(y(x),x$2)-(1+0.1*y(x))*y(x);
```

$$eq := \frac{d^2}{dx^2} y(x) - (1 + 0.1\, y(x))\, y(x)$$

```
> BCs:=(D(y)(0),D(y)(1)=1-y(1)^4);
```

$$BCs := D(y)(0), D(y)(1) = 1 - y(1)^4$$

> sol:=dsolve({eq,BCs},{y(x)},type=numeric);

Error, (in dsolve/numeric/bvp) unable to store 'Limit(0.+1.0000000000000*I, x = 0., left)' when datatype=float[8]

Maple is not able to solve this problem directly. An approximate solution can be provided to arrive at the exact solution. The approximate solution can be found using the linear boundary condition at $x = 1$. Note that the approximate solution has to be evaluated for at least eight node points.

> sola:=dsolve({eq,D(y)(0),D(y)(1)=1-y(1)},{y(x)},type=numeric, output=array([seq(i/7.,i=0..7)]));

$$sola := \begin{bmatrix} \left[x \quad y(x) \quad \frac{d}{dx}y(x) \right] \\ \begin{bmatrix} 0. & 0.357243127888639434 & 0. \\ 0.1428571429 & 0.361025571217616914 & 0.0530507700478474592 \\ 0.2857142858 & 0.372455814437752220 & 0.107264903315163834 \\ 0.4285714287 & 0.391784785487688747 & 0.163836342127088419 \\ 0.5714285716 & 0.419438037073692316 & 0.224021372243635663 \\ 0.7142857145 & 0.456027068795170632 & 0.289172960863057104 \\ 0.8571428574 & 0.502365817722847607 & 0.360779106202013922 \\ 1.000000000 & 0.559493018589254932 & 0.440506981410744569 \end{bmatrix} \end{bmatrix}$$

The numerical solution of the original boundary value problem is found as:

> sol:=dsolve({eq,BCs},{y(x)},type=numeric,approxsoln=sola);

$$sol := \mathbf{proc}(x\_bvp) \ \ldots \ \mathbf{end\ proc}$$

The derivative of the dependent variable can be plotted as:

> odeplot(sol,[x,y(x)],0..1,thickness=4,color=red,title="Figure 3.26", axes=boxed);

3.2 Nonlinear Boundary Value Problems 249

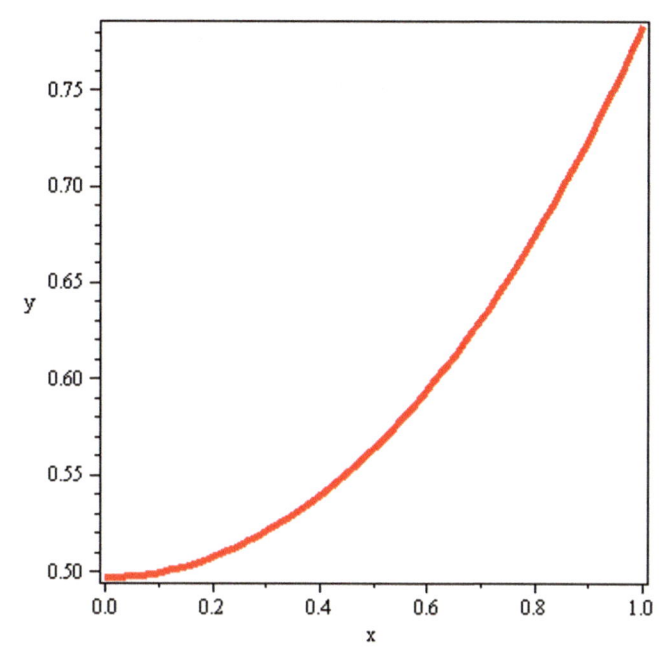

Fig. 3.26

> odeplot(sol,[x,diff(y(x),x)],0..1,color=blue,title="Figure 3.27",thickness=4, axes=boxed);

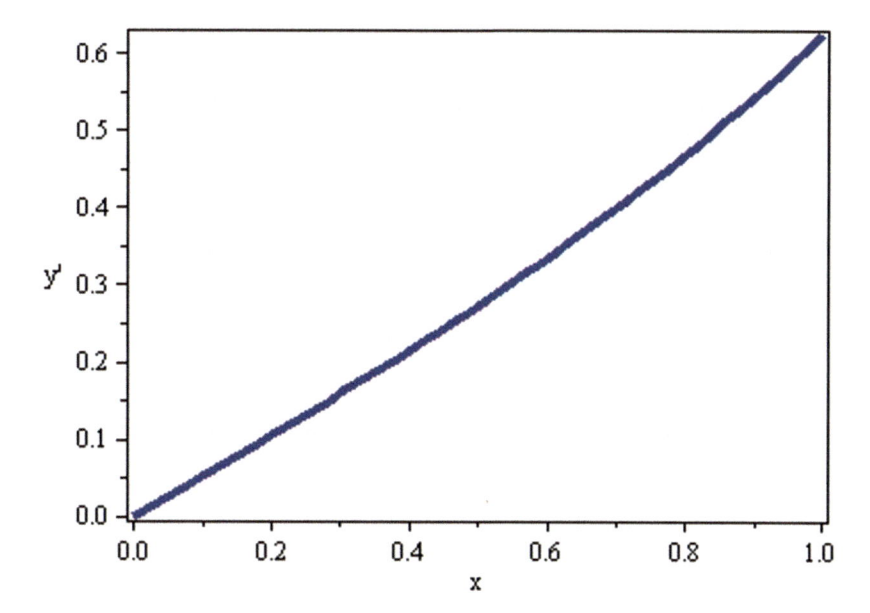

Fig. 3.27

The functions of the dependent variables can be plotted as:

```
>odeplot(sol,[x,1-y(x)^4],0..1,thickness=4,color=brown,title="Figure 3.28",
axes=boxed);
```

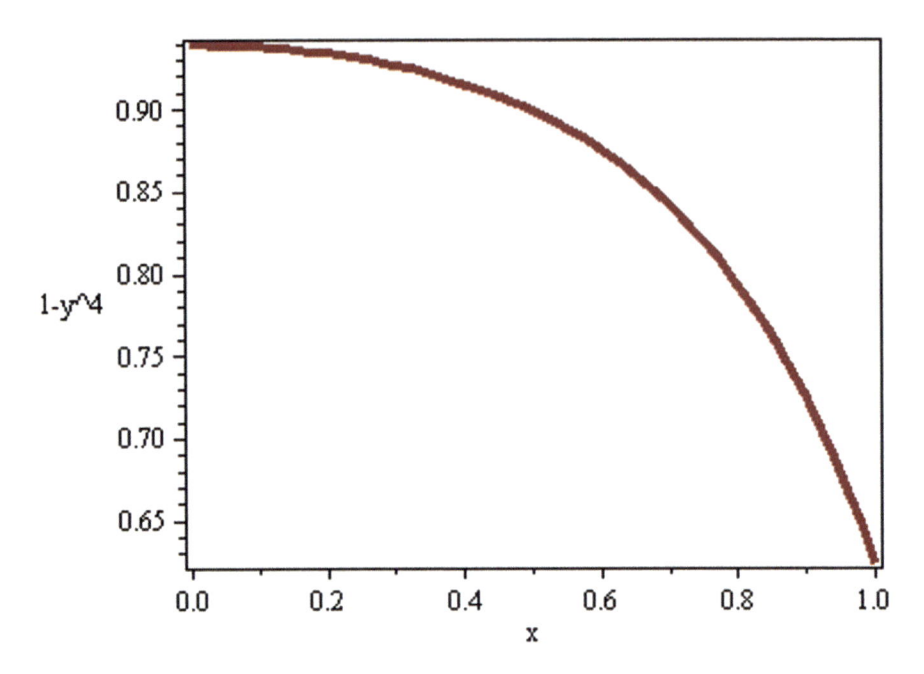

Fig. 3.28

Example 3.2.8. Diffusion of a Substrate in an Enzyme Catalyzed Reaction – BVPs with Removable Singularity

Boundary value problems in cylindrical and spherical coordinates have an inherent singularity at $x = 0$. These problems can be tackled using Maple's inbuilt midpoint methods. For example, diffusion of a substrate in an enzyme catalyzed reaction.[6] The governing equation for the dimensionless concentration is

$$\frac{1}{x^2}\frac{d}{dx}\left(x^2 f(y)\frac{dy}{dx}\right) = 10\frac{y}{y+0.1} \tag{3.49}$$

where f(y) is a dimensionless function which describes the change of diffusion coefficient as a function of concentration:

$$f(y) = 1 + \frac{\lambda}{(y+0.01)^2} \tag{3.50}$$

3.2 Nonlinear Boundary Value Problems

The boundary conditions are

$$\frac{dy}{dx}(0) = 0 \tag{3.51}$$

and

$$y(1) = 1 \tag{3.52}$$

This boundary value problem is solved in Maple below for different values of λ:

```
> restart:
> with(plots):
> f(y):=1+lambda/(y(x)+0.01)^2;
```

$$f(y) := 1 + \frac{\lambda}{(y(x) + 0.01)^2}$$

```
> Eq:=1/x^2*diff((x^2*f(y)*diff(y(x),x)),x)=10*y(x)/(y(x)+0.1);
```

$$Eq := \frac{2x\left(1 + \frac{\lambda}{(y(x)+0.01)^2}\right)\left(\frac{d}{dx}y(x)\right) - \frac{2x^2\lambda\left(\frac{d}{dx}y(x)\right)^2}{(y(x)+0.01)^3} + x^2\left(1 + \frac{\lambda}{(y(x)+0.01)^2}\right)\left(\frac{d^2}{dx^2}y(x)\right)}{x^2}$$

$$= \frac{10\,y(x)}{y(x) + 0.1}$$

```
> eq:=expand(subs(lambda=0,Eq));
```

$$eq := \frac{2\left(\frac{d}{dx}y(x)\right)}{x} + \frac{d^2}{dx^2}y(x) = \frac{10\,y(x)}{y(x) + 0.1}$$

```
> BCs:=(D(y)(0),y(1)-1);
```

$$BCs := D(y)(0), y(1) - 1$$

```
> sol:=dsolve({eq,BCs},{y(x)},type=numeric);
```

Error, (in dsolve/numeric/bvp) system is singular at left endpoint, use midpoint method instead

Maple identifies the singularity at x = 0 and suggests the midpoint method:

```
> sol:=dsolve({eq,BCs},{y(x)},type=numeric,method=bvp[midrich]);
```

$$sol := \mathbf{proc}(x\_bvp) \ \ldots \ \mathbf{end\ proc}$$

252   3 Boundary Value Problems

```
> odeplot(sol,[x,y(x)],0..1,thickness=4,title="Figure 3.29",axes=boxed);
```

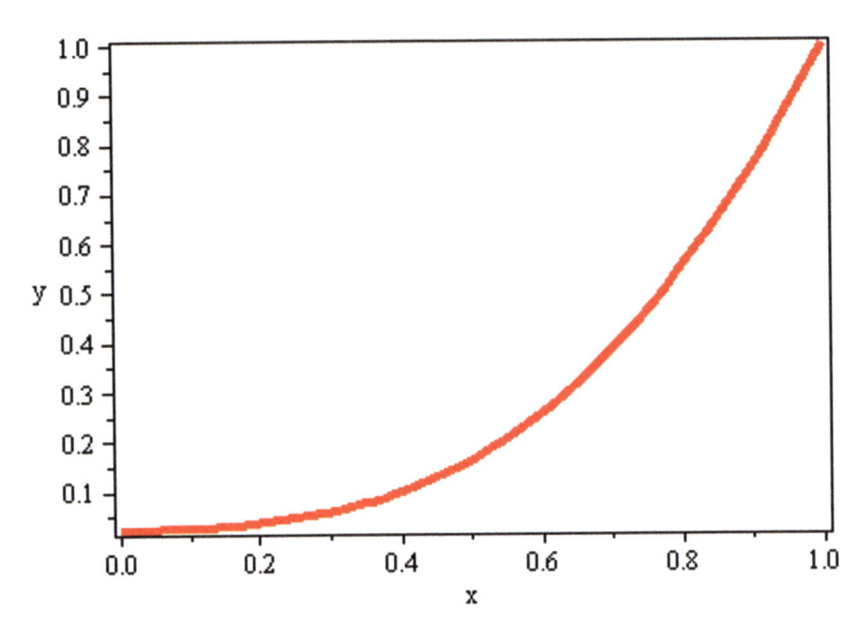

Fig. 3.29

```
> sol(0);
```

$$\left[x = 0., y(x) = 0.02279134569600320 40, \frac{d}{dx} y(x) = 0. \right]$$

```
> L:=[0,1e-2,1e-1,1,10];
```

$$L := [0, 0.01, 0.1, 1, 10]$$

```
> MM:=nops(L);
```

$$MM := 5$$

```
> clr:=[red,green,gold,blue,magenta];
```

$$clr := [red, green, gold, blue, magenta]$$

```
> for i to MM do
> eq:=expand(subs(lambda=L[i],Eq));
> sol[i]:=dsolve({eq,BCs},{y(x)},type=numeric,method=bvp[midrich]);
> p[i]:=odeplot(sol[i],[x,y(x)],0..1,thickness=4,color=clr[i]);
> od:
> pt:=textplot([seq([0.2,rhs(sol[i](0.2)[2])+0.02,lambda=L[i]],i=1..MM)]):
> display({seq(p[i],i=1..MM),pt},title="Figure 3.30",axes=boxed);
```

3.2 Nonlinear Boundary Value Problems   253

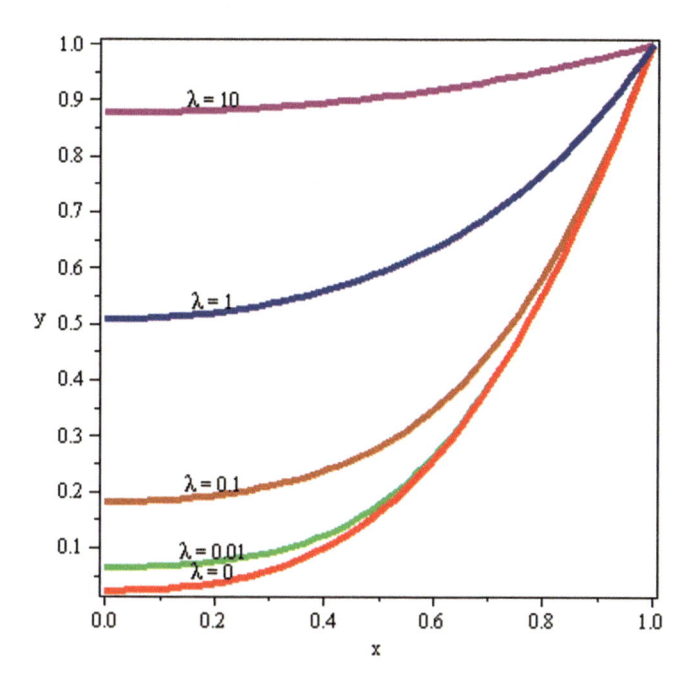

Fig. 3.30

Example 3.2.9. Multiple Steady States in a Catalyst Pellet

The catalyst pellet (example 3.2.2) is solved below using Maple's 'dsolve' command. The reaction order is taken to be second order. The governing equation becomes

$$\frac{d^2y}{dx^2} = \Phi^2 y^2 \exp\left(\gamma\beta \frac{1-y}{1+\beta(1-y)}\right) \tag{3.53}$$

The boundary conditions are the same as that of example 3.2.2. This problem is solved in Maple below:

```
> restart:
> with(plots):
> Eq:=diff(y(x),x$2)=Phi^2*y(x)^2*exp(gamma*beta*(1-y(x))/(1+beta*(1-y(x))));
```

$$Eq := \frac{d^2}{dx^2} y(x) = \Phi^2 y(x)^2 e^{\frac{\gamma\beta(1-y(x))}{1+\beta(1-y(x))}}$$

```
> eq:=subs(Phi=0.2,gamma=20,beta=0.8,Eq);
```

$$eq := \frac{d^2}{dx^2} y(x) = 0.04 y(x)^2 e^{\frac{16.0(1-y(x))}{1.8-0.8 y(x)}}$$

> BCs:=(D(y)(0),y(1)-1);

$$BCs := D(y)(0), y(1) - 1$$

> sol:=dsolve({eq,BCs},{y(x)},type=numeric);

$$sol := \mathbf{proc}(x\_bvp) \ \dots \ \mathbf{end \ proc}$$

The solution at x = 0 and 1 are found as:

> sol(0);

$$\left[x = 0., y(x) = 0.972686415959039264, \frac{d}{dx} y(x) = 0. \right]$$

> sol(1);

$$\left[x = 1., y(x) = 0.99999999999999988, \frac{d}{dx} y(x) = 0.0514954717889864398 \right]$$

> odeplot(sol,[x,y(x)],0..1,thickness=4,title="Figure 3.31",axes=boxed);

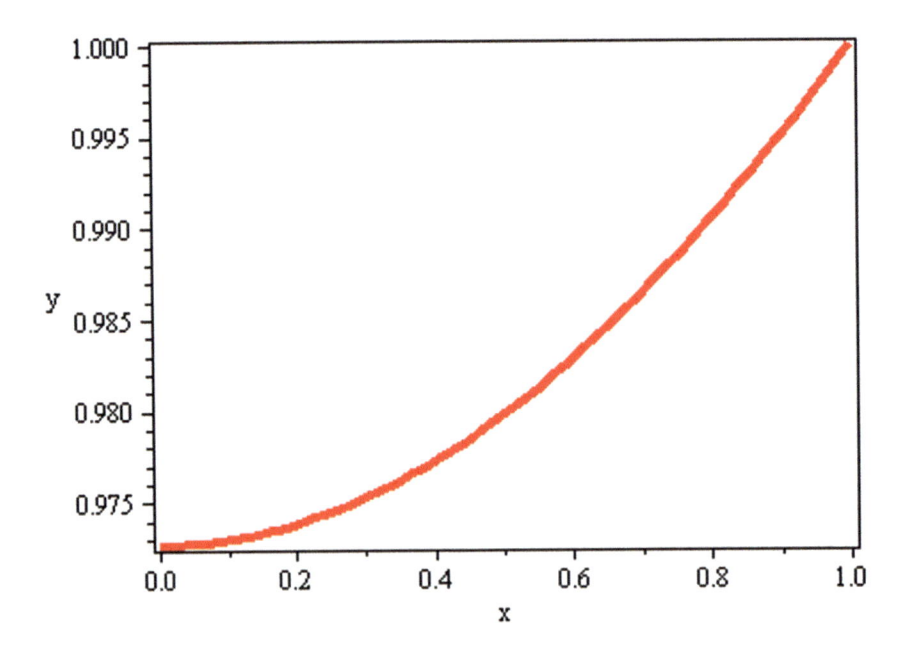

Fig. 3.31

By default Maple picks up the higher solution. The other two solutions are found by giving an initial guess. For this problem, the approximation values for y(0) are provided.

3.2 Nonlinear Boundary Value Problems

```
> p[1]:=odeplot(sol,[x,y(x)],0..1,thickness=4,color=red,axes=boxed):
> sola:=dsolve({eq,y(0)-0.7,y(1)-1},{y(x)},type=numeric,
output=array([seq(i/7.,i=0..7)]));
```

$$sola := \begin{bmatrix} \begin{bmatrix} x & y(x) & \frac{d}{dx}y(x) \end{bmatrix} \\ \begin{bmatrix} 0. & 0.699999999999999844 & -0.0257703434477653330 \\ 0.1428571429 & 0.705889467981446028 & 0.107235207080385794 \\ 0.2857142858 & 0.729936686386751910 & 0.225710290143314146 \\ 0.4285714287 & 0.769083185410072922 & 0.317328211951562800 \\ 0.5714285716 & 0.819169236540999934 & 0.379125937520926492 \\ 0.7142857145 & 0.876201812090491750 & 0.415757898501506462 \\ 0.8571428574 & 0.937130044247870386 & 0.434988921426216868 \\ 1.000000000 & 0.99999999999999978 & 0.443970056837609761 \end{bmatrix} \end{bmatrix}$$

```
> sol:=dsolve({eq,BCs},{y(x)},type=numeric,approxsoln=sola);
```

$$sol := \mathbf{proc}(x\_bvp) \ \ldots \ \mathbf{end \ proc}$$

```
> sol(0);
```

$$\left[x = 0., y(x) = 0.714650189079288346 \frac{d}{dx}y(x) = 0. \right]$$

```
> sol(1);
```

$$\left[x = 1., y(x) = 0.99999999999999966, \frac{d}{dx}y(x) = 2.04169689675851274 \right]$$

```
> p[2]:=odeplot(sol,[x,y(x)],0..1,thickness=4,color=blue,axes=boxed):
> sola:=dsolve({eq,y(0)-0.1,y(1)-1},{y(x)},type=numeric,
output=array([seq(i/7.,i=0..7)]));
```

$$sola := \begin{bmatrix} \begin{bmatrix} x & y(x) & \frac{d}{dx}y(x) \end{bmatrix} \\ \begin{bmatrix} 0. & 0.100000000000000004 & -.220113888220456938 \\ 0.1428571429 & 0.0842188120615140618 & -0.0101159634667672797 \\ 0.2857142858 & 0.0968593149070104270 & 0.194507008143603056 \\ 0.4285714287 & 0.144661678157153028 & 0.502748739515851195 \\ 0.5714285716 & 0.251406671327145725 & 1.03366444125639223 \\ 0.7142857145 & 0.446062034108055616 & 1.67179734192129702 \\ 0.8571428574 & 0.712177836637722272 & 1.98378222272051464 \\ 1.000000000 & 0.99999999999999956 & 2.02730348319667098 \end{bmatrix} \end{bmatrix}$$

> sol:=dsolve({eq,BCs},{y(x)},type=numeric,approxsoln=sola);

$$sol := \mathbf{proc}(x\_bvp) \ \ldots \ \mathbf{end\ proc}$$

> sol(0);

$$\left[x = 0., y(x) = 0.0543966732250433794, \frac{d}{dx} y(x) = 0.\right]$$

> sol(1);

$$\left[x = 1., y(x) = 0.99999999999999966, \frac{d}{dx} y(x) = 2.04169689675851274\right]$$

p[3]:=odeplot(sol,[x,y(x)],0..1,thickness=4,color=brown,title="Figure 3.32", axes=boxed):
> pt:=textplot([0.5,0.5,[Phi=0.2,gamma=20,beta=0.8]]):
> display({seq(p[i],i=1..3),pt});

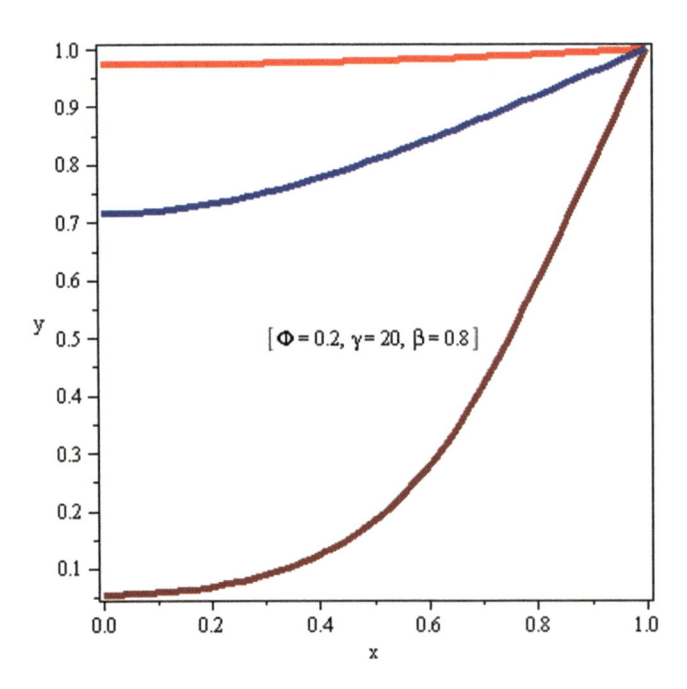

Fig. 3.32

Hence, all of the multiple states can be predicted using Maple. The solution differs slightly from the first order reaction discussed earlier.

Example 3.2.10. Blasius Equation – Infinite Domains

The Blasius problem is defined by:[22] [12]

3.2 Nonlinear Boundary Value Problems

$$f \frac{d^2 f}{d\eta^2} + 2 \frac{d^3 f}{d\eta^3} = 0 \qquad (3.54)$$

with the boundary conditions

$$f(0) = 0 \qquad (3.55)$$

$$\frac{df}{d\eta}(0) = 0 \qquad (3.56)$$

and

$$\frac{df}{d\eta}(\infty) = 1 \qquad (3.57)$$

This boundary is different from other boundary value problems discussed in this chapter because equation (3.54) is a third order ordinary differential equation with three boundary conditions. In addition, the domain is semi-infinite. This boundary value problem is solved in Maple below by replacing ∞ in equation (3.57) by 10.

```
> restart:
> with(plots):
> eq:=f(eta)*diff(f(eta),eta$2)+2*diff(f(eta),eta$3);
```

$$eq := f(\eta) \left(\frac{d^2}{d\eta^2} f(\eta) \right) + 2 \left(\frac{d^3}{d\eta^3} f(\eta) \right)$$

```
> BCs:=(f(0),D(f)(0),D(f)(10)-1);
```

$$BCs := f(0), D(f)(0), D(f)(10) - 1$$

```
> sol:=dsolve({eq,BCs},{f(eta)},type=numeric);
```

$$sol := \mathbf{proc}(x\_bvp) \ \ldots \ \mathbf{end\ proc}$$

The solution for $\eta = 0$ and $\eta = 10$ are obtained as:

```
> sol(0);
```

$$\left[\eta = 0., f(\eta) = 0., \frac{d}{d\eta} f(\eta) = 0., \frac{d^2}{d\eta^2} f(\eta) = 0.332057384255588839 \right]$$

```
> sol(10);
```

$$\left[\eta = 10., f(\eta) = 8.279212552310099274, \frac{d}{d\eta} f(\eta) = 1., \frac{d^2}{d\eta^2} f(\eta) = 8.02490660017787434 \, 10^{-9} \right]$$

The solution obtained is plotted below:

```
> odeplot(sol,[eta,f(eta)],0..10,thickness=4,color=red,title="Figure 3.33",
  axes=boxed);
```

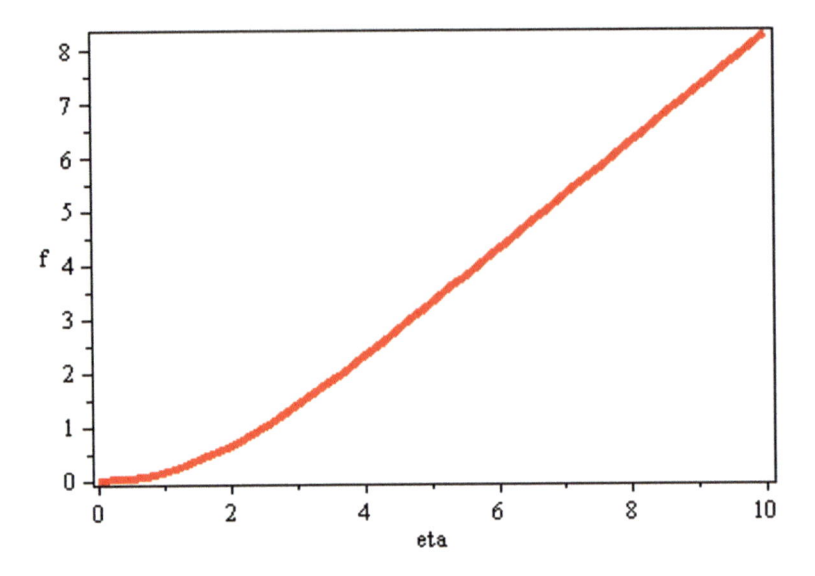

Fig. 3.33

> odeplot(sol,[eta,diff(f(eta),eta)],0..10,thickness=4,color=green,
title="Figure 3.34",axes=boxed);

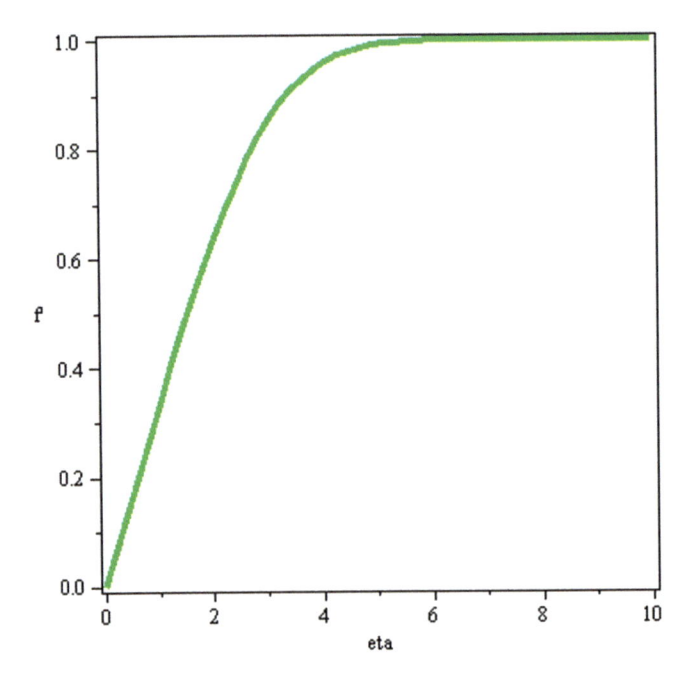

Fig. 3.34

3.2 Nonlinear Boundary Value Problems

3.2.6 Numerical Solution for Coupled BVPs Using Maple's 'dsolve' Command

Simultaneous reactions, mass/momentum transfer, heat/mass/momentum transfer are often represented by coupled boundary value problems. Coupled boundary value problems can be conveniently solved numerically using Maple's 'dsolve' command. The syntax is:

'dsolve'({"differential equations, boundary conditions"},{"dependent variables"}, type=numeric).

Example 3.2.11. Axial Conduction and Diffusion in a Tubular Reactor

Axial diffusion and conduction in an adiabatic tubular reactor can be described by: [6]

$$\frac{1}{Pe}\frac{d^2y}{dx^2} - \frac{dy}{dx} = 4y \exp\left(E\left(1-\frac{1}{\theta}\right)\right) \quad (3.58)$$

$$\frac{1}{Bo}\frac{d^2\theta}{dx^2} - \frac{d\theta}{dx} = -4\beta y \exp\left(E\left(1-\frac{1}{\theta}\right)\right) \quad (3.59)$$

with the boundary conditions

$$\frac{1}{Pe}\frac{dy}{dx}(0) = y(0) - 1 \quad (3.60)$$

$$\frac{dy}{dx}(1) = 0 \quad (3.61)$$

$$\frac{1}{Bo}\frac{d\theta}{dx}(0) = \theta(0) - 1 \quad (3.62)$$

and

$$\frac{d\theta}{dx}(1) = 0 \quad (3.63)$$

This coupled boundary value problem is solved in Maple below for the following value of parameters Pe = 10, B0 = 10, E= 18 and β = 0.05.

> with(plots):
> Eq[1]:=1/Pe*diff(y(x),x$2)-diff(y(x),x)=4*y(x)*exp(E*(1-1/theta(x)));

$$Eq_1 := \frac{\frac{d^2}{dx^2}y(x)}{Pe} - \left(\frac{d}{dx}y(x)\right) = 4y(x) e^{E\left(1-\frac{1}{\theta(x)}\right)}$$

> Eq[2]:=1/Bo*diff(theta(x),x$2)-diff(theta(x),x)=-4*beta*
  y(x)*exp(E*(1-1/theta(x)));

$$Eq_2 := \frac{\frac{d^2}{dx^2}\theta(x)}{Bo} - \left(\frac{d}{dx}\theta(x)\right) = -4\beta y(x) e^{E\left(1 - \frac{1}{\theta(x)}\right)}$$

> BCs:=[1/Pe*D(y)(0)-y(0)+1,D(y)(1),1/Bo*D(theta)(0)-theta(0)+1,D(theta)(1)];

$$BCs := \left[\frac{D(y)(0)}{Pe} - y(0) + 1, D(y)(1), \frac{D(\theta)(0)}{Bo} - \theta(0) + 1, D(\theta)(1)\right]$$

> pars:={Pe=10,Bo=10,E=18,beta=0.05};

$$pars := \{Bo = 10, E = 18, \beta = 0.05, Pe = 10\}$$

> for i to 2 do eq[i]:=subs(pars,Eq[i]);od;

$$eq_1 := \frac{1}{10}\frac{d^2}{dx^2}y(x) - \left(\frac{d}{dx}y(x)\right) = 4y(x) e^{18 - \frac{18}{\theta(x)}}$$

$$eq_2 := \frac{1}{10}\frac{d^2}{dx^2}\theta(x) - \left(\frac{d}{dx}\theta(x)\right) = -0.20 y(x) e^{18 - \frac{18}{\theta(x)}}$$

> eqs:=(eq[1],eq[2]);

$$eqs := \frac{1}{10}\frac{d^2}{dx^2}y(x) - \left(\frac{d}{dx}y(x)\right) = 4y(x) e^{18 - \frac{18}{\theta(x)}}, \frac{1}{10}\frac{d^2}{dx^2}\theta(x) - \left(\frac{d}{dx}\theta(x)\right) = -0.20 y(x) e^{18 - \frac{18}{\theta(x)}}$$

> vars:=(y(x),theta(x));

$$vars := y(x), \theta(x)$$

> bcs:=op(subs(pars,BCs));

$$bcs := \frac{1}{10}D(y)(0) - y(0) + 1, D(y)(1), \frac{1}{10}D(\theta)(0) - \theta(0) + 1, D(\theta)(1)$$

> sol:=dsolve({eqs,bcs},{vars},type=numeric);

$$sol := \mathbf{proc}(x\_bvp) \ \ldots \ \mathbf{end\ proc}$$

> sol(0);

$$\left[x = 0., \theta(x) = 1.01463595348572144, \frac{d}{dx}\theta(x) = 0.146359534857216172, y(x) = 0.707280930285567600, \frac{d}{dx}y(x)\right.$$
$$\left.= -2.92719069714432400\right]$$

> sol(1);

$$\left[x = 1., \theta(x) = 1.04981039182617586, \frac{d}{dx}\theta(x) = 0., y(x) = 0.00379216347648523786, \frac{d}{dx}y(x) = 0.\right]$$

3.2 Nonlinear Boundary Value Problems 261

> odeplot(sol,[x,y(x)],0..1,thickness=4,title="Figure 3.35",axes=boxed);

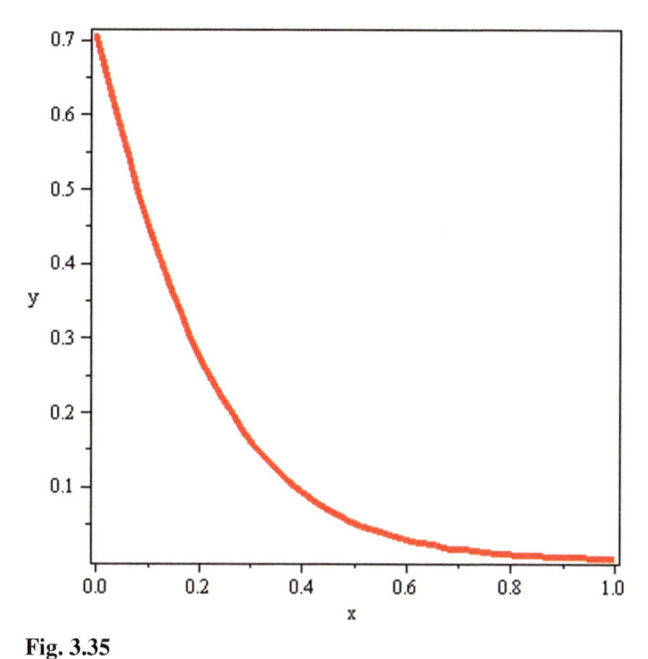

Fig. 3.35

> odeplot(sol,[x,theta(x)],0..1,thickness=4,color=green,title="Figure 3.36", axes=boxed);

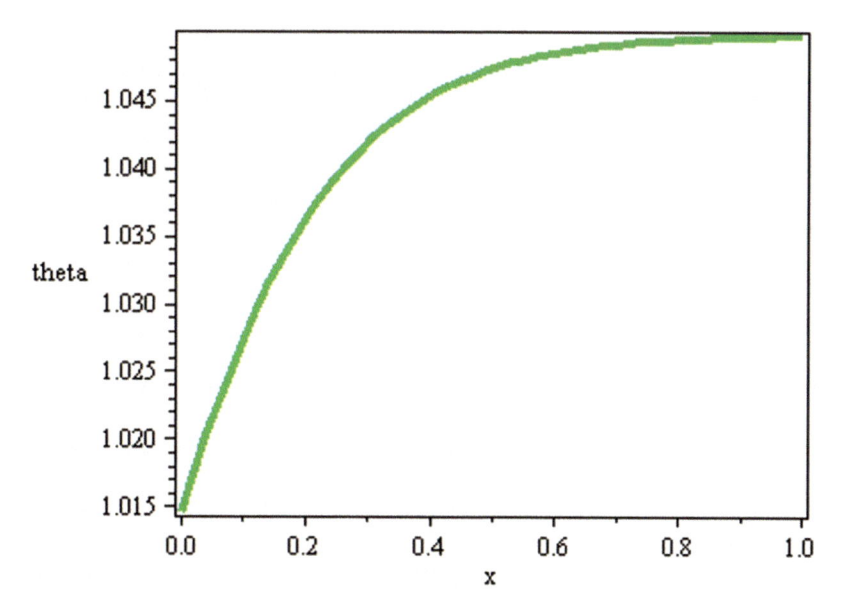

Fig. 3.36

3.2.7 Solving Boundary Value Problems and Initial Value Problems

Often boundary value problems can be cast into initial value problems for obtaining performance curves. This is best illustrated by the next example.

Example 3.2.12. Diffusion with a Second Order Reaction

Diffusion with a second order reaction (example 3.2.1) is considered here again. The dimensionless concentration is governed by:

$$\frac{d^2y}{dx^2} = \Phi^2 y^2 \qquad (3.64)$$

with the boundary conditions:

$$\frac{dy}{dx}(0) = 0 \qquad (3.65)$$

and

$$y(1) = 1 \qquad (3.66)$$

The effectiveness factor of the pellet is given by

$$\eta = \frac{1}{\Phi^2} \frac{dy}{dx}(1) \qquad (3.67)$$

For a given value of Φ, equation (3.64) can be solved for the boundary conditions ((3.65) and (3.66)). Once the numerical solution is obtained the effectiveness factor can be calculated using equation (3.67). It is of interest to plot the effectiveness factor as a function of Φ. For this purpose, one can solve the boundary value problem for different values of Φ and predict the effectiveness factor. Alternatively, this problem can be solved as an initial value problem. For convenience, the transformation $X = \Phi x$ is introduced. The BVP changes as:

$$\frac{d^2y}{dX^2} = y^2 \qquad (3.68)$$

with the boundary conditions:

$$\frac{dy}{dX}(0) = 0 \qquad (3.69)$$

and

$$y(\Phi) = 1 \qquad (3.70)$$

3.2 Nonlinear Boundary Value Problems

The effectiveness factor of the pellet is given by

$$\eta = \frac{1}{\Phi} \frac{dy}{dX}\bigg|_{X=\Phi} \tag{3.71}$$

We know that for low values of Φ, the concentration at the center y(0) is close to 1. For high values of Φ, y(0) approaches zero. We take the concentration at the center as

$$y(0) = y0 \tag{3.72}$$

where y0 varies between 0 and 1. Next equation (3.68) is solved with two initial conditions given by equations (3.69) and (3.72). This is solved as an initial value problem. The stop condition illustrated in chapter 2.2.4 is used to find the value of X for which equation (3.70) is satisfied. That is, equation (3.68) is integrated with the initial conditions given by equation (3.69) and (3.70) until y(X) becomes 1. The corresponding value of X gives the value for Φ. Once Φ is found, the effectiveness factor is found using equation (3.71). This procedure is illustrated in the following Maple program.

```
> restart:
> with(plots):
> eq:=diff(y(X),X$2)=y(X)^2;
```

$$eq := \frac{d^2}{dX^2} y(X) = y(X)^2$$

The initial conditions for y and its derivative are provided:

```
> IC:=D(y)(0)=0,y(0)=0.1;
```

$$IC := D(y)(0) = 0, y(0) = 0.1$$

```
> sol:=dsolve({eq,D(y)(0)=0,IC},{y(X)},type=numeric,abserr=1e-10,stiff=true,
  stop_cond=[y(X)-1]);
```

$$sol := \mathbf{proc}(x\_rosenbrock) \ \dots \ \mathbf{end\ proc}$$

The solution is evaluated for large values of X. The solution stops at 6.95.

```
> g1:=sol(1000);
```

Warning, cannot evaluate the solution further right of 6.9564589, stop condition #1 violated

$$g1 := \left[X = 6.95645899858277872, y(X) = 0.99999999999999988, \frac{d}{dX} y(X) = 0.816087949489618602 \right]$$

The value for Φ is obtained as:

```
> Phi1:=rhs(g1[1]);
```

$$\Phi 1 := 6.95645899858277872$$

The derivative at $X = \Phi$ is given by:

> dydX1:=rhs(g1[3]);

$$dydX1 := 0.816087949489618602$$

The effectiveness factor is calculated as:

> eta1:=dydX1/Phi1;

$$\eta 1 := 0.1173137008$$

The solution obtained is plotted as:

> odeplot(sol,[X,y(X)],0..10000,thickness=4,axes=boxed,title="Figure 3.37");
Warning, cannot evaluate the solution further right of 6.9564590, stop condition #1 violated

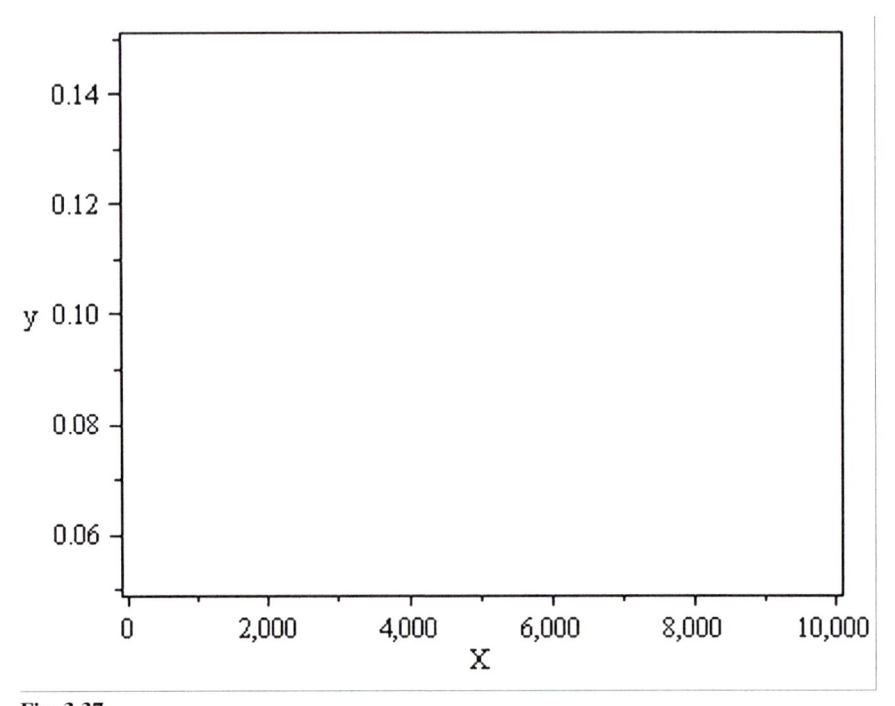

Fig. 3.37

We can ask Maple to not print the warning about the violation of stop conditions using the following command:

> _Env_dsolve_nowarnstop := true;

$$Env\_dsolve\_nowarnstop := true$$

3.2 Nonlinear Boundary Value Problems

Different initial guesses for y0 ranging from 1e - 4 to 0.999 are chosen.

> Y0:=[1e-4,1e-3,1e-2,seq(i/40.,i=1..39),0.999];

Y0 := [0.0001, 0.001, 0.01, 0.02500000000, 0.05000000000, 0.07500000000, 0.1000000000, 0.1250000000, 0.1500000000, 0.1750000000, 0.2000000000, 0.2250000000, 0.2500000000, 0.2750000000, 0.3000000000, 0.3250000000, 0.3500000000, 0.3750000000, 0.4000000000, 0.4250000000, 0.4500000000, 0.4750000000, 0.5000000000, 0.5250000000, 0.5500000000, 0.5750000000, 0.6000000000, 0.6250000000, 0.6500000000, 0.6750000000, 0.7000000000, 0.7250000000, 0.7500000000, 0.7750000000, 0.8000000000, 0.8250000000, 0.8500000000, 0.8750000000, 0.9000000000, 0.9250000000, 0.9500000000, 0.9750000000, 0.999]

> M:=nops(Y0);

$$M := 43$$

A 'for loop' is written to find the corresponding values of Φ and effectiveness factors.

> for i from 1 to M do
Sol[i]:=dsolve({eq,D(y)(0)=0,y(0)=Y0[i]},{y(X)},
type=numeric,stiff=true,abserr=1e-12,stop_cond=[y(X)-1]);;od:
> for i to M do g[i]:=Sol[i](100000):od:
> for i to M do Phi[i]:=rhs(g[i][1]);od:
> for i to M do dydX[i]:=rhs(g[i][3]);od:
> for i to M do eta[i]:=dydX[i]/Phi[i];od:

A loglogplot is made:

> loglogplot([seq([Phi[i],eta[i]],i=1..M)],axes=boxed,title="Figure 3.38",
thickness=4,labels=[Phi,eta],color=red);

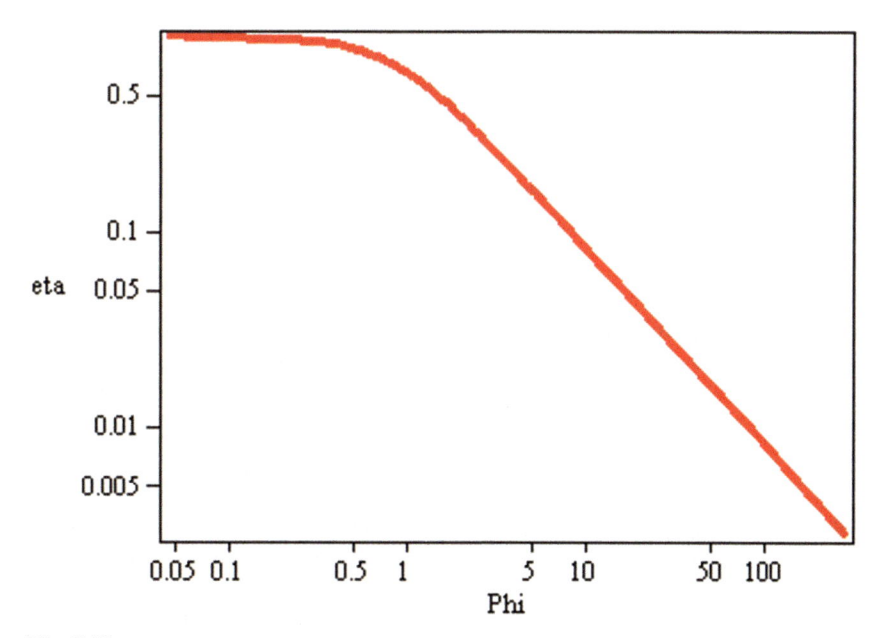

Fig. 3.38

We observe that the effectiveness factor remains close to 1 until $\Phi = 1$ and then decreases with Φ.

3.2.8 Multiple Steady States

Problems with multiple steady states are interesting to solve numerically. Computational effort required for solving these problems can be highly demanding. Multiple steady states in a rectangular catalyst pellet were analyzed in example 3.2.2, 3.2.5 and 3.2.9. One has to provide an approximate solution or a guess value to predict the three multiple solutions. It is difficult to predict the effectiveness factor of the pellet as a function of Φ or γ using the numerical approaches described earlier in this chapter. In the next example, this boundary value problem will be solved as an initial value problem.

Example 3.2.13. Multiple Steady States in a Catalyst Pellet - η vs. Φ

The non-isothermal reaction in a rectangular pellet (example 3.2.2) is again considered. The dimensionless concentration is governed by:

$$\frac{d^2y}{dx^2} = \Phi^2 y \exp\left(\gamma\beta\frac{1-y}{1+\beta(1-y)}\right) \tag{3.73}$$

with the boundary conditions:

$$\frac{dy}{dx}(0) = 0 \tag{3.74}$$

and

$$y(1) = 1 \tag{3.75}$$

The effectiveness factor of the pellet is given by

$$\eta = \frac{1}{\Phi^2}\frac{dy}{dx}(1) \tag{3.76}$$

For a given value of Φ, equation (3.63) can be solved for the boundary conditions ((3.65) and (3.66)). Once the numerical solution is obtained the effectiveness factor can be calculated using equation (3.67). It is of interest to plot the effectiveness factor as a function of Φ. For this purpose, one can solve the boundary value problem for different values of Φ and predict the effectiveness factor. However, since multiple steady states occur different initial guesses have to be used to capture all of the steady states. This boundary value problem is difficult to solve because for every value of Φ, there can 3 different ηs and, hence, 3 different values for the derivative ($\frac{dy}{dx}$) at x = 1. On the contrary, for every

3.2 Nonlinear Boundary Value Problems

value of $\frac{dy}{dx}$ (1), there is a unique value for Φ. Hence, $\frac{dy}{dx}$ (1) can be specified and Φ can be treated as an unknown, which can be solved as we know 3 boundary conditions ($\frac{dy}{dx}$ (0) = 0, y(1) = 1, $\frac{dy}{dx}$ (1) is known). The difficulty with this approach is that $\frac{dy}{dx}$(1) can take any value between 0 and ∞. Based on the previous argument, for every value of Φ, they correspond to three different values $\frac{dy}{dx}$ (1) and, hence, three different values for y(0). On the contrary, for every value of y(0), there is a unique value for Φ. The advantage of setting y(0) is that y(0) can practically vary only between 0 and 1. Now both y(0) and $\frac{dy}{dx}$ (0) are known and equation (3.73) can be solved as an initial value problem and the unknown Φ can be found be using the boundary condition y(1) = 1. Since the problem is well defined with known initial conditions for y(0) and $\frac{dy}{dx}$ (0), stiff solvers can be used to solve equation 3.73. By defining $X = \Phi x$, this boundary value problem can be transformed as:

$$\frac{d^2y}{dX^2} = y \exp\left(\gamma\beta\frac{1-y}{1+\beta(1-y)}\right) \tag{3.77}$$

with the boundary conditions:

$$\frac{dy}{dX}(0) = 0 \tag{3.78}$$

and

$$y(\Phi) = 1 \tag{3.79}$$

The effectiveness factor of the pellet is given by

$$\eta = \frac{1}{\Phi}\frac{dy}{dX}\bigg|_{y=1} \tag{3.80}$$

We know that for low values of Φ, the concentration at the center y(0) is close to 1. For high values of Φ, y(0) approaches zero. We take the concentration at the center as

$$y(0) = y0 \tag{3.81}$$

where y0 varies between 0 and 1. This system of equations is similar to example 3.2.12. The program used for example 3.2.12 is modified below to solve this example.

> restart:
> with(plots):

The values of the parameters are substituted here:

> Eq:=diff(y(X),X$2)=y(X)*exp(gamma*beta*(1-y(X))/(1+beta*(1-y(X))));

$$Eq := \frac{d^2}{dX^2} y(X) = y(X) \, e^{\frac{\gamma \beta \, (1 - y(X))}{1 + \beta \, (1 - y(X))}}$$

> eq:=subs(beta=0.8,gamma=20,Eq);

$$eq := \frac{d^2}{dX^2} y(X) = y(X) \, e^{\frac{16.0 \, (1 - y(X))}{1.8 - 0.8 \, y(X)}}$$

> IC:=D(y)(0)=0,y(0)=1e-6;

$$IC := D(y)(0) = 0, y(0) = 0.000001$$

> sol:=dsolve({eq,IC},{y(X)},type=numeric,abserr=1e-10,stiff=true,stop_cond=[y(X)-1]);

$$sol := \textbf{proc}(x\_rosenbrock) \ldots \textbf{end proc}$$

> g1:=sol(1000);
Warning, cannot evaluate the solution further right of .20038679, stop condition #1 violated

$$g1 := \left[X = 0.2003867963644 17225, y(X) = 1., \frac{d}{dX} y(X) = 19.5850020594993062 \right]$$

> Phi1:=rhs(g1[1]);

$$\Phi 1 := 0.2003867963644 17225$$

> dydX1:=rhs(g1[3]);

$$dydX1 := 19.5850020594993062$$

> eta1:=dydX1/Phi1;

$$\eta 1 := 97.73599065$$

> odeplot(sol,[X,y(X)],0..2,thickness=4,title="Figure 3.39",axes=boxed);
Warning, cannot evaluate the solution further right of .20038680, stop condition #1 violated

3.2 Nonlinear Boundary Value Problems

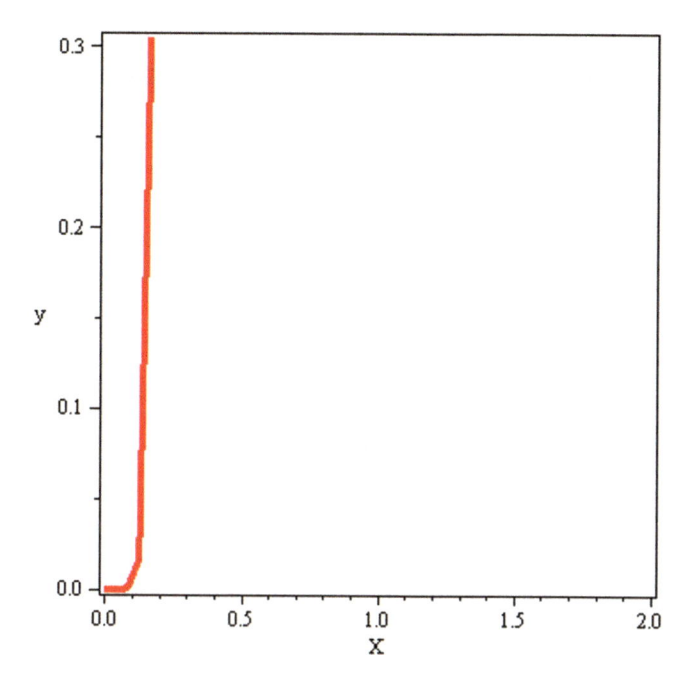

Fig. 3.39

```
> _Env_dsolve_nowarnstop := true;
```

$$\_Env\_dsolve\_nowarnstop := true$$

Initial guesses ranging from 1e - 10 to 0.9996 are provided for y(0):

```
> Y0:=[1e-10,1e-9,1e-8,1e-7,1e-6,1e-5,1e-4,1e-3,1e-2,5e-2,0.1,0.15,0.2,
0.25,0.3,0.35,0.4,0.45,0.5,0.55,0.6,0.65,0.7,0.75,0.8,0.825,0.85,0.875,0.9,
0.91,0.92,0.93,0.94,0.95,0.955,0.96,0.965,0.968,0.97,0.975,0.98,0.99,
0.995,0.996,0.997,0.998,0.999,0.9995,0.9996];
```

$Y0 := [1.\,10^{-10},\,1.\,10^{-9},\,1.\,10^{-8},\,1.\,10^{-7},\,0.000001,\,0.00001,\,0.0001,\,0.001,\,0.01,\,0.05,\,0.1,\,0.15,\,0.2,\,0.25,\,0.3,\,0.35,$
$0.4,\,0.45,\,0.5,\,0.55,\,0.6,\,0.65,\,0.7,\,0.75,\,0.8,\,0.825,\,0.85,\,0.875,\,0.9,\,0.91,\,0.92,\,0.93,\,0.94,\,0.95,\,0.955,\,0.96,\,0.965,$
$0.968,\,0.97,\,0.975,\,0.98,\,0.99,\,0.995,\,0.996,\,0.997,\,0.998,\,0.999,\,0.9995,\,0.9996]$

```
> M:=nops(Y0);
```

$$M := 49$$

```
> for i from 1 to M do
Sol[i]:=dsolve({eq,D(y)(0)=0,y(0)=Y0[i]},{y(X)},
type=numeric,stiff=true,abserr=1e-12,stop_cond=[y(X)-1]);;od:
> for i to M do g[i]:=Sol[i](100000):od:
> for i to M do Phi[i]:=rhs(g[i][1]);od:
> for i to M do dydX[i]:=rhs(g[i][3]);od:
```

```
> for i to M do eta[i]:=dydX[i]/Phi[i];od:
> loglogplot([seq([Phi[i],eta[i]],i=1..M)],axes=boxed,title="Figure 3.40",
thickness=4,labels=[phi,eta],color=green);
```

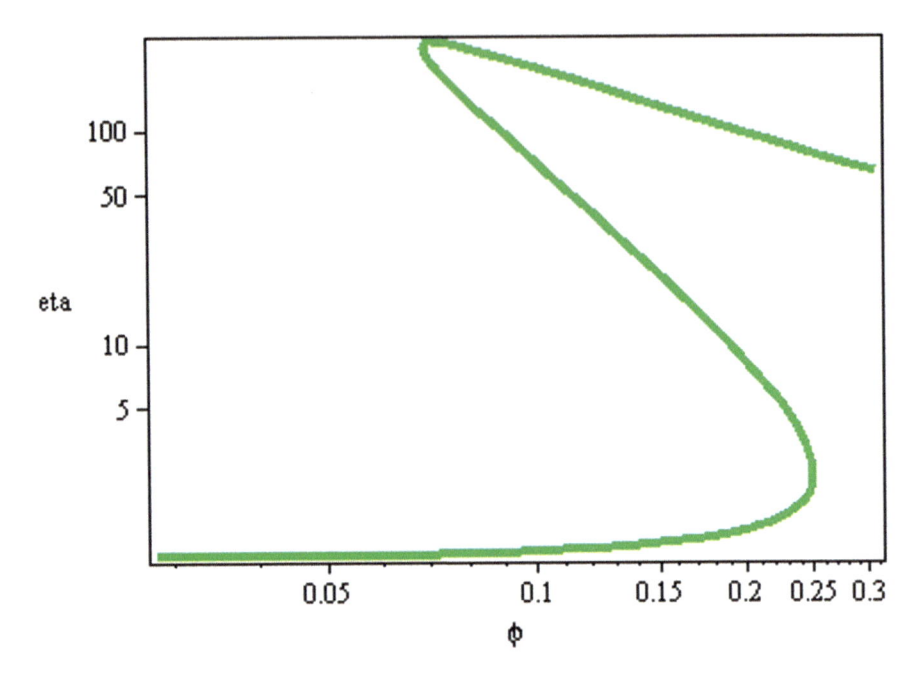

Fig. 3.40

This plot captures the multiple steady state part. However, to predict the effectiveness factor at higher values of Φ, we choose initial values ranging from 1e - 40 to 1e - 10. For these low values, one has to perform highly accurate simulations. For this purpose, absolute error (abserr) is set to 1e - 41 and the relative error (relerr) is set to 1e - 12.

```
> p1:=loglogplot([seq([Phi[i],eta[i]],i=1..M)],axes=boxed,thickness=4,
labels=[phi,eta],color=red):
> Y00:=[1e-40,1e-35,1e-30,1e-25,1e-20,1e-15,1e-10];
```

$$Y00 := [1.\ 10^{-40},\ 1.\ 10^{-35},\ 1.\ 10^{-30},\ 1.\ 10^{-25},\ 1.\ 10^{-20},\ 1.\ 10^{-15},\ 1.\ 10^{-10}]$$

```
> MM:=nops(Y00);
```

$$MM := 7$$

```
> for i from 1 to MM do
Sol0[i]:=dsolve({eq,D(y)(0)=0,y(0)=Y00[i]},{y(X)},
type=numeric,stiff=true,maxfun=1000000,abserr=1e-41,
relerr=1e-12,stop_cond=[y(X)-1]);od:
> for i to MM do g[i]:=Sol0[i](100000):od:
> for i to MM do Phi1[i]:=rhs(g[i][1]);od:
```

3.2 Nonlinear Boundary Value Problems

```
> for i to MM do dydX1[i]:=rhs(g[i][3]);od:
> for i to MM do eta1[i]:=dydX1[i]/Phi1[i];od:
> p2:=loglogplot([seq([Phi1[i],eta1[i]],i=1..MM)],axes=boxed,title="Figure 3.41",
thickness=4,labels=[phi,eta]):
> display({p1,p2});
```

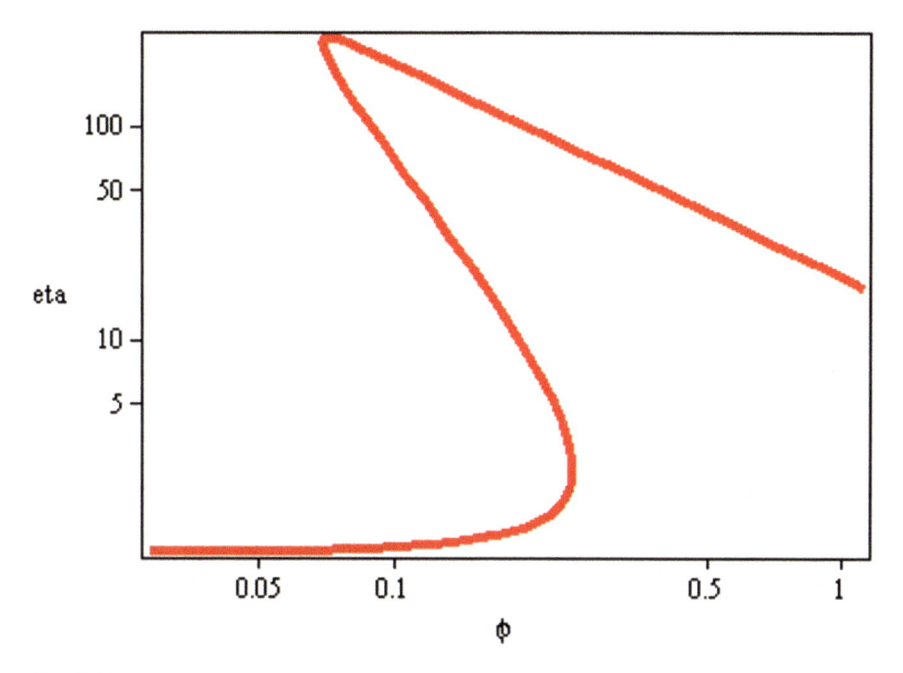

Fig. 3.41

This plot captures the multiple steady state part. However, to predict the effectiveness factor at higher values of Φ, we choose initial values ranging from 1e - 40 to 1e - 10. For these low values, one has to perform highly accurate simulations. For this purpose, absolute error (abserr) is set to 1e - 41 and the relative error (relerr) is set to 1e - 12.

```
> Eq:=diff(y(X),X$2)=y(X)*exp(gamma*beta*(1-y(X))/(1+beta*(1-y(X))));
```

$$Eq := \frac{\partial^2}{\partial X^2} y(X) = y(X) \, e^{\left(\frac{\gamma \beta (1 - y(X))}{1 + \beta (1 - y(X))}\right)}$$

```
> eq:=subs(beta=0.8,gamma=20,Eq);
```

$$eq := \frac{\partial^2}{\partial X^2} y(X) = y(X) \, e^{\left(16.0 \frac{1 - y(X)}{1.8 - .8\, y(X)}\right)}$$

The following profile is obtained:

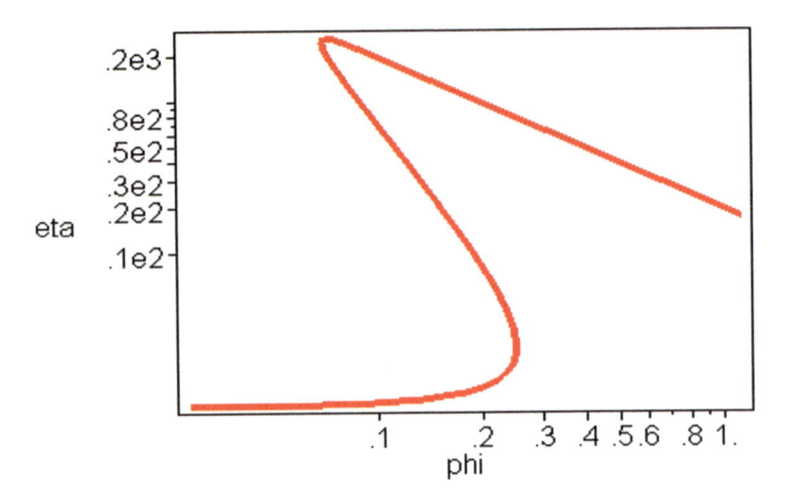

Fig. 3.42

3.2.9 Eigenvalue Problems

The separation of variables is a common technique used to solve linear PDEs. This technique will be discussed in detail in chapter 7. This technique yields ordinary differential equations for the eigenfunctions. In this section, we will present two numerical techniques for the Graetz problem.

Example 3.2.14. Graetz Problem–Finite Difference Solution

The Graetz problem (heat or mass transfer) in cylindrical coordinates with parabolic velocity profile is solved here. The governing equation for the eigenfunction is[15] [8]

$$\frac{d^2y}{dx^2} + \frac{1}{x}\frac{dy}{dx} + \lambda^2(1-x^2)y \qquad (3.82)$$

with the boundary conditions:

$$\frac{dy}{dx}(0) = 0 \qquad (3.83)$$

and

$$y(1) = 0 \qquad (3.84)$$

Equation (3.82) is a second order equation with two boundary conditions (equations (3.83) and (3.84). In equation (3.82), λ is the eigenvalue. To solve for

3.2 Nonlinear Boundary Value Problems

the eigenvalue an additional boundary condition has to be used. For this purpose, y at x = 0 is arbitrarily set to 1:

$$y(0) = 1 \tag{3.85}$$

Next, equation (3.82) is discretized using finite differences as in section 3.2.3. This yields N equations for the interior node points. The boundary conditions (equations (3.83) and (3.84)) are converted to finite difference form. This yields two equations. There are a total of N+2 node points including the boundaries. There are N+2 dependent variables ($y_{i}, i = 0..N+1$). There is an additional variable λ. The additional equation is (3.85). Hence, there are N+3 variables ($y_{i}, i = 0..N+1$ and λ) to be solved from N+3 equations. There are infinite solutions for the differential equation (3.82). Hence, there are multiple solutions for the system of finite difference equations. This example is solved in Maple below:

```
> restart:
> with(plots):
> N:=6;
```

$$N := 6$$

```
> L:=1;
```

$$L := 1$$

```
> eq:=diff(y(x),x$2)+1/x*diff(y(x),x)+lambda^2*(1-x^2)*y(x);
```

$$eq := \frac{d^2}{dx^2} y(x) + \frac{\frac{d}{dx} y(x)}{x} + \lambda^2 \left(1 - x^2\right) y(x)$$

```
> bc1:=diff(y(x),x);
```

$$bc1 := \frac{d}{dx} y(x)$$

```
> bc2:=y(x)-0;
```

$$bc2 := y(x)$$

The additional boundary condition is entered here:

```
> bc3:=y(x)-1;
```

$$bc3 := y(x) - 1$$

The central difference expressions for the second and first derivatives are

```
> d2ydx2:=(y[m+1]-2*y[m]+y[m-1])/h^2;
```

$$d2ydx2 := \frac{y_{m+1} - 2 y_m + y_{m-1}}{h^2}$$

> dydx:=(y[m+1]-y[m-1])/2/h;

$$dydx := \frac{1}{2}\frac{y_{m+1}-y_{m-1}}{h}$$

Three point forward and backward difference expressions for the derivative are:
> dydxf:=(-y[2]+4*y[1]-3*y[0])/(2*h);

$$dydxf := \frac{1}{2}\frac{-y_2+4y_1-3y_0}{h}$$

> dydxb:=(y[N-1]-4*y[N]+3*y[N+1])/(2*h);

$$dydxb := \frac{1}{2}\frac{y_5-4y_6+3y_7}{h}$$

The governing equation in finite difference form is:
> Eq[m]:=subs(diff(y(x),x$2)=d2ydx2,diff(y(x),x)=dydx,y(x)=y[m],x=m*h,eq);

$$Eq_m := \frac{y_{m+1}-2y_m+y_{m-1}}{h^2} + \frac{1}{2}\frac{y_{m+1}-y_{m-1}}{m\,h^2} + \lambda^2(1-m^2h^2)y_m$$

The boundary conditions in finite difference form are:
> Eq[0]:=subs(diff(y(x),x)=dydxf,y(x)=y[0],bc1);

$$Eq_0 := \frac{1}{2}\frac{-y_2+4y_1-3y_0}{h}$$

> Eq[N+1]:=subs(diff(y(x),x)=dydxb,y(x)=y[N+1],bc2);

$$Eq_7 := y_7$$

A 'for loop' can be written for the interior node points as
> for i to N do Eq[i]:=subs(m=i,Eq[m]);od;

$$Eq_1 := \frac{y_2-2y_1+y_0}{h^2} + \frac{1}{2}\frac{y_2-y_0}{h^2} + \lambda^2(1-h^2)y_1$$

$$Eq_2 := \frac{y_3-2y_2+y_1}{h^2} + \frac{1}{4}\frac{y_3-y_1}{h^2} + \lambda^2(1-4h^2)y_2$$

$$Eq_3 := \frac{y_4-2y_3+y_2}{h^2} + \frac{1}{6}\frac{y_4-y_2}{h^2} + \lambda^2(1-9h^2)y_3$$

$$Eq_4 := \frac{y_5-2y_4+y_3}{h^2} + \frac{1}{8}\frac{y_5-y_3}{h^2} + \lambda^2(1-16h^2)y_4$$

3.2 Nonlinear Boundary Value Problems

$$Eq_5 := \frac{y_6 - 2y_5 + y_4}{h^2} + \frac{1}{10} \frac{y_6 - y_4}{h^2} + \lambda^2 \left(1 - 25 h^2\right) y_5$$

$$Eq_6 := \frac{y_7 - 2y_6 + y_5}{h^2} + \frac{1}{12} \frac{y_7 - y_5}{h^2} + \lambda^2 \left(1 - 36 h^2\right) y_6$$

The additional equation for the eigenvalue is:

> Eqeig:=subs(diff(y(x),x)=dydxf,y(x)=y[0],bc3);

$$Eqeig := y_0 - 1$$

> h:=L/(N+1);

$$h := \frac{1}{7}$$

The finite difference equations are stored in eqs:

> eqs:=seq(eval(subs(Phi=1,Eq[i])),i=0..N+1),Eqeig;

$$eqs := -\frac{7}{2} y_2 + 14 y_1 - \frac{21}{2} y_0, \frac{147}{2} y_2 - 98 y_1 + \frac{49}{2} y_0$$
$$+ \frac{48}{49} \lambda^2 y_1, \frac{245}{4} y_3 - 98 y_2 + \frac{147}{4} y_1 + \frac{45}{49} \lambda^2 y_2, \frac{343}{6} y_4$$
$$- 98 y_3 + \frac{245}{6} y_2 + \frac{40}{49} \lambda^2 y_3, \frac{441}{8} y_5 - 98 y_4 + \frac{343}{8} y_3$$
$$+ \frac{33}{49} \lambda^2 y_4, \frac{539}{10} y_6 - 98 y_5 + \frac{441}{10} y_4 + \frac{24}{49} \lambda^2 y_5, \frac{637}{12} y_7$$
$$- 98 y_6 + \frac{539}{12} y_5 + \frac{13}{49} \lambda^2 y_6, y_7, y_0 - 1$$

The dependent variables are stored in vars:

> vars:=seq(y[i],i=0..N+1),lambda;

$$vars := y_0, y_1, y_2, y_3, y_4, y_5, y_6, y_7, \lambda$$

These variables are solved as:

> fsolve({eqs},{vars});

$\{y_7 = 0., y_0 = 1.000000000, \lambda = -2.687842398, y_2 = 0.8606034948, y_1 = 0.9651508737, y_5 = 0.3314378546, y_6$

Since there are multiple solutions, one has to provide the range for the dependent variables. The dependent variable y varies between 0 and 1 and the eigenvalue is solved in the range of 0.4.

> vars:=seq(y[i]=-1..1,i=0..N+1),lambda=0..4;

$$vars := y_0 = -1..1, y_1 = -1..1, y_2 = -1..1, y_3 = -1..1, y_4 = -1..1, y_5 = -1..1, y_6 = -1..1, y_7 = -1..1, \lambda = 0..4$$

The solution obtained in stored in sol[1] and plotted:

> sol[1]:=fsolve({eqs},{vars});

$$sol_1 := \{y_7 = 0., y_0 = 1.000000000, y_2 = 0.8606034948, y_1 = 0.9651508737, y_5 = 0.3314378546, y_6 = 0.1549393509, y_3 = 0.7046525028, y_4 = 0.5205640878, \lambda = 2.687842398\}$$

> assign(sol[1]):

The first eigenvalue is given by:

> l[1]:=lambda;

$$l_1 := 2.687842398$$

The first eigenfunction is plotted here:

> plot([seq([i*h,y[i]],i=0..N+1)],thickness=4,axes=boxed,title="Figure 3.43", labels=[x,y]);

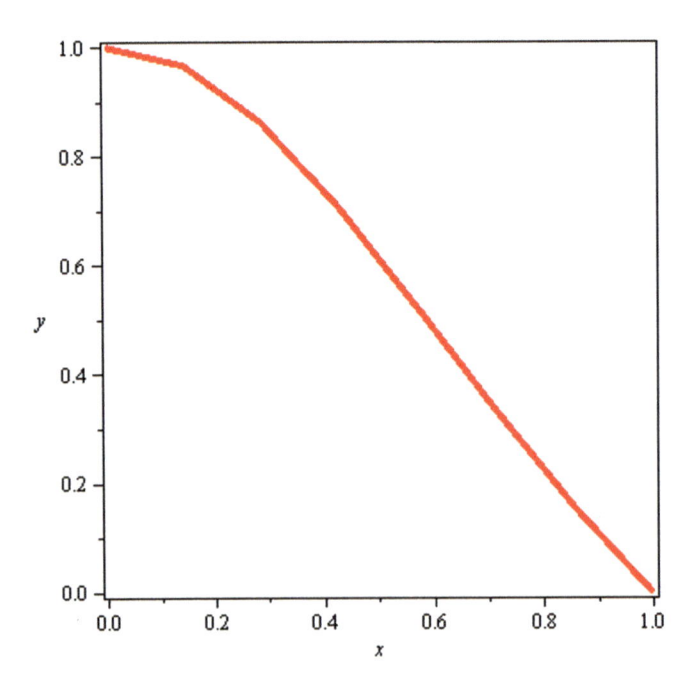

Fig. 3.43

3.2 Nonlinear Boundary Value Problems

```
>p[1]:=plot([seq([i*h,y[i]],i=0..N+1)],thickness=4,axes=boxed,labels=[x,y],
color=colorlist[1]):
> pt[1]:=textplot([0.4,0.9,'lambda=l[1]']):
> for i from 0 to N+1 do unassign('y[i]'):od:
> unassign('lambda'):
> M:=4;
```

$$M := 4$$

The next three eigenvalues are found by changing the range in increments of four:

```
> for j from 2 to M do
> vars:=seq(y[i]=-1..1,i=0..N+1),lambda=l[j-1]+1..l[j-1]+4;
> sol[j]:=fsolve({eqs},{vars});
> assign(sol[j]):
> l[j]:=lambda;
> p[j]:=plot([seq([i*h,y[i]],i=0..N+1)],thickness=4,title="Figure 3.44",
axes=boxed,labels=[x,y],color=colorlist[j]):
> for i from 0 to N+1 do unassign('y[i]'):od:
> unassign('lambda'):
> od:
```

The first four eigenvalues are:

```
> print(seq(l[i],i=1..M));
```

$$2.687842398, 6.519412138, 10.06247204, 12.97831858$$

```
> pt[2]:=textplot([0.38,0.5,lambda=l[2]]):
pt[3]:=textplot([0.87,0.4,lambda=l[3]]):
pt[4]:=textplot([0.8,-0.08,lambda=l[4]]):
```

Even after using $N = 30$ interior node points we get only two digit accuracy with the exact solution. In addition, we observe more errors for higher eigenvalues than for the lower eigenvalues.

```
> display({seq(p[i],i=1..M),seq(pt[i],i=1..M)});
```

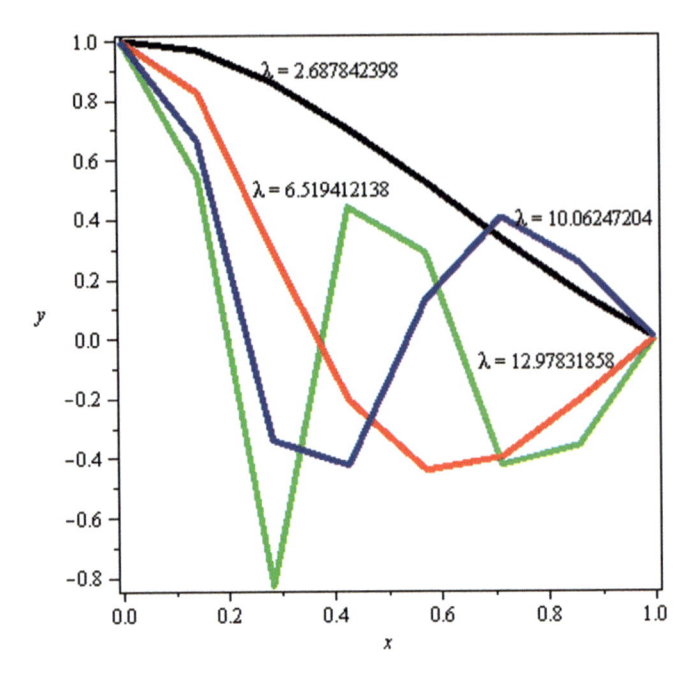

Fig. 3.44

We observe that first eigenfunction reaches zero at $x = 1$. The second eigenfunction crosses the x-axis once before reaching zero at $x = 1$. The third eigenfunction crosses the x-axis twice before reaching zero. In general, nth eigenfunction crosses the x-axis n-1 times. We observe that the eigenvalues match with the literature. However, accuracy is very poor. For accurate predictions of the eigenvalues, the shooting technique is adopted in the next example.

Example 3.2.15. Graetz Problem–Shooting Technique

The governing equation is:

$$\frac{d^2y}{dx^2} + \frac{1}{x}\frac{dy}{dx} + \lambda^2(1-x^2)y \tag{3.86}$$

with the initial conditions:

$$\frac{dy}{dx}(0) = 0 \tag{3.87}$$

and

$$y(0) = 1 \tag{3.88}$$

3.2 Nonlinear Boundary Value Problems

A guess value for λ is chosen:

$$\lambda = \lambda_{old} \tag{3.89}$$

Equation (3.86) is solved with the initial conditions (equations (3.87) and (3.89) using this assumed value of λ. Then a new value of λ is obtained using the following relationship:

$$\lambda_{new} = \lambda_{old} + \rho \left[\frac{y_{expected, x=1} - y(\lambda_{old})_{x=1}}{\left.\frac{dy(\lambda_{old})}{d\lambda}\right|_{x=1}} \right]$$

$$= \lambda_{old} + \rho \left[\frac{0 - y(\lambda_{old})_{x=1}}{\left.\frac{dy(\lambda_{old})}{d\lambda}\right|_{x=1}} \right] \tag{3.90}$$

In equation (3.90), the Jacobian $\frac{dy(\lambda_{old})}{d\lambda}$ is predicted as illustrated in 3.2.4 for the shooting technique. This Jacobian is predicted exactly by differentiating the governing equation (3.86) with respect to λ as

$$\frac{d^2}{dx^2}\left(\frac{dy}{d\lambda}\right) + \frac{1}{x}\frac{d}{dx}\left(\frac{dy}{d\lambda}\right) + \lambda^2(1-x^2)\left(\frac{dy}{d\lambda}\right) + 2\lambda(1-x^2)y \tag{3.91}$$

Next, the Jacobian is treated as the variable y2.

$$\frac{d^2 y2}{dx^2} + \frac{1}{x}\frac{dy2}{dx} + \lambda^2(1-x^2)y2 + 2\lambda(1-x^2)y \tag{3.92}$$

The initial conditions for y2 are obtained by using the differentiating equations (3.86) and (3.88) with respect to λ.

$$\frac{dy2}{dx}(0) = 0 \tag{3.93}$$

and

$$y2(0) = 0 \tag{3.94}$$

This example is solved in Maple below:

```
> restart:
> with(plots):
```

280   3 Boundary Value Problems

The governing equation is entered here:

> eq:=diff(y(x),x$2)+1/x*diff(y(x),x)+lambda^2*(1-x^2)*y(x);

$$eq := \frac{d^2}{dx^2} y(x) + \frac{\frac{d}{dx} y(x)}{x} + \lambda^2 \left(1 - x^2\right) y(x)$$

The sensitivity equation is developed by differentiating the governing equation with respect to λ:

> eqlambda:=subs(y(x)=Y(x,lambda),eq);

$$eqlambda := \frac{\partial^2}{\partial x^2} Y(x,\lambda) + \frac{\frac{\partial}{\partial x} Y(x,\lambda)}{x} + \lambda^2 \left(1 - x^2\right) Y(x,\lambda)$$

> eqlambda:=diff(eqlambda,lambda);

$$eqlambda := \frac{\partial^3}{\partial x^2 \partial \lambda} Y(x,\lambda) + \frac{\frac{\partial^2}{\partial x \partial \lambda} Y(x,\lambda)}{x} + 2\lambda \left(1 - x^2\right) Y(x,\lambda) + \lambda^2 \left(1 - x^2\right) \left(\frac{\partial}{\partial \lambda} Y(x,\lambda)\right)$$

> eqlambda:=subs(diff(Y(x,lambda),lambda)=y2(x),eqlambda);

$$eqlambda := \frac{d^2}{dx^2} y2(x) + \frac{\frac{d}{dx} y2(x)}{x} + 2\lambda \left(1 - x^2\right) Y(x,\lambda) + \lambda^2 \left(1 - x^2\right) y2(x)$$

> eqlambda:=subs(Y(x,lambda)=y(x),eqlambda);

$$eqlambda := \frac{d^2}{dx^2} y2(x) + \frac{\frac{d}{dx} y2(x)}{x} + 2\lambda \left(1 - x^2\right) y(x) + \lambda^2 \left(1 - x^2\right) y2(x)$$

The variables are stored in 'vars:'

> vars:=(y(x),y2(x));

$$vars := y(x), y2(x)$$

An initial guess for lambda is given here:

> lambda0:=1;

$$\lambda 0 := 1$$

The governing equations are stored in 'eqs:'

> eqs:=subs(lambda=lambda0,eq),subs(lambda=lambda0,eqlambda);

$$eqs := \frac{d^2}{dx^2} y(x) + \frac{\frac{d}{dx} y(x)}{x} + \left(1 - x^2\right) y(x), \frac{d^2}{dx^2} y2(x) + \frac{\frac{d}{dx} y2(x)}{x} + 2\left(1 - x^2\right) y(x) + \left(1 - x^2\right) y2(x)$$

3.2 Nonlinear Boundary Value Problems

> e:=1e-6;
$$e := 0.000001$$

The initial conditions for y and y2 are stored in ICs:

> ICs:=(D(y)(0)=0,y(0)=1,y2(0)=0,D(y2)(0)=0);
$$ICs := D(y)(0) = 0, y(0) = 1, y2(0) = 0, D(y2)(0) = 0$$

The numerical solution is obtained as:

> sol:=dsolve({eqs,ICs},{vars},type=numeric);
$$sol := \mathbf{proc}(x\_rkf45) \ \ldots \ \mathbf{end \ proc}$$

The solution is evaluated at $x = 1$:

> sol(1);
$$\left[x = 1., y(x) = 0.820278359865142193, \frac{d}{dx} y(x) = -.232215636108961648, y2(x) = -.344168805990258675, \right.$$

There is a removable singularity at $x = 0$. This is handled by replacing 0 with $e = 10^{-6}$.

> e:=1e-6;
$$e := 0.000001$$

> ICs:=(D(y)(e)=0,y(e)=1,y2(e)=0,D(y2)(e)=0);
$$ICs := D(y)(0.000001) = 0, y(0.000001) = 1, y2(0.000001) = 0, D(y2)(0.000001) = 0$$

> sol:=dsolve({eqs,ICs},{vars},type=numeric);
$$sol := \mathbf{proc}(x\_rkf45) \ \ldots \ \mathbf{end \ proc}$$

The solution at $x = 1$ is:
> sol(1);
$$\left[x = 1., y(x) = 0.820278368333642871, \frac{d}{dx} y(x) = -.232215714190456557, y2(x) = -.344168764106810410, \right.$$

The expected value of y at $x = 0$ is $y = 0$. The predicted value is:

> ypred:=rhs(sol(1)[2]);
$$ypred := 0.820278368333642871$$

The Jacobian, y2 at $x = 1$ is:

> y2pred:=rhs(sol(1)[4]);
$$y2pred := -.344168764106810410$$

The value of λ is updated as:

> lambda1:=lambda0+(0-ypred)/y2pred;

$$\lambda 1 := 3.383360879$$

The error in λ is:

> err:=lambda1-lambda0;

$$err := 2.383360879$$

> lambda0:=lambda1;

$$\lambda 0 := 3.383360879$$

> k:=1;

$$k := 1$$

The iteration is performed until the error in λ becomes less than $1e$-6:

```
> while err>1e-6 do
> eqs:=subs(lambda=lambda0,eq),subs(lambda=lambda0,eqlambda);
> sol:=dsolve({eqs,ICs},{vars},type=numeric);
> ypred:=rhs(sol(1)[2]);
> y2pred:=rhs(sol(1)[4]);
> lambda1:=lambda0+(0-ypred)/y2pred;
> err:=abs(lambda1-lambda0);
> lambda0:=lambda1;k:=k+1;
> end;
```

$$eqs := \frac{d^2}{dx^2}y(x) + \frac{\frac{d}{dx}y(x)}{x} + 11.44713084\,(1-x^2)\,y(x),\ \frac{d^2}{dx^2}y2(x) + \frac{\frac{d}{dx}y2(x)}{x} + 6.766721758\,(1-x^2)\,y(x) + 11.44713084\,(1-x^2)\,y2(x)$$

$$sol := \mathbf{proc}(x\_rkf45)\ \ldots\ \mathbf{end\ proc}$$

$$ypred := -.3012792057699952319$$

$$y2pred := -.370977885391426232$$

$$\lambda 1 := 2.571239139$$

$$err := 0.812121740$$

$$\lambda 0 := 2.571239139$$

$$k := 2$$

$$eqs := \frac{d^2}{dx^2}y(x) + \frac{\frac{d}{dx}y(x)}{x} + 6.611270710\,(1-x^2)\,y(x),\ \frac{d^2}{dx^2}y2(x) + \frac{\frac{d}{dx}y2(x)}{x} + 5.142478278\,(1-x^2)\,y(x) + 6.611270710\,(1-x^2)\,y2(x)$$

3.2 Nonlinear Boundary Value Problems

$sol := \mathbf{proc}(x\_rkf45) \ldots \mathbf{end\ proc}$

$ypred := 0.0676388401700673686$

$y2pred := -.514559657293676142$

$\lambda 1 := 2.702689087$

$err := 0.131449948$

$\lambda 0 := 2.702689087$

$k := 3$

$eqs := \frac{d^2}{dx^2} y(x) + \frac{\frac{d}{dx} y(x)}{x} + 7.304528301 \, (1 - x^2) \, y(x), \frac{d^2}{dx^2} y2(x) + \frac{\frac{d}{dx} y2(x)}{x} + 5.405378174 \, (1 - x^2) \, y(x) + 7.304528301 \, (1 - x^2) \, y2(x)$

$sol := \mathbf{proc}(x\_rkf45) \ldots \mathbf{end\ proc}$

$ypred := 0.000839192930248642546$

$y2pred := -.501097644968840461$

$\lambda 1 := 2.704363796$

$err := 0.001674709$

$\lambda 0 := 2.704363796$

$k := 4$

$eqs := \frac{d^2}{dx^2} y(x) + \frac{\frac{d}{dx} y(x)}{x} + 7.313583541 \, (1 - x^2) \, y(x), \frac{d^2}{dx^2} y2(x) + \frac{\frac{d}{dx} y2(x)}{x} + 5.408727592 \, (1 - x^2) \, y(x) + 7.313583541 \, (1 - x^2) \, y2(x)$

$sol := \mathbf{proc}(x\_rkf45) \ldots \mathbf{end\ proc}$

$ypred := 1.66400171246086464 \, 10^{-7}$

$y2pred := -.500899460370252791$

$\lambda 1 := 2.704364128$

$err := 3.32 \, 10^{-7}$

$\lambda 0 := 2.704364128$

$k := 5$

Finally, the eigenvalue obtained is:

> lambda0;

$$2.704364128$$

The number of iterations is:

> k;

$$5$$

The error in the value of λ is:

> err;

$$3.32 \; 10^{-7}$$

The first eigenfunction is plotted as:

> odeplot(sol,[x,y(x)],e..1,axes=boxed,thickness=3,title="Figure 3.45");

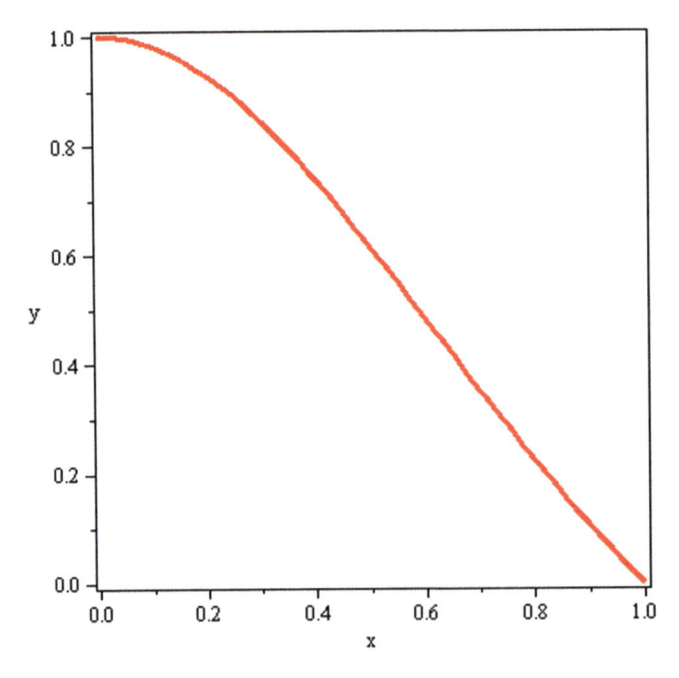

Fig. 3.45

Next, a tolerance of 1e - 10 is set. This BVP has infinite solutions. The first five eigenvalues are found by using five different initial guesses for λ. A scaling factor of $\rho = 1/3$ is introduced to ensure stability.

3.2 Nonlinear Boundary Value Problems

```
> tol:=1e-10;rho:=1/3;
```

$$tol := 1.\ 10^{-10}$$

$$\rho := \frac{1}{3}$$

```
> lambdaguess:=[1,5,9,13,17];
```

$$lambdaguess := [1, 5, 9, 13, 17]$$

```
> MM:=nops(lambdaguess);
```

$$MM := 5$$

```
> colorlist:=[black,red,blue,yellow,green];
```

$$colorlist := [black, red, blue, yellow, green]$$

```
> for i from 1 to MM do
> lambda0:=lambdaguess[i];
> k:=1;err:=1;
> while err>tol do
> eqs:=subs(lambda=lambda0,eq),subs(lambda=lambda0,eqlambda);
> sol:=dsolve({eqs,ICs},{vars},type=numeric);
> ypred:=rhs(sol(1)[2]);
> y2pred:=rhs(sol(1)[4]);
> lambda1:=lambda0+rho*(0-ypred)/y2pred;
> err:=abs(lambda1-lambda0);
> lambda0:=lambda1;k:=k+1;
> end:
> l[i]:=lambda0;
> kk[i]:=k;
> Err[i]:=err;
> p[i]:=odeplot(sol,[x,y(x)],e..1,axes=boxed,thickness=3,color=colorlist[i]);
> end:
```

The first five eigenvalues are:

```
> seq(l[i],i=1..MM);
```

$$2.704364127, 6.679031178, 10.67337916, 14.67107800, 18.66987135$$

These values match exactly with the literature and the analytical solution discussed in chapter 7 (up to six significant digits). The number of iterations required for the eigenvalue is:

```
> seq(kk[i],i=1..MM);
```

$$53, 51, 45, 44, 44$$

The error in λ is:

```
> seq(Err[i],i=1..MM);
```

$$0., 0., 0., 0., 0.$$

The first five eigenfunctions are plotted as:

```
> arw:=arrow(<0.5,0.8>,<-0.45,-0.6>,width=[1/250, relative], head_length=
[0.05, relative]):
pt:=textplot([[0.8,0.95,"Follow the arrow"],seq([0.8,0.85-(i-1)*0.09,
lambda=l[i]],i=1..MM)]);
```

$$pt := PLOT(...)$$

```
> display({seq(p[i],i=1..MM),arw,pt},title="Figure 3.46");
```

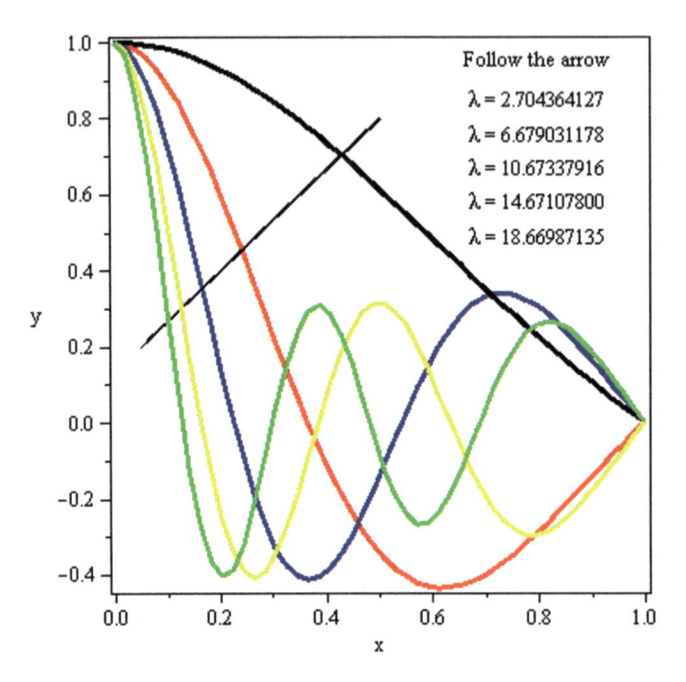

Fig. 3.46

Note that with the shooting technique we obtain exact results. The finite difference solution is not as smooth as the shooting technique solution.

3.2.10 Summary

In this chapter, nonlinear boundary value problems were solved numerically. In section 3.2.2, series solutions were derived for nonlinear boundary value problems. This is a powerful technique and is even capable of predicting multiple steady states in a catalyst pellet. However, these series solutions should be used cautiously. The convergence of the solution is not guaranteed and should be verified. This can be done by increasing the number of terms in the series and plotting the profiles.

3.2 Nonlinear Boundary Value Problems

In section 3.2.3, finite difference solutions were obtained for nonlinear boundary value problems. This is a straightforward and easy technique and can be used to obtain an initial guess for other sophisticated techniques. This technique is important because it forms the basis for the method of lines technique for solving linear and nonlinear partial differential equations (chapter 5 and 6). However, for stiff boundary value problems, this technique may not work and might demand prohibitively large number of node points. In addition, approximate initial guess should be provided for all the node points for stiff boundary value problems.

In section 3.2.4, nonlinear boundary value problems were solved using shooting technique. The given boundary value problem was converted to a system of initial value problems. The unknown initial condition was obtained using an iteration and optimization procedure. This is a very robust technique and can be used to solve stiff boundary value problems. This technique is capable of predicting multiple steady states in a catalyst pellet. However, the number of iterations required for convergence can be prohibitively large for certain boundary value problems.

In section 3.2.5, nonlinear boundary value problems were solved using Maple's 'dsolve' command. Maple's 'dsolve' command is based on finite difference coupled with Richardson interpolation. This can be used to solve a variety of B boundary value problems conveniently. For nonlinear or stiff boundary value problems, one should provide an approximate solution. This approximate solution can be arrived by providing initial conditions and solving the boundary value problem as an initial value problem. In addition, special methods must be specified for boundary value problems in cylindrical and spherical coordinates to avoid the removable singularity at the origin. If this method fails, one might resort to shooting technique. This methodology was then extended for coupled boundary value problems in section 3.2.6.

In section 3.2.7, boundary value problems were solved as initial value problems. This methodology is especially useful for predicting the performances in chemical reactors. Maple's stop condition was used in this section to obtain η vs. Φ curves. This is very useful because, it is generally easier to solve an initial value problem than a boundary value problem. This technique was then used in section 3.2.8 to predict multiple steady states in a catalyst pellet in section 3.2.8. This methodology is extremely useful for predicting the hysteresis curves in multiple steady state problems.

In section 3.2.9, eigenvalue problems were solved numerically. Two different methods were used. First, a finite difference technique was used to predict the unknown eigenvalues for the Graetz problem in cylindrical coordinates. By specifying different ranges, first five eigenvalues and eigenfunctions were obtained numerically. This technique is not robust and requires a large number of node points as the magnitude of the eigenvalue increases. Next, a sensitivity approach was used to predict the eigenvalue. The given boundary value problem was first converted to an initial value problem. This initial value problem was then solved with a guessed value of the eigenvalue. The eigenvalue was the updated based on an optimization procedure. This is a very robust technique. This technique predicts the eigenvalues accurately and the eigenfunctions obtained are very smooth.

Fifteen examples were presented in this chapter.

3.2.11 Exercise Problems

1. Consider the diffusion reaction problem discussed in example 3.2.9. Obtain the series solutions for this problem. Plot the concentration profiles for $\Phi = 0.1, 1, 2$ and 10.

2. Consider diffusion with reaction in a cylindrical pellet.[8, 18] The governing equation and boundary conditions are:

$$\frac{d^2c}{dx^2} + \frac{1}{x}\frac{dc}{dx} = \Phi^2 c^2$$

$$\frac{dc}{dx}(0) = 0 \text{ and } c(1) = 1$$

Obtain the series solutions for this problem using Maple's 'dsolve' command. (Since there is a removable singularity at x = 0, use c(1)=1 and D(c)(1) = c1 to obtain series solutions and obtain the constant c1 using the boundary condition at x = 0).

3. Consider diffusion with a reaction in a cylindrical pellet[8, 18] The governing equation and boundary conditions are:

$$\frac{d^2c}{dx^2} + \frac{2}{x}\frac{dc}{dx} = \Phi^2 c^2$$

$$\frac{dc}{dx}(0) = 0 \text{ and } c(1) = 1$$

4. Obtain series solutions for this problem using Maple's 'dsolve' command. (Since there is a removable singularity at x = 0, use c(1)=1 and D(c)(1) = c1 to obtain series solutions and obtain the constant c1 using the boundary condition at x = 0).

5. Consider the nonlinear heat transfer problem solved in example 3.2.4. Obtain series solutions for this problem and plot the profiles.

6. Consider the Blasius problem discussed in example 3.2.10. Obtain series solutions for this problem. Can you obtain physically meaningful series solutions for this problem using Maple's 'dsolve' command?

7. Consider diffusion in a slab catalyst with a highly nonlinear Hinshelwood kinetics.[18] The governing equation and boundary conditions are:

$$\frac{d^2c}{dx^2} = \Phi^2 \frac{c}{1+\delta c + \gamma c^2}$$

$$\frac{dc}{dx}(0) = 0 \text{ and } c(1) = 1$$

Obtain series solutions for this problem and plot the concentration profiles for $\Phi = \delta = \gamma = 1$.

3.2 Nonlinear Boundary Value Problems

8. Redo example 3.2.2 for a second order reaction (take the governing equation from example 3.2.9).
9. Redo problem 2 using Maple's 'dsolve' numeric command.
10. Redo problem 3 using Maple's 'dsolve' numeric command.
11. Redo problem 6 using Maple's 'dsolve' numeric command and shooting technique.
12. Solve example 3.2.2 using Maple's 'dsolve' numeric command and obtain the three steady states.
13. Consider problem 7 and solve it as an initial value problem to obtain the effectiveness factor as a function of Φ. (See examples 3.2.12 and 3.2.13).
14. Solve problem 6 as an initial value problem (see examples 3.2.12 and 3.2.13) and obtain the effectiveness factor as a function of Φ.
15. Redo example 3.2.2 using the finite difference technique illustrated in example 3.2.3. Can you obtain all the three steady states?
16. Solve the Blasius equation (example 3.2.10) using the shooting technique.
17. Consider an isothermal chemical flow reactor with dispersion.[22] The governing equation and boundary conditions are:

$$\frac{1}{Pe}\frac{d^2c}{dx^2} - \frac{dc}{dx} = \frac{1}{Pe}Dac^n$$

$$\frac{1}{Pe}\frac{dc}{dx}(0) = c(0) - 1 \text{ and } \frac{dc}{dx}(1) = 0$$

where Pe is the Peclet number, Da is the Damkohler number and n is the reaction order. Obtain an analytical solution for this problem for n = 1. Solve this problem using 'dsolve' numeric command, shooting technique, and finite difference methods for the set of parameters Pe = 1, Da = 1 and n = 2. Repeat the calculations for Pe = 50, Da = 1 and n = 2. Discuss your results.

18. Consider heat transfer associated with a boundary layer in a flat plate.[22] Velocity profile is governed by the Blasius equation and the temperature (θ) is governed by the Pohlhausen equation. The governing equations and boundary conditions are:

$$\frac{d^3f}{dx^3} + f\frac{d^2f}{dx^2} = 0$$

$$\frac{d^2\theta}{dx^2} + Prf\frac{d\theta}{dx} = 0$$

$$f(0) = 0; \quad \frac{df}{dx}(0) = 0 \text{ and } \frac{df}{dx}(5) = 1$$

$$\theta(0) = 0 \text{ and } \theta(5) = 0$$

where Pr is the Prandtl number. For mass transfer, a similar equation arises with the Prandtl number replaced by the Schmidt number. Solve this system using Maple's 'dsolve' numeric command for Pr = 2. Plot velocity and temperature profiles. Plot (0) as a function of Pr for Pr = 1..20.

19. Consider diffusion reaction problem with Langmuir-Hinshelwood kinetics (Finlayson, 1980). The governing equations and boundary conditions are:

$$\frac{d^2c}{dx^2} = \Phi^2 \frac{c}{(1+\alpha c)^2}$$

$$\frac{dc}{dx}(0) = 0 \text{ and } c(1) = 1$$

where $\alpha = 20$. Solve this problem as an initial value problem (see examples 3.2.12 and 3.2.13) and predict effectiveness factor as a function of Φ.

20. From problem 18, choose a value of Φ for which there are multiple steady states. For the chosen value of Φ, predict the multiple steady state concentration profiles using Maple's 'dsolve' numeric command and shooting technique.

21. Redo example 3.2.11 using finite difference technique.
22. Redo example 3.2.11 for Pe = Bo = 100.
23. Analyze problem 21 for multiple steady states. To do this, solve this problem using the shooting technique for the given set of parameters.
24. Consider the behavior of a thin sheet of viscous liquid emerging from a thin slot at the base of a converging channel in connection with a method of lacquer application known as "curtain coating."[6] The dimensionless governing equations and boundary conditions for the velocity are:

$$\frac{d^2y}{dx^2} - \frac{1}{y}\left(\frac{dy}{dx}\right)^2 - y\frac{dy}{dx} + 1 = 0$$

$$y(0) = 0.325 \text{ and } \frac{dy}{dx}(L) = \frac{1}{\sqrt{2L}} \text{ for large values of L}$$

25. Solve this problem using Maple's 'dsolve' numeric command, shooting technique and finite difference technique. Initially choose L = 5 and increase L to make sure that the solution has converged (i.e., change L = 6 and calculate $\frac{dy}{dx}(0)$). Compare the efficiency of the three methods for this problem.

26. Consider diffusion with a reversible reaction ($2A \Leftrightarrow B$) in a porous catalyst layer.[23] The total mass flux is given by:

$$N_A = -D\frac{dC_A}{dx} + \frac{C_A}{C_T}(N_A + N_B)$$

where D is the effective diffusion coefficient, C_T is the total concentration, N_B is the total mass flux of B and is given by $N_B = -N_A/2$. Shell balance gives:

3.2 Nonlinear Boundary Value Problems

$$-\frac{dN_A}{dx} = k\left(C_A^2 - \frac{C_T - C_A}{K}\right)$$

where k is the rate constant and K is the equilibrium constant. The boundary conditions are:

$$C_A(0) = 1 \times 10^{-5} \text{ mol/cm}^3 \text{ and } \frac{dC_A}{dx}(L) = 0$$

Use the first equation to eliminate N_A and obtain a governing equation for C_A. The values of parameters are $D = 1 \times 10^{-2}$ cm²/s, $C_T = 4 \times 10^{-5}$ mol/cm³, $L = 0.2$ cm, $k = 8 \times 10^4$ cm³/s/mol and $K = 6 \times 10^5$ cm³/mol. Solve this problem numerically (choose any appropriate numerical method) to obtain the concentration profile.

27. Consider multi-component diffusion of gases A and B through stagnant gas C (Gianakopulos, 1972; Cutlip and Shacham, 1999).[23] The governing equations for concentration of A and B are:

$$\frac{dC_A}{dz} = \frac{1}{C_T}\left(\frac{(C_A N_B - C_B N_A)}{D_{AB}} + \frac{(C_A N_C - C_C N_A)}{D_{AC}}\right)$$

$$\frac{dC_B}{dz} = \frac{1}{C_T}\left(\frac{(C_B N_A - C_A N_B)}{D_{AB}} + \frac{(C_B N_C - C_C N_B)}{D_{BC}}\right)$$

Concentration of C is given by the material balance $C_A + C_B + C_C = C_T$. Since C is stagnant N_C is zero. The boundary conditions are:
$C_A(0) = 2.229 \times 10^{-4}$ kg-mol/m³, $C_B(0) = 0$, $C_A(L) = 0$ and $C_B(L) = 2.701 \times 10^{-3}$ kg-mol/m³.

Values of the parameters are $L = 0.001$m, $D_{AB} = 1.47 \times 10^{-4}$ m²/s, $D_{AC} = 1.075 \times 10^{-4}$ m²/s, $D_{BC} = 1.245 \times 10^{-4}$ m²/s, and $C_T = 7.4309 \times 10^{-5}$ kg-mol/m³. The governing equations are first order in z and can be solved just by using the initial conditions at $z = 0$. However the values for N_A and N_B are not known. These should be found using the boundary conditions at $z = L$. Solve this problem using Maple's 'dsolve' numeric command, and finite difference technique. Plot the concentration profiles.

28. Gas A reacts with B to produce C in a finite liquid film.[23] The governing equations and boundary conditions are:

$$\frac{d^2 C_A}{dx^2} = \Phi^2 C_A C_B$$

$$\frac{d^2 C_B}{dx^2} = \beta\Phi^2 C_A C_B$$

$$C_A(0) = 1; \quad C_A(1) = 0$$

$$\frac{dC_B}{dx}(0) = 0; \quad C_B(1) = 1$$

where Φ is the Thiele modulus and β is the ratio of diffusion coefficient of B to A. Obtain the series solutions for this problem using Maple's 'dsolve' command if possible. Solve this problem using a suitable numerical method for $\Phi^2 = 3.2$ and $\beta = 0.5$.

29. Consider hydrodynamics in a rotating disk electrode.[15] [24] The governing equations for velocity distributions are:

$$2F + \frac{dH}{dx} = 0$$

$$F^2 - G^2 + H\frac{dF}{dx} = \frac{d^2F}{dx^2}$$

$$2FG + H\frac{dG}{dx} = \frac{d^2G}{dx^2}$$

$$H(0) = F(0) = 0 \text{ and } G(0) = 1$$

$$F(5) = G(5) = 0$$

Solve this problem using Maple's 'dsolve' numeric command plot for the velocity profiles. Obtain the series solutions for this problem by using unknown initial conditions calculated from the numerical solution. Is the series solution obtained convergent? For what values of x can these series solutions be safely used? Can accuracy be increased by adding more terms in the series?

30. Redo problem 27 by applying finite differences in x. How many node points are needed to obtain the converged solution?\

31. Consider a variant of Graetz problem discussed in examples 3.2.14 and 3.2.15 (Villadsen and Michelsen, 1978). The governing equation and boundary conditions are:

$$\frac{d^2y}{dx^2} + \lambda^2(1-x^2)y = 0$$

$$y(0) = 0; \quad \frac{dy}{dx}(0) = -1 \text{ and } \frac{dy}{dx}(1) = 0$$

Obtain the first five eigenvalues and eigenfunctions using the shooting technique described in example 3.2.15.

32. Consider diffusion with reaction in a pore with pore-mouth poisoning.[8] The governing equation and boundary conditions are:

$$\frac{d^2y}{dx^2} = \Phi^2 y^2 \begin{pmatrix} 1 & x \le 0.7 \\ \exp(-10^6[x-0.7]) \end{pmatrix}$$

$$\frac{dy}{dx}(0) = 0 \text{ and } y(1) = 1$$

Solve this equation using a suitable numerical method for $\Phi = 2$ and plot y and $\dfrac{dy}{dx}$ as a function of x. Note that Maple's piecewise function can be used to enter the right hand side of the governing equation.

33. Consider heat transfer in a fin with variable conductivity and nonlinear heat transfer coefficient.[25] The governing equations and boundary conditions are:

$$\dfrac{d}{dx}\left(-k_0[1+\alpha(T-T_0)]\dfrac{dT}{dx}\right) = -\dfrac{h_0\left(1-\dfrac{x^2}{L^2}\right)}{B}(T-T_a)$$

$$T(0) = T_w \text{ and } \dfrac{dT}{dx}(L) = 0$$

The values of the parameters are $h_0 = 40$ Btu/hr-ft^2-°F, $k_0 = 60$ Btu/hr-ft^2-°F, $\alpha = 0.02$ (°f)$^{-1}$, $T_{wo} = 450$°F, $T_0 = 77$°F, $T_a = 90$°F, $L = 1.5$ in and $B = 0.02$ in. Obtain series solutions for this problem using Maple's 'dsolve' command if possible. In addition, solve this problem using a suitable numerical technique.

34. Consider potential distribution in porous electrode (Newman, 1991;[15] exercise problem 7 of chapter 3.1). For nonlinear Butler-Volmer kinetics dimensionless overpotential η is governed by:

$$\dfrac{d^2\eta}{dx^2} = v^2\left(e^{0.5\eta} - e^{-0.5\eta}\right)$$

$$\dfrac{d\eta}{dx}(0) = \delta; \quad \dfrac{d\eta}{dx}(1) = -\delta\beta$$

Obtain the series solutions for this problem using Maple's 'dsolve' command if possible. Plot overpotential profiles for $v = \delta = 1$ and $\beta = -0.1$. In addition, solve this problem using a suitable numerical method.

References

1. Amundson, N.R.: Mathematical Methods in Chemical Engineering: Matrices and Their Applications. Prentice Hall, Inc., Englewood Cliffs (1966)
2. Taylor, R., Krishna, R.: Multicomponent Mass Transfer. Wiley & Sons, Chichester (1993)
3. Varma, A., Morbidelli, M.: Mathematical Methods in Chemical Engineering. Oxford University Press, Inc., Oxford (1997)
4. Subramanian, V.R., Haran, B.S., White, R.E.: Series solutions for boundary value problems using a symbolic successive substitution method. Computers & Chemical Engineering 23(3), 287–296 (1999)

5. Subramanian, V.R., White, R.E.: Symbolic solutions for boundary value problems using Maple. Computers & Chemical Engineering 24(11), 2405–2416 (2000)
6. Davis, M.E.: Numerical Methods and Modeling for Chemical Engineers. John Wiley & Sons, Chichester (1984)
7. Taylor, R.: Engineering Computing with Maple: Solution of PDEs via the Method of Lines. CACHE News 49 (Fall 1999)
8. Villadsen, J., Michelsen, M.L.: Solution of Differential Equation Models by Polynomial Approximation. Prentice Hall, Inc., Englewood Cliffs (1978)
9. Crassidis, J.L., John, L.: Optimal Estimation of Dynamic Systems. Chapman & Hall/CRC Press, Boca Raton (2004)
10. DeCarlo, R.A.: Linear Systems: A State Variable Approach with Numerical Implementation. Prentice Hall, Inc., Englewood Cliffs (1989)
11. Finlayson, B.A.: Nonlinear Analysis in Chemical Engineering. McGraw-Hill, New York (1980)
12. Schiesser, W.E., Silebi, C.A.: Dynamic Modeling of Transport Process Systems (1997)
13. Bird, R.B., Stewart, W.E., Lightfoot, E.N.: Transport Phenomena. John Wiley & Sons, Inc., Chichester (1960)
14. Aris, R.: Mathematical Modeling: A Chemical Engineer's Perspective. Academic Press, London (1999)
15. Newman, J.: Electrochemical Systems. Prentice Hall, New York (1991)
16. Newman, J., Tiedemann, W.: Porous-electrode Theory with Battery Applications. AIChE Journal 21(1), 25–41 (1975)
17. Crank, J.: The Mathematics of Diffusion, 2nd edn. Oxford University Press, Oxford (1979)
18. Rice, R.G., Do, D.D.: Applied Mathematics and Modeling for Chemical Engineers. John Wiley & Sons, Inc., Chichester (1995)
19. Ramakrishna, D., Amundson, N.R.: Linear Operator Methods in Chemical Engineering: with Applications to Transport and Chemical Reaction Systems. Prentice Hall, Inc., New York (1985)
20. White, R.E., et al.: Extension of Darby's Model of a Hydrophilic Gas Fed Porous Electrode. Journal of the Electrochemical Society 131, 268–274 (1984)
21. Constantinides, A., Mostoufi, N.: Numerical Methods for Chemical Engineers with MATLAB Applications. Prentice-Hall PTR, Englewood Cliffs (1999)
22. Hanna, O.T., Sandall, O.C.: Computational Methods in Chemical Engineering. Prentice Hall, Inc., Englewood Cliffs (1995)
23. Cutlip, M.B., Shacham, M.: Problem Solving in Chemical Engineering with Numerical Methods. Prentice Hall PTR, Englewood Cliffs (1999)
24. Levich, B.G.: Physiochemical Hydrodynamics. Prentice Hall, Inc., Englewood Cliffs (1962)
25. Riggs, J.B.: An Introduction to Numerical Methods for Chemical Engineers. Texas Tech. University Press (1988)

Chapter 4
Partial Differential Equations in Semi-infinite Domains

Mathematical modeling of mass or heat transfer in solids involves Fick's law of mass transfer or Fourier's law of heat conduction. Engineers are interested in the distribution of heat or concentration across the slab, or the material in which the experiment is performed. This process is represented by parabolic partial differential equations (unsteady state) or elliptic partial differential equations. When the length of the domain is large, it is reasonable to consider the domain as semi-infinite which simplifies the problem and helps in obtaining analytical solutions. These partial differential equations are governed by the initial condition and the boundary condition at $x = 0$. The dependent variable has to be finite at distances far ($x = \infty$) from the origin. Both parabolic and elliptic partial differential equations will be discussed in this chapter. The Laplace transform technique will be used for parabolic partial differential equations. A similarity solution technique will be used for parabolic, elliptic and nonlinear partial differential equations.

4.1 Partial Differential Equations (PDEs) in Semi-infinite Domains

Transient heat conduction or mass transfer in solids with constant physical properties (diffusion coefficient, thermal diffusivity, thermal conductivity, etc.) is usually represented by a parabolic partial differential equation. For steady state heat or mass transfer in solids, potential distribution in electrochemical cells is usually represented by elliptic partial differential equations. In this chapter, we describe how one can arrive at the analytical solutions for linear parabolic partial differential equations and elliptic partial differential equations in semi-infinite domains using the Laplace transform technique, a similarity solution technique and Maple. In addition, we describe how numerical similarity solutions can be obtained for nonlinear partial differential equations in semi-infinite domains.

4.2 Laplace Transform Technique for Parabolic PDEs

Parabolic partial differential equations are solved using the Laplace transform technique in this section. Diffusion like partial differential equations are first order in the time variable and second order in the spatial variable. This method involves applying the Laplace transform in the time variable to convert the partial differential equation to an ordinary differential equation in the Laplace domain, which becomes a boundary value problem (ordinary differential equation, ODE) in the spatial direction with s, the Laplace variable, as a parameter. The boundary conditions are converted to the Laplace domain and the differential equation in the Laplace domain is solved by using the techniques illustrated in chapter 3.1 for solving linear boundary value problems. The methodology is very similar to the technique illustrated in chapter 3.1 for solving boundary value problems (BVPs) in the semi-infinite domain. Once an analytical solution is obtained in the Laplace domain, the solution is inverted to a time domain to obtain the final analytical solution (in time and spatial coordinates), which is shown in the following examples.

Example 4.1. Heat Conduction in a rectangular slab

Consider the following transient heat conduction problem in a slab.[1-3] The governing equation is:

$$\frac{\partial u}{\partial t} = \alpha \frac{\partial^2 u}{\partial x^2}$$

$$u(x,0) = 1 \qquad (4.1)$$

$$u(0,t) = 0 \text{ and } u(\infty,t) \text{ is defined}$$

where α is the thermal diffusivity (m^2/s). Equation (4.1) is solved below using Maple:

> restart:with(linalg):with(inttrans):with(plots):

The governing equation is stored in the equation:

> eq:=diff(u(x,t),t)=alpha*diff(u(x,t),x$2);

$$eq := \frac{\partial}{\partial t} u(x,t) = \alpha \left(\frac{\partial^2}{\partial x^2} u(x,t) \right)$$

Enter the initial condition here:

> u(x,0):=1;

$$u(x,0) := 1$$

The boundary condition at $x = 0$ is entered here:

4.2 Laplace Transform Technique for Parabolic PDEs

> bc1:=u(0,t)=0;

$$bc1 := u(0, t) = 0$$

Enter the second boundary condition here:

> bc2:=u(infinity,t)=defined;

$$bc2 := u(\infty, t) = defined$$

The governing equation and the boundary condition at $x = 0$ are converted to the Laplace domain:

> eqs:=laplace(eq,t,s);

$$eqs := s \, laplace(u(x, t), t, s) - 1 = \alpha \left(\frac{\partial^2}{\partial x^2} laplace(u(x, t), t, s) \right)$$

The given partial differential equation is transformed to an ordinary differential equation in the Laplace domain since

> eqs:=subs(laplace(u(x,t),t,s)=U(x),eqs);

$$eqs := s \, U(x) - 1 = \alpha \left(\frac{d^2}{dx^2} U(x) \right)$$

where U(x) is the dependent variable in the Laplace domain:

> bc1:=laplace(bc1,t,s);

$$bc1 := laplace(u(0, t), t, s) = 0$$

> bc1:=subs(laplace(u(0,t),t,s)=U(0),bc1);

$$bc1 := U(0) = 0$$

Next, the dependent variable in the Laplace domain is solved using the 'dsolve' command since (see chapter 3.1.6):

> U(x):=rhs(dsolve(g,U(x)));

$$U(x) := e^{\frac{\sqrt{s}\,x}{\sqrt{\alpha}}} \_C2 - \frac{e^{-\frac{\sqrt{s}\,x}{\sqrt{\alpha}}} (\_C2\,s + 1)}{s} + \frac{1}{s}$$

The constant _C2 is found using the boundary condition at $x = \infty$. The dependent variable U(x) is defined at $x = \infty$. In the above expression $\left(\frac{\sqrt{s}x}{\sqrt{\alpha}} \right)$

becomes ∞ as x tends to ∞. Hence, in the above expression, the coefficient of $\left(\dfrac{\sqrt{sx}}{\sqrt{\alpha}}\right)$ is equated to zero:

> eqc:=coeff(U(x),exp(1/alpha^(1/2)*s^(1/2)*x));

$$eqc := \_C2$$

> _C2:=solve(eqc,_C2);

$$\_C2 := 0$$

This simplifies the solution since:

> U(x):=eval(U(x));

$$U(x) := -\frac{e^{-\frac{\sqrt{s}\,x}{\sqrt{\alpha}}}}{s} + \frac{1}{s}$$

The solution obtained in the Laplace domain is converted to the time domain since:

> u:=invlaplace(U(x),s,t);

$$u := -invlaplace\left(\frac{e^{-\frac{\sqrt{s}\,x}{\sqrt{\alpha}}}}{s}, s, t\right) + 1$$

This solution can be further simplified since:

> u:=convert(u,erf);

$$u := -invlaplace\left(\frac{e^{-\frac{\sqrt{s}\,x}{\sqrt{\alpha}}}}{s}, s, t\right) + 1$$

The solution obtained can be plotted for a particular value of the parameter α since:

> plot3d(subs(alpha=0.001,u),x=1..0,t=500..0,axes=boxed,title="Figure 4.1", labels =[x,t,"u"],orientation=[120,60]);

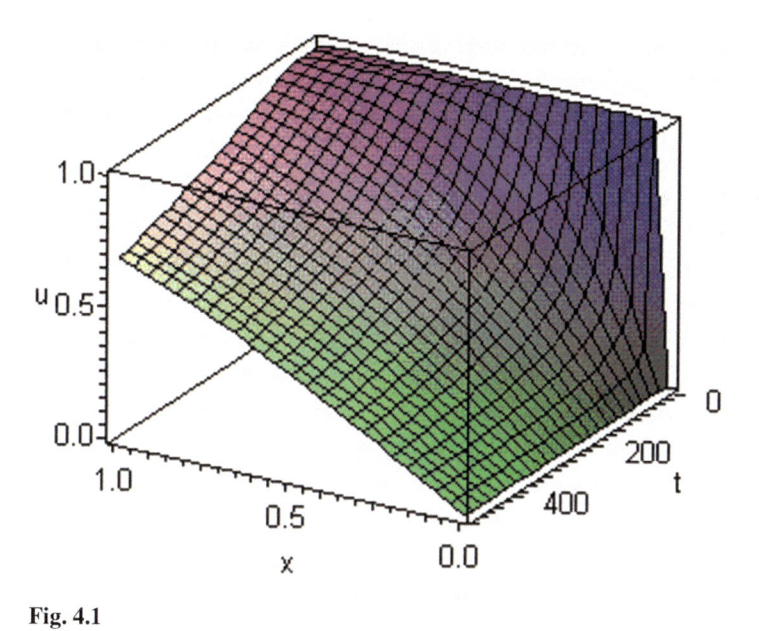

4.2 Laplace Transform Technique for Parabolic PDEs

Fig. 4.1

The same plot is made for a different value of α here:

> plot3d(subs(alpha=0.1,u),x=1..0,t=50..0,axes=boxed,title="Figure 4.2", labels =[x,t,"u"],orientation=[120,60]);

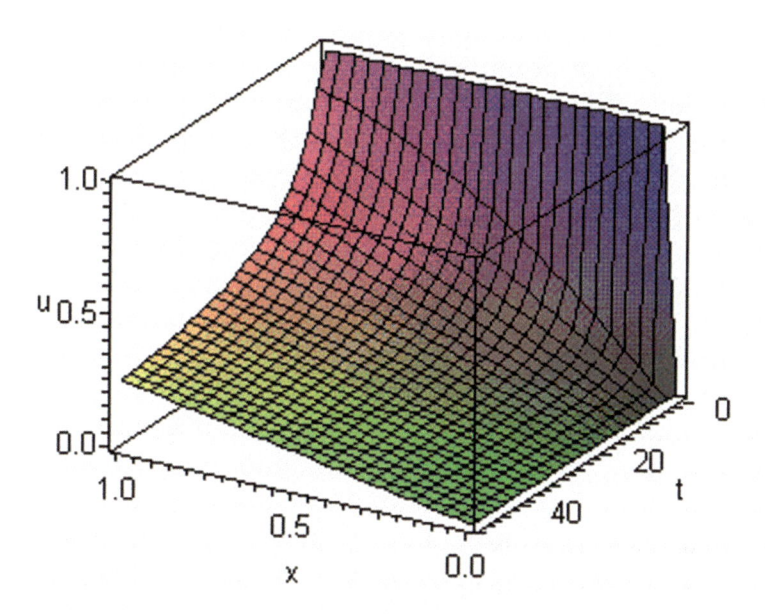

Fig. 4.2

300   4 Partial Differential Equations in Semi-infinite Domains

Next, the dimensionless temperature u is plotted versus x for different values of time as shown below:

>pfs:=plot([subs(alpha=0.001,t=1,u),subs(alpha=0.001,t=10,u), subs(alpha=0.001,t=100,u),subs(alpha=0.001,t=200,u)],x=0..1, axes=boxed,title="Figure 4.3",thickness=4,labels=[x,"u"]);

$$pfs := PLOT(...)$$

> pts:=textplot([[0.12,evalf(subs(alpha=0.001,t=1,x=0.08,u)),"t=1"], [0.25,evalf(subs(alpha=0.001,t=10,x=0.2,u)),"t=10"], [0.58,evalf(subs(alpha=0.001,t=100,x=0.5,u)),"t=100"], [0.69,evalf(subs(alpha=0.001,t=200,x=0.6,u)),"t=200"]]);

$$pts := PLOT(...)$$

> display({});

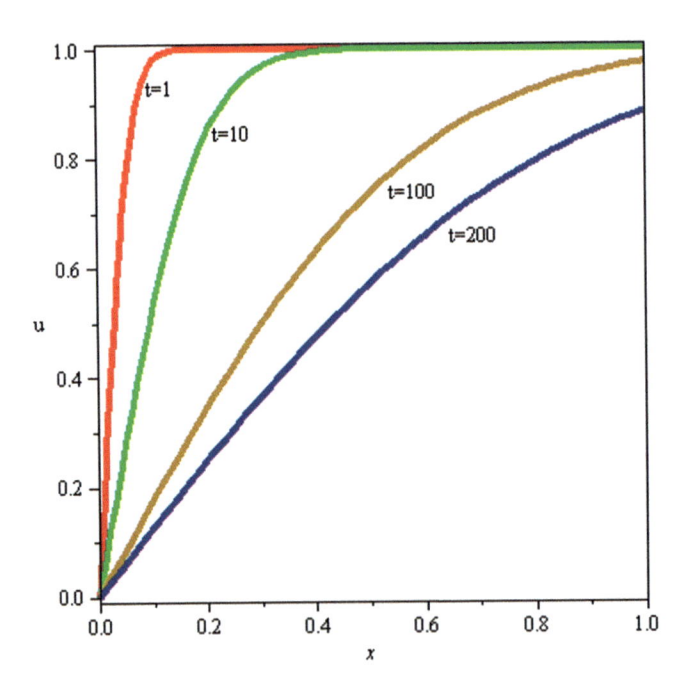

Fig. 4.3

An animation in time can be made since:

> animate(subs(alpha=0.001,u),x=0..1,t=1..500,thickness=4,title="Figure 4.4", axes =boxed,labels=[x,"u"]);

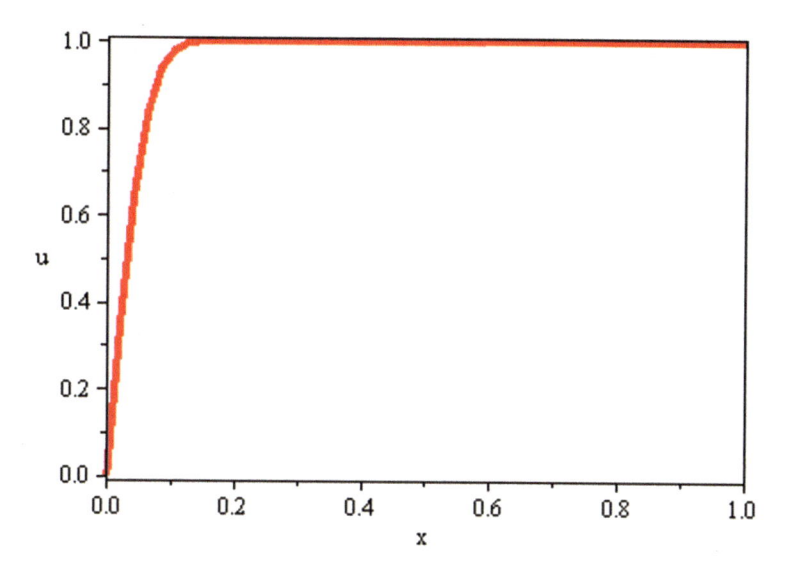

4.2 Laplace Transform Technique for Parabolic PDEs

Fig. 4.4

Example 4.2. Heat Conduction with Transient Boundary Conditions

Heat conduction with a constant boundary condition at x =0 was considered in example 4.1. The same technique can be applied for time dependent boundary conditions. Consider the transient heat conduction problem in a slab.[4] The governing equation is:

$$\frac{\partial u}{\partial t} = \alpha \frac{\partial^2 u}{\partial x^2}$$

$$u(x,0) = 0 \tag{4.2}$$

$$u(0,t) = \frac{k}{\sqrt{t}} \text{ and } u(\infty,t) \text{ is defined}$$

Equation (4.2) is solved in Maple below:

```
> restart:with(linalg):with(inttrans):with(plots):
> eq:=diff(u(x,t),t)=alpha*diff(u(x,t),x$2);
```

$$eq := \frac{\partial}{\partial t} u(x,t) = \alpha \left(\frac{\partial^2}{\partial x^2} u(x,t) \right)$$

```
> u(x,0):=0;
```
$$u(x, 0) := 0$$

```
> bc1:=u(0,t)=k/t^(1/2);
```
$$bc1 := u(0, t) = \frac{k}{\sqrt{t}}$$

```
> bc2:=u(infinity,t)=defined;
```
$$bc2 := u(\infty, t) = defined$$

```
> eqs:=laplace(eq,t,s):
```

The governing equation in the Laplace domain is:

```
> eqs:=subs(laplace(u(x,t),t,s)=U(x),eqs);
```
$$eqs := s\, U(x) = \alpha\left(\frac{d^2}{dx^2} U(x)\right)$$

```
> bc1:=laplace(bc1,t,s);
```
$$bc1 := laplace(u(0, t), t, s) = \sqrt{\frac{\pi}{s}}\, k$$

The boundary condition in the Laplace domain is:

```
> bc1:=subs(laplace(u(0,t),t,s)=U(0),bc1);
```
$$bc1 := U(0) = \sqrt{\frac{\pi}{s}}\, k$$

```
> U(x):=rhs(dsolve({eqs,bc1},U(x)));
```
$$U(x) := \left(\sqrt{\frac{\pi}{s}}\, k - \_C2\right) e^{\frac{\sqrt{s}\, x}{\sqrt{\alpha}}} + \_C2\, e^{-\frac{\sqrt{s}\, x}{\sqrt{\alpha}}}$$

```
> eqc:=coeff(U(x),exp(1/alpha^(1/2)*s^(1/2)*x));
```
$$eqc := \sqrt{\frac{\pi}{s}}\, k - \_C2$$

```
> _C2:=solve(eqc,_C2);
```
$$\_C2 := \sqrt{\frac{\pi}{s}}\, k$$

4.2 Laplace Transform Technique for Parabolic PDEs

The dimensionless temperature in the Laplace domain is:

> U(x):=eval(U(x));

$$U(x) := \sqrt{\frac{\pi}{s}}\, k\, e^{-\frac{\sqrt{s}\, x}{\sqrt{\alpha}}}$$

Next, the solution is inverted to the time domain:

> u:=invlaplace(U(x),s,t);

$$u := k\sqrt{\pi}\ \text{invlaplace}\!\left(\frac{e^{-\frac{\sqrt{s}\, x}{\sqrt{\alpha}}}}{\sqrt{s}}, s, t\right)$$

Maple is not able to invert the solution in the Laplace domain. This can be solved by using dummy variables for x and α and defining them to be positive:

> U(x):=subs(x=x1,alpha=alpha1,U(x));

$$U(x) := \sqrt{\frac{\pi}{s}}\, k\, e^{-\frac{\sqrt{s}\, x1}{\sqrt{\alpha 1}}}$$

> assume(x1>0,alpha1>0);
> u:=invlaplace(U(x),s,t);

$$u := \frac{k\, e^{-\frac{1}{4}\frac{x1\text{\textasciitilde}^2}{\alpha 1\text{\textasciitilde}\, t}}}{\sqrt{t}}$$

> u:=subs(x1=x,alpha1=alpha,u);

$$u := \frac{k\, e^{-\frac{1}{4}\frac{x^2}{\alpha\, t}}}{\sqrt{t}}$$

Hence, the final solution is:

> pars:={alpha=0.001,k=1};

$$pars := \{\alpha = 0.001, k = 1\}$$

> plot3d(subs(pars,u),x=1..0,t=300..0,axes=boxed,title="Figure 4.5", labels =[x,t,"u"],orientation=[45,45]);

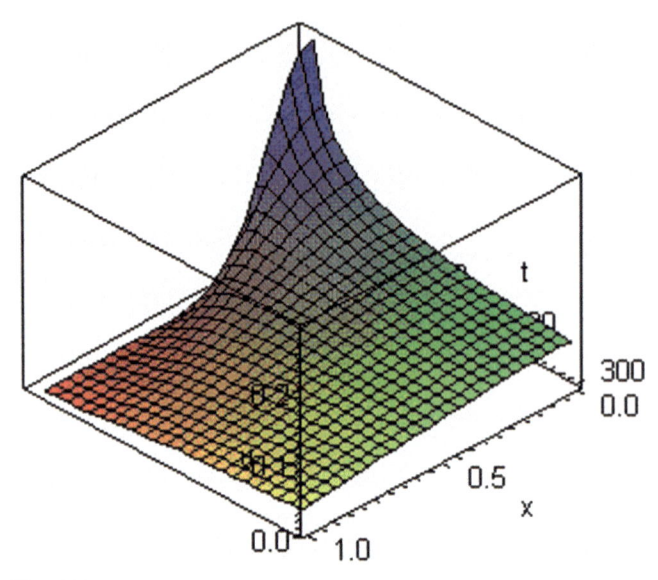

Fig. 4.5

```
> plot([subs(pars,t=1,u),subs(pars,t=10,u),subs(pars,t=100,u),
subs(pars,t=200,u)],x=0..1,axes=boxed,title="Figure 4.6",thickness=5,
labels=[x,"u"],legend=["t=1","t=10","t=100","t=200"]);
```

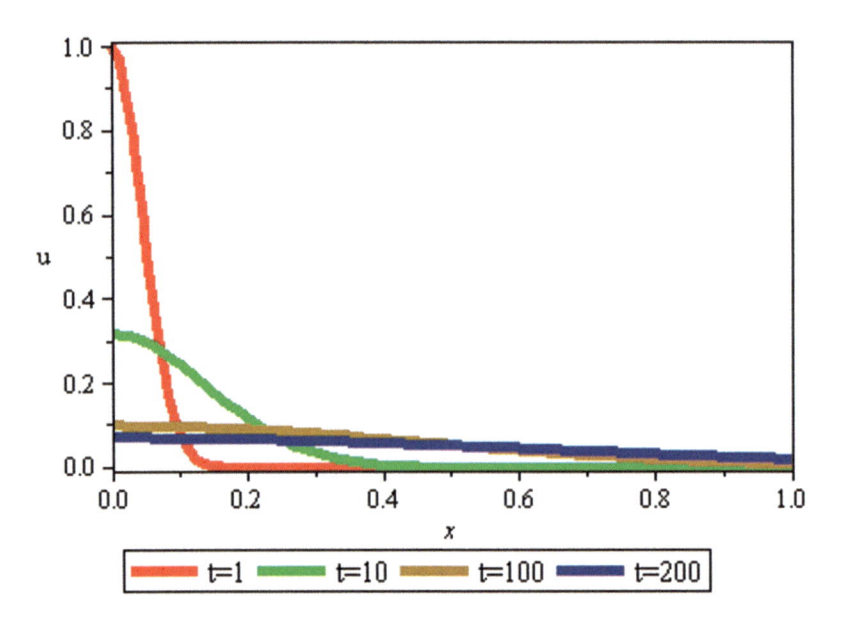

Fig. 4.6

You can make an animation using the command illustrated in example 4.1.

4.2 Laplace Transform Technique for Parabolic PDEs

Example 4.3. Heat Conduction with Flux Boundary Conditions

In the previous two examples, the temperature (dependent variable) at x = 0 was specified. The same technique can be applied for the case where the derivative of the dependent variable is known at the boundary x = 0 (flux boundary conditions). Consider the transient heat conduction problem in a slab.[4] The governing equation in dimensionless form is

$$\frac{\partial u}{\partial t} = \alpha \frac{\partial^2 u}{\partial x^2}$$

$$u(x,0) = 0 \tag{4.3}$$

$$\frac{\partial u}{\partial x}(0,t) = \text{-k and } u(\infty,t) \text{ is defined}$$

The flux boundary condition has to be considered while taking the Laplace transform. Equation (4.3) is solved in Maple below:

```
> restart:with(linalg):with(inttrans):with(plots):
> eq:=diff(u(x,t),t)=alpha*diff(u(x,t),x$2);
```

$$eq := \frac{\partial}{\partial t} u(x, t) = \alpha \left(\frac{\partial^2}{\partial x^2} u(x, t) \right)$$

```
> u(x,0):=0;
```

$$u(x, 0) := 0$$

```
> bc1:=diff(u(x,t),x)=-k;
```

$$bc1 := \frac{\partial}{\partial x} u(x, t) = -k$$

```
> bc2:=u(infinity,t)=defined;
```

$$bc2 := u(\infty, t) = defined$$

```
> eqs:=laplace(eq,t,s):
> eqs:=subs(laplace(u(x,t),t,s)=U(x),eqs);
```

$$eqs := s\, U(x) = \alpha \left(\frac{d^2}{dx^2} U(x) \right)$$

> bc1:=laplace(bc1,t,s);

$$bc1 := \frac{\partial}{\partial x} \text{laplace}(u(x,t),t,s) = -\frac{k}{s}$$

The boundary condition in the Laplace domain is:

> bc1:=subs(laplace(u(x,t),t,s)=U(x),bc1);

$$bc1 := \frac{d}{dx} U(x) = -\frac{k}{s}$$

> bc1:=subs(x=0,convert(bc1,D));

$$bc1 := D(U)(0) = -\frac{k}{s}$$

> U(x):=rhs(dsolve({eqs,bc1},U(x)));

$$U(x) := -\frac{(k\sqrt{\alpha} - \_C2\, s^{3/2})\, e^{\frac{\sqrt{s}\, x}{\sqrt{\alpha}}}}{s^{3/2}} + \_C2\, e^{-\frac{\sqrt{s}\, x}{\sqrt{\alpha}}}$$

> eqc:=coeff(U(x),exp(1/alpha^(1/2)*s^(1/2)*x));

$$eqc := -\frac{k\sqrt{\alpha} - \_C2\, s^{3/2}}{s^{3/2}}$$

> _C2:=solve(eqc,_C2);

$$\_C2 := \frac{k\sqrt{\alpha}}{s^{3/2}}$$

> U(x):=eval(U(x));

$$U(x) := \frac{k\sqrt{\alpha}\, e^{-\frac{\sqrt{s}\, x}{\sqrt{\alpha}}}}{s^{3/2}}$$

> u:=invlaplace(U(x),s,t);

$$u := k\sqrt{\alpha}\, \text{invlaplace}\left(\frac{e^{-\frac{\sqrt{s}\, x}{\sqrt{\alpha}}}}{s^{3/2}}, s, t\right)$$

4.2 Laplace Transform Technique for Parabolic PDEs

The dimensionless temperature profile is given by:

> u:=convert(u,erfc);

$$u := k\sqrt{\alpha}\ invlaplace\left(\frac{e^{-\frac{\sqrt{s}\,x}{\sqrt{\alpha}}}}{s^{3/2}}, s, t\right)$$

> pars:={alpha=0.001,k=1};

$$pars := \{\alpha = 0.001, k = 1\}$$

Plots are made for particular values of parameters:

> plot3d(subs(pars,u),x=1..0,t=300..0,axes=boxed,title="Figure 4.7",
labels=[x,t,"u"],orientation=[-60,60]);

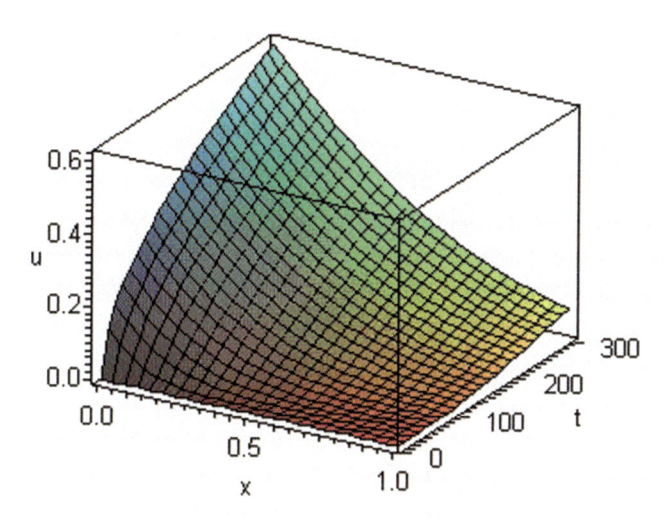

Fig. 4.7

> plot([subs(pars,t=1,u),subs(pars,t=10,u),subs(pars,t=100,u),
subs(pars,t=200,u)],x=0..1,axes=boxed,title="Figure 4.8",
thickness=5,labels=[x,"u"],legend=["t=1","t=10","t=100","t=200"]);

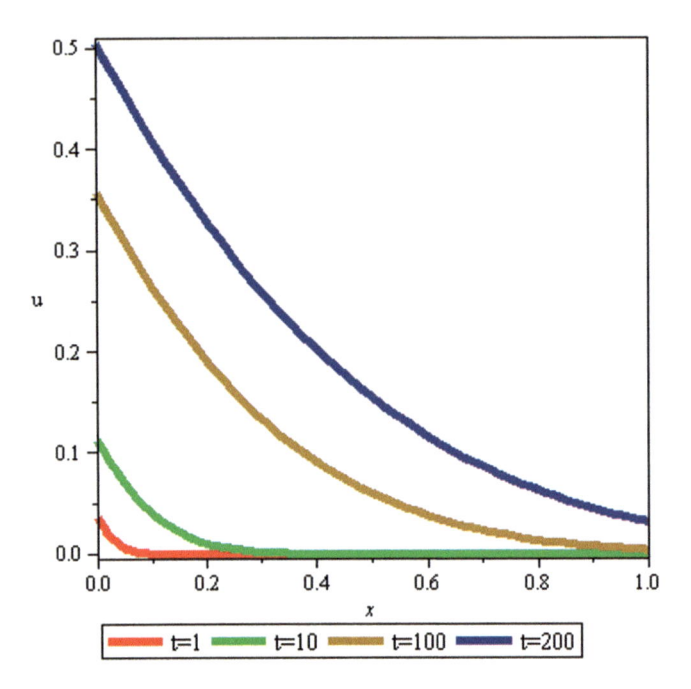

Fig. 4.8

Example 4.4. Heat Conduction with an Initial Profile

In the previous examples, the initial condition was a constant and independent of x. The same technique can be applied for the case where there is an initial temperature profile. Consider the transient heat conduction problem in a slab

$$\frac{\partial u}{\partial t} = \alpha \frac{\partial^2 u}{\partial x^2}$$

$$u(x,0) = \sin(\pi x) \tag{4.4}$$

$$u(0,t) = 0 \text{ and } u(\infty,t) \text{ is defined}$$

Equation (4.4) is solved in Maple below. The programs given for the previous examples have to be modified to solve equation (4.4) by only changing the initial condition:

```
> restart:with(linalg):with(inttrans):with(plots):
> eq:=diff(u(x,t),t)=alpha*diff(u(x,t),x$2);
```

$$eq := \frac{\partial}{\partial t} u(x,t) = \alpha \left(\frac{\partial^2}{\partial x^2} u(x,t) \right)$$

4.2 Laplace Transform Technique for Parabolic PDEs

> u(x,0):=sin(Pi*x);
$$u(x, 0) := \sin(\pi x)$$

> bc1:=u(0,t)=0;
$$bc1 := u(0, t) = 0$$

> bc2:=u(infinity,t)=defined;
$$bc2 := u(\infty, t) = defined$$

The following solution and plots are obtained:

> eqs:=laplace(eq,t,s);
$$eqs := s\, laplace(u(x, t), t, s) - \sin(\pi x) = \alpha \left(\frac{\partial^2}{\partial x^2} laplace(u(x, t), t, s) \right)$$

> eqs:=subs(laplace(u(x,t),t,s)=U(x),eqs);
$$eqs := s\, U(x) - \sin(\pi x) = \alpha \left(\frac{d^2}{dx^2} U(x) \right)$$

> bc1:=laplace(bc1,t,s);
$$bc1 := laplace(u(0, t), t, s) = 0$$

> bc1:=subs(laplace(u(0,t),t,s)=U(0),bc1);
$$bc1 := U(0) = 0$$

> U(x):=rhs(dsolve({eqs,bc1},U(x)));
$$U(x) := e^{\frac{\sqrt{s}\, x}{\sqrt{\alpha}}} \_C2 - \_C2\, e^{-\frac{\sqrt{s}\, x}{\sqrt{\alpha}}} + \frac{\sin(\pi x)}{s + \pi^2 \alpha}$$

> eqc:=coeff(U(x),exp(1/alpha^(1/2)*s^(1/2)*x));
$$eqc := \_C2$$

> _C2:=solve(eqc,_C2);
$$\_C2 := 0$$

> U(x):=eval(U(x));
$$U(x) := \frac{\sin(\pi x)}{s + \pi^2 \alpha}$$

> u:=invlaplace(U(x),s,t);
$$u := \sin(\pi x)\, e^{-\pi^2 \alpha t}$$

> pars:={alpha=0.001};
$$pars := \{\alpha = 0.001\}$$

310   4 Partial Differential Equations in Semi-infinite Domains

> plot3d(subs(pars,u),x=10..0,t=100..0,axes=boxed,title="Figure 4.9",
labels=[x,t,"u"],orientation=[-60,60]);

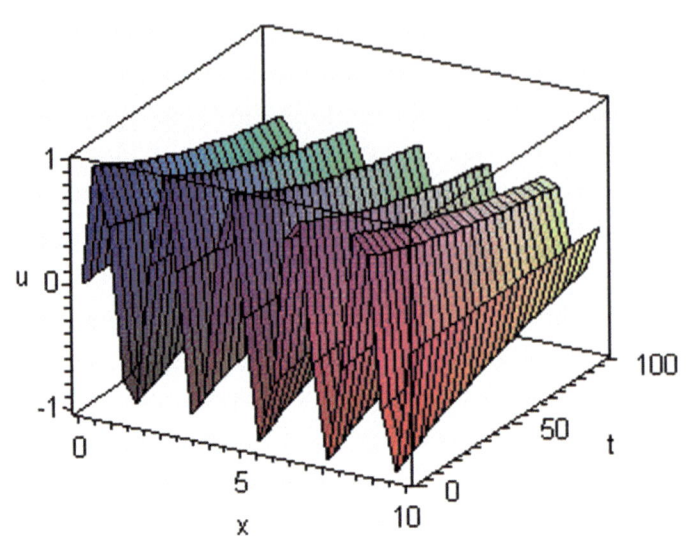

Fig. 4.9

> plot([subs(pars,t=1,u),subs(pars,t=10,u),subs(pars,t=100,u),
subs(pars,t=200,u)],x=0..1,axes=boxed,title="Figure 4.10",
thickness=5,labels=[x,"u"]);

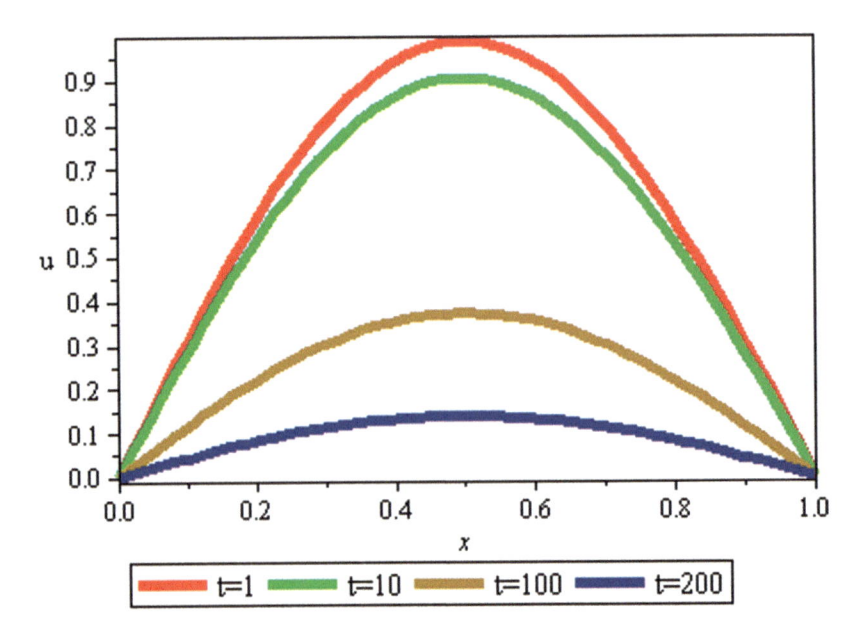

Fig. 4.10

4.2 Laplace Transform Technique for Parabolic PDEs

Example 4.5. Heat Conduction with a Source Term

The technique illustrated in the previous examples can be applied for the case where there is a source term (this source term can be a function of x and t). Consider the transient heat conduction problem in a slab[5]

$$\frac{\partial u}{\partial t} = \alpha \frac{\partial^2 u}{\partial x^2} + \sin(x)e^{-t}$$

$$u(x,0) = 0 \qquad (4.5)$$

$$u(0,t) = 0 \text{ and } u(\infty,t) \text{ is defined}$$

Equation (4.5) is solved in Maple below. The programs given for the previous example can be modified to solve equation (4.5). Only the governing equation has to be changed since:

> restart:with(linalg):with(inttrans):with(plots):

Only the governing equation has to be changed since:

> eq:=diff(u(x,t),t)=alpha*diff(u(x,t),x$2)+sin(x)*exp(-t);

$$eq := \frac{\partial}{\partial t} u(x,t) = \alpha \left(\frac{\partial^2}{\partial x^2} u(x,t) \right) + \sin(x)\, e^{-t}$$

> u(x,0):=0;

$$u(x,0) := 0$$

> bc1:=u(0,t)=0;

$$bc1 := u(0,t) = 0$$

> bc2:=u(infinity,t)=defined;

$$bc2 := u(\infty, t) = defined$$

The following solution and plots are obtained:

> eqs:=laplace(eq,t,s);

$$eqs := s\, laplace(u(x,t), t, s) = \alpha \left(\frac{\partial^2}{\partial x^2} laplace(u(x,t), t, s) \right) + \frac{\sin(x)}{1+s}$$

> eqs:=subs(laplace(u(x,t),t,s)=U(x),eqs);

$$eqs := s\,U(x) = \alpha\left(\frac{d^2}{dx^2}U(x)\right) + \frac{\sin(x)}{1+s}$$

> bc1:=laplace(bc1,t,s);

$$bc1 := laplace\,(u(0,t),t,s) = 0$$

> bc1:=subs(laplace(u(0,t),t,s)=U(0),bc1);

$$bc1 := U(0) = 0$$

> U(x):=rhs(dsolve({eqs,bc1},U(x)));

$$U(x) := e^{\frac{\sqrt{s}\,x}{\sqrt{\alpha}}}\_C2 - \_C2\,e^{-\frac{\sqrt{s}\,x}{\sqrt{\alpha}}} + \frac{\sin(x)}{(1+s)(s+\alpha)}$$

> eqc:=coeff(U(x),exp(1/alpha^(1/2)*s^(1/2)*x));

$$eqc := \_C2$$

> _C2:=solve(eqc,_C2);

$$\_C2 := 0$$

> U(x):=eval(U(x));

$$U(x) := \frac{\sin(x)}{(1+s)(s+\alpha)}$$

> u:=invlaplace(U(x),s,t);

$$u := \frac{\sin(x)\left(e^{-t} - e^{-\alpha t}\right)}{-1+\alpha}$$

> pars:={alpha=1/5};

$$pars := \left\{\alpha = \frac{1}{5}\right\}$$

> plot3d(subs(pars,u),t=5..0,x=10..0,axes=boxed,title="Figure 4.11", labels=[x,t,"u"]);

4.2 Laplace Transform Technique for Parabolic PDEs   313

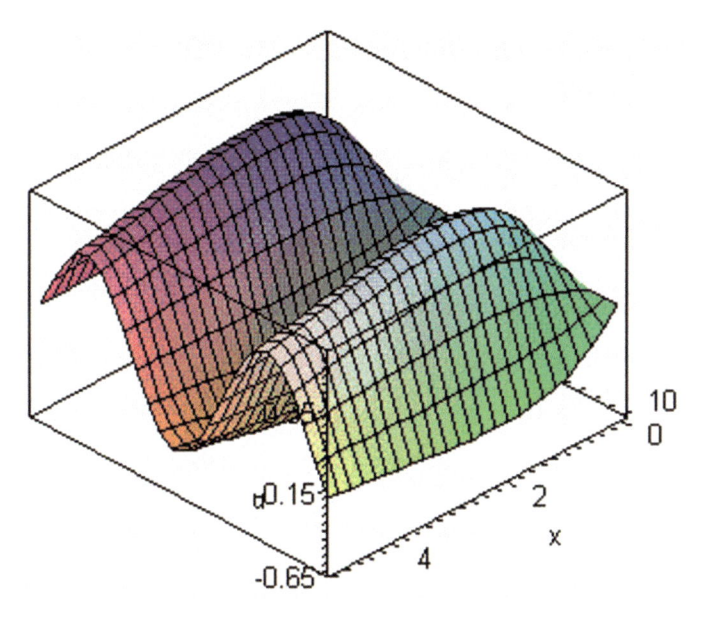

Fig. 4.11

> plot([subs(pars,t=0,u),subs(pars,t=0.5,u),subs(pars,t=1,u),
subs(pars,t=2,u)],x=0..10,axes=boxed,title="Figure 4.12",
thickness=5,labels=[x,"u"]);

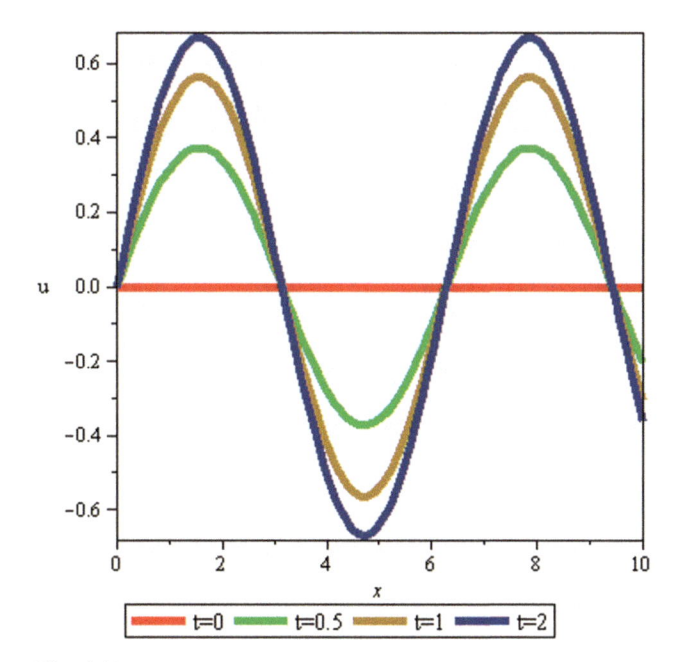

Fig. 4.12

4.3 Laplace Transform Technique for Parabolic PDEs – Advanced Problems

For some complicated problems, Maple cannot find the inverse Laplace transform. In these cases, one can split use standard Laplace transform formulae to simplify the expressions. By manipulating the expressions, Maple can be used to find the inverse Laplace transform. This is best illustrated with the following examples.

Example 4.6. Heat Conduction with Radiation at the Surface

Consider the transient heat conduction problem in a slab.[4] The governing equation in dimensionless form is

$$\frac{\partial u}{\partial t} = \alpha \frac{\partial^2 u}{\partial x^2}$$

$$u(x,0) = 1 \tag{4.6}$$

$$\frac{\partial u}{\partial x}(0,t) = hu(0,t) \text{ and } u(\infty,t) \text{ is defined}$$

where α is the thermal diffusivity $\left(m^2/s\right)$ and h is the heat transfer coefficient $(m-1)$. Carslaw and Jaeger[4] presented solutions for this problem after transforming the governing equation and boundary conditions to a form convenient for similarity transformation. Equation (4.6) is solved in Maple below using the Laplace transform technique (note that the transformation is not necessary with this approach):

> restart:with(linalg):with(inttrans):with(plots):

The governing equation is entered here:

> eq:=diff(u(x,t),t)=alpha*diff(u(x,t),x$2);

$$eq := \frac{\partial}{\partial t} u(x,t) = \alpha \left(\frac{\partial^2}{\partial x^2} u(x,t) \right)$$

> u(x,0):=1;

$$u(x,0) := 1$$

Enter the boundary condition here:

> bc1:=diff(u(x,t),x)-h*u(x,t)=0;

4.3 Laplace Transform Technique for Parabolic PDEs – Advanced Problems

$$bc1 := \frac{\partial}{\partial x} u(x,t) - h\, u(x,t) = 0$$

> bc2:=u(infinity,t)=defined;

$$bc2 := u(\infty, t) = defined$$

> eqs:=laplace(eq,t,s);

$$eqs := s\, laplace(u(x,t), t, s) - 1 = \alpha \left(\frac{\partial^2}{\partial x^2} laplace(u(x,t), t, s) \right)$$

> eqs:=subs(laplace(u(x,t),t,s)=U(x),eqs);

$$eqs := s\, U(x) - 1 = \alpha \left(\frac{d^2}{dx^2} U(x) \right)$$

> bc1:=laplace(bc1,t,s);

$$bc1 := \frac{\partial}{\partial x} laplace(u(x,t), t, s) - h\, laplace(u(x,t), t, s) = 0$$

> bc1:=subs(laplace(u(x,t),t,s)=U(x),bc1);

$$bc1 := \frac{d}{dx} U(x) - h\, U(x) = 0$$

> bc1:=convert(bc1,D);

$$bc1 := D(U)(x) - h\, U(x) = 0$$

The boundary condition in the Laplace domain is:

> bc1:=subs(x=0,bc1);

$$bc1 := D(U)(0) - h\, U(0) = 0$$

U is solved as:

> U(x):=rhs(dsolve({eqs,bc1}));

$$U(x) := \frac{1}{2} \frac{e^{\frac{\sqrt{s}\, x}{\sqrt{\alpha}}} \left(\sqrt{s}\, h\, \_C1\, \sqrt{\alpha} + \_C1\, s - 1 \right)}{s} + \frac{1}{2} \frac{e^{-\frac{\sqrt{s}\, x}{\sqrt{\alpha}}} \left(\_C1\, s - \sqrt{s}\, h\, \_C1\, \sqrt{\alpha} - 1 \right)}{s} + \frac{1}{s}$$

> eqc:=coeff(U(x),exp(1/alpha^(1/2)*s^(1/2)*x));

$$eqc := \frac{1}{2} \frac{\sqrt{s}\, h\, \_C1\, \sqrt{\alpha} + \_C1\, s - 1}{s}$$

> coef:=solve(eqc,{_C1});

$$coef := \left\{ \_C1 = \frac{1}{\sqrt{s}\, h\, \sqrt{\alpha} + s} \right\}$$

The temperature profile in the Laplace domain is:

> U(x):=evala(subs(coef,U(x)));

$$U(x) := -\frac{e^{-\frac{\sqrt{s}\,x}{\sqrt{\alpha}}}\,\alpha\,h}{\left(h\,\alpha + \sqrt{\alpha}\,\sqrt{s}\right)s} + \frac{1}{s}$$

Maple cannot find the inverse of the Laplace transform:

> invlaplace(U(x),s,t);

$$-\text{invlaplace}\left(\frac{e^{-\frac{\sqrt{s}\,x}{\sqrt{\alpha}}}}{s},s,t\right) + \frac{\text{invlaplace}\left(\frac{e^{-\frac{\sqrt{s}\,x}{\sqrt{\alpha}}}}{\sqrt{s}},s,t\right)}{h\sqrt{\alpha}} - \text{invlaplace}\left(\frac{e^{-\frac{\sqrt{s}\,x}{\sqrt{\alpha}}}}{h\,\alpha + \sqrt{\alpha}\,\sqrt{s}},s,t\right) + 1$$

The above expression is split into two terms U1 and U2 as shown below:

> U1:=1/(s);

$$U1 := \frac{1}{s}$$

> u1:=invlaplace(U1,s,t);

$$u1 := 1$$

> U2:=U(x)-U1;

$$U2 := -\frac{e^{-\frac{\sqrt{s}\,x}{\sqrt{\alpha}}}\,\alpha\,h}{\left(h\,\alpha + \sqrt{\alpha}\,\sqrt{s}\right)s}$$

The inverse of U2 is:

> U2:=subs({x=x1,alpha=alpha1},U2);

$$U2 := -\frac{e^{-\frac{\sqrt{s}\,x1}{\sqrt{\alpha 1}}}\,\alpha 1\,h}{\left(h\,\alpha 1 + \sqrt{\alpha 1}\,\sqrt{s}\right)s}$$

> assume(x1>0,alpha1>0);
> u2:=invlaplace(U2,s,t);

$$u2 := -\text{erfc}\left(\frac{1}{2}\frac{x1\sim}{\sqrt{\alpha 1\sim t}}\right) + e^{h\,(h\,\alpha 1\sim t + x1\sim)}\,\text{erfc}\left(\frac{1}{2}\frac{x1\sim + 2\,h\,\alpha 1\sim t}{\sqrt{\alpha 1\sim t}}\right)$$

4.3 Laplace Transform Technique for Parabolic PDEs – Advanced Problems

> u2:=subs({x1=x,alpha1=alpha},u2);

$$u2 := -\text{erfc}\left(\frac{1}{2}\frac{x}{\sqrt{\alpha t}}\right) + e^{h(h\alpha t + x)} \text{erfc}\left(\frac{1}{2}\frac{x + 2h\alpha t}{\sqrt{\alpha t}}\right)$$

Finally, the temperature distribution is given by:

> u:=u1+u2;

$$u := 1 - \text{erfc}\left(\frac{1}{2}\frac{x}{\sqrt{\alpha t}}\right) + e^{h(h\alpha t + x)} \text{erfc}\left(\frac{1}{2}\frac{x + 2h\alpha t}{\sqrt{\alpha t}}\right)$$

> pars:={alpha=1e-3,h=0.01};

$$pars := \{h = 0.01, \alpha = 0.001\}$$

> plot3d(subs(pars,u),x=1..0,t=500..0,axes=boxed,title="Figure 4.13",
labels=[x,t,"u"],orientation=[110,60]);

Fig. 4.13

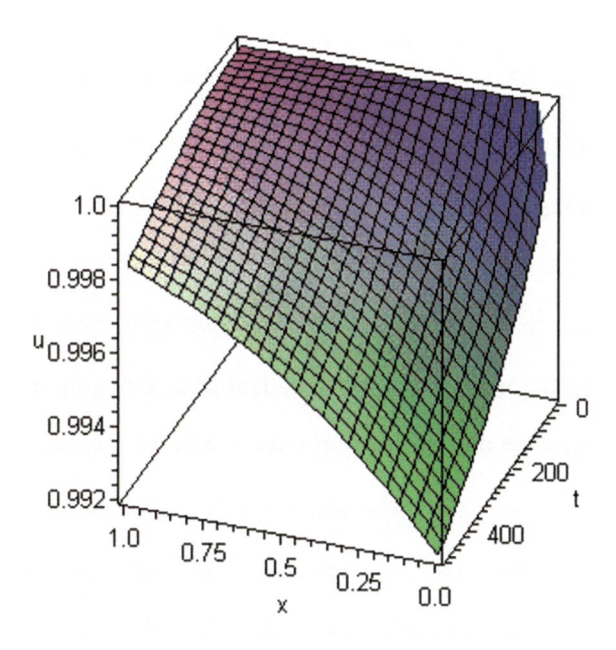

> plot([subs(pars,t=10,u),subs(pars,t=100,u),subs(pars,t=200,u),
subs(pars,t=500,u)],x=0..1,axes=boxed,title="Figure 4.14",
thickness=5,labels=[x,"u"]);

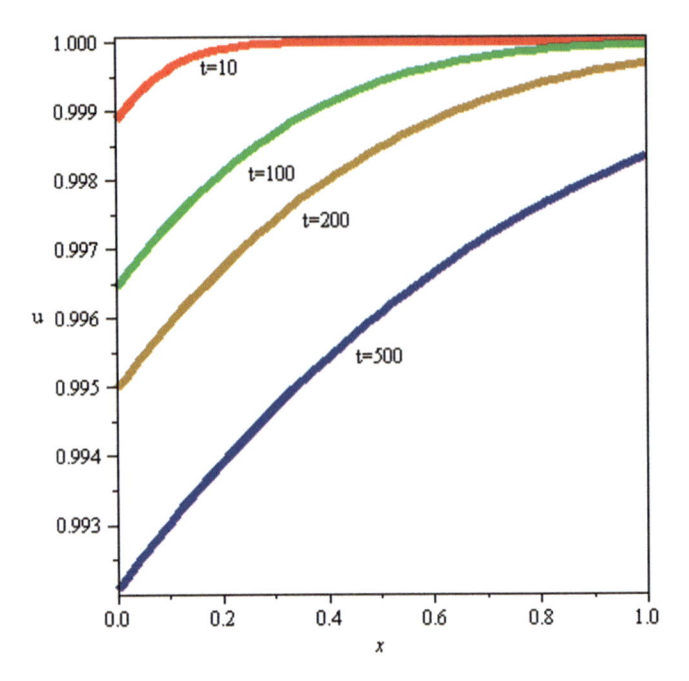

Fig. 4.14

```
> us:=eval(subs(x=0,u));
```

$$us := e^{h^2 \alpha t} \operatorname{erfc}\left(\frac{h \alpha t}{\sqrt{\alpha t}}\right)$$

Example 4.7. Unsteady State Diffusion with a First-Order Reaction

Consider the transient diffusion problem.[6] The governing equation is

$$\frac{\partial u}{\partial t} = D \frac{\partial^2 u}{\partial x^2} - ku$$

$$u(x,0) = 0 \tag{4.7}$$

$$u(0,t) = 1 \text{ and } u(\infty,t) \text{ is defined}$$

where D is the diffusivity (m²/s) and k is the rate constant $\left(s^{-1}\right)$. Equation (4.7) is solved in Maple below using the Laplace transform technique:

4.3 Laplace Transform Technique for Parabolic PDEs – Advanced Problems

```
> restart:with(linalg):with(inttrans):with(plots):
> eq:=diff(u(x,t),t)=D*diff(u(x,t),x$2)-k*u(x,t);
```

$$eq := \frac{\partial}{\partial t} u(x,t) = D\left(\frac{\partial^2}{\partial x^2} u(x,t)\right) - k\, u(x,t)$$

```
> u(x,0):=0;
```

$$u(x,0) := 0$$

```
> bc1:=u(0,t)=1;
```

$$bc1 := u(0,t) = 1$$

```
> bc2:=u(infinity,t)=defined;
```

$$bc2 := u(\infty, t) = \textit{defined}$$

```
> eqs:=laplace(eq,t,s);
```

$$eqs := s\,\textit{laplace}(u(x,t),t,s) = D\left(\frac{\partial^2}{\partial x^2}\textit{laplace}(u(x,t),t,s)\right) - k\,\textit{laplace}(u(x,t),t,s)$$

```
> eqs:=subs(laplace(u(x,t),t,s)=U(x),eqs);
```

$$eqs := s\,U(x) = D\left(\frac{d^2}{dx^2} U(x)\right) - k\,U(x)$$

```
> bc1:=laplace(bc1,t,s);
```

$$bc1 := \textit{laplace}(u(0,t), t, s) = \frac{1}{s}$$

```
> bc1:=subs(laplace(u(0,t),t,s)=U(0),bc1);
```

$$bc1 := U(0) = \frac{1}{s}$$

```
> U(x):=rhs(dsolve({eqs,bc1}));
```

$$U(x) := -\frac{(-1 + \_C2\,s)\,e^{\frac{\sqrt{s+k}\,x}{\sqrt{D}}}}{s} + \_C2\,e^{-\frac{\sqrt{s+k}\,x}{\sqrt{D}}}$$

```
> eqc:=coeff(U(x),exp(1/D^(1/2)*(s+k)^(1/2)*x));
```

$$eqc := -\frac{-1 + \_C2\,s}{s}$$

```
> _C2:=solve(eqc,_C2);
```

$$\_C2 := \frac{1}{s}$$

> U(x):=eval(U(x));

$$U(x) := \frac{e^{-\frac{\sqrt{s+k}\ x}{\sqrt{D}}}}{s}$$

Here again Maple cannot find the inverse Laplace transform:

> invlaplace(U(x),s,t);

$$invlaplace\left(\frac{e^{-\frac{\sqrt{s+k}\ x}{\sqrt{D}}}}{s}, s, t\right)$$

From the property of the Laplace transform,[7] we know that

$$\text{if } L^{-1}F(s) = f(t)$$
$$\text{then } L^{-1}\left(\frac{F(s)}{s}\right) = \int_0^t f(t) \quad (4.8)$$

> U1:=U(x)*s;

$$U1 := e^{-\frac{\sqrt{s+k}\ x}{\sqrt{D}}}$$

Again, Maple cannot invert U1 directly:

> invlaplace(U1,s,t);

$$invlaplace\left(e^{-\frac{\sqrt{s+k}\ x}{\sqrt{D}}}, s, t\right)$$

Another formula[7] is used:

$$L^{-1}F(s+a) = e^{-at}f(t) \quad (4.9)$$

> U2:=subs(s=s-k,U1);

$$U2 := e^{-\frac{\sqrt{s}\ x}{\sqrt{D}}}$$

The inverse transform for U2 is:

> U2:=subs({D=D1,x=x1},U2);

$$U2 := e^{-\frac{\sqrt{s}\ x1}{\sqrt{D1}}}$$

4.3 Laplace Transform Technique for Parabolic PDEs – Advanced Problems

```
> assume(D1>0,x1>0);
> u2:=invlaplace(U2,s,t);
```

$$u2 := \frac{1}{2} \frac{x1\sim e^{-\frac{1}{4}\frac{x1\sim^2}{D1\sim t}}}{\sqrt{D1\sim \pi}\, t^{3/2}}$$

```
> u2:=subs({D1=D,x1=x},u2);
```

$$u2 := \frac{1}{2} \frac{x\, e^{-\frac{1}{4}\frac{x^2}{D\,t}}}{\sqrt{D\pi}\, t^{3/2}}$$

The inverse transform for U1 is:

```
> u1:=exp(-k*t)*u2;
```

$$u1 := \frac{1}{2} \frac{e^{-kt}\, x\, e^{-\frac{1}{4}\frac{x^2}{D\,t}}}{\sqrt{D\pi}\, t^{3/2}}$$

```
> I1:=int(u1,t);
```

$$I1 := \int \frac{1}{2} \frac{e^{-kt}\, x\, e^{-\frac{1}{4}\frac{x^2}{D\,t}}}{\sqrt{D\pi}\, t^{3/2}}\, dt$$

```
> with(student):
> I1:=simplify(changevar(t=x^2/4/D/T^2,I1,T));
```

$$I1 := -\frac{2x \left(\int \dfrac{e^{-\frac{1}{4}\frac{kx^2 + 4T^4 D}{D T^2}}}{T\sqrt{\dfrac{x^2}{D T^2}}}\, dT \right)}{\sqrt{\pi}\, \sqrt{D}}$$

```
> I1:=subs(T=T1,x=x1,D=D1,I1);
```

$$I1 := -\frac{2x1\sim \left(\int \dfrac{e^{-\frac{1}{4}\frac{k\,x1\sim^2 + 4T1^4 D1\sim}{D1\sim T1^2}}}{T1\sqrt{\dfrac{x1\sim^2}{D1\sim T1^2}}}\, dT1 \right)}{\sqrt{\pi}\, \sqrt{D1\sim}}$$

> assume(T1>0);
> I1:=simplify(I1);

$$I1 := -\frac{1}{2} e^{\frac{x1\sim \sqrt{k}}{\sqrt{D1\sim}}} \operatorname{erf}\left(\frac{1}{2} \frac{2 D1\sim T1\sim^2 + x1\sim \sqrt{D1\sim}\sqrt{k}}{D1\sim T1\sim}\right)$$

$$+ \frac{1}{2} e^{-\frac{x1\sim \sqrt{k}}{\sqrt{D1\sim}}} \operatorname{erf}\left(\frac{1}{2} \frac{-2 D1\sim T1\sim^2 + x1\sim \sqrt{D1\sim}\sqrt{k}}{D1\sim T1\sim}\right)$$

> I1:=subs({x1=x,T1=T,D1=D},I1);

$$I1 := -\frac{1}{2} e^{\frac{x\sqrt{k}}{\sqrt{D}}} \operatorname{erf}\left(\frac{1}{2}\frac{2 D T^2 + x\sqrt{D}\sqrt{k}}{D T}\right) + \frac{1}{2} e^{-\frac{x\sqrt{k}}{\sqrt{D}}} \operatorname{erf}\left(\frac{1}{2}\frac{-2 D T^2 + x\sqrt{D}\sqrt{k}}{D T}\right)$$

> I1:=expand(I1);

$$I1 := -\frac{1}{2} e^{\frac{x\sqrt{k}}{\sqrt{D}}} \operatorname{erf}\left(T + \frac{1}{2}\frac{x\sqrt{k}}{\sqrt{D}\,T}\right) - \frac{1}{2} \frac{\operatorname{erf}\left(T - \frac{1}{2}\frac{x\sqrt{k}}{\sqrt{D}\,T}\right)}{e^{\frac{x\sqrt{k}}{\sqrt{D}}}}$$

> I2:=eval(subs(T=infinity,I1));

$$I2 := -\frac{1}{2} e^{\frac{x\sqrt{k}}{\sqrt{D}}} - \frac{1}{2 e^{\frac{x\sqrt{k}}{\sqrt{D}}}}$$

> u:=I1-I2;

$$u := -\frac{1}{2} e^{\frac{x\sqrt{k}}{\sqrt{D}}} \operatorname{erf}\left(T + \frac{1}{2}\frac{x\sqrt{k}}{\sqrt{D}\,T}\right) - \frac{1}{2} \frac{\operatorname{erf}\left(T - \frac{1}{2}\frac{x\sqrt{k}}{\sqrt{D}\,T}\right)}{e^{\frac{x\sqrt{k}}{\sqrt{D}}}} + \frac{1}{2} e^{\frac{x\sqrt{k}}{\sqrt{D}}} + \frac{1}{2 e^{\frac{x\sqrt{k}}{\sqrt{D}}}}$$

> u:=collect(u,exp(x/D^(1/2)*k^(1/2)));

$$u := \left(-\frac{1}{2}\operatorname{erf}\left(T + \frac{1}{2}\frac{x\sqrt{k}}{\sqrt{D}\,T}\right) + \frac{1}{2}\right) e^{\frac{x\sqrt{k}}{\sqrt{D}}} + \frac{-\frac{1}{2}\operatorname{erf}\left(T - \frac{1}{2}\frac{x\sqrt{k}}{\sqrt{D}\,T}\right) + \frac{1}{2}}{e^{\frac{x\sqrt{k}}{\sqrt{D}}}}$$

> u:=convert(u,erfc);

$$u := \frac{1}{2}\operatorname{erfc}\left(T + \frac{1}{2}\frac{x\sqrt{k}}{\sqrt{D}\,T}\right) e^{\frac{x\sqrt{k}}{\sqrt{D}}} + \frac{1}{2}\frac{\operatorname{erfc}\left(T - \frac{1}{2}\frac{x\sqrt{k}}{\sqrt{D}\,T}\right)}{e^{\frac{x\sqrt{k}}{\sqrt{D}}}}$$

4.3 Laplace Transform Technique for Parabolic PDEs – Advanced Problems

> u:=subs(T=x/2/(D*t)^(1/2),u);

$$u := \frac{1}{2}\operatorname{erfc}\left(\frac{1}{2}\frac{x}{\sqrt{Dt}} + \frac{\sqrt{Dt}\sqrt{k}}{\sqrt{D}}\right)e^{\frac{x\sqrt{k}}{\sqrt{D}}} + \frac{1}{2}\frac{\operatorname{erfc}\left(\frac{1}{2}\frac{x}{\sqrt{Dt}} - \frac{\sqrt{Dt}\sqrt{k}}{\sqrt{D}}\right)}{e^{\frac{x\sqrt{k}}{\sqrt{D}}}}$$

> convert(expand(simplify(eval(subs(k=0,u)))),erfc);

$$\operatorname{erfc}\left(\frac{1}{2}\frac{x}{\sqrt{Dt}}\right)$$

> expand(D*(expand(-eval(subs(x=0,diff(u,x))))));

$$\frac{D}{\sqrt{\pi}\,e^{kt}\sqrt{Dt}} + \sqrt{D}\,\sqrt{k}\,\operatorname{erf}\left(\frac{\sqrt{Dt}\sqrt{k}}{\sqrt{D}}\right)$$

> pars:={D=1e-6,k=0.1};

$$pars := \{D = 0.000001, k = 0.1\}$$

> plot3d(subs(pars,u),x=1e-3..0,t=10..0,axes=boxed,labels=[x,t,"u"],
orientation=[-60,60]);

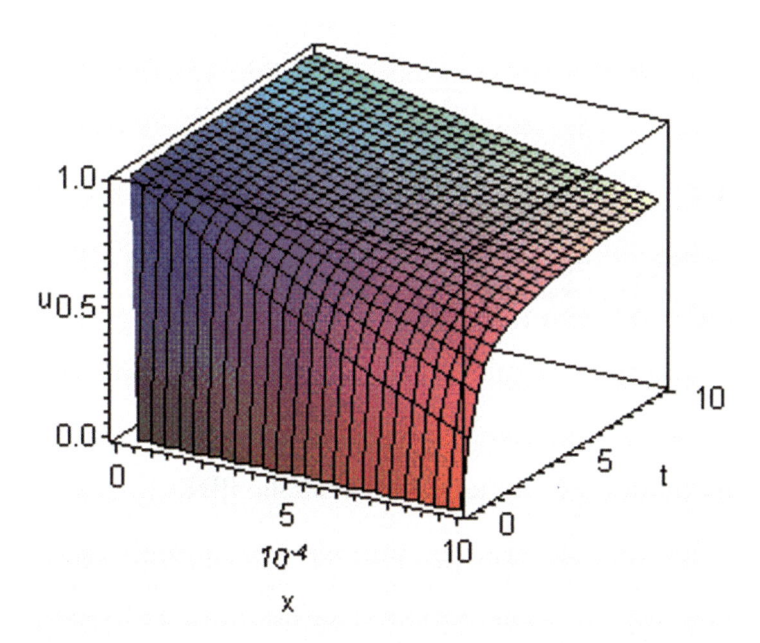

Fig. 4.15

> pfs:=plot([subs(pars,t=0.01,u),subs(pars,t=0.1,u),subs(pars,t=1,u),
subs(pars,t=2,u)],x=0..1e-3,axes=boxed,thickness=5,labels=[x,"u"]);

pfs := *PLOT*(...)

> pts:=textplot([[0.00016,evalf(subs(pars,t=0.01,x=0.0001,u)),"t=0.01"],
[0.00036,evalf(subs(pars,t=0.1,x=0.0003,u)),"t=0.1"],[0.00053,
evalf(subs(pars,t=1,x=0.0006,u)),"t=1"],[0.00069,
evalf(subs(pars,t=2,x=0.0006,u)),"t=2"]]);

pts := *PLOT*(...)

> display({pfs,pts});

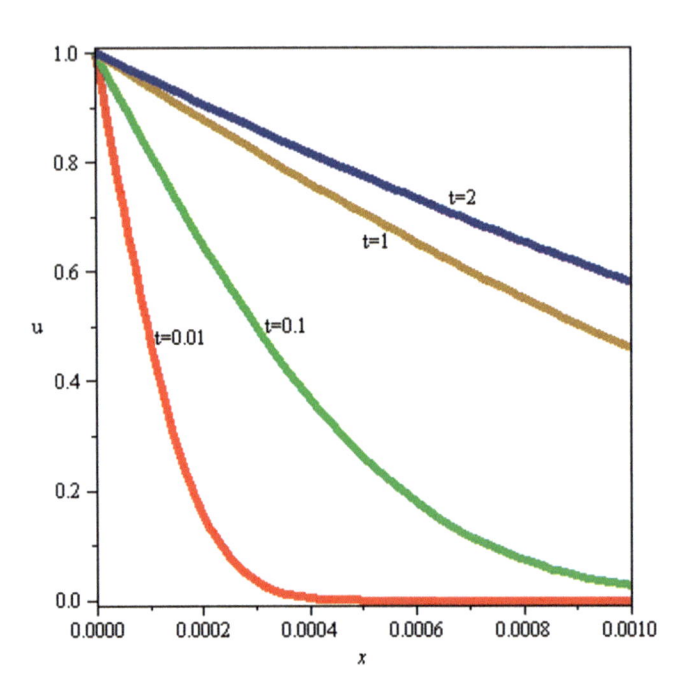

Fig. 4.16

4.4 Similarity Solution Technique for Parabolic PDEs

Parabolic partial differential equations are solved using the similarity solution technique in this section. This method involves combining the two independent variables (x and t) as one (η). For this purpose, the original initial and boundary conditions should become two boundary conditions in the new combined variable (η). The methodology involves converting the governing equation (PDE) to an ordinary differential equation (ODE) in the combined variable (η). This variable transformation is very difficult to do by hand. In this chapter, we will show how

4.4 Similarity Solution Technique for Parabolic PDEs

this variable transformation can be done using Maple. The original problem becomes a boundary value problem (ODE) in the new combined variable (η). The original initial and boundary conditions are converted to boundary conditions in the combined variable. This boundary value problem can then be solved using the techniques illustrated in chapter 3.1 for solving boundary value problems. Unlike the Laplace transform technique, there is no need for inversion.

Example 4.8. Heat Conduction in a Rectangular Slab

Example 4.1 is solved here with the boundary and initial conditions switched.[4] The governing equation is

$$\frac{\partial u}{\partial t} = \alpha \frac{\partial^2 u}{\partial x^2}$$

$$u(x,0) = 0 \tag{4.10}$$

$$u(0,t) = 1 \text{ and } u(\infty,t) \text{ is defined}$$

The following transformation is used to combine the variable:[7]

$$\eta = \frac{x}{2\sqrt{\alpha t}} \tag{4.11}$$

The variable u in the new coordinate η is represented by U. The governing equation (ODE) for U is obtained by converting the time and spatial derivative in equation (4.10) (PDE) to derivatives in the η coordinate. The boundary conditions for U are:

$$U(0) = 1$$
$$U(\infty) = 0 \tag{4.12}$$

The governing equation for U is then solved with the above boundary conditions to obtain the final solution. Example 4.8 is solved in Maple below:

> restart:

The with(student) package is called to facilitate variable transformations:

> with(student):

The governing equation is entered here:

> eq:=diff(u(x,t),t)-alpha*diff(u(x,t),x$2);

$$eq := \frac{\partial}{\partial t} u(x,t) - \alpha \left(\frac{\partial^2}{\partial x^2} u(x,t) \right)$$

First, u(x,t) is transformed to U(η(x,t)):

> eq1:=changevar(u(x,t)=U(eta(x,t)),eq);

$$eq1 := D(U)(\eta(x,t)) \left(\frac{\partial}{\partial t} \eta(x,t) \right) - \alpha \left(D^{(2)}(U)(\eta(x,t)) \left(\frac{\partial}{\partial x} \eta(x,t) \right)^2 + D(U)(\eta(x,t)) \left(\frac{\partial^2}{\partial x^2} \eta(x,t) \right) \right)$$

The transformation for η is substituted here:

> eq2:=expand(simplify(subs(eta(x,t)=x/2/(alpha*t)^(1/2),eq1)));

$$eq2 := -\frac{1}{4} \frac{D(U)\left(\frac{1}{2} \frac{x}{\sqrt{\alpha t}} \right) x}{t\sqrt{\alpha t}} - \frac{1}{4} \frac{D^{(2)}(U)\left(\frac{1}{2} \frac{x}{\sqrt{\alpha t}} \right)}{t}$$

The governing equation is further simplified here:

> eq2:=expand(eq2*t);

$$eq2 := -\frac{1}{4} \frac{D(U)\left(\frac{1}{2} \frac{x}{\sqrt{\alpha t}} \right) x}{\sqrt{\alpha t}} - \frac{1}{4} D^{(2)}(U)\left(\frac{1}{2} \frac{x}{\sqrt{\alpha t}} \right)$$

> eq2:=subs(x=eta*2*(alpha*t)^(1/2),eq2);

$$eq2 := -\frac{1}{2} D(U)(\eta) \eta - \frac{1}{4} D^{(2)}(U)(\eta)$$

> eq2:=convert(eq2,diff);

$$eq2 := -\frac{1}{2} \left(\frac{d}{d\eta} U(\eta) \right) \eta - \frac{1}{4} \frac{d^2}{d\eta^2} U(\eta)$$

The final form for the governing equation is:

> eq2:=expand(-2*eq2);

$$eq2 := \left(\frac{d}{d\eta} U(\eta) \right) \eta + \frac{1}{2} \frac{d^2}{d\eta^2} U(\eta)$$

Enter the boundary condition here:

> bc1:=U(0)=1;

$$bc1 := U(0) = 1$$

4.4 Similarity Solution Technique for Parabolic PDEs

> bc2:=U(infinity)=0;

$$bc2 := U(\infty) = 0$$

U is solved as:

> U:=rhs(dsolve({eq2,bc1,bc2},U(eta)));

$$U := 1 - \text{erf}(\eta)$$

> U:=convert(U,erfc);

$$U := \text{erfc}(\eta)$$

Next, u as a function of x and t is obtained as:

> u:=subs(eta=x/2/(alpha*t)^(1/2),U);

$$u := \text{erfc}\left(\frac{1}{2}\frac{x}{\sqrt{\alpha t}}\right)$$

The solution obtained can be plotted:

> plot3d(subs(alpha=0.001,u),x=1..0,t=500..0,axes=boxed,title="Figure 4.17", labels =[x,t,"u"],orientation=[-60,60]);

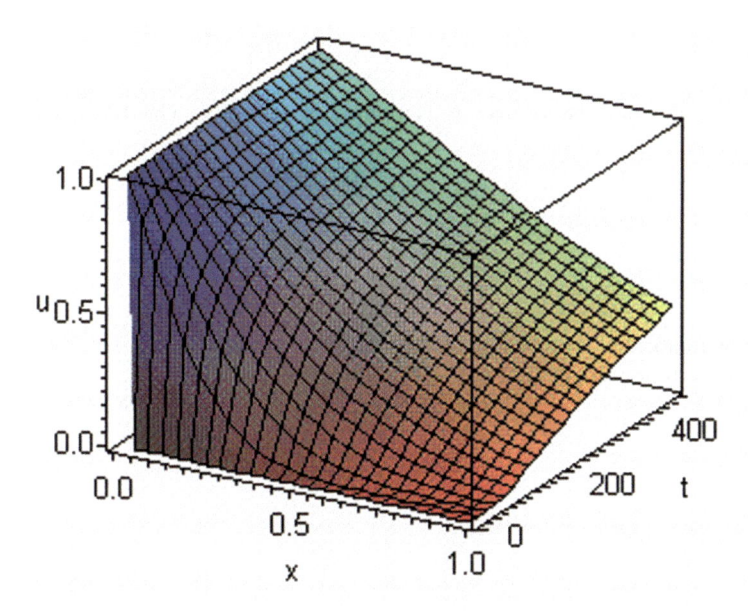

Fig. 4.17

Example 4.9. Laminar Flow in a CVD Reactor

Consider the laminar flow in a CVD reactor. The governing equation is

$$2v_{max} \frac{x}{B} \frac{\partial u}{\partial z} = D \frac{\partial^2 u}{\partial x^2}$$

$$u(x,0) = 1 \qquad (4.13)$$

$$u(0,z) = 0 \text{ and } u(\infty, z) = 1$$

where v_{max} is the average velocity (cm/s), B is the half-width of the reactor (cm) and D is the diffusion coefficient (cm²/s). Next, the transformation $V = \dfrac{2v_{max}}{BD}$ is used to simplify the governing equation as:

$$Vx \frac{\partial u}{\partial z} = \frac{\partial^2 u}{\partial x^2} \qquad (4.14)$$

The following transformation is used to combine the variable [7]

$$\eta = \frac{x}{\left(\dfrac{9z}{V}\right)^{\frac{1}{3}}} \qquad (4.15)$$

The variable u in the new coordinate η is represented by U. The boundary conditions for U are:

$$U(0) = 0$$
$$U(\infty) = 1 \qquad (4.16)$$

The governing equation for U is then solved with the above boundary conditions to obtain the final solution. Example 4.9 is solved in Maple below:

```
> restart:
> with(student):
> with(plots):
> eq:=V*x*diff(u(x,z),z)-diff(u(x,z),x$2);
```

$$eq := Vx\left(\frac{\partial}{\partial z} u(x,z)\right) - \left(\frac{\partial^2}{\partial x^2} u(x,z)\right)$$

4.4 Similarity Solution Technique for Parabolic PDEs

```
> eq1:=changevar(u(x,z)=U(eta(x,z)),eq):
> eq2:=expand(simplify(subs(eta(x,z)=x/(9*z/V)^(1/3),eq1))):
> eq2:=subs(x=eta*(9*z/V)^(1/3),eq2):
> eq2:=expand(eq2*z):
> eq2:=simplify(eq2/(z/V)^(1/3)):
> eq2:=eq2*9/3^(2/3)/V:
> eq2:=convert(eq2,diff):
```

The governing equation in the combined variable is:

```
> eq2:=-eq2;
```

$$eq2 := 3\eta^2 \left(\frac{d}{d\eta} U(\eta) \right) + \frac{d^2}{d\eta^2} U(\eta)$$

The boundary conditions for U are entered here:

```
> bc1:=U(0)=0;
```

$$bc1 := U(0) = 0$$

```
> bc2:=U(infinity)=1;
```

$$bc2 := U(\infty) = 1$$

```
> rhs(dsolve({eq2,bc1,bc2},U(eta)));
```

$$\lim_{\_a \to \infty} \left(\frac{e^{-\frac{1}{2}\eta^3} \left(3\eta^3\, \text{WhittakerM}\left(\frac{1}{6}, \frac{2}{3}, \eta^3\right) + 4\, \text{WhittakerM}\left(\frac{7}{6}, \frac{2}{3}, \eta^3\right) \right) \_a^{5/2}}{\eta^{5/2} e^{-\frac{1}{2}\_a^3} \left(3\_a^3\, \text{WhittakerM}\left(\frac{1}{6}, \frac{2}{3}, \_a^3\right) + 4\, \text{WhittakerM}\left(\frac{7}{6}, \frac{2}{3}, \_a^3\right) \right)} \right)$$

Maple cannot find the limit as $\eta \to \infty$. Alternatively, the governing equation is solved using the first boundary condition (bc1) only:

```
> U:=rhs(dsolve({eq2,bc1},U(eta)));
```

$$U := \frac{e^{-\frac{1}{2}\eta^3} \left(3\eta^3\, \text{WhittakerM}\left(\frac{1}{6}, \frac{2}{3}, \eta^3\right) + 4\, \text{WhittakerM}\left(\frac{7}{6}, \frac{2}{3}, \eta^3\right) \right) \_C2}{\eta^{5/2}}$$

The solution is a combination of exponential and Whittaker functions. In the literature this problem is usually left in terms of integrals (Gamma functions). However, Maple is able to solve the differential equation explicitly. Next, the constant _C2 is found using the boundary condition bc2.

```
> eval(limit(U,eta=infinity));
```

$$\lim_{\eta \to \infty} \left(\frac{e^{-\frac{1}{2}\eta^3} \left(3\eta^3\, \text{WhittakerM}\left(\frac{1}{6}, \frac{2}{3}, \eta^3\right) + 4\, \text{WhittakerM}\left(\frac{7}{6}, \frac{2}{3}, \eta^3\right) \right) \_C2}{\eta^{5/2}} \right)$$

Since the limit does not exit, U/_C2 is plotted until η=10.

> plot(U/_C2,eta=0..10,axes=boxed,thickness=4,title="Figure 4.18");

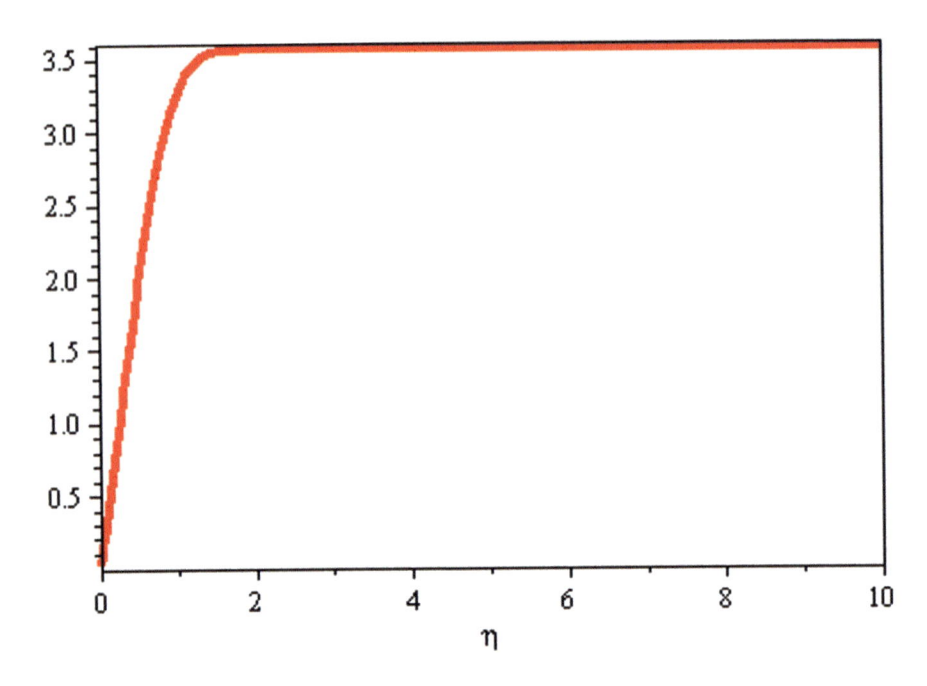

Fig. 4.18

An initial guess L = 1 is used to replace x = ∞.

> evalf(subs(eta=1.,U/_C2));

$$3.230044728$$

> L:=1;

$$L := 1$$

> err:=1;

$$err := 1$$

The value of U/C2 is found at x = L:

> c0:=evalf(subs(eta=1.,U/_C2));

$$c0 := 3.230044728$$

4.4 Similarity Solution Technique for Parabolic PDEs

Next, the length L is increased until the U/_C2(x =) becomes a constant:

> while err>1e-6 do L:=L+1; c1:=evalf(subs(eta=L,U/_C2));err:=abs(c1-c0); c0:=c1;od;\

$$L := 2$$
$$c1 := 3.571814055$$
$$err := 0.341769327$$
$$c0 := 3.571814055$$
$$L := 3$$
$$c1 := 3.571918047$$
$$err := 0.000103992$$
$$c0 := 3.571918047$$
$$L := 4$$
$$c1 := 3.571918047$$
$$err := 0.$$
$$c0 := 3.571918047$$

The length and the constant have converged to 10 digit accuracy. The final length is

> L;

$$4$$

The constant _C2 is found using the boundary condition U(L) = 1.

> _C2:=1/c0;

$$\_C2 := 0.2799616304$$

The solution to the transformed coordinate is:

> U;

$$\frac{0.2799616304 \, e^{-\frac{1}{2}\eta^3} \left(3\eta^3 \, \text{WhittakerM}\left(\frac{1}{6}, \frac{2}{3}, \eta^3\right) + 4\, \text{WhittakerM}\left(\frac{7}{6}, \frac{2}{3}, \eta^3\right)\right)}{\eta^{5/2}}$$

> plot(U,eta=0..10.,thickness=4,axes=boxed,title="Figure 4.19", labels=[eta,'u']);

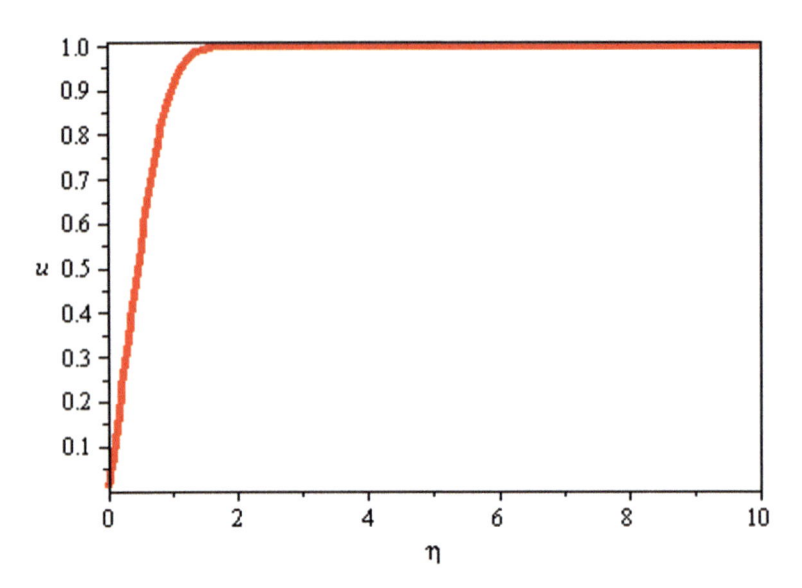

Fig. 4.19

The solution in the original coordinates is obtained as:

> u:=subs(eta=x/(9*z/V)^(1/3),U);

$$u := \frac{1}{\left(\dfrac{x\,9^{2/3}}{\left(\dfrac{z}{V}\right)^{1/3}}\right)^{5/2}} \left(22.67689206\,\sqrt{9}\; e^{-\frac{1}{18}\frac{x^3 V}{z}} \left(\frac{1}{3}\, \frac{x^3\,V\,\text{WhittakerM}\left(\frac{1}{6},\frac{2}{3},\frac{1}{9}\frac{x^3 V}{z}\right)}{z} \right) \right)$$

> pars:={V=0.001};

$$pars := \{V = 0.001\}$$

A plot was made by specifying a value for V.

> plot3d(subs(pars,u),x=0..10,z=0..10,axes=boxed,title="Figure 4.20", orientation=[120,60],labels=[x,y,"u"]);

4.5 Similarity Solution Technique for Elliptic Partial Differential Equations

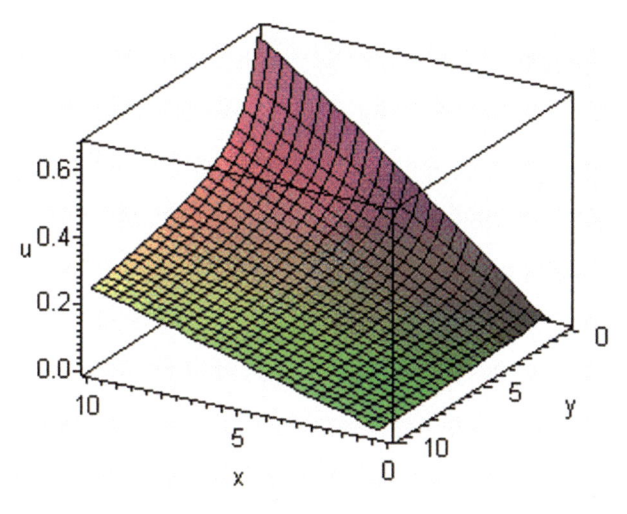

Fig. 4.20

```
> u;
```

$$22.67689206\sqrt{9}\, e^{-\frac{1}{18}\frac{x^3 V}{z}} \left(\frac{1}{3} \frac{x^3 V \text{WhittakerM}\left(\frac{1}{6}, \frac{2}{3}, \frac{1}{9}\frac{x^3 V}{z}\right)}{z} + 4\, \text{WhittakerM}\left(\frac{7}{6}, \frac{2}{3}, \frac{1}{9}\frac{x^3 V}{z}\right) \right) \bigg/ \left(\frac{x\, 9^{2/3}}{\left(\frac{z}{V}\right)^{1/3}} \right)^{5/2}$$

```
>
```

4.5 Similarity Solution Technique for Elliptic Partial Differential Equations

Elliptic partial differential equations are solved using the similarity solution technique in this section. The method described in section 4.4 is also valid for elliptic partial differential equations. The methodology involves converting the governing equation (PDE) to an ordinary differential equation (ODE) in the combined variable (η). This variable transformation is very difficult to do by hand. In this section, we show how this variable transformation can be done using Maple. The original problem becomes a boundary value problem (ODE) in the new combined variable (η). This is best illustrated with the following examples.

Example 4.10. Steady State Heat Conduction in a Plate

Consider steady state conduction in a semi-infinite rectangular strip. The governing equation in dimensionless form is

334   4 Partial Differential Equations in Semi-infinite Domains

$$\frac{\partial^2 u}{\partial x^2} + \frac{\partial^2 u}{\partial y^2} = 0$$

$$u(x,0) = u(\infty, y) = 1 \tag{4.17}$$

$$u(0,y) = u(x,\infty) = 0$$

The following transformation is used to combine the variable:

$$\eta = \frac{y}{x} \tag{4.18}$$

The variable u in the new coordinate η is represented by U. The governing equation (ODE) for U is obtained by converting the spatial derivatives (x and y) in equation (4.17) (PDE) to derivatives in the η coordinate. The boundary conditions for U are:

$$U(0) = 1$$
$$\tag{4.19}$$
$$U(\infty) = 0$$

Example 4.10 is solved in Maple below:

```
> restart:
> with(student):
> with(plots):
```

The governing equation is entered here:

```
> eq:=diff(u(x,y),x$2)+diff(u(x,y),y$2);
```

$$eq := \frac{\partial^2}{\partial x^2} u(x,y) + \frac{\partial^2}{\partial y^2} u(x,y)$$

```
> eq:=changevar(u(x,y)=U(eta(x,y)),eq):
> eq1:=(simplify(subs(eta(x,y)=y/x,eq))):
> eq1:=subs(y=eta*x,eq1):
> eq1:=simplify(eq1*x^2):
```

The governing equation in the combined variable is:

```
> eq2:=convert(eq1,diff);
```

$$eq2 := \left(\frac{d^2}{d\eta^2} U(\eta)\right) \eta^2 + 2\left(\frac{d}{d\eta} U(\eta)\right) \eta + \frac{d^2}{d\eta^2} U(\eta)$$

4.5 Similarity Solution Technique for Elliptic Partial Differential Equations

> bc1:=U(0)=1;

$$bc1 := U(0) = 1$$

> bc2:=U(infinity)=0;

$$bc2 := U(\infty) = 0$$

> U:=rhs(dsolve({eq2,bc1,bc2},U(eta)));

$$U := \frac{\pi - 2\arctan(\eta)}{\pi}$$

The dimensionless temperature U is given by:

> U:=expand(U);

$$U := 1 - \frac{2\arctan(\eta)}{\pi}$$

> plot(U,eta=0..10,thickness=5,title="Figure 4.21",axes=boxed);

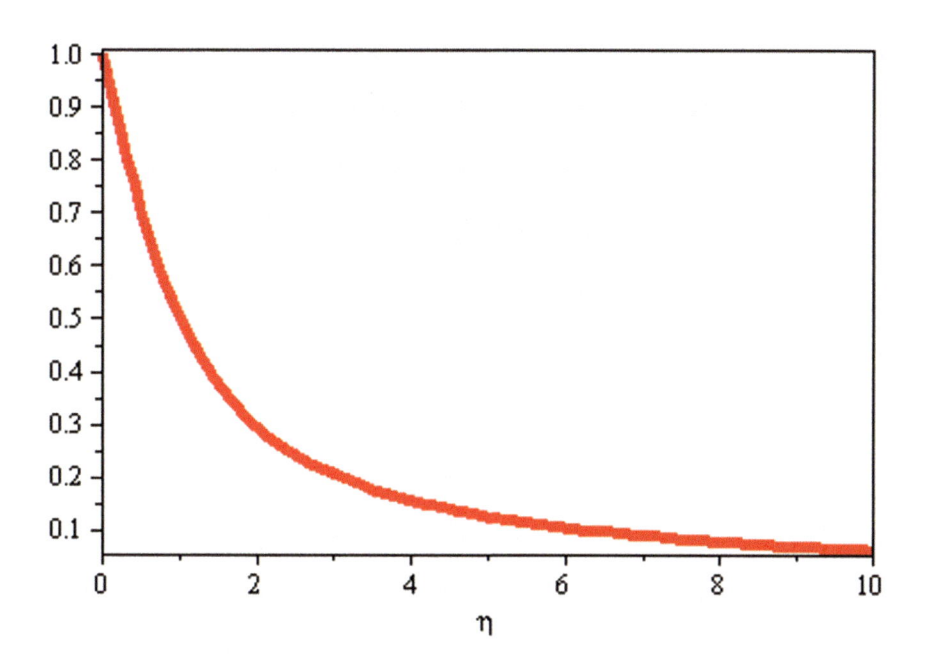

Fig. 4.21

The dimensionless temperature in the original coordinate is:

> u:=expand(subs(eta=y/x,U));

$$u := 1 - \frac{2 \arctan\left(\dfrac{y}{x}\right)}{\pi}$$

The solution obtained is plotted:

> plot3d(u,x=0..50,y=0..50,axes=boxed,title="Figure 4.22", orientation=[120,60],labels=[x,y,"u"]);

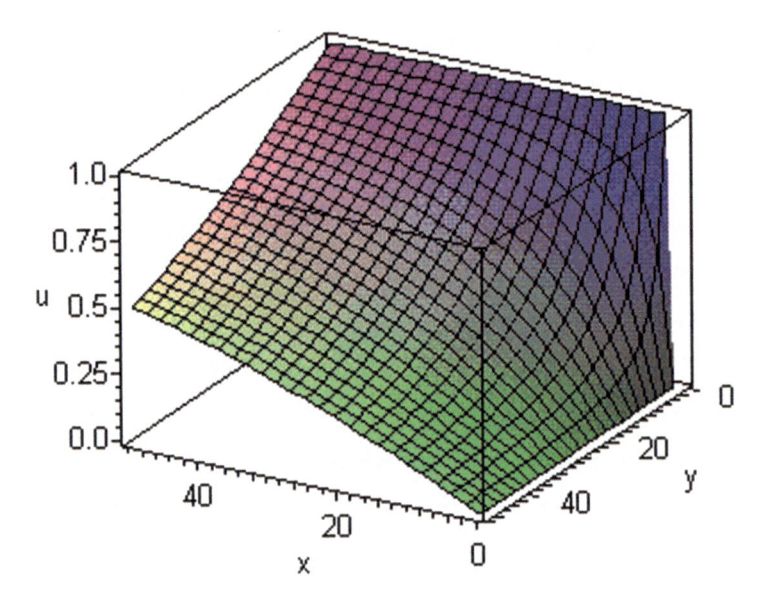

Fig. 4.22

The dimensionless heat flux at $y = 0$ is given by:

> flux:=subs(y=0,-diff(u,y));

$$flux := \frac{2}{\pi x}$$

Example 4.11. Current Distribution in an Electrochemical Cell

Primary and secondary current distributions in electrochemical cells are governed by the Laplace equation.[8] Consider a rectangular geometry governed by the following equation[9]

4.5 Similarity Solution Technique for Elliptic Partial Differential Equations

$$\frac{\partial^2 u}{\partial x^2} + \frac{\partial^2 u}{\partial y^2} = 0$$

$$u(x,0) = 1$$

(4.20)

$$u(L,y) = u(x,\infty) = 0$$

Note that this geometry is of finite dimension in x (L) and semi-infinite in y. The following transformation is used to combine the variable:

$$\eta = \frac{y}{L-x} \tag{4.21}$$

The variable u in the new coordinate η is represented by U. The boundary conditions for U are:

$$U(0) = 1$$

(4.22)

$$U(\infty) = 0$$

Example 4.11 is solved in Maple below. The program used for example 4.10 can be modified to solve this example. Only the variable transformation (equation (4.21)) has to be modified. The following results are obtained:

```
> restart:
> with(student):
> with(plots):
> eq:=diff(u(x,y),x$2)+diff(u(x,y),y$2);
```

$$eq := \frac{\partial^2}{\partial x^2} u(x,y) + \frac{\partial^2}{\partial y^2} u(x,y)$$

```
> eq:=changevar(u(x,y)=U(eta(x,y)),eq):
> eq1:=(simplify(subs(eta(x,y)=y/(L-x),eq))):
> eq1:=subs(y=eta*(L-x),eq1):
> eq1:=simplify(eq1*(L-x)^2):
> eq2:=convert(eq1,diff);
```

$$eq2 := \left(\frac{d^2}{d\eta^2} U(\eta)\right)\eta^2 + 2\left(\frac{d}{d\eta} U(\eta)\right)\eta + \frac{d^2}{d\eta^2} U(\eta)$$

```
> bc1:=U(0)=1;
```

$$bc1 := U(0) = 1$$

> bc2:=U(infinity)=0;

$$bc2 := U(\infty) = 0$$

> U:=rhs(dsolve({eq2,bc1,bc2},U(eta)));

$$U := \frac{\pi - 2\arctan(\eta)}{\pi}$$

> U:=expand(U);

$$U := 1 - \frac{2\arctan(\eta)}{\pi}$$

> plot(U,eta=0..10,thickness=4,title="Figure 4.23",axes=boxed);

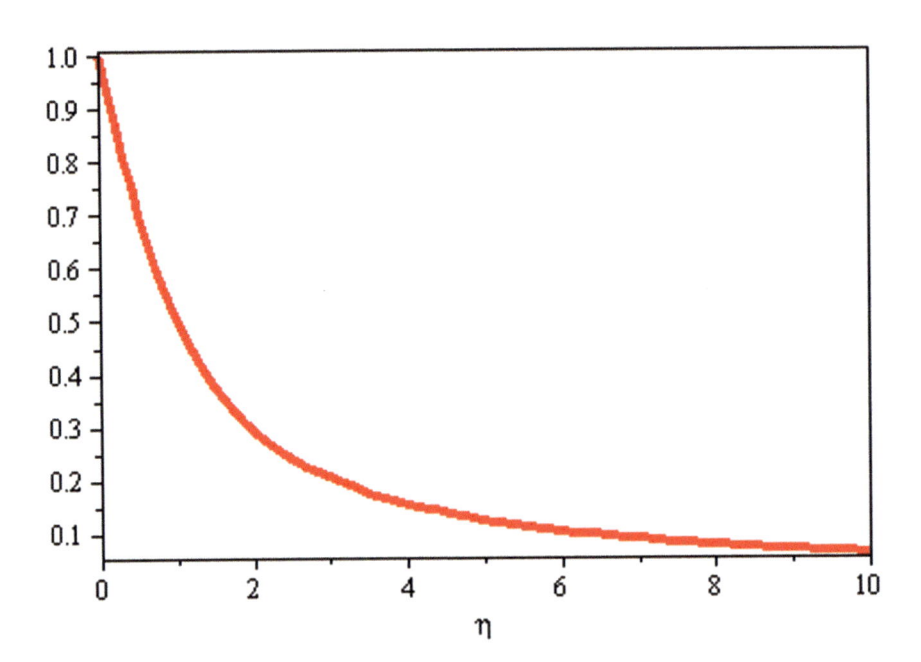

Fig. 4.23

> u:=expand(subs(eta=y/(L-x),U));

$$u := 1 - \frac{2\arctan\left(\dfrac{y}{L-x}\right)}{\pi}$$

> plot3d(subs(L=1,u),x=0..1,y=0..2,axes=boxed,title="Figure 4.24", orientation=[30,60],labels=[x,y,"u"]);

4.6 Similarity Solution Technique for Nonlinear Partial Differential Equations

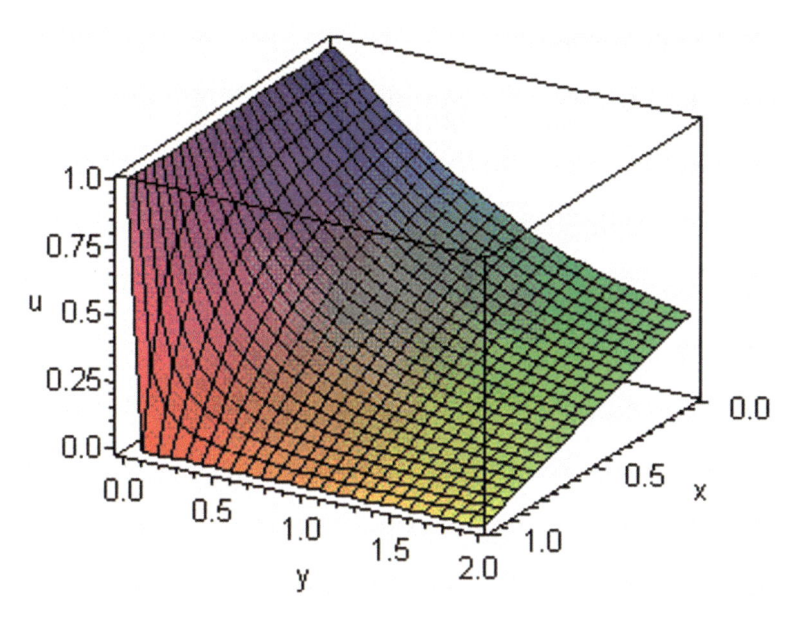

Fig. 4.24

The current distribution at the electrode $(y=0)$ is given by:

> curr:=subs(y=0,-diff(u,y));

$$curr := \frac{2}{\pi(L-x)}$$

>

4.6 Similarity Solution Technique for Nonlinear Partial Differential Equations

Nonlinear parabolic and elliptic partial differential equations are solved using the similarity solution technique in this section. The methods described in section 4.4 and sections 4.5 are valid for nonlinear partial differential equations, also. The methodology involves converting the governing equation (PDE) to an ordinary differential equation in the combined variable (η). This variable transformation is very difficult to do by hand. In this section, we will show how this variable transformation can be done using Maple. The original problem becomes a nonlinear boundary value problem (ODE) in the new combined variable (η). This is best illustrated with the following examples.

Example 4.12. Variable Diffusivity

Consider the transient diffusion in a rectangle in which the diffusivity varies linearly as a function of concentration.[10] The governing equation is:

$$\frac{\partial u}{\partial t} = \frac{\partial}{\partial x}\left((1+u)\frac{\partial u}{\partial x}\right)$$

$$u(x,0) = 0 \tag{4.23}$$

$$u(0,t) = 1 \text{ and } u(\infty,t) \text{ is defined}$$

The following transformation is used to combine the variable:[7]

$$\eta = \frac{x}{2\sqrt{t}} \tag{4.24}$$

The variable u in the new coordinate η is represented by U. The governing equation (ODE) for U is obtained by converting the time and spatial derivative in equation(4.23) (PDE) to derivatives in the η coordinate. The boundary conditions for U are:

$$U(0) = 1$$
$$\tag{4.25}$$
$$U(\infty) = 0$$

The governing equation for U is then solved with the above boundary conditions to obtain the final solution. Example 4.12 is solved in Maple below:

```
> restart:
> with(student):
> with(plots):
```

The governing equation is entered here:

```
> eq:=diff(u(x,t),t)-diff(((1+u(x,t))*diff(u(x,t),x),x));
```

$$eq := \frac{\partial}{\partial t}u(x,t) - \left(\frac{\partial}{\partial x}u(x,t)\right)^2 - (1+u(x,t))\left(\frac{\partial^2}{\partial x^2}u(x,t)\right)$$

```
> eq1:=changevar(u(x,t)=U(eta(x,t)),eq):
> eq2:=expand(simplify(subs(eta(x,t)=x/2/(t)^(1/2),eq1))):
> eq2:=expand(eq2*t):
> eq2:=subs(x=eta*2*(t)^(1/2),eq2):
> eq2:=convert(eq2,diff):
```

4.6 Similarity Solution Technique for Nonlinear Partial Differential Equations

The governing equation in the combined variable is:
```
> eq2:=expand(-2*eq2);
```

$$eq2 := \left(\frac{d}{d\eta} U(\eta)\right)\eta + \frac{1}{2}\left(\frac{d}{d\eta} U(\eta)\right)^2 + \frac{1}{2}\frac{d^2}{d\eta^2} U(\eta)$$
$$+ \frac{1}{2}\left(\frac{d^2}{d\eta^2} U(\eta)\right) U(\eta)$$

```
> bc1:=U(0)=1;
```

$$bc1 := U(0) = 1$$

The length of the domain is taken to be 5:
```
> bc2:=U(5)=0;
```

$$bc2 := U(5) = 0$$

The nonlinear equation is solved numerically and plotted:
```
> sol:=dsolve({eq2,bc1,bc2},U(eta),type=numeric);
```

$$sol := \textbf{proc}(x\_bvp) \; ... \; \textbf{end proc}$$

```
> odeplot(sol,[eta,U(eta)],0..5,axes=boxed,title="Figure 4.25",thickness=4);
```

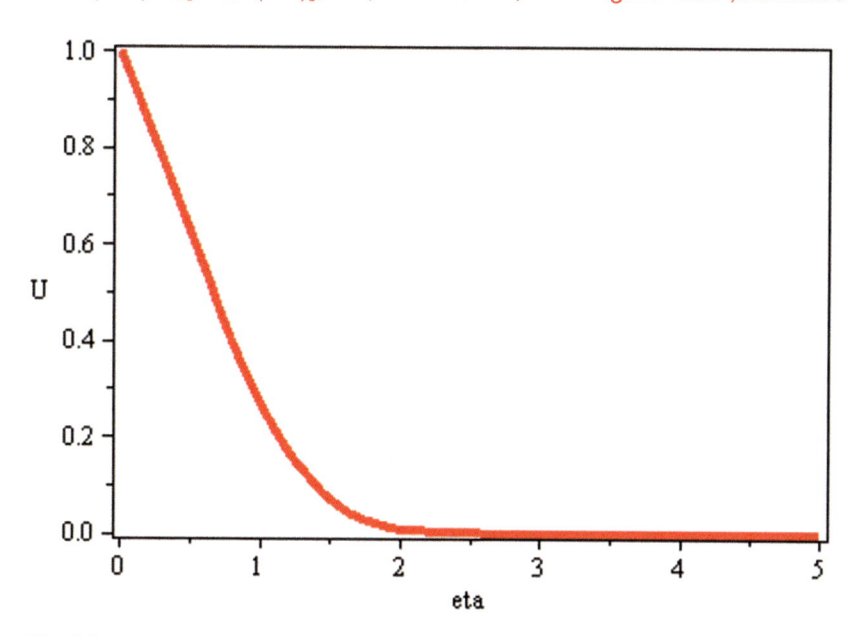

Fig. 4.25

```
>
```

Example 4.13. Plane Flow Past a Flat Plate – Blassius Equation

The velocity distribution in the boundary layer is given by:[6]

$$\frac{\partial u}{\partial x} + \frac{\partial v}{\partial y} = 0$$

$$u\frac{\partial u}{\partial x} + v\frac{\partial u}{\partial y} = \frac{\partial^2 u}{\partial y^2}$$

$$u(0,y) = 1 \tag{4.26}$$

$$u(x,0) = 0 \text{ and } u(x,\infty) = 1$$

$$v(x,0) = 0$$

where u and v are the x and y components of the velocity. Next, the stream function is introduced:

$$u = \frac{\partial \psi}{\partial y} \text{ and } v = -\frac{\partial \psi}{\partial x} \tag{4.27}$$

By definition (4.27), the stream function ψ satisfies the first equation in equation (4.26). The boundary conditions for ψ are

$$\frac{\partial \psi}{\partial y}(0,y) = 1$$

$$\frac{\partial \psi}{\partial y}(x,0) = 0 \text{ and } \frac{\partial \psi}{\partial y}(x,\infty) = 1 \tag{4.28}$$

$$-\frac{\partial \psi}{\partial x}(x,0) = 0$$

The following transformation is used to combine the independent variables:

$$\eta = \frac{y}{\sqrt{x}} \tag{4.29}$$

Next, the following transformation is introduced:

$$\psi = \sqrt{x}\ f(\eta) \tag{4.30}$$

The boundary conditions for $f(\eta)$ are:

$$f(0) = 0 \text{ and } \frac{df}{d\eta}(0) = 0$$

$$\frac{df}{d\eta}(\infty) = 1 \tag{4.31}$$

4.6 Similarity Solution Technique for Nonlinear Partial Differential Equations

Using Maple the transformation involved in the governing equation and boundary conditions in example 4.13 is solved below:

```
> restart:
> with(student):
> with(plots):
```

Enter the governing equation:

```
> eq:=u(x,y)*diff(u(x,y),x)+v(x,y)*diff(u(x,y),y)-diff(u(x,y),y$2);
```

$$eq := u(x,y) \left(\frac{\partial}{\partial x} u(x,y) \right) + v(x,y) \left(\frac{\partial}{\partial y} u(x,y) \right) - \left(\frac{\partial^2}{\partial y^2} u(x,y) \right)$$

The stream function is introduced:

```
> vars:={u(x,y)=diff(psi(x,y),y),v(x,y)=-diff(psi(x,y),x)};
```

$$vars := \left\{ u(x,y) = \frac{\partial}{\partial y} \psi(x,y), v(x,y) = -\left(\frac{\partial}{\partial x} \psi(x,y) \right) \right\}$$

The governing equation for the stream function is:

```
> eq:=subs(vars,eq);
```

$$eq := \left(\frac{\partial}{\partial y} \psi(x,y) \right) \left(\frac{\partial^2}{\partial x \partial y} \psi(x,y) \right) - \left(\frac{\partial}{\partial x} \psi(x,y) \right) \left(\frac{\partial^2}{\partial y^2} \psi(x,y) \right) - \left(\frac{\partial^3}{\partial y^3} \psi(x,y) \right)$$

Next, the transformation defined in equation (4.30) is introduced:

```
> eq:=changevar(psi(x,y)=x^(1/2)*f(eta(x,y)),eq);
```

$$eq := \frac{1}{2} D(f)(\eta(x,y)) \left(\frac{\partial}{\partial y} \eta(x,y) \right) \left(D(f)(\eta(x,y)) \left(\frac{\partial}{\partial y} \eta(x,y) \right) + 2 x D^{(2)}(f)(\eta(x,y)) \left(\frac{\partial}{\partial y} \eta(x,y) \right) \left(\frac{\partial}{\partial x} \eta(x,y) \right) + 2 x D(f)(\eta(x,y)) \left(\frac{\partial^2}{\partial y \partial x} \eta(x,y) \right) \right) - \frac{1}{2} \left(f(\eta(x,y)) + 2 x D(f)(\eta(x,y)) \left(\frac{\partial}{\partial x} \eta(x,y) \right) \right) \left(D^{(2)}(f)(\eta(x,y)) \left(\frac{\partial}{\partial y} \eta(x,y) \right)^2 + D(f)(\eta(x,y)) \left(\frac{\partial^2}{\partial y^2} \eta(x,y) \right) \right) - \sqrt{x} D^{(3)}(f)(\eta(x,y)) \left(\frac{\partial}{\partial y} \eta(x,y) \right)^3 - 3 \sqrt{x} D^{(2)}(f)(\eta(x,y)) \left(\frac{\partial}{\partial y} \eta(x,y) \right) \left(\frac{\partial^2}{\partial y^2} \eta(x,y) \right) - \sqrt{x} D(f)(\eta(x,y)) \left(\frac{\partial^3}{\partial y^3} \eta(x,y) \right)$$

The independent variables are combined using the transformation defined in equation (4.29):

```
> eq1:=(simplify(subs(eta(x,y)=y/x^(1/2),eq))):
> eq1:=subs(y=eta*x^(1/2),eq1):
> eq1:=simplify(eq1*x):
```

The governing equation for f in the combined variable is:

```
> eq2:=convert(-eq1,diff);
```

$$eq2 := \frac{1}{2}\left(\frac{d^2}{d\eta^2}f(\eta)\right)f(\eta) + \frac{d^3}{d\eta^3}f(\eta)$$

Next, the velocity variables u and v (i.e., derivatives of the stream function) are expressed in terms of the combined variable and f:

> v(eta):=-diff(psi(x,y),x);

$$v(\eta) := -\left(\frac{\partial}{\partial x}\psi(x,y)\right)$$

> v(eta):=changevar(psi(x,y)=x^(1/2)*f(eta(x,y)),v(eta)):
> v(eta):=expand(subs(eta(x,y)=y/x^(1/2),v(eta))):
> v(eta):=subs(y=eta*x^(1/2),v(eta)):
> v(eta):=factor(v(eta));

$$v(\eta) := \frac{1}{2}\frac{-f(\eta) + D(f)(\eta)\eta}{\sqrt{x}}$$

> u(eta):=diff(psi(x,y),y);

$$u(\eta) := \frac{\partial}{\partial y}\psi(x,y)$$

> u(eta):=changevar(psi(x,y)=x^(1/2)*f(eta(x,y)),u(eta)):
> u(eta):=expand(subs(eta(x,y)=y/x^(1/2),u(eta))):
> u(eta):=subs(y=eta*x^(1/2),u(eta));

$$u(\eta) := D(f)(\eta)$$

Next, the boundary conditions are expressed in terms of f:

> bc1:=subs(eta=0,v(eta))=0;

$$bc1 := -\frac{1}{2}\frac{f(0)}{\sqrt{x}} = 0$$

> bc1:=-bc1*2*x^(1/2);

$$bc1 := f(0) = 0$$

> bc2:=subs(eta=0,u(eta))=0;

$$bc2 := D(f)(0) = 0$$

> bc3:=subs(eta=infinity,u(eta))=1;

$$bc3 := D(f)(\infty) = 1$$

The length of the domain is taken to be 5:

> bc3:=subs(infinity=5,bc3);

$$bc3 := D(f)(5) = 1$$

4.6 Similarity Solution Technique for Nonlinear Partial Differential Equations

The numerical solution for the Blassius equation is obtained as:

> sol:=dsolve({eq2,bc1,bc2,bc3},f(eta),type=numeric);

$$sol := \mathbf{proc}(x\_bvp) \ ... \ \mathbf{end \ proc}$$

> odeplot(sol,[eta,f(eta)],0..5,thickness=3,title="Figure 4.26",axes=boxed);

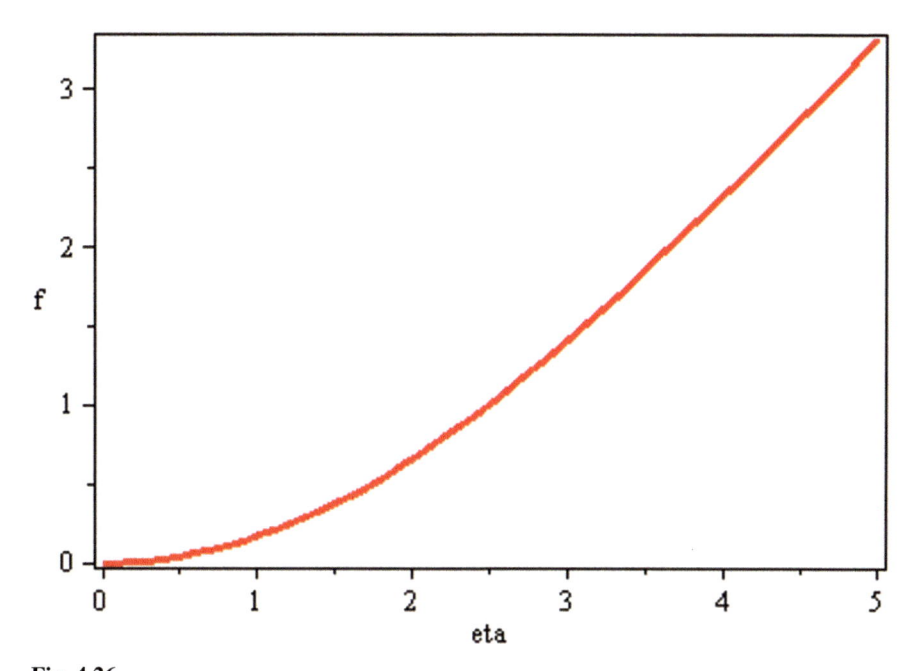

Fig. 4.26

Next, the velocity profiles are obtained by converting the corresponding expression to 'diff' form:

> u(eta):=convert(u(eta),diff);

$$u(\eta) := \frac{d}{d\eta} f(\eta)$$

> v(eta):=convert(v(eta),diff);

$$v(\eta) := \frac{1}{2} \frac{-f(\eta) + \left(\frac{d}{d\eta} f(\eta)\right) \eta}{\sqrt{x}}$$

Since v is a function of x, $v*x^{1/2}$ is plotted:

> odeplot(sol,[eta,u(eta)],0..5,thickness=4,color=blue,title="Figure 4.27", axes=boxed,labels=[eta,u]);

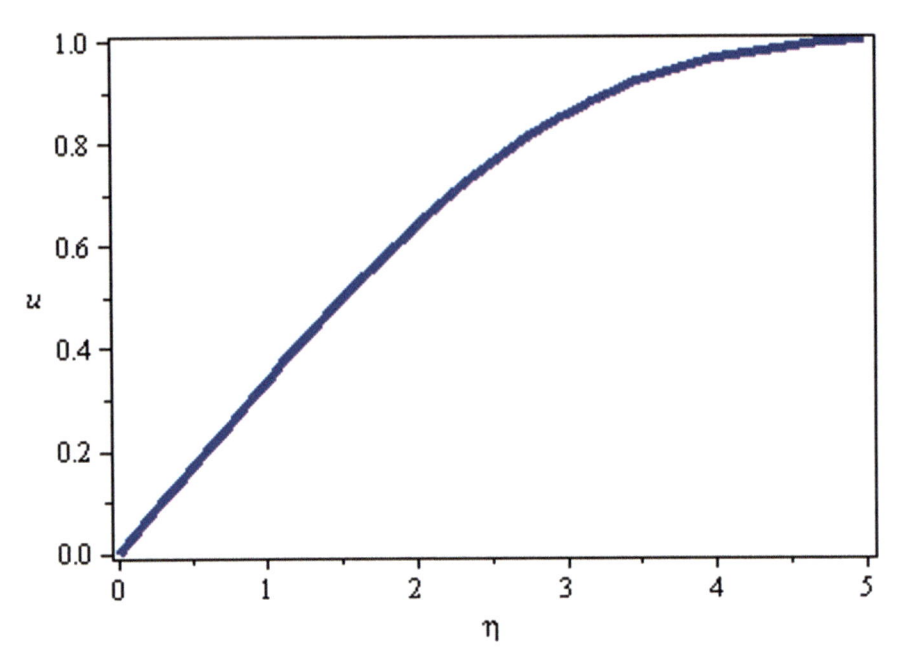

Fig. 4.27

> odeplot(sol,[eta,v(eta)*x^(1/2)],0..5,thickness=4,color=brown,
title="Figure 4.28",axes=boxed,labels=[eta,"v*x^(1/2)"]);

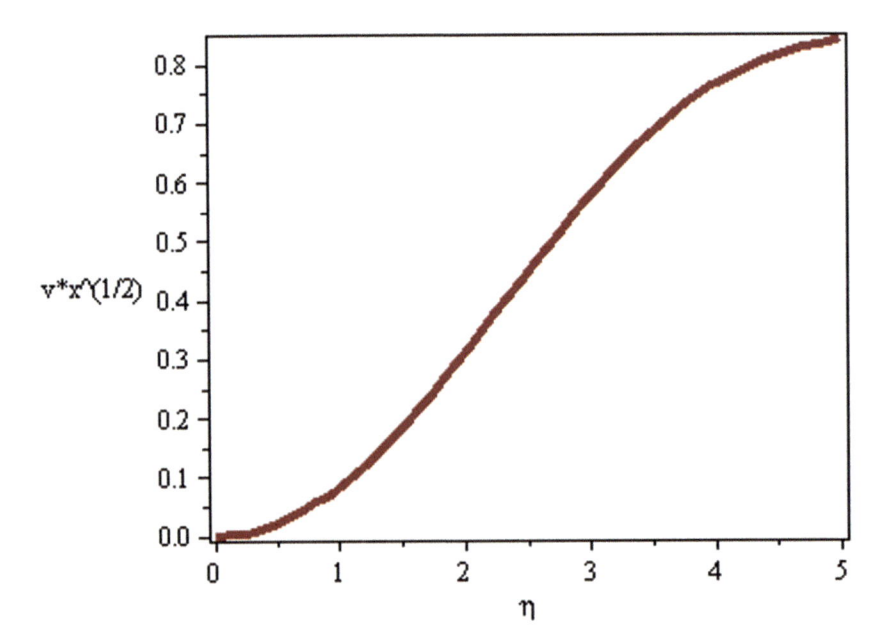

Fig. 4.28

4.6 Similarity Solution Technique for Nonlinear Partial Differential Equations

The solution at $\eta = 0$ is obtained as:

> sol(0);

$$\left[\eta = 0., f(\eta) = 0., \frac{d}{d\eta} f(\eta) = 0., \frac{d^2}{d\eta^2} f(\eta) = 0.336152378983945622 \right]$$

Stress is related to the Reynolds number (re) and the flux at $y = 0$:

> S:=re*diff(u(x,y),y);

$$S := re \left(\frac{\partial}{\partial y} u(x, y) \right)$$

The velocity gradient is terms of the stream function is:

> subs(u(x,y)=diff(psi(x,y),y),S);

$$re \left(\frac{\partial^2}{\partial y^2} \psi(x, y) \right)$$

The second derivative of the stream function is expressed in terms of f and η.

> d:=diff(psi(x,y),y$2);

$$d := \frac{\partial^2}{\partial y^2} \psi(x, y)$$

> d:=changevar(psi(x,y)=x^(1/2)*f(eta(x,y)),d):
> d:=expand(subs(eta(x,y)=y/x^(1/2),d)):
> d:=subs(y=eta*x^(1/2),d):
> d:=convert(d,diff);

$$d := \frac{\frac{d^2}{d\eta^2} f(\eta)}{\sqrt{x}}$$

> S:=re*d;

$$S := \frac{re \left(\frac{d^2}{d\eta^2} f(\eta) \right)}{\sqrt{x}}$$

The second derivative of f is found from the numerical solution.

> eqd3:=sol(0)[4];

$$eqd3 := \frac{d^2}{d\eta^2} f(\eta) = 0.33615237898394562$$

Hence, the stress Reynolds number relationship becomes:

> S:=subs(diff(f(eta),`$`(eta,2))=rhs(eqd3),S);

$$S := \frac{0.3361523789839456 2 e}{\sqrt{x}}$$

4.7 Summary

In this chapter, analytical solutions were obtained for parabolic and elliptic partial differential equations in semi-infinite domains. In section 4.2, the given linear parabolic partial differential equations were converted to an ordinary differential equation boundary value problem in the Laplace domain. The dependent variable was then solved in the Laplace domain using Maple's 'dsolve' command. The solution obtained in the Laplace domain was then converted to the time domain using Maple's inverse Laplace transform technique. Maple is not capable of inverting complicated functions. Two such examples were illustrated in section 4.3. As shown in section 4.3, even when Maple fails, one can arrive at the transient solution by simplifying the integrals using standard Laplace transform formulae.

In section 4.4, the given linear parabolic partial differential equation in semi-infinite domain was solved by combining the independent variables (similarity solution). This technique is capable of providing special function solutions as shown in example 4.9. In section 4.5, elliptic partial differential equations were solved using the similarity solution technique. In section 4.6, similarity solution was extended for nonlinear parabolic and elliptic partial differential equations.

Both the Laplace transform and the similarity solution techniques are powerful techniques for partial differential equations in semi-infinite domains. The Laplace transform technique can be used for all linear partial differential equations with all possible boundary conditions. The similarity solution can be used only if the independent variables can be combined and if the boundary conditions in x and t can be converted to boundary conditions in the combined variable. In addition, unlike the Laplace transform technique, the similarity solution technique cannot handle partial differential equations in which the dependent variable appears explicitly. The Laplace transform cannot handle elliptic or nonlinear partial differential equations. The similarity solution can be used for elliptic and for a few nonlinear partial differential equations as shown in section 4.6. There are thirteen examples in this chapter.

4.8 Exercise Problems

1. Redo example 4.2 if the boundary condition at x = 0 is replaced by
$$u(0,t) = \frac{k}{t\sqrt{t}}.$$

4.8 Exercise Problems

2. Redo example 4.3 if the boundary condition at x = 0 is replaced by
$$\frac{\partial u}{\partial x}(0,t) = -k\exp(-t).$$

3. Complete the details missing in example 4.4 (*i.e.*, complete the Maple program).

4. Complete the details missing in example 4.11.

5. Consider heat transfer in a semi-infinite solid with heat generated within it.[4] The governing equations and the boundary/initial conditions are:
$$\frac{\partial u}{\partial t} = \alpha \frac{\partial^2 u}{\partial x^2} - \frac{A\alpha}{\kappa}$$

$$u(x,0) = 1$$

$$u(0,t) = 0 \text{ and } u(\infty,t) \text{ is defined}$$

where α is the thermal diffusivity, κ is the thermal conductivity, and A is the heat produced per second per unit volume. Obtain an analytical solution for this problem using the Laplace transform technique.

6. Obtain an analytical solution for problem 5 using the similarity solution technique. Hint: define $w = u + \frac{Ax^2}{2\kappa}$ and solve for w instead of u.

7. Consider heat or mass transfer in a region internally bounded by a sphere of radius R initially at zero temperature/concentration. The governing equations and boundary conditions are:
$$\frac{\partial u}{\partial t} = \alpha \left(\frac{\partial^2 u}{\partial x^2} + \frac{2}{x}\frac{\partial u}{\partial x} \right) \qquad R \leq x < \infty$$

$$u(x,0) = 0$$

$$u(R,t) = 1 \text{ and } u(\infty,t) \text{ is defined}$$

Obtain an analytical solution for this problem using the Laplace transform technique.

8. Obtain an analytical solution for problem 7 using a similarity solution technique. Hint: define w = u/x and derive an equation for w. Define X = x − R and solve for w as a function of X ant t.

9. Consider a region internally bounded by a sphere of radius R initially at zero temperature/concentration with a specified flux at x = R. The governing equations and boundary conditions are:

$$\frac{\partial u}{\partial t} = \alpha\left(\frac{\partial^2 u}{\partial x^2} + \frac{2}{x}\frac{\partial u}{\partial x}\right) \qquad R \leq x < \infty$$

$$u(x,0) = 0$$

$$\frac{\partial u}{\partial x}(R,t) = F \text{ and } u(\infty,t) \text{ is defined}$$

Obtain an analytical solution for this problem using the Laplace transform technique.

10. Obtain an analytical solution for problem 9 using a similarity solution technique. (See problem 8 for a hint).

11. Consider a variant of example 4.12:

$$\frac{\partial u}{\partial t} = \frac{\partial}{\partial x}\left((1+0.1u^2)\frac{\partial u}{\partial x}\right)$$

$$u(x,0) = 0$$

$$u(0,t) = 1 \text{ and } u(\infty,t) \text{ is defined}$$

Obtain a similarity solution for this problem.

12. Example 4.9 is sometimes solved in terms of dimensionless independent variables for temperature distribution in a boundary layer (see Slattery, 1999) as:

$$2X\frac{\partial u}{\partial Z} = \frac{\partial^2 u}{\partial X^2}$$

$$u(X,0) = 0$$

$$u(0,Z) = 1 \text{ and } u(\infty, Z) = 0$$

Define $\eta = \dfrac{X}{\left(\dfrac{9Z}{2}\right)^{\frac{1}{3}}}$ and obtain a similarity solution for this problem.

13. Consider problem 12 with a flux boundary condition at the surface.[6] The governing equations and boundary conditions are:

4.8 Exercise Problems

$$2X\frac{\partial u}{\partial Z} = \frac{\partial^2 u}{\partial X^2}$$

$$u(X,0) = 0$$

$$\frac{\partial u}{\partial X}(0,Z) = -1 \text{ and } u(\infty, Z) = 0$$

Obtain a similarity solution for this problem. Hint: define $w = \dfrac{\partial u}{\partial X}$ and solve for w and then obtain u.

14. Consider plane flow past a flat plate discussed in example 4.13. In the example discussed, a numerical solution was obtained for f as a function of η. Obtain a series solutions for f as a function of η. Hint: see chapter 3 for more information on obtaining series solutions for nonlinear boundary value problems. How many terms will be required in the series for convergence?

15. Consider steady state flow in a convergent channel.[6] The governing equations are:

$$\frac{\partial u}{\partial x} + \frac{\partial v}{\partial y} = 0$$

$$u\frac{\partial u}{\partial x} + v\frac{\partial u}{\partial y} = \bar{u}\frac{d\bar{u}}{dx} + \frac{\partial^2 u}{\partial y^2}$$

$$u(x,0) = 0 \text{ and } u(x,\infty) = -\frac{1}{x}$$

$$v(x,0) = 0 \text{ and } \frac{\partial v}{\partial y}(x,\infty) = 0$$

$$\bar{u} = -\frac{1}{x}$$

Convert u and v to stream functions and rewrite the governing equations and boundary conditions as in example 4.13. Next define $\psi = -f(\eta); \eta = \dfrac{y}{x}$ and rewrite the governing equations and boundary conditions in terms of f and

η. Define $p = \dfrac{df}{d\eta}$ and solve for p. Obtain analytical, series, and numerical solutions for this problem and plot $\dfrac{u}{\bar{u}}$ as a function of η.

16. Consider natural convection in flow past a flat plate.[6] The velocity distribution is governed as in example 4.13. In addition, the temperature T is governed by:

$$u\frac{\partial T}{\partial x} + v\frac{\partial T}{\partial y} = \frac{1}{Pr}\frac{\partial^2 T}{\partial y^2} + \frac{Br}{Pr}\left(\frac{\partial u}{\partial y}\right)^2$$

$T(x,0) = 1$ and $T(x,\infty) = 0$

$T(0,y) = 0$

where Pr is the Prandtl number and Br is the Brinkman number. Obtain a similarity solution for the velocity and temperature distribution. Plot the temperature profiles for Pr = 0.7 with different values of Br ranging from –3 to 3.

17. Redo problem 16 if Br = 0.
18. In problem 17 what happens when Pr is 1? How is the temperature distribution different from the velocity distribution?

References

1. Carslaw, H.S., Jaeger, J.C.: Conduction of Heat in Solids. Oxford University Press, Oxford (1973)
2. Crank, J.: The Mathematics of Diffusion, 2nd edn. Oxford University Press, Oxford (1979)
3. Schiesser, W.E., Silebi, C.A.: Dynamic Modeling of Transport Process Systems (1997)
4. Carslaw, H.S., Jaeger, J.C.: Conduction of Heat in Solids. Oxford University Press, Oxford (1972)
5. Articolo, G.A.: Partial Differential Equations and Boundary Value Problems with Maple V. Harcourt Brace/Academic Press (1999)
6. Slattery, J.: Advanced Transport Phenomena. Cambride University Press, NY (1999)
7. Rice, R.G., Do, D.D.: Applied Mathematics and Modeling for Chemical Engineers. John Wiley & Sons, Inc., Chichester (1995)
8. Newman, J.: Electrochemical Systems. Prentice Hall, New York (1991)
9. West, A.C., Newman, J.: Modern Aspects of Electrochemistry. In: Conway, B.E., Bockris, J.O'M., White, R.E. (eds.) Determination of Current Distribtuions Governed by Laplace's Equation, vol. 23. Plenum Press (1992)
10. Finlayson, B.A.: Nonlinear Analysis in Chemical Engineering. McGraw-Hill, New York (1980)

Chapter 5
Method of Lines for Parabolic Partial Differential Equations

Mathematical modeling of mass or heat transfer in solids involves Fick's law of mass transfer or Fourier's law of heat conduction. Engineers are interested in the distribution of heat or concentration across the slab or the material in which the experiment is performed. This process is usually time varying and eventually reaches a steady state. This process is represented by parabolic partial differential equations with known initial conditions and boundary conditions at two ends. Both linear and nonlinear parabolic partial differential equations will be discussed in this chapter. We will present semianalytical solutions for linear parabolic partial differential equations and numerical solutions for nonlinear parabolic partial differential equations based on the numerical method of lines.

5.1 Semianalytical Method for Parabolic Partial Differential Equations (PDEs)

5.1.1 Introduction

Transient heat conduction or mass transfer in solids with constant physical properties (diffusion coefficient, thermal diffusivity, thermal conductivity, etc.) is usually represented by a linear parabolic partial differential equation. In this chapter, we describe how one can arrive at the semianalytical solutions (solutions are analytical in the time variable and numerical in the spatial dimension) for linear parabolic partial differential equations using Maple, the method of lines and the matrix exponential.

5.1.2 Semianalytical Method for Homogeneous PDEs

Consider a general linear homogeneous parabolic partial differential equation in dimensionless form

$$\frac{\partial u}{\partial t} = a_0(x)\frac{\partial^2 u}{\partial x^2} + a_1(x)\frac{\partial u}{\partial x} + a_2(x)u \qquad (5.1)$$

with a known initial condition

$$u(x,0) = 1 \qquad (5.2)$$

and linear homogeneous boundary conditions at both of the boundaries

$$\alpha_1 u(0,t) + \beta_1 \frac{\partial u}{\partial x}(0,t) = 0 \qquad (5.3)$$

and

$$\alpha_2 u(1,t) + \beta_2 \frac{\partial u}{\partial x}(1,t) = 0 \qquad (5.4)$$

where α_1, β_1, α_2, and β_2 are constants.

The method of lines involves converting the governing equation (equation (5.1)) to a system of coupled ordinary differential equations in time by applying finite difference approximations for the spatial derivatives. The governing equation (equation (5.1)) can be converted to its finite difference form as follows:

$$\frac{du_i}{dt} = a_0(x_i)\frac{u_{i+1} - 2u_i + u_{i-1}}{h^2} + a_1(x_i)\frac{u_{i+1} - u_{i-1}}{2h} + a_2(x_i)u_i \qquad (5.5)$$

where i is the node number, N is the number of interior node points, and h is the node spacing defined as

$$h = \frac{L}{N+1} \qquad (5.6)$$

where L = 1 is the length of the domain of interest. Thus, x = 0 corresponds to the node point i = 0; x = 1 corresponds to the node point i = N+1 and x = xi = ih is the value of x at the node point i. The variable u_i corresponds to the dependent variable at node point i. Equation (5.6) is a system of N linear coupled ODEs for N dependent variables ($u_i, i=1..N$). The boundary values u_0 and u_{N+1} are eliminated using the boundary conditions. The boundary conditions (equations (5.3) and (5.4)) can be written in finite difference form as

$$\alpha_1 u_0 + \beta_1 \frac{-3u_0 + 4u_1 - u_2}{2h} = 0 \qquad (5.7)$$

and

$$\alpha_2 u_{N+1} + \beta_2 \frac{3u_{N+1} - 4u_N + u_{N-1}}{2h} = 0 \qquad (5.8)$$

5.1 Semianalytical Method for Parabolic Partial Differential Equations (PDEs)

Using the boundary conditions (equations (5.7) and (5.8)) the boundary values u_0 and u_{N+1} can be eliminated. Hence, the method of lines technique reduces the linear parabolic ODE partial differential equation (equation (5.1)) to a linear system of N coupled first order ordinary differential equations (equation (5.5)). Traditionally this linear system of ordinary differential equations is integrated numerically in time.[1] [2] [3] [4] However, since the governing equation (equation (5.5)) is linear, it can be written as a matrix differential equation (see section 2.1.2):

$$\frac{dY}{dt} = AY \qquad (5.9)$$

where the dependent variable vector is

$$Y = \begin{bmatrix} u_1 & u_2 & \ldots & u_{N-1} & u_N \end{bmatrix}^T \qquad (5.10)$$

the coefficient matrix A depends on both the governing equation (equation (5.5)) and the boundary conditions (equations (5.7) and (5.8)). The solution for equation (5.9) is obtained by finding the exponential matrix:

$$Y = \exp(At)Y_0 \qquad (5.11)$$

where Y0 is the initial condition vector. Hence, the dependent variables at all the node points are obtained as an analytical solution of time t. We call this a semianalytical solution since the solution is analytical in time and numerical in x.[5] A procedure for using Maple to solve linear parabolic partial differential equations with homogeneous boundary conditions can be summarized as follows:

1. Start the Maple worksheet with a 'restart' command to clear all variables.
2. Call 'with(linalg)' and 'with(plots)' commands.
3. Enter the governing equation.
4. Store the boundary conditions in bc1 (x = 0) and bc2 (x = 1).
5. Store the initial condition in Y0.
6. Enter the number of interior node points, N.
7. Enter the length of the domain, L.
8. Convert the governing equations to a finite difference form by using central difference expression accurate to the order h^2 for first and second derivatives.
9. Convert the boundary conditions to a finite difference form by using the 3-point forward and backward differences (accurate to the order h^2), respectively, for bc1 and bc2.
10. Eliminate the boundary values ($u_0(t)$ and $u_{N+1}(t)$) using the boundary conditions.
11. Store the right hand side of the finite difference equations in eqs.
12. Store the dependent variables, u_i, i = 1..N in Y.
13. Generate **A** matrix using Maple's 'genmatrix' command.
14. Find the exponential matrix ($\exp(\mathbf{A}t)$) by using Maple's 'exponential(A,t)' command and store it in mat.
15. Store the initial conditions in Y0 vector.
16. Find the solution Y by multiplying mat times Y0.
17. Once the semianalytical solution is obtained plots can be made.

Example 5.1. Heat Conduction in a Rectangular Slab

Consider the heat conduction problem in a slab.[6] The governing equation in dimensionless form is

$$\frac{\partial u}{\partial t} = \frac{\partial^2 u}{\partial x^2}$$

(5.12)

$$u(x,0) = 1$$

$$u(0,t) = 0 \text{ and } u(1,t) = 0$$

This equation is solved in Maple below using the procedure described above.

> restart;

> with(linalg):with(plots):

Enter the governing equation here:

> ge:=diff(u(x,t),t)=diff(u(x,t),x$2);

$$ge := \frac{\partial}{\partial t} u(x,t) = \frac{\partial^2}{\partial x^2} u(x,t)$$

Enter the boundary condition at x = 0:

> bc1:=u(x,t);

$$bc1 := u(x,t)$$

Enter the boundary condition at x = L:

> bc2:=u(x,t);

$$bc2 := u(x,t)$$

Enter the initial condition:

> IC:=u(x,0)=1;

$$IC := u(x,0) = 1$$

Enter the number of interior node points:

> N:=4;

$$N := 4$$

Enter the length of the domain:

> L:=1;

$$L := 1$$

5.1 Semianalytical Method for Parabolic Partial Differential Equations (PDEs)

Enter the three point backward difference expression (accurate to the order h^2) for the first derivative at $x = L$:

> dydxf:=1/2*(-u[2](t)-3*u[0](t)+4*u[1](t))/h;

$$dydxf := \frac{1}{2} \frac{-u_2(t) - 3\,u_0(t) + 4\,u_1(t)}{h}$$

Enter the three point central difference expression (accurate to the order h^2) for the second derivative at $x = L$:

> dydxb:=1/2*(u[N-1](t)+3*u[N+1](t)-4*u[N](t))/h;

$$dydxb := \frac{1}{2} \frac{u_3(t) + 3\,u_5(t) - 4\,u_4(t)}{h}$$

Convert the boundary conditions to the finite difference form:

> dydx:=1/2/h*(u[m+1](t)-u[m-1](t));

$$dydx := \frac{1}{2} \frac{u_{m+1}(t) - u_{m-1}(t)}{h}$$

> d2ydx2:=1/h^2*(u[m-1](t)-2*u[m](t)+u[m+1](t));

$$d2ydx2 := \frac{u_{m-1}(t) - 2\,u_m(t) + u_{m+1}(t)}{h^2}$$

> bc1:=subs(diff(u(x,t),x)=dydxf,u(x,t)=u[0](t),x=0,bc1);

$$bc1 := u_0(t)$$

> bc2:=subs(diff(u(x,t),x)=dydxb,u(x,t)=u[N+1](t),x=1,bc2);

$$bc2 := u_5(t)$$

The boundary conditions are stored in eq[0] and eq[N+1].

> eq[0]:=bc1;

$$eq_0 := u_0(t)$$

> eq[N+1]:=bc2;

$$eq_5 := u_5(t)$$

The governing equations are converted to the finite difference form:

> for i from 1 to N do eq[i]:=diff(u[i](t),t)= subs(diff(u(x,t),x$2) = subs(m=i,d2ydx2),diff(u(x,t),x) = subs(m=i,dydx),u(x,t)=u[i](t), x=i*h,rhs(ge));od;

$$eq_1 := \frac{d}{dt} u_1(t) = \frac{u_0(t) - 2u_1(t) + u_2(t)}{h^2}$$

$$eq_2 := \frac{d}{dt} u_2(t) = \frac{u_1(t) - 2u_2(t) + u_3(t)}{h^2}$$

$$eq_3 := \frac{d}{dt} u_3(t) = \frac{u_2(t) - 2u_3(t) + u_4(t)}{h^2}$$

$$eq_4 := \frac{d}{dt} u_4(t) = \frac{u_3(t) - 2u_4(t) + u_5(t)}{h^2}$$

The boundary values u[0](t) and u[N+0](t) are eliminated:

> u[0](t):=(solve(eq[0],u[0](t)));

$$u_0(t) := 0$$

> u[N+1](t):=solve(eq[N+1],u[N+1](t));

$$u_5(t) := 0$$

The governing equations are simplified as:

> for i from 1 to N do eq[i]:=eval(eq[i]);od;

$$eq_1 := \frac{d}{dt} u_1(t) = \frac{-2u_1(t) + u_2(t)}{h^2}$$

$$eq_2 := \frac{d}{dt} u_2(t) = \frac{u_1(t) - 2u_2(t) + u_3(t)}{h^2}$$

$$eq_3 := \frac{d}{dt} u_3(t) = \frac{u_2(t) - 2u_3(t) + u_4(t)}{h^2}$$

$$eq_4 := \frac{d}{dt} u_4(t) = \frac{u_3(t) - 2u_4(t)}{h^2}$$

The coefficient matrix (**A**) is generated using Maple's 'genmatrix' command.

> eqs:=[seq(rhs(eq[j]),j=1..N)];

$$eqs := \left[\frac{-2u_1(t) + u_2(t)}{h^2}, \frac{u_1(t) - 2u_2(t) + u_3(t)}{h^2}, \frac{u_2(t) - 2u_3(t) + u_4(t)}{h^2}, \frac{u_3(t) - 2u_4(t)}{h^2} \right]$$

> Y:=[seq(u[i](t),i=1..N)];

$$Y := [u_1(t), u_2(t), u_3(t), u_4(t)]$$

5.1 Semianalytical Method for Parabolic Partial Differential Equations (PDEs)

> A:=genmatrix(eqs,Y);

$$A := \begin{bmatrix} -\dfrac{2}{h^2} & \dfrac{1}{h^2} & 0 & 0 \\ \dfrac{1}{h^2} & -\dfrac{2}{h^2} & \dfrac{1}{h^2} & 0 \\ 0 & \dfrac{1}{h^2} & -\dfrac{2}{h^2} & \dfrac{1}{h^2} \\ 0 & 0 & \dfrac{1}{h^2} & -\dfrac{2}{h^2} \end{bmatrix}$$

The node spacing, h, is evaluated here:

> h:=eval(1/(N+1));

$$h := \frac{1}{5}$$

The **A** matrix is simplified as:
> A:=map(eval,A);

$$A := \begin{bmatrix} -50 & 25 & 0 & 0 \\ 25 & -50 & 25 & 0 \\ 0 & 25 & -50 & 25 \\ 0 & 0 & 25 & -50 \end{bmatrix}$$

When more than four node points are used, the entries of the **A** matrix should be decimals. This problem can be handled by using the 'map(A,evalf)' command as given below:

> if N > 4 then A:=map(evalf,A);end:
> evalm(A);

$$\begin{bmatrix} -50 & 25 & 0 & 0 \\ 25 & -50 & 25 & 0 \\ 0 & 25 & -50 & 25 \\ 0 & 0 & 25 & -50 \end{bmatrix}$$

The solution is found using the exponential matrix:

> mat:=exponential(A,t):

> mat:=map(evalf,mat):

360   5 Method of Lines for Parabolic Partial Differential Equations

The initial condition is stored in the Y0 vector.

> Y0:=matrix(N,1):for i from 1 to N do Y0[i,1]:=evalf(subs(x=i*h,rhs(IC)));
od:evalm(Y0);

$$\begin{bmatrix} 1. \\ 1. \\ 1. \\ 1. \end{bmatrix}$$

> Y:=evalm(mat&*Y0):

> Y:=map(simplify,Y);

$$Y := \begin{bmatrix} 0.7236067982\, e^{-9.549150288\, t} + 0.2763932022\, e^{-65.45084971\, t} \\ -0.1708203934\, e^{-65.45084971\, t} + 1.170820394\, e^{-9.549150288\, t} \\ -0.1708203934\, e^{-65.45084971\, t} + 1.170820394\, e^{-9.549150288\, t} \\ 0.7236067982\, e^{-9.549150288\, t} + 0.2763932022\, e^{-65.45084971\, t} \end{bmatrix}$$

Next, the dependent variables can be stored in $u_i(t), i = 0..N+1$.

> for i from 1 to N do u[i](t):=evalf((Y[i,1]));od:
> for i from 0 to N+1 do u[i](t):=eval(u[i](t));od;

$$u_0(t) := 0$$
$$u_1(t) := 0.7236067982\, e^{-9.549150288\, t} + 0.2763932022\, e^{-65.45084971\, t}$$
$$u_2(t) := -0.1708203934\, e^{-65.45084971\, t} + 1.170820394\, e^{-9.549150288\, t}$$
$$u_3(t) := -0.1708203934\, e^{-65.45084971\, t} + 1.170820394\, e^{-9.549150288\, t}$$
$$u_4(t) := 0.7236067982\, e^{-9.549150288\, t} + 0.2763932022\, e^{-65.45084971\, t}$$
$$u_5(t) := 0$$

Hence, an analytical solution in time is obtained for the dependent variables at all of the node points. One can plot the concentration profiles by using 'seq' to assign color automatically for every curve.

> pp:=plot([seq(u[i](t),i=0..N+1)],t=0..0.4);

$$pp := PLOT(...)$$

> pt:=textplot([[0.05,0.05,typeset(u[0],"(t), ",u[5],"(t)")],[0.1,0.2, typeset(u[1],"(t), ",u[4],"(t",u, u[4],"(t)")],[0.15,0.4, typeset(u[2],"(t), ",u[3],"(t)")]]);

$$pt := PLOT(...)$$

> display({pp,pt},axes=boxed,thickness=4,title="Figure 5.1",labels=[t,"u"]);

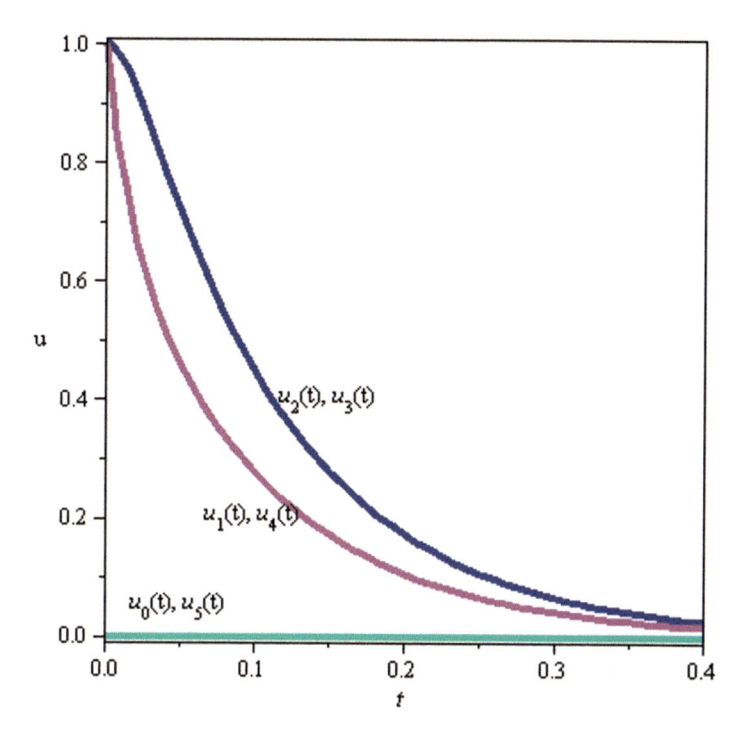

5.1 Semianalytical Method for Parabolic Partial Differential Equations (PDEs)

Fig. 5.1

A three dimensional plot can be made by storing the solution in a matrix (PP). Enter the time that you want to use to plot your profiles (this time can be changed depending on the problem):

> tf:=0.1;

$$tf := 0.1$$

Enter the number of time steps (excluding 0):

> M:=30;

$$M := 30$$

The time intervals are stored in T1:

> T1:=[seq(tf*i/M,i=0..M)];

$T1 := [0., 0.003333333333, 0.006666666667, 0.01000000000, 0.01333333333, 0.01666666667, 0.02000000000,$
$0.02333333333, 0.02666666667, 0.03000000000, 0.03333333333, 0.03666666667, 0.04000000000,$
$0.04333333333, 0.04666666667, 0.05000000000, 0.05333333333, 0.05666666667, 0.06000000000,$
$0.06333333333, 0.06666666667, 0.07000000000, 0.07333333333, 0.07666666667, 0.08000000000,$
$0.08333333333, 0.08666666667, 0.09000000000, 0.09333333333, 0.09666666667, 0.1000000000]$

> PP:=matrix(N+2,M+1);

$$PP := array(1..6, 1..31, [\ ])$$

The first column of PP is filled by using the initial condition:

> for i from 1 to N+2 do PP[i,1]:=evalf(subs(x=(i-1)*h,rhs(IC)));od:

The remaining columns are filled in by using the solution obtained:

> for i from 1 to N+2 do for j from 2 to M+1 do PP[i,j]:= evalf(subs(t=T1[j],u[i-1](t)));od;od:

Next, data points are stored in plotdata for obtaining a 3D plot using Maple's 'surfdata' command.

> plotdata := [seq([seq([(i-1)*h,T1[j],PP[i,j]], i=1..N+2)], j=1..M+1)]:

> surfdata(plotdata, axes=boxed, title="Figure 5.2", labels=[x,t,u],orientation=[45,60]);

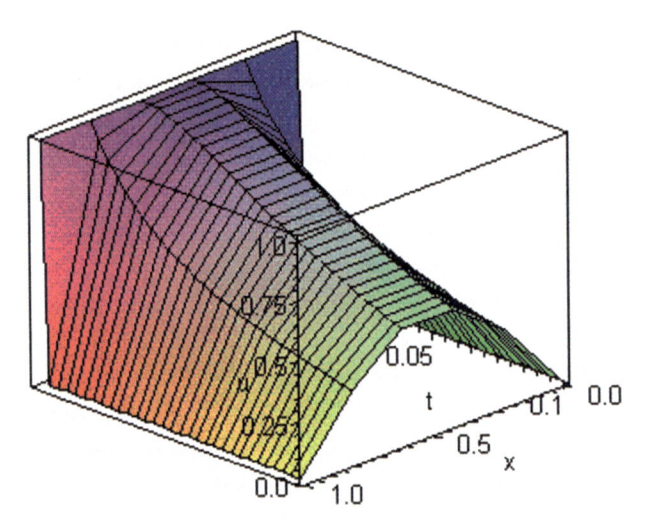

Fig. 5.2

>

Accuracy can be increased by increasing the number of interior node points. Using the program above by increasing N = 10, the following plots are obtained.

> pp:=plot([seq(u[i](t),i=0..N+1)],t=0..0.4,thickness=4);

$$pp := PLOT(...)$$

> arw:=arrow(<0.10,0.8>,<-0.09,-0.6>,width=[1/500,relative], head_length=[1/20,relative]):

5.1 Semianalytical Method for Parabolic Partial Differential Equations (PDEs)

```
pt:=textplot([[0.2,0.9,"Follow the arrow:"],seq([0.18,0.9-i*0.06,
typeset(u[5-i+1],"(t), ",u[6+i-1],"(t)")],i=1..6)]):
```
> `display({pp,arw,pt},axes=boxed,title="Figure 5.3",labels=[t,"u"]);`

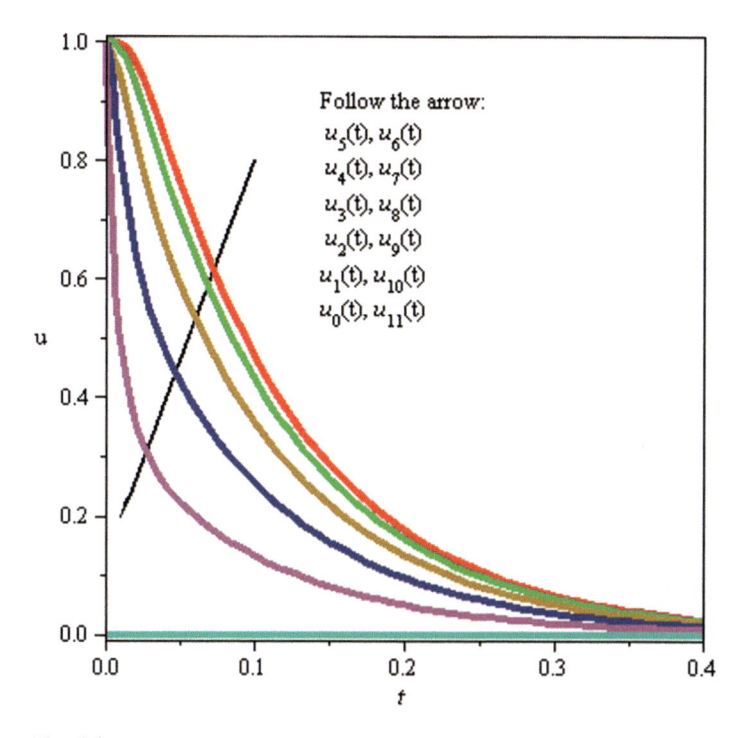

Fig. 5.3

A three dimensional plot can be made by storing the solution in a matrix (PP). Enter the time that you want to use to plot your profiles (this time can be changed depending on the problem):

> `tf:=0.1;`

$$tf := 0.1$$

Enter the number of time steps (excluding 0):

> `M:=30;`

$$M := 30$$

The time intervals are stored in T1:

> `T1:=[seq(tf*i/M,i=0..M)];`

$T1 := [0., 0.003333333333, 0.006666666667, 0.01000000000, 0.01333333333, 0.01666666667, 0.02000000000, 0.02333333333, 0.02666666667, 0.03000000000, 0.03333333333, 0.03666666667, 0.04000000000,$

0.04333333333, 0.04666666667, 0.05000000000, 0.05333333333, 0.05666666667, 0.06000000000, 0.06333333333, 0.06666666667, 0.07000000000, 0.07333333333, 0.07666666667, 0.08000000000, 0.08333333333, 0.08666666667, 0.09000000000, 0.09333333333, 0.09666666667, 0.1000000000]

> PP:=matrix(N+2,M+1);

$$PP := array(1..12, 1..31, [\,])$$

The first column of PP is filled by using the initial condition:

> for i from 1 to N+2 do PP[i,1]:=evalf(subs(x=(i-1)*h,rhs(IC)));od:

The remaining columns are filled in by using the solution obtained:

> for i from 1 to N+2 do for j from 2 to M+1 do PP[i,j]:= evalf(subs(t=T1[j],u[i-1](t)));od;od:

Next, data points are stored in plotdata for obtaining a 3D plot using Maple's 'surfdata' command.

> plotdata := [seq([seq([(i-1)*h,T1[j],PP[i,j]], i=1..N+2)], j=1..M+1)]:

> surfdata(plotdata, axes=boxed, title="Figure 5.4", labels=[x,t,u],orientation=[45,60]);

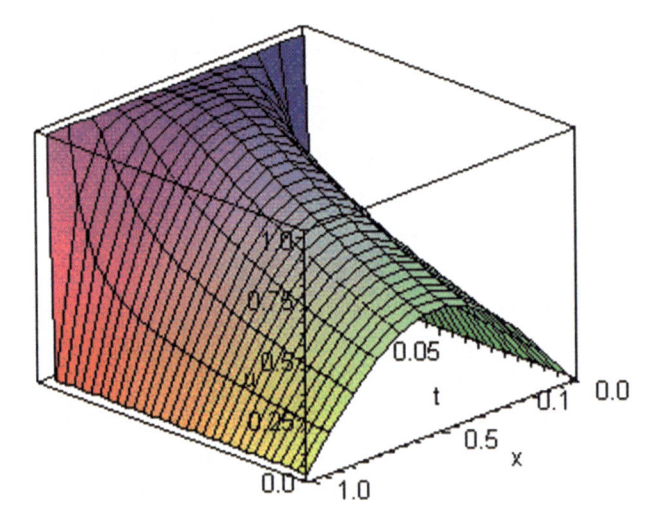

Fig. 5.4

>

We observe that the process slowly approaches the steady state. We note that the profiles are symmetrical about x = 0.5 and, hence, we obtain u[0] = u[N+1], u[1] = u[N], etc.

5.1.3 Semianalytical Method for Nonhomogeneous PDEs

Consider a general parabolic PDE with a source term where γ is the source term

$$\frac{\partial u}{\partial t} = a_0(x)\frac{\partial^2 u}{\partial x^2} + a_1(x)\frac{\partial u}{\partial x} + a_2(x)u + \gamma \qquad (5.13)$$

independent of u. In equation (5.13) γ can be functions of both t and x or just a constant. The following nonhomogeneous boundary conditions are considered:

$$\alpha_1 u(0,t) + \beta_1 \frac{\partial u}{\partial x}(0,t) = \gamma_1 \qquad (5.14)$$

$$\alpha_2 u(1,t) + \beta_2 \frac{\partial u}{\partial x}(1,t) = \gamma_2 \qquad (5.15)$$

where γ_1, and γ_2 can either be functions of time or constants. When the governing equation (5.13) and the boundary conditions (5.14) and (5.15) are converted to the finite difference form, N coupled ODEs arise which can be written in the matrix form

$$\frac{d\mathbf{Y}}{dt} = \mathbf{AY} + \mathbf{b}(t) \qquad (5.16)$$

The solution for equation (5.16) is obtained by adding the nonhomogeneous solution to the matrix exponential (Amundson, 1966;[7] Taylor and Krishna, 1993;[8] Subramanian and White, 2000;[5] see section 2.1.3):

$$\mathbf{Y} = \exp(\mathbf{A}t)\mathbf{Y}_0 + \int_0^t \exp[-\mathbf{A}(\tau-t)]\,\mathbf{b}(\tau)\,d\tau \qquad (5.17)$$

where τ is a dummy variable of integration. When **b** is a constant vector equation (5.17) reduces to

$$\mathbf{Y} = \exp(\mathbf{A}t)(\mathbf{Y}_0 + \mathbf{A}^{-1}\mathbf{b}) - \mathbf{A}^{-1}\mathbf{b} \qquad (5.18)$$

Hence, we obtain a semianalytical solution, i.e., the dependent variables at all the node points are obtained as an analytical solution of time t. The procedure for solving linear parabolic partial differential equations with nonhomogeneous boundary conditions can be summarized as follows:

1. Start the Maple program with a 'restart' command to clear all variables.
2. Call 'with(linalg)' and 'with(plots)' commands.
3. Enter the governing equation.
4. Store the boundary conditions in bc1 (x = 0) and bc2 (x = 1).
5. Store the initial condition in Y0.
6. Enter the number of interior node points, N.

7. Enter the length of the domain, L.
8. Convert the governing equations to the finite difference form by a using central difference expression accurate to the order h^2 for first and second derivatives.
9. Convert the boundary conditions to the finite difference form by using the 3-point forward and backward differences (accurate to the order h^2), respectively, for bc1 and bc2.
10. Eliminate the boundary values ($u_0(t)$ and $u_{N+1}(t)$) using the boundary conditions.
11. Store the right hand side of the finite difference equations in eqs.
12. Store the dependent variables, u_i, i = 1..N in vars.
13. Generate **A** matrix and **b** vector using Maple's 'genmatrix' command.
14. Find the exponential matrix (exp(**A**t)) by using Maple's 'exponential(A,t)' command and store it in mat.
15. Store the initial conditions in the Y0 vector.
16. Find the solution Y by multiplying mat times Y0 and adding the nonhomogeneous part according to equation (5.17).
17. Once the semianalytical solution is obtained, plots can be made.

Example 5.2

Consider the heat conduction/mass transfer problem in a cylinder.[6] [9] [10] The governing equation in dimensionless form is

$$\frac{\partial u}{\partial t} = \frac{\partial^2 u}{\partial x^2} + \frac{1}{x}\frac{\partial u}{\partial x}$$

$$u(x,0) = 0 \tag{5.19}$$

$$\frac{\partial u}{\partial x}(0,t) = 0 \text{ and } u(1,t) = 1$$

The nonhomogeneous boundary condition at x = 1 contributes to the forcing function vector **b**. However, in this case the **b** vector is a constant and, hence, equation (5.18) can be used. (Note that equation (5.17) is valid even when the **b** vector is a constant). When the governing equation is applied at x = 0 in cylindrical or spherical coordinates, we have singularity at x = 0.[11] [12] This singularity is avoided in our semianalytical technique as we use the boundary condition at x = 0 (symmetry boundary condition) to eliminate the dependent variable. Equation (5.19) is solved in Maple below using the procedure described above.

> restart;

> with(linalg):with(plots):

> ge:=diff(u(x,t),t)=diff(u(x,t),x$2)+1/x*diff(u(x,t),x);

5.1 Semianalytical Method for Parabolic Partial Differential Equations (PDEs)

$$ge := \frac{\partial}{\partial t} u(x,t) = \frac{\partial^2}{\partial x^2} u(x,t) + \frac{\frac{\partial}{\partial x} u(x,t)}{x}$$

> bc1:=diff(u(x,t),x);

$$bc1 := \frac{\partial}{\partial x} u(x,t)$$

> bc2:=u(x,t)-1;

$$bc2 := u(x,t) - 1$$

> IC:=u(x,0)=0;

$$IC := u(x,0) = 0$$

> N:=10;

$$N := 10$$

> L:=1;

$$L := 1$$

> dydxf:=1/2*(-u[2](t)-3*u[0](t)+4*u[1](t))/h:
> dydxb:=1/2*(u[N-1](t)+3*u[N+1](t)-4*u[N](t))/h:
> dydx:=1/2/h*(u[m+1](t)-u[m-1](t)):
> d2ydx2:=1/h^2*(u[m-1](t)-2*u[m](t)+u[m+1](t)):
> bc1:=subs(diff(u(x,t),x)=dydxf,u(x,t)=u[0](t),x=0,bc1):
> bc2:=subs(diff(u(x,t),x)=dydxb,u(x,t)=u[N+1](t),x=1,bc2):
> eq[0]:=bc1;

$$eq_0 := \frac{1}{2} \frac{-u_2(t) - 3u_0(t) + 4u_1(t)}{h}$$

> eq[N+1]:=bc2;

$$eq_{11} := u_{11}(t) - 1$$

> for i from 1 to N do eq[i]:=diff(u[i](t),t)= subs(diff(u(x,t),x$2) = subs(m=i,d2ydx2),diff(u(x,t),x) = subs(m=i,dydx),u(x,t)=u[i](t),x=i*h,rhs(ge));od:
> u[0](t):=(solve(eq[0],u[0](t)));

$$u_0(t) := -\frac{1}{3} u_2(t) + \frac{4}{3} u_1(t)$$

```
> u[N+1](t):=solve(eq[N+1],u[N+1](t));
```

$$u_{11}(t) := 1$$

```
> for i from 1 to N do eq[i]:=eval(eq[i]);od;
```

$$eq_1 := \frac{d}{dt}u_1(t) = \frac{\frac{2}{3}u_2(t) - \frac{2}{3}u_1(t)}{h^2} + \frac{1}{2}\frac{\frac{4}{3}u_2(t) - \frac{4}{3}u_1(t)}{h^2}$$

$$eq_2 := \frac{d}{dt}u_2(t) = \frac{u_1(t) - 2u_2(t) + u_3(t)}{h^2} + \frac{1}{4}\frac{u_3(t) - u_1(t)}{h^2}$$

$$eq_3 := \frac{d}{dt}u_3(t) = \frac{u_2(t) - 2u_3(t) + u_4(t)}{h^2} + \frac{1}{6}\frac{u_4(t) - u_2(t)}{h^2}$$

$$eq_4 := \frac{d}{dt}u_4(t) = \frac{u_3(t) - 2u_4(t) + u_5(t)}{h^2} + \frac{1}{8}\frac{u_5(t) - u_3(t)}{h^2}$$

$$eq_5 := \frac{d}{dt}u_5(t) = \frac{u_4(t) - 2u_5(t) + u_6(t)}{h^2} + \frac{1}{10}\frac{u_6(t) - u_4(t)}{h^2}$$

$$eq_6 := \frac{d}{dt}u_6(t) = \frac{u_5(t) - 2u_6(t) + u_7(t)}{h^2} + \frac{1}{12}\frac{u_7(t) - u_5(t)}{h^2}$$

$$eq_7 := \frac{d}{dt}u_7(t) = \frac{u_6(t) - 2u_7(t) + u_8(t)}{h^2} + \frac{1}{14}\frac{u_8(t) - u_6(t)}{h^2}$$

$$eq_8 := \frac{d}{dt}u_8(t) = \frac{u_7(t) - 2u_8(t) + u_9(t)}{h^2} + \frac{1}{16}\frac{u_9(t) - u_7(t)}{h^2}$$

$$eq_9 := \frac{d}{dt}u_9(t) = \frac{u_8(t) - 2u_9(t) + u_{10}(t)}{h^2} + \frac{1}{18}\frac{u_{10}(t) - u_8(t)}{h^2}$$

$$eq_{10} := \frac{d}{dt}u_{10}(t) = \frac{u_9(t) - 2u_{10}(t) + 1}{h^2} + \frac{1}{20}\frac{1 - u_9(t)}{h^2}$$

5.1 Semianalytical Method for Parabolic Partial Differential Equations (PDEs)

```
> eqs:=[seq(rhs(eq[j]),j=1..N)]:
> Y:=[seq(u[i](t),i=1..N)];
```

$$Y := \left[u_1(t), u_2(t), u_3(t), u_4(t), u_5(t), u_6(t), u_7(t), u_8(t), u_9(t), u_{10}(t) \right]$$

```
> A:=genmatrix(eqs,Y,'b1'):
> b:=matrix(N,1):for i to N do b[i,1]:=-eval(b1[i]);od:evalm(b);
```

$$\begin{bmatrix} 0 \\ 0 \\ 0 \\ 0 \\ 0 \\ 0 \\ 0 \\ 0 \\ 0 \\ \dfrac{21}{20\,h^2} \end{bmatrix}$$

```
> h:=eval(L/(N+1));
```

$$h := \frac{1}{11}$$

```
> A:=map(eval,A):
> if N > 4 then A:=map(evalf,A);end:
> evalm(A);
```

$$\begin{bmatrix} -80.66666667 & 80.66666667 & 0. & 0. & 0. & 0. & 0. & 0. & 0. & 0. \\ 121. & -242. & 121. & 0. & 0. & 0. & 0. & 0. & 0. & 0. \\ 0. & 121. & -242. & 121. & 0. & 0. & 0. & 0. & 0. & 0. \\ 0. & 0. & 121. & -242. & 121. & 0. & 0. & 0. & 0. & 0. \\ 0. & 0. & 0. & 121. & -242. & 121. & 0. & 0. & 0. & 0. \\ 0. & 0. & 0. & 0. & 121. & -242. & 121. & 0. & 0. & 0. \\ 0. & 0. & 0. & 0. & 0. & 121. & -242. & 121. & 0. & 0. \\ 0. & 0. & 0. & 0. & 0. & 0. & 121. & -242. & 121. & 0. \\ 0. & 0. & 0. & 0. & 0. & 0. & 0. & 121. & -242. & 121. \\ 0. & 0. & 0. & 0. & 0. & 0. & 0. & 0. & 80.66666667 & -80.66666667 \end{bmatrix}$$

> det(A);

> mat:=exponential(A,t):

> mat:=map(evalf,mat):

> Y0:=matrix(N,1):for i from 1 to N do Y0[i,1]:=
evalf(subs(x=i*h,rhs(IC)));od:evalm(Y0);

$$\begin{bmatrix} 0. \\ 0. \\ 0. \\ 0. \\ 0. \\ 0. \\ 0. \\ 0. \\ 0. \\ 0. \end{bmatrix}$$

The nonhomogeneous solution (equation(5.17)) is found and added to the homogeneous solution.

> s1:=evalm(Y0+inverse(A)&*b):

> Y:=evalm(mat&*s1-inverse(A)&*b):

> Y:=map(simplify,Y):

The number of digits is decreased to five for brevity. However, for accuracy, one has to use a minimum of 10 digits, which is the default in Maple.

> Digits:=5;

$$Digits := 5$$

> for i from 1 to N do u[i](t):=evalf((Y[i,1]));od:

> for i from 0 to N+1 do u[i](t):=eval(u[i](t));od;

$u_0(t) := 0.015562\, e^{-474.73\,t} - 0.059845\, e^{-444.17\,t} + 0.12905\, e^{-395.52\,t} - 0.22028\, e^{-333.54\,t} + 0.33059\, e^{-264.02\,t}$
$- 0.45799\, e^{-193.21\,t} + 0.60388\, e^{-127.20\,t} - 0.78001\, e^{-71.337\,t} + 1.0329\, e^{-29.851\,t} - 1.5938\, e^{-5.7568\,t} + 0.99997$

$u_1(t) := 0.0078565\, e^{-474.73\,t} - 0.031207\, e^{-444.17\,t} + 0.071018\, e^{-395.52\,t} - 0.13041\, e^{-333.54\,t} + 0.21391\, e^{-264.02\,t}$
$- 0.32732\, e^{-193.21\,t} + 0.47820\, e^{-127.20\,t} - 0.67982\, e^{-71.337\,t} + 0.97287\, e^{-29.851\,t} - 1.5751\, e^{-5.7568\,t} + 1.0000$

$u_2(t) := -0.015262\, e^{-474.73\,t} + 0.054709\, e^{-444.17\,t} - 0.10309\, e^{-395.52\,t} + 0.13920\, e^{-333.54\,t} - 0.13614\, e^{-264.02\,t}$
$+ 0.064669\, e^{-193.21\,t} + 0.10117\, e^{-127.20\,t} - 0.37923\, e^{-71.337\,t} + 0.79286\, e^{-29.851\,t} - 1.5189\, e^{-5.7568\,t} + 1.0000$

5.1 Semianalytical Method for Parabolic Partial Differential Equations (PDEs)

$u_3(t) := 0.018770\, e^{-474.73\,t} - 0.054402\, e^{-444.17\,t} + 0.062025\, e^{-395.52\,t} - 0.0060035\, e^{-333.54\,t} - 0.10853\, e^{-264.02\,t}$
$+ 0.21726\, e^{-193.21\,t} - 0.21013\, e^{-127.20\,t} - 0.020008\, e^{-71.337\,t} + 0.52837\, e^{-29.851\,t} - 1.4274\, e^{-5.7568\,t} + 1.0000$

$u_4(t) := -0.020044\, e^{-474.73\,t} + 0.038831\, e^{-444.17\,t} + 0.0061803\, e^{-395.52\,t} - 0.095535\, e^{-333.54\,t} + 0.11417\, e^{-264.02\,t}$
$+ 0.028901\, e^{-193.21\,t} - 0.24315\, e^{-127.20\,t} + 0.24669\, e^{-71.337\,t} + 0.22772\, e^{-29.851\,t} - 1.3038\, e^{-5.7568\,t} + 1.0000$

$u_5(t) := 0.019671\, e^{-474.73\,t} - 0.015357\, e^{-444.17\,t} - 0.055213\, e^{-395.52\,t} + 0.068916\, e^{-333.54\,t} + 0.065942\, e^{-264.02\,t}$
$- 0.15862\, e^{-193.21\,t} - 0.041621\, e^{-127.20\,t} + 0.32484\, e^{-71.337\,t} - 0.056055\, e^{-29.851\,t} - 1.1525\, e^{-5.7568\,t}$
$+ 1.0000$

$u_6(t) := -0.017996\, e^{-474.73\,t} - 0.0084442\, e^{-444.17\,t} + 0.058627\, e^{-395.52\,t} + 0.030767\, e^{-333.54\,t} - 0.10432\, e^{-264.02\,t}$
$- 0.081793\, e^{-193.21\,t} + 0.16304\, e^{-127.20\,t} + 0.21468\, e^{-71.337\,t} - 0.27566\, e^{-29.851\,t} - 0.97890\, e^{-5.7568\,t}$
$+ 1.0000$

$u_7(t) := 0.015306\, e^{-474.73\,t} + 0.026019\, e^{-444.17\,t} - 0.021944\, e^{-395.52\,t} - 0.079799\, e^{-333.54\,t} - 0.038275\, e^{-264.02\,t}$
$+ 0.10377\, e^{-193.21\,t} + 0.17801\, e^{-127.20\,t} + 0.0046386\, e^{-71.337\,t} - 0.39871\, e^{-29.851\,t} - 0.78901\, e^{-5.7568\,t}$
$+ 1.0000$

$u_8(t) := -0.011882\, e^{-474.73\,t} - 0.033254\, e^{-444.17\,t} - 0.024824\, e^{-395.52\,t} + 0.029683\, e^{-333.54\,t} + 0.096913\, e^{-264.02\,t}$
$+ 0.10994\, e^{-193.21\,t} + 0.016322\, e^{-127.20\,t} - 0.17995\, e^{-71.337\,t} - 0.41355\, e^{-29.851\,t} - 0.58940\, e^{-5.7568\,t}$
$+ 1.0000$

$u_9(t) := 0.0080034\, e^{-474.73\,t} + 0.029336\, e^{-444.17\,t} + 0.049007\, e^{-395.52\,t} + 0.049276\, e^{-333.54\,t} + 0.017176\, e^{-264.02\,t}$
$- 0.049836\, e^{-193.21\,t} - 0.14249\, e^{-127.20\,t} - 0.24297\, e^{-71.337\,t} - 0.33061\, e^{-29.851\,t} - 0.38689\, e^{-5.7568\,t}$
$+ 1.0000$

$u_{10}(t) := -0.0039531\, e^{-474.73\,t} - 0.016680\, e^{-444.17\,t} - 0.036693\, e^{-395.52\,t} - 0.061876\, e^{-333.54\,t}$
$- 0.089672\, e^{-264.02\,t} - 0.11741\, e^{-193.21\,t} - 0.14268\, e^{-127.20\,t} - 0.16365\, e^{-71.337\,t} - 0.17914\, e^{-29.851\,t}$
$- 0.18825\, e^{-5.7568\,t} + 1.0000$

$$u_{11}(t) := 1$$

Hence, we have obtained a semianalytical solution (analytical in time and numerical in x). Next, plots are made:

> for i from 0 to N+1 do p[i]:=plot(u[i](t),t=0..0.4,
color=COLOR(HUE,i/(N+2)), thickness=3):end do:

> arw:=arrow(<0.26,0.6>,<-0.2,0.3>,width=[1/300,relative],
head_width=[5,relative],head_length=[1/20,relative]):

pt:=textplot([[0.3,0.55,"Follow the arrow:"],seq([0.26,0.5-i*0.05,
typeset(u[i],"(t)")],i=0..2),[0.26,0.35,"..."],[0.265,0.3,typeset(u[11],"(t)")]]):

> display([seq(p[i],i=0..11),arw,pt],axes=boxed,labels=[t,"u"],title="Figure 5.5");

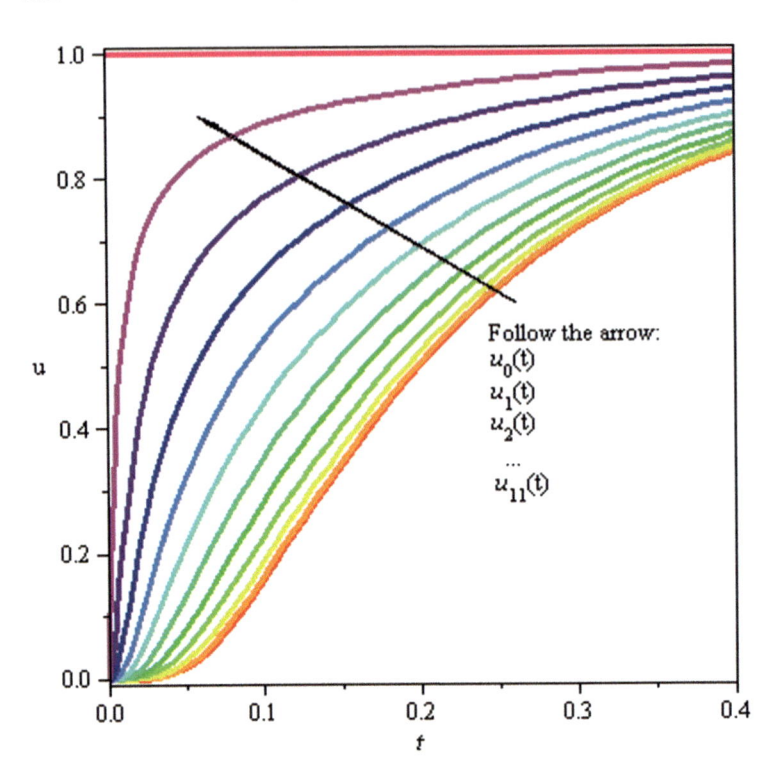

Fig. 5.5

> tf:=0.2;

$$tf := 0.2$$

> M:=30;

$$M := 30$$

> T1:=[seq(tf*i/M,i=0..M)];

$T1 := [0., 0.0066667, 0.013333, 0.020000, 0.026667, 0.033333, 0.040000, 0.046667, 0.053333, 0.060000, 0.066667, 0.073333, 0.080000, 0.086667, 0.093333, 0.10000, 0.10667, 0.11333, 0.12000, 0.12667, 0.13333, 0.14000, 0.14667, 0.15333, 0.16000, 0.16667, 0.17333, 0.18000, 0.18667, 0.19333, 0.20000]$

> PP:=matrix(N+2,M+1);

$$PP := array(1..12, 1..31, [\,])$$

> for i from 1 to N+2 do PP[i,1]:=evalf(subs(x=(i-1)*h,rhs(IC)));od:
> for i from 1 to N+2 do for j from 2 to M+1 do
PP[i,j]:=evalf(subs(t=T1[j],u[i-1](t)));od;od:

5.1 Semianalytical Method for Parabolic Partial Differential Equations (PDEs)

```
> plotdata := [seq([ seq([(i-1)*h,T1[j],PP[i,j]], i=1..N+2)], j=1..M+1)]:
> surfdata( plotdata, axes=boxed, title="Figure 5.6",labels=[x,t,u],
orientation=[-145,45]);
```

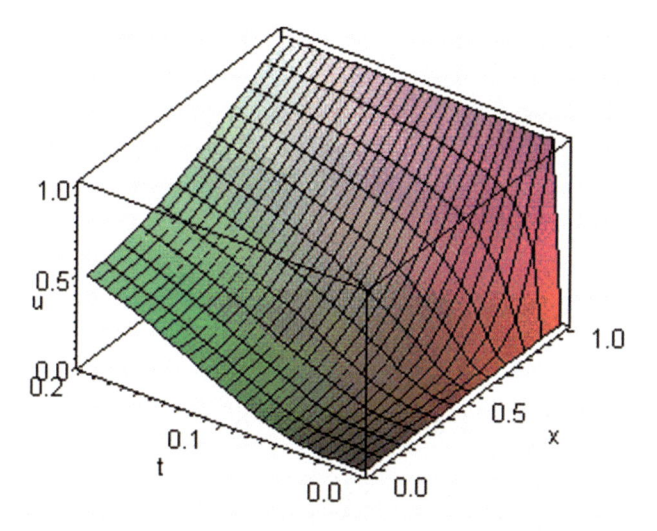

Fig. 5.6

Sometimes it is important to know how the temperature or concentration at the center of the cylinder changes with time. Analytical solutions for cylinders involve Bessel functions and an infinite series. With our semianalytical method, we can find how the temperature varies analytically with time. The time dependent variable at the center of the cylinder varies with time as:

```
> u[0](t);
```

$$0.015562\, e^{-474.73\, t} - 0.059845\, e^{-444.17\, t} + 0.12905\, e^{-395.52\, t} - 0.22028\, e^{-333.54\, t} + 0.33059\, e^{-264.02\, t}$$
$$- 0.45799\, e^{-193.21\, t} + 0.60388\, e^{-127.20\, t} - 0.78001\, e^{-71.337\, t} + 1.0329\, e^{-29.851\, t} - 1.5938\, e^{-5.7568\, t} + 0.99997$$

We can find the time taken for the center of the cylinder to reach 0.5 (i.e., 50% of the steady state):

```
> fsolve(u[0](t)-0.5,t=0..2);
```

$$0.20048$$

In this example temperature (or concentration) at the surface is fixed ($u(1,t) = 1$). With our semianalytical solution we can find how the flux at the surface varies as a function of time. This flux is given by a three point backward difference at $x = 1$ $x = 1$.

> flux:=dydxb;

$$flux := 0.13098\,e^{-474.73\,t} + 0.52830\,e^{-444.17\,t} + 1.0768\,e^{-395.52\,t} + 1.6323\,e^{-333.54\,t} + 2.0673\,e^{-264.02\,t}$$
$$+ 2.3089\,e^{-193.21\,t} + 2.3552\,e^{-127.20\,t} + 2.2640\,e^{-71.337\,t} + 2.1227\,e^{-29.851\,t} + 2.0136\,e^{-5.7568\,t}$$

Example 5.3

Consider the electrochemical discharge of a planar electrode.[5]

$$\frac{\partial u}{\partial t} = \frac{\partial^2 u}{\partial x^2}$$

$$u(x,0) = 1 \qquad (5.20)$$

$$\frac{\partial u}{\partial x}(0,t) = 0 \text{ and } \frac{\partial u}{\partial x}(1,t) = -\delta$$

where u is the dimensionless concentration and δ is the dimensionless applied current density at the surface. The electrochemical performance of the electrode depends on the concentration (u) at the surface. When this problem is cast into finite differences we arrive at the matrix differential equation (equation (5.16) with a constant b vector. Hence, we can use equation (5.18) to arrive at the semianalytical solution. However, because of flux boundary conditions at both the ends, the A matrix becomes singular. Hence, A cannot be inverted and equation (5.18) cannot be used. It should be noted that if A is singular that does not mean that equation (5.16) does not have a solution. One can prove that the solution (equation (5.18)) is independent of A-1 by using the series expansion for the exponential matrix[7][8]

$$\begin{aligned} \mathbf{Y} &= \exp(\mathbf{A}t)(\mathbf{Y}_0 + \mathbf{A}^{-1}\mathbf{b}) - \mathbf{A}^{-1}\mathbf{b} \\ &= \exp(\mathbf{A}t)\mathbf{Y}_0 + \exp(\mathbf{A}t)\mathbf{A}^{-1}\mathbf{b} - \mathbf{A}^{-1}\mathbf{b} \end{aligned} \qquad (5.21)$$

Now the exponential matrix in the second term is represented as an infinite series:

$$\mathbf{Y} = \exp(\mathbf{A}t)\mathbf{Y}_0 + \left(\mathbf{I} + \mathbf{A}t + \frac{\mathbf{A}^2 t^2}{2!} + \frac{\mathbf{A}^3 t^3}{3!} + \dots + \frac{\mathbf{A}^n t^n}{n!} + \dots\right)(\mathbf{A}^{-1}\mathbf{b}) - \mathbf{A}^{-1}\mathbf{b}$$

(5.22)

where **I** is the identity matrix of order N x N. Equation (5.22) can be further simplified by factoring out $\mathbf{A}^{-1}\mathbf{b}$ from the second and third terms.

5.1 Semianalytical Method for Parabolic Partial Differential Equations (PDEs)

$$Y = \exp(At)Y_0 + \left(I + At + \frac{A^2t^2}{2!} + \frac{A^3t^3}{3!} + ... \frac{A^nt^n}{n!} + ... -I\right)(A^{-1}b) \quad (5.23)$$

The identity matrices inside the parenthesis get cancelled. Next, A^{-1} outside the parenthesis is taken inside the parenthesis to obtain:

$$Y = \exp(At)Y_0 + \left(It + \frac{At^2}{2!} + \frac{A^2t^3}{3!} + ... \frac{A^{n-1}t^n}{n!} + ...\right)b \quad (5.24)$$

We observe that equation (5.24) is independent of A^{-1} and, hence, we can obtain a solution for equation (5.16) even when A is singular. The infinite series in equation (5.23) is difficult to calculate. Alternatively, we use equation (5.17), which is valid even when A is singular to obtain the semianalytical solutions. Equation (5.19) is solved below in Maple using the procedure described above with equation (5.17) as the nonhomogeneous part.

```
> restart;
> with(linalg):with(plots):
> ge:=diff(u(x,t),t)=diff(u(x,t),x$2);
```

$$ge := \frac{\partial}{\partial t} u(x,t) = \frac{\partial^2}{\partial x^2} u(x,t)$$

```
> bc1:=diff(u(x,t),x);
```

$$bc1 := \frac{\partial}{\partial x} u(x,t)$$

```
> bc2:=diff(u(x,t),x)+delta;
```

$$bc2 := \frac{\partial}{\partial x} u(x,t) + \delta$$

```
> #Digits:=50;
> IC:=u(x,0)=1;
```

$$IC := u(x,0) = 1$$

```
> N:=4;
```

$$N := 4$$

```
> L:=1;
```

$$L := 1$$

```
> dydxf:=1/2*(-u[2](t)-3*u[0](t)+4*u[1](t))/h:
> dydxb:=1/2*(u[N-1](t)+3*u[N+1](t)-4*u[N](t))/h:
> dydx:=1/2/h*(u[m+1](t)-u[m-1](t)):
> d2ydx2:=1/h^2*(u[m-1](t)-2*u[m](t)+u[m+1](t)):
> bc1:=subs(diff(u(x,t),x)=dydxf,u(x,t)=u[0](t),x=0,bc1):
> bc2:=subs(diff(u(x,t),x)=dydxb,u(x,t)=u[N+1](t),x=1,bc2):
> eq[0]:=bc1;
```

$$eq_0 := \frac{1}{2} \frac{-u_2(t) - 3 u_0(t) + 4 u_1(t)}{h}$$

```
> eq[N+1]:=bc2;
```

$$eq_5 := \frac{1}{2} \frac{u_3(t) + 3 u_5(t) - 4 u_4(t)}{h} + \delta$$

```
> for i from 1 to N do eq[i]:=diff(u[i](t),t)= subs(diff(u(x,t),x$2) =
subs(m=i,d2ydx2),diff(u(x,t),x) = subs(m=i,dydx),u(x,t)=u[i](t),
x=i*h,rhs(ge));od:
> u[0](t):=(solve(eq[0],u[0](t)));
```

$$u_0(t) := -\frac{1}{3} u_2(t) + \frac{4}{3} u_1(t)$$

```
> u[N+1](t):=solve(eq[N+1],u[N+1](t));
```

$$u_5(t) := -\frac{1}{3} u_3(t) + \frac{4}{3} u_4(t) - \frac{2}{3} \delta h$$

```
> for i from 1 to N do eq[i]:=eval(eq[i]);od;
```

$$eq_1 := \frac{d}{dt} u_1(t) = \frac{\frac{2}{3} u_2(t) - \frac{2}{3} u_1(t)}{h^2}$$

$$eq_2 := \frac{d}{dt} u_2(t) = \frac{u_1(t) - 2 u_2(t) + u_3(t)}{h^2}$$

$$eq_3 := \frac{d}{dt} u_3(t) = \frac{u_2(t) - 2 u_3(t) + u_4(t)}{h^2}$$

$$eq_4 := \frac{d}{dt} u_4(t) = \frac{\frac{2}{3} u_3(t) - \frac{2}{3} u_4(t) - \frac{2}{3} \delta h}{h^2}$$

5.1 Semianalytical Method for Parabolic Partial Differential Equations (PDEs)

```
> eqs:=[seq(rhs(eq[j]),j=1..N)]:
> Y:=[seq(u[i](t),i=1..N)];
```
$$Y := \left[u_1(t), u_2(t), u_3(t), u_4(t) \right]$$

```
> A:=genmatrix(eqs,Y,'b1'):
> b:=matrix(N,1):for i to N do b[i,1]:=-eval(b1[i]);od:evalm(b);
```

$$\begin{bmatrix} 0 \\ 0 \\ 0 \\ -\dfrac{2}{3}\dfrac{\delta}{h} \end{bmatrix}$$

```
> h:=eval(L/(N+1));
```

$$h := \frac{1}{5}$$

```
> A:=map(eval,A):
> if N > 4 then A:=map(evalf,A);end:
> evalm(A);
```

$$\begin{bmatrix} -\dfrac{50}{3} & \dfrac{50}{3} & 0 & 0 \\ 25 & -50 & 25 & 0 \\ 0 & 25 & -50 & 25 \\ 0 & 0 & \dfrac{50}{3} & -\dfrac{50}{3} \end{bmatrix}$$

```
> det(A);
```

$$0$$

We observe that the **A** matrix is singular.

```
> mat:=exponential(A,t):
> mat:=map(evalf,mat):
> mat:=map(simplify,mat):
> Y0:=matrix(N,1):for i from 1 to N do
Y0[i,1]:=evalf(subs(x=i*h,rhs(IC)));od:evalm(Y0);
```

$$\begin{bmatrix} 1. \\ 1. \\ 1. \\ 1. \end{bmatrix}$$

> b2:=subs(t=tau,evalm(b));

$$b2 := \begin{bmatrix} 0 \\ 0 \\ 0 \\ -\dfrac{2}{3}\dfrac{\delta}{h} \end{bmatrix}$$

> mat2:=subs(t=tau,evalm(mat)):

> mat3:=evalm(mat2&*b2):

> mat4:=map(int,mat3,tau=0..t):

> Y:=evalm(mat&*Y0+mat4):

> #Y:=map(simplify,Y):

> #Digits:=20;

> for i from 1 to N do u[i](t):=evalf((Y[i,1]));od:

> for i from 0 to N+1 do u[i](t):=eval(u[i](t));od;

$u_0(t) := 1.000000001 + 0.1540000000\,\delta + 0.02933333336\,\delta\,e^{-41.66666667\,t} - 0.004861149165\,\delta\,e^{-81.43334892\,t}$

$- 0.1784721841\,\delta\,e^{-10.23331773\,t} - 1.000000001\,\delta\,t$

$u_1(t) := 1.000000001 + 0.1340000000\,\delta - 0.001849283003\,\delta\,e^{-81.43334892\,t} - 0.1481507170\,\delta\,e^{-10.23331773\,t}$

$+ 0.01600000001\,\delta\,e^{-41.66666667\,t} - 1.000000001\,\delta\,t$

$u_2(t) := 1.000000001 + 0.07400000001\,\delta - 0.02400000002\,\delta\,e^{-41.66666667\,t} + 0.007186315481\,\delta\,e^{-81.43334892\,t}$

$- 0.05718631547\,\delta\,e^{-10.23331773\,t} - 1.000000001\,\delta\,t$

$u_3(t) := 1.000000001 - 0.02599999997\,\delta - 0.02400000002\,\delta\,e^{-41.66666667\,t} - 0.007186315481\,\delta\,e^{-81.43334892\,t}$

$+ 0.05718631547\,\delta\,e^{-10.23331773\,t} - 1.000000001\,\delta\,t$

$u_4(t) := 1.000000001 - 0.1660000000\,\delta - 1.000000001\,\delta\,t + 0.01600000001\,\delta\,e^{-41.66666667\,t}$

$+ 0.001849283003\,\delta\,e^{-81.43334892\,t} + 0.1481507170\,\delta\,e^{-10.23331773\,t}$

5.1 Semianalytical Method for Parabolic Partial Differential Equations (PDEs)

$u_5(t) := 1.000000001 - 0.3459999999\,\delta + 0.02933333336\,\delta e^{-41.66666667\,t} + 0.004861149165\,\delta e^{-81.43334892\,t}$

$+ 0.1784721841\,\delta e^{-10.23331773\,t} - 1.000000001\,\delta t$

We obtain the semianalytical solution for the concentration profile as a function of dimensionless current density, δ. Concentration profiles can be plotted specifying values for δ.

> pp:=plot([seq(subs(delta=1,u[i](t)),i=0..N+1)],t=0..0.4);

$$pp := PLOT(...)$$

> pt:=textplot([[0.25,evalf(subs(t=0.25,delta=1,u[0](t))),typeset(u[0],"(t)"),align=above],seq([0.25,evalf(subs(t=0.25,delta=1,u[i](t))),typeset(u[i],"(t)"),align=below],i=1..N+1)]);

$$pt := PLOT(...)$$

> display([pp,pt],thickness=4,title="Figure 5.7",axes=boxed,labels=[t,"u"]);

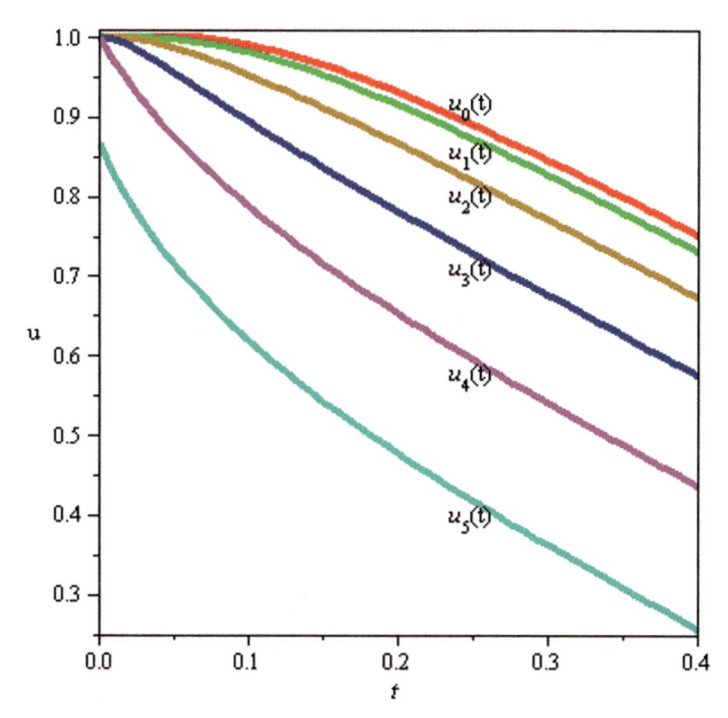

Fig. 5.7

> pp:=plot([seq(subs(delta=0.1,u[i](t)),i=0..N+1)],t=0..0.4);

$$pp := PLOT(...)$$

380   5 Method of Lines for Parabolic Partial Differential Equations

```
> pt:=textplot([[0.25,evalf(subs(t=0.25,delta=0.1,u[0](t))),typeset(u[0],"(t)"),
  align=above],seq([0.25,evalf(subs(t=0.25,delta=0.1,u[i](t))),typeset(u[i],"(t)"),
  align=below],i=1..N+1)]);
```

$$pt := PLOT(\ldots)$$

```
> display([pp,pt],thickness=4,title="Figure 5.8",axes=boxed,labels=[t,"u"]);
```

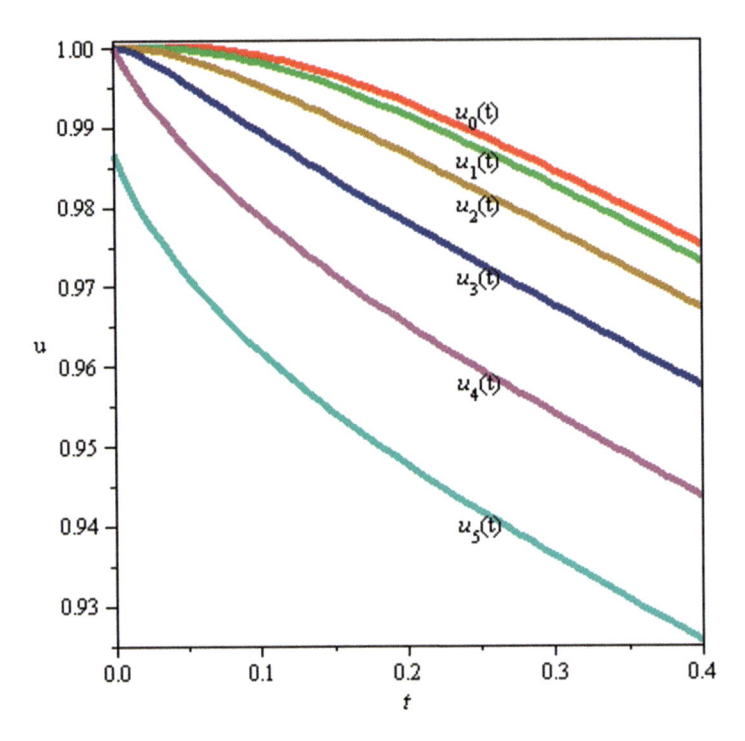

Fig. 5.8

```
> tf:=0.4;
```

$$tf := 0.4$$

We observe that as δ decreases, the time taken for discharge (the concentration to decrease from 1 to 0) increases. A three dimensional plot for the concentration profile can be made as follows:

```
> M:=30;
```

$$M := 30$$

5.1 Semianalytical Method for Parabolic Partial Differential Equations (PDEs)

> T1:=[seq(tf*i/M,i=0..M)];

T1 := [0., 0.01333333333, 0.02666666667, 0.04000000000, 0.05333333333, 0.06666666667, 0.08000000000, 0.09333333333, 0.1066666667, 0.1200000000, 0.1333333333, 0.1466666667, 0.1600000000, 0.1733333333, 0.1866666667, 0.2000000000, 0.2133333333, 0.2266666667, 0.2400000000, 0.2533333333, 0.2666666667, 0.2800000000, 0.2933333333, 0.3066666667, 0.3200000000, 0.3333333333, 0.3466666667, 0.3600000000, 0.3733333333, 0.3866666667, 0.4000000000]

> PP:=matrix(N+2,M+1);

$$PP := array(1..6, 1..31, [\,])$$

> for i from 1 to N+2 do PP[i,1]:=evalf(subs(x=(i-1)*h,rhs(IC)));od:

> for i from 1 to N+2 do for j from 2 to M+1 do
PP[i,j]:=evalf(subs(t=T1[j],subs(delta=1,u[i-1](t))));od;od:

> plotdata := [seq([seq([(i-1)*h,T1[j],PP[i,j]], i=1..N+2)], j=1..M+1)]:

> surfdata(plotdata, axes=boxed, title="Figure 5.9",
labels=[x,t,u],orientation=[45,45]);

Fig. 5.9

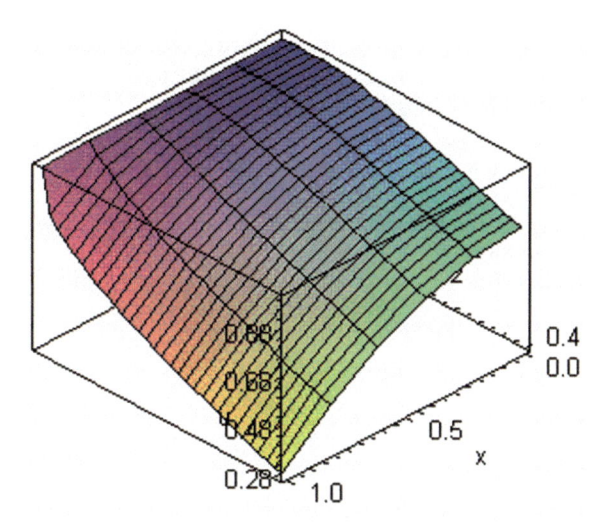

Electrochemical behavior of the electrode depends on the surface concentration given by:

> u[N+1](t);

$1.000000001 - 0.3459999999\,\delta + 0.02933333336\,\delta\,e^{-41.66666667\,t} + 0.004861149165\,\delta\,e^{-81.43334892\,t}$
$+ 0.1784721841\,\delta\,e^{-10.23331773\,t} - 1.000000001\,\delta\,t$

```
> plot3d(u[N+1](t),delta=0..1,t=0..1,axes=boxed,title="Figure 5.10",
view=[0..5,0..1,0..1],orientation=[0,90],labels=[delta,t,"us"]);
```

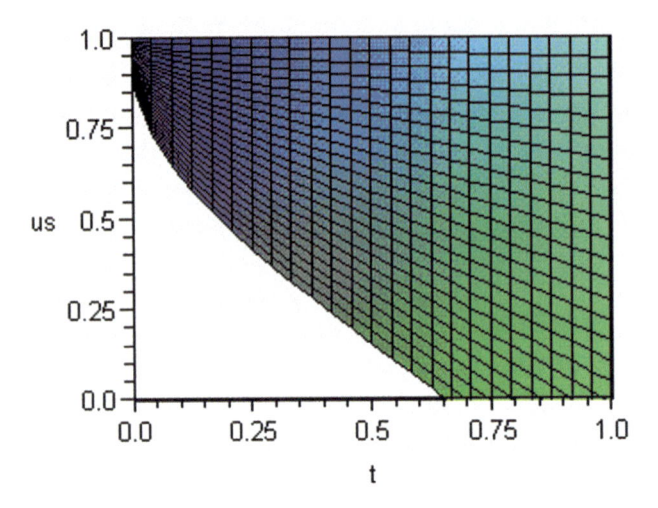

Fig. 5.10

For low values of δ, the surface concentration remains close to 1, and as δ increases, the surface concentration depletes faster. Accuracy can be increased by increasing the number of node points.

Example 5.4

Consider the following heat/mass transfer problem with a time dependent boundary condition,

$$\frac{\partial u}{\partial t} = \frac{\partial^2 u}{\partial x^2}$$

$$u(x,0) = 0 \tag{5.25}$$

$$\frac{\partial u}{\partial x}(0,t) = 0 \text{ and } u(1,t) = 1 - e^{-t}$$

This BVP is solved below in Maple by following the procedure described earlier. In this case, the forcing function vector, **b**(t) is a function of time and hence equation (5.17)a is used to obtain the semianalytical solution. The program used

5.1 Semianalytical Method for Parabolic Partial Differential Equations (PDEs)

for example 5.3 can be used to solve this example by just modifying the boundary conditions:

> restart;

> with(linalg):with(plots):

> ge:=diff(u(x,t),t)=diff(u(x,t),x$2);

$$ge := \frac{\partial}{\partial t} u(x,t) = \frac{\partial^2}{\partial x^2} u(x,t)$$

> bc1:=diff(u(x,t),x);

$$bc1 := \frac{\partial}{\partial x} u(x,t)$$

> bc2:=u(x,t)-1+exp(-t);

$$bc2 := u(x,t) - 1 + e^{-t}$$

> IC:=u(x,0)=0;

$$IC := u(x,0) = 0$$

> N:=10;

$$N := 10$$

> L:=1;

$$L := 1$$

> dydxf:=1/2*(-u[2](t)-3*u[0](t)+4*u[1](t))/h:
> dydxb:=1/2*(u[N-1](t)+3*u[N+1](t)-4*u[N](t))/h:
> dydx:=1/2/h*(u[m+1](t)-u[m-1](t)):
> d2ydx2:=1/h^2*(u[m-1](t)-2*u[m](t)+u[m+1](t)):
> bc1:=subs(diff(u(x,t),x)=dydxf,u(x,t)=u[0](t),x=0,bc1):
> bc2:=subs(diff(u(x,t),x)=dydxb,u(x,t)=u[N+1](t),x=1,bc2):
> eq[0]:=bc1;

$$eq_0 := \frac{1}{2} \frac{-u_2(t) - 3u_0(t) + 4u_1(t)}{h}$$

> eq[N+1]:=bc2;

$$eq_{11} := u_{11}(t) - 1 + e^{-t}$$

```
> for i from 1 to N do eq[i]:=diff(u[i](t),t)= subs(diff(u(x,t),x$2) =
subs(m=i,d2ydx2),diff(u(x,t),x) = subs(m=i,dydx),u(x,t)=u[i](t),
x=i*h,rhs(ge));od;
```

$$eq_1 := \frac{d}{dt} u_1(t) = \frac{u_0(t) - 2u_1(t) + u_2(t)}{h^2}$$

$$eq_2 := \frac{d}{dt} u_2(t) = \frac{u_1(t) - 2u_2(t) + u_3(t)}{h^2}$$

$$eq_3 := \frac{d}{dt} u_3(t) = \frac{u_2(t) - 2u_3(t) + u_4(t)}{h^2}$$

$$eq_4 := \frac{d}{dt} u_4(t) = \frac{u_3(t) - 2u_4(t) + u_5(t)}{h^2}$$

$$eq_5 := \frac{d}{dt} u_5(t) = \frac{u_4(t) - 2u_5(t) + u_6(t)}{h^2}$$

$$eq_6 := \frac{d}{dt} u_6(t) = \frac{u_5(t) - 2u_6(t) + u_7(t)}{h^2}$$

$$eq_7 := \frac{d}{dt} u_7(t) = \frac{u_6(t) - 2u_7(t) + u_8(t)}{h^2}$$

$$eq_8 := \frac{d}{dt} u_8(t) = \frac{u_7(t) - 2u_8(t) + u_9(t)}{h^2}$$

$$eq_9 := \frac{d}{dt} u_9(t) = \frac{u_8(t) - 2u_9(t) + u_{10}(t)}{h^2}$$

$$eq_{10} := \frac{d}{dt} u_{10}(t) = \frac{u_9(t) - 2u_{10}(t) + u_{11}(t)}{h^2}$$

```
> eqs:=[seq(rhs(eq[j]),j=1..N)]:
> Y:=[seq(u[i](t),i=1..N)];
```

$$Y := \left[u_1(t), u_2(t), u_3(t), u_4(t), u_5(t), u_6(t), u_7(t), u_8(t), u_9(t), u_{10}(t) \right]$$

```
> A:=genmatrix(eqs,Y,'b1'):
> b:=matrix(N,1):for i to N do b[i,1]:=-eval(b1[i]);od:evalm(b);
```

5.1 Semianalytical Method for Parabolic Partial Differential Equations (PDEs) 385

$$\begin{bmatrix} 0 \\ 0 \\ 0 \\ 0 \\ 0 \\ 0 \\ 0 \\ 0 \\ 0 \\ \dfrac{1 - e^{-t}}{h^2} \end{bmatrix}$$

> h:=eval(L/(N+1));

$$h := \frac{1}{11}$$

> A:=map(eval,A):

> if N > 4 then A:=map(evalf,A);end:

> evalm(A);

$$\begin{bmatrix} -80.66666667 & 80.66666667 & 0. & 0. & 0. & 0. & 0. & 0. & 0. & 0. \\ 121. & -242. & 121. & 0. & 0. & 0. & 0. & 0. & 0. & 0. \\ 0. & 121. & -242. & 121. & 0. & 0. & 0. & 0. & 0. & 0. \\ 0. & 0. & 121. & -242. & 121. & 0. & 0. & 0. & 0. & 0. \\ 0. & 0. & 0. & 121. & -242. & 121. & 0. & 0. & 0. & 0. \\ 0. & 0. & 0. & 0. & 121. & -242. & 121. & 0. & 0. & 0. \\ 0. & 0. & 0. & 0. & 0. & 121. & -242. & 121. & 0. & 0. \\ 0. & 0. & 0. & 0. & 0. & 0. & 121. & -242. & 121. & 0. \\ 0. & 0. & 0. & 0. & 0. & 0. & 0. & 121. & -242. & 121. \\ 0. & 0. & 0. & 0. & 0. & 0. & 0. & 0. & 121. & -242. \end{bmatrix}$$

> det(A);

$$4.484999965 \cdot 10^{20}$$

> mat:=exponential(A,t):

> mat:=map(evalf,mat):

> mat:=map(simplify,mat):

> Y0:=matrix(N,1):for i from 1 to N do
Y0[i,1]:=evalf(subs(x=i*h,rhs(IC)));od:evalm(Y0);

386   5 Method of Lines for Parabolic Partial Differential Equations

$$\begin{bmatrix} 0. \\ 0. \\ 0. \\ 0. \\ 0. \\ 0. \\ 0. \\ 0. \\ 0. \\ 0. \end{bmatrix}$$

> b2:=subs(t=tau,evalm(b));

$$b2 := \begin{bmatrix} 0 \\ 0 \\ 0 \\ 0 \\ 0 \\ 0 \\ 0 \\ 0 \\ 0 \\ \dfrac{1-e^{-\tau}}{h^2} \end{bmatrix}$$

> mat2:=subs(t=t-tau,evalm(mat)):
> mat3:=evalm(mat2&*b2):
> mat4:=map(int,mat3,tau=0..t):
> Y:=evalm(mat&*Y0+mat4):
> Y:=map(simplify,Y):
> for i from 1 to N do u[i](t):=evalf((Y[i,1]));od:
> for i from 0 to N+1 do u[i](t):=eval(u[i](t));od;

$u_1(t) := 0.8593820229\, e^{-2.465470971\, t} - 0.01833306308\, e^{-22.03592529\, t} + 0.003336802648\, e^{-60.19228428\, t}$
$\quad - 0.0009999822868\, e^{-114.4543404\, t} + 0.0003749437317\, e^{-180.5698886\, t} - 0.0001577109432\, e^{-252.7906017\, t}$
$\quad + 0.00006938999532\, e^{-324.4981258\, t} - 0.00002970655764\, e^{-388.9460231\, t} + 0.00001095911825\, e^{-439.9888576\, t}$
$\quad - 0.000002457989159\, e^{-472.7251489\, t} + 1.000000001 - 1.843651199\, e^{-1.t}$

5.1 Semianalytical Method for Parabolic Partial Differential Equations (PDEs)

$u_2(t) := 0.8331161339\, e^{-2.465470971\,t} - 0.01332497207\, e^{-22.03592529\,t} + 0.0008469294252\, e^{-60.19228428\,t}$
$+ 0.0004188480425\, e^{-114.4543404\,t} - 0.0004643564486\, e^{-180.5698886\,t} + 0.0003365185313\, e^{-252.7906017\,t}$
$- 0.0002097454192\, e^{-324.4981258\,t} + 0.0001135279147\, e^{-388.9460231\,t} - 0.000004881637660\, e^{-439.9888576\,t}$
$+ 0.00001194639051\, e^{-472.7251489\,t} + 0.9999999965 - 1.820796010\, e^{-1.\,t}$

$u_3(t) := 0.7898748320\, e^{-2.465470971\,t} - 0.005890202896\, e^{-22.03592529\,t} - 0.002064254676\, e^{-60.19228428\,t}$
$+ 0.001441488470\, e^{-114.4543404\,t} - 0.0006106913971\, e^{-180.5698886\,t} + 0.0001277007230\, e^{-252.7906017\,t}$
$+ 0.00007361499674\, e^{-324.4981258\,t} - 0.0001081651419\, e^{-388.9460231\,t} + 0.00006891773000\, e^{-439.9888576\,t}$
$- 0.00002032162006\, e^{-472.7251489\,t} + 0.9999999718 - 1.782892890\, e^{-1.\,t}$

$u_4(t) := 0.7305391818\, e^{-2.465470971\,t} + 0.002617261358\, e^{-22.03592529\,t} - 0.003948561064\, e^{-60.19228428\,t}$
$+ 0.001100619638\, e^{-114.4543404\,t} + 0.0001543164672\, e^{-180.5698886\,t} - 0.0003479066720\, e^{-252.7906017\,t}$
$+ 0.0001595545185\, e^{-324.4981258\,t} + 0.00001783106908\, e^{-388.9460231\,t} - 0.00006395174382\, e^{-439.9888576\,t}$
$+ 0.00002680326919\, e^{-472.7251489\,t} + 0.9999999183 - 1.730255067\, e^{-1.\,t}$

$u_5(t) := 0.6563181841\, e^{-2.465470971\,t} + 0.01064808398\, e^{-22.03592529\,t} - 0.003868628354\, e^{-60.19228428\,t}$
$- 0.0002813295096\, e^{-114.4543404\,t} + 0.0006890359068\, e^{-180.5698886\,t} - 0.00009667487201\, e^{-252.7906017\,t}$
$- 0.0001823996905\, e^{-324.4981258\,t} + 0.00008651053831\, e^{-388.9460231\,t} + 0.00003572468919\, e^{-439.9888576\,t}$
$- 0.00003078737469\, e^{-472.7251489\,t} + 0.9999998421 - 1.663317561\, e^{-1.\,t}$

$u_6(t) := 0.5687241527\, e^{-2.465470971\,t} + 0.01673973203\, e^{-22.03592529\,t} - 0.001864219414\, e^{-60.19228428\,t}$
$- 0.001397168387\, e^{-114.4543404\,t} + 0.0001954981461\, e^{-180.5698886\,t} + 0.0003565280771\, e^{-252.7906017\,t}$
$- 0.00003519390704\, e^{-324.4981258\,t} - 0.0001228921263\, e^{-388.9460231\,t} + 0.000005496445607\, e^{-439.9888576\,t}$
$+ 0.00003190269183\, e^{-472.7251489\,t} + 0.9999997705 - 1.582633607\, e^{-1.\,t}$

$u_8(t) := 0.3607924401\, e^{-2.465470971\,t} + 0.01922316709\, e^{-22.03592529\,t} + 0.003468273597\, e^{-60.19228428\,t}$
$+ 0.0001412951673\, e^{-114.4543404\,t} - 0.0004949233948\, e^{-180.5698886\,t} - 0.0003623138704\, e^{-252.7906017\,t}$
$- 0.0001055267207\, e^{-324.4981258\,t} + 0.00004670678991\, e^{-388.9460231\,t} + 0.00006767495220\, e^{-439.9888576\,t}$
$+ 0.00002538826557\, e^{-472.7251489\,t} + 0.9999998251 - 1.382802007\, e^{-1.\,t}$

$u_9(t) := 0.2446916090\, e^{-2.465470971\,t} + 0.01516267965\, e^{-22.03592529\,t} + 0.004143672356\, e^{-60.19228428\,t}$
$+ 0.001340359231\, e^{-114.4543404\,t} + 0.0003385180311\, e^{-180.5698886\,t} - 0.00003256957870\, e^{-252.7906017\,t}$

$- 0.0001344465310\, e^{-324.4981258\, t} - 0.0001194556882\, e^{-388.9460231\, t} - 0.00006601621273\, e^{-439.9888576\, t}$

$- 0.00001836555023\, e^{-472.7251489\, t} + 0.9999999483 - 1.265305933\, e^{-1.\, t}$

$u_{10}(t) := 0.000009631500831\, e^{-472.7251489\, t} + 0.1236050601\, e^{-2.465470971\, t} + 0.008340836747\, e^{-22.03592529\, t}$

$+ 0.002757772835\, e^{-60.19228428\, t} + 0.001271572114\, e^{-114.4543404\, t} + 0.0006667847976\, e^{-180.5698886\, t}$

$+ 0.0003652184998\, e^{-252.7906017\, t} + 0.0001971928238\, e^{-324.4981258\, t} + 0.00009836338367\, e^{-388.9460231\, t}$

$+ 0.00004034554277\, e^{-439.9888576\, t} + 1.000000010 - 1.137352789\, e^{-1.\, t}$

$$u_{11}(t) := 1 - e^{-t}$$

The following plots are obtained:

> for i from 0 to N+1 do p[i]:=plot(subs(delta=1,u[i](t)),t=0..0.4, thickness=4,color=COLOR(HUE,i/(N+2)));end do:

> arw:=arrow(<0.3,0.02>,<- 0.15,0.11>,width=[1/1000,relative=false], head_width=[1/200,relative=false],head_length=[1/20,relative]):

pt:=textplot([[0.12,0.17,typeset("Follow the arrow: ",u[0],"(t), ..., ", u[N+1],"(t).")]]):

> display([seq(p[i],i=1..N),p[0],pt,arw],title="Figure 5.11", axes=boxed,labels=[t,"u"]);

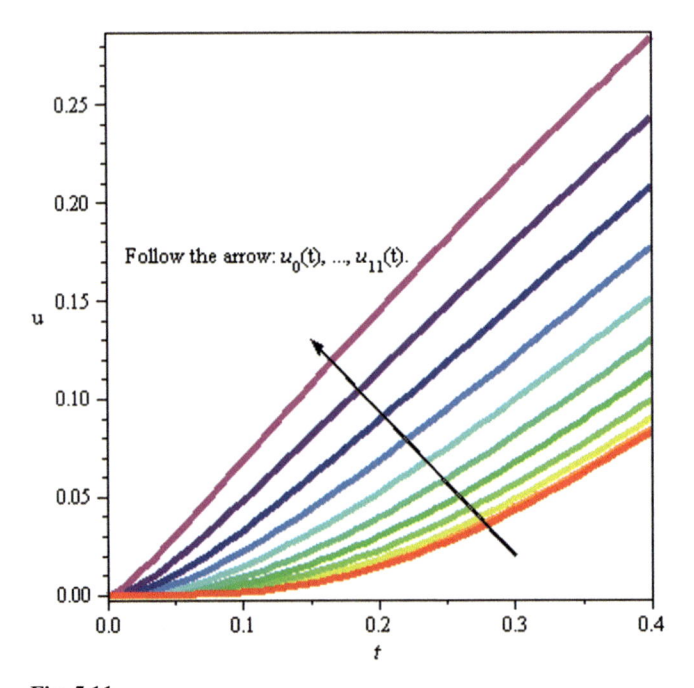

Fig. 5.11

5.1 Semianalytical Method for Parabolic Partial Differential Equations (PDEs)

```
> for i from 0 to N+1 do
p[i]:=plot(subs(delta=0.1,u[i](t)),t=0..0.4,thickness=3);od:

> tf:=0.4;
```

$$tf := 0.4$$

```
> M:=30;
```

$$M := 30$$

```
> T1:=[seq(tf*i/M,i=0..M)];

> for i from 0 to N+1 do
p[i]:=plot(subs(delta=0.1,u[i](t)),t=0..0.4,thickness=3);od:

> tf:=0.4;
```

$$tf := 0.4$$

```
> M:=30;
```

$$M := 30$$

```
> T1:=[seq(tf*i/M,i=0..M)];
```

$T1 := [0., 0.01333333333, 0.02666666667, 0.04000000000, 0.05333333333, 0.06666666667, 0.08000000000,$
$0.09333333333, 0.1066666667, 0.1200000000, 0.1333333333, 0.1466666667, 0.1600000000, 0.1733333333,$
$0.1866666667, 0.2000000000, 0.2133333333, 0.2266666667, 0.2400000000, 0.2533333333, 0.2666666667,$
$0.2800000000, 0.2933333333, 0.3066666667, 0.3200000000, 0.3333333333, 0.3466666667, 0.3600000000,$
$0.3733333333, 0.3866666667, 0.4000000000]$

```
> PP:=matrix(N+2,M+1);
```

$$PP := array(1..12, 1..31, [\ ])$$

```
> for i from 1 to N+2 do PP[i,1]:=evalf(subs(x=(i-1)*h,rhs(IC)));od:

> for i from 1 to N+2 do for j from 2 to M+1 do
PP[i,j]:=evalf(subs(t=T1[j],subs(delta=1,u[i-1](t))));od;od:

> plotdata := [seq([ seq([(i-1)*h,T1[j],PP[i,j]], i=1..N+2)], j=1..M+1)]:

> surfdata( plotdata, axes=boxed,title="Figure 5.12",
labels=[x,t,u],orientation=[-145,45]);
```

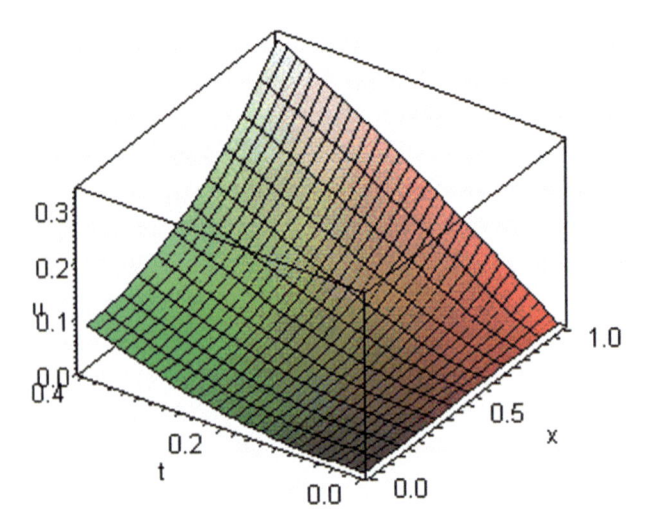

Fig. 5.12

Example 5.5

Consider the diffusion of a gas (A) through a stagnant liquid (B) in a container.[2] A reacts with B according to the irreversible reaction $A + B \xrightarrow{k} C$. The governing equation for this problem is,

$$\frac{\partial c_A}{\partial t_1} = D_{AB}\frac{\partial^2 c_A}{\partial z^2} - kc_A$$

$$c_A(z,0) = 0 \tag{5.26}$$

$$c_A(0,t_1) = c_{A0} \text{ and } \frac{\partial c_A}{\partial z}(L,t_1) = 0$$

where c_A is the concentration of A (mol/m^3), z is the distance (m), t_1 is the time variable (s), D_{AB} is the diffusion coefficient of A in B (2×10^{-9} m^2/s), L is the height of the container (10 cm). The concentration of A at z = 0 is c_{A0} = 0.01mol/m^3 and k is the first order rate constant (2×10^{-7} s^{-1}). The following dimensionless variables are introduced for convenience,

$$u = \frac{c_A}{c_{A0}}; \ x = \frac{z}{L}; \ \text{and } t = \frac{D_{AB}t_1}{L^2} \tag{5.27}$$

5.1 Semianalytical Method for Parabolic Partial Differential Equations (PDEs)

Substituting equation (5.26) into equation (5.25) we get:

$$\frac{\partial u}{\partial t} = \frac{\partial^2 u}{\partial x^2} - \Phi^2 u$$

$$u(x,0) = 0 \qquad (5.28)$$

$$u(0,t) = 1 \text{ and } \frac{\partial u}{\partial x}(1,t) = 0$$

where $\Phi = \sqrt{\frac{kL^2}{D_{AB}}}$ is the Thiele modulus. Once the solution is obtained, one can find the mass transfer flux at $x = 0$ and find the time taken for the flux to reach a steady state. This BVP is solved below in Maple by following the procedure described earlier for example 5.2 as we have a constant **b** vector. Both dimensionless and dimensional plots are obtained.

> restart;

> with(linalg):with(plots):

> ge:=diff(u(x,t),t)=diff(u(x,t),x$2)+Phi^2*u(x,t);

$$ge := \frac{\partial}{\partial t} u(x,t) = \frac{\partial^2}{\partial x^2} u(x,t) + \Phi^2 u(x,t)$$

> k:=2e-7;Dab:=2e-9;ca0:=0.01;Lc:=10e-2;

$$k := 2.\,10^{-7}$$

$$Dab := 2.\,10^{-9}$$

$$ca0 := 0.01$$

$$Lc := 0.10$$

> Phi:=sqrt(k*Lc^2/Dab);

$$\Phi := 1.000000000$$

> bc1:=u(x,t)-1;

$$bc1 := u(x,t) - 1$$

> bc2:=diff(u(x,t),x);

$$bc2 := \frac{\partial}{\partial x} u(x,t)$$

> IC:=u(x,0)=0;
$$IC := u(x, 0) = 0$$

> N:=10;
$$N := 10$$

> L:=1;
$$L := 1$$

> dydxf:=1/2*(-u[2](t)-3*u[0](t)+4*u[1](t))/h:
> dydxb:=1/2*(u[N-1](t)+3*u[N+1](t)-4*u[N](t))/h:
> dydx:=1/2/h*(u[m+1](t)-u[m-1](t)):
> d2ydx2:=1/h^2*(u[m-1](t)-2*u[m](t)+u[m+1](t)):
> bc1:=subs(diff(u(x,t),x)=dydxf,u(x,t)=u[0](t),x=0,bc1):
> bc2:=subs(diff(u(x,t),x)=dydxb,u(x,t)=u[N+1](t),x=1,bc2):
> eq[0]:=bc1;
$$eq_0 := u_0(t) - 1$$

> eq[N+1]:=bc2;
$$eq_{11} := \frac{1}{2} \frac{u_9(t) + 3 u_{11}(t) - 4 u_{10}(t)}{h}$$

> for i from 1 to N do eq[i]:=diff(u[i](t),t)= subs(diff(u(x,t),x$2) = subs(m=i,d2ydx2),diff(u(x,t),x) = subs(m=i,dydx),u(x,t)=u[i](t), x=i*h,rhs(ge));od:

> u[0](t):=(solve(eq[0],u[0](t)));
$$u_0(t) := 1$$

> u[N+1](t):=solve(eq[N+1],u[N+1](t));
$$u_{11}(t) := -\frac{1}{3} u_9(t) + \frac{4}{3} u_{10}(t)$$

> for i from 1 to N do eq[i]:=eval(eq[i]);od:
> eqs:=[seq(rhs(eq[j]),j=1..N)]:
> Y:=[seq(u[i](t),i=1..N)];
$$Y := [u_1(t), u_2(t), u_3(t), u_4(t), u_5(t), u_6(t), u_7(t), u_8(t), u_9(t), u_{10}(t)]$$

5.1 Semianalytical Method for Parabolic Partial Differential Equations (PDEs)

```
> A:=genmatrix(eqs,Y,'b1'):
> b:=matrix(N,1):for i to N do b[i,1]:=-eval(b1[i]);od:evalm(b);
```

$$\begin{bmatrix} \dfrac{1.}{h^2} \\ 0 \\ 0 \\ 0 \\ 0 \\ 0 \\ 0 \\ 0 \\ 0 \\ 0. \end{bmatrix}$$

```
> h:=eval(L/(N+1));
```

$$h := \frac{1}{11}$$

```
> A:=map(eval,A):
> if N > 4 then A:=map(evalf,A);end:
> evalm(A);
```

$$[[-241.0000000, 121., 0., 0., 0., 0., 0., 0., 0., 0.],$$
$$[121., -241.0000000, 121., 0., 0., 0., 0., 0., 0., 0.],$$
$$[0., 121., -241.0000000, 121., 0., 0., 0., 0., 0., 0.],$$
$$[0., 0., 121., -241.0000000, 121., 0., 0., 0., 0., 0.],$$
$$[0., 0., 0., 121., -241.0000000, 121., 0., 0., 0., 0.],$$
$$[0., 0., 0., 0., 121., -241.0000000, 121., 0., 0., 0.],$$
$$[0., 0., 0., 0., 0., 121., -241.0000000, 121., 0., 0.],$$
$$[0., 0., 0., 0., 0., 0., 121., -241.0000000, 121., 0.],$$
$$[0., 0., 0., 0., 0., 0., 0., 121., -241.0000000, 121.],$$
$$[0., 0., 0., 0., 0., 0., 0., 0., 80.66666667, -79.66666667]]$$

```
> mat:=exponential(A,t):
> mat:=map(evalf,mat):
```

> Y0:=matrix(N,1):for i from 1 to N do
Y0[i,1]:=evalf(subs(x=i*h,rhs(IC)));od:evalm(Y0);

$$\begin{bmatrix} 0. \\ 0. \\ 0. \\ 0. \\ 0. \\ 0. \\ 0. \\ 0. \\ 0. \\ 0. \end{bmatrix}$$

> s1:=evalm(Y0+inverse(A)&*b):

> Y:=evalm(mat&*s1-inverse(A)&*b):

> Y:=map(simplify,Y):

> Digits:=5;

$$Digits := 5$$

> for i from 1 to N do u[i](t):=evalf((Y[i,1]));od:

> for i from 0 to N+1 do u[i](t):=eval(u[i](t));od;

$$u_0(t) := 1$$

$u_1(t) := -0.0045530\, e^{-471.73\, t} - 0.017752\, e^{-438.99\, t} - 0.038258\, e^{-387.95\, t} - 0.063989\, e^{-323.50\, t} - 0.092324\, e^{-251.79\, t}$

$- 0.12040\, e^{-179.57\, t} - 0.14554\, e^{-113.45\, t} - 0.16600\, e^{-59.192\, t} - 0.18380\, e^{-21.036\, t} - 0.30475\, e^{-1.4655\, t} + 1.1374$

$u_2(t) := 0.0086819\, e^{-471.73\, t} + 0.029046\, e^{-438.99\, t} + 0.046462\, e^{-387.95\, t} + 0.043627\, e^{-323.50\, t} + 0.0082334\, e^{-251.79\, t}$

$- 0.061127\, e^{-179.57\, t} - 0.15341\, e^{-113.45\, t} - 0.24942\, e^{-59.192\, t} - 0.33412\, e^{-21.036\, t} - 0.60328\, e^{-1.4655\, t}$

$+ 1.2653$

$u_3(t) := -0.012002\, e^{-471.73\, t} - 0.029776\, e^{-438.99\, t} - 0.018166\, e^{-387.95\, t} + 0.034243\, e^{-323.50\, t} + 0.091590\, e^{-251.79\, t}$

$+ 0.089368\, e^{-179.57\, t} - 0.016171\, e^{-113.45\, t} - 0.20876\, e^{-59.192\, t} - 0.42360\, e^{-21.036\, t} - 0.88952\, e^{-1.4655\, t}$

$+ 1.3828$

$u_4(t) := 0.014203\, e^{-471.73\, t} + 0.019676\, e^{-438.99\, t} - 0.024399\, e^{-387.95\, t} - 0.066975\, e^{-323.50\, t} - 0.016401\, e^{-251.79\, t}$

$+ 0.10650\, e^{-179.57\, t} + 0.13636\, e^{-113.45\, t} - 0.064260\, e^{-59.192\, t} - 0.43593\, e^{-21.036\, t} - 1.1576\, e^{-1.4655\, t} + 1.4889$

5.1 Semianalytical Method for Parabolic Partial Differential Equations (PDEs)

$u_5(t) := -0.015081\, e^{-471.73\, t} - 0.0024177\, e^{-438.99\, t} + 0.047799\, e^{-387.95\, t} + 0.011420\, e^{-323.50\, t} - 0.090127\, e^{-251.79\, t}$

$- 0.035302\, e^{-179.57\, t} + 0.15991\, e^{-113.45\, t} + 0.11221\, e^{-59.192\, t} - 0.36888\, e^{-21.036\, t} - 1.4022\, e^{-1.4655\, t} + 1.5826$

$u_6(t) := 0.014554\, e^{-471.73\, t} - 0.015718\, e^{-438.99\, t} - 0.033647\, e^{-387.95\, t} + 0.059188\, e^{-323.50\, t} + 0.024439\, e^{-251.79\, t}$

$- 0.12442\, e^{-179.57\, t} + 0.032200\, e^{-113.45\, t} + 0.23286\, e^{-59.192\, t} - 0.23464\, e^{-21.036\, t} - 1.6181\, e^{-1.4655\, t} + 1.6633$

$u_7(t) := -0.012671\, e^{-471.73\, t} + 0.028139\, e^{-438.99\, t} - 0.0069347\, e^{-387.95\, t} - 0.051776\, e^{-323.50\, t} + 0.087947\, e^{-251.79\, t}$

$- 0.027866\, e^{-179.57\, t} - 0.12597\, e^{-113.45\, t} + 0.23767\, e^{-59.192\, t} - 0.057674\, e^{-21.036\, t} - 1.8011\, e^{-1.4655\, t}$

$+ 1.7303$

$u_8(t) := 0.0096066\, e^{-471.73\, t} - 0.030323\, e^{-438.99\, t} + 0.042071\, e^{-387.95\, t} - 0.023888\, e^{-323.50\, t} - 0.032282\, e^{-251.79\, t}$

$+ 0.11027\, e^{-179.57\, t} - 0.16498\, e^{-113.45\, t} + 0.12425\, e^{-59.192\, t} + 0.12980\, e^{-21.036\, t} - 1.9474\, e^{-1.4655\, t} + 1.7829$

$u_9(t) := -0.0056473\, e^{-471.73\, t} + 0.021479\, e^{-438.99\, t} - 0.044157\, e^{-387.95\, t} + 0.068062\, e^{-323.50\, t} - 0.085069\, e^{-251.79\, t}$

$+ 0.083848\, e^{-179.57\, t} - 0.047939\, e^{-113.45\, t} - 0.050980\, e^{-59.192\, t} + 0.29363\, e^{-21.036\, t} - 2.0540\, e^{-1.4655\, t}$

$+ 1.8208$

$u_{10}(t) := 0.0011620\, e^{-471.73\, t} - 0.0048222\, e^{-438.99\, t} + 0.011554\, e^{-387.95\, t} - 0.022517\, e^{-323.50\, t}$

$+ 0.039868\, e^{-251.79\, t} - 0.067704\, e^{-179.57\, t} + 0.11445\, e^{-113.45\, t} - 0.20085\, e^{-59.192\, t} + 0.40399\, e^{-21.036\, t}$

$- 2.1188\, e^{-1.4655\, t} + 1.8437$

$u_{11}(t) := 0.0034317\, e^{-471.73\, t} - 0.013589\, e^{-438.99\, t} + 0.030124\, e^{-387.95\, t} - 0.052710\, e^{-323.50\, t} + 0.081513\, e^{-251.79\, t}$

$- 0.11822\, e^{-179.57\, t} + 0.16858\, e^{-113.45\, t} - 0.25081\, e^{-59.192\, t} + 0.44077\, e^{-21.036\, t} - 2.1404\, e^{-1.4655\, t} + 1.8514$

Semianalytical solutions are obtained in dimensionless form. The dimensionless concentration profiles are plotted as:

> setcolors(["Red", "Blue", "LimeGreen", "Goldenrod", "maroon", "DarkTurquoise", "coral", "aquamarine", "magenta", "khaki", "sienna", "orange", "yellow", "gray"]);

["Red", "LimeGreen", "Goldenrod", "Blue", "MediumOrchid",

"DarkTurquoise"]

> pp:=plot([seq(u[i](t),i=0..N+1)],t=0..0.4,thickness=4);

$$pp := PLOT(\ldots)$$

> arw:=arrow(<0.1,1.01>,<0.1,-0.8>,width=[1/600,relative=false], head_width=[1/200,relative=false],head_length=[1/30,relative=false]):

```
pt:=textplot([[0.28,0.15,typeset("Follow the arrow: ",u[0],"(t), ..., ",
u[N+1],"(t).")]]):
```
> display([pp,pt,arw],title="Figure 5.13",axes=boxed,labels=[t,"u"]);

Follow the arrow: $u_0(t), ..., u_{11}(t)$.

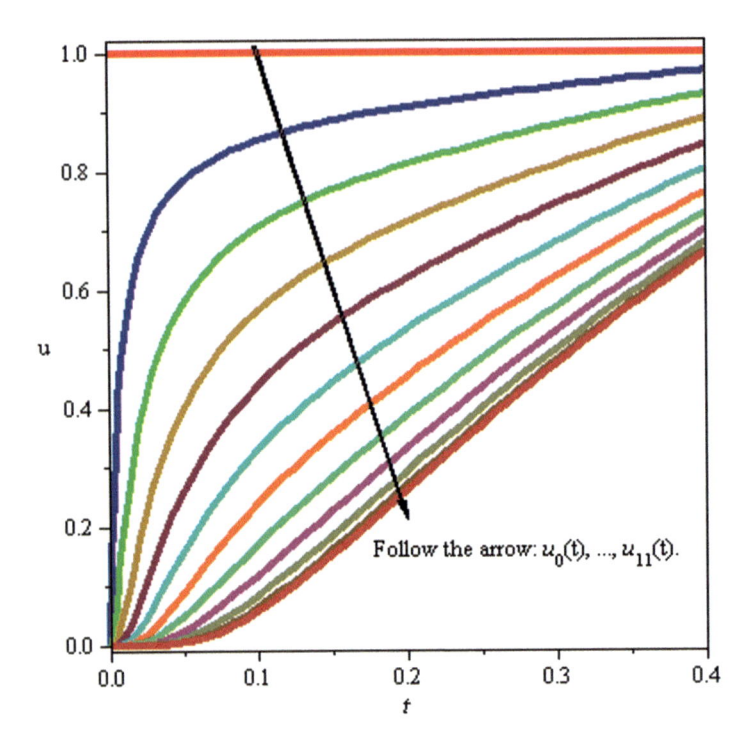

Fig. 5.13

> tf:=0.5;

$$tf := 0.5$$

> M:=30;

$$M := 30$$

> T1:=[seq(tf*i/M,i=0..M)];

$T1 := [0., 0.016667, 0.033333, 0.050000, 0.066667, 0.083333, 0.10000, 0.11667, 0.13333, 0.15000, 0.16667, 0.18333,$

$0.20000, 0.21667, 0.23333, 0.25000, 0.26667, 0.28333, 0.30000, 0.31667, 0.33333, 0.35000, 0.36667, 0.38333,$

$0.40000, 0.41667, 0.43333, 0.45000, 0.46667, 0.48333, 0.50000]$

5.1 Semianalytical Method for Parabolic Partial Differential Equations (PDEs)

```
> PP:=matrix(N+2,M+1);
```

$$PP := array(1..12, 1..31, [\,])$$

```
> for i from 1 to N+2 do PP[i,1]:=evalf(subs(x=(i-1)*h,rhs(IC)));od:
> for i from 1 to N+2 do for j from 2 to M+1 do
PP[i,j]:=evalf(subs(t=T1[j],u[i-1](t)));od;od:
> plotdata := [seq([ seq([(i-1)*h,T1[j],PP[i,j]], i=1..N+2)], j=1..M+1)]:
> surfdata( plotdata, axes=boxed, title="Figure 5.14",
labels=[x,t,u],orientation=[-75,75]);
```

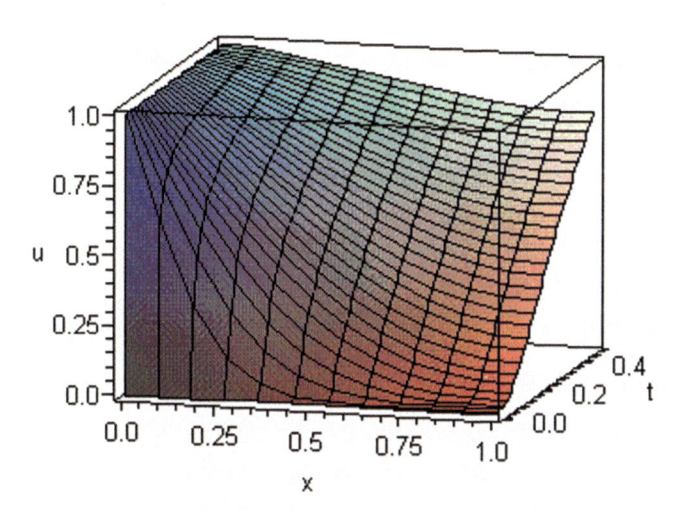

Fig. 5.14

The mass transafer flux at $x = 0$ in dimensionless form is given by:

```
> flux:=-dydxf;
```

$$flux := 0.14792\,e^{-471.73\,t} + 0.55030\,e^{-438.99\,t} + 1.0972\,e^{-387.95\,t} + 1.6478\,e^{-323.50\,t} + 2.0764\,e^{-251.79\,t}$$
$$+ 2.3126\,e^{-179.57\,t} + 2.3581\,e^{-113.45\,t} + 2.2802\,e^{-59.192\,t} + 2.2060\,e^{-21.036\,t} + 3.3864\,e^{-1.4655\,t} - 1.5636$$

```
> plot(flux,t=0..0.5,thickness=4,title="Figure 5.15",
axes=boxed,labels=[t,"flux"]);
```

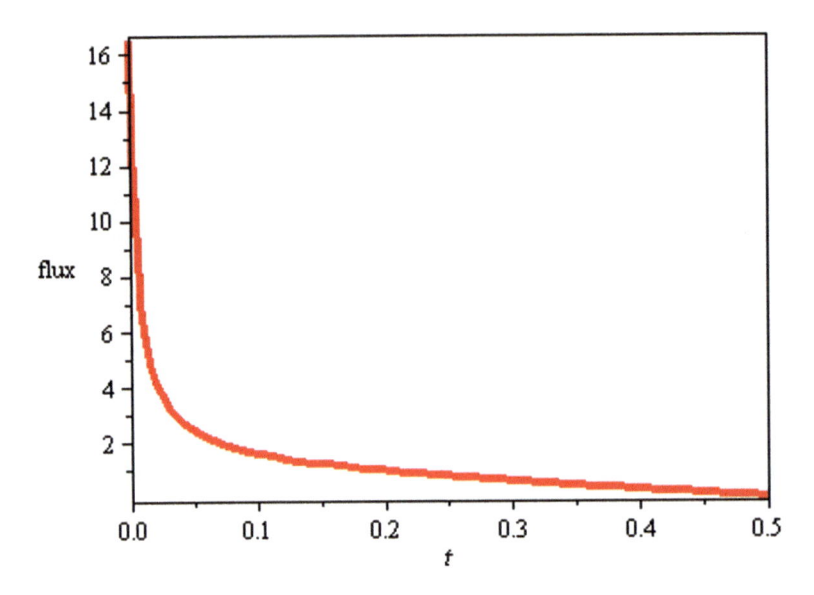

398   5 Method of Lines for Parabolic Partial Differential Equations

Fig. 5.15

Next, the concentration profiles are converted to dimensionless form as:

> for i from 0 to N+1 do ca[i](t):=u[i](t)*ca0;od:

> for i from 0 to N+1 do ca[i](t1):=subs(t=t1*Dab/Lc^2,ca[i](t));od:

> for i from 0 to N+1 do p[i]:=plot(ca[i](t1),t1=0..2e6,thickness=3);od:

> pp:=plot([seq(ca[i](t1),i=0..N+1)],t1=0..2e6,thickness=4);

$$pp := PLOT(...)$$

> arw:=arrow(<0.1e6,0.0105>,<1.0e6,-0.0085>,width=[1/1.0e5,relative=false], head_width=[1/5.0e3,relative=false],head_length=[1/30,relative=true]):

> pt:=textplot([[1.4e6,0.0015,typeset("Follow the arrow: ",ca[0],"(t), ..., ", ca[N+1],"(t).")]]):

> display([pp,pt,arw],title="Figure 5.16.",axes=boxed,labels=[t1,"ca"]);

5.1 Semianalytical Method for Parabolic Partial Differential Equations (PDEs) 399

Follow the arrow: $ca_0(t), ..., ca_{11}(t)$.

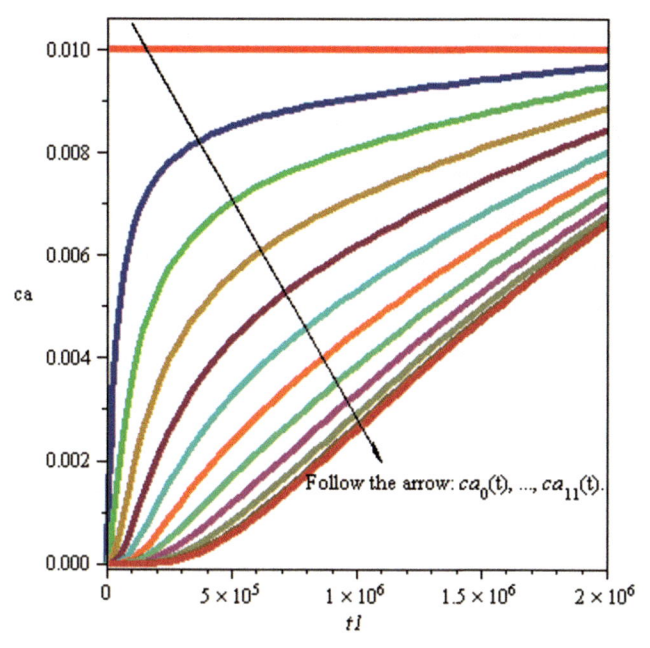

Fig. 5.16

> tf:=2e6;

$$tf := 2.\, 10^6$$

> M:=30;

$$M := 30$$

> T1:=[seq(tf*i/M,i=0..M)];

$T1 := [0., 66667., 1.3333\, 10^5, 2.0000\, 10^5, 2.6667\, 10^5, 3.3333\, 10^5, 4.0000\, 10^5, 4.6667\, 10^5, 5.3333\, 10^5, 6.0000\, 10^5,$
$6.6667\, 10^5, 7.3333\, 10^5, 8.0000\, 10^5, 8.6667\, 10^5, 9.3333\, 10^5, 1.0000\, 10^6, 1.0667\, 10^6, 1.1333\, 10^6, 1.2000\, 10^6,$
$1.2667\, 10^6, 1.3333\, 10^6, 1.4000\, 10^6, 1.4667\, 10^6, 1.5333\, 10^6, 1.6000\, 10^6, 1.6667\, 10^6, 1.7333\, 10^6, 1.8000\, 10^6,$
$1.8667\, 10^6, 1.9333\, 10^6, 2.0000\, 10^6]$

> PP:=matrix(N+2,M+1);

$$PP := array\,(1..12,\, 1..31,\, [\,])$$

> for i from 1 to N+2 do PP[i,1]:=evalf(subs(x=(i-1)*h*Lc,0));od:

> for i from 1 to N+2 do for j from 2 to M+1 do PP[i,j]:=evalf(subs(t1=T1[j],ca[i-1](t1)));od;od:

> plotdata := [seq([seq([(i-1)*h*Lc,T1[j],PP[i,j]], i=1..N+2)], j=1..M+1)]:

> surfdata(plotdata,title="Figure 5.17",axes=boxed, labels=[z,t1,"ca"],orientation=[-75,75]);

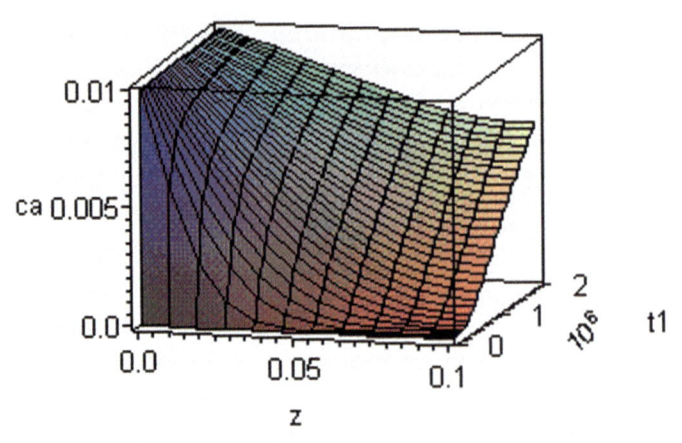

400   5 Method of Lines for Parabolic Partial Differential Equations

Fig. 5.17

Next, the mass transfer flux is converted to dimensionless form as:

> Flux:=Dab*ca0/Lc*flux:

> Flux:=subs(t=t1*Dab/Lc^2,Flux);

$$Flux := 2.9584 \, 10^{-11} \, e^{-0.000094346 \, t1} + 1.1006 \, 10^{-10} \, e^{-0.000087798 \, t1} + 2.1944 \, 10^{-10} \, e^{-0.000077590 \, t1}$$
$$+ 3.2956 \, 10^{-10} \, e^{-0.000064700 \, t1} + 4.1528 \, 10^{-10} \, e^{-0.000050358 \, t1} + 4.6252 \, 10^{-10} \, e^{-0.000035914 \, t1}$$
$$+ 4.7162 \, 10^{-10} \, e^{-0.000022690 \, t1} + 4.5604 \, 10^{-10} \, e^{-0.000011838 \, t1} + 4.4120 \, 10^{-10} \, e^{-0.0000042072 \, t1}$$
$$+ 6.7728 \, 10^{-10} \, e^{-2.9310 \, 10^{-7} \, t1} - 3.1272 \, 10^{-10}$$

> plot(Flux,t1=0..2e6,thickness=4,title="Figure 5.18",
axes=boxed,labels=[t,"Flux"]);

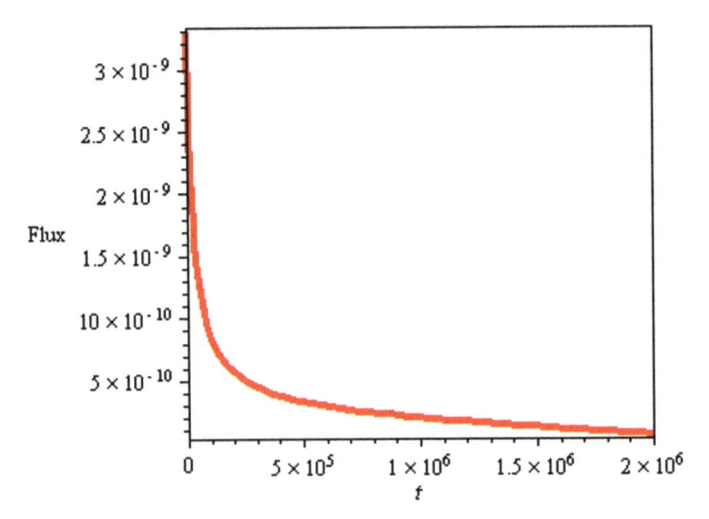

Fig. 5.18

5.1 Semianalytical Method for Parabolic Partial Differential Equations (PDEs)

Example 5.6. Semianalytical Method for the Graetz Problem

Consider the classical Graetz problem,[1]

$$2Pe(1-x^2)\frac{\partial u}{\partial z} = \frac{\partial^2 u}{\partial x^2} + \frac{1}{x}\frac{\partial u}{\partial x}$$

$$u(x,0) = 0 \tag{5.29}$$

$$\frac{\partial u}{\partial x}(0,z) = 0 \text{ and } u(1,z) = 1$$

This PDE is first order in z. The z variable can be treated as a time variable. The temperature profiles depend on the Peclet number Pe. For convenience we introduce the variable transformation z = 2Pet, which converts equation (5.29) to

$$\frac{\partial u}{\partial t} = \frac{1}{1-x^2}\frac{\partial^2 u}{\partial x^2} + \frac{1}{1-x^2}\frac{1}{x}\frac{\partial u}{\partial x}$$

$$u(x,0) = 0 \tag{5.30}$$

$$\frac{\partial u}{\partial x}(0,t) = 0 \text{ and } u(1,t) = 1$$

Equation (5.30) is solved below in Maple using the Maple program developed for example 5.2 by making very few changes as:

> restart;

> with(linalg):with(plots):

> ge:=diff(u(x,t),t)=1/(1-x^2)*diff(u(x,t),x$2)+1/(1-x^2)/x*diff(u(x,t),x);

$$ge := \frac{\partial}{\partial t}u(x,t) = \frac{\frac{\partial^2}{\partial x^2}u(x,t)}{1-x^2} + \frac{\frac{\partial}{\partial x}u(x,t)}{(1-x^2)x}$$

> bc1:=diff(u(x,t),x);

$$bc1 := \frac{\partial}{\partial x}u(x,t)$$

> bc2:=u(x,t)-1;

$$bc2 := u(x,t) - 1$$

> IC:=u(x,0)=0;

$$IC := u(x, 0) = 0$$

> N:=10;

$$N := 10$$

> L:=1;

$$L := 1$$

> dydxf:=1/2*(-u[2](t)-3*u[0](t)+4*u[1](t))/h:
> dydxb:=1/2*(u[N-1](t)+3*u[N+1](t)-4*u[N](t))/h:
> dydx:=1/2/h*(u[m+1](t)-u[m-1](t)):
> d2ydx2:=1/h^2*(u[m-1](t)-2*u[m](t)+u[m+1](t)):
> bc1:=subs(diff(u(x,t),x)=dydxf,u(x,t)=u[0](t),x=0,bc1):
> bc2:=subs(diff(u(x,t),x)=dydxb,u(x,t)=u[N+1](t),x=1,bc2):
> eq[0]:=bc1;

$$eq_0 := \frac{1}{2} \frac{-u_2(t) - 3u_0(t) + 4u_1(t)}{h}$$

> eq[N+1]:=bc2;

$$eq_{11} := u_{11}(t) - 1$$

> for i from 1 to N do eq[i]:=diff(u[i](t),t)= subs(diff(u(x,t),x$2) = subs(m=i,d2ydx2),diff(u(x,t),x) = subs(m=i,dydx),u(x,t)=u[i](t), x=i*h,rhs(ge));od;

$$eq_1 := \frac{d}{dt} u_1(t) = \frac{u_0(t) - 2u_1(t) + u_2(t)}{(1-h^2)h^2} + \frac{1}{2}\frac{u_2(t) - u_0(t)}{(1-h^2)h^2}$$

$$eq_2 := \frac{d}{dt} u_2(t) = \frac{u_1(t) - 2u_2(t) + u_3(t)}{(1-4h^2)h^2} + \frac{1}{4}\frac{u_3(t) - u_1(t)}{(1-4h^2)h^2}$$

$$eq_3 := \frac{d}{dt} u_3(t) = \frac{u_2(t) - 2u_3(t) + u_4(t)}{(1-9h^2)h^2} + \frac{1}{6}\frac{u_4(t) - u_2(t)}{(1-9h^2)h^2}$$

$$eq_4 := \frac{d}{dt} u_4(t) = \frac{u_3(t) - 2u_4(t) + u_5(t)}{(1-16h^2)h^2} + \frac{1}{8}\frac{u_5(t) - u_3(t)}{(1-16h^2)h^2}$$

$$eq_5 := \frac{d}{dt} u_5(t) = \frac{u_4(t) - 2u_5(t) + u_6(t)}{(1-25h^2)h^2} + \frac{1}{10}\frac{u_6(t) - u_4(t)}{(1-25h^2)h^2}$$

5.1 Semianalytical Method for Parabolic Partial Differential Equations (PDEs)

$$eq_6 := \frac{d}{dt} u_6(t) = \frac{u_5(t) - 2u_6(t) + u_7(t)}{(1 - 36h^2)h^2} + \frac{1}{12} \frac{u_7(t) - u_5(t)}{(1 - 36h^2)h^2}$$

$$eq_7 := \frac{d}{dt} u_7(t) = \frac{u_6(t) - 2u_7(t) + u_8(t)}{(1 - 49h^2)h^2} + \frac{1}{14} \frac{u_8(t) - u_6(t)}{(1 - 49h^2)h^2}$$

$$eq_8 := \frac{d}{dt} u_8(t) = \frac{u_7(t) - 2u_8(t) + u_9(t)}{(1 - 64h^2)h^2} + \frac{1}{16} \frac{u_9(t) - u_7(t)}{(1 - 64h^2)h^2}$$

$$eq_9 := \frac{d}{dt} u_9(t) = \frac{u_8(t) - 2u_9(t) + u_{10}(t)}{(1 - 81h^2)h^2} + \frac{1}{18} \frac{u_{10}(t) - u_8(t)}{(1 - 81h^2)h^2}$$

$$eq_{10} := \frac{d}{dt} u_{10}(t) = \frac{u_9(t) - 2u_{10}(t) + u_{11}(t)}{(1 - 100h^2)h^2} + \frac{1}{20} \frac{u_{11}(t) - u_9(t)}{(1 - 100h^2)h^2}$$

> u[0](t):=(solve(eq[0],u[0](t)));

$$u_0(t) := -\frac{1}{3} u_2(t) + \frac{4}{3} u_1(t)$$

> u[N+1](t):=solve(eq[N+1],u[N+1](t));

$$u_{11}(t) := 1$$

> for i from 1 to N do eq[i]:=eval(eq[i]);od;

$$eq_1 := \frac{d}{dt} u_1(t) = \frac{\frac{2}{3} u_2(t) - \frac{2}{3} u_1(t)}{(1 - h^2)h^2} + \frac{1}{2} \frac{\frac{4}{3} u_2(t) - \frac{4}{3} u_1(t)}{(1 - h^2)h^2}$$

$$eq_2 := \frac{d}{dt} u_2(t) = \frac{u_1(t) - 2u_2(t) + u_3(t)}{(1 - 4h^2)h^2} + \frac{1}{4} \frac{u_3(t) - u_1(t)}{(1 - 4h^2)h^2}$$

$$eq_3 := \frac{d}{dt} u_3(t) = \frac{u_2(t) - 2u_3(t) + u_4(t)}{(1 - 9h^2)h^2} + \frac{1}{6} \frac{u_4(t) - u_2(t)}{(1 - 9h^2)h^2}$$

$$eq_4 := \frac{d}{dt} u_4(t) = \frac{u_3(t) - 2u_4(t) + u_5(t)}{(1 - 16h^2)h^2} + \frac{1}{8} \frac{u_5(t) - u_3(t)}{(1 - 16h^2)h^2}$$

$$eq_5 := \frac{d}{dt} u_5(t) = \frac{u_4(t) - 2u_5(t) + u_6(t)}{(1 - 25h^2)h^2} + \frac{1}{10} \frac{u_6(t) - u_4(t)}{(1 - 25h^2)h^2}$$

$$eq_6 := \frac{d}{dt} u_6(t) = \frac{u_5(t) - 2u_6(t) + u_7(t)}{(1-36h^2)h^2} + \frac{1}{12} \frac{u_7(t) - u_5(t)}{(1-36h^2)h^2}$$

$$eq_7 := \frac{d}{dt} u_7(t) = \frac{u_6(t) - 2u_7(t) + u_8(t)}{(1-49h^2)h^2} + \frac{1}{14} \frac{u_8(t) - u_6(t)}{(1-49h^2)h^2}$$

$$eq_8 := \frac{d}{dt} u_8(t) = \frac{u_7(t) - 2u_8(t) + u_9(t)}{(1-64h^2)h^2} + \frac{1}{16} \frac{u_9(t) - u_7(t)}{(1-64h^2)h^2}$$

$$eq_9 := \frac{d}{dt} u_9(t) = \frac{u_8(t) - 2u_9(t) + u_{10}(t)}{(1-81h^2)h^2} + \frac{1}{18} \frac{u_{10}(t) - u_8(t)}{(1-81h^2)h^2}$$

$$eq_{10} := \frac{d}{dt} u_{10}(t) = \frac{u_9(t) - 2u_{10}(t) + 1}{(1-100h^2)h^2} + \frac{1}{20} \frac{1 - u_9(t)}{(1-100h^2)h^2}$$

```
> eqs:=[seq(rhs(eq[j]),j=1..N)]:
> Y:=[seq(u[i](t),i=1..N)];
```

$$Y := [u_1(t), u_2(t), u_3(t), u_4(t), u_5(t), u_6(t), u_7(t), u_8(t), u_9(t), u_{10}(t)]$$

```
> A:=genmatrix(eqs,Y,'b1'):
> b:=matrix(N,1):for i to N do b[i,1]:=-eval(b1[i]);od:evalm(b);
```

$$\begin{bmatrix} 0 \\ 0 \\ 0 \\ 0 \\ 0 \\ 0 \\ 0 \\ 0 \\ 0 \\ \dfrac{21}{20(1-100h^2)h^2} \end{bmatrix}$$

```
> h:=eval(L/(N+1));
```

$$h := \frac{1}{11}$$

5.1 Semianalytical Method for Parabolic Partial Differential Equations (PDEs) 405

```
> A:=map(eval,A):
> if N > 4 then A:=map(evalf,A);end:
> evalm(A);
```

$$\begin{bmatrix} [-162.6777778, 162.6777778, 0., 0., 0., 0., 0., 0., 0., 0.] \\ [93.85256410, -250.2735043, 156.4209402, 0., 0., 0., 0., 0., 0., 0.] \\ [0., 108.9360119, -261.4464286, 152.5104167, 0., 0., 0., 0., 0., 0.] \\ [0., 0., 122.0083333, -278.8761905, 156.8678571, 0., 0., 0., 0., 0.] \\ [0., 0., 0., 137.2593750, -305.0208333, 167.7614583, 0., 0., 0., 0.] \\ [0., 0., 0., 0., 157.8931373, -344.4941176, 186.6009804, 0., 0., 0.] \\ [0., 0., 0., 0., 0., 188.8224206, -406.6944444, 217.8720238, 0., 0.] \\ [0., 0., 0., 0., 0., 0., 240.8059211, -513.7192982, 272.9133772, 0.] \\ [0., 0., 0., 0., 0., 0., 0., 345.6902778, -732.0500000, 386.3597222] \\ [0., 0., 0., 0., 0., 0., 0., 0., 662.3309524, -1394.380952] \end{bmatrix}$$

```
> mat:=exponential(A,t):
> mat:=map(evalf,mat):
> Y0:=matrix(N,1):for i from 1 to N do
Y0[i,1]:=evalf(subs(x=i*h,rhs(IC)));od:evalm(Y0);
```

$$\begin{bmatrix} 0. \\ 0. \\ 0. \\ 0. \\ 0. \\ 0. \\ 0. \\ 0. \\ 0. \\ 0. \end{bmatrix}$$

```
> s1:=evalm(Y0+inverse(A)&*b):
> Y:=evalm(mat&*s1-inverse(A)&*b):
> Y:=map(simplify,Y):
> Digits:=10;
```

$$Digits := 10$$

```
> for i from 1 to N do u[i](t):=evalf((Y[i,1]));od:
> for i from 0 to N+1 do u[i](t):=eval(u[i](t));od;
```

$$u_0(t) := 6.396258436\,10^{-8}\,e^{-1687.725405\,t} - 0.0002046970356\,e^{-834.0458065\,t} + 0.01195590815\,e^{-592.7186781\,t}$$
$$- 0.07536306752\,e^{-486.6117320\,t} + 0.1813271970\,e^{-398.2866134\,t} - 0.2906086458\,e^{-296.1252973\,t}$$
$$+ 0.4092192622\,e^{-195.1433199\,t} - 0.5577027862\,e^{-108.0790702\,t} + 0.7964509565\,e^{-43.62287680\,t}$$
$$- 1.475074105\,e^{-7.274747081\,t} + 1.000000001$$

$$u_1(t) := 1.453919354\,10^{-8}\,e^{-1687.725405\,t} - 0.0000755639912\,e^{-834.0458065\,t} + 0.005398922725\,e^{-592.7186781\,t}$$
$$- 0.03773647616\,e^{-486.6117320\,t} + 0.09984394132\,e^{-398.2866134\,t} - 0.1808647670\,e^{-296.1252973\,t}$$
$$+ 0.2923294177\,e^{-195.1433199\,t} - 0.4565876363\,e^{-108.0790702\,t} + 0.7311014315\,e^{-43.62287680\,t}$$
$$- 1.453409223\,e^{-7.274747081\,t} + 1.000000001$$

$$u_2(t) := -1.337309789\,10^{-7}\,e^{-1687.725405\,t} + 0.000311835142\,e^{-834.0458065\,t} - 0.01427203354\,e^{-592.7186781\,t}$$
$$+ 0.0751432979\,e^{-486.6117320\,t} - 0.1446058256\,e^{-398.2866134\,t} + 0.1483668692\,e^{-296.1252973\,t}$$
$$- 0.05834011584\,e^{-195.1433199\,t} - 0.1532421868\,e^{-108.0790702\,t} + 0.5350528562\,e^{-43.62287680\,t}$$
$$- 1.388414576\,e^{-7.274747081\,t} + 1.000000001$$

$$u_3(t) := 0.000001239372059\,e^{-1687.725405\,t} - 0.001118464477\,e^{-834.0458065\,t} + 0.02800583789\,e^{-592.7186781\,t}$$
$$- 0.0908929085\,e^{-486.6117320\,t} + 0.07692633264\,e^{-398.2866134\,t} + 0.06502818339\,e^{-296.1252973\,t}$$
$$- 0.1959595423\,e^{-195.1433199\,t} + 0.1346477576\,e^{-108.0790702\,t} + 0.2682074831\,e^{-43.62287680\,t}$$
$$- 1.284846092\,e^{-7.274747081\,t} + 1.000000001$$

$$u_4(t) := -0.00001147897273\,e^{-1687.725405\,t} + 0.003976521410\,e^{-834.0458065\,t} - 0.05063790949\,e^{-592.7186781\,t}$$
$$+ 0.0805199277\,e^{-486.6117320\,t} + 0.03426686589\,e^{-398.2866134\,t} - 0.1207627439\,e^{-296.1252973\,t}$$
$$- 0.04352101267\,e^{-195.1433199\,t} + 0.2448628472\,e^{-108.0790702\,t} + 0.0008876810364\,e^{-43.62287680\,t}$$
$$- 1.149581428\,e^{-7.274747081\,t} + 1.000000001$$

$$u_5(t) := 0.0001021378324\,e^{-1687.725405\,t} - 0.01320336041\,e^{-834.0458065\,t} + 0.0795280937\,e^{-592.7186781\,t}$$
$$- 0.03593552194\,e^{-486.6117320\,t} - 0.08591726287\,e^{-398.2866134\,t} - 0.03729876452\,e^{-296.1252973\,t}$$
$$+ 0.1291825819\,e^{-195.1433199\,t} + 0.1618785802\,e^{-108.0790702\,t} - 0.2072749858\,e^{-43.62287680\,t}$$
$$- 0.9910638176\,e^{-7.274747081\,t} + 1.000000002$$

$$u_6(t) := -0.0008324323948\,e^{-1687.725405\,t} + 0.03838244329\,e^{-834.0458065\,t} - 0.0949531851\,e^{-592.7186781\,t}$$
$$- 0.02698165080\,e^{-486.6117320\,t} + 0.01973244921\,e^{-398.2866134\,t} + 0.0968236728\,e^{-296.1252973\,t}$$
$$+ 0.1202213700\,e^{-195.1433199\,t} - 0.01030786394\,e^{-108.0790702\,t} - 0.3236964873\,e^{-43.62287680\,t}$$
$$- 0.8183912023\,e^{-7.274747081\,t} + 1.000000002$$

5.1 Semianalytical Method for Parabolic Partial Differential Equations (PDEs)

$u_7(t) := 0.005905778525\, e^{-1687.725405\,t} - 0.08952503648\, e^{-834.0458065\,t} + 0.0590173413\, e^{-592.7186781\,t}$
$+ 0.05094503871\, e^{-486.6117320\,t} + 0.06704438659\, e^{-398.2866134\,t} + 0.05664106821\, e^{-296.1252973\,t}$
$- 0.01308675884\, e^{-195.1433199\,t} - 0.1500210013\, e^{-108.0790702\,t} - 0.3465435354\, e^{-43.62287680\,t}$
$- 0.6403737811\, e^{-7.274747081\,t} + 1.000000002$

$u_8(t) := -0.03400299773\, e^{-1687.725405\,t} + 0.1423367722\, e^{-834.0458065\,t} + 0.03190253306\, e^{-592.7186781\,t}$
$+ 0.004672299517\, e^{-486.6117320\,t} - 0.01448728294\, e^{-398.2866134\,t} - 0.05513595735\, e^{-296.1252973\,t}$
$- 0.1169316676\, e^{-195.1433199\,t} - 0.1966484843\, e^{-108.0790702\,t} - 0.2969618136\, e^{-43.62287680\,t}$
$- 0.4647094462\, e^{-7.274747081\,t} + 1.000000002$

$u_9(t) := 0.1410615395\, e^{-1687.725405\,t} - 0.08807283196\, e^{-834.0458065\,t} - 0.06129252106\, e^{-592.7186781\,t}$
$- 0.04465692535\, e^{-486.6117320\,t} - 0.06522043432\, e^{-398.2866134\,t} - 0.09424743043\, e^{-296.1252973\,t}$
$- 0.1248596239\, e^{-195.1433199\,t} - 0.1601360322\, e^{-108.0790702\,t} - 0.2055804919\, e^{-43.62287680\,t}$
$- 0.2973207223\, e^{-7.274747081\,t} + 1.000000001$

$u_{10}(t) := -0.3184973256\, e^{-1687.725405\,t} - 0.1041036052\, e^{-834.0458065\,t} - 0.05065880317\, e^{-592.7186781\,t}$
$- 0.03244292688\, e^{-486.6117320\,t} - 0.04361420333\, e^{-398.2866134\,t} - 0.05670111642\, e^{-296.1252973\,t}$
$- 0.06927980334\, e^{-195.1433199\,t} - 0.08222994379\, e^{-108.0790702\,t} - 0.1010983264\, e^{-43.62287680\,t}$
$- 0.1419351022\, e^{-7.274747081\,t} + 1.000000001$

$$u_{11}(t) := 1$$

Set the colors for the curves:

> setcolors(["Red", "Blue", "LimeGreen", "Goldenrod", "maroon", "DarkTurquoise", "coral", "aquamarine", "magenta", "khaki", "sienna", "orange", "yellow", "gray"]);

["Red", "LimeGreen", "Goldenrod", "Blue", "MediumOrchid", "DarkTurquoise"]

> pp:=plot([seq(u[i](t),i=0..N+1)],t=0..0.4,thickness=3);

$$pp := PLOT(\ldots)$$

408   5 Method of Lines for Parabolic Partial Differential Equations

Plot the texts for the corresponding curves:

> arw:=arrow(<0.15,0.4>,<-0.05,0.63>,width=[1/700,relative=true],
head_width= head_length=[1/30,relative=false]):pt:=textplot([[0.25,0.35,
typeset("Follow the arrow: ",u[0],"(t), ..., ",u[N+1]", u[N+1],"(t).")]]):

> display([pp,pt,arw],axes=boxed,title="Figure 5.19",labels=[t,"u"]);

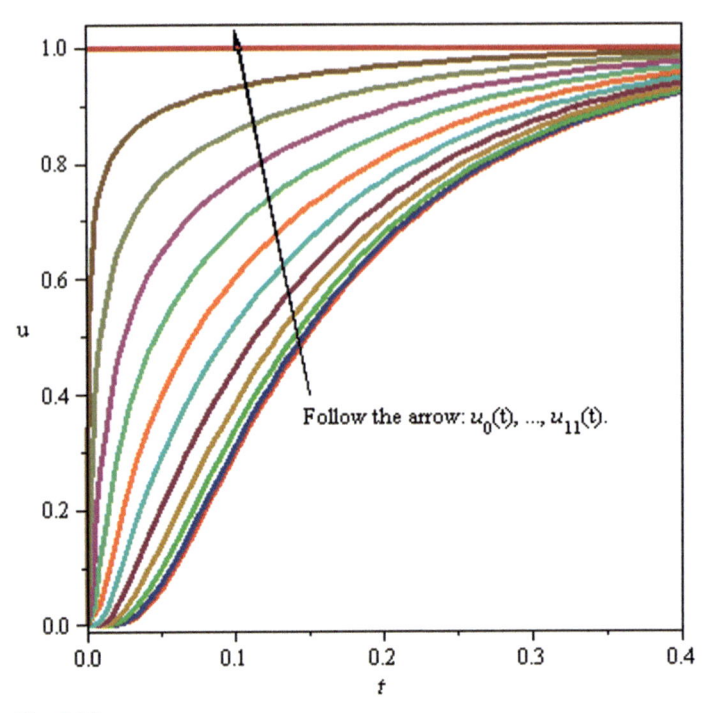

Fig. 5.19

> tf:=0.2;

$$tf := 0.2$$

> M:=30;

$$M := 30$$

> T1:=[seq(tf*i/M,i=0..M)];

$T1 := [0., 0.03333333333, 0.06666666667, 0.1000000000, 0.1333333333, 0.1666666667, 0.2000000000,$
$0.2333333333, 0.2666666667, 0.3000000000, 0.3333333333, 0.3666666667, 0.4000000000, 0.4333333333,$
$0.4666666667, 0.5000000000, 0.5333333333, 0.5666666667, 0.6000000000, 0.6333333333, 0.6666666667,$
$0.7000000000, 0.7333333333, 0.7666666667, 0.8000000000, 0.8333333333, 0.8666666667, 0.9000000000,$
$0.9333333333, 0.9666666667, 1.000000000]$

5.1 Semianalytical Method for Parabolic Partial Differential Equations (PDEs) 409

```
> PP:=matrix(N+2,M+1);
```

$$PP := array(1..12, 1..31, [\,])$$

```
> for i from 1 to N+2 do PP[i,1]:=evalf(subs(x=(i-1)*h,rhs(IC)));od:
> for i from 1 to N+2 do for j from 2 to M+1 do
PP[i,j]:=evalf(subs(t=T1[j],u[i-1](t)));od;od:
> plotdata := [seq([ seq([(i-1)*h,T1[j],PP[i,j]], i=1..N+2)], j=1..M+1)]:
> surfdata(plotdata,axes=boxed,title="Figure 5.20",
labels=[x,t,u],orientation=[-135,45]);
```

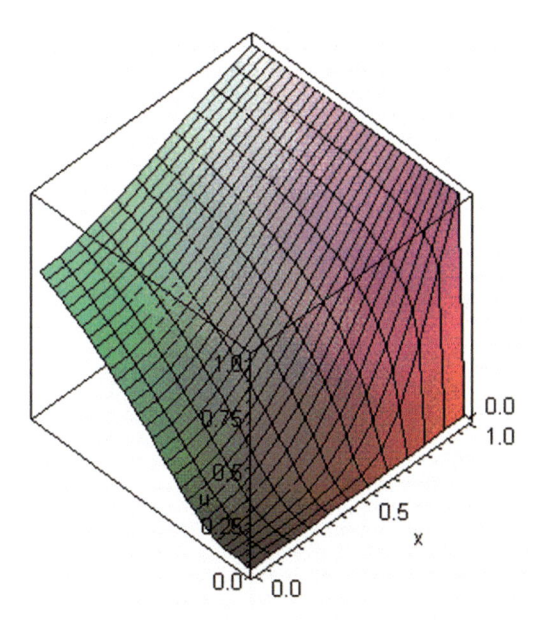

Fig. 5.20

```
> for i from 0 to N+1 do u[i](z):=subs(t=z/2/Pe,u[i](t));od:
>   pp:=plot([seq(subs(Pe=1,u[i](z)),i=0..N+1)],z=0..1,thickness=3;
```

$$pp := PLOT(...)$$

Plot the texts for the corresponding curves:

```
> arw:=arrow(<0.3,0.4>,<-0.15,0.63>,width=[1/700,relative=true,
head_width=[1/150,relative=false,head_length=[1/30,relative=false]):
> pt:=textplot([[0.55,0.35,typeset("Follow the arrow:",u[0], "(z), ..., ",
u[N+1],"(z).")]]):
> display([pp,pt,arw],axes=boxed,title="Figure 5.21",labels=[z,"u"]);
```

410   5 Method of Lines for Parabolic Partial Differential Equations

Follow the arrow: $u_0(z), ..., u_{11}(z)$.

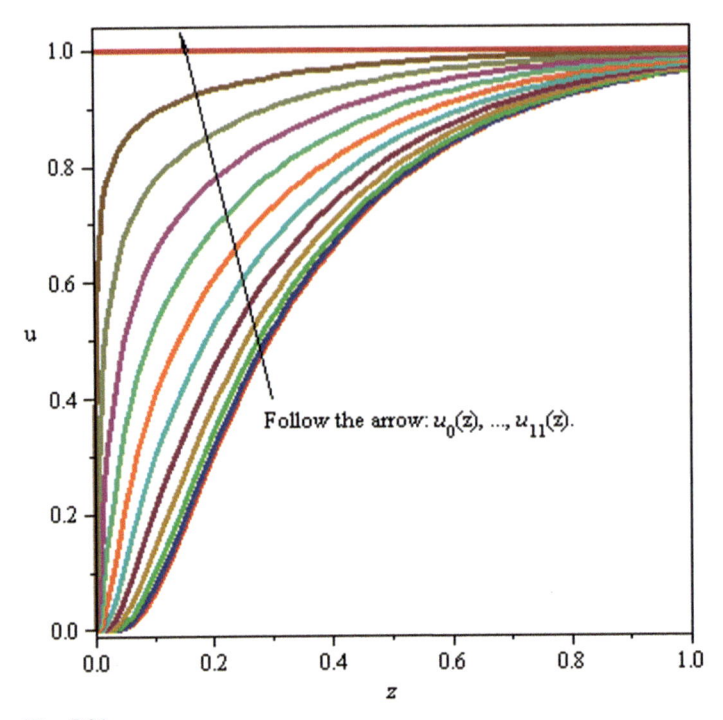

Fig. 5.21

> tf:=1.;

$$tf := 1.$$

> M:=30;

$$M := 30$$

> T1:=[seq(tf*i/M,i=0..M)];

$T1 := [0., 0.006666666667, 0.01333333333, 0.02000000000, 0.02666666667, 0.03333333333, 0.04000000000,$

$0.04666666667, 0.05333333333, 0.06000000000, 0.06666666667, 0.07333333333, 0.08000000000,$

$0.08666666667, 0.09333333333, 0.1000000000, 0.1066666667, 0.1133333333, 0.1200000000, 0.1266666667,$

$0.1333333333, 0.1400000000, 0.1466666667, 0.1533333333, 0.1600000000, 0.1666666667, 0.1733333333,$

$0.1800000000, 0.1866666667, 0.1933333333, 0.2000000000]$

> PP:=matrix(N+2,M+1);

$$PP := array(1..12, 1..31, [\,])$$

> for i from 1 to N+2 do PP[i,1]:=evalf(subs(x=(i-1)*h,rhs(IC)));od:
> for i from 1 to N+2 do for j from 2 to M+1 do
PP[i,j]:=evalf(subs(z=T1[j],Pe=1.,u[i-1](z)));od;od:

5.1 Semianalytical Method for Parabolic Partial Differential Equations (PDEs) 411

```
> plotdata := [seq([ seq([(i-1)*h,T1[j],PP[i,j]], i=1..N+2)], j=1..M+1)]:
> surfdata(plotdata,axes=boxed,title="Figure 5.22",
labels=[x,z,u],orientation=[-135,45]);
```

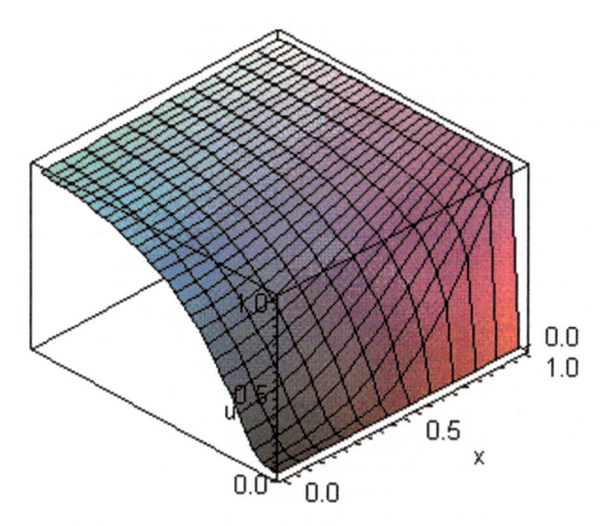

Fig. 5.22

We observe that as the Peclet number increases, we observe the penetration depth (the distance required in the z direction to reach the steady state value 1) increases. The analytical solution for this problem involves transcendental equations and infinite series. (Jacob, 1949)[1] According to the analytical solution, the dimensionless temperature (u) at x = 0 varies as:

$$ua(0,z) = 1-1.477\exp\left(-3.658\frac{z}{Pe}\right)+0.810\exp\left(-22.178\frac{z}{Pe}\right)-\exp\left(-53.05\frac{z}{Pe}\right)+...$$
(5.31)

We observe that at z = 0, equation (5.31) needs infinite terms for convergence. However, equation (5.31) converges rapidly for high values of the ratio z/Pe. Hence, our semianalytical solution is compared with the exact solution for different values of the ratio z/Pe.

```
> pp:=plot([seq(subs(Pe=10,u[i](z)),i=0..N+1)],z=0..1,thickness=3);
```

$$pp := PLOT(...)$$

```
> display([pp],axes=boxed,title="Figure 5.23", labels=[z,"u"],
caption=typeset("From the bottom to the top of the figure: ",u[0],"(z), ..., ",
u[N+1],"(z)"));
```

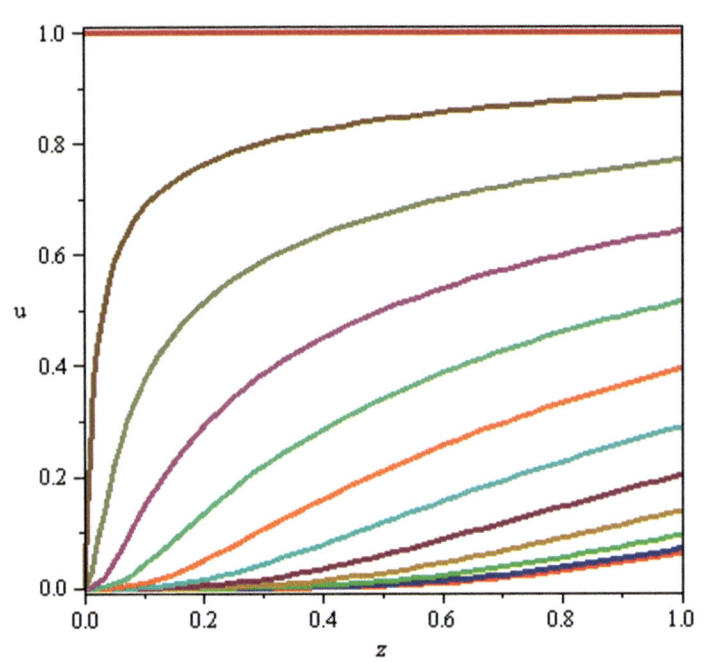

412　　　　　　　　　　5 Method of Lines for Parabolic Partial Differential Equations

From the bottom to the top of the figure: $u_0(z), ..., u_{11}(z)$

Fig. 5.23

> tf:=1.;

$$tf := 1.$$

> M:=30;

$$M := 30$$

> T1:=[seq(tf*i/M,i=0..M)];

$T1 := [0., 0.03333333333, 0.06666666667, 0.1000000000, 0.1333333333, 0.1666666667, 0.2000000000,$

$0.2333333333, 0.2666666667, 0.3000000000, 0.3333333333, 0.3666666667, 0.4000000000, 0.4333333333,$

$0.4666666667, 0.5000000000, 0.5333333333, 0.5666666667, 0.6000000000, 0.6333333333, 0.6666666667,$

$0.7000000000, 0.7333333333, 0.7666666667, 0.8000000000, 0.8333333333, 0.8666666667, 0.9000000000,$

$0.9333333333, 0.9666666667, 1.000000000\,]$

> PP:=matrix(N+2,M+1);

$$PP := array\,(1..12,\,1..31,\,[\,\,])$$

> for i from 1 to N+2 do PP[i,1]:=evalf(subs(x=(i-1)*h,rhs(IC)));od:
> for i from 1 to N+2 do for j from 2 to M+1 do
PP[i,j]:=evalf(subs(z=T1[j],Pe=10.,u[i-1](z)));od;od:

5.1 Semianalytical Method for Parabolic Partial Differential Equations (PDEs)

```
> plotdata:= [seq([ seq([(i-1)*h,T1[j],PP[i,j]], i=1..N+2)], j=1..M+1)]:
> surfdata(plotdata,axes=boxed,title="Figure 5.24",
labels=[x,z,u],orientation=[-135,45]);
```

Fig. 5.24

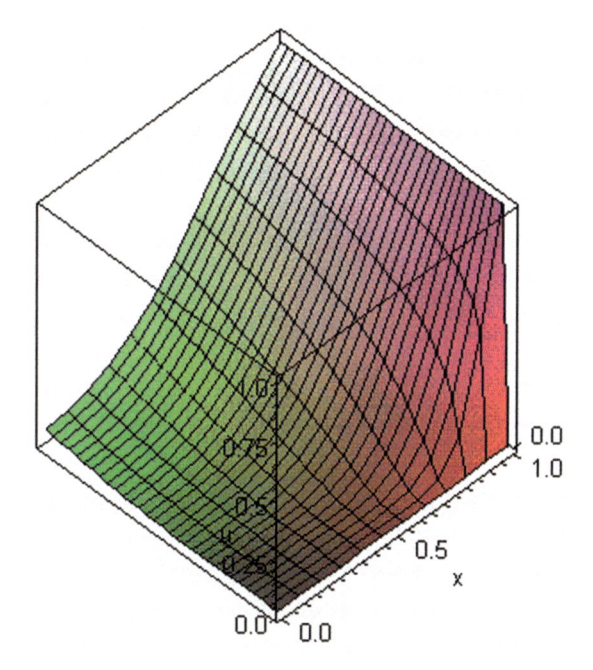

```
> ua0:=1-1.477*exp(-3.658*z/Pe)+0.81*exp(-22.178*z/Pe)-0.385*exp
(-53.05*z/Pe);
```

$$ua0 := 1 - 1.477\,e^{-\frac{3.658\,z}{Pe}} + 0.81\,e^{-\frac{22.178\,z}{Pe}} - 0.385\,e^{-\frac{53.05\,z}{Pe}}$$

```
> u[0](z);
```

$$6.396258436\,10^{-8}\,e^{-\frac{843.8627025\,z}{Pe}} - 0.0002046970356\,e^{-\frac{417.0229032\,z}{Pe}} + 0.01195590815\,e^{-\frac{296.3593390\,z}{Pe}}$$
$$- 0.07536306752\,e^{-\frac{243.3058660\,z}{Pe}} + 0.1813271970\,e^{-\frac{199.1433067\,z}{Pe}} - 0.2906086458\,e^{-\frac{148.0626486\,z}{Pe}}$$
$$+ 0.4092192622\,e^{-\frac{97.57165995\,z}{Pe}} - 0.5577027862\,e^{-\frac{54.03953510\,z}{Pe}} + 0.7964509565\,e^{-\frac{21.81143840\,z}{Pe}}$$
$$- 1.475074105\,e^{-\frac{3.637373540\,z}{Pe}} + 1.000000001$$

Note that with our semianalytical technique the temperature at $x = 0$ is obtained as an analytical function of z and Pe.

414   5 Method of Lines for Parabolic Partial Differential Equations

> pars:=[0.1,0.2,0.25,0.5,1,2];

$$pars := [0.1, 0.2, 0.25, 0.5, 1, 2]$$

> M:=nops(pars);

$$M := 6$$

> seq(evalf(subs(z=Pe*pars[i],ua0)),i=1..M);

0.06174643482, 0.2989457048, 0.4113066251, 0.7628444874, 0.9619169890, 0.9990180665

> seq(evalf(subs(z=Pe*pars[i],z=10,u[0](z))),i=1..M);

0.0621584604, 0.2974939700, 0.4092682142, 0.7607005605, 0.9611740074, 0.9989780471

>

We obtain reasonable accuracy with the semianalytical solution. Accuracy can be increased by increasing the number of node points.

Example 5.7. Semianalytical Method for PDEs with Known Initial Profiles

In all the previous examples, initial conditions were constants. The initial condition can be a function of x. For example, consider the heat transfer problem:[6]

$$\frac{\partial u}{\partial t} = D_s \frac{\partial^2 u}{\partial x^2}$$

$$u(x,0) = \cos\left(\frac{\pi x}{2L}\right) \quad (5.32)$$

$$\frac{\partial u}{\partial x}(0,t) = 0 \text{ and } u(L,t) = 0$$

This BVP is solved for a given value of $D_s = 1e^{-5}$ cm^2/s and $L = 0.02$ cm below. Note that this BVP is solved directly in dimensionless form. The program used above for example 5.1 is used here, as the boundary conditions are homogeneous. Only the following statements need to be changed.

> restart;
> with(linalg):with(plots):
> ge:=diff(u(x,t),t)=Ds*diff(u(x,t),x$2);

$$ge := \frac{\partial}{\partial t} u(x,t) = Ds \left(\frac{\partial^2}{\partial x^2} u(x,t) \right)$$

5.1 Semianalytical Method for Parabolic Partial Differential Equations (PDEs)

```
> bc1:=diff(u(x,t),x);
```

$$bc1 := \frac{\partial}{\partial x} u(x, t)$$

```
> bc2:=u(x,t);
```

$$bc2 := u(x, t)$$

```
> IC:=u(x,0)=cos(Pi*x/2/L);
```

$$IC := u(x, 0) = \cos\left(\frac{1}{2} \frac{\pi x}{L}\right)$$

```
> L:=0.02;
```

$$L := 0.02$$

```
> Ds:=2e-5;
```

$$Ds := 0.00002$$

Note that since the problem is solved in dimensionless form, we plot the profiles until tf = 20 seconds. The following plots are obtained.

```
> N:=10;
```

$$N := 10$$

```
> dydxf:=1/2*(-u[2](t)-3*u[0](t)+4*u[1](t))/h;
```

$$dydxf := \frac{1}{2} \frac{-u_2(t) - 3 u_0(t) + 4 u_1(t)}{h}$$

```
> dydxb:=1/2*(u[N-1](t)+3*u[N+1](t)-4*u[N](t))/h;
```

$$dydxb := \frac{1}{2} \frac{u_9(t) + 3 u_{11}(t) - 4 u_{10}(t)}{h}$$

```
> dydx:=1/2/h*(u[m+1](t)-u[m-1](t));
```

$$dydx := \frac{1}{2} \frac{u_{m+1}(t) - u_{m-1}(t)}{h}$$

```
> d2ydx2:=1/h^2*(u[m-1](t)-2*u[m](t)+u[m+1](t));
```

$$d2ydx2 := \frac{u_{m-1}(t) - 2 u_m(t) + u_{m+1}(t)}{h^2}$$

```
> bc1:=subs(diff(u(x,t),x)=dydxf,u(x,t)=u[0](t),x=0,bc1);
```

$$bc1 := \frac{1}{2} \frac{-u_2(t) - 3 u_0(t) + 4 u_1(t)}{h}$$

```
> bc2:=subs(diff(u(x,t),x)=dydxb,u(x,t)=u[N+1](t),x=1,bc2);
```

$$bc2 := u_{11}(t)$$

```
> eq[0]:=bc1;
```

$$eq_0 := \frac{1}{2} \frac{-u_2(t) - 3u_0(t) + 4u_1(t)}{h}$$

```
> eq[N+1]:=bc2;
```

$$eq_{11} := u_{11}(t)$$

```
> for i from 1 to N do eq[i]:=diff(u[i](t),t)= subs(diff(u(x,t),x$2) =
subs(m=i,d2ydx2),diff(u(x,t),x) = subs(m=i,dydx),
u(x,t)=u[i](t),x=i*h,rhs(ge));od;
```

$$eq_1 := \frac{d}{dt} u_1(t) = \frac{0.00002 \left(u_0(t) - 2 u_1(t) + u_2(t)\right)}{h^2}$$

$$eq_2 := \frac{d}{dt} u_2(t) = \frac{0.00002 \left(u_1(t) - 2 u_2(t) + u_3(t)\right)}{h^2}$$

$$eq_3 := \frac{d}{dt} u_3(t) = \frac{0.00002 \left(u_2(t) - 2 u_3(t) + u_4(t)\right)}{h^2}$$

$$eq_4 := \frac{d}{dt} u_4(t) = \frac{0.00002 \left(u_3(t) - 2 u_4(t) + u_5(t)\right)}{h^2}$$

$$eq_5 := \frac{d}{dt} u_5(t) = \frac{0.00002 \left(u_4(t) - 2 u_5(t) + u_6(t)\right)}{h^2}$$

$$eq_6 := \frac{d}{dt} u_6(t) = \frac{0.00002 \left(u_5(t) - 2 u_6(t) + u_7(t)\right)}{h^2}$$

$$eq_7 := \frac{d}{dt} u_7(t) = \frac{0.00002 \left(u_6(t) - 2 u_7(t) + u_8(t)\right)}{h^2}$$

$$eq_8 := \frac{d}{dt} u_8(t) = \frac{0.00002 \left(u_7(t) - 2 u_8(t) + u_9(t)\right)}{h^2}$$

$$eq_9 := \frac{d}{dt} u_9(t) = \frac{0.00002 \left(u_8(t) - 2 u_9(t) + u_{10}(t)\right)}{h^2}$$

$$eq_{10} := \frac{d}{dt} u_{10}(t) = \frac{0.00002 \left(u_9(t) - 2 u_{10}(t) + u_{11}(t)\right)}{h^2}$$

5.1 Semianalytical Method for Parabolic Partial Differential Equations (PDEs) 417

```
> u[0](t):=(solve(eq[0],u[0](t)));
```

$$u_0(t) := -\frac{1}{3} u_2(t) + \frac{4}{3} u_1(t)$$

```
> u[N+1](t):=solve(eq[N+1],u[N+1](t));
```

$$u_{11}(t) := 0$$

```
> for i from 1 to N do eq[i]:=eval(eq[i]);od;
```

$$eq_1 := \frac{d}{dt} u_1(t) = \frac{0.00002 \left(\frac{2}{3} u_2(t) - \frac{2}{3} u_1(t) \right)}{h^2}$$

$$eq_2 := \frac{d}{dt} u_2(t) = \frac{0.00002 \left(u_1(t) - 2 u_2(t) + u_3(t) \right)}{h^2}$$

$$eq_3 := \frac{d}{dt} u_3(t) = \frac{0.00002 \left(u_2(t) - 2 u_3(t) + u_4(t) \right)}{h^2}$$

$$eq_4 := \frac{d}{dt} u_4(t) = \frac{0.00002 \left(u_3(t) - 2 u_4(t) + u_5(t) \right)}{h^2}$$

$$eq_5 := \frac{d}{dt} u_5(t) = \frac{0.00002 \left(u_4(t) - 2 u_5(t) + u_6(t) \right)}{h^2}$$

$$eq_6 := \frac{d}{dt} u_6(t) = \frac{0.00002 \left(u_5(t) - 2 u_6(t) + u_7(t) \right)}{h^2}$$

$$eq_7 := \frac{d}{dt} u_7(t) = \frac{0.00002 \left(u_6(t) - 2 u_7(t) + u_8(t) \right)}{h^2}$$

$$eq_8 := \frac{d}{dt} u_8(t) = \frac{0.00002 \left(u_7(t) - 2 u_8(t) + u_9(t) \right)}{h^2}$$

$$eq_9 := \frac{d}{dt} u_9(t) = \frac{0.00002 \left(u_8(t) - 2 u_9(t) + u_{10}(t) \right)}{h^2}$$

$$eq_{10} := \frac{d}{dt} u_{10}(t) = \frac{0.00002 \left(u_9(t) - 2 u_{10}(t) \right)}{h^2}$$

> eqs:=[seq(rhs(eq[j]),j=1..N)];

$$eqs := \left[\frac{0.00002\left(\frac{2}{3}u_2(t) - \frac{2}{3}u_1(t)\right)}{h^2}, \frac{0.00002\left(u_1(t) - 2u_2(t) + u_3(t)\right)}{h^2}, \right.$$
$$\frac{0.00002\left(u_2(t) - 2u_3(t) + u_4(t)\right)}{h^2}, \frac{0.00002\left(u_3(t) - 2u_4(t) + u_5(t)\right)}{h^2},$$
$$\frac{0.00002\left(u_4(t) - 2u_5(t) + u_6(t)\right)}{h^2}, \frac{0.00002\left(u_5(t) - 2u_6(t) + u_7(t)\right)}{h^2},$$
$$\frac{0.00002\left(u_6(t) - 2u_7(t) + u_8(t)\right)}{h^2}, \frac{0.00002\left(u_7(t) - 2u_8(t) + u_9(t)\right)}{h^2},$$
$$\left. \frac{0.00002\left(u_8(t) - 2u_9(t) + u_{10}(t)\right)}{h^2}, \frac{0.00002\left(u_9(t) - 2u_{10}(t)\right)}{h^2} \right]$$

> Y:=[seq(u[i](t),i=1..N)];

$$Y := \left[u_1(t), u_2(t), u_3(t), u_4(t), u_5(t), u_6(t), u_7(t), u_8(t), u_9(t), u_{10}(t) \right]$$

> A:=genmatrix(eqs,Y);

$$A := \left[\left[-\frac{0.00001333333333}{h^2}, \frac{0.00001333333333}{h^2}, 0, 0, 0, 0, 0, 0, 0, 0 \right], \right.$$
$$\left[\frac{0.00002}{h^2}, -\frac{0.00004}{h^2}, \frac{0.00002}{h^2}, 0, 0, 0, 0, 0, 0, 0 \right],$$
$$\left[0, \frac{0.00002}{h^2}, -\frac{0.00004}{h^2}, \frac{0.00002}{h^2}, 0, 0, 0, 0, 0, 0 \right],$$
$$\left[0, 0, \frac{0.00002}{h^2}, -\frac{0.00004}{h^2}, \frac{0.00002}{h^2}, 0, 0, 0, 0, 0 \right],$$
$$\left[0, 0, 0, \frac{0.00002}{h^2}, -\frac{0.00004}{h^2}, \frac{0.00002}{h^2}, 0, 0, 0, 0 \right],$$
$$\left[0, 0, 0, 0, \frac{0.00002}{h^2}, -\frac{0.00004}{h^2}, \frac{0.00002}{h^2}, 0, 0, 0 \right],$$
$$\left[0, 0, 0, 0, 0, \frac{0.00002}{h^2}, -\frac{0.00004}{h^2}, \frac{0.00002}{h^2}, 0, 0 \right],$$
$$\left[0, 0, 0, 0, 0, 0, \frac{0.00002}{h^2}, -\frac{0.00004}{h^2}, \frac{0.00002}{h^2}, 0 \right],$$
$$\left[0, 0, 0, 0, 0, 0, 0, \frac{0.00002}{h^2}, -\frac{0.00004}{h^2}, \frac{0.00002}{h^2} \right],$$
$$\left. \left[0, 0, 0, 0, 0, 0, 0, 0, \frac{0.00002}{h^2}, -\frac{0.00004}{h^2} \right] \right]$$

5.1 Semianalytical Method for Parabolic Partial Differential Equations (PDEs)

```
> h:=eval(L/(N+1));
```

$$h := 0.001818181818$$

```
> A:=map(eval,A);
```

$A := [[-4.033333334, 4.033333334, 0, 0, 0, 0, 0, 0, 0, 0]$,
$[6.050000002, -12.10000000, 6.050000002, 0, 0, 0, 0, 0, 0, 0]$,
$[0, 6.050000002, -12.10000000, 6.050000002, 0, 0, 0, 0, 0, 0]$,
$[0, 0, 6.050000002, -12.10000000, 6.050000002, 0, 0, 0, 0, 0]$,
$[0, 0, 0, 6.050000002, -12.10000000, 6.050000002, 0, 0, 0, 0]$,
$[0, 0, 0, 0, 6.050000002, -12.10000000, 6.050000002, 0, 0, 0]$,
$[0, 0, 0, 0, 0, 6.050000002, -12.10000000, 6.050000002, 0, 0]$,
$[0, 0, 0, 0, 0, 0, 6.050000002, -12.10000000, 6.050000002, 0]$,
$[0, 0, 0, 0, 0, 0, 0, 6.050000002, -12.10000000, 6.050000002]$,
$[0, 0, 0, 0, 0, 0, 0, 0, 6.050000002, -12.10000000]]$

```
> if N > 4 then A:=map(evalf,A);end:
> evalm(A);
```

$[[-4.033333334, 4.033333334, 0., 0., 0., 0., 0., 0., 0., 0.]$,
$[6.050000002, -12.10000000, 6.050000002, 0., 0., 0., 0., 0., 0., 0.]$,
$[0., 6.050000002, -12.10000000, 6.050000002, 0., 0., 0., 0., 0., 0.]$,
$[0., 0., 6.050000002, -12.10000000, 6.050000002, 0., 0., 0., 0., 0.]$,
$[0., 0., 0., 6.050000002, -12.10000000, 6.050000002, 0., 0., 0., 0.]$,
$[0., 0., 0., 0., 6.050000002, -12.10000000, 6.050000002, 0., 0., 0.]$,
$[0., 0., 0., 0., 0., 6.050000002, -12.10000000, 6.050000002, 0., 0.]$,
$[0., 0., 0., 0., 0., 0., 6.050000002, -12.10000000, 6.050000002, 0.]$,
$[0., 0., 0., 0., 0., 0., 0., 6.050000002, -12.10000000, 6.050000002]$,
$[0., 0., 0., 0., 0., 0., 0., 0., 6.050000002, -12.10000000]]$

```
> mat:=exponential(A,t):
> mat:=map(evalf,mat):
> Y0:=matrix(N,1):for i from 1 to N do
Y0[i,1]:=evalf(subs(x=i*h,rhs(IC)));od:evalm(Y0);
```

$$\begin{bmatrix} 0.9898214419 \\ 0.9594929736 \\ 0.9096319953 \\ 0.8412535328 \\ 0.7557495745 \\ 0.6548607337 \\ 0.5406408173 \\ 0.4154150135 \\ 0.2817325565 \\ 0.1423148380 \end{bmatrix}$$

> Y:=evalm(mat&*Y0):
> Y:=map(simplify,Y);

$Y := [[3.874\ 10^{-8}\ e^{-23.63625745\ t} - 2.6083\ 10^{-7}\ e^{-21.99944289\ t} + 4.3738\ 10^{-7}\ e^{-19.44730116\ t}$

$+ 8.3849\ 10^{-7}\ e^{-16.22490629\ t} + 0.00000203523\ e^{-12.63953008\ t} + 0.00000401314\ e^{-9.028494431\ t}$

$+ 0.0000097097\ e^{-5.722717017\ t} + 0.00002568336\ e^{-3.009614211\ t} + 0.000098799274\ e^{-1.101796262\ t}$

$+ 0.9896801601\ e^{-0.1232735457\ t}]$

$[-1.3283\ 10^{-7}\ e^{-23.63625745\ t} - 0.00000106632\ e^{-21.99944289\ t} - 0.00000127240\ e^{-19.44730116\ t}$

$- 0.00000256339\ e^{-16.22490629\ t} - 0.000003544791\ e^{-12.63953008\ t} - 0.00000480002\ e^{-9.028494431\ t}$

$- 0.00000361535\ e^{-5.722717017\ t} + 0.00000634806\ e^{-3.009614211\ t} + 0.00007172044\ e^{-1.101796262\ t}$

$+ 0.9594319078\ e^{-0.1232735457\ t}]$

$[2.8098\ 10^{-7}\ e^{-23.63625745\ t} + 3.463\ 10^{-7}\ e^{-21.99944289\ t} + 0.0000014775\ e^{-19.44730116\ t}$

$+ 6.7690\ 10^{-7}\ e^{-16.22490629\ t} - 0.000001077036\ e^{-12.63953008\ t} - 0.00000624224\ e^{-9.028494431\ t}$

$- 0.0000131827\ e^{-5.722717017\ t} - 0.00001627944\ e^{-3.009614211\ t} + 0.00003153605\ e^{-1.101796262\ t}$

$+ 0.9096344779\ e^{-0.1232735457\ t}]$

$[-3.0377\ 10^{-7}\ e^{-23.63625745\ t} - 0.0000012950\ e^{-21.99944289\ t} - 4.2666\ 10^{-8}\ e^{-19.44730116\ t}$

$+ 0.00000156056\ e^{-16.22490629\ t} + 0.0000044547\ e^{-12.63953008\ t} + 0.00000198327\ e^{-9.028494431\ t}$

$- 0.000009902638\ e^{-5.722717017\ t} - 0.0000309767\ e^{-3.009614211\ t} - 0.000014416716\ e^{-1.101796262\ t}$

$+ 0.8413025228\ e^{-0.1232735457\ t}]$

$[4.1045\ 10^{-7}\ e^{-23.63625745\ t} + 7.714\ 10^{-9}\ e^{-21.99944289\ t} - 8.9017\ 10^{-7}\ e^{-19.44730116\ t}$

$- 0.00000259910\ e^{-16.22490629\ t} + 0.00000168819\ e^{-12.63953008\ t} + 0.0000077293\ e^{-9.028494431\ t}$

$+ 0.00000315510\ e^{-5.722717017\ t} - 0.00003044509\ e^{-3.009614211\ t} - 0.00005775664\ e^{-1.101796262\ t}$

$+ 0.7558283513\ e^{-0.1232735457\ t}]$

$[-3.838\ 10^{-7}\ e^{-23.63625745\ t} - 2.8438\ 10^{-7}\ e^{-21.99944289\ t} + 0.00000164389\ e^{-19.44730116\ t}$

$- 8.68593\ 10^{-7}\ e^{-16.22490629\ t} - 0.00000345417\ e^{-12.63953008\ t} + 0.00000250530\ e^{-9.028494431\ t}$

$+ 0.0000136316\ e^{-5.722717017\ t} - 0.00001496036\ e^{-3.009614211\ t} - 0.00009059476\ e^{-1.101796262\ t}$

$+ 0.6549535629\ e^{-0.1232735457\ t}]$

$[3.78327\ 10^{-7}\ e^{-23.63625745\ t} - 8.381\ 10^{-7}\ e^{-21.99944289\ t} - 6.5485\ 10^{-7}\ e^{-19.44730116\ t}$

5.1 Semianalytical Method for Parabolic Partial Differential Equations (PDEs)

$+ 0.0000020400 \, e^{-16.22490629 \, t} - 2.163191 \, 10^{-7} \, e^{-12.63953008 \, t} - 0.0000058573 \, e^{-9.028494431 \, t}$

$+ 0.00001158066 \, e^{-5.722717017 \, t} + 0.00000777327 \, e^{-3.009614211 \, t} - 0.0001069645 \, e^{-1.101796262 \, t}$

$+ 0.5407335620 \, e^{-0.1232735457 \, t}]$

$[-3.3037 \, 10^{-7} \, e^{-23.63625745 \, t} + 6.151 \, 10^{-7} \, e^{-21.99944289 \, t} - 4.8967 \, 10^{-7} \, e^{-19.44730116 \, t}$

$- 0.00000159382 \, e^{-16.22490629 \, t} + 0.0000045009 \, e^{-12.63953008 \, t} - 0.0000049180 \, e^{-9.028494431 \, t}$

$- 0.00000113087 \, e^{-5.722717017 \, t} + 0.0000264453 \, e^{-3.009614211 \, t} - 0.0001038854 \, e^{-1.101796262 \, t}$

$+ 0.4154956788 \, e^{-0.1232735457 \, t}]$

$[2.1380 \, 10^{-7} \, e^{-23.63625745 \, t} - 9.872 \, 10^{-7} \, e^{-21.99944289 \, t} + 0.00000151806 \, e^{-19.44730116 \, t}$

$- 0.00000183848 \, e^{-16.22490629 \, t} + 5.9985 \, 10^{-7} \, e^{-12.63953008 \, t} + 0.00000382424 \, e^{-9.028494431 \, t}$

$- 0.0000125561 \, e^{-5.722717017 \, t} + 0.0000317937 \, e^{-3.009614211 \, t} - 0.00008191467 \, e^{-1.101796262 \, t}$

$+ 0.2817917411 \, e^{-0.1232735457 \, t}]$

$[-1.3112 \, 10^{-7} \, e^{-23.63625745 \, t} + 4.1057 \, 10^{-7} \, e^{-21.99944289 \, t} - 0.00000117061 \, e^{-19.44730116 \, t}$

$+ 0.00000221781 \, e^{-16.22490629 \, t} - 0.0000040503 \, e^{-12.63953008 \, t} + 0.00000718977 \, e^{-9.028494431 \, t}$

$- 0.00001197020 \, e^{-5.722717017 \, t} + 0.00002120177 \, e^{-3.009614211 \, t} - 0.00004505144 \, e^{-1.101796262 \, t}$

$+ 0.1423460760 \, e^{-0.1232735457 \, t}]]$

```
> for i from 1 to N do u[i](t):=evalf((Y[i,1]));od:
> for i from 0 to N+1 do u[i](t):=eval(u[i](t));od;
```

$u_0(t) := 9.593000000 \, 10^{-8} \, e^{-23.63625745 \, t} + 7.6666667 \, 10^{-9} \, e^{-21.99944289 \, t} + 0.000001007306667 \, e^{-19.44730116 \, t}$

$+ 0.000001972450000 \, e^{-16.22490629 \, t} + 0.000003895237000 \, e^{-12.63953008 \, t}$

$+ 0.000006950860000 \, e^{-9.028494431 \, t} + 0.00001415138334 \, e^{-5.722717017 \, t} + 0.00003212846000 \, e^{-3.009614211 \, t}$

$+ 0.0001078255520 \, e^{-1.101796262 \, t} + 0.9997629111 \, e^{-0.1232735457 \, t}$

$u_1(t) := 3.874 \, 10^{-8} \, e^{-23.63625745 \, t} - 2.6083 \, 10^{-7} \, e^{-21.99944289 \, t} + 4.3738 \, 10^{-7} \, e^{-19.44730116 \, t}$

$+ 8.3849 \, 10^{-7} \, e^{-16.22490629 \, t} + 0.00000203523 \, e^{-12.63953008 \, t} + 0.00000401314 \, e^{-9.028494431 \, t}$

$+ 0.0000097097 \, e^{-5.722717017 \, t} + 0.00002568336 \, e^{-3.009614211 \, t} + 0.000098799274 \, e^{-1.101796262 \, t}$

$+ 0.9896801601 \, e^{-0.1232735457 \, t}$

$$u_2(t) := -1.3283 \cdot 10^{-7} e^{-23.63625745\,t} - 0.00000106632\, e^{-21.99944289\,t} - 0.00000127240\, e^{-19.44730116\,t}$$

$$- 0.00000256339\, e^{-16.22490629\,t} - 0.000003544791\, e^{-12.63953008\,t} - 0.00000480002\, e^{-9.028494431\,t}$$

$$- 0.00000361535\, e^{-5.722717017\,t} + 0.00000634806\, e^{-3.009614211\,t} + 0.00007172044\, e^{-1.101796262\,t}$$

$$+ 0.9594319078\, e^{-0.1232735457\,t}$$

$$u_3(t) := 2.8098 \cdot 10^{-7} e^{-23.63625745\,t} + 3.463 \cdot 10^{-7} e^{-21.99944289\,t} + 0.0000014775\, e^{-19.44730116\,t}$$

$$+ 6.7690 \cdot 10^{-7} e^{-16.22490629\,t} - 0.000001077036\, e^{-12.63953008\,t} - 0.00000624224\, e^{-9.028494431\,t}$$

$$- 0.0000131827\, e^{-5.722717017\,t} - 0.00001627944\, e^{-3.009614211\,t} + 0.00003153605\, e^{-1.101796262\,t}$$

$$+ 0.9096344779\, e^{-0.1232735457\,t}$$

$$u_4(t) := -3.0377 \cdot 10^{-7} e^{-23.63625745\,t} - 0.0000012950\, e^{-21.99944289\,t} - 4.2666 \cdot 10^{-8} e^{-19.44730116\,t}$$

$$+ 0.00000156056\, e^{-16.22490629\,t} + 0.0000044547\, e^{-12.63953008\,t} + 0.00000198327\, e^{-9.028494431\,t}$$

$$- 0.000009902638\, e^{-5.722717017\,t} - 0.0000309767\, e^{-3.009614211\,t} - 0.000014416716\, e^{-1.101796262\,t}$$

$$+ 0.8413025228\, e^{-0.1232735457\,t}$$

$$u_5(t) := 4.1045 \cdot 10^{-7} e^{-23.63625745\,t} + 7.714 \cdot 10^{-9} e^{-21.99944289\,t} - 8.9017 \cdot 10^{-7} e^{-19.44730116\,t}$$

$$- 0.00000259910\, e^{-16.22490629\,t} + 0.00000168819\, e^{-12.63953008\,t} + 0.0000077293\, e^{-9.028494431\,t}$$

$$+ 0.00000315510\, e^{-5.722717017\,t} - 0.00003044509\, e^{-3.009614211\,t} - 0.00005775664\, e^{-1.101796262\,t}$$

$$+ 0.7558283513\, e^{-0.1232735457\,t}$$

$$u_6(t) := -3.838 \cdot 10^{-7} e^{-23.63625745\,t} - 2.8438 \cdot 10^{-7} e^{-21.99944289\,t} + 0.00000164389\, e^{-19.44730116\,t}$$

$$- 8.68593 \cdot 10^{-7} e^{-16.22490629\,t} - 0.00000345417\, e^{-12.63953008\,t} + 0.00000250530\, e^{-9.028494431\,t}$$

$$+ 0.0000136316\, e^{-5.722717017\,t} - 0.00001496036\, e^{-3.009614211\,t} - 0.00009059476\, e^{-1.101796262\,t}$$

$$+ 0.6549535629\, e^{-0.1232735457\,t}$$

$$u_7(t) := 3.78327 \cdot 10^{-7} e^{-23.63625745\,t} - 8.381 \cdot 10^{-7} e^{-21.99944289\,t} - 6.5485 \cdot 10^{-7} e^{-19.44730116\,t}$$

$$+ 0.0000020400\, e^{-16.22490629\,t} - 2.163191 \cdot 10^{-7} e^{-12.63953008\,t} - 0.0000058573\, e^{-9.028494431\,t}$$

$$+ 0.00001158066\, e^{-5.722717017\,t} + 0.00000777327\, e^{-3.009614211\,t} - 0.0001069645\, e^{-1.101796262\,t}$$

$$+ 0.5407335620\, e^{-0.1232735457\,t}$$

5.1 Semianalytical Method for Parabolic Partial Differential Equations (PDEs) 423

$u_8(t) := -3.3037\ 10^{-7}\ e^{-23.63625745\,t} + 6.151\ 10^{-7}\ e^{-21.99944289\,t} - 4.8967\ 10^{-7}\ e^{-19.44730116\,t}$

$- 0.00000159382\ e^{-16.22490629\,t} + 0.0000045009\ e^{-12.63953008\,t} - 0.0000049180\ e^{-9.028494431\,t}$

$- 0.00000113087\ e^{-5.722717017\,t} + 0.0000264453\ e^{-3.009614211\,t} - 0.0001038854\ e^{-1.101796262\,t}$

$+ 0.4154956788\ e^{-0.1232735457\,t}$

$u_9(t) := 2.1380\ 10^{-7}\ e^{-23.63625745\,t} - 9.872\ 10^{-7}\ e^{-21.99944289\,t} + 0.00000151806\ e^{-19.44730116\,t}$

$- 0.00000183848\ e^{-16.22490629\,t} + 5.9985\ 10^{-7}\ e^{-12.63953008\,t} + 0.00000382424\ e^{-9.028494431\,t}$

$- 0.0000125561\ e^{-5.722717017\,t} + 0.0000317937\ e^{-3.009614211\,t} - 0.00008191467\ e^{-1.101796262\,t}$

$+ 0.2817917411\ e^{-0.1232735457\,t}$

$u_{10}(t) := -1.3112\ 10^{-7}\ e^{-23.63625745\,t} + 4.1057\ 10^{-7}\ e^{-21.99944289\,t} - 0.00000117061\ e^{-19.44730116\,t}$

$+ 0.00000221781\ e^{-16.22490629\,t} - 0.0000040503\ e^{-12.63953008\,t} + 0.00000718977\ e^{-9.028494431\,t}$

$- 0.00001197020\ e^{-5.722717017\,t} + 0.00002120177\ e^{-3.009614211\,t} - 0.00004505144\ e^{-1.101796262\,t}$

$+ 0.1423460760\ e^{-0.1232735457\,t}$

$$u_{11}(t) := 0$$

> tf:=20;

$$tf := 20$$

Set the color list:

> setcolors(["Red", "Blue", "LimeGreen", "Goldenrod", "maroon", "DarkTurquoise", "coral", "aquamarine", "magenta", "khaki", "sienna", "orange", "yellow", "gray"]);

["Red", "LimeGreen", "Goldenrod", "Blue", "MediumOrchid",

"DarkTurquoise"]

> pp:=plot([seq(u[i](t),i=0..N+1)],t=0..tf,thickness=3);

$$pp := PLOT(\ldots)$$

Plot the texts:

> arw:=arrow(<6.0,0.55>,<-5.0,-0.57>,width=[1/1000,relative=true], head_width=[1/50,relative=false],head_length=[1/5,relative=false]):

 pt:=textplot([[11,0.6,typeset("Follow the arrow: ",u[0],"(t), ..., ", u[N+1],"(t).")]]):

> display([pp,pt,arw],title="Figure 5.25",axes=boxed,labels=[t,"u"]);

424   5 Method of Lines for Parabolic Partial Differential Equations

Follow the arrow: $u_0(t), ..., u_{11}(t)$.

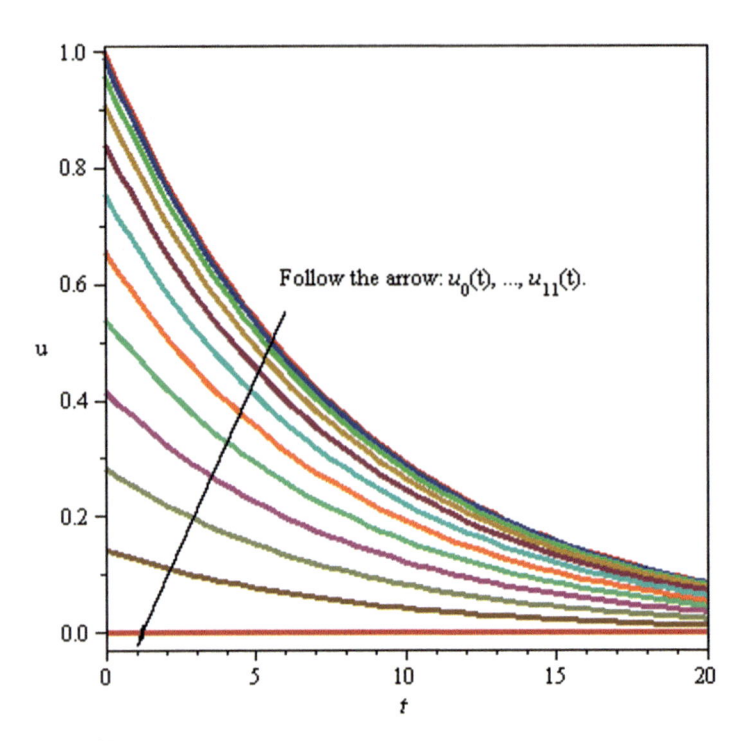

Fig. 5.25

> tf:=20;

$$tf := 20$$

> M:=30;

$$M := 30$$

> T1:=[seq(tf*i/M,i=0..M)];

$$T1 := \left[0, \frac{2}{3}, \frac{4}{3}, 2, \frac{8}{3}, \frac{10}{3}, 4, \frac{14}{3}, \frac{16}{3}, 6, \frac{20}{3}, \frac{22}{3}, 8, \frac{26}{3}, \frac{28}{3}, 10, \frac{32}{3}, \frac{34}{3}, 12, \frac{38}{3}, \frac{40}{3}, 14, \frac{44}{3}, \frac{46}{3}, 16, \frac{50}{3}, \frac{52}{3}, 18, \frac{56}{3}, \frac{58}{3}, 20\right]$$

> PP:=matrix(N+2,M+1);

$$PP := array(1..12, 1..31, [\,])$$

> for i from 1 to N+2 do PP[i,1]:=evalf(subs(x=(i-1)*h,rhs(IC)));od:
> for i from 1 to N+2 do for j from 2 to M+1 do
PP[i,j]:=evalf(subs(t=T1[j],u[i-1](t)));od;od:
> plotdata:=[seq([seq([(i-1)*h,T1[j],PP[i,j]], i=1..N+2)], j=1..M+1)]:
> surfdata(plotdata,axes=boxed,title="Figure 5.26",
labels=[x,t,u],orientation=[45,60]);

5.1 Semianalytical Method for Parabolic Partial Differential Equations (PDEs) 425

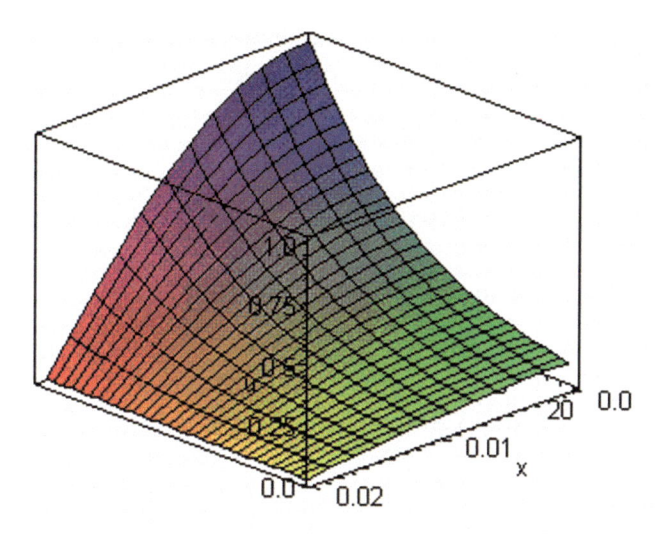

Fig. 5.26

5.1.4 Semianalytical Method for PDEs in Composite Domains

The semianalytical method developed earlier can be used to solve partial differential equations in composite domains also. Mass or heat transfer in composite domains involves two different diffusion coefficients or thermal conductivities in the two layers of the composite material.[6] In addition, even in case of solids with a single domain and constant physical properties, the reaction may take place mainly near the surface. This leads to the formation of boundary layer near one of the boundaries. In this section, the semianalytical method developed earlier is extended to composite domains.

Example 5.8

For example, in the diffusion reaction problem solved in example 5.5 for higher values of Thiele modulus ($\Phi > 10$), concentration depletes very close to the surface ($x = 0$). Since we are interested in the flux at $x = 0$ it makes more sense to choose more node points near $x = 0$. Equation (5.28) can be rewritten as

$$\frac{\partial u_1}{\partial t} = \frac{\partial^2 u_1}{\partial x^2} - \Phi^2 u_1, \quad 0 < x < \alpha \text{ (Region 1)}$$

$$\frac{\partial u_2}{\partial t} = \frac{\partial^2 u_2}{\partial x^2} - \Phi^2 u_2, \quad \alpha < x < 1 \text{ (Region 2)} \tag{5.33}$$

$$u_1(x,0) = u_2(x,0) = 0$$

where u1 and u2 are the dependent variables in region 1 ($0 < x < \alpha$) and region 2 ($\alpha < x < 1$), respectively. The boundary conditions at $x = 0$ and $x = 1$ are:

$$u_1(0,t) = 1 \tag{5.34}$$

$$\frac{\partial u_2}{\partial x}(1,t) = 0 \tag{5.35}$$

Both the dependent variable and its derivative are continuous at $x = \alpha$:

$$u_1(\alpha,t) = u_2(\alpha,t) \tag{5.36}$$

$$\frac{\partial u_1}{\partial x}(\alpha,x) = \frac{\partial u_2}{\partial x}(\alpha,x) \tag{5.37}$$

When mass or heat transfer in composite domains are modeled, different thermal diffusivities or diffusion coefficients enter in the governing equation for each region and the mass/heat flux is continuous at $x = \alpha$. Equation (5.33) can be converted to the finite difference form as:

$$\frac{\partial u_i}{\partial t} = \frac{u_{i+1} - 2u_i + u_{i-1}}{h_1^2} - \Phi^2 u_i \quad i = 1..N \text{ [Region 1]}$$

$$\tag{5.38}$$

$$\frac{\partial u_i}{\partial t} = \frac{u_{i+1} - 2u_i + u_{i-1}}{h_2^2} - \Phi^2 u_i \quad i = N+2..N+M+1 \text{ [Region 2]}$$

where N and M are the number of interior node points used in region 1 and 2, respectively, and the node spacing in each region are defined by:

$$h_1 = \frac{\alpha}{N+1}; \text{ and } h_2 = \frac{1-\alpha}{M+1} \tag{5.39}$$

We are using the same dependent variable u_i at interior node points, for both u_1 and u_2 in equation (5.38) for convenience. This satisfies the continuity of dependent variable at $x = \alpha$ (equation (5.36)) by default. The initial conditions are

$$u_i(0) = 0 \quad i = 1..N, \, i = N+2..N+M+1 \tag{5.40}$$

Equation (5.36) can be written in finite difference form as:

$$\frac{1}{2}\frac{3u_{N+1} - 4u_N + u_{N-1}}{h_1} = \frac{1}{2}\frac{-3u_{N+1} + 4u_{N+2} - u_{N+3}}{h_2} \tag{5.41}$$

The procedure involved in solving PDEs in a composite domain can be summarized as follows:

5.1 Semianalytical Method for Parabolic Partial Differential Equations (PDEs)

1. Start the Maple program with a 'restart' command to clear all variables.
2. Call 'with(linalg)' and 'with(plots)' commands.
3. Enter the governing equations in two regions.
4. Enter the value for α.
5. Enter the initial conditions for the both the regions.
6. Enter the boundary conditions at $x = 0$ and $x = 1$.
7. Enter the number of interior node points N, M for region 1 and region 2, respectively.
8. Enter the boundary condition at $x = \alpha$ (equation (5.36)).
9. Convert the governing equations in both the regions to the finite difference form.
10. Convert all the boundary conditions to the finite difference form.
11. Eliminate the boundary values ($u_0(t)$ and $u_{N+M+2}(t)$) using the boundary conditions at $x = 0$ and $x = 1$, respectively.
12. Eliminate $u_{N+1}(t)$ i.e., the dependent variable at $x = \alpha$ using the boundary condition at $x = \alpha$ (equation 5.36 and 5.40).
13. Store the dependent variables, u_i, $i = 1..N, N+2..N+M+2$ in vars.
14. Generate **A** matrix and **b** vector using Maple's 'genmatrix'.
15. Find the semianalytical solution and plot your results.
16. Equation 5.37 is solved below in Maple for $\Phi = 5$ using $\alpha = 0.5$:

> restart;

> with(linalg):with(plots):

> ge1:=diff(u1(x,t),t)=diff(u1(x,t),x$2)-Phi^2*u1(x,t);

$$ge1 := \frac{\partial}{\partial t} u1(x,t) = \frac{\partial^2}{\partial x^2} u1(x,t) - \Phi^2 u1(x,t)$$

> ge2:=diff(u2(x,t),t)=diff(u2(x,t),x$2)-Phi^2*u2(x,t);

$$ge2 := \frac{\partial}{\partial t} u2(x,t) = \frac{\partial^2}{\partial x^2} u2(x,t) - \Phi^2 u2(x,t)$$

> bc1:=u1(x,t)-1;

$$bc1 := u1(x,t) - 1$$

> bcalpha:=diff(u1(x,t),x)=diff(u2(x,t),x);

$$bcalpha := \frac{\partial}{\partial x} u1(x,t) = \frac{\partial}{\partial x} u2(x,t)$$

> bc2:=diff(u2(x,t),x);

$$bc2 := \frac{\partial}{\partial x} u2(x,t)$$

> IC1:=u1(x,0)=0;

$$IC1 := u1(x, 0) = 0$$

> IC2:=u2(x,0)=0;

$$IC2 := u2(x, 0) = 0$$

The boundary condition u1(alpha,t) = u2(alpha,t) is satisfied by default. Let N and M be the number of node points in region 1 and region 2, respectively.

> N:=5;

$$N := 5$$

> M:=5;

$$M := 5$$

> alpha:=0.25;

$$\alpha := 0.25$$

> Phi:=5;

$$\Phi := 5$$

> dydxf:=1/2*(-u[2](t)-3*u[0](t)+4*u[1](t))/h1;

$$dydxf := \frac{1}{2} \frac{-u_2(t) - 3 u_0(t) + 4 u_1(t)}{h1}$$

> dydxb:=1/2*(u[N+M](t)+3*u[N+M+2](t)-4*u[N+M+1](t))/h2;

$$dydxb := \frac{1}{2} \frac{u_{10}(t) + 3 u_{12}(t) - 4 u_{11}(t)}{h2}$$

> dydxb2:=1/2*(u[N-1](t)+3*u[N+1](t)-4*u[N](t))/h1;

$$dydxb2 := \frac{1}{2} \frac{u_4(t) + 3 u_6(t) - 4 u_5(t)}{h1}$$

> dydxf2:=1/2*(-u[N+3](t)-3*u[N+1](t)+4*u[N+2](t))/h2;

$$dydxf2 := \frac{1}{2} \frac{-u_8(t) - 3 u_6(t) + 4 u_7(t)}{h2}$$

The first and second derivatives in the governing equation are converted to finite difference form using the following finite difference approximations:

> dydx:=piecewise(i<N+1,1/2/h1*(u[m+1](t)-u[m-1](t)),i>N+1,
1/2/h2*(u[m+1](t)-u[m-1](t)));

5.1 Semianalytical Method for Parabolic Partial Differential Equations (PDEs)

$$dydx := \begin{cases} \dfrac{1}{2} \dfrac{u_{m+1}(t) - u_{m-1}(t)}{h1} & i < 6 \\ \dfrac{1}{2} \dfrac{u_{m+1}(t) - u_{m-1}(t)}{h2} & 6 < i \end{cases}$$

Maple's 'piecewise' command is useful in defining the first derivative and derivatives in different regions:

> d2ydx2:=piecewise(i<N+1,1/h1^2*(u[m-1](t)-2*u[m](t)+u[m+1](t)),i>N+1,1/h2^2*(u[m-1](t)-2*u[m](t)+u[m+1](t)));

$$d2ydx2 := \begin{cases} \dfrac{u_{m-1}(t) - 2u_m(t) + u_{m+1}(t)}{h1^2} & i < 6 \\ \dfrac{u_{m-1}(t) - 2u_m(t) + u_{m+1}(t)}{h2^2} & 6 < i \end{cases}$$

> bc1:=subs(diff(u1(x,t),x)=dydxf,u1(x,t)=u[0](t),x=0,bc1);

$$bc1 := u_0(t) - 1$$

> bc2:=subs(diff(u2(x,t),x)=dydxb,u2(x,t)=u[N+M+2](t),bc2);

$$bc2 := \dfrac{1}{2} \dfrac{u_{10}(t) + 3u_{12}(t) - 4u_{11}(t)}{h2}$$

The boundary condition at the interface, $x = \alpha$ is:

> bcalpha:=subs(diff(u1(x,t),x)=dydxb2,diff(u2(x,t),x)=dydxf2,u1(x,t)=u[N+1](t),u2(x,t)=u[N+1](t),bcalpha);

$$bcalpha := \dfrac{1}{2} \dfrac{u_4(t) + 3u_6(t) - 4u_5(t)}{h1} = \dfrac{1}{2} \dfrac{-u_8(t) - 3u_6(t) + 4u_7(t)}{h2}$$

> eq[0]:=bc1;

$$eq_0 := u_0(t) - 1$$

> eq[N+1]:=bcalpha;

$$eq_6 := \dfrac{1}{2} \dfrac{u_4(t) + 3u_6(t) - 4u_5(t)}{h1} = \dfrac{1}{2} \dfrac{-u_8(t) - 3u_6(t) + 4u_7(t)}{h2}$$

> eq[N+M+2]:=bc2;

$$eq_{12} := \dfrac{1}{2} \dfrac{u_{10}(t) + 3u_{12}(t) - 4u_{11}(t)}{h2}$$

For the example given, the governing equation does not depend on x explicitly. However, for a general case (example 5.6), x appears in the governing equation explicitly. Hence, when the governing equation is converted to finite difference form the independent variable, x, has to be expressed in the finite difference form appropriately.

```
> for i from 1 to N do eq[i]:=diff(u[i](t),t)= subs(diff(u1(x,t),x$2) =
subs(m=i,d2ydx2),diff(u1(x,t),x) = subs(m=i,dydx),u1(x,t)=u[i](t),
x=i*h1,rhs(ge1));od;
```

$$eq_1 := \frac{d}{dt} u_1(t) = \frac{u_0(t) - 2u_1(t) + u_2(t)}{h1^2} - 25 u_1(t)$$

$$eq_2 := \frac{d}{dt} u_2(t) = \frac{u_1(t) - 2u_2(t) + u_3(t)}{h1^2} - 25 u_2(t)$$

$$eq_3 := \frac{d}{dt} u_3(t) = \frac{u_2(t) - 2u_3(t) + u_4(t)}{h1^2} - 25 u_3(t)$$

$$eq_4 := \frac{d}{dt} u_4(t) = \frac{u_3(t) - 2u_4(t) + u_5(t)}{h1^2} - 25 u_4(t)$$

$$eq_5 := \frac{d}{dt} u_5(t) = \frac{u_4(t) - 2u_5(t) + u_6(t)}{h1^2} - 25 u_5(t)$$

Eq[N+1] is given the boundary condition at the interface between region 1 and region 2. The finite difference equations for region 2 are:

```
> for i from N+2 to N+M+1 do eq[i]:=diff(u[i](t),t)= subs(diff(u2(x,t),x$2) =
subs(m=i,d2ydx2),diff(u2(x,t),x) = subs(m=i,dydx),u2(x,t)=u[i](t),
x=alpha+(i-N-1)*h2,rhs(ge2));od;
```

$$eq_7 := \frac{d}{dt} u_7(t) = \frac{u_6(t) - 2u_7(t) + u_8(t)}{h2^2} - 25 u_7(t)$$

$$eq_8 := \frac{d}{dt} u_8(t) = \frac{u_7(t) - 2u_8(t) + u_9(t)}{h2^2} - 25 u_8(t)$$

$$eq_9 := \frac{d}{dt} u_9(t) = \frac{u_8(t) - 2u_9(t) + u_{10}(t)}{h2^2} - 25 u_9(t)$$

$$eq_{10} := \frac{d}{dt} u_{10}(t) = \frac{u_9(t) - 2u_{10}(t) + u_{11}(t)}{h2^2} - 25 u_{10}(t)$$

$$eq_{11} := \frac{d}{dt} u_{11}(t) = \frac{u_{10}(t) - 2u_{11}(t) + u_{12}(t)}{h2^2} - 25 u_{11}(t)$$

5.1 Semianalytical Method for Parabolic Partial Differential Equations (PDEs)

The dependent variable at the boundary $x = 0$ is eliminated as:

> u[0](t):=(solve(eq[0],u[0](t)));

$$u_0(t) := 1$$

The dependent variable at the boundary $x = \alpha$ is eliminated as:

> u[N+1](t):=solve(eq[N+1],u[N+1](t));

$$u_6(t) := -\frac{1}{3} \frac{h2\, u_4(t) - 4\, h2\, u_5(t) + h1\, u_8(t) - 4\, h1\, u_7(t)}{h2 + h1}$$

The dependent variable at the boundary $x = 1$ is eliminated as:

> u[N+M+2](t):=solve(eq[N+M+2],u[N+M+2](t));

$$u_{12}(t) := -\frac{1}{3} u_{10}(t) + \frac{4}{3} u_{11}(t)$$

> for i from 1 to N do eq[i]:=eval(eq[i]):od:for i from N+2 to N+M+1 do eq[i]:=eval(eq[i]);od:
> eqs:=[seq(rhs(eq[j]),j=1..N),seq(rhs(eq[N+1+j]),j=1..M)]:
> Y:=[seq(u[i](t),i=1..N),seq(u[N+1+i](t),i=1..M)];

$$Y := [u_1(t), u_2(t), u_3(t), u_4(t), u_5(t), u_7(t), u_8(t), u_9(t), u_{10}(t), u_{11}(t)]$$

> A:=genmatrix(eqs,Y,'b1'):
> b:=matrix(N+M,1):for i to N+M do b[i,1]:=-eval(b1[i]);od:evalm(b);

$$\begin{bmatrix} \frac{1}{h1^2} \\ 0 \\ 0 \\ 0 \\ 0 \\ 0 \\ 0 \\ 0 \\ 0 \\ 0 \end{bmatrix}$$

> h1:=eval(alpha/(N+1));h2:=eval((1-alpha)/(M+1));

$$h1 := 0.04166666667$$

$$h2 := 0.1250000000$$

```
> A:=map(eval,A):
> if N+M >4 then A:=map(evalf,A);end:
> mat:=exponential(A,t):
> mat:=map(evalf,mat):
> Y0:=matrix(N+M,1):for i from 1 to N+M do Y0[i,1]:=0;od:evalm(Y0):
> s1:=evalm(Y0+inverse(A)&*b):
> Y:=evalm(mat&*s1-inverse(A)&*b):
> Y:=map(simplify,Y):
> for i from 1 to N do u[i](t):=evalf((Y[i,1]));od:
> for i from N+2 to N+M+1 do u[i](t):=evalf((Y[i-1,1]));od:
> for i from 0 to N+M+2 do u[i](t):=eval(u[i](t));od;
```

$$u_0(t) := 1$$

$u_1(t) := -0.03091209639\,e^{-2135.706266\,t} - 0.1095917467\,e^{-1627.882051\,t} - 0.1971320872\,e^{-992.4793230\,t}$
$- 0.2317232980\,e^{-453.8126765\,t} - 0.008252005333\,e^{-254.7872600\,t} - 0.04401991117\,e^{-196.4066095\,t}$
$- 0.08906562450\,e^{-142.3665004\,t} - 0.05737130039\,e^{-89.25945415\,t} - 0.03704878998\,e^{-47.17920866\,t}$
$- 0.007427847916\,e^{-27.45398338\,t} + 0.8136647751$

$u_2(t) := 0.05145073177\,e^{-2135.706266\,t} + 0.08578636168\,e^{-1627.882051\,t} - 0.06315086679\,e^{-992.4793230\,t}$
$- 0.2909403193\,e^{-453.8126765\,t} - 0.01310035887\,e^{-254.7872600\,t} - 0.07563469087\,e^{-196.4066095\,t}$
$- 0.1595724959\,e^{-142.3665004\,t} - 0.1079881092\,e^{-89.25945415\,t} - 0.07041383529\,e^{-47.17920866\,t}$
$- 0.01474271868\,e^{-27.45398338\,t} + 0.6626448621$

$u_3(t) := -0.0547235584\,e^{-2135.706266\,t} + 0.04243976573\,e^{-1627.882051\,t} + 0.1769018551\,e^{-992.4793230\,t}$
$- 0.1335596821\,e^{-453.8126765\,t} - 0.01294580961\,e^{-254.7872600\,t} - 0.08394325344\,e^{-196.4066095\,t}$
$- 0.1997386700\,e^{-142.3665004\,t} - 0.1469493354\,e^{-89.25945415\,t} - 0.1039265117\,e^{-47.17920866\,t}$
$- 0.02186419352\,e^{-27.45398338\,t} + 0.5403855767$

$u_4(t) := 0.03963228879\,e^{-2135.706266\,t} - 0.1190074386\,e^{-1627.882051\,t} + 0.1198211774\,e^{-992.4793230\,t}$
$+ 0.1232485900\,e^{-453.8126765\,t} - 0.007366069283\,e^{-254.7872600\,t} - 0.06774836806\,e^{-196.4066095\,t}$
$- 0.1956017917\,e^{-142.3665004\,t} - 0.1698921694\,e^{-89.25945415\,t} - 0.1311603877\,e^{-47.17920866\,t}$
$- 0.02979082442\,e^{-27.45398338\,t} + 0.4415805267$

$u_5(t) := -0.01124123932\,e^{-2135.706266\,t} + 0.05071702281\,e^{-1627.882051\,t} - 0.1385173191\,e^{-992.4793230\,t}$
$+ 0.2882975361\,e^{-453.8126765\,t} + 0.0009494631225\,e^{-254.7872600\,t} - 0.03036675844\,e^{-196.4066095\,t}$
$- 0.1530628463\,e^{-142.3665004\,t} - 0.1728641835\,e^{-89.25945415\,t} - 0.1533983382\,e^{-47.17920866\,t}$
$- 0.03664785705\,e^{-27.45398338\,t} + 0.3619412984$

5.1 Semianalytical Method for Parabolic Partial Differential Equations (PDEs) 433

$u_6(t) := -0.02092214010\, e^{-2135.706266\, t} + 0.07930709060\, e^{-1627.882051\, t} - 0.1641949933\, e^{-992.4793230\, t}$

$+ 0.2387179606\, e^{-453.8126765\, t} + 0.009086019223\, e^{-254.7872600\, t} + 0.01575746081\, e^{-196.4066095\, t}$

$- 0.07961245357\, e^{-142.3665004\, t} - 0.1572161399\, e^{-89.25945415\, t} - 0.1703969609\, e^{-47.17920866\, t}$

$- 0.04356613220\, e^{-27.45398338\, t} + 0.2980113278$

$u_7(t) := 0.000676052988\, e^{-2135.706266\, t} - 0.003447900844\, e^{-1627.882051\, t} + 0.01259146322\, e^{-992.4793230\, t}$

$- 0.05332465560\, e^{-453.8126765\, t} + 0.01185488948\, e^{-254.7872600\, t} + 0.07177020865\, e^{-196.4066095\, t}$

$+ 0.09802573165\, e^{-142.3665004\, t} - 0.05426857189\, e^{-89.25945415\, t} - 0.1817569683\, e^{-47.17920866\, t}$

$- 0.06313235592\, e^{-27.45398338\, t} + 0.1612865893$

$u_8(t) := -0.00002184511233\, e^{-2135.706266\, t} + 0.0001498983846\, e^{-1627.882051\, t} - 0.0009655896652\, e^{-992.4793230\, t}$

$+ 0.01191051274\, e^{-453.8126765\, t} - 0.02812090745\, e^{-254.7872600\, t} - 0.06316469229\, e^{-196.4066095\, t}$

$+ 0.09750358869\, e^{-142.3665004\, t} + 0.1048256976\, e^{-89.25945415\, t} - 0.1295632390\, e^{-47.17920866\, t}$

$- 0.08013764866\, e^{-27.45398338\, t} + 0.08756442500$

$u_9(t) := 7.056595043\, 10^{-7}\, e^{-2135.706266\, t} - 0.000006516968117\, e^{-1627.882051\, t} + 0.00007404630868\, e^{-992.4793230\, t}$

$- 0.002657225756\, e^{-453.8126765\, t} + 0.03283163899\, e^{-254.7872600\, t} - 0.02864891624\, e^{-196.4066095\, t}$

$- 0.08194666516\, e^{-142.3665004\, t} + 0.1589529044\, e^{-89.25945415\, t} - 0.03252068870\, e^{-47.17920866\, t}$

$- 0.09412738437\, e^{-27.45398338\, t} + 0.04804711418$

$u_{10}(t) := -2.287947486\, 10^{-8}\, e^{-2135.706266\, t} + 2.831369768\, 10^{-7}\, e^{-1627.882051\, t} - 0.000005664996920\, e^{-992.4793230\, t}$

$+ 0.0005789528829\, e^{-453.8126765\, t} - 0.02409834130\, e^{-254.7872600\, t} + 0.08260776949\, e^{-196.4066095\, t}$

$- 0.1111397707\, e^{-142.3665004\, t} + 0.05348524739\, e^{-89.25945415\, t} + 0.07578564973\, e^{-47.17920866\, t}$

$- 0.1045123074\, e^{-27.45398338\, t} + 0.02729820735$

$u_{11}(t) := 4.487798877\, 10^{-10}\, e^{-2135.706266\, t} - 7.738918553\, 10^{-9}\, e^{-1627.882051\, t} + 2.613680068\, 10^{-7}\, e^{-992.4793230\, t}$

$- 0.00006397067255\, e^{-453.8126765\, t} + 0.005494828496\, e^{-254.7872600\, t} - 0.02737764444\, e^{-196.4066095\, t}$

$+ 0.06348022895\, e^{-142.3665004\, t} - 0.1056853134\, e^{-89.25945415\, t} + 0.1578291256\, e^{-47.17920866\, t}$

$- 0.1108901702\, e^{-27.45398338\, t} + 0.01721266276$

$u_{12}(t) := 8.224864804\, 10^{-9}\, e^{-2135.706266\, t} - 1.046975503\, 10^{-7}\, e^{-1627.882051\, t} + 0.000002236822983\, e^{-992.4793230\, t}$

$- 0.0002782785244\, e^{-453.8126765\, t} + 0.01535921843\, e^{-254.7872600\, t} - 0.06403944908\, e^{-196.4066095\, t}$

$+ 0.1216868955\, e^{-142.3665004\, t} - 0.1587421670\, e^{-89.25945415\, t} + 0.1851769509\, e^{-47.17920866\, t}$

$- 0.1130161245\, e^{-27.45398338\, t} + 0.01385081456$

> setcolors(["Red", "Blue", "LimeGreen", "Goldenrod", "maroon", "DarkTurquoise", "coral", "aquamarine", "magenta", "khaki", "sienna", "orange", "yellow", "gray"]);

["Red", "LimeGreen", "Goldenrod", "Blue", "MediumOrchid", "DarkTurquoise"]

434   5 Method of Lines for Parabolic Partial Differential Equations

> pp:=plot([seq(u[i](t),i=0..N+M+2)],t=0..0.1);

$$pp := PLOT(...)$$

> pt:=textplot([[0.015,1.0,typeset(u[0],"(t)"),align=below],seq([0.02+i*0.005, evalf(subs(t=0.02+i*0.005,u[i](t))),typeset(u[i],"(t)"),align=above], i=1..N+M-2)]);

$$pt := PLOT(...)$$

> display([pp,pt],title="Figure 5.27",thickness=3,axes=boxed,labels=[t,"u"]);

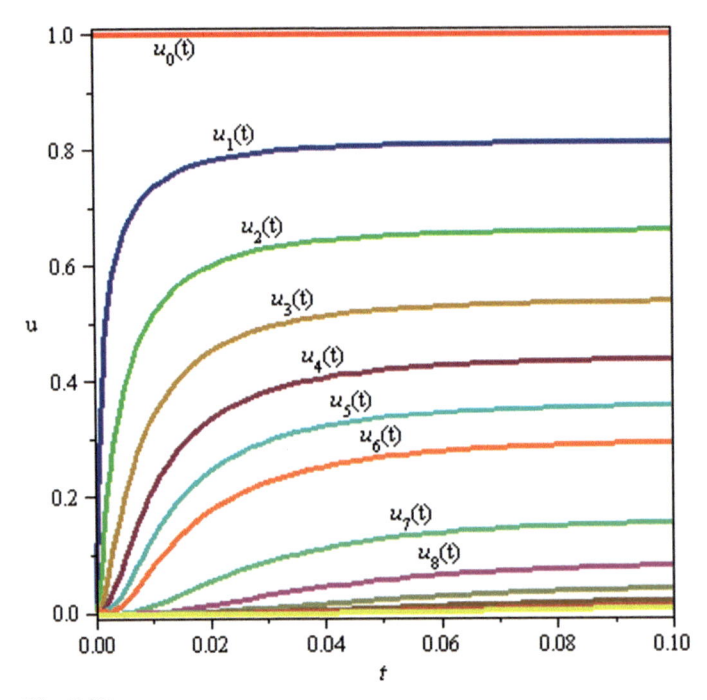

Fig. 5.27

> tf:=0.1;

$$tf := 0.1$$

> MM:=30;

$$MM := 30$$

> T1:=[seq(tf*i/MM,i=0..MM)];

$T1 :=$ [0., 0.003333333333, 0.006666666667, 0.01000000000, 0.01333333333, 0.01666666667, 0.02000000000, 0.02333333333, 0.02666666667, 0.03000000000, 0.03333333333, 0.03666666667, 0.04000000000, 0.04333333333, 0.04666666667, 0.05000000000, 0.05333333333, 0.05666666667, 0.06000000000, 0.06333333333, 0.06666666667, 0.07000000000, 0.07333333333, 0.07666666667, 0.08000000000, 0.08333333333, 0.08666666667, 0.09000000000, 0.09333333333, 0.09666666667, 0.1000000000]

5.1 Semianalytical Method for Parabolic Partial Differential Equations (PDEs)

> PP:=matrix(N+M+3,MM+1);

$$PP := array(1..13, 1..31, [\,])$$

> for i from 1 to N+M+3 do PP[i,1]:=0;od:

> for i from 1 to N+M+3 do for j from 2 to MM+1 do PP[i,j]:=evalf(subs(t=T1[j],u[i-1](t)));od;od:

> evalm(PP):

> i:='i';

$$i := i$$

For making 3-D plots the x coordinate has to be defined appropriately.

> X:=piecewise(i<N+3,(i-1)*h1,i>N+2,alpha+(i-N-2)*h2);

$$X := \begin{cases} 0.04166666667i - 0.04166666667 & i < 8 \\ -0.6250000000 + 0.1250000000i & 7 < i \end{cases}$$

> plot(X,i=1..N+M+3,thickness=4,title="Figure 5.28", axes=boxed,labels=['i',"X"]);

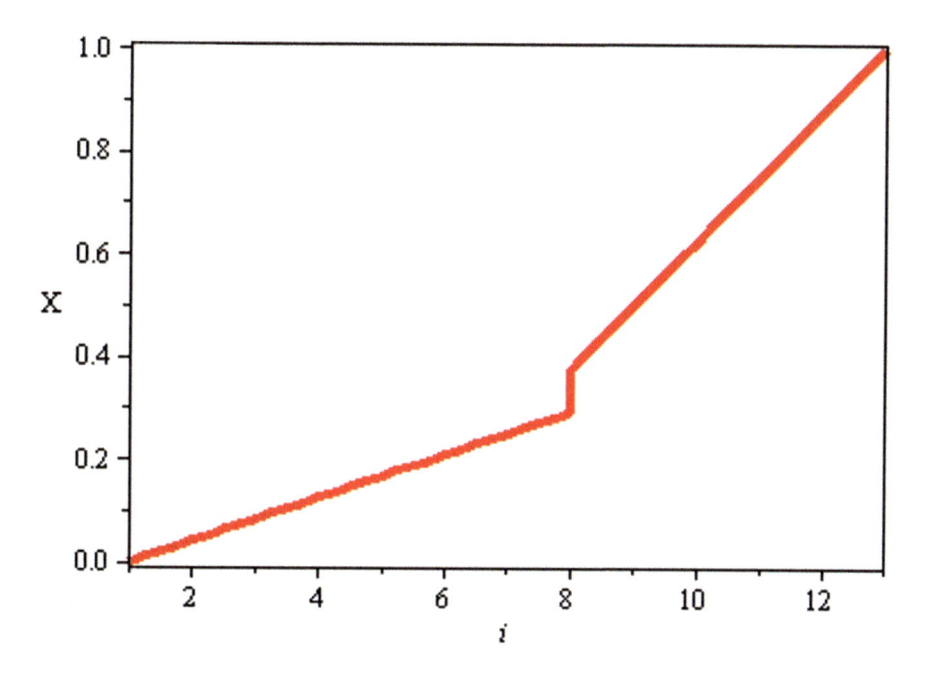

Fig. 5.28

> plotdata:=[seq([seq([eval(X),T1[j],PP[i,j]], i=1..N+M+3)], j=1..MM+1)]:
> surfdata(plotdata,axes=boxed,title="Figure 5.29",
labels=[x,t,u],orientation=[-75,60]);

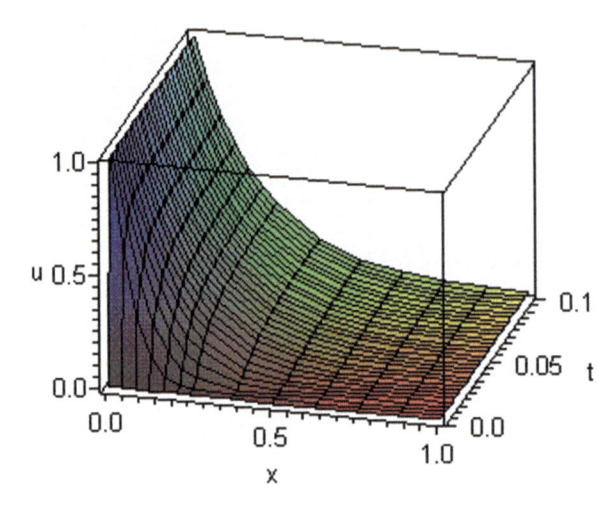

Fig. 5.29

We observe that most of the reaction takes place near the surface, $x = 0$. An analytical solution for the steady state distribution can be obtained as:

> eq:=diff(U(x),x$2)-Phi^2*U(x);

$$eq := \frac{d^2}{dx^2} U(x) - 25\, U(x)$$

> Ua:=rhs(dsolve({eq,U(0)=1,D(U)(1)=0},U(x)));

$$Ua := \frac{e^{-5}\, e^{5x}}{e^5 + e^{-5}} + \frac{e^5\, e^{-5x}}{e^5 + e^{-5}}$$

A more compact analytical solution can be obtained by using the matrix exponential method described in section 2.1.2. The dimensionless mass flux at $x = 0$ is given by:

> Flux:=-evalf(subs(x=0.,diff(Ua,x)));

$$Flux := 4.999546023$$

5.1 Semianalytical Method for Parabolic Partial Differential Equations (PDEs)

```
> flux:=-dydxf;
```

$$flux := 2.101189409\, e^{-2135.706266\, t} + 6.289840180\, e^{-1627.882051\, t} + 8.704529785\, e^{-992.4793230\, t}$$
$$+ 7.631434470\, e^{-453.8126765\, t} + 0.2388919495\, e^{-254.7872600\, t} + 1.205339446\, e^{-196.4066095\, t}$$
$$+ 2.360280025\, e^{-142.3665004\, t} + 1.457965109\, e^{-89.25945415\, t} + 0.9333758955\, e^{-47.17920866\, t}$$
$$+ 0.1796240758\, e^{-27.45398338\, t} + 4.895829144$$

Steady state flux is obtained as:

```
> flux:=limit(flux,t=infinity);Flux;
```

$$flux := 4.895829144$$

$$4.999546023$$

We obtain almost a 5% error with our semianalytical solution. The computation time taken for the semianalytical method depends on the total number of interior node points (N+M). By using $\alpha = 0.25$ without changing N and M, we obtain the following results:

```
> Err:=abs(flux-Flux)/Flux*100;
```

$$Err := 2.074525937$$

```
>
```

The error has reduced significantly from 5% to 2%. Hence, we conclude that stiff PDEs with boundary layers can be solved efficiently by dividing the region as two composite domains.

5.1.5 Expediting the Calculation of Exponential Matrix

We have used Maple to calculate the exponential matrix in all of the above examples. When N increases, the time taken by Maple for calculating the exponential matrix increases drastically. For N = 10, the matrix order is 10 x 10. For this matrix, Maple takes around 1 minute to calculate the exponential matrix in a 2.6 GHz processor with 2GB RAM. For a particular problem, one can derive an analytical expression for the exponential matrix by calculating the eigenvalues and eigenvectors analytically.[13] However, these expressions are valid for a particular problem only. If the governing equation or the boundary conditions change, one has to redo all the steps. This involves tedious algebra. To avoid this, when all the eigenvalues of the coefficient matrix A are distinct, A matrix is converted to canonical form as

$$A = PDP-1 \qquad (5.42)$$

Where **D** is the diagonal matrix of order NxN with the N distinct eigenvalues (λ_k, i = 1..N) as its diagonal elements. **P** is the eigenvector matrix defined as

$$\mathbf{P} = [X_1, X_2, ... X_N] \qquad (5.43)$$

where P_k is the eigenvector corresponding to the eigenvalue λ_k. One of the main advantages of equation (5.42) is that, it simplifies the calculation of exponential matrix as:

$$\exp(\mathbf{A}t) = \mathbf{P}\exp(\mathbf{D}t)\,\mathbf{P}^{-1} \tag{5.44}$$

Since $\mathbf{D}$ is a diagonal matrix, the exponential matrix of $\mathbf{D}$ is easily obtained as:

$$\exp(\mathbf{D}t) = \begin{bmatrix} e^{\lambda_1 t} & 0 & \cdots & 0 & 0 \\ 0 & e^{\lambda_2 t} & \cdots & 0 & 0 \\ \cdots & \cdots & \cdots & \cdots & \cdots \\ 0 & 0 & \cdots & e^{\lambda_{N-1} t} & 0 \\ 0 & 0 & \cdots & 0 & e^{\lambda_N t} \end{bmatrix} \tag{5.45}$$

Maple can be used to obtain the eigenvalues and eigenvector matrix (P). Maple takes only a few seconds to calculate the eigenvalues (for a 20 x 20 matrix it takes less than a second). However, Maple takes a long time to calculate the eigenvector matrix. To overcome this problem, we can obtain the particular eigenvector Xk using the equation

$$\mathbf{G} = (\mathbf{A} - \lambda_k \mathbf{U})\mathbf{X}_k = \mathbf{0} \tag{5.46}$$

where $\mathbf{U}$ is the diagonal matrix of order NxN. We define X_k as

$$X_k = [\beta_1, \beta_2, \ldots \beta_N]^T \tag{5.47}$$

On substituting equation (5.47) into equation (5.46), we obtain equations for β_i, i = 1..N. Next by arbitrarily choosing $\beta_1 = 1$, and by using rows 1 to N-1 of equation (5.46) we can solve β_i, 1 = 2..N and obtain X_k.

The following procedure in Maple can be used to obtain exponential matrix for any matrix with distinct eigenvalues.

Example 5.9

> restart;
> with(linalg):with(plots):
> UseHardwareFloats := true;

$$UseHardwareFloats := true$$

> ge:=diff(u(x,t),t)=diff(u(x,t),x$2);

$$ge := \frac{\partial}{\partial t} u(x,t) = \frac{\partial^2}{\partial x^2} u(x,t)$$

5.1 Semianalytical Method for Parabolic Partial Differential Equations (PDEs) 439

> bc1:=u(x,t);

$$bc1 := u(x, t)$$

> bc2:=u(x,t);

$$bc2 := u(x, t)$$

> IC:=u(x,0)=1;

$$IC := u(x, 0) = 1$$

> N:=40;

$$N := 40$$

> L:=1;

$$L := 1$$

> dydxf:=1/2*(-u[2](t)-3*u[0](t)+4*u[1](t))/h:
> dydxb:=1/2*(u[N-1](t)+3*u[N+1](t)-4*u[N](t))/h:
> dydx:=1/2/h*(u[m+1](t)-u[m-1](t)):
> d2ydx2:=1/h^2*(u[m-1](t)-2*u[m](t)+u[m+1](t)):
> bc1:=subs(diff(u(x,t),x)=dydxf,u(x,t)=u[0](t),x=0,bc1):
> bc2:=subs(diff(u(x,t),x)=dydxb,u(x,t)=u[N+1](t),x=1,bc2):
> eq[0]:=bc1:
> eq[N+1]:=bc2:
> for i from 1 to N do eq[i]:=diff(u[i](t),t)= subs(diff(u(x,t),x$2) = subs(m=i,d2ydx2),diff(u(x,t),x) = subs(m=i,dydx),u(x,t)=u[i](t), x=i*h,rhs(ge));od:
> u[0](t):=(solve(eq[0],u[0](t)));

$$u_0(t) := 0$$

> u[N+1](t):=solve(eq[N+1],u[N+1](t));

$$u_{41}(t) := 0$$

> for i from 1 to N do eq[i]:=eval(eq[i]);od:
> eqs:=[seq(rhs(eq[j]),j=1..N)]:
> Y:=[seq(u[i](t),i=1..N)]:
> A:=genmatrix(eqs,Y,'b1'):
> b:=matrix(N,1):for i to N do b[i,1]:=-eval(b1[i]);od:evalm(b):
> h:=eval(L/(N+1)):
> A:=map(eval,A):

```
> if N > 4 then A:=map(evalf,A);end:
> #evalm(A);
> Nrow:=rowdim(A):
> l:=evalf(eigenvalues(A)):

> for i to Nrow do lambda[i]:=l[i];od:
> Id:=Matrix(Nrow,Nrow,shape=identity):
> X:=matrix(Nrow,1,[seq(beta[i],i=1..Nrow)]):
> for k to Nrow do:
> G:=evalm((A-lambda[k]*Id)&*X):
> eqx[1]:=beta[1]=1:for i from 2 to Nrow do eqx[i]:=G[i-1,1]:od:
> for i to Nrow do beta[i]:=fsolve(eqx[i],beta[i]);od:
> XX[k]:=map(eval,evalm(X)):
> for i to Nrow do unassign('beta[i]'):od:od:

> P:=Matrix([seq(evalm(XX[i]),i=1..Nrow)]):
> expD1:=Matrix(1..Nrow,1..Nrow,shape=diagonal):
> for i to Nrow do expD1[i,i]:=exp(lambda[i]*t):od:
> mat:=evalm(P&*expD1&*inverse(P)):
```

When the above procedure was used to calculate the exponential matrix for example 7.1, the time taken for N=40 interior node points was less than 30 seconds. For the same number of node points, Maple takes more than 5 minutes to calculate the exponential matrix in a 2.6Ghz processor with 2 GB RAM. The 3-D plot obtained for example 7.1 with N=40 node points is given below:

```
> Y0:=matrix(N,1):for i from 1 to N do
Y0[i,1]:=evalf(subs(x=i*h,rhs(IC)));od:evalm(Y0):
> b2:=subs(t=tau,evalm(b)):
> mat2:=subs(t=t-tau,evalm(mat)):
> mat3:=evalm(mat2&*b2):
> mat4:=map(int,mat3,tau=0..t):
> Y:=evalm(mat&*Y0+mat4):
> for i from 1 to N do u[i](t):=evalf((Y[i,1]));od:
> for i from 0 to N+1 do u[i](t):=eval(u[i](t));od:
```

5.1 Semianalytical Method for Parabolic Partial Differential Equations (PDEs)

```
> setcolors(["Red", "Blue", "LimeGreen", "Goldenrod", "maroon",
  "DarkTurquoise", "coral", "aquamarine", "magenta", "khaki", "sienna",
  "orange", "yellow", "gray"]):
```

```
> pp:=plot([seq(subs(delta=1,u[i](t)),i=0..N+1,5)],t=0..0.4,thickness=3);
```

$$pp := PLOT(\ldots)$$

```
> pt:=textplot([seq([0.05+i*0.002,evalf(subs(delta=1,t=0.05+i*0.002,u[i](t))),
  typeset(u[i]),align=right],i=0..N+1,5)]);
```

$$pt := PLOT(\ldots)$$

```
> display([pp,pt],title="Figure 5.30",axes=boxed,labels=[t,"u"]);
```

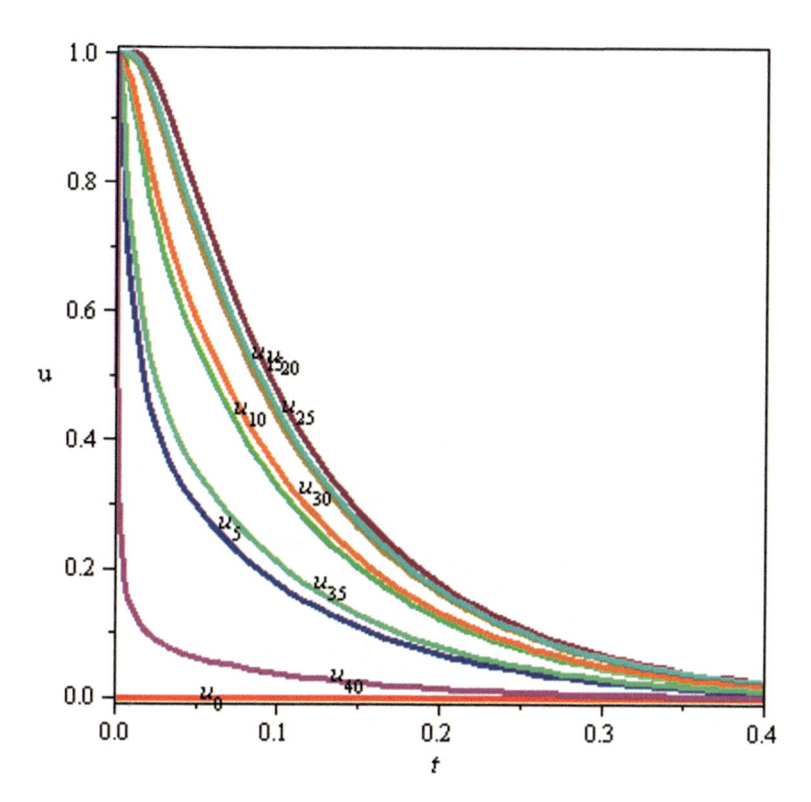

Fig. 5.30

```
> for i from 0 to N+1 do
  p[i]:=plot(subs(delta=0.1,u[i](t)),t=0..0.4,thickness=3);od:
```

```
> tf:=0.1;
```

$$tf := 0.1$$

> M:=30;

$$M := 30$$

> T1:=[seq(tf*i/M,i=0..M)];

$T1 := [0., 0.003333333333, 0.006666666667, 0.01000000000, 0.01333333333, 0.01666666667, 0.02000000000,$
$0.02333333333, 0.02666666667, 0.03000000000, 0.03333333333, 0.03666666667, 0.04000000000,$
$0.04333333333, 0.04666666667, 0.05000000000, 0.05333333333, 0.05666666667, 0.06000000000,$
$0.06333333333, 0.06666666667, 0.07000000000, 0.07333333333, 0.07666666667, 0.08000000000,$
$0.08333333333, 0.08666666667, 0.09000000000, 0.09333333333, 0.09666666667, 0.1000000000]$

> PP:=matrix(N+2,M+1);

$$PP := array(1..42, 1..31, [\ ])$$

> for i from 1 to N+2 do PP[i,1]:=evalf(subs(x=(i-1)*h,rhs(IC)));od:

> for i from 1 to N+2 do for j from 2 to M+1 do
PP[i,j]:=evalf(subs(t=T1[j],subs(delta=1,u[i-1](t))));od;od:

> plotdata:= [seq([seq([(i-1)*h,T1[j],PP[i,j]], i=1..N+2)], j=1..M+1)]:

> surfdata(plotdata,axes=boxed,title="Figure 5.31",
labels=[x,t,u],orientation=[45,45]);

Fig. 5.31

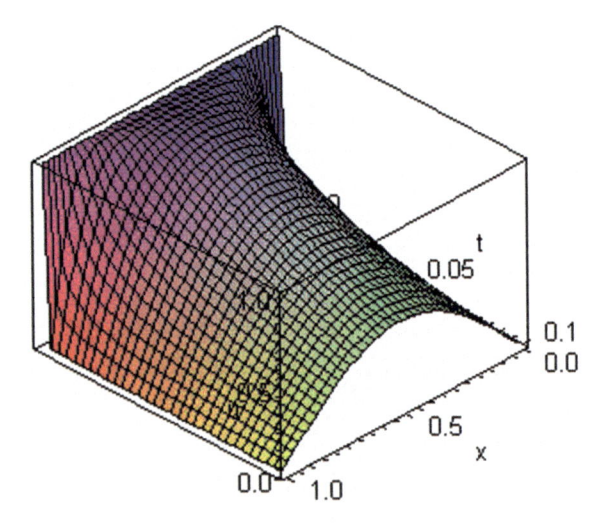

Example 5.10

> restart;

> with(linalg):with(plots):

5.1 Semianalytical Method for Parabolic Partial Differential Equations (PDEs)

> ge:=diff(u(x,t),t)=-diff(u(x,t),x);

$$ge := \frac{\partial}{\partial t} u(x,t) = -\left(\frac{\partial}{\partial x} u(x,t)\right)$$

> bc1:=u(x,t)-1;

$$bc1 := u(x,t) - 1$$

> bc2:=diff(u(x,t),t)=-diff(u(x,t),x);

$$bc2 := \frac{\partial}{\partial t} u(x,t) = -\left(\frac{\partial}{\partial x} u(x,t)\right)$$

> IC:=u(x,0)=0;

$$IC := u(x,0) = 0$$

> N:=2;

$$N := 2$$

> L:=1;

$$L := 1$$

> dydxf:=1/2*(-u[2](t)-3*u[0](t)+4*u[1](t))/h:
> dydxb:=1/2*(u[N-1](t)+3*u[N+1](t)-4*u[N](t))/h:
> dydxb:=(u[N+1](t)-u[N](t))/h:
> dydx:=1/h*(u[m](t)-u[m-1](t)):
> d2ydx2:=1/h^2*(u[m-1](t)-2*u[m](t)+u[m+1](t)):
> bc1:=subs(diff(u(x,t),x)=dydxf,u(x,t)=u[0](t),x=0,bc1):
> bc2:=subs(diff(u(x,t),x)=dydxb,u(x,t)=u[N+1](t),x=1,bc2):
> eq[0]:=bc1;

$$eq_0 := u_0(t) - 1$$

> eq[N+1]:=bc2;

$$eq_3 := \frac{d}{dt} u_3(t) = -\frac{u_3(t) - u_2(t)}{h}$$

> for i from 1 to N do eq[i]:=diff(u[i](t),t)= subs(diff(u(x,t),x$2) = subs(m=i,d2ydx2),diff(u(x,t),x) = subs(m=i,dydx), u(x,t)=u[i](t),x=i*h,rhs(ge));od:
> u[0](t):=(solve(eq[0],u[0](t)));

$$u_0(t) := 1$$

> #u[N+1](t):=solve(eq[N+1],u[N+1](t));
> for i from 1 to N+1 do eq[i]:=eval(eq[i]);od;

$$eq_1 := \frac{d}{dt} u_1(t) = -\frac{u_1(t) - 1}{h}$$

$$eq_2 := \frac{d}{dt} u_2(t) = -\frac{u_2(t) - u_1(t)}{h}$$

$$eq_3 := \frac{d}{dt} u_3(t) = -\frac{u_3(t) - u_2(t)}{h}$$

> eqs:=[seq(rhs(eq[j]),j=1..N+1)]:
> Y:=[seq(u[i](t),i=1..N+1)]:
> A:=genmatrix(eqs,Y,'b1'):
> b:=matrix(N+1,1):for i to N+1 do b[i,1]:=-eval(b1[i]);od:evalm(b):
> h:=eval(L/(N+1)):
> A:=map(eval,A):
> if N > 4 then A:=map(evalf,A);end:
> evalm(A);

$$\begin{bmatrix} -3 & 0 & 0 \\ 3 & -3 & 0 \\ 0 & 3 & -3 \end{bmatrix}$$

> J:=jordan(A,S);

$$J := \begin{bmatrix} -3 & 1 & 0 \\ 0 & -3 & 1 \\ 0 & 0 & -3 \end{bmatrix}$$

> evalm(S);

$$\begin{bmatrix} 0 & 0 & 1 \\ 0 & 3 & 0 \\ 9 & 0 & 0 \end{bmatrix}$$

5.1 Semianalytical Method for Parabolic Partial Differential Equations (PDEs) 445

> mat:=evalm(S&*exponential(J,t)&*inverse(S));

$$mat := \begin{bmatrix} e^{-3t} & 0 & 0 \\ 3te^{-3t} & e^{-3t} & 0 \\ \frac{9}{2}t^2 e^{-3t} & 3te^{-3t} & e^{-3t} \end{bmatrix}$$

> Nrow:=rowdim(A):
> l:=evalf(eigenvalues(A));

$$l := -3., -3., -3.$$

> for i to Nrow do lambda[i]:=l[i];od:
> Id:=Matrix(Nrow,Nrow,shape=identity):
> X:=matrix(Nrow,1,[seq(beta[i],i=1..Nrow)]):
> for k to Nrow do:
> G:=evalm((A-lambda[k]*Id)&*X):
> eqx[1]:=beta[1]=1:for i from 2 to Nrow do eqx[i]:=G[i-1,1]:od:
> for i to Nrow do beta[i]:=solve(eqx[i],beta[i]);od:
> XX[k]:=map(eval,evalm(X)):
> for i to Nrow do unassign('beta[i]'):od:od:

> P:=Matrix([seq(evalm(XX[i]),i=1..Nrow)]);

$$P := \begin{bmatrix} 1 & 1 & 1 \\ \beta_2 & \beta_2 & \beta_2 \\ NULL & NULL & NULL \end{bmatrix}$$

> expD1:=Matrix(1..Nrow,1..Nrow,shape=diagonal):
> for i to Nrow do expD1[i,i]:=exp(lambda[i]*t):od:
> expD1:=map(convert,expD1,trig):
> mat:=evalm(P&*expD1&*inverse(P));

Error, (in evalm) unnamed vector or array with undefined entries.

> mat:=map(expand,mat):
> mat:=map(simplify,mat):
> Y0:=matrix(N+1,1):for i from 1 to N+1 do Y0[i,1]:=evalf(subs(x=i*h,rhs(IC)));od:evalm(Y0):

```
> Y:=evalm(mat&*(Y0+inverse(A)&*b)-inverse(A)&*b):
> #b2:=subs(t=tau,evalm(b)):
> #mat2:=subs(t=t-tau,evalm(mat)):
> #mat3:=evalm(mat2&*b2):
> #mat4:=map(int,mat3,tau=0..t):
> #Y:=evalm(mat&*Y0+mat4):
> for i from 1 to N+1 do u[i](t):=evalf((Y[i,1]));od:
> for i from 0 to N+1 do u[i](t):=eval(u[i](t));od:
> for i from 0 to N+1 do u[i](t):=subs(l=0,u[i](t));od:
> setcolors(["Red", "Blue", "LimeGreen", "Goldenrod", "maroon",
"DarkTurquoise", "coral", "aquamarine", "magenta", "khaki", "sienna",
"orange", "yellow", "gray"]):
> pp:=plot([seq(subs(delta=1,u[i](t)),i=0..N+1)],t=0..1,thickness=4);
```

$$pp := PLOT(...)$$

```
> pt:=textplot([seq([0.4,evalf(subs(delta=1,t=0.4,u[i](t))),typeset(u[i],"(t)"),
align={below,right}],i=0..N+1)]);
```

$$pt := PLOT(...)$$

```
> display([pp,pt],title="Figure 5.32",axes=boxed,labels=[t,"u"]);
```

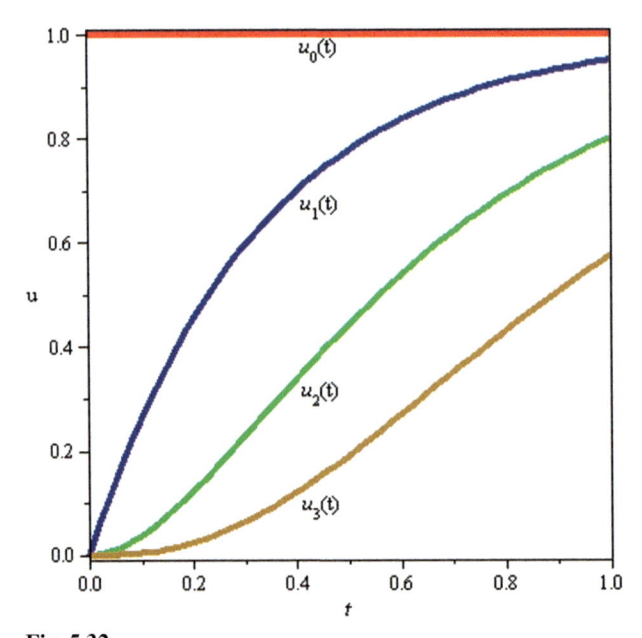

Fig. 5.32

5.1 Semianalytical Method for Parabolic Partial Differential Equations (PDEs)

```
> for i from 0 to N+1 do p[i]:=plot(subs(delta=0.1,u[i](t)),
t=0..0.4,thickness=3);od:
> tf:=1;
```
$$tf := 1$$
```
> M:=30;
```
$$M := 30$$
```
> T1:=[seq(tf*i/M,i=0..M)];
```

$$T1 := \left[0, \frac{1}{30}, \frac{1}{15}, \frac{1}{10}, \frac{2}{15}, \frac{1}{6}, \frac{1}{5}, \frac{7}{30}, \frac{4}{15}, \frac{3}{10}, \frac{1}{3}, \frac{11}{30}, \frac{2}{5}, \frac{13}{30},\right.$$
$$\frac{7}{15}, \frac{1}{2}, \frac{8}{15}, \frac{17}{30}, \frac{3}{5}, \frac{19}{30}, \frac{2}{3}, \frac{7}{10}, \frac{11}{15}, \frac{23}{30}, \frac{4}{5}, \frac{5}{6}, \frac{13}{15}, \frac{9}{10},$$
$$\left.\frac{14}{15}, \frac{29}{30}, 1\right]$$

```
> PP:=matrix(N+2,M+1);
```
$$PP := array(1..4, 1..31, [\,])$$
```
> for i from 1 to N+2 do PP[i,1]:=evalf(subs(x=(i-1)*h,rhs(IC)));od:
> for i from 1 to N+2 do for j from 2 to M+1 do
PP[i,j]:=evalf(subs(t=T1[j],subs(delta=1,u[i-1](t))));od;od:
> plotdata := [seq([ seq([(i-1)*h,T1[j],PP[i,j]], i=1..N+2)], j=1..M+1)]:
> surfdata( plotdata, title="Figure 5.33",axes=boxed,
labels=[x,t,u],orientation=[45,45]);
```

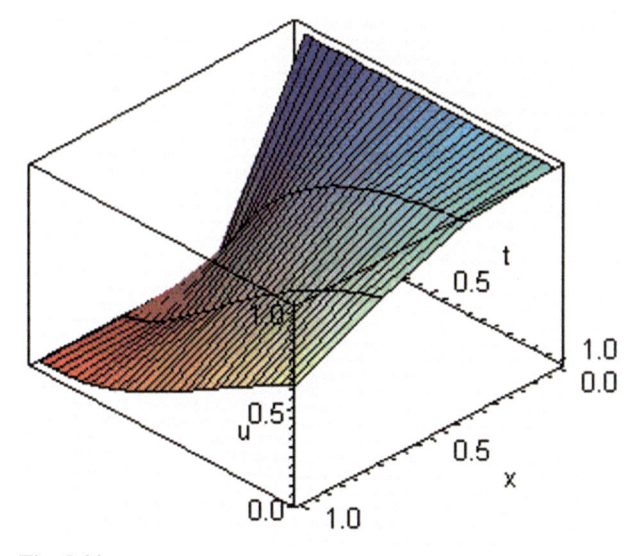

Fig. 5.33

```
>
```

Example 5.11

> restart;
> with(linalg):with(plots):
> ge:=diff(u(x,t),t)=-diff(u(x,t),x);

$$ge := \frac{\partial}{\partial t} u(x,t) = -\left(\frac{\partial}{\partial x} u(x,t)\right)$$

> bc1:=u(x,t)-1;

$$bc1 := u(x,t) - 1$$

> bc2:=diff(u(x,t),t)=-diff(u(x,t),x);

$$bc2 := \frac{\partial}{\partial t} u(x,t) = -\left(\frac{\partial}{\partial x} u(x,t)\right)$$

> IC:=u(x,0)=0;N:=200;

$$N := 200$$

> L:=1;

$$L := 1$$

> dydxf:=1/2*(-u[2](t)-3*u[0](t)+4*u[1](t))/h:
> #dydxb:=1/2*(u[N-1](t)+3*u[N+1](t)-4*u[N](t))/h:
> dydxb:=(u[N+1](t)-u[N](t))/h:
> dydx:=1/h*(u[m](t)-u[m-1](t)):
> #dydx:=1/2/h*(u[m+1](t)-u[m-1](t)):
> d2ydx2:=1/h^2*(u[m-1](t)-2*u[m](t)+u[m+1](t)):
> bc1:=subs(diff(u(x,t),x)=dydxf,u(x,t)=u[0](t),x=0,bc1):
> bc2:=subs(diff(u(x,t),x)=dydxb,u(x,t)=u[N+1](t),x=1,bc2):
> eq[0]:=bc1;

$$eq_0 := u_0(t) - 1$$

> eq[N+1]:=bc2;

$$eq_{201} := \frac{d}{dt} u_{201}(t) = -\frac{u_{201}(t) - u_{200}(t)}{h}$$

5.1 Semianalytical Method for Parabolic Partial Differential Equations (PDEs) 449

```
> for i from 1 to N do eq[i]:=diff(u[i](t),t)= subs(diff(u(x,t),x$2) =
subs(m=i,d2ydx2),diff(u(x,t),x) = subs(m=i,dydx),u(x,t)=u[i](t),
x=i*h,rhs(ge));od:

> u[0](t):=(solve(eq[0],u[0](t)));
```

$$u_0(t) := 1$$

```
> for i from 1 to N+1 do eq[i]:=eval(eq[i]);od:

> eqs:=[seq(rhs(eq[j]),j=1..N+1)]:

> Y:=[seq(u[i](t),i=1..N+1)]:

> A:=genmatrix(eqs,Y,'b1'):

> b:=matrix(N+1,1):for i to N+1 do b[i,1]:=-eval(b1[i]);od:evalm(b):

> h:=eval(L/(N+1)):

> A:=map(eval,A):

> if N > 4 then A:=map(evalf,A);end:

> Nrow:=rowdim(A);
```

$$Nrow := 201$$

```
> #exponential(A,t);

> #seq([seq(exp(-Nrow*t)*(Nrow*t)^(i-j),j=1..i)],i=1..Nrow);

> mat:=Matrix(Nrow,[seq([seq(exp(-Nrow*t)*(Nrow*t)^(i-j)/factorial(i-j),
j=1..i)],i=1..Nrow)],shape=triangular[lower]):

> Y0:=matrix(N+1,1):for i from 1 to N+1 do
Y0[i,1]:=evalf(subs(x=i*h,rhs(IC)));od:evalm(Y0):

> Y:=evalm(mat&*(Y0+inverse(A)&*b)-inverse(A)&*b):

> #b2:=subs(t=tau,evalm(b)):

> #mat2:=subs(t=t-tau,evalm(mat)):

> #mat3:=evalm(mat2&*b2):

> #mat4:=map(int,mat3,tau=0..t):

> #Y:=evalm(mat&*Y0+mat4):

> for i from 1 to N+1 do u[i](t):=evalf((Y[i,1]));od:

> for i from 0 to N+1 do u[i](t):=eval(u[i](t));od:

> for i from 0 to N+1 do u[i](t):=subs(l=0,u[i](t));od:

> setcolors(["Red", "Blue", "LimeGreen", "Goldenrod", "maroon",
"DarkTurquoise", "coral", "aquamarine", "magenta", "khaki", "sienna",
"orange", "yellow", "gray"]):
```

450   5 Method of Lines for Parabolic Partial Differential Equations

> pp:=plot([seq(subs(delta=1,u[i](t)),i=0..N+1,20)],t=0..1,thickness=3);

$$pp := PLOT(\ldots)$$

> pt:=textplot([seq([0.05+i/20*0.09,evalf(subs(delta=1,t=0.05+i/20*0.09,u[i](t))),typeset(u[i]),align={below,right}],i=0..N+1,20)]);

$$pt := PLOT(\ldots)$$

> display([pp,pt],title="Figure 5.34",axes=boxed,labels=[t,"u"]);

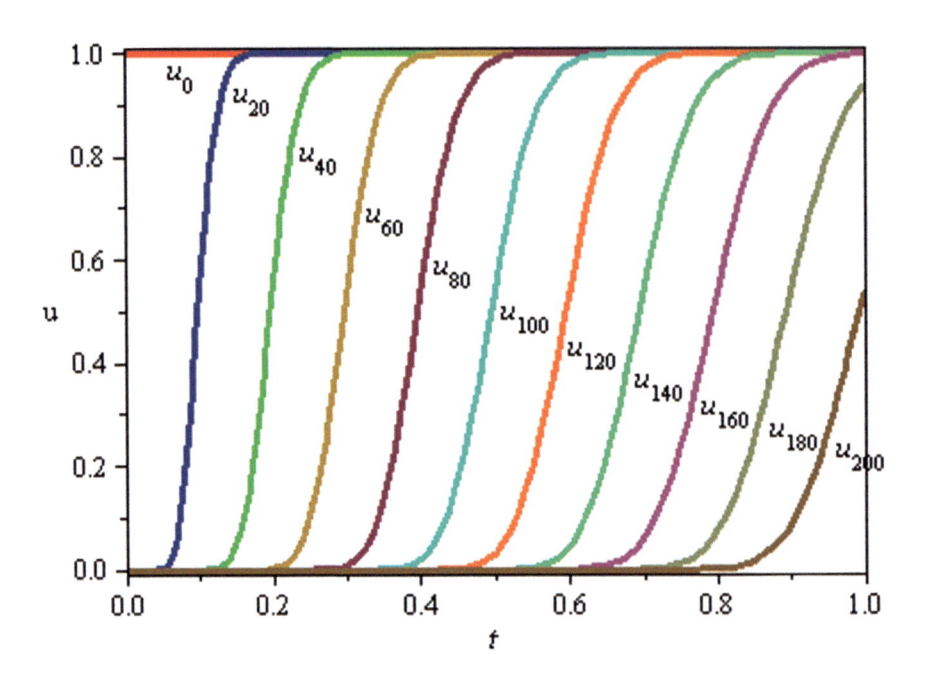

Fig. 5.34

> for i from 0 to N+1 do
p[i]:=plot(subs(delta=0.1,u[i](t)),t=0..0.4,thickness=3);od:

> tf:=1;

$$tf := 1$$

> M:=30;

$$M := 30$$

> T1:=[seq(tf*i/M,i=0..M)]:

> PP:=matrix(N+2,M+1);

$$PP := array(1..202, 1..31, [\,])$$

5.1 Semianalytical Method for Parabolic Partial Differential Equations (PDEs)

```
> for i from 1 to N+2 do PP[i,1]:=evalf(subs(x=(i-1)*h,rhs(IC)));od:
> for i from 1 to N+2 do for j from 2 to M+1 do
PP[i,j]:=evalf(subs(t=T1[j],subs(delta=1,u[i-1](t))));od;od:
> plotdata := [seq([ seq([(i-1)*h,T1[j],PP[i,j]], i=1..N+2)], j=1..M+1)]:
> surfdata(plotdata,title="Figure 5.35",axes=boxed,
labels=[x,t,u],orientation=[-45,45]);
```

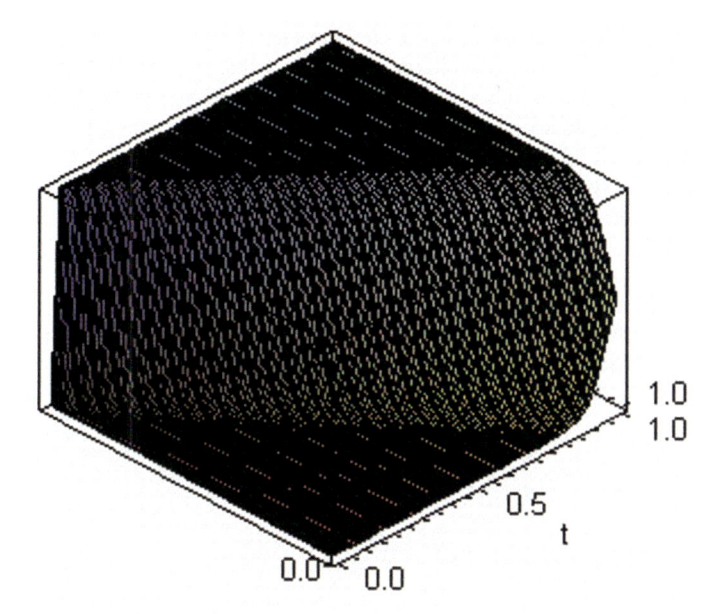

Fig. 5.35

```
>
```

It is recommended that one check the exponential matrix obtained using this expedited procedure with the exponential matrix obtained using Maple's exponential matrix command for at least two values of interior node points (for e.g., N = 2, 4 etc). Once it is verified that the above procedure works for a particular problem, one can use the procedure for obtaining the exponential matrix efficiently for high values of N.

5.1.6 Summary

In this chapter semianalytical solutions (solutions analytical in t and numerical in x) were obtained for parabolic PDEs. In section 5.1.2, the given homogeneous parabolic PDE was converted to matrix form by applying finite differences in the spatial direction. The resulting matrix differential equation was then integrated analytically in time using Maple's matrix exponential. This methodology helps us solve the dependent variables at different node points as an analytical function of time. This is a powerful technique and is valid for all linear parabolic PDEs. This technique was then extended to nonhomogeneous parabolic PDEs in section 5.1.3 by adding the nonhomogeneous part to the homogeneous solution.

In section 5.1.4, the Graetz problem was solved using the semianalytical technique. The solution obtained is numerical in x and analytical in z. The solution is obtained as a function of the Peclet number. The solution obtained compares well with the analytical solution reported in the literature. Our technique avoids calculation of special functions and at the same time provides solutions explicit in the Peclet number. In section 5.1.5, the semianalytical technique developed earlier was extended to the case when the initial condition is a function of x.

In section 5.1.6, semianalytical technique was extended to composite domains. Many chemical problems with mass transfer or kinetics limitations form a boundary layer near the surface. These problems can be handled conveniently by splitting the domain to composite domain as illustrated in this section. In addition, composite solids (composite electrodes) with different physical properties can be handled by this technique. Eight examples were given in this chapter.

In section 5.1.7, a procedure to expedite the calculation of exponential matrix was developed. This procedure is valid as long as all the eigenvalues are distinct.

5.1.7 Exercise Problems

1. Consider diffusion with convection in a coated wall reactor, where the reaction takes place at the wall.[9] The governing equation and boundary conditions for concentration are:

$$v_0 \frac{\partial c}{\partial z} = D\left(\frac{\partial^2 c}{\partial r^2} + \frac{1}{r}\frac{\partial c}{\partial r}\right)$$

$$\frac{\partial c}{\partial r}(0,z) = 0 \text{ and } -D\frac{\partial c}{\partial r}(R,z) = k\, c(R,z)$$

$$c(r,0) = c_0$$

where v_0 is the velocity, D is the diffusion coefficient, R is the radius, c_0 is the inlet concentration and k is the rate constant. Use the dimensionless variables $u = \frac{c}{c_0}$; $x = \frac{r}{R}$; $Z = \frac{zD}{v_0 R^2}$ to obtain the dimensionless governing equation and boundary/initial conditions:

$$\frac{\partial u}{\partial Z} = \frac{\partial^2 u}{\partial x^2} + \frac{1}{x}\frac{\partial u}{\partial x}$$

$$\frac{\partial u}{\partial x}(0,Z) = 0 \text{ and } \frac{\partial u}{\partial x}(1,Z) + Ha\, u(1,Z) = 0$$

$$u(x,0) = 1$$

where $Ha = \frac{kR}{D}$ is the Hatta number. Obtain semianalytical solutions for this problem and plot the profiles for Ha = 0.1, 1, 2 and 10. (Hint: for

5.1 Semianalytical Method for Parabolic Partial Differential Equations (PDEs)

convenience, use t instead of Z so that the program given in the chapter can be used directly). Solve this problem using Maple's exponential matrix and also using the code given for expediting the calculation of exponential matrix. How much computation time is saved for N = 20 node points?Redo example 5.2 with the expedited code for the exponential matrix. How much computation time is saved?

2. Redo example 5.4 with the expedited code for the exponential matrix. How much computation time is saved?
3. Redo example 5.5 with the expedited code for the exponential matrix. How much computation time is saved?
4. Redo example 5.6 with the expedited code for the exponential matrix. How much computation time is saved?
5. Redo example 5.7 with the expedited code for the exponential matrix. How much computation time is saved?
6. Consider cooling of spherical nuclear pellets.[9] The dimensionless temperature distribution is governed by:

$$\frac{\partial u}{\partial t} = \frac{\partial^2 u}{\partial x^2} + \frac{2}{x}\frac{\partial u}{\partial x} + Q$$

$$\frac{\partial u}{\partial x}(0,t) = 0 \text{ and } \frac{\partial u}{\partial x}(1,t) + Bi\ u(1,t) = 0$$

$$u(x,0) = 1$$

where Q is the ratio of heat generation to heat conduction and Bi is the Biot number. Solve this problem for the set of parameters Q = 1, Bi = 0.2 and Q = 1, Bi = 10.

7. Consider the electrochemical discharge of a planar electrode (example 5.3) again. After applying the finite differences, and eliminating the boundary values, the coefficient matrix for the node points i = 1..N, A is singular because of flux-boundary conditions at both the ends. One cannot use the code given in the chapter for expediting the calculation of exponential matrix because A is singular. The governing equations at the interior node points are:

$$\frac{du_1}{dt} = \frac{2}{3}\frac{u_2 - u_1}{h^2}$$

$$\frac{du_2}{dt} = \frac{u_1 - 2u_2 + u_3}{h^2}$$

.....

$$\frac{du_{N-1}}{dt} = \frac{u_{N-2} - 2u_{N-1} + u_N}{h^2}$$

$$\frac{du_N}{dt} = \frac{2}{3}\frac{u_{N-1} - u_N}{h^2} - \frac{2}{3}\frac{\delta}{h}$$

Multiplying the first and the last equation by 3/2 and adding up all the equations we get:

$$\frac{d\left(\frac{3}{2}[u_1 + u_N] + \sum_{i=2}^{N-1} u_i\right)}{dt} = -\frac{\delta}{h}$$

8. This differential equation can be integrated using the initial condition as:

$$\frac{3}{2}[u_1 + u_N] + \sum_{i=2}^{N-1} u_i - (N+1) = -\frac{\delta t}{h}$$

This equation can be used to eliminate u_N completely and we need to solve only the following $N-1$ ODEs:

$$\frac{du_1}{dt} = \frac{2}{3}\frac{u_2 - u_1}{h^2}$$

$$\frac{du_2}{dt} = \frac{u_1 - 2u_2 + u_3}{h^2}$$

.....

$$\frac{du_{N-1}}{dt} = \frac{u_{N-2} - 2u_{N-1} - u_1 - \frac{2}{3}\sum_{i=2}^{N-1} u_i}{h^2} + \frac{2\left(N+1-\frac{\delta t}{h}\right)}{3 h^2}$$

This is a system of N-1 first-order nonhomogeneous ODEs, which can be solved as illustrated in section 5.1.3 (example 5.4). The resulting coefficient matrix is non-singular and has distinct N-1 eigenvalues. The exponential matrix can be obtained efficiently using the expedition-procedure given in the section 5.1.7. Modify the program given in the chapter to obtain semianalytical solutions. Plot the dimensionless concentration profiles for $\delta = 1$.

9. Consider dispersion of a linear kinematic wave in dimensionless form.[14] The governing equation and boundary/initial conditions are:

$$\frac{\partial u}{\partial t} = \frac{\partial^2 u}{\partial x^2} - Pe\frac{\partial u}{\partial x}$$

$$u(0,t) = 1; \quad u(1,t) = 0$$

$$u(x,0) = 0$$

Obtain a semianalytical solution for this problem for Pe = 1, 10 and 50. What is the computation time saved for Pe = 10 and 50? Hint: Maple's exponential matrix command may not work for this problem for all

5.1 Semianalytical Method for Parabolic Partial Differential Equations (PDEs)

values of Pe and N. In addition, different finite difference approximations might have to be used (see section 3.1.3).

10. Consider the fluid-flow problem:[12]

$$\frac{\partial u}{\partial t} = \frac{\partial^2 u}{\partial x^2} + \frac{1}{x}\frac{\partial u}{\partial x} + 4$$

$$\frac{\partial u}{\partial x}(0,t) = 0 \text{ and } u(1,t) = 0$$

$$u(x,0) = 0$$

Obtain the semianalytical solution and plot the dimensionless velocity profiles.

11. Consider the Graetz problem discussed in example 5.6. The same problem in planar geometry is:

$$2\text{Pe}(1-x^2)\frac{\partial u}{\partial z} = \frac{\partial^2 u}{\partial x^2}$$

$$u(x,0) = 0$$

$$\frac{\partial u}{\partial x}(0,z) = 0 \text{ and } u(1,z) = 1$$

Solve this problem and plot the profiles for different values of Peclet number.

12. Consider heat conduction in a slab with radiation at both ends.[6] The dimensionless governing equations and boundary/initial conditions are:

$$\frac{\partial u}{\partial t} = \frac{\partial^2 u}{\partial x^2}$$

$$-\frac{\partial u}{\partial x}(0,t) + Hu(0,t) = 0 \text{ and } \frac{\partial u}{\partial x}(1,t) + Hu(1,t) = 0$$

$$u(x,0) = 1$$

where H is the dimensionless heat transfer coefficient. Obtain the semianalytical solutions for H = 1 and plot the profiles.

13. Solve problem 20, chapter 7 using the semianalytical method.
14. Solve problem 18, chapter 8 using the semianalytical method.
15. Solve problem 19, chapter 8 using the semianalytical method.
16. Solve example 7.4 using the semianalytical method.

5.2 Numerical Method of Lines for Parabolic Partial Differential Equations (PDEs)

5.2.1 Introduction

Transient heat conduction or mass transfer in solids with varying physical properties (diffusion coefficient, thermal diffusivity, thermal conductivity, etc.) with nonlinear chemical reaction or heat source term is usually represented by a nonlinear parabolic partial differential equation. The semianalytical or analytical method of lines developed in chapter 5.1 cannot be used for nonlinear parabolic PDEs. In this chapter, we describe how one can arrive at the numerical solution by applying numerical method of lines for nonlinear parabolic PDEs by discretizing the spatial derivatives using finite differences and integrating numerically using Maple's numerical IVP solver (Runge-Kutta, Gear and Rosenbrock solver).

5.2.2 Numerical Method of Lines for Parabolic PDEs with Linear Boundary

Conditions

Consider a general nonlinear parabolic partial differential equation in dimensionless form

$$\frac{\partial u}{\partial t} = a_0(x)\frac{\partial^2 u}{\partial x^2} + a_1(x)\frac{\partial u}{\partial x} + a_2(x)u + f(u) \qquad (5.48)$$

with a known initial condition

$$u(x,0) = 1 \qquad (5.49)$$

and linear boundary conditions at both the boundaries:

$$\alpha_1 u(0,t) + \beta_1 \frac{\partial u}{\partial x}(0,t) = \gamma_1 \qquad (5.50)$$

and

$$\alpha_2 u(1,t) + \beta_2 \frac{\partial u}{\partial x}(1,t) = \gamma_2 \qquad (5.51)$$

where α_1, β_1, α_2, and β_2 are constants and γ_1, and γ_2 can be functions of time.

The numerical method of lines[1] [3] [4] [2] (Schiesser and Silebi, 1997; Cutlip and Shacham, 1999; Taylor; 1999; Constantinides and Mostoufi, 1999) involves converting the governing equation (equation (5.48)) to a system of coupled ODEs in time by applying finite difference approximations for the spatial derivatives

5.2 Numerical Method of Lines for Parabolic Partial Differential Equations (PDEs)

(see chapter 5.1). The governing equation (equation (5.48)) can be converted to finite difference form as follows:

$$\frac{du_i}{dt} = a_0(x_i)\frac{u_{i+1} - 2u_i + u_{i-1}}{h^2} + a_1(x_i)\frac{u_{i+1} - u_{i-1}}{2h} + a_2(x_i)u_i + f(u_i) \tag{5.52}$$

where i is the node number, N is the number of interior node points, and h is the node spacing defined as

$$h = \frac{L}{N+1} \tag{5.53}$$

where L is the length of the domain of interest. Thus, x = 0 corresponds to the node point i = 0, x = 1 corresponds to the node point i = N+1 and x = xi = ih is the value of x at the node point i. The variable ui corresponds to the dependent variable at node point i. Equation (5.53) is a system of N nonlinear coupled ODEs for N dependent variables $(u_i, i=1..N)$. The boundary values u_0 and u_{N+1} are eliminated using the boundary conditions. The boundary conditions (equations (5.50) and (5.51)) can be written in finite difference form as

$$\alpha_1 u_0 + \beta_1 \frac{-3u_0 + 4u_1 + u_2}{2h} = 0 \tag{5.54}$$

and

$$\alpha_2 u_{N+1} + \beta_2 \frac{3u_{N+1} - 4u_N + u_{N-1}}{2h} = 0 \tag{5.55}$$

Using the boundary conditions (equations (5.54) and (5.55)) the boundary values u_0 and u_{N+1} can be eliminated. Hence, the method of lines technique reduces the nonlinear parabolic PDE (equation (5.48)) to a nonlinear system of N coupled first order ODEs (equation (5.52)). This nonlinear system of ODEs is integrated numerically in time using Maple's numerical ODE solver (Runge-Kutta, Gear, and Rosenbrock for stiff ODEs; see chapter 2.2.5). The procedure for using Maple to solve nonlinear parabolic partial differential equations with linear boundary conditions can be summarized as follows:

1. Start the Maple worksheet with a 'restart' command to clear all variables.
2. Call 'with(linalg)' and 'with(plots)' commands.
3. Enter the governing equation.
4. Store the boundary conditions in bc1 (x = 0) and bc2 (x = 1).
5. Enter the number of interior node points, N.
6. Enter the length of the domain, L.
7. Convert the governing equations to the finite difference form by using central the difference expression accurate to the order h^2 for the first and second derivatives.

8. Convert the boundary conditions to the finite difference form by using the 3-point forward and backward differences (accurate to the order h^2), respectively, for bc1 and bc2.
9. Eliminate the boundary values ($u_0(t)$ and $u_{N+1}(t)$) using the boundary conditions.
10. Store the finite difference equations in eqs.
11. Store the dependent variables, u_i, $i = 1..N$ in Y.
12. Store the initial conditions for the dependent variables in ICs.

Find the numerical solution using Maple's 'dsolve' command. The syntax is:

1. "dsolve({eqs,ICs},{Y},type=numeric)". Maple's default numerical ODE solver, Runge-Kutta method accurate to the order Δt^6, is used in this chapter.
2. Once the numerical solution is obtained, plots can be made.

Example 5.2.1. Diffusion with Second Order Reaction

Consider diffusion in a slab with a second order reaction. The governing equation in dimensionless form is (see example 5.1.5)

$$\frac{\partial u}{\partial t} = \frac{\partial^2 u}{\partial x^2} - \Phi^2 u^2$$

$$u(x,0) = 0 \tag{5.56}$$

$$u(0,t) = 1 \text{ and } \frac{\partial u}{\partial x}(1,t) = 0$$

where Φ is the Thiele modulus. This equation is solved in Maple below using the procedure described above.

> restart;

> with(linalg):with(plots):

Enter the governing equation:

> ge:=diff(u(x,t),t)=diff(u(x,t),x$2)-Phi^2*u(x,t)^2;

$$ge := \frac{\partial}{\partial t} u(x,t) = \frac{\partial^2}{\partial x^2} u(x,t) - \Phi^2 u(x,t)^2$$

Enter the boundary conditions:

> bc1:=u(x,t)-1;

$$bc1 := u(x,t) - 1$$

> bc2:=diff(u(x,t),x);

$$bc2 := \frac{\partial}{\partial x} u(x,t)$$

5.2 Numerical Method of Lines for Parabolic Partial Differential Equations (PDEs)

Enter the initial condition:

> IC:=u(x,0)=0;

$$IC := u(x, 0) = 0$$

Enter the number of interior node points:

> N:=10;

$$N := 10$$

> L:=1;

$$L := 1$$

Enter the value of the parameters:

> Phi:=1;

$$\Phi := 1$$

> dydxf:=1/2*(-u[2](t)-3*u[0](t)+4*u[1](t))/h;

$$dydxf := \frac{1}{2} \frac{-u_2(t) - 3 u_0(t) + 4 u_1(t)}{h}$$

> dydxb:=1/2*(u[N-1](t)+3*u[N+1](t)-4*u[N](t))/h;

$$dydxb := \frac{1}{2} \frac{u_9(t) + 3 u_{11}(t) - 4 u_{10}(t)}{h}$$

> dydx:=1/2/h*(u[m+1](t)-u[m-1](t));

$$dydx := \frac{1}{2} \frac{u_{m+1}(t) - u_{m-1}(t)}{h}$$

> d2ydx2:=1/h^2*(u[m-1](t)-2*u[m](t)+u[m+1](t));

$$d2ydx2 := \frac{u_{m-1}(t) - 2 u_m(t) + u_{m+1}(t)}{h^2}$$

Convert the boundary conditions to the finite difference form:

> bc1:=subs(diff(u(x,t),x)=dydxf,u(x,t)=u[0](t),x=0,bc1);

$$bc1 := u_0(t) - 1$$

> bc2:=subs(diff(u(x,t),x)=dydxb,u(x,t)=u[N+1](t),x=1,bc2);

$$bc2 := \frac{1}{2} \frac{u_9(t) + 3 u_{11}(t) - 4 u_{10}(t)}{h}$$

> eq[0]:=bc1;

$$eq_0 := u_0(t) - 1$$

> eq[N+1]:=bc2;

$$eq_{11} := \frac{1}{2} \frac{u_9(t) + 3u_{11}(t) - 4u_{10}(t)}{h}$$

Convert the governing equations to the finite difference form:

> for i from 1 to N do eq[i]:=diff(u[i](t),t)= subs(diff(u(x,t),x$2) = subs(m=i,d2ydx2),diff(u(x,t),x) = subs(m=i,dydx),u(x,t)=u[i](t), x=i*h,rhs(ge)));od;

$$eq_1 := \frac{d}{dt} u_1(t) = \frac{u_0(t) - 2u_1(t) + u_2(t)}{h^2} - u_1(t)^2$$

$$eq_2 := \frac{d}{dt} u_2(t) = \frac{u_1(t) - 2u_2(t) + u_3(t)}{h^2} - u_2(t)^2$$

$$eq_3 := \frac{d}{dt} u_3(t) = \frac{u_2(t) - 2u_3(t) + u_4(t)}{h^2} - u_3(t)^2$$

$$eq_4 := \frac{d}{dt} u_4(t) = \frac{u_3(t) - 2u_4(t) + u_5(t)}{h^2} - u_4(t)^2$$

$$eq_5 := \frac{d}{dt} u_5(t) = \frac{u_4(t) - 2u_5(t) + u_6(t)}{h^2} - u_5(t)^2$$

$$eq_6 := \frac{d}{dt} u_6(t) = \frac{u_5(t) - 2u_6(t) + u_7(t)}{h^2} - u_6(t)^2$$

$$eq_7 := \frac{d}{dt} u_7(t) = \frac{u_6(t) - 2u_7(t) + u_8(t)}{h^2} - u_7(t)^2$$

$$eq_8 := \frac{d}{dt} u_8(t) = \frac{u_7(t) - 2u_8(t) + u_9(t)}{h^2} - u_8(t)^2$$

$$eq_9 := \frac{d}{dt} u_9(t) = \frac{u_8(t) - 2u_9(t) + u_{10}(t)}{h^2} - u_9(t)^2$$

$$eq_{10} := \frac{d}{dt} u_{10}(t) = \frac{u_9(t) - 2u_{10}(t) + u_{11}(t)}{h^2} - u_{10}(t)^2$$

The boundary values are eliminated using the boundary conditions:

> u[0](t):=(solve(eq[0],u[0](t)));

$$u_0(t) := 1$$

> u[N+1](t):=solve(eq[N+1],u[N+1](t));

$$u_{11}(t) := -\frac{1}{3} u_9(t) + \frac{4}{3} u_{10}(t)$$

5.2 Numerical Method of Lines for Parabolic Partial Differential Equations (PDEs)

> h:=L/(N+1);

$$h := \frac{1}{11}$$

> for i from 1 to N do eq[i]:=eval(eq[i]);od;

$$eq_1 := \frac{d}{dt} u_1(t) = 121 - 242\, u_1(t) + 121\, u_2(t) - u_1(t)^2$$

$$eq_2 := \frac{d}{dt} u_2(t) = 121\, u_1(t) - 242\, u_2(t) + 121\, u_3(t) - u_2(t)^2$$

$$eq_3 := \frac{d}{dt} u_3(t) = 121\, u_2(t) - 242\, u_3(t) + 121\, u_4(t) - u_3(t)^2$$

$$eq_4 := \frac{d}{dt} u_4(t) = 121\, u_3(t) - 242\, u_4(t) + 121\, u_5(t) - u_4(t)^2$$

$$eq_5 := \frac{d}{dt} u_5(t) = 121\, u_4(t) - 242\, u_5(t) + 121\, u_6(t) - u_5(t)^2$$

$$eq_6 := \frac{d}{dt} u_6(t) = 121\, u_5(t) - 242\, u_6(t) + 121\, u_7(t) - u_6(t)^2$$

$$eq_7 := \frac{d}{dt} u_7(t) = 121\, u_6(t) - 242\, u_7(t) + 121\, u_8(t) - u_7(t)^2$$

$$eq_8 := \frac{d}{dt} u_8(t) = 121\, u_7(t) - 242\, u_8(t) + 121\, u_9(t) - u_8(t)^2$$

$$eq_9 := \frac{d}{dt} u_9(t) = 121\, u_8(t) - 242\, u_9(t) + 121\, u_{10}(t) - u_9(t)^2$$

$$eq_{10} := \frac{d}{dt} u_{10}(t) = \frac{242}{3} u_9(t) - \frac{242}{3} u_{10}(t) - u_{10}(t)^2$$

> eqs:=seq((eq[j]),j=1..N);

$$eqs := \frac{d}{dt} u_1(t) = 121 - 242\, u_1(t) + 121\, u_2(t) - u_1(t)^2,\ \frac{d}{dt} u_2(t) = 121\, u_1(t) - 242\, u_2(t) + 121\, u_3(t) - u_2(t)^2,$$

$$\frac{d}{dt} u_3(t) = 121\, u_2(t) - 242\, u_3(t) + 121\, u_4(t) - u_3(t)^2,\ \frac{d}{dt} u_4(t) = 121\, u_3(t) - 242\, u_4(t) + 121\, u_5(t)$$

$$- u_4(t)^2,\ \frac{d}{dt} u_5(t) = 121\, u_4(t) - 242\, u_5(t) + 121\, u_6(t) - u_5(t)^2,\ \frac{d}{dt} u_6(t) = 121\, u_5(t) - 242\, u_6(t)$$

$$+ 121\, u_7(t) - u_6(t)^2,\ \frac{d}{dt} u_7(t) = 121\, u_6(t) - 242\, u_7(t) + 121\, u_8(t) - u_7(t)^2,\ \frac{d}{dt} u_8(t) = 121\, u_7(t)$$

$$- 242\, u_8(t) + 121\, u_9(t) - u_8(t)^2,\ \frac{d}{dt} u_9(t) = 121\, u_8(t) - 242\, u_9(t) + 121\, u_{10}(t) - u_9(t)^2,\ \frac{d}{dt} u_{10}(t)$$

$$= \frac{242}{3} u_9(t) - \frac{242}{3} u_{10}(t) - u_{10}(t)^2$$

> Y:=seq(u[i](t),i=1..N);

$$Y := u_1(t), u_2(t), u_3(t), u_4(t), u_5(t), u_6(t), u_7(t), u_8(t), u_9(t), u_{10}(t)$$

> ICs:=seq(u[i](0)=rhs(IC),i=1..N);

$$ICs := u_1(0) = 0, u_2(0) = 0, u_3(0) = 0, u_4(0) = 0, u_5(0) = 0, u_6(0) = 0, u_7(0) = 0, u_8(0) = 0, u_9(0) = 0, u_{10}(0) = 0$$

Solve the equations numerically and store the numerical solutions in U[i].

> sol:=dsolve({eqs,ICs},{Y},type=numeric,output=listprocedure);

$sol := [t = \text{proc}(t) \ldots \text{end proc}, u_1(t) = \text{proc}(t) \ldots \text{end proc}, u_2(t) = \text{proc}(t) \ldots \text{end proc}, u_3(t) = \text{proc}(t)$
$\text{end proc}, u_4(t) = \text{proc}(t) \ldots \text{end proc}, u_5(t) = \text{proc}(t) \ldots \text{end proc}, u_6(t) = \text{proc}(t) \ldots \text{end proc}, u_7(t) = \text{proc}(t)$
$\text{end proc}, u_8(t) = \text{proc}(t) \ldots \text{end proc}, u_9(t) = \text{proc}(t) \ldots \text{end proc}, u_{10}(t) = \text{proc}(t) \ldots \text{end proc}]$

> for i to N do U[i]:=subs(sol,u[i](t));od:
> U[0]:=subs(u[1](t)=U[1],u[2](t)=U[2],u[0](t));

$$U_0 := 1$$

> U[N+1]:=subs(u[N](t)=U[N],u[N-1](t)=U[N-1],u[N+1](t));

$$U_{11} := -\frac{1}{3} U_9 + \frac{4}{3} U_{10}$$

Plot the numerical solutions:

> for i from 0 to N+1 do p[i]:=plot(U[i](t),t=0..0.4,thickness=3);od:
> pp:=plot([seq(U[i](t),i=0..N+1)],t=0..0.4);

$$pp := PLOT(\ldots)$$

> display(pp,title="Figure 5.36",axes=boxed,thickness=4,labels=[t,"u"]);

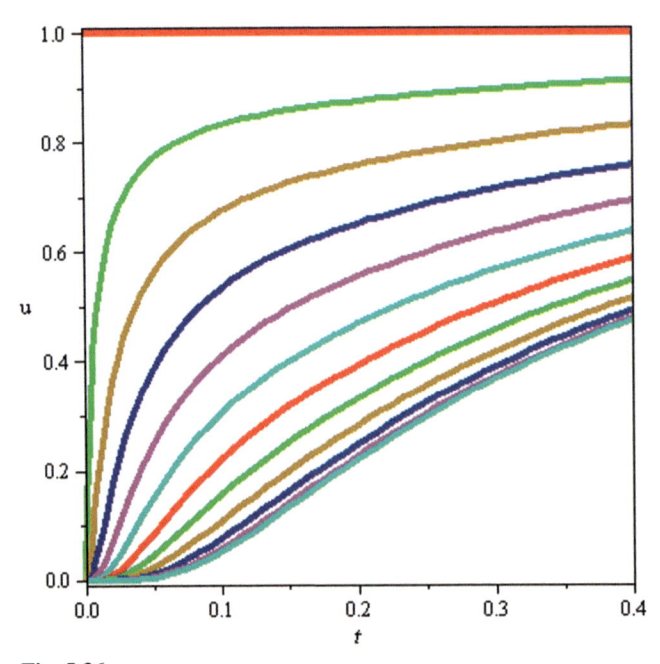

Fig. 5.36

5.2 Numerical Method of Lines for Parabolic Partial Differential Equations (PDEs)   463

> tf:=0.4;

$$tf := 0.4$$

> M:=30;

$$M := 30$$

> T1:=[seq(tf*i/M,i=0..M)];

$T1 := [0., 0.01333333333, 0.02666666667, 0.04000000000, 0.05333333333, 0.06666666667, 0.08000000000,$
$0.09333333333, 0.1066666667, 0.1200000000, 0.1333333333, 0.1466666667, 0.1600000000, 0.1733333333,$
$0.1866666667, 0.2000000000, 0.2133333333, 0.2266666667, 0.2400000000, 0.2533333333, 0.2666666667,$
$0.2800000000, 0.2933333333, 0.3066666667, 0.3200000000, 0.3333333333, 0.3466666667, 0.3600000000,$
$0.3733333333, 0.3866666667, 0.4000000000]$

> PP:=matrix(N+2,M+1);

$$PP := array(1..12, 1..31, [\,])$$

> for i from 1 to N+2 do PP[i,1]:=evalf(subs(x=(i-1)*h,rhs(IC)));od:

> for i from 1 to N+2 do for j from 2 to M+1 do
PP[i,j]:=evalf(subs(t=T1[j],U[i-1](t)));od;od:

> plotdata := [seq([seq([(i-1)*h,T1[j],PP[i,j]], i=1..N+2)], j=1..M+1)]:

> surfdata(plotdata,axes=boxed,title="Figure 5.37",
labels=[x,t,u],orientation=[-45,60]);

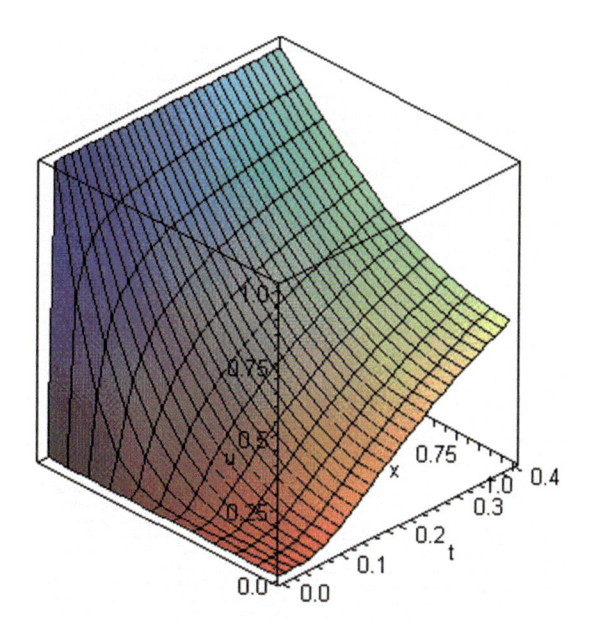

Fig. 5.37

>

Example 5.2.2. Variable Diffusivity

Consider diffusion in a slab in which the diffusion coefficient varies as a function of concentration.[15] The governing equation in dimensionless form is

$$\frac{\partial u}{\partial t} = \frac{\partial}{\partial x}\left((1+\alpha u)\frac{\partial u}{\partial x}\right)$$

$$u(x,0) = 0 \qquad (5.57)$$

$$u(0,t) = 1 \text{ and } \frac{\partial u}{\partial x}(1,t) = 0$$

This equation (for $\alpha = 1$) is solved below using the Maple program given for example 5.2.1 by just modifying the governing equation:

```
> restart;
> with(linalg):with(plots):
> ge:=diff(u(x,t),t)=diff((1+alpha*u(x,t))*diff(u(x,t),x),x);
```

$$ge := \frac{\partial}{\partial t}u(x,t) = \alpha\left(\frac{\partial}{\partial x}u(x,t)\right)^2 + (1+\alpha u(x,t))\left(\frac{\partial^2}{\partial x^2}u(x,t)\right)$$

```
> bc1:=u(x,t)-1;
```

$$bc1 := u(x,t) - 1$$

```
> bc2:=diff(u(x,t),x);
```

$$bc2 := \frac{\partial}{\partial x}u(x,t)$$

```
> IC:=u(x,0)=0;
```

$$IC := u(x,0) = 0$$

```
> N:=10;
```

$$N := 10$$

```
> L:=1;
```

$$L := 1$$

```
> alpha:=1;
```

$$\alpha := 1$$

5.2 Numerical Method of Lines for Parabolic Partial Differential Equations (PDEs) 465

> dydxf:=1/2*(-u[2](t)-3*u[0](t)+4*u[1](t))/h;

$$dydxf := \frac{1}{2} \frac{-u_2(t) - 3 u_0(t) + 4 u_1(t)}{h}$$

> dydxb:=1/2*(u[N-1](t)+3*u[N+1](t)-4*u[N](t))/h;

$$dydxb := \frac{1}{2} \frac{u_9(t) + 3 u_{11}(t) - 4 u_{10}(t)}{h}$$

> dydx:=1/2/h*(u[m+1](t)-u[m-1](t));

$$dydx := \frac{1}{2} \frac{u_{m+1}(t) - u_{m-1}(t)}{h}$$

> d2ydx2:=1/h^2*(u[m-1](t)-2*u[m](t)+u[m+1](t));

$$d2ydx2 := \frac{u_{m-1}(t) - 2 u_m(t) + u_{m+1}(t)}{h^2}$$

> bc1:=subs(diff(u(x,t),x)=dydxf,u(x,t)=u[0](t),x=0,bc1);

$$bc1 := u_0(t) - 1$$

> bc2:=subs(diff(u(x,t),x)=dydxb,u(x,t)=u[N+1](t),x=1,bc2);

$$bc2 := \frac{1}{2} \frac{u_9(t) + 3 u_{11}(t) - 4 u_{10}(t)}{h}$$

> eq[0]:=bc1;

$$eq_0 := u_0(t) - 1$$

> eq[N+1]:=bc2;

$$eq_{11} := \frac{1}{2} \frac{u_9(t) + 3 u_{11}(t) - 4 u_{10}(t)}{h}$$

> for i from 1 to N do eq[i]:=diff(u[i](t),t)= subs(diff(u(x,t),x$2) = subs(m=i,d2ydx2),diff(u(x,t),x) = subs(m=i,dydx),u(x,t)=u[i](t), x=i*h,rhs(ge));od;

$$eq_1 := \frac{d}{dt} u_1(t) = \frac{1}{4} \frac{(u_2(t) - u_0(t))^2}{h^2} + \frac{(1 + u_1(t))(u_0(t) - 2 u_1(t) + u_2(t))}{h^2}$$

$$eq_2 := \frac{d}{dt} u_2(t) = \frac{1}{4} \frac{(u_3(t) - u_1(t))^2}{h^2} + \frac{(1 + u_2(t))(u_1(t) - 2 u_2(t) + u_3(t))}{h^2}$$

$$eq_3 := \frac{d}{dt} u_3(t) = \frac{1}{4} \frac{(u_4(t) - u_2(t))^2}{h^2} + \frac{(1 + u_3(t))(u_2(t) - 2 u_3(t) + u_4(t))}{h^2}$$

$$eq_4 := \frac{d}{dt} u_4(t) = \frac{1}{4} \frac{(u_5(t) - u_3(t))^2}{h^2} + \frac{(1 + u_4(t))(u_3(t) - 2u_4(t) + u_5(t))}{h^2}$$

$$eq_5 := \frac{d}{dt} u_5(t) = \frac{1}{4} \frac{(u_6(t) - u_4(t))^2}{h^2} + \frac{(1 + u_5(t))(u_4(t) - 2u_5(t) + u_6(t))}{h^2}$$

$$eq_6 := \frac{d}{dt} u_6(t) = \frac{1}{4} \frac{(u_7(t) - u_5(t))^2}{h^2} + \frac{(1 + u_6(t))(u_5(t) - 2u_6(t) + u_7(t))}{h^2}$$

$$eq_7 := \frac{d}{dt} u_7(t) = \frac{1}{4} \frac{(u_8(t) - u_6(t))^2}{h^2} + \frac{(1 + u_7(t))(u_6(t) - 2u_7(t) + u_8(t))}{h^2}$$

$$eq_8 := \frac{d}{dt} u_8(t) = \frac{1}{4} \frac{(u_9(t) - u_7(t))^2}{h^2} + \frac{(1 + u_8(t))(u_7(t) - 2u_8(t) + u_9(t))}{h^2}$$

$$eq_9 := \frac{d}{dt} u_9(t) = \frac{1}{4} \frac{(u_{10}(t) - u_8(t))^2}{h^2} + \frac{(1 + u_9(t))(u_8(t) - 2u_9(t) + u_{10}(t))}{h^2}$$

$$eq_{10} := \frac{d}{dt} u_{10}(t) = \frac{1}{4} \frac{(u_{11}(t) - u_9(t))^2}{h^2} + \frac{(1 + u_{10}(t))(u_9(t) - 2u_{10}(t) + u_{11}(t))}{h^2}$$

> u[0](t):=(solve(eq[0],u[0](t)));

$$u_0(t) := 1$$

> u[N+1](t):=solve(eq[N+1],u[N+1](t));

$$u_{11}(t) := -\frac{1}{3} u_9(t) + \frac{4}{3} u_{10}(t)$$

> h:=L/(N+1);

$$h := \frac{1}{11}$$

> for i from 1 to N do eq[i]:=eval(eq[i]);od;

$$eq_1 := \frac{d}{dt} u_1(t) = \frac{121}{4} (u_2(t) - 1)^2 + 121 (1 + u_1(t))(1 - 2u_1(t) + u_2(t))$$

$$eq_2 := \frac{d}{dt} u_2(t) = \frac{121}{4} (u_3(t) - u_1(t))^2 + 121 (1 + u_2(t))(u_1(t) - 2u_2(t) + u_3(t))$$

$$eq_3 := \frac{d}{dt} u_3(t) = \frac{121}{4} (u_4(t) - u_2(t))^2 + 121 (1 + u_3(t))(u_2(t) - 2u_3(t) + u_4(t))$$

$$eq_4 := \frac{d}{dt} u_4(t) = \frac{121}{4} (u_5(t) - u_3(t))^2 + 121 (1 + u_4(t))(u_3(t) - 2u_4(t) + u_5(t))$$

$$eq_5 := \frac{d}{dt} u_5(t) = \frac{121}{4} (u_6(t) - u_4(t))^2 + 121 (1 + u_5(t))(u_4(t) - 2u_5(t) + u_6(t))$$

$$eq_6 := \frac{d}{dt} u_6(t) = \frac{121}{4} (u_7(t) - u_5(t))^2 + 121 (1 + u_6(t))(u_5(t) - 2u_6(t) + u_7(t))$$

$$eq_7 := \frac{d}{dt} u_7(t) = \frac{121}{4} (u_8(t) - u_6(t))^2 + 121 (1 + u_7(t))(u_6(t) - 2u_7(t) + u_8(t))$$

$$eq_8 := \frac{d}{dt} u_8(t) = \frac{121}{4} (u_9(t) - u_7(t))^2 + 121 (1 + u_8(t))(u_7(t) - 2u_8(t) + u_9(t))$$

5.2 Numerical Method of Lines for Parabolic Partial Differential Equations (PDEs) 467

$$eq_9 := \frac{d}{dt} u_9(t) = \frac{121}{4} (u_{10}(t) - u_8(t))^2 + 121 (1 + u_9(t))(u_8(t) - 2u_9(t) + u_{10}(t))$$

$$eq_{10} := \frac{d}{dt} u_{10}(t) = \frac{121}{4} \left(-\frac{4}{3} u_9(t) + \frac{4}{3} u_{10}(t)\right)^2 + 121 (1 + u_{10}(t)) \left(\frac{2}{3} u_9(t) - \frac{2}{3} u_{10}(t)\right)$$

> eqs:=seq((eq[j]),j=1..N);

$$eqs := \frac{d}{dt} u_1(t) = \frac{121}{4} (u_2(t) - 1)^2 + 121 (1 + u_1(t))(1 - 2u_1(t) + u_2(t)), \frac{d}{dt} u_2(t) = \frac{121}{4} (u_3(t) - u_1(t))^2 + 121 (1 + u_2(t))(u_1(t) - 2u_2(t) + u_3(t)), \frac{d}{dt} u_3(t) = \frac{121}{4} (u_4(t) - u_2(t))^2 + 121 (1 + u_3(t))(u_2(t) - 2u_3(t) + u_4(t)), \frac{d}{dt} u_4(t) = \frac{121}{4} (u_5(t) - u_3(t))^2 + 121 (1 + u_4(t))(u_3(t) - 2u_4(t) + u_5(t)), \frac{d}{dt} u_5(t) = \frac{121}{4} (u_6(t) - u_4(t))^2 + 121 (1 + u_5(t))(u_4(t) - 2u_5(t) + u_6(t)), \frac{d}{dt} u_6(t) = \frac{121}{4} (u_7(t) - u_5(t))^2 + 121 (1 + u_6(t))(u_5(t) - 2u_6(t) + u_7(t)), \frac{d}{dt} u_7(t) = \frac{121}{4} (u_8(t) - u_6(t))^2 + 121 (1 + u_7(t))(u_6(t) - 2u_7(t) + u_8(t)), \frac{d}{dt} u_8(t) = \frac{121}{4} (u_9(t) - u_7(t))^2 + 121 (1 + u_8(t))(u_7(t) - 2u_8(t) + u_9(t)), \frac{d}{dt} u_9(t) = \frac{121}{4} (u_{10}(t) - u_8(t))^2 + 121 (1 + u_9(t))(u_8(t) - 2u_9(t) + u_{10}(t)), \frac{d}{dt} u_{10}(t) = \frac{121}{4} \left(-\frac{4}{3} u_9(t) + \frac{4}{3} u_{10}(t)\right)^2 + 121 (1 + u_{10}(t)) \left(\frac{2}{3} u_9(t) - \frac{2}{3} u_{10}(t)\right)$$

> Y:=seq(u[i](t),i=1..N);

$$Y := u_1(t), u_2(t), u_3(t), u_4(t), u_5(t), u_6(t), u_7(t), u_8(t), u_9(t), u_{10}(t)$$

> ICs:=seq(u[i](0)=rhs(IC),i=1..N);

$ICs := u_1(0) = 0, u_2(0) = 0, u_3(0) = 0, u_4(0) = 0, u_5(0) = 0, u_6(0) = 0, u_7(0) = 0, u_8(0) = 0, u_9(0) = 0, u_{10}(0) = 0$

> sol:=dsolve({eqs,ICs},{Y},type=numeric,output=listprocedure);

$sol := [t = \mathbf{proc}(t) \ldots \mathbf{end\ proc}, u_1(t) = \mathbf{proc}(t) \ldots \mathbf{end\ proc}, u_2(t) = \mathbf{proc}(t) \ldots \mathbf{end\ proc}, u_3(t) = \mathbf{proc}(t)$

$\ldots$

$\mathbf{end\ proc}, u_4(t) = \mathbf{proc}(t) \ldots \mathbf{end\ proc}, u_5(t) = \mathbf{proc}(t) \ldots \mathbf{end\ proc}, u_6(t) = \mathbf{proc}(t) \ldots \mathbf{end\ proc}, u_7(t) = \mathbf{proc}(t)$

$\ldots$

$\mathbf{end\ proc}, u_8(t) = \mathbf{proc}(t) \ldots \mathbf{end\ proc}, u_9(t) = \mathbf{proc}(t) \ldots \mathbf{end\ proc}, u_{10}(t) = \mathbf{proc}(t) \ldots \mathbf{end\ proc}]$

> for i to N do U[i]:=subs(sol,u[i](t));od:
> U[0]:=subs(u[1](t)=U[1],u[2](t)=U[2],u[0](t));

$$U_0 := 1$$

> U[N+1]:=subs(u[N](t)=U[N],u[N-1](t)=U[N-1],u[N+1](t));

$$U_{11} := -\frac{1}{3} U_9 + \frac{4}{3} U_{10}$$

```
> for i from 0 to N+1 do p[i]:=plot(U[i](t),t=0..1,thickness=3);od:
> pp:=plot([seq(U[i](t),i=0..N+1)],t=0..0.4);
```

$$pp := PLOT(...)$$

```
> display(pp,axes=boxed,title="Figure 5.38",thickness=3,labels=[t,"u"]);
```

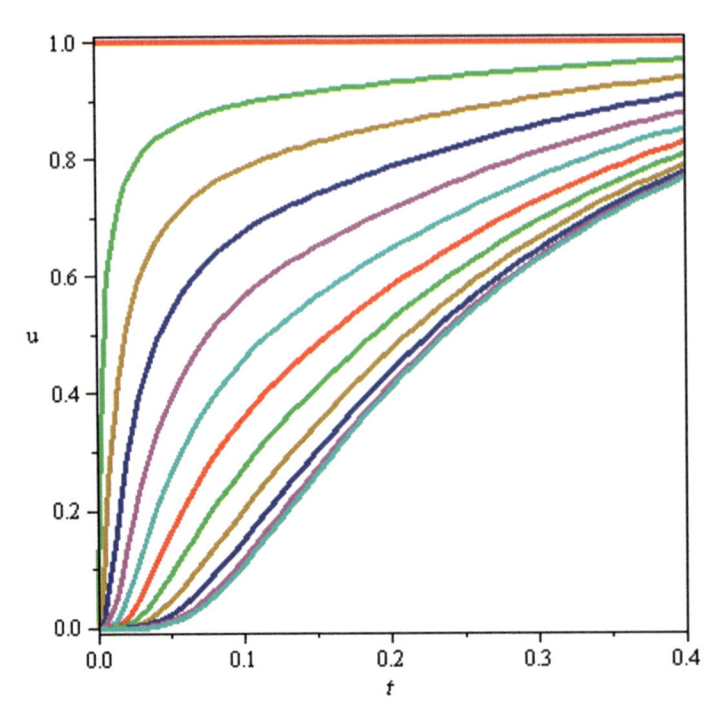

Fig. 5.38

```
> tf:=1.;
```

$$tf := 1.$$

```
> M:=30;
```

$$M := 30$$

```
> T1:=[seq(tf*i/M,i=0..M)];
```

$T1 := [0., 0.03333333333, 0.06666666667, 0.1000000000, 0.1333333333, 0.1666666667, 0.2000000000,$

$0.2333333333, 0.2666666667, 0.3000000000, 0.3333333333, 0.3666666667, 0.4000000000, 0.4333333333,$

$0.4666666667, 0.5000000000, 0.5333333333, 0.5666666667, 0.6000000000, 0.6333333333, 0.6666666667,$

$0.7000000000, 0.7333333333, 0.7666666667, 0.8000000000, 0.8333333333, 0.8666666667, 0.9000000000,$

$0.9333333333, 0.9666666667, 1.000000000]$

5.2 Numerical Method of Lines for Parabolic Partial Differential Equations (PDEs)

```
> PP:=matrix(N+2,M+1);
```

$$PP := array(1..12, 1..31, [\ ])$$

```
> for i from 1 to N+2 do PP[i,1]:=evalf(subs(x=(i-1)*h,rhs(IC)));od:
> for i from 1 to N+2 do for j from 2 to M+1 do
PP[i,j]:=evalf(subs(t=T1[j],U[i-1](t)));od;od:
> plotdata := [seq([ seq([(i-1)*h,T1[j],PP[i,j]], i=1..N+2)], j=1..M+1)]:
> surfdata(plotdata,axes=boxed,title="Figure 5.39",
labels=[x,t,u],orientation=[-45,60]);
```

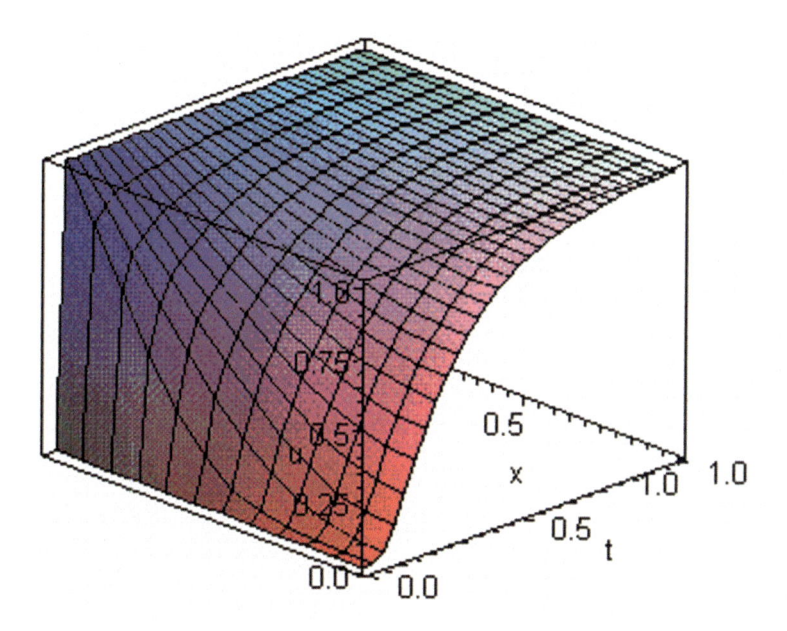

Fig. 5.39

5.2.3 Numerical Method of Lines for Parabolic PDEs with Nonlinear Boundary

Conditions

When the boundary conditions are nonlinear, the procedure described in section 5.2.2 cannot be used because the boundary values cannot be eliminated because of the nonlinear boundary conditions. This is handled by differentiating the finite difference form of the boundary condition with respect to t. This yields two additional nonlinear ODEs in time (see section 2.2.6 on DAEs), which are then solved simultaneously with N nonlinear ODEs arising from the discretization of the governing equation at N interior node points. This methodology is illustrated in the next example.

Example 5.2.3. Nonlinear Radiation at the Surface

Consider heat transfer in a slab with a nonlinear fourth order radiation boundary condition at the surface.[16] (Schiesser, 1991). The governing equation in dimensionless form is

$$\frac{\partial u}{\partial t} = \frac{\partial^2 u}{\partial x^2}$$

$$u(x,0) = 0 \tag{5.58}$$

$$u(0,t) = 1 \text{ and } \frac{\partial u}{\partial x}(1,t) = 1-u(1,t)^4$$

This equation is solved below using the Maple program given for example 5.2.1 by differentiating the boundary conditions:

> restart;

> with(linalg):with(plots):

> ge:=diff(u(x,t),t)=diff(u(x,t),x$2);

$$ge := \frac{\partial}{\partial t} u(x,t) = \frac{\partial^2}{\partial x^2} u(x,t)$$

> bc1:=u(x,t)-1;

$$bc1 := u(x,t) - 1$$

> bc2:=diff(u(x,t),x)-1+(u(x,t))^4;

$$bc2 := \frac{\partial}{\partial x} u(x,t) - 1 + u(x,t)^4$$

> IC:=u(x,0)=0;

$$IC := u(x,0) = 0$$

> N:=4;

$$N := 4$$

> L:=1;

$$L := 1$$

5.2 Numerical Method of Lines for Parabolic Partial Differential Equations (PDEs)

```
> dydxf:=1/2*(-u[2](t)-3*u[0](t)+4*u[1](t))/h;
```

$$dydxf := \frac{1}{2} \frac{-u_2(t) - 3u_0(t) + 4u_1(t)}{h}$$

```
> dydxb:=1/2*(u[N-1](t)+3*u[N+1](t)-4*u[N](t))/h;
```

$$dydxb := \frac{1}{2} \frac{u_3(t) + 3u_5(t) - 4u_4(t)}{h}$$

```
> dydx:=1/2/h*(u[m+1](t)-u[m-1](t));
```

$$dydx := \frac{1}{2} \frac{u_{m+1}(t) - u_{m-1}(t)}{h}$$

```
> d2ydx2:=1/h^2*(u[m-1](t)-2*u[m](t)+u[m+1](t));
```

$$d2ydx2 := \frac{u_{m-1}(t) - 2u_m(t) + u_{m+1}(t)}{h^2}$$

```
> bc1:=subs(diff(u(x,t),x)=dydxf,u(x,t)=u[0](t),x=0,bc1);
```

$$bc1 := u_0(t) - 1$$

```
> bc2:=subs(diff(u(x,t),x)=dydxb,u(x,t)=u[N+1](t),x=1,bc2);
```

$$bc2 := \frac{1}{2} \frac{u_3(t) + 3u_5(t) - 4u_4(t)}{h} - 1 + u_5(t)^4$$

Differentiate the boundary conditions:
```
> eq[0]:=diff(bc1,t);
```

$$eq_0 := \frac{d}{dt} u_0(t)$$

```
> eq[N+1]:=diff(bc2,t);
```

$$eq_5 := \frac{1}{2} \frac{\frac{d}{dt} u_3(t) + 3\left(\frac{d}{dt} u_5(t)\right) - 4\left(\frac{d}{dt} u_4(t)\right)}{h} + 4u_5(t)^3 \left(\frac{d}{dt} u_5(t)\right)$$

```
> for i from 1 to N do eq[i]:=diff(u[i](t),t)= subs(diff(u(x,t),x$2) =
subs(m=i,d2ydx2),diff(u(x,t),x) = subs(m=i,dydx),
u(x,t)=u[i](t),x=i*h,rhs(ge));od;
```

$$eq_1 := \frac{d}{dt} u_1(t) = \frac{u_0(t) - 2u_1(t) + u_2(t)}{h^2}$$

$$eq_2 := \frac{d}{dt} u_2(t) = \frac{u_1(t) - 2u_2(t) + u_3(t)}{h^2}$$

$$eq_3 := \frac{d}{dt} u_3(t) = \frac{u_2(t) - 2u_3(t) + u_4(t)}{h^2}$$

$$eq_4 := \frac{d}{dt} u_4(t) = \frac{u_3(t) - 2u_4(t) + u_5(t)}{h^2}$$

> h:=L/(N+1);

$$h := \frac{1}{5}$$

> for i from 0 to N+1 do eq[i]:=eval(eq[i]);od;

$$eq_0 := \frac{d}{dt} u_0(t)$$

$$eq_1 := \frac{d}{dt} u_1(t) = 25 u_0(t) - 50 u_1(t) + 25 u_2(t)$$

$$eq_2 := \frac{d}{dt} u_2(t) = 25 u_1(t) - 50 u_2(t) + 25 u_3(t)$$

$$eq_3 := \frac{d}{dt} u_3(t) = 25 u_2(t) - 50 u_3(t) + 25 u_4(t)$$

$$eq_4 := \frac{d}{dt} u_4(t) = 25 u_3(t) - 50 u_4(t) + 25 u_5(t)$$

$$eq_5 := \frac{5}{2} \frac{d}{dt} u_3(t) + \frac{15}{2} \frac{d}{dt} u_5(t) - 10 \left(\frac{d}{dt} u_4(t)\right) + 4 u_5(t)^3 \left(\frac{d}{dt} u_5(t)\right)$$

> eqs:=seq((eq[j]),j=0..N+1);

$$eqs := \frac{d}{dt} u_0(t), \frac{d}{dt} u_1(t) = 25 u_0(t) - 50 u_1(t) + 25 u_2(t), \frac{d}{dt} u_2(t) = 25 u_1(t) - 50 u_2(t) + 25 u_3(t), \frac{d}{dt} u_3(t)$$
$$= 25 u_2(t) - 50 u_3(t) + 25 u_4(t), \frac{d}{dt} u_4(t) = 25 u_3(t) - 50 u_4(t) + 25 u_5(t), \frac{5}{2} \frac{d}{dt} u_3(t) + \frac{15}{2} \frac{d}{dt} u_5(t)$$
$$- 10 \left(\frac{d}{dt} u_4(t)\right) + 4 u_5(t)^3 \left(\frac{d}{dt} u_5(t)\right)$$

Enter the initial conditions separately for the boundary values consistent with the boundary conditions:

> Y:=seq(u[i](t),i=0..N+1);

$$Y := u_0(t), u_1(t), u_2(t), u_3(t), u_4(t), u_5(t)$$

> ICs:=u[0](0)=1,seq(u[i](0)=rhs(IC),i=1..N),u[N+1](0)=0;

$$ICs := u_0(0) = 1, u_1(0) = 0, u_2(0) = 0, u_3(0) = 0, u_4(0) = 0, u_5(0) = 0$$

> sol:=dsolve({eqs,ICs},{Y},type=numeric,output=listprocedure);

$$sol := [t = \mathbf{proc}(t) \ldots \mathbf{end\ proc}, u_0(t) = \mathbf{proc}(t) \ldots \mathbf{end\ proc}, u_1(t) = \mathbf{proc}(t) \ldots \mathbf{end\ proc}, u_2(t) = \mathbf{proc}(t)$$
$$\mathbf{end\ proc}, u_3(t) = \mathbf{proc}(t) \ldots \mathbf{end\ proc}, u_4(t) = \mathbf{proc}(t) \ldots \mathbf{end\ proc}, u_5(t) = \mathbf{proc}(t) \ldots \mathbf{end\ proc}]$$

5.2 Numerical Method of Lines for Parabolic Partial Differential Equations (PDEs)   473

> for i from 0 to N+1 do U[i]:=subs(sol,u[i](t));od:

> pp:=plot([seq(U[i](t),i=0..N+1)],t=0..1);

$$pp := PLOT(...)$$

> display(pp,axes=boxed,title="Figure 5.40",thickness=4,labels=[t,"u"]);

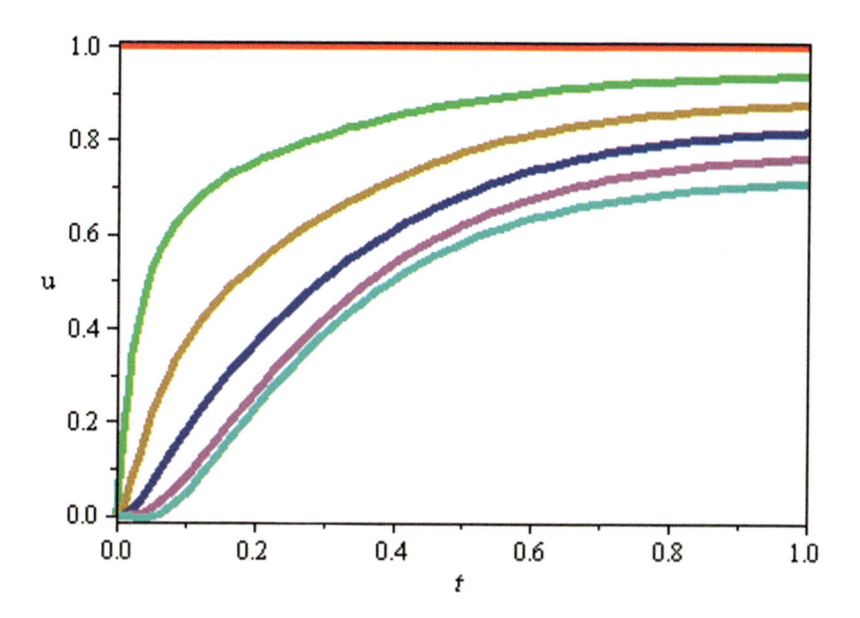

Fig. 5.40

> tf:=.2;

$$tf := 0.2$$

> M:=30;

$$M := 30$$

> T1:=[seq(tf*i/M,i=0..M)];

$T1 := [0., 0.006666666667, 0.01333333333, 0.02000000000, 0.02666666667, 0.03333333333, 0.04000000000,$
$0.04666666667, 0.05333333333, 0.06000000000, 0.06666666667, 0.07333333333, 0.08000000000,$
$0.08666666667, 0.09333333333, 0.1000000000, 0.1066666667, 0.1133333333, 0.1200000000, 0.1266666667,$
$0.1333333333, 0.1400000000, 0.1466666667, 0.1533333333, 0.1600000000, 0.1666666667, 0.1733333333,$
$0.1800000000, 0.1866666667, 0.1933333333, 0.2000000000]$

> PP:=matrix(N+2,M+1);

$$PP := array(1..6, 1..31, [\,])$$

```
> for i from 1 to N+2 do PP[i,1]:=evalf(subs(x=(i-1)*h,rhs(IC)));od:

> for i from 1 to N+2 do for j from 2 to M+1 do
PP[i,j]:=evalf(subs(t=T1[j],U[i-1](t)));od;od:

> plotdata := [seq([ seq([(i-1)*h,T1[j],PP[i,j]], i=1..N+2)], j=1..M+1)]:

> surfdata(plotdata,axes=boxed,title="Figure 5.41",
labels=[x,t,u],orientation=[-90,90]);
```

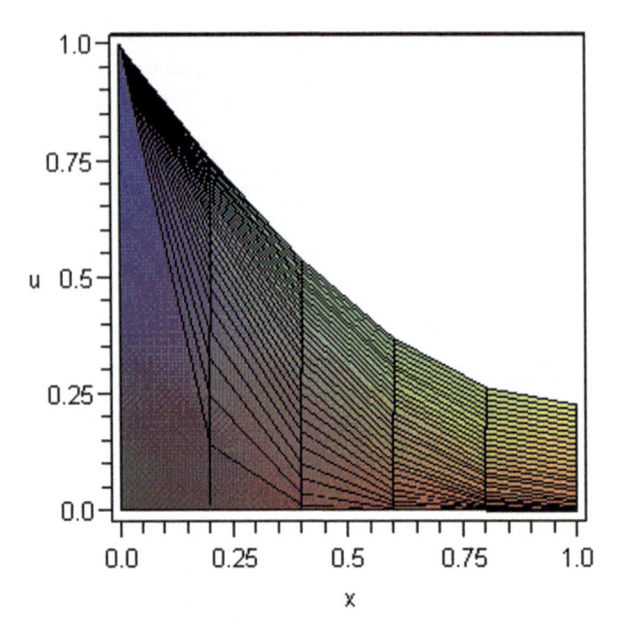

Fig. 5.41

Accuracy can be increased by increasing the number of node points.

5.2.4 Numerical Method of Lines for Stiff Nonlinear PDEs

Stiff nonlinear PDEs cannot be solved using the Runge-Kutta subroutine (see chapter 2.2.5). Maple's stiff solver can be used to solve stiff nonlinear PDEs efficiently.

Example 5.2.4. Exothermal Reaction in a Sphere

Consider heat transfer in a slab with a nonlinear fourth order radiation boundary condition at the surface.[16] (Schiesser, 1991) The governing equation in dimensionless form is

5.2 Numerical Method of Lines for Parabolic Partial Differential Equations (PDEs)

$$\frac{\partial u}{\partial t} = \frac{\partial^2 u}{\partial x^2} + \frac{2}{x}\frac{\partial u}{\partial x} + \beta\exp\left(-\frac{E}{R(u+273.16)}\right)$$

$$u(x,0) = 25 \tag{5.59}$$

$$\frac{\partial u}{\partial x}(0,t) = 0 \text{ and } u(1,t) = 158$$

where E = 30800, R = 1.987 and β the ratio of reaction rate to diffusion rate is 6.699x1017. This equation is solved below using the Maple program given for example 5.2.1 and by calling the stiff solver:

> restart;

> with(linalg):with(plots):

> ge:=diff(u(x,t),t)=diff(u(x,t),x$2)+2/x*diff(u(x,t),x)+ beta*exp(-E/R/(u(x,t)+273.16));;

$$ge := \frac{\partial}{\partial t}u(x,t) = \frac{\partial^2}{\partial x^2}u(x,t) + \frac{2\left(\frac{\partial}{\partial x}u(x,t)\right)}{x} + \beta\, e^{-\frac{E}{R(u(x,t)+273.16)}}$$

> bc1:=diff(u(x,t),x);;

$$bc1 := \frac{\partial}{\partial x}u(x,t)$$

> bc2:=u(x,t)-158;

$$bc2 := u(x,t) - 158$$

> IC:=u(x,0)=25;

$$IC := u(x,0) = 25$$

> N:=10;

$$N := 10$$

> L:=1;

$$L := 1$$

> beta:=6.699e17;E:=30800;R:=1.987;

$$\beta := 6.699\ 10^{17}$$
$$E := 30800$$
$$R := 1.987$$

> dydxf:=1/2*(-u[2](t)-3*u[0](t)+4*u[1](t))/h;

$$dydxf := \frac{1}{2}\frac{-u_2(t) - 3u_0(t) + 4u_1(t)}{h}$$

> dydxb:=1/2*(u[N-1](t)+3*u[N+1](t)-4*u[N](t))/h;

$$dydxb := \frac{1}{2}\frac{u_9(t) + 3u_{11}(t) - 4u_{10}(t)}{h}$$

> dydx:=1/2/h*(u[m+1](t)-u[m-1](t));

$$dydx := \frac{1}{2}\frac{u_{m+1}(t) - u_{m-1}(t)}{h}$$

> d2ydx2:=1/h^2*(u[m-1](t)-2*u[m](t)+u[m+1](t));

$$d2ydx2 := \frac{u_{m-1}(t) - 2u_m(t) + u_{m+1}(t)}{h^2}$$

> bc1:=subs(diff(u(x,t),x)=dydxf,u(x,t)=u[0](t),x=0,bc1);

$$bc1 := \frac{1}{2}\frac{-u_2(t) - 3u_0(t) + 4u_1(t)}{h}$$

> bc2:=subs(diff(u(x,t),x)=dydxb,u(x,t)=u[N+1](t),x=1,bc2);

$$bc2 := u_{11}(t) - 158$$

> eq[0]:=bc1;

$$eq_0 := \frac{1}{2}\frac{-u_2(t) - 3u_0(t) + 4u_1(t)}{h}$$

> eq[N+1]:=bc2;

$$eq_{11} := u_{11}(t) - 158$$

> for i from 1 to N do eq[i]:=diff(u[i](t),t)= subs(diff(u(x,t),x$2) = subs(m=i,d2ydx2),diff(u(x,t),x) = subs(m=i,dydx),u(x,t)=u[i](t), x=i*h,rhs(ge));od:

> u[0](t):=(solve(eq[0],u[0](t)));

5.2 Numerical Method of Lines for Parabolic Partial Differential Equations (PDEs) 477

$$u_0(t) := -\frac{1}{3} u_2(t) + \frac{4}{3} u_1(t)$$

> u[N+1](t):=solve(eq[N+1],u[N+1](t));

$$eq_{11} := u_{11}(t) - 158$$

> h:=L/(N+1);

$$h := \frac{1}{11}$$

> for i from 1 to N do eq[i]:=eval(eq[i]);od:
> eqs:=seq((eq[j]),j=1..N):
> Y:=seq(u[i](t),i=1..N);

$$Y := u_1(t), u_2(t), u_3(t), u_4(t), u_5(t), u_6(t), u_7(t), u_8(t), u_9(t), u_{10}(t)$$

> ICs:=seq(u[i](0)=rhs(IC),i=1..N);

$ICs := u_1(0) = 25, u_2(0) = 25, u_3(0) = 25, u_4(0) = 25, u_5(0) = 25, u_6(0) = 25, u_7(0) = 25, u_8(0) = 25, u_9(0) = 25, u_{10}(0) = 25$

Maple's stiff solver (Rosenbrock algorithm) is used by setting the option stiff = true:

> sol:=dsolve({eqs,ICs},{Y},type=numeric,stiff=true,output=listprocedure);

$sol := [t = \textbf{proc}(t) \ ... \ \textbf{end proc}, u_1(t) = \textbf{proc}(t) \ ... \ \textbf{end proc}, u_2(t) = \textbf{proc}(t) \ ... \ \textbf{end proc}, u_3(t) = \textbf{proc}(t)$
$\textbf{end proc}, u_4(t) = \textbf{proc}(t) \ ... \ \textbf{end proc}, u_5(t) = \textbf{proc}(t) \ ... \ \textbf{end proc}, u_6(t) = \textbf{proc}(t) \ ... \ \textbf{end proc}, u_7(t) = \textbf{proc}(t)$
$\textbf{end proc}, u_8(t) = \textbf{proc}(t) \ ... \ \textbf{end proc}, u_9(t) = \textbf{proc}(t) \ ... \ \textbf{end proc}, u_{10}(t) = \textbf{proc}(t) \ ... \ \textbf{end proc}]$

> for i to N do U[i]:=subs(sol,u[i](t));od:
> U[0]:=subs(u[1](t)=U[1],u[2](t)=U[2],u[0](t));

$$U_0 := -\frac{1}{3} U_2 + \frac{4}{3} U_1$$

> U[N+1]:=subs(u[N](t)=U[N],u[N-1](t)=U[N-1],u[N+1](t));

$$U_{11} := 158$$

> for i from 0 to N+1 do p[i]:=plot(U[i](t),t=0..0.4,thickness=3);od:
> display({seq(p[i],i=0..N+1)},axes=boxed,title="Figure 5.42",labels=[t,"u"]);

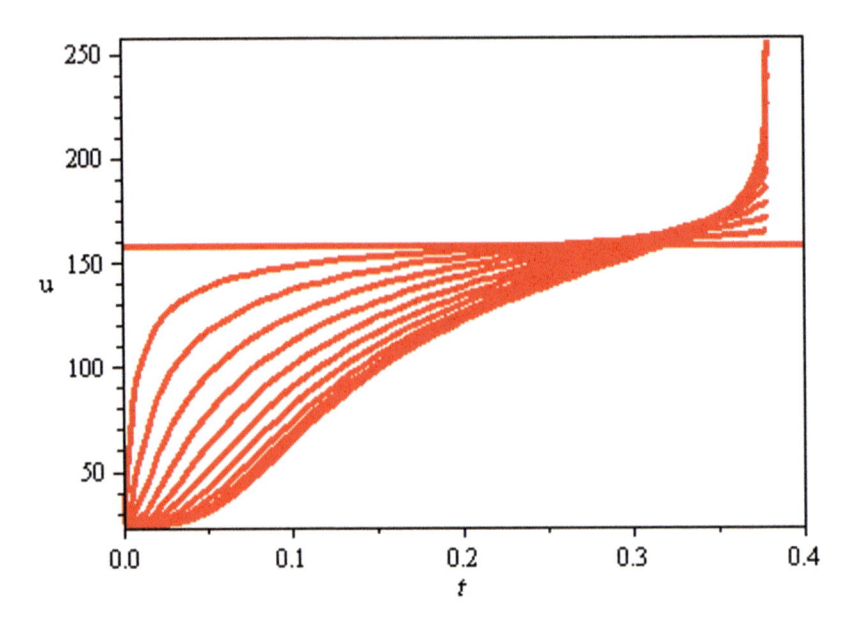

478   5 Method of Lines for Parabolic Partial Differential Equations

Fig. 5.42

> tf:=0.3784;

$$tf := 0.3784$$

> M:=30;

$$M := 30$$

> T1:=[seq(tf*i/M,i=0..M)];

$T1 := [0., 0.01261333333, 0.02522666667, 0.03784000000, 0.05045333333, 0.06306666667, 0.07568000000, 0.08829333333, 0.1009066667, 0.1135200000, 0.1261333333, 0.1387466667, 0.1513600000, 0.1639733333, 0.1765866667, 0.1892000000, 0.2018133333, 0.2144266667, 0.2270400000, 0.2396533333, 0.2522666667, 0.2648800000, 0.2774933333, 0.2901066667, 0.3027200000, 0.3153333333, 0.3279466667, 0.3405600000, 0.3531733333, 0.3657866667, 0.3784000000]$

> PP:=matrix(N+2,M+1);

$$PP := array(1..12, 1..31, [\,])$$

> for i from 1 to N+2 do PP[i,1]:=evalf(subs(x=(i-1)*h,rhs(IC)));od:

> for i from 1 to N+2 do for j from 2 to M+1 do PP[i,j]:=evalf(subs(t=T1[j],U[i-1](t)));od;od:

> plotdata := [seq([seq([(i-1)*h,T1[j],PP[i,j]], i=1..N+2)], j=1..M+1)]:

> surfdata(plotdata,axes=boxed,title="Figure 5.43", labels=[x,t,u],orientation=[-45,60]);

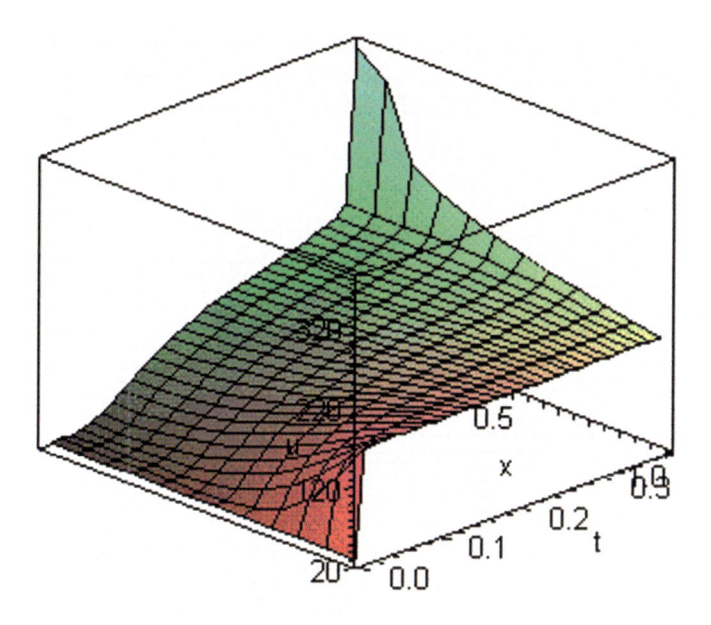

5.2 Numerical Method of Lines for Parabolic Partial Differential Equations (PDEs)   479

Fig. 5.43

The center of the sphere explodes after t = 0.3784 and the numerical calculations stop:

> plot(U[0](t),t=0..1,thickness=3,title="Figure 5.44",axes=boxed,
labels=[t,"u[0]"]);

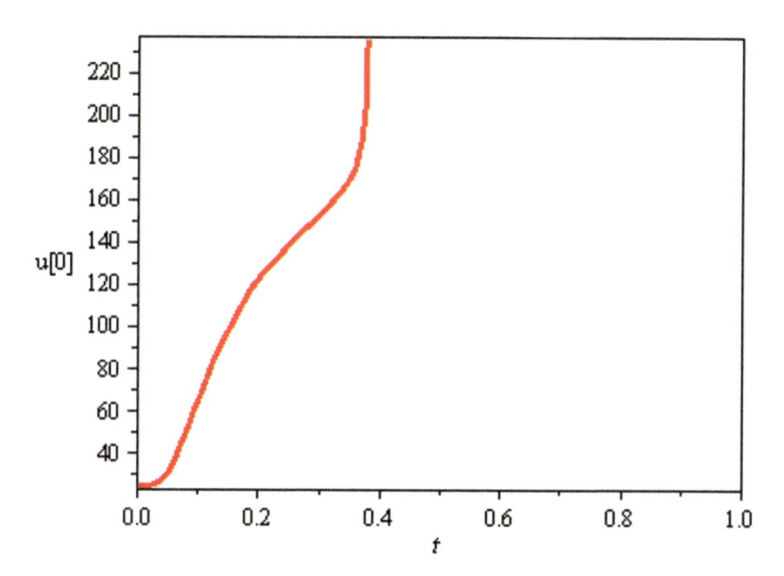

Fig. 5.44

A plot of u versus x is made. The temperature inside the sphere is less than the surface temperature of 158 ^{0}C. Because of the exothermal reaction, the temperature inside the sphere increases after t=0.36 and the process becomes unstable.

> px[0]:=plot([seq([i*h,subs(x=i*h,rhs(IC))],i=0..N+1)],thickness=3):

> for j from 2 to M+1 do
px[j]:=plot([seq([i*h,U[i](T1[j])],i=0..N+1)],thickness=3):od:;

> display({seq(px[j*2],j=0..(M+1)/2)},title="Figure 5.45",axes=boxed, labels=[x,u]);

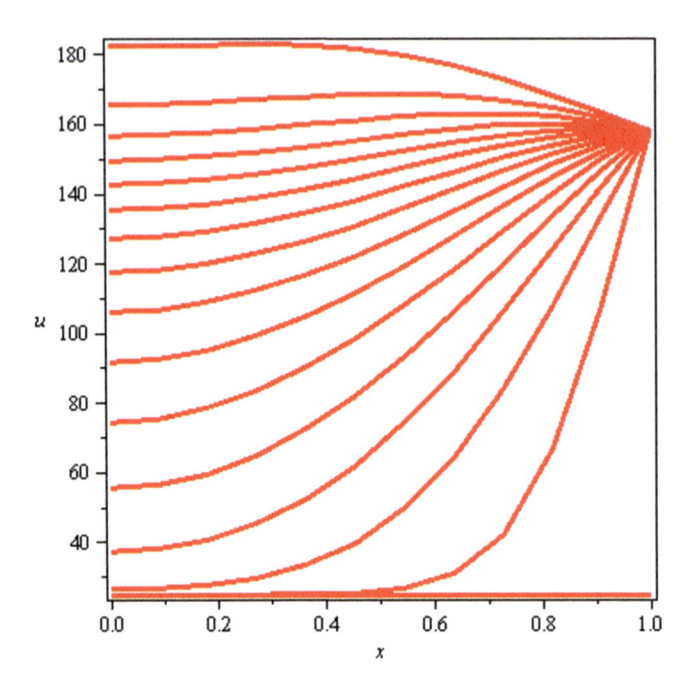

Fig. 5.45

5.2.5 Numerical Method of Lines for Nonlinear Coupled PDEs

The procedure developed for a single nonlinear PDE can be extended to solve coupled PDEs. Numerical method of lines provides an efficient way to solve nonlinear coupled PDEs.

Example 5.2.5. Two Coupled PDEs

Consider the following highly coupled nonlinear PDEs[16]

5.2 Numerical Method of Lines for Parabolic Partial Differential Equations (PDEs)

$$\frac{\partial u}{\partial t} = \frac{\partial}{\partial x}\left((v-1)\frac{\partial u}{\partial x}\right) + (16xt - 2t - 16(v-1))(u-1) + 10x\exp(-4x)$$

$$\frac{\partial v}{\partial t} = \frac{\partial^2 v}{\partial x^2} + \frac{\partial u}{\partial x} + 4(u-1) + x^2 - 2t - 10t\exp(-4x)$$

$$u(x,0) = v(x,0) = 1$$

$$u(0,t) = v(0,t) = 1$$

$$\frac{\partial u}{\partial x}(1,t) + 3(u(1,t)-1) = 0$$

$$5\frac{\partial v}{\partial x}(1,t) - \exp(4)(u(1,t)-1) = 0$$

(5.60)

Equation (5.59) is chosen for illustration, because these equations are highly nonlinear, coupled, and also have an analytical solution:

$$u = 1 + 10xt\exp(-4x) \text{ and } v = 1 + x2t \qquad (5.61)$$

Equation (5.59) is solved using the general procedure for nonlinear-coupled PDEs given below:

1. Start the Maple worksheet with a 'restart' command to clear all variables.
2. Call 'with(linalg)' and 'with(plots)' commands.
3. Enter the governing equations (ge[1] and ge[2]).
4. Store the boundary conditions in bc1[1]; bc1[2] (x = 0) and bc2[1]; bc2[2] (x = 1).
5. Enter the number of interior node points, N.
6. Enter the length of the domain, L.
7. Convert the governing equations to the finite difference form by using a central difference expression accurate to the order h^2 for the first and second derivatives.
8. u[i,1], i = 0..N+1 corresponds to the first dependent variable u and u[i,2] corresponds to the a, second dependent variable, v.
9. Convert the boundary conditions to the finite difference form by using the 3-point forward and backward differences (accurate to the order h^2).
10. If the boundary conditions are linear, eliminate the boundary values using the boundary conditions and solve the equations as in section 5.2.2.

11. If the boundary conditions are nonlinear, differentiate the finite difference form of the boundary conditions and solve the equations as in section 5.2.3.
12. Once the numerical solution is obtained, plots can be made.

Equation (5.59) is solved in Maple below using this procedure:

> restart;

> with(linalg):with(plots):

Enter the governing equations:

> ge[1]:=diff(u[1](x,t),t)=diff((u[2](x,t)-1)*diff(u[1](x,t),x),x)+
(16*x*t-2*t-16*(u[2](x,t)-1))*(u[1](x,t)-1)+10*x*exp(-4*x);

$$ge_1 := \frac{\partial}{\partial t} u_1(x,t) = \left(\frac{\partial}{\partial x} u_2(x,t)\right)\left(\frac{\partial}{\partial x} u_1(x,t)\right) + (u_2(x,t)-1)\left(\frac{\partial^2}{\partial x^2} u_1(x,t)\right) + (16xt - 2t - 16u_2(x,t) + 16)(u_1(x,t) - 1) + 10xe^{-4x}$$

> ge[2]:=diff(u[2](x,t),t)=diff(u[2](x,t),x$2)+diff(u[1](x,t),x)+
4*(u[1](x,t)-1)+x^2-2*t-10*t*exp(-4*x);

$$ge_2 := \frac{\partial}{\partial t} u_2(x,t) = \frac{\partial^2}{\partial x^2} u_2(x,t) + \frac{\partial}{\partial x} u_1(x,t) + 4u_1(x,t) - 4 + x^2 - 2t - 10te^{-4x}$$

Enter the boundary conditions at x = 0:

> bc1[1]:=u[1](x,t)-1;

$$bc1_1 := u_1(x,t) - 1$$

> bc1[2]:=u[2](x,t)-1;

$$bc1_2 := u_2(x,t) - 1$$

Enter the boundary conditions at x = 1:

> bc2[1]:=3*u[1](x,t)+diff(u[1](x,t),x)-3;

$$bc2_1 := 3u_1(x,t) + \frac{\partial}{\partial x} u_1(x,t) - 3$$

> bc2[2]:=5*diff(u[2](x,t),x)-evalf(exp(4))*(u[1](x,t)-1);

$$bc2_2 := 5\left(\frac{\partial}{\partial x} u_2(x,t)\right) - 54.59815003 u_1(x,t) + 54.5981500$$

Enter the initial conditions:

> IC[1]:=u[1](x,0)=1;

$$IC_1 := u_1(x,0) = 1$$

> IC[2]:=u[2](x,0)=1;

$$IC_2 := u_2(x,0) = 1$$

5.2 Numerical Method of Lines for Parabolic Partial Differential Equations (PDEs)

Enter the number of governing equations:

> NN:=2;

$$NN := 2$$

> N:=2;

$$N := 2$$

> L:=1;

$$L := 1$$

Develop finite difference expressions for the first and second derivatives for the given two dependent dependent variables:

> for i to NN do
> dydxf[i]:=1/2*(-u[2,i](t)-3*u[0,i](t)+4*u[1,i](t))/h;
> dydxb[i]:=1/2*(u[N-1,i](t)+3*u[N+1,i](t)-4*u[N,i](t))/h;
> dydx[i]:=1/2/h*(u[m+1,i](t)-u[m-1,i](t));
> d2ydx2[i]:=1/h^2*(u[m-1,i](t)-2*u[m,i](t)+u[m+1,i](t));od;

$$dydxf_1 := \frac{1}{2} \frac{-u_{2,1}(t) - 3u_{0,1}(t) + 4u_{1,1}(t)}{h}$$

$$dydxb_1 := \frac{1}{2} \frac{u_{1,1}(t) + 3u_{3,1}(t) - 4u_{2,1}(t)}{h}$$

$$dydx_1 := \frac{1}{2} \frac{u_{m+1,1}(t) - u_{m-1,1}(t)}{h}$$

$$d2ydx2_1 := \frac{u_{m-1,1}(t) - 2u_{m,1}(t) + u_{m+1,1}(t)}{h^2}$$

$$dydxf_2 := \frac{1}{2} \frac{-u_{2,2}(t) - 3u_{0,2}(t) + 4u_{1,2}(t)}{h}$$

$$dydxb_2 := \frac{1}{2} \frac{u_{1,2}(t) + 3u_{3,2}(t) - 4u_{2,2}(t)}{h}$$

$$dydx_2 := \frac{1}{2} \frac{u_{m+1,2}(t) - u_{m-1,2}(t)}{h}$$

$$d2ydx2_2 := \frac{u_{m-1,2}(t) - 2u_{m,2}(t) + u_{m+1,2}(t)}{h^2}$$

Convert the boundary conditions to the finite difference form:

> for i to NN do bc1[i]:=subs(diff(u[1](x,t),x)=dydxf[1],
diff(u[2](x,t),x)=dydxf[2],u[1](x,t)

=u[0,1](t),u[2](x,t)=u[0,2](t),x=0,bc1[i]);od;

$$bc1_1 := u_{0,1}(t) - 1$$

$$bc1_2 := u_{0,2}(t) - 1$$

> for i to NN do bc2[i]:=subs(diff(u[1](x,t),x)=dydxb[1],
diff(u[2](x,t),x)=dydxb[2],u[1](x,t)

=u[N+1,1](t),u[2](x,t)=u[N+1,2](t),x=L,bc2[i]);od;

$$bc2_1 := 3\,u_{3,1}(t) + \frac{1}{2}\frac{u_{1,1}(t) + 3\,u_{3,1}(t) - 4\,u_{2,1}(t)}{h} - 3$$

$$bc2_2 := \frac{5}{2}\frac{u_{1,2}(t) + 3\,u_{3,2}(t) - 4\,u_{2,2}(t)}{h} - 54.59815003\,u_{3,1}(t) + 54.59815003$$

> for i to NN do eq[0,i]:=bc1[i];eq[N+1,i]:=bc2[i];od;

$$eq_{0,1} := u_{0,1}(t) - 1$$

$$eq_{3,1} := 3\,u_{3,1}(t) + \frac{1}{2}\frac{u_{1,1}(t) + 3\,u_{3,1}(t) - 4\,u_{2,1}(t)}{h} - 3$$

$$eq_{0,2} := u_{0,2}(t) - 1$$

$$eq_{3,2} := \frac{5}{2}\frac{u_{1,2}(t) + 3\,u_{3,2}(t) - 4\,u_{2,2}(t)}{h} - 54.59815003\,u_{3,1}(t) + 54.59815003$$

Convert the first governing equation to the finite difference form:

> for i from 1 to N do eq[i,1]:=diff(u[i,1](t),t)= subs(diff(u[1](x,t),x$2) =
subs(m=i,d2ydx2[1]),

diff(u[2](x,t),x$2) = subs(m=i,d2ydx2[2]),diff(u[1](x,t),x) =
subs(m=i,dydx[1]),diff(u[2](x,t),x) = subs(m=i,dydx[2]),u[1](x,t)=u[i,1](t),
u[2](x,t)=u[i,2](t),x=i*h,rhs(ge[1]));od;

$$eq_{1,1} := \frac{d}{dt}u_{1,1}(t) = \frac{1}{4}\frac{(u_{2,2}(t) - u_{0,2}(t))(u_{2,1}(t) - u_{0,1}(t))}{h^2}$$

$$+ \frac{(u_{1,2}(t) - 1)(u_{0,1}(t) - 2u_{1,1}(t) + u_{2,1}(t))}{h^2} + (16\,h\,t - 2\,t - 16\,u_{1,2}(t) + 16)(u_{1,1}(t) - 1)$$

$$+ 10\,h\,e^{-4h}$$

5.2 Numerical Method of Lines for Parabolic Partial Differential Equations (PDEs)   485

$$eq_{2,1} := \frac{d}{dt} u_{2,1}(t) = \frac{1}{4} \frac{(u_{3,2}(t) - u_{1,2}(t))(u_{3,1}(t) - u_{1,1}(t))}{h^2}$$

$$+ \frac{(u_{2,2}(t) - 1)(u_{1,1}(t) - 2 u_{2,1}(t) + u_{3,1}(t))}{h^2} + (32 h t - 2 t - 16 u_{2,2}(t) + 16)(u_{2,1}(t) - 1)$$

$$+ 20 h e^{-8h}$$

Convert the second governing equation to the finite difference form:

> for i from 1 to N do eq[i,2]:=diff(u[i,2](t),t)= subs(diff(u[1](x,t),x$2) = subs(m=i,d2ydx2[1]),

diff(u[2](x,t),x$2) = subs(m=i,d2ydx2[2]),diff(u[1](x,t),x) = subs(m=i,dydx[1]),diff(u[2](x,t),x) = subs(m=i,dydx[2]),u[1](x,t)=u[i,1](t), u[2](x,t)=u[i,2](t),x=i*h,rhs(ge[2]));od;

$$eq_{1,2} := \frac{d}{dt} u_{1,2}(t) = \frac{u_{0,2}(t) - 2 u_{1,2}(t) + u_{2,2}(t)}{h^2} + \frac{1}{2} \frac{u_{2,1}(t) - u_{0,1}(t)}{h} + 4 u_{1,1}(t) - 4 + h^2 - 2 t$$

$$- 10 t e^{-4h}$$

$$eq_{2,2} := \frac{d}{dt} u_{2,2}(t) = \frac{u_{1,2}(t) - 2 u_{2,2}(t) + u_{3,2}(t)}{h^2} + \frac{1}{2} \frac{u_{3,1}(t) - u_{1,1}(t)}{h} + 4 u_{2,1}(t) - 4 + 4 h^2 - 2 t$$

$$- 10 t e^{-8h}$$

> for i to NN do u[0,i](t):=(solve(eq[0,i],u[0,i](t)));od;

$$u_{0,1}(t) := 1$$

$$u_{0,2}(t) := 1$$

Eliminate the boundary values:

> for i to NN do u[N+1,i](t):=(solve(eq[N+1,i],u[N+1,i](t)));od;

$$u_{3,1}(t) := \frac{1}{3} \frac{-u_{1,1}(t) + 4 u_{2,1}(t) + 6 h}{2 h + 1}$$

$$u_{3,2}(t) := -\frac{1}{2. h + 1.} (1.333333333\ 10^{-9} (5.00000000\ 10^8 u_{1,2}(t) h + 2.50000000\ 10^8 u_{1,2}(t)$$
$$- 2.000000000\ 10^9 u_{2,2}(t) h - 1.000000000\ 10^9 u_{2,2}(t) + 1.819938334\ 10^9 h u_{1,1}(t)$$
$$- 7.279753336\ 10^9 h u_{2,1}(t) + 2.\ h^2 + 5.459815003\ 10^9 h))$$

> h:=L/(N+1);

$$h := \frac{1}{3}$$

> for i from 1 to N do eq[i,1]:=eval(eq[i,1]);od;

$$eq_{1,1} := \frac{d}{dt} u_{1,1}(t) = \frac{9}{4} (u_{2,2}(t) - 1)(u_{2,1}(t) - 1) + 9 (u_{1,2}(t) - 1)(1 - 2 u_{1,1}(t) + u_{2,1}(t)) + \left(\frac{10}{3} t\right.$$

$$\left. - 16 u_{1,2}(t) + 16\right)(u_{1,1}(t) - 1) + \frac{10}{3} e^{-\frac{4}{3}}$$

$$eq_{2,1} := \frac{d}{dt} u_{2,1}(t) = \frac{9}{4} (-1.333333333 u_{1,2}(t) + 1.333333333 u_{2,2}(t) - 0.4853168888 u_{1,1}(t)$$

$$+ 1.941267556\, u_{2,1}(t) - 1.455950666) \left(-\frac{6}{5} u_{1,1}(t) + \frac{4}{5} u_{2,1}(t) + \frac{2}{5}\right) + 9\left(u_{2,2}(t) - 1\right)\left(\frac{4}{5} u_{1,1}(t)\right.$$

$$\left.- \frac{6}{5} u_{2,1}(t) + \frac{2}{5}\right) + \left(\frac{26}{3} t - 16\, u_{2,2}(t) + 16\right)\left(u_{2,1}(t) - 1\right) + \frac{20}{3} e^{-\frac{8}{3}}$$

> for i from 1 to N do eq[i,2]:=eval(eq[i,2]);od;

$$eq_{1,2} := \frac{d}{dt} u_{1,2}(t) = \frac{65}{18} - 18\, u_{1,2}(t) + 9\, u_{2,2}(t) + \frac{3}{2} u_{2,1}(t) + 4\, u_{1,1}(t) - 2t - 10\, t\, e^{-\frac{4}{3}}$$

$$eq_{2,2} := \frac{d}{dt} u_{2,2}(t) = 6.000000000\, u_{1,2}(t) - 6.00000000\, u_{2,2}(t) - 6.167852000\, u_{1,1}(t) + 22.67140800\, u_{2,1}(t)$$

$$- 16.05911156 - 2t - 10\, t\, e^{-\frac{8}{3}}$$

Store the governing equations in eqs:

> eqs:=seq(seq((eq[i,j]),i=1..N),j=1..NN);

$$eqs := \frac{d}{dt} u_{1,1}(t) = \frac{9}{4}\left(u_{2,2}(t) - 1\right)\left(u_{2,1}(t) - 1\right) + 9\left(u_{1,2}(t) - 1\right)\left(1 - 2\, u_{1,1}(t) + u_{2,1}(t)\right) + \left(\frac{10}{3} t\right.$$

$$\left.- 16\, u_{1,2}(t) + 16\right)\left(u_{1,1}(t) - 1\right) + \frac{10}{3} e^{-\frac{4}{3}},\ \frac{d}{dt} u_{2,1}(t) = \frac{9}{4}\left(-1.333333333\, u_{1,2}(t)\right.$$

$$\left.+ 1.333333333\, u_{2,2}(t) - 0.4853168888\, u_{1,1}(t) + 1.941267556\, u_{2,1}(t) - 1.455950666\right)\left(-\frac{6}{5} u_{1,1}(t)\right.$$

$$\left.+ 16\right)\left(u_{2,1}(t) - 1\right) + \frac{20}{3} e^{-\frac{8}{3}},\ \frac{d}{dt} u_{1,2}(t) = \frac{65}{18} - 18\, u_{1,2}(t) + 9\, u_{2,2}(t) + \frac{3}{2} u_{2,1}(t) + 4\, u_{1,1}(t) - 2t$$

$$+ \frac{4}{5} u_{2,1}(t) + \frac{2}{5}) + 9\left(u_{2,2}(t) - 1\right)\left(\frac{4}{5} u_{1,1}(t) - \frac{6}{5} u_{2,1}(t) + \frac{2}{5}\right) + \left(\frac{26}{3} t - 16\, u_{2,2}(t)\right.$$

$$\left.+ 16\right)\left(u_{2,1}(t) - 1\right) + \frac{20}{3} e^{-\frac{8}{3}},\ \frac{d}{dt} u_{1,2}(t) = \frac{65}{18} - 18\, u_{1,2}(t) + 9\, u_{2,2}(t) + \frac{3}{2} u_{2,1}(t) + 4\, u_{1,1}(t) - 2t$$

$$- 10\, t\, e^{-\frac{4}{3}},\ \frac{d}{dt} u_{2,2}(t) = 6.000000000\, u_{1,2}(t) - 6.00000000\, u_{2,2}(t) - 6.167852000\, u_{1,1}(t)$$

$$+ 22.67140800\, u_{2,1}(t) - 16.05911156 - 2t - 10\, t\, e^{-\frac{8}{3}}$$

Store the dependent variables in Y:

> Y:=seq(seq(u[i,j](t),i=1..N),j=1..NN);

$$Y := u_{1,1}(t), u_{2,1}(t), u_{1,2}(t), u_{2,2}(t)$$

Store the initial conditions in ICs:

> ICs:=seq(u[i,1](0)=rhs(IC[1]),i=1..N),seq(u[i,2](0)=rhs(IC[2]),i=1..N);

$$ICs := u_{1,1}(0) = 1, u_{2,1}(0) = 1, u_{1,2}(0) = 1, u_{2,2}(0) = 1$$

> sol:=dsolve({eqs,ICs},{Y},type=numeric,stiff=true,maxfun=1000000, abserr=1e-6,relerr=1e-5,output=listprocedure);

$sol := [t = \mathbf{proc}(t) \ \ldots\ \mathbf{end\ proc}, u_{1,1}(t) = \mathbf{proc}(t) \ \ldots\ \mathbf{end\ proc}, u_{1,2}(t) = \mathbf{proc}(t) \ \ldots\ \mathbf{end\ proc}, u_{2,1}(t) = \mathbf{proc}(t)$

$\mathbf{end\ proc}, u_{2,2}(t) = \mathbf{proc}(t) \ \ldots\ \mathbf{end\ proc}\,]$

5.2 Numerical Method of Lines for Parabolic Partial Differential Equations (PDEs)

```
> for j to NN do for i to N do U[i,j]:=subs(sol,u[i,j](t));od:od;
> for i to NN do U[0,i]:=subs(u[1,1](t)=U[1,1],u[1,2](t)=U[1,2],
  u[2,1](t)=U[2,1],u[2,2](t)=U[2,2],u[0,i](t));od;
```

$$U_{0,1} := 1$$

$$U_{0,2} := 1$$

```
> for i to NN do U[N+1,i]:=eval(subs(u[N,1](t)=U[N,1],u[N,2](t)=U[N,2],
  u[N-1,1](t)=U[N-1,1],u[N-1,2](t)=U[N-1,2],u[N+1,i](t)));od;
```

$$U_{3,1} := -\frac{1}{5} U_{1,1} + \frac{4}{5} U_{2,1} + \frac{2}{5}$$

$$U_{3,2} := -0.3333333332\, U_{1,2} + 1.333333333\, U_{2,2} - 0.4853168888\, U_{1,1} + 1.941267556\, U_{2,1} - 1.455950666$$

The numerical solution obtained is compared with the exact analytical solution at x = 1:

```
> ua:=1+10*x*t*exp(-4*x);
```

$$ua := 1 + 10\, x\, t\, e^{-4x}$$

```
> va:=1+x^2*t;
```

$$va := 1 + x^2 t$$

```
> U[N+1,1](1);evalf(subs(x=1.,t=1.,ua));
```

$$1.169488249$$
$$1.183156389$$

```
> evalf(U[N+1,2](1));evalf(subs(x=1.,t=1.,va));
```

$$1.961089198$$

$$2.$$

```
> U[N+1,1](2);evalf(subs(x=1.,t=2.,ua));
```

$$1.341573468$$
$$1.366312778$$

```
> evalf(U[N+1,2](2));evalf(subs(x=1.,t=2.,va));
```

$$2.920892289$$

$$3.$$

488   5 Method of Lines for Parabolic Partial Differential Equations

We obtain reasonable results even with N=2 node points.

> tf:=1.;

$$tf := 1.$$

> M:=30;

$$M := 30$$

> T1:=[seq(tf*i/M,i=0..M)];

T1 := [0., 0.03333333333, 0.06666666667, 0.1000000000, 0.1333333333, 0.1666666667, 0.2000000000, 0.2333333333, 0.2666666667, 0.3000000000, 0.3333333333, 0.3666666667, 0.4000000000, 0.4333333333, 0.4666666667, 0.5000000000, 0.5333333333, 0.5666666667, 0.6000000000, 0.6333333333, 0.6666666667, 0.7000000000, 0.7333333333, 0.7666666667, 0.8000000000, 0.8333333333, 0.8666666667, 0.9000000000, 0.9333333333, 0.9666666667, 1.000000000]

> PP:=matrix(N+2,M+1);

$$PP := array(1..4, 1..31, [\,])$$

> for i from 1 to N+2 do PP[i,1]:=evalf(subs(x=(i-1)*h,rhs(IC[1])));od:

> for i from 1 to N+2 do for j from 2 to M+1 do PP[i,j]:=evalf(subs(t=T1[j],U[i-1,1](t)));od;od:

> plotdata := [seq([seq([(i-1)*h,T1[j],PP[i,j]], i=1..N+2)], j=1..M+1)]:

> surfdata(plotdata,axes=boxed,title="Figure 5.46", labels=[x,t,"u"],orientation=[-45,60]);

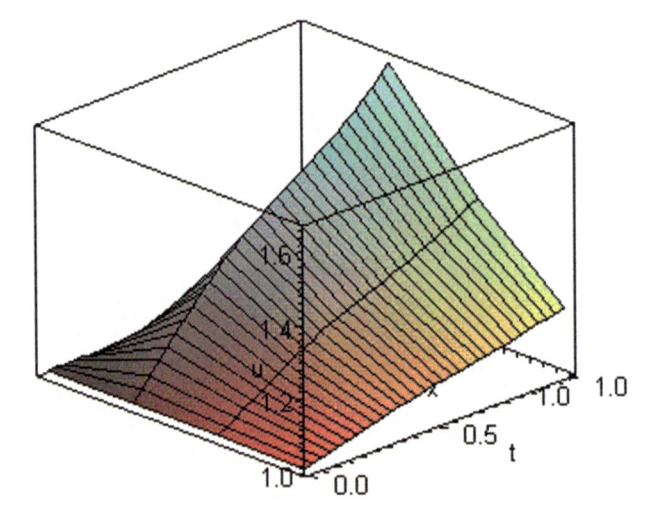

Fig. 5.46

5.2 Numerical Method of Lines for Parabolic Partial Differential Equations (PDEs)   489

> plot3d(ua,x=0..1,t=0..1,axes=boxed,title="Figure 5.47",
labels=[x,t,"ua"],orientation=[-45,60]);

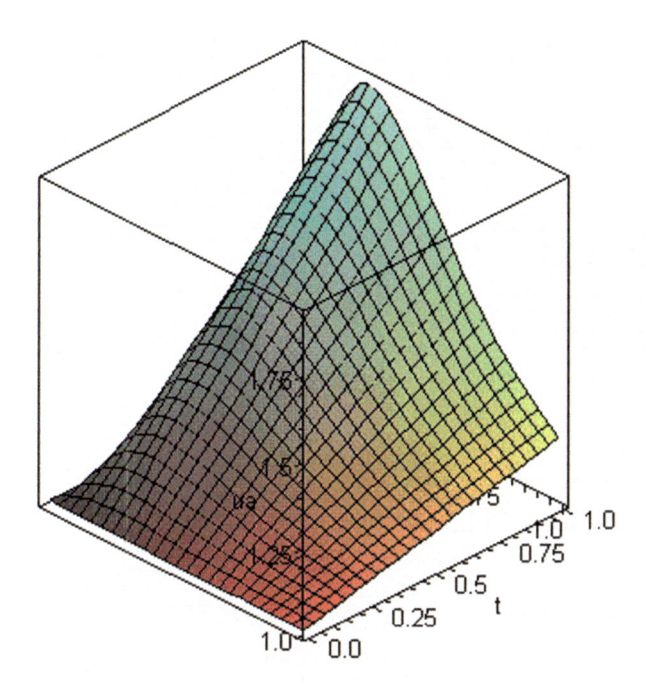

Fig. 5.47

> PP:=matrix(N+2,M+1);

$$PP := array(1..4, 1..31, [\ ])$$

> for i from 1 to N+2 do PP[i,1]:=evalf(subs(x=(i-1)*h,rhs(IC[2])));od:

> for i from 1 to N+2 do for j from 2 to M+1 do
PP[i,j]:=evalf(subs(t=T1[j],U[i-1,2](t)));od;od:

> plotdata := [seq([seq([(i-1)*h,T1[j],PP[i,j]], i=1..N+2)], j=1..M+1)]:

> surfdata(plotdata,axes=boxed,title="Figure 5.48",
labels=[x,t,"v"],orientation=[-45,60]);

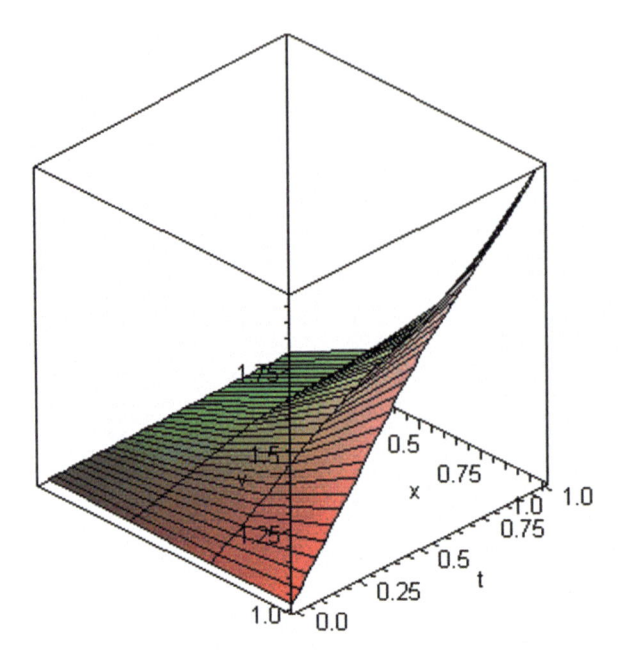

490   5 Method of Lines for Parabolic Partial Differential Equations

Fig. 5.48

> plot3d(va,x=0..1,t=0..1,labels=[x,t,"va"],axes=boxed,title="Figure 5.49", orientation=[-45,60]);

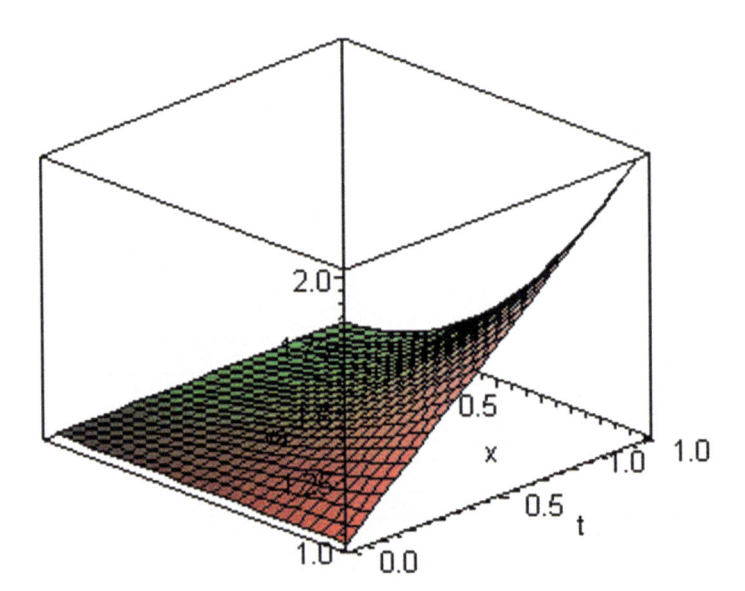

Fig. 5.49

>

5.2 Numerical Method of Lines for Parabolic Partial Differential Equations (PDEs)

The accuracy can be increased by increasing the number of interior node points. The results obtained with N = 10 are:

> U[N+1,1](1);evalf(subs(x=1.,t=1.,ua));

$$1.180862704$$
$$1.183156389$$

> evalf(U[N+1,2](1));evalf(subs(x=1.,t=1.,va));

$$1.989643591$$
$$2.$$

> U[N+1,1](2);evalf(subs(x=1.,t=2.,ua));

$$1.362175566$$
$$1.366312778$$

> evalf(U[N+1,2](2));evalf(subs(x=1.,t=2.,va));

$$2.978132652$$
$$3.$$

5.2.6 Numerical Method of Lines for Moving Boundary Problems

The procedure developed for nonlinear PDEs can be extended to solve PDEs with moving boundaries. Analytical solutions for moving problems are restricted to linear models and pseudo-steady state solutions. The numerical method of lines provides an efficient way to solve nonlinear PDEs with moving boundaries.

Example 5.2.6. The Shrinking Core Model for Catalyst Regeneration

Catalyst regeneration in a spherical particle (burning coal particle) can be represented by the following dimensionless equations[17]

$$\frac{\partial u}{\partial t} = \frac{\partial^2 u}{\partial x^2} + \frac{2}{x}\frac{\partial u}{\partial x}$$

$$u(x,0) = 0$$

$$u(x_c,t) = 0 \tag{5.62}$$

$$u(1,t) = 1$$

where x_c, the dimensionless shrinking core is governed by the flux at the shrinking interface:

$$\frac{dx_c}{dt} = -k \frac{\partial u}{\partial x}\bigg|_{x=x_c}$$

$$x_c(0) = 1 \tag{5.63}$$

$$k = \frac{c_{A0}}{\rho\phi}$$

where c_{A0} is the dimensional concentration at the surface of the particle (moles/m^3), ρ is the molar density (moles/m^3) and ϕ is the volume fraction (dimensionless). In equation (5.61) t is the dimensionless time, x is the dimensionless distance and at any particular time t, varies from x_c to 1. Even though equation (5.61) is linear, the finite difference form of the equation (5.61) involves node space h, which varies as a function of time as:

$$h = \frac{1 - x_c}{N+1} \tag{5.64}$$

Since h changes as a function of time (t), the finite difference form of equation (5.18) (5.61) becomes nonlinear. Equation (5.61) is solved in Maple below using the program developed for example 5.2.1 by solving the finite difference form of the moving boundary equation (equation (5.62)) simultaneously with the governing equations for the concentration profiles:

> restart;

> with(linalg):with(plots):

Enter the governing equation for the dimensionless concentration with the boundary and initial conditions:

> ge:=diff(u(x,t),t)=diff(u(x,t),x$2)+2/x*diff(u(x,t),x);

$$ge := \frac{\partial}{\partial t} u(x,t) = \frac{\partial^2}{\partial x^2} u(x,t) + \frac{2\left(\frac{\partial}{\partial x} u(x,t)\right)}{x}$$

> bc1:=u(x,t)-0;

$$bc1 := u(x,t)$$

> bc2:=u(x,t)-1;

$$bc2 := u(x,t) - 1$$

5.2 Numerical Method of Lines for Parabolic Partial Differential Equations (PDEs)   493

> IC:=u(x,0)=0;

$$IC := u(x, 0) = 0$$

Enter the governing equation for the shrinking interface:

> ge2:=diff(xc[1](t),t)=-k*diff(u(x,t),x);

$$ge2 := \frac{d}{dt} xc_1(t) = -k \left(\frac{\partial}{\partial x} u(x, t) \right)$$

Note that xc(t) is entered as xc[1](t) as Maple cannot handle a nonindexed variable xc with an indexed entrey u[i](t). Enter the parameter values:

> k:=0.1;

$$k := 0.1$$

> N:=10;

$$N := 10$$

> L:=1;

$$L := 1$$

> dydxf:=1/2*(-u[2](t)-3*u[0](t)+4*u[1](t))/h;

$$dydxf := \frac{1}{2} \frac{-u_2(t) - 3 u_0(t) + 4 u_1(t)}{h}$$

> dydxb:=1/2*(u[N-1](t)+3*u[N+1](t)-4*u[N](t))/h;

$$dydxb := \frac{1}{2} \frac{u_9(t) + 3 u_{11}(t) - 4 u_{10}(t)}{h}$$

> dydx:=1/2/h*(u[m+1](t)-u[m-1](t));

$$dydx := \frac{1}{2} \frac{u_{m+1}(t) - u_{m-1}(t)}{h}$$

> d2ydx2:=1/h^2*(u[m-1](t)-2*u[m](t)+u[m+1](t));

$$d2ydx2 := \frac{u_{m-1}(t) - 2 u_m(t) + u_{m+1}(t)}{h^2}$$

> bc1:=subs(diff(u(x,t),x)=dydxf,u(x,t)=u[0](t),bc1);

$$bc1 := u_0(t)$$

> bc2:=subs(diff(u(x,t),x)=dydxb,u(x,t)=u[N+1](t),bc2);

$$bc2 := u_{11}(t) - 1$$

Convert the moving boundary equation to the finite difference form:

> eqX:=subs(diff(u(x,t),x)=dydxf,u(x,t)=u[0](t),ge2);

$$eqX := \frac{d}{dt} xc_1(t) = -\frac{0.05000000000(-u_2(t) - 3u_0(t) + 4u_1(t))}{h}$$

> eq[0]:=bc1;

$$eq_0 := u_0(t)$$

> eq[N+1]:=bc2;

$$eq_{11} := u_{11}(t) - 1$$

> for i from 1 to N do eq[i]:=diff(u[i](t),t)= subs(diff(u(x,t),x$2) = subs(m=i,d2ydx2),diff(u(x,t),x) = subs(m=i,dydx),u(x,t)=u[i](t), x=xc[1](t)+i*h,rhs(ge));od;

$$eq_1 := \frac{d}{dt} u_1(t) = \frac{u_0(t) - 2u_1(t) + u_2(t)}{h^2} + \frac{u_2(t) - u_0(t)}{(xc_1(t) + h)h}$$

$$eq_2 := \frac{d}{dt} u_2(t) = \frac{u_1(t) - 2u_2(t) + u_3(t)}{h^2} + \frac{u_3(t) - u_1(t)}{(xc_1(t) + 2h)h}$$

$$eq_3 := \frac{d}{dt} u_3(t) = \frac{u_2(t) - 2u_3(t) + u_4(t)}{h^2} + \frac{u_4(t) - u_2(t)}{(xc_1(t) + 3h)h}$$

$$eq_4 := \frac{d}{dt} u_4(t) = \frac{u_3(t) - 2u_4(t) + u_5(t)}{h^2} + \frac{u_5(t) - u_3(t)}{(xc_1(t) + 4h)h}$$

$$eq_5 := \frac{d}{dt} u_5(t) = \frac{u_4(t) - 2u_5(t) + u_6(t)}{h^2} + \frac{u_6(t) - u_4(t)}{(xc_1(t) + 5h)h}$$

$$eq_6 := \frac{d}{dt} u_6(t) = \frac{u_5(t) - 2u_6(t) + u_7(t)}{h^2} + \frac{u_7(t) - u_5(t)}{(xc_1(t) + 6h)h}$$

$$eq_7 := \frac{d}{dt} u_7(t) = \frac{u_6(t) - 2u_7(t) + u_8(t)}{h^2} + \frac{u_8(t) - u_6(t)}{(xc_1(t) + 7h)h}$$

$$eq_8 := \frac{d}{dt} u_8(t) = \frac{u_7(t) - 2u_8(t) + u_9(t)}{h^2} + \frac{u_9(t) - u_7(t)}{(xc_1(t) + 8h)h}$$

$$eq_9 := \frac{d}{dt} u_9(t) = \frac{u_8(t) - 2u_9(t) + u_{10}(t)}{h^2} + \frac{u_{10}(t) - u_8(t)}{(xc_1(t) + 9h)h}$$

$$eq_{10} := \frac{d}{dt} u_{10}(t) = \frac{u_9(t) - 2u_{10}(t) + u_{11}(t)}{h^2} + \frac{u_{11}(t) - u_9(t)}{(xc_1(t) + 10h)h}$$

5.2 Numerical Method of Lines for Parabolic Partial Differential Equations (PDEs)

```
> u[0](t):=(solve(eq[0],u[0](t)));
```

$$u_0(t) := 0$$

```
> u[N+1](t):=solve(eq[N+1],u[N+1](t));
```

$$u_{11}(t) := 1$$

The node spacing varies as a function time as:

```
> h:=(1-xc[1](t))/(N+1);
```

$$h := \frac{1}{11} - \frac{1}{11} xc_1(t)$$

```
> for i from 1 to N do eq[i]:=eval(eq[i]);od;
```

$$eq_1 := \frac{d}{dt} u_1(t) = \frac{-2 u_1(t) + u_2(t)}{\left(\frac{1}{11} - \frac{1}{11} xc_1(t)\right)^2} + \frac{u_2(t)}{\left(\frac{10}{11} xc_1(t) + \frac{1}{11}\right)\left(\frac{1}{11} - \frac{1}{11} xc_1(t)\right)}$$

$$eq_2 := \frac{d}{dt} u_2(t) = \frac{u_1(t) - 2 u_2(t) + u_3(t)}{\left(\frac{1}{11} - \frac{1}{11} xc_1(t)\right)^2} + \frac{u_3(t) - u_1(t)}{\left(\frac{9}{11} xc_1(t) + \frac{2}{11}\right)\left(\frac{1}{11} - \frac{1}{11} xc_1(t)\right)}$$

$$eq_3 := \frac{d}{dt} u_3(t) = \frac{u_2(t) - 2 u_3(t) + u_4(t)}{\left(\frac{1}{11} - \frac{1}{11} xc_1(t)\right)^2} + \frac{u_4(t) - u_2(t)}{\left(\frac{8}{11} xc_1(t) + \frac{3}{11}\right)\left(\frac{1}{11} - \frac{1}{11} xc_1(t)\right)}$$

$$eq_4 := \frac{d}{dt} u_4(t) = \frac{u_3(t) - 2 u_4(t) + u_5(t)}{\left(\frac{1}{11} - \frac{1}{11} xc_1(t)\right)^2} + \frac{u_5(t) - u_3(t)}{\left(\frac{7}{11} xc_1(t) + \frac{4}{11}\right)\left(\frac{1}{11} - \frac{1}{11} xc_1(t)\right)}$$

$$eq_5 := \frac{d}{dt} u_5(t) = \frac{u_4(t) - 2 u_5(t) + u_6(t)}{\left(\frac{1}{11} - \frac{1}{11} xc_1(t)\right)^2} + \frac{u_6(t) - u_4(t)}{\left(\frac{6}{11} xc_1(t) + \frac{5}{11}\right)\left(\frac{1}{11} - \frac{1}{11} xc_1(t)\right)}$$

$$eq_6 := \frac{d}{dt} u_6(t) = \frac{u_5(t) - 2 u_6(t) + u_7(t)}{\left(\frac{1}{11} - \frac{1}{11} xc_1(t)\right)^2} + \frac{u_7(t) - u_5(t)}{\left(\frac{5}{11} xc_1(t) + \frac{6}{11}\right)\left(\frac{1}{11} - \frac{1}{11} xc_1(t)\right)}$$

$$eq_7 := \frac{d}{dt} u_7(t) = \frac{u_6(t) - 2 u_7(t) + u_8(t)}{\left(\frac{1}{11} - \frac{1}{11} xc_1(t)\right)^2} + \frac{u_8(t) - u_6(t)}{\left(\frac{4}{11} xc_1(t) + \frac{7}{11}\right)\left(\frac{1}{11} - \frac{1}{11} xc_1(t)\right)}$$

$$eq_8 := \frac{d}{dt} u_8(t) = \frac{u_7(t) - 2 u_8(t) + u_9(t)}{\left(\frac{1}{11} - \frac{1}{11} xc_1(t)\right)^2} + \frac{u_9(t) - u_7(t)}{\left(\frac{3}{11} xc_1(t) + \frac{8}{11}\right)\left(\frac{1}{11} - \frac{1}{11} xc_1(t)\right)}$$

$$eq_9 := \frac{d}{dt} u_9(t) = \frac{u_8(t) - 2u_9(t) + u_{10}(t)}{\left(\frac{1}{11} - \frac{1}{11} xc_1(t)\right)^2} + \frac{u_{10}(t) - u_8(t)}{\left(\frac{2}{11} xc_1(t) + \frac{9}{11}\right)\left(\frac{1}{11} - \frac{1}{11} xc_1(t)\right)}$$

$$eq_{10} := \frac{d}{dt} u_{10}(t) = \frac{u_9(t) - 2u_{10}(t) + 1}{\left(\frac{1}{11} - \frac{1}{11} xc_1(t)\right)^2} + \frac{1 - u_9(t)}{\left(\frac{1}{11} xc_1(t) + \frac{10}{11}\right)\left(\frac{1}{11} - \frac{1}{11} xc_1(t)\right)}$$

Now eqX is solved simultaneously with the finite difference governing equations (eq[i], i = 1..N)

> eqX:=eval(eqX);

$$eqX := \frac{d}{dt} xc_1(t) = -\frac{0.05000000000 \left(-u_2(t) + 4u_1(t)\right)}{\frac{1}{11} - \frac{1}{11} xc_1(t)}$$

> eqs:=seq((eq[j]),j=1..N),eqX;

$$eqs := \frac{d}{dt} u_1(t) = \frac{-2u_1(t) + u_2(t)}{\left(\frac{1}{11} - \frac{1}{11} xc_1(t)\right)^2} + \frac{u_2(t)}{\left(\frac{10}{11} xc_1(t) + \frac{1}{11}\right)\left(\frac{1}{11} - \frac{1}{11} xc_1(t)\right)}, \frac{d}{dt} u_2(t)$$

$$= \frac{u_1(t) - 2u_2(t) + u_3(t)}{\left(\frac{1}{11} - \frac{1}{11} xc_1(t)\right)^2} + \frac{u_3(t) - u_1(t)}{\left(\frac{9}{11} xc_1(t) + \frac{2}{11}\right)\left(\frac{1}{11} - \frac{1}{11} xc_1(t)\right)}, \frac{d}{dt} u_3(t)$$

$$= \frac{u_2(t) - 2u_3(t) + u_4(t)}{\left(\frac{1}{11} - \frac{1}{11} xc_1(t)\right)^2} + \frac{u_4(t) - u_2(t)}{\left(\frac{8}{11} xc_1(t) + \frac{3}{11}\right)\left(\frac{1}{11} - \frac{1}{11} xc_1(t)\right)}, \frac{d}{dt} u_4(t)$$

$$= \frac{u_3(t) - 2u_4(t) + u_5(t)}{\left(\frac{1}{11} - \frac{1}{11} xc_1(t)\right)^2} + \frac{u_5(t) - u_3(t)}{\left(\frac{7}{11} xc_1(t) + \frac{4}{11}\right)\left(\frac{1}{11} - \frac{1}{11} xc_1(t)\right)}, \frac{d}{dt} u_5(t)$$

$$= \frac{u_4(t) - 2u_5(t) + u_6(t)}{\left(\frac{1}{11} - \frac{1}{11} xc_1(t)\right)^2} + \frac{u_6(t) - u_4(t)}{\left(\frac{6}{11} xc_1(t) + \frac{5}{11}\right)\left(\frac{1}{11} - \frac{1}{11} xc_1(t)\right)}, \frac{d}{dt} u_6(t)$$

$$= \frac{u_5(t) - 2u_6(t) + u_7(t)}{\left(\frac{1}{11} - \frac{1}{11} xc_1(t)\right)^2} + \frac{u_7(t) - u_5(t)}{\left(\frac{5}{11} xc_1(t) + \frac{6}{11}\right)\left(\frac{1}{11} - \frac{1}{11} xc_1(t)\right)}, \frac{d}{dt} u_7(t)$$

$$= \frac{u_6(t) - 2u_7(t) + u_8(t)}{\left(\frac{1}{11} - \frac{1}{11} xc_1(t)\right)^2} + \frac{u_8(t) - u_6(t)}{\left(\frac{4}{11} xc_1(t) + \frac{7}{11}\right)\left(\frac{1}{11} - \frac{1}{11} xc_1(t)\right)}, \frac{d}{dt} u_8(t)$$

$$= \frac{u_7(t) - 2u_8(t) + u_9(t)}{\left(\frac{1}{11} - \frac{1}{11} xc_1(t)\right)^2} + \frac{u_9(t) - u_7(t)}{\left(\frac{3}{11} xc_1(t) + \frac{8}{11}\right)\left(\frac{1}{11} - \frac{1}{11} xc_1(t)\right)}, \frac{d}{dt} u_9(t)$$

$$= \frac{u_8(t) - 2u_9(t) + u_{10}(t)}{\left(\frac{1}{11} - \frac{1}{11} xc_1(t)\right)^2} + \frac{u_{10}(t) - u_8(t)}{\left(\frac{2}{11} xc_1(t) + \frac{9}{11}\right)\left(\frac{1}{11} - \frac{1}{11} xc_1(t)\right)}, \frac{d}{dt} u_{10}(t)$$

5.2 Numerical Method of Lines for Parabolic Partial Differential Equations (PDEs)   497

$$= \frac{u_9(t) - 2u_{10}(t) + 1}{\left(\frac{1}{11} - \frac{1}{11}xc_1(t)\right)^2} + \frac{1 - u_9(t)}{\left(\frac{1}{11}xc_1(t) + \frac{10}{11}\right)\left(\frac{1}{11} - \frac{1}{11}xc_1(t)\right)}, \frac{d}{dt}xc_1(t) =$$

$$-\frac{0.05000000000\,(-u_2(t) + 4u_1(t))}{\frac{1}{11} - \frac{1}{11}xc_1(t)}$$

> Y:=seq(u[i](t),i=1..N),xc[1](t);

$$Y := u_1(t), u_2(t), u_3(t), u_4(t), u_5(t), u_6(t), u_7(t), u_8(t), u_9(t), u_{10}(t), xc_1(t)$$

The initial condition for xc[1](T) is taken to be 0.999999 instead of 1 to avoid singularity in the governeing equations:

> ICs:=seq(u[i](0)=rhs(IC),i=1..N),xc[1](0)=0.999999;

$ICs := u_1(0) = 0, u_2(0) = 0, u_3(0) = 0, u_4(0) = 0, u_5(0) = 0, u_6(0) = 0, u_7(0) = 0, u_8(0) = 0, u_9(0) = 0, u_{10}(0)$

$= 0, xc_1(0) = 0.999999$

> sol:=dsolve({eqs,ICs},{Y},type=numeric,stiff=true,abserr=1e-20,
stop_cond=[u[2](t)-1,xc[1](t)],output=listprocedure);

$sol := [t = \mathbf{proc}(t)\ ...\ \mathbf{end\ proc}, u_1(t) = \mathbf{proc}(t)\ ...\ \mathbf{end\ proc}, u_2(t) = \mathbf{proc}(t)\ ...\ \mathbf{end\ proc}, u_3(t) = \mathbf{proc}(t)$

...

$\mathbf{end\ proc}, u_4(t) = \mathbf{proc}(t)\ ...\ \mathbf{end\ proc}, u_5(t) = \mathbf{proc}(t)\ ...\ \mathbf{end\ proc}, u_6(t) = \mathbf{proc}(t)\ ...\ \mathbf{end\ proc}, u_7(t) = \mathbf{proc}(t)$

...

$\mathbf{end\ proc}, u_8(t) = \mathbf{proc}(t)\ ...\ \mathbf{end\ proc}, u_9(t) = \mathbf{proc}(t)\ ...\ \mathbf{end\ proc}, u_{10}(t) = \mathbf{proc}(t)\ ...\ \mathbf{end\ proc}, xc_1(t) = \mathbf{proc}(t)$

...

$\mathbf{end\ proc}\,]$

Maple's stop condition is used to halt the computation when u[2](t) becomes 1 or xc[1](t) becomes zero.

> for i to N do U[i]:=subs(sol,u[i](t));od:

> U[0]:=subs(u[1](t)=U[1],u[2](t)=U[2],u[0](t));

$$U_0 := 0$$

> U[N+1]:=subs(u[N](t)=U[N],u[N-1](t)=U[N-1],u[N+1](t));

$$U_{11} := 1$$

> Xc:=subs(sol,xc[1](t));

$$Xc := \mathbf{proc}(t)\ ...\ \mathbf{end\ proc}$$

> sol(2.);

Warning, cannot evaluate the solution further right of 1.8213372, stop condition #2 violated
Warning, cannot evaluate the solution further right of 1.8213372, stop condition #2 violated
Warning, cannot evaluate the solution further right of 1.8213372, stop condition #2 violated
Warning, cannot evaluate the solution further right of 1.8213372, stop condition #2 violated
Warning, cannot evaluate the solution further right of 1.8213372, stop condition #2 violated
Warning, cannot evaluate the solution further right of 1.8213372, stop condition #2 violated
Warning, cannot evaluate the solution further right of 1.8213372, stop condition #2 violated
Warning, cannot evaluate the solution further right of 1.8213372, stop condition #2 violated
Warning, cannot evaluate the solution further right of 1.8213372, stop condition #2 violated
Warning, cannot evaluate the solution further right of 1.8213372, stop condition #2 violated
Warning, cannot evaluate the solution further right of 1.8213372, stop condition #2 violated
Warning, cannot evaluate the solution further right of 1.8213372, stop condition #2 violated

$$[t(2.) = 1.82133726317519740, u_1(t)(2.) = 0.802350250154852618, u_2(t)(2.) = 0.835678564920583566, u_3(t)(2.)$$
$$= 0.864423811797214281, u_4(t)(2.) = 0.889601891633213126, u_5(t)(2.) = 0.911842776442204461, u_6(t)(2.)$$
$$= 0.931564836408688590, u_7(t)(2.) = 0.949057252908217296, u_8(t)(2.) = 0.964524022333724584, u_9(t)(2.)$$
$$= 0.978109356914030226, u_{10}(t)(2.) = 0.989913060274005052, xc_1(t)(2.) = -7.030373462210692821 \cdot 10^{-17}]$$

Stop condition 2 has been violated which means that the shrinking interface xc has shrunk to zero.

> tf:=1.8213373*0.99;

$$\mathit{tf} := 1.803123927$$

The shrinking interace is plotted as a function of time as:

> plot(Xc(t),t=0..tf,axes=boxed,title="Figure 5.50",
thickness=3,labels=[t,"xc"]);

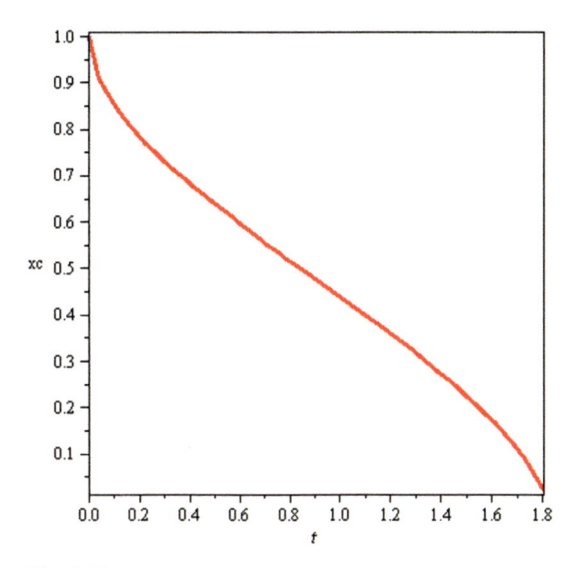

5.2 Numerical Method of Lines for Parabolic Partial Differential Equations (PDEs)   499

Fig. 5.50

Concentration profiles are plotted below:

> pp:=plot([seq(U[i](t),i=0..N+1)],t=0..tf);

$$pp := PLOT(\dots)$$

> display(pp,axes=boxed,title="Figure 5.51",thickness=3,labels=[t,"u"]);

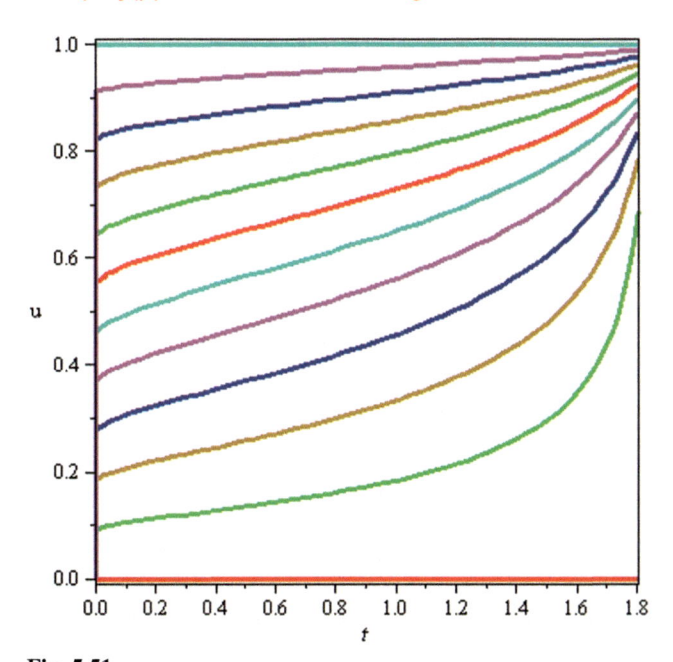

Fig. 5.51

500   5 Method of Lines for Parabolic Partial Differential Equations

> h:=subs(xc[1]=Xc,h);

$$h := \frac{1}{11} - \frac{1}{11} Xc(t)$$

> for j from 0 to 20 do
p[j]:=plot([seq([evalf(subs(t=tf*j/20,Xc(t)+i*h)),evalf(subs(t=tf*j/20,
U[i](t)))],i=0..N+1)]):od:
> display({seq(p[j],j=0..20)},thickness=4,axes=boxed,title="Figure 5.52",
labels=[x,"u"]);

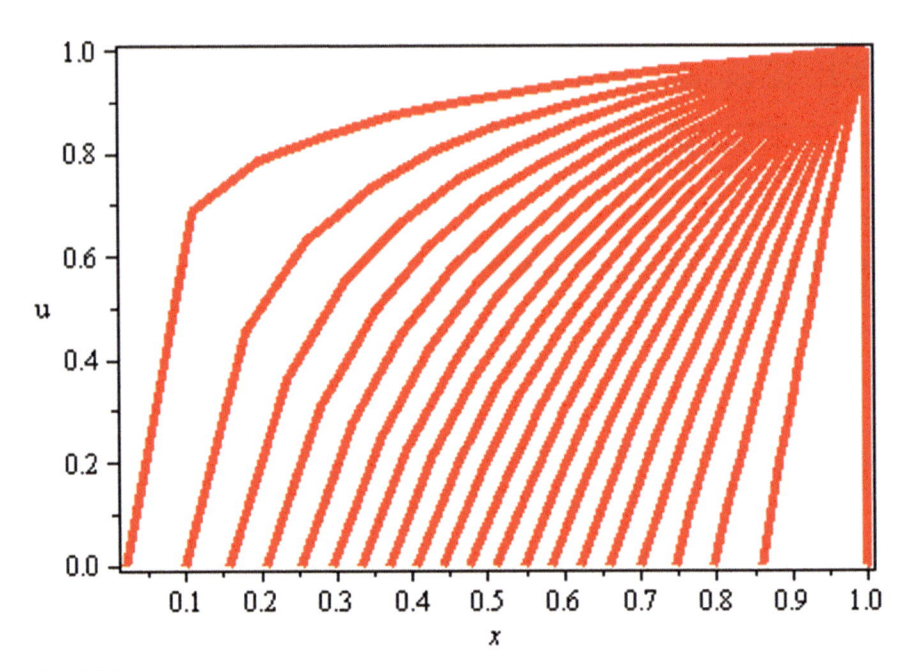

Fig. 5.52

> M:=30;

$$M := 30$$

> T1:=[seq(tf*i/M,i=0..M)];

$T1 := $ [0., 0.06010413090, 0.1202082618, 0.1803123927, 0.2404165236, 0.3005206545, 0.3606247853,

0.4207289163, 0.4808330473, 0.5409371780, 0.6010413090, 0.6611454400, 0.7212495707, 0.7813537017,

0.8414578327, 0.9015619633, 0.9616660943, 1.021770225, 1.081874356, 1.141978487, 1.202082618,

1.262186749, 1.322290880, 1.382395011, 1.442499142, 1.502603273, 1.562707403, 1.622811534, 1.682915665,

1.743019796, 1.803123927]

5.2 Numerical Method of Lines for Parabolic Partial Differential Equations (PDEs)

```
> PP:=matrix(N+2,M+1);
```

$$PP := array(1..12, 1..31, [\ ])$$

```
> for i from 1 to N+2 do PP[i,1]:=evalf(subs(x=(i-1)*L/(N+1),rhs(IC)));od:
> for i from 1 to N+2 do for j from 1 to M+1 do
PP[i,j]:=evalf(subs(t=T1[j],U[i-1](t)));od;od:
> plotdata := [seq([ seq([eval(subs(t=T1[j],Xc(t)+(i-1)*h)),T1[j],PP[i,j]],
i=1..N+2)], j=1..M+1)]:
> surfdata(plotdata,axes=boxed,title="Figure 5.53",
labels=[x,t,u],orientation=[-90,0]);
```

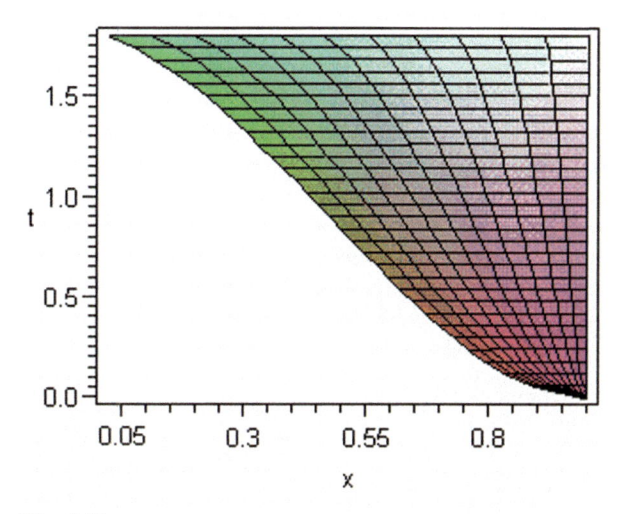

Fig. 5.53

The preceding figure shows how the radius of the particles shrinks with time.

5.2.7 Summary

In this chapter nonlinear parabolic PDEs were solved numerically using numerical method of lines. In section 5.2.2, the given nonlinear parabolic PDE with linear boundary conditions was converted to a system of nonlinear ODEs in time by applying finite differences in the spatial direction. The resulting system of nonlinear ODEs was then integrated numerically in time using Maple's 'dsolve' command. This methodology solves the dependent variables at different node points numerically in time. This is a powerful technique and is valid for all parabolic PDEs. This technique was then extended to parabolic PDEs with nonlinear boundary conditions in section 5.2.3 by differentiating the finite difference form of the boundary conditions. The numerical method of lines developed in this chapter is a powerful technique capable of handling most of the parabolic PDEs in the literature.

In section 5.2.4, a stiff nonlinear PDE was solved using numerical method of lines. This stiff problem was handled by calling Maple's stiff solver. The temperature explodes after a certain time. The numerical method of lines (NMOL) technique was then extended to coupled nonlinear parabolic PDEs in section 5.2.5. By comparing with the analytical solution, we observed that NMOL predicts the behavior accurately.

In section 5.2.6, NMOL was extended to moving boundary problems. For moving boundary problems, the length of the domain changes with time. The finite difference equations for the PDEs were solved simultaneously with the governing equation for the moving boundary. For this purpose, the moving boundary equation is converted to finite difference form. NMOL can be used solve the moving boundary problem accurately and efficiently. A total of six examples were solved in this chapter.

5.2.8 Exercise Problems

1. Complete the details missing in example 5.2.2.
2. Consider chapter 5.1, exercise problem 9.

$$\frac{\partial u}{\partial t} = \frac{\partial^2 u}{\partial x^2} - Pe \frac{\partial u}{\partial x}$$

$$u(0,t) = 1; \quad u(1,t) = 0$$

$$u(x,0) = 0$$

Solve this linear problem using numerical method of lines for Pe = 1, 10. How many node points are needed for obtaining three digits accuracy if average concentration at t = 1 is used to verify convergence?

3. Consider the Graetz problem discussed in example 5.6. Solve this problem using numerical method of lines for Pe = 1, 10 and 20.
4. Consider the Graetz problem discussed in problem 11 of chapter 5.1. Solve this problem using numerical method of lines for Pe = 1, 10 and 20.

Material and energy balances for a spherical catalyst can be written as:[12] (Davis, 1984)

$$\frac{\partial u}{\partial t} = \frac{\partial^2 u}{\partial x^2} + \frac{2}{x}\frac{\partial u}{\partial x} - \Phi^2 u \left(\exp\left[\gamma\left(1 - \frac{1}{\theta}\right)\right]\right)$$

$$Le \frac{\partial \theta}{\partial t} = \frac{\partial^2 \theta}{\partial x^2} + \frac{2}{x}\frac{\partial \theta}{\partial x} + \beta\Phi^2 u \left(\exp\left[\gamma\left(1 - \frac{1}{\theta}\right)\right]\right)$$

$$u(x,0) = 0; \quad \theta(x,0) = 1$$

$$\frac{\partial u}{\partial x}(0,t) = \frac{\partial \theta}{\partial x}(0,t) = 0$$

$$u(1,t) = \theta(1,t) = 1$$

where u is the dimensionless concentration and θ is the dimensionless temperature. Values of the parameters are $\Phi = 1$, $\gamma = 18$, $\beta = 0.04$. Le is the ratio of molecular diffusivity to thermal diffusivity. Solve these equations and plot the profiles using numerical method of lines for Le= 1 and Le = 10.

5. Consider adsorption in a pore[12]. (Davis, 1984) The dimensionless concentration u and the fraction of coverage, f are governed by:

$$\frac{\partial u}{\partial t} = \frac{\partial^2 u}{\partial x^2} - \left[k_a(1-f)u - k_d f\right]$$

$$\frac{\partial f}{\partial t} = \left[k_a(1-f)u - k_d f\right]$$

$$u(x,0) = 0; \; f(x,0) = 1$$

$$u(0,t) = 1$$

$$\frac{\partial u}{\partial x}(1,t) = 0$$

Equation for f as the boundary conditions for f at x = 0 and x = 1.

6. Consider Burger's equation in one-dimension: [18] [1]

$$\frac{\partial u}{\partial t} = \mu \frac{\partial^2 u}{\partial x^2} - u \frac{\partial u}{\partial x}$$

$$u(0,t) = \frac{1}{1 + \exp\left(-\frac{t}{4\mu}\right)}; \; u(1,t) = \frac{1}{1 + \exp\left(\frac{1}{2\mu} - \frac{t}{4\mu}\right)}$$

$$u(x,0) = \frac{1}{1 + \exp\left(\frac{x}{2\mu}\right)}$$

Solve this problem using numerical method of lines and compare with the exact solution $ua = \dfrac{1}{1 + \exp\left(\dfrac{x}{2\mu} - \dfrac{t}{4\mu}\right)}$ [18](Byrne and Hinmarsh, 1987) for $\mu = 1$ and $\mu = 0.1$.

7. Redo problem 7 for the boundary and initial conditions given in problem 2.
8. Solve problem 20, chapter 7 using numerical method of lines.
9. Solve problem 18, chapter 8 using numerical method of lines.
10. Solve problem 19, chapter 8 using numerical method of lines.
11. Consider the transient version of multiple steady state problem discussed in example 3.2.2:

$$\frac{\partial u}{\partial t} = \frac{\partial^2 u}{\partial x^2} - \Phi^2 u \exp\left(\frac{\gamma\beta(1-u)}{1+\beta(1-u)}\right)$$

$$\frac{\partial u}{\partial x}(0,t) = 0 \text{ and } u(1,t) = 1$$

In chapter 3.2 we obtained multiple steady states (three states) for this problem for the values of the parameters $\Phi = 0.2$, $\beta = 0.8$ and $\gamma = 20$. Solve this transient problem using numerical method of lines for two different initial conditions $u(x,0) = 1$ and $u(x,0) = 0$? What do you observe? Can you obtain all the three steady states discussed in example 3.2.2

12. Consider the shrinking core problem discussed in example 5.2.6. Redo this problem if the particle is rectangular instead of spherical. The governing equations are:

$$\frac{\partial u}{\partial t} = \frac{\partial^2 u}{\partial x^2}$$

$$u(x,0) = 0;\ u(x_c,t) = 0$$

$$u(1,t) = 1$$

$$\frac{dx_c}{dt} = -k\frac{\partial u}{\partial x}\bigg|_{x=x_c}$$

$$x_c(0) = 1$$

$$k = \frac{c_{A0}}{\rho\phi}$$

13. Redo problem 13, if the particle is cylindrical instead of rectangular.

14. Metal hydride electrodes involve change or shrinking of phases during discharge. Diffusion of hydrogen atoms inside a metal hydride particle can be modeled in dimensionless form:[19]

$$\frac{\partial u}{\partial t} = \frac{\partial^2 u}{\partial x^2} + \frac{2}{x}\frac{\partial u}{\partial x}$$

$$u(x,0) = u_0;\ u(x_c,t) = 1$$

$$\frac{\partial u}{\partial x}(1,t) = -\delta$$

$$\frac{dx_c}{dt} = \frac{1}{u_0-1}\frac{\partial u}{\partial x}\bigg|_{x=x_c}$$

$$x_c(0) = 1$$

where δ is the dimensionless current density and u_0 is the dimensionless initial concentration. Solve this problem for the parameters δ = 0.1 and u_0 = 9.

References

1. Schiesser, W.E., Silebi, C.A.: Dynamic Modeling of Transport Process Systems (1997)
2. Constantinides, A., Mostoufi, N.: Numerical Methods for Chemical Engineers with MATLAB Applications. Prentice-Hall PTR, Englewood Cliffs (1999)
3. Cutlip, M.B., Shacham, M.: Problem Solving in Chemical Engineering with Numerical Methods. Prentice Hall PTR, Englewood Cliffs (1999)
4. Taylor, R.: Engineering Computing with Maple: Solution of PDEs via the Method of Lines. CACHE News, 49 (Fall 1999)
5. Subramanian, V.R., White, R.E.: Solving Differential Equations with Maple. Chemical Engineering Education, 328–336 (Fall 2000)
6. Carslaw, H.S., Jaeger, J.C.: Conduction of Heat in Solids. Oxford University Press, Oxford (1972)
7. Amundson, N.R.: Mathematical Methods in Chemical Engineering: Matrices and Their Applications. Prentice Hall, Inc., Englewood Cliffs (1966)
8. Taylor, R., Krishna, R.: Multicomponent Mass Transfer. Wiley & Sons, Chichester (1993)
9. Rice, R.G., Do, D.D.: Applied Mathematics and Modeling for Chemical Engineers. John Wiley & Sons, Inc., Chichester (1995)
10. Crank, J.: The Mathematics of Diffusion, 2nd edn. Oxford University Press, Oxford (1979)
11. Riggs, J.B.: An Introduction to Numerical Methods for Chemical Engineers. Texas Tech University Press (1988)
12. Davis, M.E.: Numerical Methods and Modeling for Chemical Engineers. John Wiley & Sons, Chichester (1984)
13. Varma, A., Morbidelli, M.: Mathematical Methods in Chemical Engineering. Oxford University Press, Inc., Oxford (1997)
14. Aris, R.: Mathematical Modeling: A Chemical Engineer's Perspective. Academic Press, London (1999)
15. Finlayson, B.A.: Nonlinear Analysis in Chemical Engineering. McGraw-Hill, New York (1980)
16. Schiesser, W.E.: The Numerical Methods of Lines. Academic Press, Inc., New York (1991)
17. Fogler, H.S.: Elements of Chemical Reaction Engineering, 3rd edn. Prentice Hall, Englewood Cliffs (1998)
18. Byrne, G.D., Hindmarsh, A.C.: Stiff ODE Solvers: A Review of Current and Coming Attractions. Journal of Computational Physics 70(1), 1–62 (1987)
19. Subramanian, V.R., Ploehn, H.J., White, R.E.: Shrinking Core Model for the Discharge of a Metal Hydride Electrode. Journal of the Electrochemical Society 147, 2868–2873 (2000)

Chapter 6
Method of Lines for Elliptic Partial Differential Equations

6.1 Semianalytical and Numerical Method of Lines for Elliptic PDEs

6.1.1 Introduction

Steady state mass or heat transfer in solids and current distribution in electrochemical systems involve solving elliptic partial differential equations. The method of lines has not been used for elliptic partial differential equations to our knowledge. Schiesser and Silebi (1997)[1] added a time derivative to the steady state elliptic partial differential equation and applied finite differences in both x and y directions and then arrived at the steady state solution by waiting for the process to reach steady state.[2] When finite differences are applied only in the x direction, we arrive at a system of second order ordinary differential equations in y. Unfortunately, this is a coupled system of boundary value problems in y (boundary conditions defined at y = 0 and y = 1) and, hence, initial value problem solvers cannot be used to solve these boundary value problems directly. In this chapter, we introduce two methods to solve this system of boundary value problems. Both linear and nonlinear elliptic partial differential equations will be discussed in this chapter. We will present semianalytical solutions for linear elliptic partial differential equations and numerical solutions for nonlinear elliptic partial differential equations based on method of lines.

6.1.2 Semianalytical Method for Elliptic PDEs in Rectangular Coordinates

Steady state heat conduction or mass transfer in solids with constant physical properties (diffusion coefficient, thermal diffusivity, thermal conductivity, *etc.*) is usually represented by a linear elliptic partial differential equation. For linear parabolic partial differential equations, finite differences can be used to convert to any given partial differential equation to system of linear first order ordinary differential equations in time. In chapter 5.1, we showed how an exponential matrix method [3] [4] [5] could be used to integrate these simultaneous equations analytically in time. This exponential matrix method is extended to solve elliptic equations in this chapter (Subramanian & White, 2000).[6] This method involves applying finite differences in the x direction and analytically integrating in y. The dependent variable and its derivative are solved simultaneously. The unknown initial condition for either the variable or its derivative is found by using the boundary condition at the second boundary (e.g., y = 1). An important aspect of our technique is that the solution obtained is semianalytical (analytical in y, finite differences in x). A useful aspect of our technique is that the solution obtained is valid for both nonlinear and linear boundary conditions at y = 0 and y = 1.

In this chapter, we describe how one can arrive at the semianalytical solutions (solutions are analytical in the y variable and numerical in the spatial dimension) for linear elliptic partial differential equations using Maple and the matrix exponential method.

Example 6.1. Heat Transfer in a Rectangle

The methodology is illustrated using a Laplace equation for heat transfer in a rectangle[7] [8] using length L and height H. The governing equation for the temperature in dimensionless form can be written as[8]

$$\varepsilon^2 \frac{\partial^2 u}{\partial x^2} + \frac{\partial^2 u}{\partial y^2} = 0 \qquad (6.1)$$

where ε = H/L is the aspect ratio. For simplicity, the following boundary conditions are considered

$$u(0,y) = 0 \text{ for } 0 \le y \le 1 \qquad (6.2)$$

$$u(1,y) = 0 \text{ for } 0 \le y \le 1 \qquad (6.3)$$

$$u(x,0) = 0 \text{ for } 0 \le x \le 1 \qquad (6.4)$$

and

$$u(x,1) = 1 \text{ for } 0 \le x \le 1 \qquad (6.5)$$

Now finite differences are used to replace $\frac{\partial^2}{\partial x^2}$ in equation 6.1 to give

$$\frac{d^2 u_i}{dy^2} = -\varepsilon^2 \frac{u_{i+1} - 2u_i + u_{i-1}}{h^2} \qquad i = 1..N \qquad (6.6)$$

where N is the number of interior node points used in discretization and h = 1/(N+1) is the node spacing. Note that a central difference accurate to the order of h^2 is used in equation 6.1. Note that u_i denotes the temperature at point i on the line at x = ih. The boundary conditions at x = 0 and x = 1 (equations (6.2) and (6.3) are transformed as

6.1 Semianalytical and Numerical Method of Lines for Elliptic PDEs

$$u_0 = 0 \tag{6.7}$$

$$u_{N+1} = 0 \tag{6.8}$$

The boundary conditions in y are transformed as

$$u_i(y=0) = 0 \quad i = 1..N \tag{6.9}$$

$$u_i(y=1) = 1 \quad i = 1..N \tag{6.10}$$

For convenience, let $\zeta = \dfrac{y\varepsilon}{h}$. This converts the governing equation (equation 6.6) and boundary conditions (equations (6.7) – (6.10)) to

$$\begin{aligned}
\frac{d^2 u_i}{d\zeta^2} &= -u_{i+1} + 2u_i - u_{i-1} \quad i = 1..N \\
u_0 &= 0 \\
u_{N+1} &= 0 \\
u_i(\zeta = 0) &= 0 \quad i = 1..N \\
u_i(\zeta = \frac{\varepsilon}{h}) &= 1 \quad i = 1..N
\end{aligned} \tag{6.11}$$

In equation (6.11), there are N second order equations. These are converted to 2N first order equations as follows (Subramanian & White, 2000b;[9] Rice and Do, 1995;[10] see chapter 2.1.2):

$$\begin{aligned}
\frac{du_i}{d\zeta} &= u_{N+1+i} \quad i = 1..N \\
\frac{du_{N+1+i}}{d\zeta} &= -u_{i+1} + 2u_i - u_{i-1} \quad i = 1..N
\end{aligned} \tag{6.12}$$

with $u_0 = 0$ and $u_{N+1} = 0$. The initial conditions for these 2N differential equations are

$$u_i(\zeta = 0) = 0 \quad i = 1..N \tag{6.13}$$

and

$$u_{N+1+i}(\zeta = 0) = c_i \quad i = 1..N \tag{6.14}$$

In equation (6.14), the unknown constants ci, i = 1..N are found after integrating the equations in (6.12) and by using the boundary conditions at y = 1.

$$u_i(\zeta = \frac{\varepsilon}{h}) = 1 \tag{6.15}$$

Equation (6.12) is a system of 2N linear first order differential equations and can be written in matrix form as

$$\frac{d\mathbf{Y}}{d\zeta} = \mathbf{AY} + \mathbf{b}(x) \tag{6.16}$$

where

$$\mathbf{Y} = [u_1, u_2, ..u_N, u_{N+2}, u_{N+3}, ..u_{2N+1}]^T \tag{6.17}$$

and **A** is the 2N x 2N coefficient matrix defined by

$$\mathbf{A} = \begin{bmatrix} \mathbf{0} & \mathbf{I} \\ \mathbf{a} & \mathbf{0} \end{bmatrix} \tag{6.18}$$

where **0** is the zero matrix of order N x N, **I** is the identity matrix of order N x N and **a** is a N x N matrix given by

$$\mathbf{a} = \begin{bmatrix} 2 & -1 & 0 & 0 & 0 & ... & 0 \\ -1 & 2 & -1 & 0 & 0 & ... & 0 \\ 0 & -1 & 2 & -1 & 0 & ... & 0 \\ ... & ... & ... & ... & ... & ... & ... \\ 0 & ... & 0 & -1 & 2 & -1 & 0 \\ 0 & ... & 0 & 0 & -1 & 2 & -1 \\ 0 & ... & 0 & 0 & 0 & -1 & 2 \end{bmatrix} \tag{6.19}$$

The constant vector $\mathbf{b}(\zeta)$ is a column vector of order 2N x 1. Equation (6.16) can be integrated analytically by finding the exponential matrix [3] [4] [5] (Subramanian & White, 2000a; Varma & Morbidelli, 1997; Taylor & Krishna, 1993; Amundson, 1966)

$$\mathbf{Y} = \exp(\mathbf{A}\zeta)\mathbf{Y}_0 + \int_0^\zeta \exp[\mathbf{A}(\zeta - \lambda)] \mathbf{b}(\lambda) \, d\lambda \tag{6.20}$$

where λ is a dummy variable. For the example chosen, both u0 and uN+1 are zero and, hence, the forcing function b (ζ) is a 2N x 1 zero vector. However, if the boundary conditions are functions of y in equations 6.2 and 6.3, then b(ζ) is a function of ζ and the integral in equation (6.20) has to be evaluated. We call this a semianalytical solution since the solution obtained is analytical in ζ (or y). In our previous publication, the exponential matrix was found as a function of y.

6.1 Semianalytical and Numerical Method of Lines for Elliptic PDEs

However, since the **a** matrix has the elements $-2/h^2$ and $1/h^2$ compared to the identity matrix **I** in equation 6.18, the resultant **A** matrix becomes unstable and the ratio of the largest eigenvalue to the smallest eigenvalue becomes very large for some cases.

The procedure involved in solving a linear steady state elliptic PDE is summarized as follows:

1. Start the Maple worksheet with a 'restart' command to clear all variables.
2. Call 'with(linalg)' and 'with(plots)' commands.
3. Enter the governing equation.
4. Store the 'x' boundary conditions in bc1 (x = 0) and bc2 (x = 1).
5. Store the 'y' boundary conditions in bc3 (y = 0) and bc4 (y = 1).
6. Enter the number of interior node points, N.
7. Enter the length of the domain, L.
8. Transform the elliptic PDE from 'y' coordinate to 'ζ' coordinate using the variable transformation $\zeta = y\varepsilon/h$.
9. Convert the boundary conditions in x (bc1, and bc2) to finite difference form by using 3-point forward and backward differences (accurate to the order h^2), respectively, for bc1 and bc2.
10. Convert the governing equation to finite difference form by using central difference expression accurate to the order h^2 for the first and second derivatives in the spatial variable, x (equation (6.11)). This gives raise to N second order linear ODEs in ζ. This system of second order equations is converted to 2N first order linear ODEs in ζ as described in equation (6.12).
11. The variable $u_i(\zeta)$, i = 0..N+1 corresponds to the dependent variable, u_i at node point i.
12. The variable $u_{N+1+i}(\zeta)$, i = 1..N corresponds to the derivative of the dependent variable, $\dfrac{du_i}{d\zeta}$ at node point i.
13. Eliminate the boundary values ($u_0(\zeta)$ and $u_{N+1}(\zeta)$) using the boundary conditions.
14. Store the right hand side of the finite difference equations in eqs.
15. Store the dependent variables, u_i, i = 1..N, I = N+2..2N+1 in Y.
16. Generate **A** matrix using Maple's 'genmatrix' command. Find the exponential matrix using the expedited matrix procedure given in chapter 5. (Note that this can be done only if the eigenvalues are distinct and nonzero).
17. Find the solution by adding the non-homogeneous part according to equation 6.20. The first N rows of Y vector correspond to the dependent variable $u_i(\zeta)$ and the second N rows of Y vector correspond to the derivative $\dfrac{du_i}{d\zeta}$ at node point i.
18. Take initial condition as Y0 = g21

19. g^T. There are 2N dependent variables and hence 2N unknown constants (initial conditions). These constants are found out using the boundary conditions in y (bc3 and bc4). Note that in example 6.1 the dependent variable is known at y = ζ = 0 is known and hence the constants $p_1, p_2,..p_N$ are known before hand. However, the boundary condition at y = 0 may be of mixed type.
20. Once the constants are solved, the solution obtained is converted to 'y' variable and the plots are made.

Example 6.1 is solved below in Maple using this procedure. For illustration, N = 2 node points are used.

>restart;with(plottools):with(linalg):with(plots):

Enter the governing equation:

> ge:=diff(u(x,y),y$2)=-epsilon^2*diff(u(x,y),x$2);

$$ge := \frac{\partial^2}{\partial y^2} u(x,y) = -\varepsilon^2 \left(\frac{\partial^2}{\partial x^2} u(x,y) \right)$$

Enter the boundary conditions:

> bc1:=u(x,y)-0;

$$bc1 := u(x,y)$$

> bc2:=u(x,y)-0;

$$bc2 := u(x,y)$$

> bc3:=u(x,y)-0;

$$bc3 := u(x,y)$$

> bc4:=u(x,y)-1;

$$bc4 := u(x,y) - 1$$

> epsilon:=1;

$$\varepsilon := 1$$

Enter the finite difference approximations for the derivatives:

> dydxf:=1/2/h*(-u[m+2](zeta)-3*u[m](zeta)+4*u[m+1](zeta));

$$dydxf := \frac{1}{2} \frac{-u_{m+2}(\zeta) - 3 u_m(\zeta) + 4 u_{m+1}(\zeta)}{h}$$

> dydxb:=1/2/h*(u[m-2](zeta)+3*u[m](zeta)-4*u[m-1](zeta));

$$dydxb := \frac{1}{2} \frac{u_{m-2}(\zeta) + 3 u_m(\zeta) - 4 u_{m-1}(\zeta)}{h}$$

6.1 Semianalytical and Numerical Method of Lines for Elliptic PDEs

```
> dydx:=1/2/h*(u[m+1](zeta)-u[m-1](zeta));
```

$$dydx := \frac{1}{2} \frac{u_{m+1}(\zeta) - u_{m-1}(\zeta)}{h}$$

```
> d2ydx2:=1/h^2*(u[m-1](zeta)-2*u[m](zeta)+u[m+1](zeta));
```

$$d2ydx2 := \frac{u_{m-1}(\zeta) - 2u_m(\zeta) + u_{m+1}(\zeta)}{h^2}$$

Convert the boundary conditions to finite difference form:

```
> bc1:=subs(diff(u(x,y),x)=subs(m=0,dydxf),u(x,y)=u[0](zeta),x=0,bc1);
```

$$bc1 := u_0(\zeta)$$

```
> bc2:=subs(diff(u(x,y),x)=subs(m=N+1,dydxb),u(x,y)=u[N+1](zeta),x=1,bc2);
```

$$bc2 := u_{N+1}(\zeta)$$

Enter the number of interior node points:

```
> N:=2;
```

$$N := 2$$

```
> eq[0]:=bc1;
```

$$eq_0 := u_0(\zeta)$$

```
> eq[N+1]:=bc2;
```

$$eq_3 := u_3(\zeta)$$

Convert the governing equation to finite difference from (equation 6.12):

```
> for i from 1 to N do eq[N+1+i]:=diff(u[N+1+i](zeta),zeta)=
subs(diff(u(x,y),x$2) = subs(m=i,d2ydx2),diff(u(x,y),x) =
subs(m=i,dydx),u(x,y)=u[i](zeta),
x=i*h,rhs(h^2/epsilon^2*ge));od;
```

$$eq_4 := \frac{d}{d\zeta} u_4(\zeta) = -u_0(\zeta) + 2u_1(\zeta) - u_2(\zeta)$$

$$eq_5 := \frac{d}{d\zeta} u_5(\zeta) = -u_1(\zeta) + 2u_2(\zeta) - u_3(\zeta)$$

Enter the boundary values:

```
> u[0](zeta):=(solve(eq[0],u[0](zeta)));
```

$$u_0(\zeta) := 0$$

> u[N+1](zeta):=solve(eq[N+1],u[N+1](zeta));

$$u_3(\zeta) := 0$$

> for i from 1 to N do eq[i]:=diff(u[i](zeta),zeta)=u[N+1+i](zeta);od;

$$eq_1 := \frac{d}{d\zeta} u_1(\zeta) = u_4(\zeta)$$

$$eq_2 := \frac{d}{d\zeta} u_2(\zeta) = u_5(\zeta)$$

> for i from 1 to N do eq[i]:=eval(eq[i]);od;for i from 1 to N do eq[N+1+i]:=eval(eq[N+1+i]);od;

$$eq_1 := \frac{d}{d\zeta} u_1(\zeta) = u_4(\zeta)$$

$$eq_2 := \frac{d}{d\zeta} u_2(\zeta) = u_5(\zeta)$$

$$eq_4 := \frac{d}{d\zeta} u_4(\zeta) = 2 u_1(\zeta) - u_2(\zeta)$$

$$eq_5 := \frac{d}{d\zeta} u_5(\zeta) = -u_1(\zeta) + 2 u_2(\zeta)$$

Generate the **A** matrix using the governing equations and the dependent variables.

> eqns:=[seq(rhs(eq[j]),j=1..N),seq(rhs(eq[N+1+j]),j=1..N)];

$$eqns := \left[u_4(\zeta), u_5(\zeta), 2 u_1(\zeta) - u_2(\zeta), -u_1(\zeta) + 2 u_2(\zeta) \right]$$

> Y:=[seq(u[i](zeta),i=1..N),seq(u[N+1+i](zeta),i=1..N)];

$$Y := \left[u_1(\zeta), u_2(\zeta), u_4(\zeta), u_5(\zeta) \right]$$

> A:=genmatrix(eqns,Y,'b1');

$$A := \begin{bmatrix} 0 & 0 & 1 & 0 \\ 0 & 0 & 0 & 1 \\ 2 & -1 & 0 & 0 \\ -1 & 2 & 0 & 0 \end{bmatrix}$$

6.1 Semianalytical and Numerical Method of Lines for Elliptic PDEs

Convert the entries of the **A** matrix as decimals if N is greater than two (as shown in chapter 5.1):

> if N>2 then A:=map(evalf,A):end;

A Maple procedure is written to expedite the calculation for the exponential matrix (see chapter 5.1). First the eigenvalues are found:

>NRow:=rowdim(A);

$$NRow := 4$$

> L:=evalf(eigenvalues(A));

$$L := 1., -1., 1.732050808, -1.732050808$$

Note that this procedure can be used only if all the eigenvalues are real and distinct. Also, for obtaining the eigenvectors (equation 5.26) since βs are coupled, all of the equations are solved simultaneously. In chapter 5.1, the equations for βs were solved individually one by one.

> evalm(A);

$$\begin{bmatrix} 0 & 0 & 1 & 0 \\ 0 & 0 & 0 & 1 \\ 2 & -1 & 0 & 0 \\ -1 & 2 & 0 & 0 \end{bmatrix}$$

> b:=matrix(2*N,1):for i from 1 to 2*N do b[i,1]:=-b1[i];od:evalm(b);

$$\begin{bmatrix} 0 \\ 0 \\ 0 \\ 0 \end{bmatrix}$$

Note that for the example given the **b** vector is zero. However, depending on the boundary conditions, bc3 and bc3, the **b** vector can be a function of ζ or a constant vector.

> h:=eval(1/(N+1));

$$h := \frac{1}{3}$$

> J:=jordan(A,S);

$$J := \begin{bmatrix} -1 & 0 & 0 & 0 \\ 0 & 1 & 0 & 0 \\ 0 & 0 & \sqrt{3} & 0 \\ 0 & 0 & 0 & -\sqrt{3} \end{bmatrix}$$

> mat:=evalm(S&*exponential(J,zeta)&*inverse(S)):

> mat1:=evalm(subs(zeta=zeta-zeta1,evalm(mat))):

> b2:=evalm(subs(zeta=zeta1,evalm(b))):

> mat2:=evalm(mat1&*b2):

> mat2:=map(expand,mat2):

> mat3:=map(int,mat2,zeta1=0..zeta):

The initial condition vector is defined here.

> Y0:=matrix(2*N,1);

$$Y0 := array(1..4, 1..1, [\,])$$

> for i to N do Y0[i,1]:=p[i];od:

> for i to N do Y0[N+i,1]:=c[i]:od:

> evalm(Y0);

$$\begin{bmatrix} p_1 \\ p_2 \\ c_1 \\ c_2 \end{bmatrix}$$

The solution is found by adding the nonhomogeneous part to the homogeneous part.

> Y:=evalm(mat&*Y0+mat3):

The solution at y = 0 and y = 1 is stored in sol0 and sol1 to calculate the unknown constants.

> sol0:=map(eval,evalm(subs(zeta=0,evalm(Y)))):

> sol1:=map(eval,evalm(subs(zeta=epsilon/h,evalm(Y)))):

6.1 Semianalytical and Numerical Method of Lines for Elliptic PDEs

Now the boundary conditions bc3 and bc4 are applied.

> for i to N do Eq[i]:=subs(diff(u(x,y),y)=epsilon/h*c[i],u(x,y)=p[i],x=i*h,bc3);od;

$$Eq_1 := p_1$$
$$Eq_2 := p_2$$

> for i to N do Eq[N+i]:=evalf(subs(diff(u(x,y),y)=epsilon/h*sol1[N+i,1],u(x,y)=sol1[i,1],bc4));od;

$$Eq_3 := 50.17924626 p_1 - 40.11158426 p_2 + 31.07205648 c_1 - 21.05418156 c_2 - 1.$$

$$Eq_4 := -40.11158426 p_1 + 50.17924626 p_2 - 21.05418156 c_1 + 31.07205648 c_2 - 1.$$

The unknown constants are solved as:

> csol:=solve({seq(Eq[i],i=1..2*N)},{seq(c[i],i=1..N),seq(p[i],i=1..N)});

$$csol := \{p_2 = 0., c_2 = 0.09982156974, c_1 = 0.09982156974, p_1 = 0.\}$$

> assign(csol);

> Y:=map(eval,Y):

> for i from 1 to N do u[i](zeta):=eval((Y[i,1]));od:

> for i from 0 to N+1 do u[i](zeta):=eval(u[i](zeta));od:

> for i from 0 to N+1 do u[i](y):=eval(subs(zeta=epsilon*y/h,u[i](zeta)));od;

$$u_0(y) := 0$$

$$u_1(y) := -\frac{0.09982156974 e^{-3y}}{(1+\sqrt{3})(\sqrt{3}-1)} + 0.04991078488 e^{3y}$$

$$u_2(y) := -\frac{0.09982156974 e^{-3y}}{(1+\sqrt{3})(\sqrt{3}-1)} + 0.04991078488 e^{3y}$$

$$u_3(y) := 0$$

Hence, the semianalytical solution is obtained for temperature distribution. The plots obtained for N=10 node points are given below:

518   6 Method of Lines for Elliptic Partial Differential Equations

```
> for i from 0 to N+1 do pl[i]:=line([0.3,0.98-abs(i-5.25)*0.14],
[0.6,evalf(subs(y=0.6,u[i](y)))],thickness=1,linestyle=dot);
pt[i]:=textplot([0.3,0.98-abs(i-5.25)*0.14,typeset(u[i],"(y)")],
align=left):end do:
> pp:=plot([seq(u[i](y),i=0..N+1)],y=0..1);
```

$$pp := PLOT(...)$$

```
> display([pp,seq(pl[i],i=0..N+1),seq(pt[i],i=0..N+1)],axes=boxed,thickness=3,
title="Figure 6.1",labels=[y,"u"]);
```

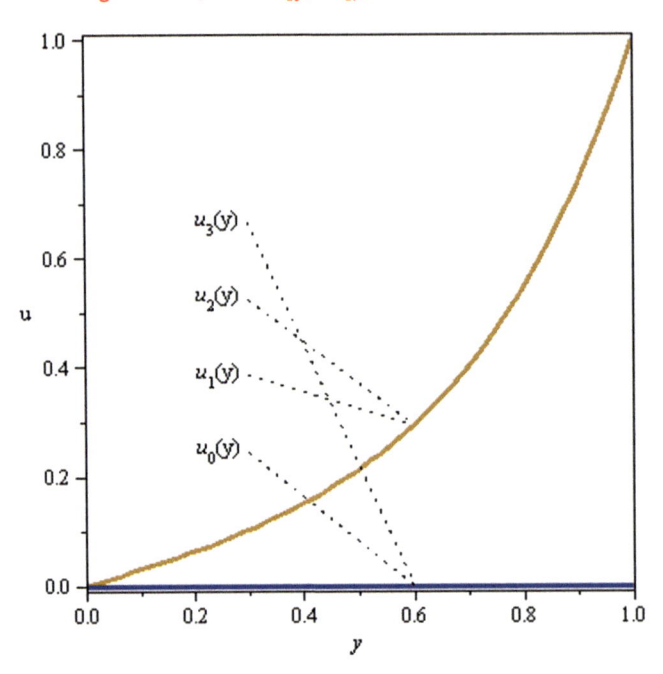

Fig. 6.1

```
> M:=10;
```

$$M := 10$$

```
> T1:=[seq(evalf(i/M),i=0..M)];
```

$$T1 := [0., 0.1000000000, 0.2000000000, 0.3000000000, 0.4000000000,$$
$$0.5000000000, 0.6000000000, 0.7000000000, 0.8000000000,$$
$$0.9000000000, 1.]$$

```
> for j from 1 to M do
P[j]:=plot([seq([h*i,evalf(subs(y=T1[j],evalf(u[i](y))))],i=0..N+1)],

style=line,thickness=3,axes=boxed,view=[0..1,0..1.1]):od:
> P[M+1]:=plot([seq([h*i,evalf(subs(x=i*h,1))],i=0..N+1)],style=line,
thickness=3,title="Figure 6.2",axes=boxed):
```

6.1 Semianalytical and Numerical Method of Lines for Elliptic PDEs

```
> for j from 1 to M+1 do
pt[j]:=textplot([0.5,evalf(subs(y=T1[j],u[1](y))),typeset(y,sprintf("=%4.2f",
T1[j]))],align=above);od:
> display({seq(P[i],i=1..M+1),seq(pt[j],j=1..M+1)},labels=[x,u]);
```

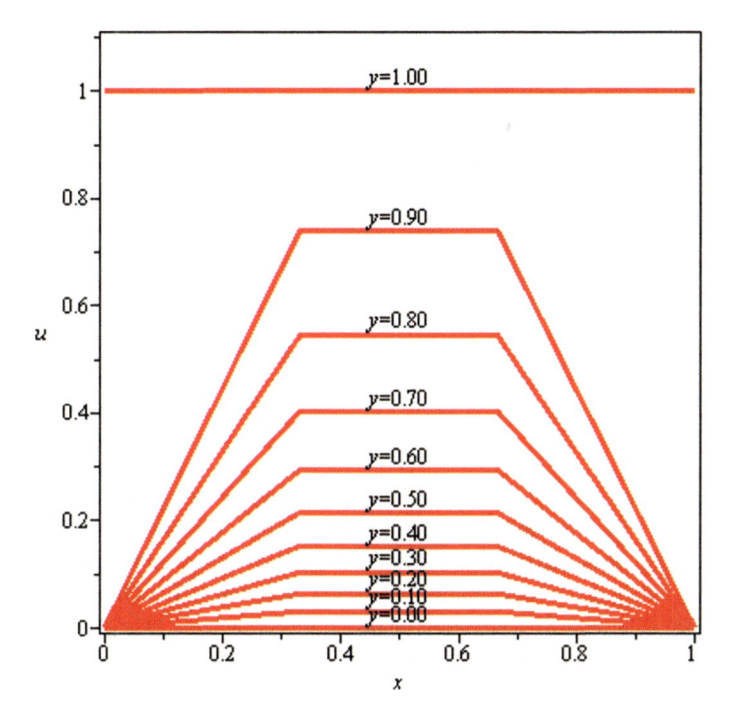

Fig. 6.2

```
> Ny:=30;
```

$$Ny := 30$$

```
> PP:=matrix(N+2,Ny);
```

$$PP := array(1..4, 1..30, [\,])$$

```
> for i to Ny do PP[1,i]:=0;PP[N+2,i]:=0;od:
> for i to N+2 do PP[i,1]:=0;PP[i,Ny]:=1;od:
> for i from 2 to N+1 do for j from 2 to Ny-1 do
PP[i,j]:=evalf(subs(y=(j-1)/(Ny-1),u[i-1](y)));od;od:
> plotdata := [seq([ seq([(i-1)/(N+1),(j-1)/(Ny-1),PP[i,j]], i=1..N+2)],
j=1..Ny)]:
> surfdata(plotdata,axes=boxed,title="Figure 6.3",
labels=[x,y,u],orientation=[-120,60] );
```

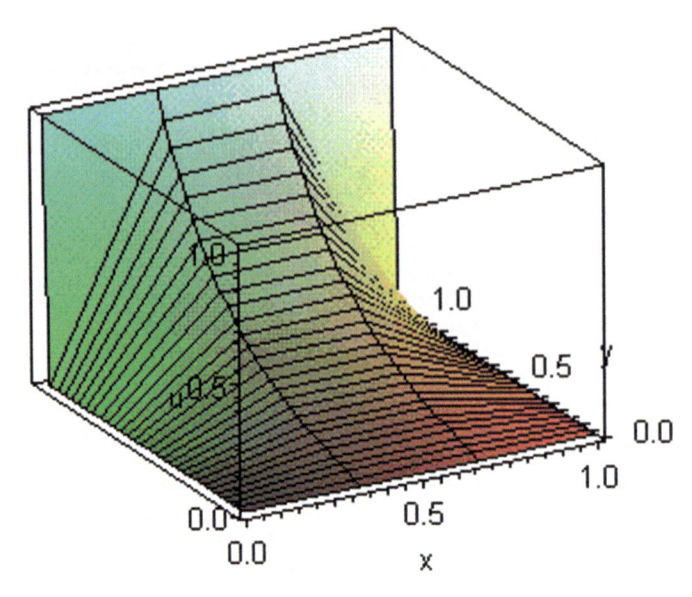

520   6 Method of Lines for Elliptic Partial Differential Equations

Fig. 6.3

Because of symmetry, we get $u_1 = u_N$, $u_2 = u_{N-1}$, etc. Note that **A** matrix depends only on the governing equation and the boundary conditions at $x = 0$ and $x = 1$. Once the exponential matrix $(\exp(A\zeta))$ is found, the exponential matrix can be used for a different set of boundary conditions at $y = 0$ and $y = 1$. This is true because the solution obtained is analytical in the y direction and valid for any boundary conditions in y as illustrated in the next example.

Example 6.2

For example, consider the following boundary value problem

$$\frac{\partial^2 u}{\partial x^2} + \frac{\partial^2 u}{\partial y^2} = 0$$

$$u(0, y) = 0 \quad 0 \le y \le 1$$

$$u(1, y) = 0 \quad 0 \le y \le 1 \quad (6.21)$$

$$\frac{\partial u}{\partial y}(x, 0) - u(x, 0) \quad 0 \le x \le 1$$

$$u(x, 1) = 1 \quad 0 < x < 1$$

For solving equation (6.21) there is no need to find the exponential matrix again. Since the boundary conditions at $x = 0$ and $x = 1$ do not change, the complete solution can be obtained using the exponential matrix obtained for the previous example by just recalculating the constants as described below:

6.1 Semianalytical and Numerical Method of Lines for Elliptic PDEs

> restart;with(plottools):with(linalg):with(plots):
> Digits:=12;

$$Digits := 12$$

> ge:=diff(u(x,y),y$2)=-epsilon^2*diff(u(x,y),x$2);

$$ge := \frac{\partial^2}{\partial y^2} u(x,y) = -\varepsilon^2 \left(\frac{\partial^2}{\partial x^2} u(x,y) \right)$$

> bc1:=u(x,y)-0;

$$bc1 := u(x,y)$$

> bc2:=u(x,y)-0;

$$bc2 := u(x,y)$$

Now, the boundary condition at y = 0 is redefined.

> bc3:=diff(u(x,y),y)-u(x,y);

$$bc3 := \frac{\partial}{\partial y} u(x,y) - u(x,y)$$

> bc4:=u(x,y)-1;

$$bc4 := u(x,y) - 1$$

> epsilon:=1;

$$\varepsilon := 1$$

> dydxf:=1/2/h*(-u[m+2](zeta)-3*u[m](zeta)+4*u[m+1](zeta));

$$dydxf := \frac{1}{2} \frac{-u_{m+2}(\zeta) - 3 u_m(\zeta) + 4 u_{m+1}(\zeta)}{h}$$

> dydxb:=1/2/h*(u[m-2](zeta)+3*u[m](zeta)-4*u[m-1](zeta));

$$dydxb := \frac{1}{2} \frac{u_{m-2}(\zeta) + 3 u_m(\zeta) - 4 u_{m-1}(\zeta)}{h}$$

> dydx:=1/2/h*(u[m+1](zeta)-u[m-1](zeta));

$$dydx := \frac{1}{2} \frac{u_{m+1}(\zeta) - u_{m-1}(\zeta)}{h}$$

> d2ydx2:=1/h^2*(u[m-1](zeta)-2*u[m](zeta)+u[m+1](zeta));

$$d2ydx2 := \frac{u_{m-1}(\zeta) - 2 u_m(\zeta) + u_{m+1}(\zeta)}{h^2}$$

> bc1:=subs(diff(u(x,y),x)=subs(m=0,dydxf),u(x,y)=u[0](zeta),bc1);

$$bc1 := u_0(\zeta)$$

> bc2:=subs(diff(u(x,y),x)=subs(m=N+1,dydxb),u(x,y)=u[N+1](zeta),bc2);

$$bc2 := u_{N+1}(\zeta)$$

> N:=10;

$$N := 10$$

> eq[0]:=bc1;

$$eq_0 := u_0(\zeta)$$

> eq[N+1]:=bc2;

$$eq_{11} := u_{11}(\zeta)$$

> for i from 1 to N do eq[N+1+i]:=diff(u[N+1+i](zeta),zeta)= subs(diff(u(x,y),x$2) = subs(m=i,d2ydx2),diff(u(x,y),x) = subs(m=i,dydx),u(x,y)=u[i](zeta),

x=i*h,rhs(h^2/epsilon^2*ge));od;

$$eq_{12} := \frac{d}{d\zeta} u_{12}(\zeta) = -u_0(\zeta) + 2u_1(\zeta) - u_2(\zeta)$$

$$eq_{13} := \frac{d}{d\zeta} u_{13}(\zeta) = -u_1(\zeta) + 2u_2(\zeta) - u_3(\zeta)$$

$$eq_{14} := \frac{d}{d\zeta} u_{14}(\zeta) = -u_2(\zeta) + 2u_3(\zeta) - u_4(\zeta)$$

$$eq_{15} := \frac{d}{d\zeta} u_{15}(\zeta) = -u_3(\zeta) + 2u_4(\zeta) - u_5(\zeta)$$

$$eq_{16} := \frac{d}{d\zeta} u_{16}(\zeta) = -u_4(\zeta) + 2u_5(\zeta) - u_6(\zeta)$$

$$eq_{17} := \frac{d}{d\zeta} u_{17}(\zeta) = -u_5(\zeta) + 2u_6(\zeta) - u_7(\zeta)$$

$$eq_{18} := \frac{d}{d\zeta} u_{18}(\zeta) = -u_6(\zeta) + 2u_7(\zeta) - u_8(\zeta)$$

$$eq_{19} := \frac{d}{d\zeta} u_{19}(\zeta) = -u_7(\zeta) + 2u_8(\zeta) - u_9(\zeta)$$

$$eq_{20} := \frac{d}{d\zeta} u_{20}(\zeta) = -u_8(\zeta) + 2u_9(\zeta) - u_{10}(\zeta)$$

$$eq_{21} := \frac{d}{d\zeta} u_{21}(\zeta) = -u_9(\zeta) + 2u_{10}(\zeta) - u_{11}(\zeta)$$

6.1 Semianalytical and Numerical Method of Lines for Elliptic PDEs

```
> u[0](zeta):=(solve(eq[0],u[0](zeta)));
```

$$u_0(\zeta) := 0$$

```
> u[N+1](zeta):=solve(eq[N+1],u[N+1](zeta));
```

$$u_{11}(\zeta) := 0$$

```
> for i from 1 to N do eq[i]:=diff(u[i](zeta),zeta)= u[N+1+i](zeta);od;
```

$$eq_1 := \frac{d}{d\zeta} u_1(\zeta) = u_{12}(\zeta)$$

$$eq_2 := \frac{d}{d\zeta} u_2(\zeta) = u_{13}(\zeta)$$

$$eq_3 := \frac{d}{d\zeta} u_3(\zeta) = u_{14}(\zeta)$$

$$eq_4 := \frac{d}{d\zeta} u_4(\zeta) = u_{15}(\zeta)$$

$$eq_5 := \frac{d}{d\zeta} u_5(\zeta) = u_{16}(\zeta)$$

$$eq_6 := \frac{d}{d\zeta} u_6(\zeta) = u_{17}(\zeta)$$

$$eq_7 := \frac{d}{d\zeta} u_7(\zeta) = u_{18}(\zeta)$$

$$eq_8 := \frac{d}{d\zeta} u_8(\zeta) = u_{19}(\zeta)$$

$$eq_9 := \frac{d}{d\zeta} u_9(\zeta) = u_{20}(\zeta)$$

$$eq_{10} := \frac{d}{d\zeta} u_{10}(\zeta) = u_{21}(\zeta)$$

```
> for i from 1 to N do eq[i]:=eval(eq[i]);od;for i from 1 to N do
eq[N+1+i]:=eval(eq[N+1+i]);od;
```

$$eq_1 := \frac{d}{d\zeta} u_1(\zeta) = u_{12}(\zeta)$$

$$eq_2 := \frac{d}{d\zeta} u_2(\zeta) = u_{13}(\zeta)$$

$$eq_3 := \frac{d}{d\zeta} u_3(\zeta) = u_{14}(\zeta)$$

$$eq_4 := \frac{d}{d\zeta} u_4(\zeta) = u_{15}(\zeta)$$

$$eq_5 := \frac{d}{d\zeta} u_5(\zeta) = u_{16}(\zeta)$$

$$eq_6 := \frac{d}{d\zeta} u_6(\zeta) = u_{17}(\zeta)$$

$$eq_7 := \frac{d}{d\zeta} u_7(\zeta) = u_{18}(\zeta)$$

$$eq_8 := \frac{d}{d\zeta} u_8(\zeta) = u_{19}(\zeta)$$

$$eq_9 := \frac{d}{d\zeta} u_9(\zeta) = u_{20}(\zeta)$$

$$eq_{10} := \frac{d}{d\zeta} u_{10}(\zeta) = u_{21}(\zeta)$$

$$eq_{12} := \frac{d}{d\zeta} u_{12}(\zeta) = 2 u_1(\zeta) - u_2(\zeta)$$

$$eq_{13} := \frac{d}{d\zeta} u_{13}(\zeta) = -u_1(\zeta) + 2 u_2(\zeta) - u_3(\zeta)$$

$$eq_{14} := \frac{d}{d\zeta} u_{14}(\zeta) = -u_2(\zeta) + 2 u_3(\zeta) - u_4(\zeta)$$

$$eq_{15} := \frac{d}{d\zeta} u_{15}(\zeta) = -u_3(\zeta) + 2 u_4(\zeta) - u_5(\zeta)$$

$$eq_{16} := \frac{d}{d\zeta} u_{16}(\zeta) = -u_4(\zeta) + 2 u_5(\zeta) - u_6(\zeta)$$

$$eq_{17} := \frac{d}{d\zeta} u_{17}(\zeta) = -u_5(\zeta) + 2 u_6(\zeta) - u_7(\zeta)$$

$$eq_{18} := \frac{d}{d\zeta} u_{18}(\zeta) = -u_6(\zeta) + 2 u_7(\zeta) - u_8(\zeta)$$

$$eq_{19} := \frac{d}{d\zeta} u_{19}(\zeta) = -u_7(\zeta) + 2 u_8(\zeta) - u_9(\zeta)$$

6.1 Semianalytical and Numerical Method of Lines for Elliptic PDEs

$$eq_{20} := \frac{d}{d\zeta} u_{20}(\zeta) = -u_8(\zeta) + 2 u_9(\zeta) - u_{10}(\zeta)$$

$$eq_{21} := \frac{d}{d\zeta} u_{21}(\zeta) = -u_9(\zeta) + 2 u_{10}(\zeta)$$

> eqns:=[seq(rhs(eq[j]),j=1..N),seq(rhs(eq[N+1+j]),j=1..N)];

$eqns := [u_{12}(\zeta), u_{13}(\zeta), u_{14}(\zeta), u_{15}(\zeta), u_{16}(\zeta), u_{17}(\zeta), u_{18}(\zeta), u_{19}(\zeta), u_{20}(\zeta), u_{21}(\zeta), 2 u_1(\zeta) - u_2(\zeta),$
$-u_1(\zeta) + 2 u_2(\zeta) - u_3(\zeta), -u_2(\zeta) + 2 u_3(\zeta) - u_4(\zeta), -u_3(\zeta) + 2 u_4(\zeta) - u_5(\zeta), -u_4(\zeta) + 2 u_5(\zeta)$
$- u_6(\zeta), -u_5(\zeta) + 2 u_6(\zeta) - u_7(\zeta), -u_6(\zeta) + 2 u_7(\zeta) - u_8(\zeta), -u_7(\zeta) + 2 u_8(\zeta) - u_9(\zeta), -u_8(\zeta)$
$+ 2 u_9(\zeta) - u_{10}(\zeta), -u_9(\zeta) + 2 u_{10}(\zeta)]$

> Y:=[seq(u[i](zeta),i=1..N),seq(u[N+1+i](zeta),i=1..N)];

$Y := [u_1(\zeta), u_2(\zeta), u_3(\zeta), u_4(\zeta), u_5(\zeta), u_6(\zeta), u_7(\zeta), u_8(\zeta), u_9(\zeta), u_{10}(\zeta), u_{12}(\zeta), u_{13}(\zeta), u_{14}(\zeta), u_{15}(\zeta),$
$u_{16}(\zeta), u_{17}(\zeta), u_{18}(\zeta), u_{19}(\zeta), u_{20}(\zeta), u_{21}(\zeta)]$

> A:=genmatrix(eqns,Y,'b1');

$$A := \begin{bmatrix}
0. & 0. & 0. & 0. & 0. & 0. & 0. & 0. & 0. & 0. & 1. & 0. & 0. & 0. & 0. & 0. & 0. & 0. & 0. & 0. \\
0. & 0. & 0. & 0. & 0. & 0. & 0. & 0. & 0. & 0. & 0. & 1. & 0. & 0. & 0. & 0. & 0. & 0. & 0. & 0. \\
0. & 0. & 0. & 0. & 0. & 0. & 0. & 0. & 0. & 0. & 0. & 0. & 1. & 0. & 0. & 0. & 0. & 0. & 0. & 0. \\
0. & 0. & 0. & 0. & 0. & 0. & 0. & 0. & 0. & 0. & 0. & 0. & 0. & 1. & 0. & 0. & 0. & 0. & 0. & 0. \\
0. & 0. & 0. & 0. & 0. & 0. & 0. & 0. & 0. & 0. & 0. & 0. & 0. & 0. & 1. & 0. & 0. & 0. & 0. & 0. \\
0. & 0. & 0. & 0. & 0. & 0. & 0. & 0. & 0. & 0. & 0. & 0. & 0. & 0. & 0. & 1. & 0. & 0. & 0. & 0. \\
0. & 0. & 0. & 0. & 0. & 0. & 0. & 0. & 0. & 0. & 0. & 0. & 0. & 0. & 0. & 0. & 1. & 0. & 0. & 0. \\
0. & 0. & 0. & 0. & 0. & 0. & 0. & 0. & 0. & 0. & 0. & 0. & 0. & 0. & 0. & 0. & 0. & 1. & 0. & 0. \\
0. & 0. & 0. & 0. & 0. & 0. & 0. & 0. & 0. & 0. & 0. & 0. & 0. & 0. & 0. & 0. & 0. & 0. & 1. & 0. \\
0. & 0. & 0. & 0. & 0. & 0. & 0. & 0. & 0. & 0. & 0. & 0. & 0. & 0. & 0. & 0. & 0. & 0. & 0. & 1. \\
2. & -1. & 0. & 0. & 0. & 0. & 0. & 0. & 0. & 0. & 0. & 0. & 0. & 0. & 0. & 0. & 0. & 0. & 0. & 0. \\
-1. & 2. & -1. & 0. & 0. & 0. & 0. & 0. & 0. & 0. & 0. & 0. & 0. & 0. & 0. & 0. & 0. & 0. & 0. & 0. \\
0. & -1. & 2. & -1. & 0. & 0. & 0. & 0. & 0. & 0. & 0. & 0. & 0. & 0. & 0. & 0. & 0. & 0. & 0. & 0. \\
0. & 0. & -1. & 2. & -1. & 0. & 0. & 0. & 0. & 0. & 0. & 0. & 0. & 0. & 0. & 0. & 0. & 0. & 0. & 0. \\
0. & 0. & 0. & -1. & 2. & -1. & 0. & 0. & 0. & 0. & 0. & 0. & 0. & 0. & 0. & 0. & 0. & 0. & 0. & 0. \\
0. & 0. & 0. & 0. & -1. & 2. & -1. & 0. & 0. & 0. & 0. & 0. & 0. & 0. & 0. & 0. & 0. & 0. & 0. & 0. \\
0. & 0. & 0. & 0. & 0. & -1. & 2. & -1. & 0. & 0. & 0. & 0. & 0. & 0. & 0. & 0. & 0. & 0. & 0. & 0. \\
0. & 0. & 0. & 0. & 0. & 0. & -1. & 2. & -1. & 0. & 0. & 0. & 0. & 0. & 0. & 0. & 0. & 0. & 0. & 0. \\
0. & 0. & 0. & 0. & 0. & 0. & 0. & -1. & 2. & -1. & 0. & 0. & 0. & 0. & 0. & 0. & 0. & 0. & 0. & 0. \\
0. & 0. & 0. & 0. & 0. & 0. & 0. & 0. & -1. & 2. & 0. & 0. & 0. & 0. & 0. & 0. & 0. & 0. & 0. & 0.
\end{bmatrix}$$

526   6 Method of Lines for Elliptic Partial Differential Equations

> if N>2 then A:=map(evalf,A):end;

$$A := \begin{bmatrix} 0. & 0. & 0. & 0. & 0. & 0. & 0. & 0. & 0. & 0. & 1. & 0. & 0. & 0. & 0. & 0. & 0. & 0. & 0. & 0. \\ 0. & 0. & 0. & 0. & 0. & 0. & 0. & 0. & 0. & 0. & 0. & 1. & 0. & 0. & 0. & 0. & 0. & 0. & 0. & 0. \\ 0. & 0. & 0. & 0. & 0. & 0. & 0. & 0. & 0. & 0. & 0. & 0. & 1. & 0. & 0. & 0. & 0. & 0. & 0. & 0. \\ 0. & 0. & 0. & 0. & 0. & 0. & 0. & 0. & 0. & 0. & 0. & 0. & 0. & 1. & 0. & 0. & 0. & 0. & 0. & 0. \\ 0. & 0. & 0. & 0. & 0. & 0. & 0. & 0. & 0. & 0. & 0. & 0. & 0. & 0. & 1. & 0. & 0. & 0. & 0. & 0. \\ 0. & 0. & 0. & 0. & 0. & 0. & 0. & 0. & 0. & 0. & 0. & 0. & 0. & 0. & 0. & 1. & 0. & 0. & 0. & 0. \\ 0. & 0. & 0. & 0. & 0. & 0. & 0. & 0. & 0. & 0. & 0. & 0. & 0. & 0. & 0. & 0. & 1. & 0. & 0. & 0. \\ 0. & 0. & 0. & 0. & 0. & 0. & 0. & 0. & 0. & 0. & 0. & 0. & 0. & 0. & 0. & 0. & 0. & 1. & 0. & 0. \\ 0. & 0. & 0. & 0. & 0. & 0. & 0. & 0. & 0. & 0. & 0. & 0. & 0. & 0. & 0. & 0. & 0. & 0. & 1. & 0. \\ 0. & 0. & 0. & 0. & 0. & 0. & 0. & 0. & 0. & 0. & 0. & 0. & 0. & 0. & 0. & 0. & 0. & 0. & 0. & 1. \\ 2. & -1. & 0. & 0. & 0. & 0. & 0. & 0. & 0. & 0. & 0. & 0. & 0. & 0. & 0. & 0. & 0. & 0. & 0. & 0. \\ -1. & 2. & -1. & 0. & 0. & 0. & 0. & 0. & 0. & 0. & 0. & 0. & 0. & 0. & 0. & 0. & 0. & 0. & 0. & 0. \\ 0. & -1. & 2. & -1. & 0. & 0. & 0. & 0. & 0. & 0. & 0. & 0. & 0. & 0. & 0. & 0. & 0. & 0. & 0. & 0. \\ 0. & 0. & -1. & 2. & -1. & 0. & 0. & 0. & 0. & 0. & 0. & 0. & 0. & 0. & 0. & 0. & 0. & 0. & 0. & 0. \\ 0. & 0. & 0. & -1. & 2. & -1. & 0. & 0. & 0. & 0. & 0. & 0. & 0. & 0. & 0. & 0. & 0. & 0. & 0. & 0. \\ 0. & 0. & 0. & 0. & -1. & 2. & -1. & 0. & 0. & 0. & 0. & 0. & 0. & 0. & 0. & 0. & 0. & 0. & 0. & 0. \\ 0. & 0. & 0. & 0. & 0. & -1. & 2. & -1. & 0. & 0. & 0. & 0. & 0. & 0. & 0. & 0. & 0. & 0. & 0. & 0. \\ 0. & 0. & 0. & 0. & 0. & 0. & -1. & 2. & -1. & 0. & 0. & 0. & 0. & 0. & 0. & 0. & 0. & 0. & 0. & 0. \\ 0. & 0. & 0. & 0. & 0. & 0. & 0. & -1. & 2. & -1. & 0. & 0. & 0. & 0. & 0. & 0. & 0. & 0. & 0. & 0. \\ 0. & 0. & 0. & 0. & 0. & 0. & 0. & 0. & -1. & 2. & 0. & 0. & 0. & 0. & 0. & 0. & 0. & 0. & 0. & 0. \end{bmatrix}$$

> evalm(A);

$$\begin{bmatrix} 0. & 0. & 0. & 0. & 0. & 0. & 0. & 0. & 0. & 0. & 1. & 0. & 0. & 0. & 0. & 0. & 0. & 0. & 0. & 0. \\ 0. & 0. & 0. & 0. & 0. & 0. & 0. & 0. & 0. & 0. & 0. & 1. & 0. & 0. & 0. & 0. & 0. & 0. & 0. & 0. \\ 0. & 0. & 0. & 0. & 0. & 0. & 0. & 0. & 0. & 0. & 0. & 0. & 1. & 0. & 0. & 0. & 0. & 0. & 0. & 0. \\ 0. & 0. & 0. & 0. & 0. & 0. & 0. & 0. & 0. & 0. & 0. & 0. & 0. & 1. & 0. & 0. & 0. & 0. & 0. & 0. \\ 0. & 0. & 0. & 0. & 0. & 0. & 0. & 0. & 0. & 0. & 0. & 0. & 0. & 0. & 1. & 0. & 0. & 0. & 0. & 0. \\ 0. & 0. & 0. & 0. & 0. & 0. & 0. & 0. & 0. & 0. & 0. & 0. & 0. & 0. & 0. & 1. & 0. & 0. & 0. & 0. \\ 0. & 0. & 0. & 0. & 0. & 0. & 0. & 0. & 0. & 0. & 0. & 0. & 0. & 0. & 0. & 0. & 1. & 0. & 0. & 0. \\ 0. & 0. & 0. & 0. & 0. & 0. & 0. & 0. & 0. & 0. & 0. & 0. & 0. & 0. & 0. & 0. & 0. & 1. & 0. & 0. \\ 0. & 0. & 0. & 0. & 0. & 0. & 0. & 0. & 0. & 0. & 0. & 0. & 0. & 0. & 0. & 0. & 0. & 0. & 1. & 0. \\ 0. & 0. & 0. & 0. & 0. & 0. & 0. & 0. & 0. & 0. & 0. & 0. & 0. & 0. & 0. & 0. & 0. & 0. & 0. & 1. \\ 2. & -1. & 0. & 0. & 0. & 0. & 0. & 0. & 0. & 0. & 0. & 0. & 0. & 0. & 0. & 0. & 0. & 0. & 0. & 0. \\ -1. & 2. & -1. & 0. & 0. & 0. & 0. & 0. & 0. & 0. & 0. & 0. & 0. & 0. & 0. & 0. & 0. & 0. & 0. & 0. \\ 0. & -1. & 2. & -1. & 0. & 0. & 0. & 0. & 0. & 0. & 0. & 0. & 0. & 0. & 0. & 0. & 0. & 0. & 0. & 0. \\ 0. & 0. & -1. & 2. & -1. & 0. & 0. & 0. & 0. & 0. & 0. & 0. & 0. & 0. & 0. & 0. & 0. & 0. & 0. & 0. \\ 0. & 0. & 0. & -1. & 2. & -1. & 0. & 0. & 0. & 0. & 0. & 0. & 0. & 0. & 0. & 0. & 0. & 0. & 0. & 0. \\ 0. & 0. & 0. & 0. & -1. & 2. & -1. & 0. & 0. & 0. & 0. & 0. & 0. & 0. & 0. & 0. & 0. & 0. & 0. & 0. \\ 0. & 0. & 0. & 0. & 0. & -1. & 2. & -1. & 0. & 0. & 0. & 0. & 0. & 0. & 0. & 0. & 0. & 0. & 0. & 0. \\ 0. & 0. & 0. & 0. & 0. & 0. & -1. & 2. & -1. & 0. & 0. & 0. & 0. & 0. & 0. & 0. & 0. & 0. & 0. & 0. \\ 0. & 0. & 0. & 0. & 0. & 0. & 0. & -1. & 2. & -1. & 0. & 0. & 0. & 0. & 0. & 0. & 0. & 0. & 0. & 0. \\ 0. & 0. & 0. & 0. & 0. & 0. & 0. & 0. & -1. & 2. & 0. & 0. & 0. & 0. & 0. & 0. & 0. & 0. & 0. & 0. \end{bmatrix}$$

6.1 Semianalytical and Numerical Method of Lines for Elliptic PDEs   527

> b:=matrix(2*N,1):for i from 1 to 2*N do b[i,1]:=-b1[i];od:evalm(b);

$$\begin{bmatrix} 0 \\ 0 \\ 0 \\ 0 \\ 0 \\ 0 \\ 0 \\ 0 \\ 0 \\ 0 \\ 0 \\ 0 \\ 0 \\ 0 \\ 0 \\ 0 \\ 0 \\ 0 \\ 0 \\ 0 \end{bmatrix}$$

> h:=eval(1/(N+1));

$$h := \frac{1}{11}$$

> J:=jordan(A,S);

$J := [\,[0.284629676547, 0, 0, 0, 0, 0, 0, 0, 0, 0, 0, 0, 0, 0, 0, 0, 0, 0, 0, 0\,],$

$[0, -.284629676547, 0, 0, 0, 0, 0, 0, 0, 0, 0, 0, 0, 0, 0, 0, 0, 0, 0, 0\,],$

$[0, 0, -.563465113678, 0, 0, 0, 0, 0, 0, 0, 0, 0, 0, 0, 0, 0, 0, 0, 0, 0\,],$

$[0, 0, 0, 0.563465113678, 0, 0, 0, 0, 0, 0, 0, 0, 0, 0, 0, 0, 0, 0, 0, 0\,],$

$[0, 0, 0, 0, 1.08128163361, 0, 0, 0, 0, 0, 0, 0, 0, 0, 0, 0, 0, 0, 0, 0\,],$

$[0, 0, 0, 0, 0, -1.08128163361, 0, 0, 0, 0, 0, 0, 0, 0, 0, 0, 0, 0, 0, 0\,],$

$[0, 0, 0, 0, 0, 0, 0.830830026146, 0, 0, 0, 0, 0, 0, 0, 0, 0, 0, 0, 0, 0\,],$

$$[0, 0, 0, 0, 0, 0, 0, -.830830026146, 0, 0, 0, 0, 0, 0, 0, 0, 0, 0, 0, 0],$$
$$[0, 0, 0, 0, 0, 0, 0, 0, 1.68250709437, 0, 0, 0, 0, 0, 0, 0, 0, 0, 0, 0],$$
$$[0, 0, 0, 0, 0, 0, 0, 0, 0, -1.68250709437, 0, 0, 0, 0, 0, 0, 0, 0, 0, 0],$$
$$[0, 0, 0, 0, 0, 0, 0, 0, 0, 0, 1.51149913265, 0, 0, 0, 0, 0, 0, 0, 0, 0],$$
$$[0, 0, 0, 0, 0, 0, 0, 0, 0, 0, 0, -1.51149913265, 0, 0, 0, 0, 0, 0, 0, 0],$$
$$[0, 0, 0, 0, 0, 0, 0, 0, 0, 0, 0, 0, 1.30972147378, 0, 0, 0, 0, 0, 0, 0],$$
$$[0, 0, 0, 0, 0, 0, 0, 0, 0, 0, 0, 0, 0, -1.30972147378, 0, 0, 0, 0, 0, 0],$$
$$[0, 0, 0, 0, 0, 0, 0, 0, 0, 0, 0, 0, 0, 0, 1.81926395646, 0, 0, 0, 0, 0],$$
$$[0, 0, 0, 0, 0, 0, 0, 0, 0, 0, 0, 0, 0, 0, 0, -1.81926395646, 0, 0, 0, 0],$$
$$[0, 0, 0, 0, 0, 0, 0, 0, 0, 0, 0, 0, 0, 0, 0, 0, 1.91898597269, 0, 0, 0],$$
$$[0, 0, 0, 0, 0, 0, 0, 0, 0, 0, 0, 0, 0, 0, 0, 0, 0, -1.91898597269, 0, 0],$$
$$[0, 0, 0, 0, 0, 0, 0, 0, 0, 0, 0, 0, 0, 0, 0, 0, 0, 0, 1.97964287516, 0],$$
$$[0, 0, 0, 0, 0, 0, 0, 0, 0, 0, 0, 0, 0, 0, 0, 0, 0, 0, 0, -1.97964287516]]$$

```
> mat:=evalm(S&*exponential(J,zeta)&*inverse(S)):
> mat1:=evalm(subs(zeta=zeta-zeta1,evalm(mat))):
> b2:=evalm(subs(zeta=zeta1,evalm(b))):
> mat2:=evalm(mat1&*b2):
> mat2:=map(expand,mat2):
> mat3:=map(int,mat2,zeta1=0..zeta):
> Y0:=matrix(2*N,1);
```

$$Y0 := array(1..20, 1..1, [\ ])$$

```
> for i to N do Y0[i,1]:=p[i];od:
> for i to N do Y0[N+i,1]:=c[i]:od:
> evalm(Y0);
```

6.1 Semianalytical and Numerical Method of Lines for Elliptic PDEs

$$\begin{bmatrix} p_1 \\ p_2 \\ p_3 \\ p_4 \\ p_5 \\ p_6 \\ p_7 \\ p_8 \\ p_9 \\ p_{10} \\ c_1 \\ c_2 \\ c_3 \\ c_4 \\ c_5 \\ c_6 \\ c_7 \\ c_8 \\ c_9 \\ c_{10} \end{bmatrix}$$

> Y:=evalm(mat&*Y0+mat3):

> sol0:=map(eval,evalm(subs(zeta=0,evalm(Y)))):

> sol1:=map(eval,evalm(subs(zeta=epsilon/h,evalm(Y)))):

>

> bc3:=diff(u(x,y),y)-u(x,y);

$$bc3 := \frac{\partial}{\partial y} u(x,y) - u(x,y)$$

> for i to N do Eq[i]:=subs(diff(u(x,y),y)=epsilon/h*c[i],u(x,y)=p[i], x=i*h,bc3);od;

$$Eq_1 := 11\, c_1 - p_1$$

$$Eq_2 := 11\, c_2 - p_2$$

$$Eq_3 := 11\, c_3 - p_3$$

$$Eq_4 := 11\, c_4 - p_4$$

530   6 Method of Lines for Elliptic Partial Differential Equations

$$Eq_5 := 11\,c_5 - p_5$$

$$Eq_6 := 11\,c_6 - p_6$$

$$Eq_7 := 11\,c_7 - p_7$$

$$Eq_8 := 11\,c_8 - p_8$$

$$Eq_9 := 11\,c_9 - p_9$$

$$Eq_{10} := 11\,c_{10} - p_{10}$$

> for i to N do Eq[N+i]:=evalf(subs(diff(u(x,y),y)=epsilon/h*sol1[N+i,1], u(x,y)=sol1[i,1],bc4));od:

The new sets of constants are:

> csol:=solve({seq(Eq[i],i=1..2*N)},{seq(c[i],i=1..N),seq(p[i],i=1..N)});

$csol := \{p_6 = 0.0827183801643, p_4 = 0.0760595917920, p_9 = 0.0452952522478, p_1 = 0.0236225323780, p_8$

$= 0.0632520997147, p_5 = 0.0827183825965, p_2 = 0.0452954269030, c_9 = 0.00411775020434, c_1$

$= 0.00214750294346, c_{10} = 0.00214749901393, p_3 = 0.0632521702481, p_{10} = 0.0236224891532, p_7$

$= 0.0760593939720, c_7 = 0.00691449036109, c_8 = 0.00575019088316, c_3 = 0.00575019729528, c_2$

$= 0.00411776608209, c_6 = 0.00751985274221, c_4 = 0.00691450834473, c_5 = 0.00751985296332\}$

> assign(csol);
> YY:=map(eval,Y):
> for i from 1 to N do u[i](zeta):=eval((YY[i,1]));od:
> for i from 0 to N+1 do u[i](zeta):=eval(u[i](zeta));od:
> for i from 0 to N+1 do u[i](y):=eval(subs(zeta=epsilon*y/h,u[i](zeta)));od;

$$u_0(y) := 0$$

$u_1(y) := 1.9593\,10^{-11}\,e^{21.1088456996\,y} + 2.1334560\,10^{-8}\,e^{-6.19811625046\,y} + 0.0155451316190\,e^{3.13092644202\,y}$

$+ 2.1876687\,10^{-8}\,e^{6.19811625046\,y} + 0.000032315018793\,e^{9.13913028761\,y} + 2.00999\,10^{-10}\,e^{11.8940979697\,y}$

$+ 1.14976135\,10^{-7}\,e^{14.4069362116\,y} - 5.82\,10^{-13}\,e^{16.6264904592\,y} + 9.74298\,10^{-10}\,e^{18.5075780381\,y}$

$- 7.1\,10^{-15}\,e^{21.7760716268\,y} - 2.14334431\,10^{-8}\,e^{-21.7760716268\,y} + 3.0105881\,10^{-8}\,e^{-21.1088456996\,y}$

$- 3.0012705\,10^{-8}\,e^{-20.0119035211\,y} + 1.77594567\,10^{-8}\,e^{-18.5075780381\,y} - 2.0913750\,10^{-8}\,e^{-16.6264904592\,y}$

$+ 1.0091299\,10^{-7}\,e^{-14.4069362116\,y} - 8.731597\,10^{-9}\,e^{-11.8940979697\,y} + 0.000025940185863\,e^{-9.13913028761\,y}$

$+ 0.00801889983580\,e^{-3.13092644202\,y}$

6.1 Semianalytical and Numerical Method of Lines for Elliptic PDEs 531

$u_2(y) := -3.2979 \cdot 10^{-11} e^{21.1088456996\,y} + 3.5876453 \cdot 10^{-8} e^{-6.19811625046\,y} + 0.0298308131162 \, e^{3.13092644202\,y}$

$+ 3.6807637 \cdot 10^{-8} e^{6.19811625046\,y} + 0.000042323655695 \, e^{9.13913028761\,y} + 1.66997 \cdot 10^{-10} e^{11.8940979697\,y}$

$+ 3.2725652 \cdot 10^{-8} e^{14.4069362116\,y} + 1.587 \cdot 10^{-13} e^{16.6264904592\,y} - 8.09466 \cdot 10^{-10} e^{18.5075780381\,y}$

$+ 2. \cdot 10^{-14} e^{20.0119035211\,y} - 1.2 \cdot 10^{-14} e^{21.7760716268\,y} + 5.492818 \cdot 10^{-9} e^{-21.7760716268\,y}$

$+ 4.3788651 \cdot 10^{-8} e^{-21.1088456996\,y} - 3.1910413 \cdot 10^{-8} e^{-20.0119035211\,y} + 2.6044584 \cdot 10^{-9} e^{-18.5075780381\,y}$

$- 4.79686813 \cdot 10^{-8} e^{-16.6264904592\,y} + 3.8507362 \cdot 10^{-8} e^{-14.4069362116\,y} - 8.595199 \cdot 10^{-9} e^{-11.8940979697\,y}$

$+ 0.000033974397602 \, e^{-9.13913028761\,y} + 0.0153881734157 \, e^{-3.13092644202\,y}$

$u_3(y) := 3.5887 \cdot 10^{-11} e^{21.1088456996\,y} + 3.8991288 \cdot 10^{-8} e^{-6.19811625046\,y} + 0.0416998777148 \, e^{3.13092644202\,y}$

$+ 4.0052526 \cdot 10^{-8} e^{6.19811625046\,y} + 0.000023117177749 \, e^{9.13913028761\,y} - 6.22399 \cdot 10^{-11} e^{11.8940979697\,y}$

$- 1.05661538 \cdot 10^{-7} e^{14.4069362116\,y} + 5.14 \cdot 10^{-13} e^{16.6264904592\,y} - 3.017766 \cdot 10^{-10} e^{18.5075780381\,y}$

$- 1.9 \cdot 10^{-14} e^{20.0119035211\,y} + 1.8 \cdot 10^{-14} e^{21.7760716268\,y} - 5.7443959 \cdot 10^{-8} e^{-21.7760716268\,y}$

$+ 7.7360245 \cdot 10^{-8} e^{-21.1088456996\,y} - 6.4679902 \cdot 10^{-8} e^{-20.0119035211\,y} + 1.34542234 \cdot 10^{-8} e^{-18.5075780381\,y}$

$- 6.8858452 \cdot 10^{-8} e^{-16.6264904592\,y} - 7.119263 \cdot 10^{-8} e^{-14.4069362116\,y} - 9.739646 \cdot 10^{-10} e^{-11.8940979697\,y}$

$+ 0.000018556848442 \, e^{-9.13913028761\,y} + 0.0215107681571 \, e^{-3.13092644202\,y}$

$u_4(y) := -2.7393 \cdot 10^{-11} e^{21.1088456996\,y} + 2.9673853 \cdot 10^{-8} e^{-6.19811625046\,y} + 0.0501906103509 \, e^{3.13092644202\,y}$

$+ 3.0580871 \cdot 10^{-8} e^{6.19811625046\,y} - 0.0000120465751253 \, e^{9.13913028761\,y} - 2.18733 \cdot 10^{-10} e^{11.8940979697\,y}$

$- 6.2800093 \cdot 10^{-8} e^{14.4069362116\,y} - 3.22 \cdot 10^{-13} e^{16.6264904592\,y} + 1.060139 \cdot 10^{-9} e^{18.5075780381\,y}$

$+ 9. \cdot 10^{-15} e^{20.0119035211\,y} - 1.1 \cdot 10^{-14} e^{21.7760716268\,y} + 9.310310 \cdot 10^{-9} e^{-21.7760716268\,y}$

$+ 7.9090704 \cdot 10^{-8} e^{-21.1088456996\,y} - 7.38125616 \cdot 10^{-8} e^{-20.0119035211\,y} + 3.29810352 \cdot 10^{-8} e^{-18.5075780381\,y}$

$- 7.69171600 \cdot 10^{-8} e^{-16.6264904592\,y} - 3.257667 \cdot 10^{-8} e^{-14.4069362116\,y} + 4.217362 \cdot 10^{-9} e^{-11.8940979697\,y}$

$- 0.000009670142361 \, e^{-9.13913028761\,y} + 0.0258906981071 \, e^{-3.13092644202\,y}$

$u_5(y) := 1.02164 \cdot 10^{-11} e^{21.1088456996\,y} + 1.0870448 \cdot 10^{-8} e^{-6.19811625046\,y} + 0.0546152278097 \, e^{3.13092644202\,y}$

$+ 1.14001170 \cdot 10^{-8} e^{6.19811625046\,y} - 0.000038894850445 \, e^{9.13913028761\,y} - 1.19455 \cdot 10^{-10} e^{11.8940979697\,y}$

$+ 8.7786730 \cdot 10^{-8} e^{14.4069362116\,y} - 4.37 \cdot 10^{-13} e^{16.6264904592\,y} - 5.790855 \cdot 10^{-10} e^{18.5075780381\,y}$

$$+\,2.\,10^{-14}\,e^{20.0119035211\,y} + 9.\,10^{-15}\,e^{21.7760716268\,y} - 7.5177505\,10^{-8}\,e^{-21.7760716268\,y}$$

$$+\,9.39308486\,10^{-8}\,e^{-21.1088456996\,y} - 6.7436090\,10^{-8}\,e^{-20.0119035211\,y} + 1.55311800\,10^{-8}\,e^{-18.5075780381\,y}$$

$$-\,8.30504774\,10^{-8}\,e^{-16.6264904592\,y} + 9.341526\,10^{-8}\,e^{-14.4069362116\,y} + 2.00482\,10^{-10}\,e^{-11.8940979697\,y}$$

$$-\,0.000031222033546\,e^{-9.13913028761\,y} + 0.0281731203119\,e^{-3.13092644202\,y}$$

$$u_6(y) := 1.02016\,10^{-11}\,e^{21.1088456996\,y} - 1.1455664\,10^{-8}\,e^{-6.19811625046\,y} + 0.0546152200540\,e^{3.13092644202\,y}$$

$$-\,1.14001109\,10^{-8}\,e^{6.19811625046\,y} - 0.000038894851086\,e^{9.13913028761\,y} + 1.19468\,10^{-10}\,e^{11.8940979697\,y}$$

$$+\,8.7786738\,10^{-8}\,e^{14.4069362116\,y} + 4.49\,10^{-13}\,e^{16.6264904592\,y} - 5.790813\,10^{-10}\,e^{18.5075780381\,y}$$

$$-\,4.\,10^{-14}\,e^{20.0119035211\,y} + 1.\,10^{-15}\,e^{21.7760716268\,y} + 1.0153019\,10^{-8}\,e^{-21.7760716268\,y}$$

$$+\,9.39307729\,10^{-8}\,e^{-21.1088456996\,y} - 8.6537338\,10^{-8}\,e^{-20.0119035211\,y} + 1.55311228\,10^{-8}\,e^{-18.5075780381\,y}$$

$$-\,8.85003472\,10^{-8}\,e^{-16.6264904592\,y} + 9.341521\,10^{-8}\,e^{-14.4069362116\,y} - 8.693832\,10^{-9}\,e^{-11.8940979697\,y}$$

$$-\,0.000031222040580\,e^{-9.13913028761\,y} + 0.0281731186450\,e^{-3.13092644202\,y}$$

$$u_7(y) := -2.7407\,10^{-11}\,e^{21.1088456996\,y} - 3.0216059\,10^{-8}\,e^{-6.19811625046\,y} + 0.0501906228028\,e^{3.13092644202\,y}$$

$$-\,3.0580900\,10^{-8}\,e^{6.19811625046\,y} - 0.0000120465688538\,e^{9.13913028761\,y} + 2.18735\,10^{-10}\,e^{11.8940979697\,y}$$

$$-\,6.2800047\,10^{-8}\,e^{14.4069362116\,y} + 3.21\,10^{-13}\,e^{16.6264904592\,y} + 1.060191\,10^{-9}\,e^{18.5075780381\,y}$$

$$+\,1.1\,10^{-14}\,e^{20.0119035211\,y} + 1.2\,10^{-14}\,e^{21.7760716268\,y} - 6.9107107\,10^{-8}\,e^{-21.7760716268\,y}$$

$$+\,7.9090482\,10^{-8}\,e^{-21.1088456996\,y} - 6.78960073\,10^{-8}\,e^{-20.0119035211\,y} + 3.29805631\,10^{-8}\,e^{-18.5075780381\,y}$$

$$-\,8.08157188\,10^{-8}\,e^{-16.6264904592\,y} - 3.257672\,10^{-8}\,e^{-14.4069362116\,y} - 1.2066683\,10^{-8}\,e^{-11.8940979697\,y}$$

$$-\,0.000009670134784\,e^{-9.13913028761\,y} + 0.0258907011190\,e^{-3.13092644202\,y}$$

$$u_8(y) := 3.5905\,10^{-11}\,e^{21.1088456996\,y} - 3.9447760\,10^{-8}\,e^{-6.19811625046\,y} + 0.0416998809893\,e^{3.13092644202\,y}$$

$$-\,4.0052498\,10^{-8}\,e^{6.19811625046\,y} + 0.000023117204054\,e^{9.13913028761\,y} + 6.22353\,10^{-11}\,e^{11.8940979697\,y}$$

$$-\,1.05661490\,10^{-7}\,e^{14.4069362116\,y} - 5.10\,10^{-13}\,e^{16.6264904592\,y} - 3.017775\,10^{-10}\,e^{18.5075780381\,y}$$

$$+\,4.0\,10^{-14}\,e^{20.0119035211\,y} - 1.7\,10^{-14}\,e^{21.7760716268\,y} + 7.707786\,10^{-9}\,e^{-21.7760716268\,y}$$

$$+\,7.7359978\,10^{-8}\,e^{-21.1088456996\,y} - 5.3326352\,10^{-8}\,e^{-20.0119035211\,y} + 1.34538118\,10^{-8}\,e^{-18.5075780381\,y}$$

6.1 Semianalytical and Numerical Method of Lines for Elliptic PDEs

$- 6.2298433 \ 10^{-8} \ e^{-16.6264904592 y} - 7.119268 \ 10^{-8} \ e^{-14.4069362116 y} - 5.6088259 \ 10^{-9} \ e^{-11.8940979697 y}$

$+ 0.000018556859563 \ e^{-9.13913028761 y} + 0.0215107743026 \ e^{-3.13092644202 y}$

$u_9(y) := -3.2982 \ 10^{-11} \ e^{21.1088456996 y} - 3.6207855 \ 10^{-8} \ e^{-6.19811625046 y} + 0.0298308467346 \ e^{3.13092644202 y}$

$- 3.6807645 \ 10^{-8} \ e^{6.19811625046 y} + 0.000042323646752 \ e^{9.13913028761 y} - 1.66997 \ 10^{-10} \ e^{11.8940979697 y}$

$+ 3.2725625 \ 10^{-8} \ e^{14.4069362116 y} - 1.656 \ 10^{-13} \ e^{16.6264904592 y} - 8.09477 \ 10^{-10} \ e^{18.5075780381 y}$

$- 1. \ 10^{-14} \ e^{20.0119035211 y} + 1. \ 10^{-15} \ e^{21.7760716268 y} - 4.1114618 \ 10^{-8} \ e^{-21.7760716268 y}$

$+ 4.3788523 \ 10^{-8} \ e^{-21.1088456996 y} - 5.2695428 \ 10^{-8} \ e^{-20.0119035211 y} + 2.6041325 \ 10^{-9} \ e^{-18.5075780381 y}$

$- 4.59367072 \ 10^{-8} \ e^{-16.6264904592 y} + 3.8507335 \ 10^{-8} \ e^{-14.4069362116 y} + 3.838072 \ 10^{-9} \ e^{-11.8940979697 y}$

$+ 0.000033974435420 \ e^{-9.13913028761 y} + 0.0153881640751 \ e^{-3.13092644202 y}$

$u_{10}(y) := 1.9622 \ 10^{-11} \ e^{21.1088456996 y} - 2.1508854 \ 10^{-8} \ e^{-6.19811625046 y} + 0.0155451185550 \ e^{3.13092644202 y}$

$- 2.1876658 \ 10^{-8} \ e^{6.19811625046 y} + 0.000032315015175 \ e^{9.13913028761 y} - 2.01007 \ 10^{-10} \ e^{11.8940979697 y}$

$+ 1.14976241 \ 10^{-7} \ e^{14.4069362116 y} + 5.44 \ 10^{-13} \ e^{16.6264904592 y} + 9.74288 \ 10^{-10} \ e^{18.5075780381 y}$

$+ 2. \ 10^{-14} \ e^{20.0119035211 y} + 6.3 \ 10^{-15} \ e^{21.7760716268 y} + 2.8542449 \ 10^{-9} \ e^{-21.7760716268 y}$

$+ 3.0105726 \ 10^{-8} \ e^{-21.1088456996 y} - 1.4142247 \ 10^{-8} \ e^{-20.0119035211 y} + 1.77592352 \ 10^{-8} \ e^{-18.5075780381 y}$

$- 2.8051518 \ 10^{-8} \ e^{-16.6264904592 y} + 1.0091286 \ 10^{-7} \ e^{-14.4069362116 y} + 6.233241 \ 10^{-9} \ e^{-11.8940979697 y}$

$+ 0.000025940212368 \ e^{-9.13913028761 y} + 0.00801890866497 \ e^{-3.13092644202 y}$

$u_{11}(y) := 0$

Using the new values for constants the semianalytical solution is recalculated, the following plots are obtained, and the constants are unassigned.

> for i to N do unassign('c[i]'):unassign('p[i]'):od:

> for i from 0 to N+1 do

pl[i]:=line([0.3,0.98-abs(i-5.25)*0.14],[0.6,evalf(subs(y=0.6,u[i](y)))], thickness=1,linestyle=dot);

pt[i]:=textplot([0.3,0.98-abs(i-5.25)*0.14,typeset(u[i],"(y)")],align=left):

end do:

534   6 Method of Lines for Elliptic Partial Differential Equations

```
> for i from 0 to N+1 do p[i]:=plot(u[i](y),y=0..1,thickness=3);od:

> pp:=plot([seq(u[i](y),i=0..N+1)],y=0..1,thickness=3):

> display([pp,seq(pl[i],i=0..N+1),seq(pt[i],i=0..N+1)],title="Figure 6.4",
axes=boxed,labels=[y,"u"]);
```

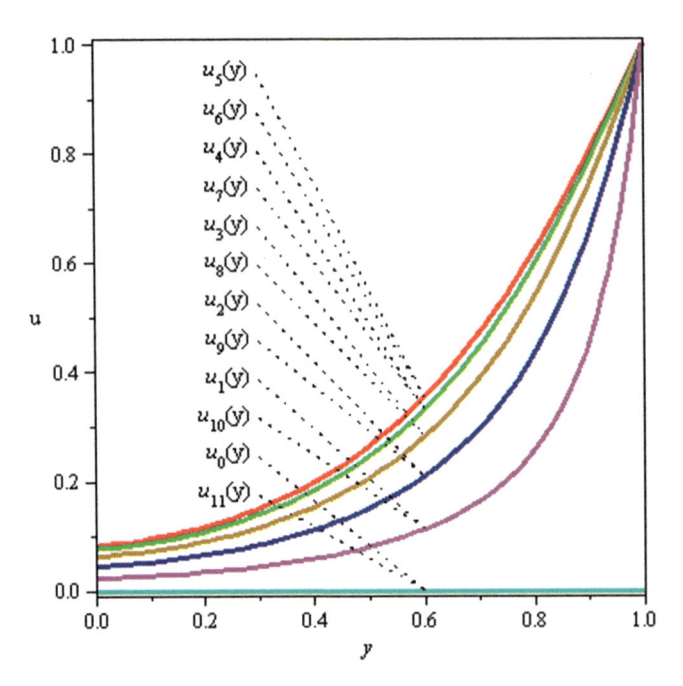

Fig. 6.4

```
> M:=5;
```

$$M := 5$$

```
> T1:=[seq(evalf(i/M),i=0..M)];
```

$T1 := [0., 0.200000000000, 0.400000000000, 0.600000000000, 0.800000000000, 1.]$

```
> for j from 1 to M do
P[j]:=plot([seq([h*i,evalf(subs(y=T1[j],evalf(u[i](y))))],i=0..N+1)],style=line,
thickness=3,axes=boxed,view=[0..1,0..1.1]):od:

> P[M+1]:=plot([seq([h*i,evalf(subs(x=i*h,1))],i=0..N+1)],
style=line,thickness=3,title="Figure 6.5",axes=boxed):

> for j from 1 to M+1 do
pt[j]:=textplot([0.5,evalf(subs(y=T1[j],u[5](y))),typeset(y,sprintf("=%4.2f",
T1[j]))],align=above);od:
```

6.1 Semianalytical and Numerical Method of Lines for Elliptic PDEs   535

> display({seq(P[i],i=1..M+1),seq(pt[i],i=1..M+1)},labels=[x,u]);

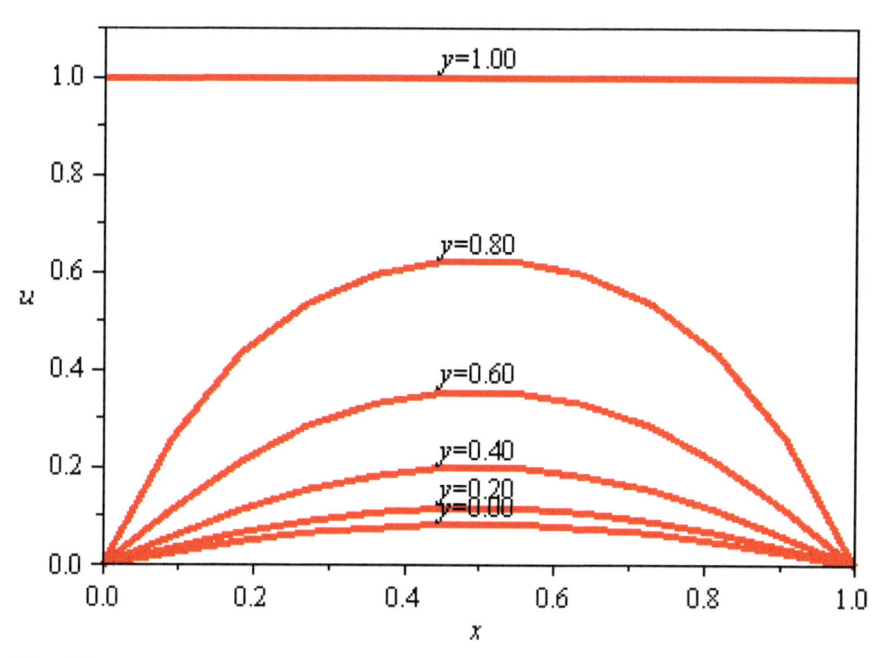

Fig. 6.5

> Ny:=20;

$$Ny := 20$$

> PP:=matrix(N+2,Ny);

$$PP := array(1..12, 1..20, [\,])$$

For the three dimensional plot, first the boundaries, x = 0, x = 1, and y = 1 are defined.

> for i to Ny do PP[1,i]:=0;PP[N+2,i]:=0;od:

> for i to N+2 do PP[i,Ny]:=1;od:

The temperature inside the rectangle is obtained using the semianalytical solution.

> for i from 2 to N+1 do for j from 1 to Ny-1 do PP[i,j]:=evalf(subs(y=(j-1)/(Ny-1),u[i-1](y)));od;od:

> plotdata := [seq([seq([(i-1)/(N+1),(j-1)/(Ny-1),PP[i,j]], i=1..N+2)], j=1..Ny)]:

> surfdata(plotdata,axes=boxed,title="Figure 6.6", labels=[x,y,u],orientation=[-120,60]);

Fig. 6.6

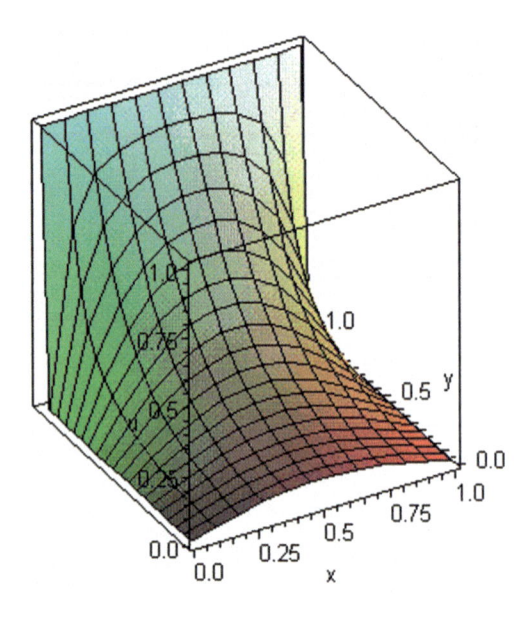

Note that a semianalytical solution in x can be obtained instead of y in the previous two examples by discretizing the spatial derivatives in the y derivatives.

6.1.3 Semianalytical Method for Elliptic PDEs in Cylindrical Coordinates – Graetz Problem

Example 6.3. Graetz Problem with a Fixed Wall Temperature

As an aside, it is worth mentioning, that the technique described earlier can also be used for solving partial differential equations in cylindrical coordinates. For example, consider the Graetz problem,[1]

$$2\text{Pe}(1-x^2)\frac{\partial u}{\partial y} = \frac{\partial^2 u}{\partial x^2} + \frac{1}{x}\frac{\partial u}{\partial x} + \frac{\partial^2 u}{\partial y^2}$$

$$\frac{\partial u}{\partial x}(0,y) = 0 \text{ for } 0 \leq y \leq H$$

$$u(1,y) = 1 \text{ for } 0 < y \leq H \quad (6.22)$$

$$u(x,0) = 0 \text{ for } 0 \leq x \leq 1$$

$$\frac{\partial u}{\partial x}(x,H) \text{ for } 0 \leq x \leq 1$$

Schiesser and Silebi (1997)[1] solved this problem using the numerical method of lines by adding a time derivative for u and waiting for the steady state. However,

6.1 Semianalytical and Numerical Method of Lines for Elliptic PDEs

our method directly yields the steady state solution (semianalytical in y and numerical in x) for the temperature profiles. Note that the semianalytical solution can be obtained only in the y direction and not in the x direction for cylindrical coordinate problems because the coefficient matrix, A, in equation (6.16) becomes a function of x and the solution should be found using the matrizant instead of exponential matrix. The calculation of the matrizant is time consuming. See chapter 3.1.4 for additional information. Note that a low value of the Peclet number, Pe = 1, is chosen so that the effect of the axial conduction can be seen. The Maple program developed for example 6.1 can be used for this example by making minor changes as follows:

> restart;with(linalg):with(plots):

The governing equation is entered in the following form:

> ge:=diff(u(x,y),y$2)=2*Pe*(1-x^2)*diff(u(x,y),y)-diff(u(x,y),x$2)-1/x*diff(u(x,y),x);

$$ge := \frac{\partial^2}{\partial y^2} u(x,y) = 2\, Pe\, (1-x^2) \left(\frac{\partial}{\partial y} u(x,y) \right) - \left(\frac{\partial^2}{\partial x^2} u(x,y) \right) - \frac{\frac{\partial}{\partial x} u(x,y)}{x}$$

> Digits:=30;

$$Digits := 30$$

For this example, 'Digits' has to be set to 30 for accurate predictions. The boundary conditions are entered as:

> bc1:=diff(u(x,y),x);

$$bc1 := \frac{\partial}{\partial x} u(x,y)$$

> bc2:=u(x,y)-1;

$$bc2 := u(x,y) - 1$$

> bc3:=u(x,y)-0;

$$bc3 := u(x,y)$$

> bc4:=diff(u(x,y),y);

$$bc4 := \frac{\partial}{\partial y} u(x,y)$$

Parameters are entered here:

> Pe:=1.0;

$$Pe := 1.0$$

> epsilon:=1;

$$\varepsilon := 1$$

Note that epsilon is given as 1 for this example since L and H are taken care of separately.

> L:=1;

$$L := 1$$

> H:=2;

$$H := 2$$

> dydxf:=1/2/h*(-u[m+2](zeta)-3*u[m](zeta)+4*u[m+1](zeta)):
> dydxb:=1/2/h*(u[m-2](zeta)+3*u[m](zeta)-4*u[m-1](zeta)):
> dydx:=1/2/h*(u[m+1](zeta)-u[m-1](zeta)):
> d2ydx2:=1/h^2*(u[m-1](zeta)-2*u[m](zeta)+u[m+1](zeta)):
> bc1:=subs(diff(u(x,y),x)=subs(m=0,dydxf),u(x,y)=u[0](zeta),x=0,bc1):
> bc2:=subs(diff(u(x,y),x)=subs(m=N+1,dydxb),u(x,y)=u[N+1](zeta),x=1,bc2):
> N:=10;

$$N := 10$$

> eq[0]:=bc1;

$$eq_0 := \frac{1}{2} \frac{-u_2(\zeta) - 3 u_0(\zeta) + 4 u_1(\zeta)}{h}$$

> eq[N+1]:=bc2;

$$eq_{11} := u_{11}(\zeta) - 1$$

The governing equation is converted to finite difference form here. Note that the first derivative with respect to 'y' is replaced by u[N+1+i], i= 1..N.

> for i from 1 to N do eq[N+1+i]:=diff(u[N+1+i](zeta),zeta)=
subs(diff(u(x,y),x$2) = subs(m=i,d2ydx2),diff(u(x,y),x) =
subs(m=i,dydx),diff(u(x,y),y)=epsilon/h*u[N+1+i](zeta),u(x,y)=u[i](zeta),
x=i*h,-rhs(h^2/epsilon^2*ge));od;

$$eq_{12} := \frac{d}{d\zeta} u_{12}(\zeta) = -h^2 \left(\frac{2.0 (1 - h^2) u_{12}(\zeta)}{h} - \frac{u_0(\zeta) - 2 u_1(\zeta) + u_2(\zeta)}{h^2} - \frac{1}{2} \frac{u_2(\zeta) - u_0(\zeta)}{h^2} \right)$$

$$eq_{13} := \frac{d}{d\zeta} u_{13}(\zeta) = -h^2 \left(\frac{2.0 (1 - 4h^2) u_{13}(\zeta)}{h} - \frac{u_1(\zeta) - 2 u_2(\zeta) + u_3(\zeta)}{h^2} - \frac{1}{4} \frac{u_3(\zeta) - u_1(\zeta)}{h^2} \right)$$

$$eq_{14} := \frac{d}{d\zeta} u_{14}(\zeta) = -h^2 \left(\frac{2.0 (1 - 9h^2) u_{14}(\zeta)}{h} - \frac{u_2(\zeta) - 2 u_3(\zeta) + u_4(\zeta)}{h^2} - \frac{1}{6} \frac{u_4(\zeta) - u_2(\zeta)}{h^2} \right)$$

$$eq_{15} := \frac{d}{d\zeta} u_{15}(\zeta) = -h^2 \left(\frac{2.0 (1 - 16 h^2) u_{15}(\zeta)}{h} - \frac{u_3(\zeta) - 2 u_4(\zeta) + u_5(\zeta)}{h^2} - \frac{1}{8} \frac{u_5(\zeta) - u_3(\zeta)}{h^2} \right)$$

6.1 Semianalytical and Numerical Method of Lines for Elliptic PDEs

$$eq_{16} := \frac{d}{d\zeta} u_{16}(\zeta) = -h^2 \left(\frac{2.0\,(1 - 25\,h^2)\,u_{16}(\zeta)}{h} - \frac{u_4(\zeta) - 2\,u_5(\zeta) + u_6(\zeta)}{h^2} - \frac{1}{10}\,\frac{u_6(\zeta) - u_4(\zeta)}{h^2} \right)$$

$$eq_{17} := \frac{d}{d\zeta} u_{17}(\zeta) = -h^2 \left(\frac{2.0\,(1 - 36\,h^2)\,u_{17}(\zeta)}{h} - \frac{u_5(\zeta) - 2\,u_6(\zeta) + u_7(\zeta)}{h^2} - \frac{1}{12}\,\frac{u_7(\zeta) - u_5(\zeta)}{h^2} \right)$$

$$eq_{18} := \frac{d}{d\zeta} u_{18}(\zeta) = -h^2 \left(\frac{2.0\,(1 - 49\,h^2)\,u_{18}(\zeta)}{h} - \frac{u_6(\zeta) - 2\,u_7(\zeta) + u_8(\zeta)}{h^2} - \frac{1}{14}\,\frac{u_8(\zeta) - u_6(\zeta)}{h^2} \right)$$

$$eq_{19} := \frac{d}{d\zeta} u_{19}(\zeta) = -h^2 \left(\frac{2.0\,(1 - 64\,h^2)\,u_{19}(\zeta)}{h} - \frac{u_7(\zeta) - 2\,u_8(\zeta) + u_9(\zeta)}{h^2} - \frac{1}{16}\,\frac{u_9(\zeta) - u_7(\zeta)}{h^2} \right)$$

$$eq_{20} := \frac{d}{d\zeta} u_{20}(\zeta) = -h^2 \left(\frac{2.0\,(1 - 81\,h^2)\,u_{20}(\zeta)}{h} - \frac{u_8(\zeta) - 2\,u_9(\zeta) + u_{10}(\zeta)}{h^2} - \frac{1}{18}\,\frac{u_{10}(\zeta) - u_8(\zeta)}{h^2} \right)$$

$$eq_{21} := \frac{d}{d\zeta} u_{21}(\zeta) = -h^2 \left(\frac{2.0\,(1 - 100\,h^2)\,u_{21}(\zeta)}{h} - \frac{u_9(\zeta) - 2\,u_{10}(\zeta) + u_{11}(\zeta)}{h^2} \right.$$
$$\left. - \frac{1}{20}\,\frac{u_{11}(\zeta) - u_9(\zeta)}{h^2} \right)$$

> u[0](zeta):=(solve(eq[0],u[0](zeta)));

$$u_0(\zeta) := -\frac{1}{3}\,u_2(\zeta) + \frac{4}{3}\,u_1(\zeta)$$

> u[N+1](zeta):=solve(eq[N+1],u[N+1](zeta));

$$u_{11}(\zeta) := 1$$

> for i from 1 to N do eq[i]:=diff(u[i](zeta),zeta)= -u[N+1+i](zeta);od;

$$eq_1 := \frac{d}{d\zeta} u_1(\zeta) = -u_{12}(\zeta)$$

$$eq_2 := \frac{d}{d\zeta} u_2(\zeta) = -u_{13}(\zeta)$$

$$eq_3 := \frac{d}{d\zeta} u_3(\zeta) = -u_{14}(\zeta)$$

$$eq_4 := \frac{d}{d\zeta} u_4(\zeta) = -u_{15}(\zeta)$$

$$eq_5 := \frac{d}{d\zeta} u_5(\zeta) = -u_{16}(\zeta)$$

$$eq_6 := \frac{d}{d\zeta} u_6(\zeta) = -u_{17}(\zeta)$$

$$eq_7 := \frac{d}{d\zeta} u_7(\zeta) = -u_{18}(\zeta)$$

$$eq_8 := \frac{d}{d\zeta} u_8(\zeta) = -u_{19}(\zeta)$$

$$eq_9 := \frac{d}{d\zeta} u_9(\zeta) = -u_{20}(\zeta)$$

$$eq_{10} := \frac{d}{d\zeta} u_{10}(\zeta) = -u_{21}(\zeta)$$

> h:=L/(N+1);

$$h := \frac{1}{11}$$

> for i from 1 to N do eq[i]:=eval(eq[i]);od;for i from 1 to N do eq[N+1+i]:=eval(eq[N+1+i]);od;

$$eq_1 := \frac{d}{d\zeta} u_1(\zeta) = -u_{12}(\zeta)$$

$$eq_2 := \frac{d}{d\zeta} u_2(\zeta) = -u_{13}(\zeta)$$

$$eq_3 := \frac{d}{d\zeta} u_3(\zeta) = -u_{14}(\zeta)$$

$$eq_4 := \frac{d}{d\zeta} u_4(\zeta) = -u_{15}(\zeta)$$

$$eq_5 := \frac{d}{d\zeta} u_5(\zeta) = -u_{16}(\zeta)$$

$$eq_6 := \frac{d}{d\zeta} u_6(\zeta) = -u_{17}(\zeta)$$

$$eq_7 := \frac{d}{d\zeta} u_7(\zeta) = -u_{18}(\zeta)$$

$$eq_8 := \frac{d}{d\zeta} u_8(\zeta) = -u_{19}(\zeta)$$

$$eq_9 := \frac{d}{d\zeta} u_9(\zeta) = -u_{20}(\zeta)$$

$$eq_{10} := \frac{d}{d\zeta} u_{10}(\zeta) = -u_{21}(\zeta)$$

$$eq_{12} := \frac{d}{d\zeta} u_{12}(\zeta) = -0.18031555221637866265965439 5192\, u_{12}(\zeta) + \frac{4}{3} u_2(\zeta) - \frac{4}{3} u_1(\zeta)$$

$$eq_{13} := \frac{d}{d\zeta} u_{13}(\zeta) = -0.17580766341096919609316303 5312\, u_{13}(\zeta) + \frac{3}{4} u_1(\zeta) - 2 u_2(\zeta) + \frac{5}{4} u_3(\zeta)$$

$$eq_{14} := \frac{d}{d\zeta} u_{14}(\zeta) = -0.16829451540195341848234410 2179\, u_{14}(\zeta) + \frac{5}{6} u_2(\zeta) - 2 u_3(\zeta) + \frac{7}{6} u_4(\zeta)$$

$$eq_{15} := \frac{d}{d\zeta} u_{15}(\zeta) = -0.15777610818933132982719759 5793\, u_{15}(\zeta) + \frac{7}{8} u_3(\zeta) - 2 u_4(\zeta) + \frac{9}{8} u_5(\zeta)$$

$$eq_{16} := \frac{d}{d\zeta} u_{16}(\zeta) = -0.14425244177310293012772351 6153\, u_{16}(\zeta) + \frac{9}{10} u_4(\zeta) - 2 u_5(\zeta) + \frac{11}{10} u_6(\zeta)$$

6.1 Semianalytical and Numerical Method of Lines for Elliptic PDEs 541

$eq_{17} := \frac{d}{d\zeta} u_{17}(\zeta) = -0.127723516153268219383921863260\, u_{17}(\zeta) + \frac{11}{12} u_5(\zeta) - 2 u_6(\zeta) + \frac{13}{12} u_7(\zeta)$

$eq_{18} := \frac{d}{d\zeta} u_{18}(\zeta) = -0.108189331329827197595792637115\, u_{18}(\zeta) + \frac{13}{14} u_6(\zeta) - 2 u_7(\zeta) + \frac{15}{14} u_8(\zeta)$

$eq_{19} := \frac{d}{d\zeta} u_{19}(\zeta) = -0.085649887302779864763335837715 7\, u_{19}(\zeta) + \frac{15}{16} u_7(\zeta) - 2 u_8(\zeta) + \frac{17}{16} u_9(\zeta)$

$eq_{20} := \frac{d}{d\zeta} u_{20}(\zeta) = -0.060105184072126220886551465063 9\, u_{20}(\zeta) + \frac{17}{18} u_8(\zeta) - 2 u_9(\zeta) + \frac{19}{18} u_{10}(\zeta)$

$eq_{21} := \frac{d}{d\zeta} u_{21}(\zeta) = -0.031555221637866265965439519158 5\, u_{21}(\zeta) + \frac{19}{20} u_9(\zeta) - 2 u_{10}(\zeta) + \frac{21}{20}$

```
> eqns:=[seq(rhs(eq[j]),j=1..N),seq(rhs(eq[N+1+j]),j=1..N)]:
> Y:=[seq(u[i](zeta),i=1..N),seq(u[N+1+i](zeta),i=1..N)];
```

$Y := [u_1(\zeta), u_2(\zeta), u_3(\zeta), u_4(\zeta), u_5(\zeta), u_6(\zeta), u_7(\zeta), u_8(\zeta), u_9(\zeta), u_{10}(\zeta), u_{12}(\zeta), u_{13}(\zeta), u_{14}(\zeta), u_{15}(\zeta),$
$u_{16}(\zeta), u_{17}(\zeta), u_{18}(\zeta), u_{19}(\zeta), u_{20}(\zeta), u_{21}(\zeta)]$

```
> A:=genmatrix(eqns,Y,'b1'):
> Nrow:=rowdim(A);
```

$$Nrow := 20$$

```
> ll:=eigenvalues(A):

> for i to Nrow do

 lambda[i]:=ll[i];

end do:

> Id:=Matrix(Nrow,Nrow,shape=identity):

> X:=matrix(Nrow,1,[seq(beta[i],i=1..Nrow)]):

> for k to Nrow do

 G:=evalm((A-lambda[k]*Id)&*X);

 eqx[1]:=beta[1]=1:

 for i from 2 to Nrow do

  eqx[i]:=G[i-1,1]:

 end do:
```

```
cons:=fsolve({seq(eqx[i],i=1..Nrow)},{seq(beta[i],i=1..Nrow)}):

assign(cons):

XX[k]:=map(eval,evalm(X)):

for i to Nrow do

  unassign('beta[i]'):

 end do:

end do:
> PV:=Matrix(Nrow,Nrow,[seq(evalm(XX[i]),i=1..Nrow)]):
> expD1:=Matrix(1..Nrow,1..Nrow,shape=diagonal):
> for i to Nrow do

  expD1[i,i]:=exp(lambda[i]*zeta):

end do:
> mat:=evalm(PV&*expD1&*inverse(PV)):
> if N>2 then A:=map(evalf,A):end:
> b:=matrix(2*N,1):for i from 1 to 2*N do b[i,1]:=-b1[i];od:evalm(b):
> h:=eval(1/(N+1));
```

$$h := \frac{1}{11}$$

```
> mat1:=evalm(subs(zeta=zeta-zeta1,evalm(mat))):
> b2:=evalm(subs(zeta=zeta1,evalm(b))):
> mat2:=evalm(mat1&*b2):
> mat2:=map(expand,mat2):
> mat3:=map(int,mat2,zeta1=0..zeta):
> Y0:=matrix(2*N,1);
```

$$Y0 := array\,(1..20,\,1..1,\,[\,])$$

6.1 Semianalytical and Numerical Method of Lines for Elliptic PDEs

```
> for i to N do Y0[i,1]:=p[i];od:
> for i to N do Y0[N+i,1]:=c[i]:od:
> evalm(Y0);
```

$$\begin{bmatrix} p_1 \\ p_2 \\ p_3 \\ p_4 \\ p_5 \\ p_6 \\ p_7 \\ p_8 \\ p_9 \\ p_{10} \\ c_1 \\ c_2 \\ c_3 \\ c_4 \\ c_5 \\ c_6 \\ c_7 \\ c_8 \\ c_9 \\ c_{10} \end{bmatrix}$$

```
> Y:=evalm(mat&*Y0+mat3):
```

The solution should be evaluated at $y = H$ to find the constants.

```
> sol0:=map(eval,evalm(subs(zeta=0,evalm(Y)))):
> sol1:=map(eval,evalm(subs(zeta=epsilon*H/h,evalm(Y)))):
> for i to N do Eq[i]:=subs(diff(u(x,y),y)=epsilon/h*c[i],
u(x,y)=p[i],x=i*h,bc3);od;
```

$$Eq_1 := p_1$$
$$Eq_2 := p_2$$
$$Eq_3 := p_3$$
$$Eq_4 := p_4$$
$$Eq_5 := p_5$$
$$Eq_6 := p_6$$
$$Eq_7 := p_7$$
$$Eq_8 := p_8$$
$$Eq_9 := p_9$$
$$Eq_{10} := p_{10}$$

> for i to N do Eq[N+i]:=evalf(subs(diff(u(x,y),y)=epsilon/h*sol1[N+i,1], u(x,y)=sol1[i,1],bc4));od:

Constants are found as:

> csol:=solve({seq(Eq[i],i=1..2*N)},{seq(c[i],i=1..N),seq(p[i],i=1..N)});

$csol := \{p_9 = 0., p_4 = 0., p_8 = 0., c_1 = -.2195259805418114603425758145, c_6$

$= -.2409918767221317893513018348333, c_7 = -.2623172864214406469777245519681, c_5$

$= -.2299807153670956344258406866330, c_8 = -.3062158206724079774129412111681, c_{10}$

$= -.91491270422522078166606182203, c_9 = -.4141358218661252966883304741120, c_3$

$= -.2214149182354713229205662079687, c_4 = -.2242720866445001506277711480982, c_2$

$= -.2200813478777443684791684891161, p_3 = 0., p_{10} = 0., p_2 = 0., p_5 = 0., p_1 = 0., p_6 = 0., p_7 = 0.\}$

> assign(csol);

> Y:=map(eval,Y):

> for i from 1 to N do u[i](zeta):=eval((Y[i,1]));od:

> for i from 0 to N+1 do u[i](zeta):=eval(u[i](zeta));od:

> for i from 0 to N+1 do u[i](y):=eval(subs(zeta=epsilon*y/h,u[i](zeta)));od:

The following plots are obtained using $N = 10$ interior node points. Digits = 30 is required for $N = 10$ interior node points. For $N = 3$ node points, the default number of Digits =10 is enough.

6.1 Semianalytical and Numerical Method of Lines for Elliptic PDEs   545

> for i from 0 to N+1 by 2 do p[i]:=plot(u[i](y),y=0..H,thickness=3);od:

> pp:=plot([seq(u[i](y),i=0..N+1,2)],y=0..H,thickness=3,
legend=[seq(typeset(u[i],"(y)"),i=0..N+1,2)]);

$$pp := PLOT(...)$$

> display(pp,axes=boxed,title="Figure 6.7",labels=[y,"u"]);

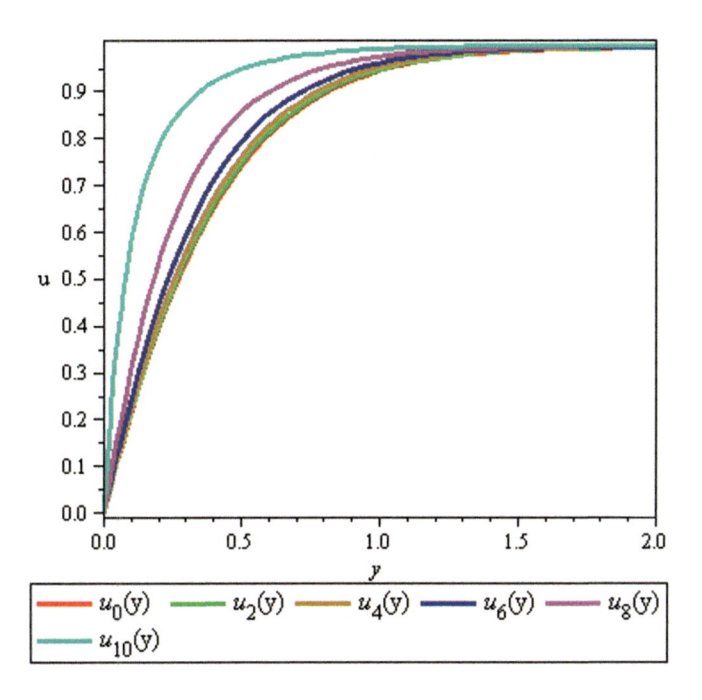

Fig. 6.7

> M:=5;

$$M := 5$$

> T1:=[seq(evalf(i*H/M),i=0..M)]:

> P[1]:=plot([seq([h*i,0.],i=0..N+1)],style=line,thickness=3,axes=boxed):

> for j from 2 to M+1 do
P[j]:=plot([seq([h*i,evalf(subs(y=T1[j],evalf(u[i](y))))],i=0..N+1)],

style=line,thickness=3,title="Figure 6.8",axes=boxed):od:

> for j from 1 to M+1 do
pt[j]:=textplot([0.5,evalf(subs(y=T1[j],u[5](y))),typeset(y,sprint

("=%4.2f",T1[j]))],align={above,left});od:

> display({seq(P[i],i=1..M+1),seq(pt[i],i=1..M+1)},labels=[x,u]);

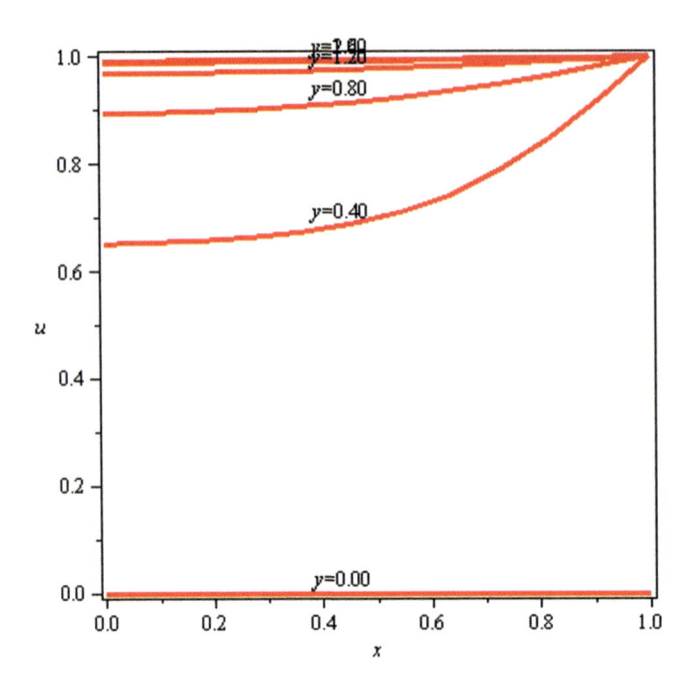

Fig. 6.8

> a:=convert(T1[2],string);

$$a := ".4000000000000000000000000000000"$$

> a1:=sprintf("%4.2f",T1[2]);

$$a1 := "0.40"$$

> Ny:=30;

$$Ny := 30$$

> PP:=matrix(N+2,Ny);

$$PP := array(1..12, 1..30, [\,])$$

> for i from 1 to N+2 do for j from 1 to Ny do
PP[i,j]:=evalf(subs(y=(j-1)*H/(Ny-1),u[i-1](y)));od;od:

> PP[N+2,1]:=0;

$$PP_{12,1} := 0$$

6.1 Semianalytical and Numerical Method of Lines for Elliptic PDEs

```
> plotdata := [seq([ seq([(i-1)/(N+1),(j-1)*H/(Ny-1),PP[i,j]], i=1..N+2)],
j=1..Ny)]:
> surfdata(plotdata,axes=boxed,title="Figure 6.9",
labels=[x,y,u],orientation=[-150,45] );
```

Fig. 6.9

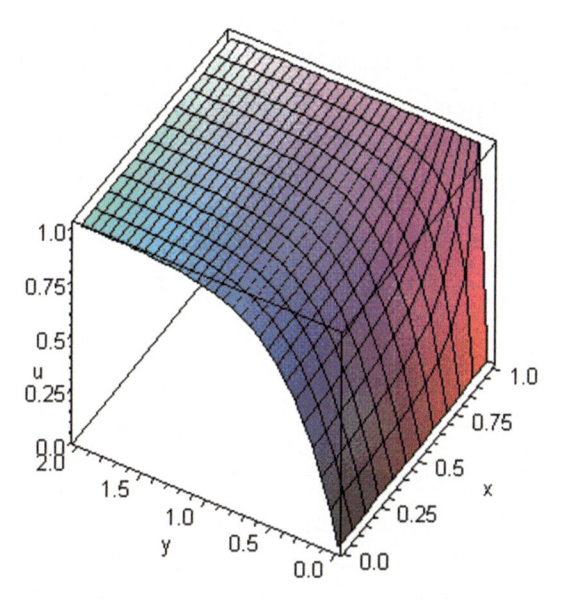

The program developed for example 6.3 is very general and can be used for any linear elliptic partial differential equation with linear boundary conditions.

6.1.4 Semianalytical Method for Elliptic PDEs with Nonlinear Boundary Conditions

Example 6.4. Nonlinear Radiation Boundary Condition

Consider the following boundary value problem with a nonlinear radiation boundary condition at y = 0.

$$\frac{\partial^2 u}{\partial x^2} + \frac{\partial^2 u}{\partial y^2} = 0$$

$$u(0, y) = 0 \quad 0 \le y < 1$$

$$\frac{\partial u}{\partial x}(1, y) = 0 \quad 0 \le y \le 1 \quad (6.23)$$

$$\frac{\partial u}{\partial y}(x, 0) - u(x,0)^4 \quad 0 \le x \le 1$$

$$u(x, H) = 1 \quad 0 \le x \le 1$$

6 Method of Lines for Elliptic Partial Differential Equations

This equation is solved below in Maple using the program developed for example 6.3. The semianalytical method developed earlier is valid for nonlinear boundary conditions also. This is true because the vector equation (6.6) is linear as both the governing equation and the boundary conditions in x are linear. The nonlinear boundary condition comes into the picture only for solving the constants. This is illustrated in the following program.

> restart;with(linalg):with(plots):

> ge:=diff(u(x,y),y$2)=-epsilon^2*diff(u(x,y),x$2);

$$ge := \frac{\partial^2}{\partial y^2} u(x,y) = -\varepsilon^2 \left(\frac{\partial^2}{\partial x^2} u(x,y) \right)$$

> bc1:=u(x,y)-0;

$$bc1 := u(x,y)$$

> bc2:=diff(u(x,y),x);

$$bc2 := \frac{\partial}{\partial x} u(x,y)$$

The nonlinear boundary condition at y = 0 is entered:

> bc3:=diff(u(x,y),y)-u(x,y)^4;

$$bc3 := \frac{\partial}{\partial y} u(x,y) - u(x,y)^4$$

> bc4:=u(x,y)-1;

$$bc4 := u(x,y) - 1$$

> H:=0.5;

$$H := 0.5$$

> epsilon:=1;

$$\varepsilon := 1$$

> dydxf:=1/2/h*(-u[m+2](zeta)-3*u[m](zeta)+4*u[m+1](zeta)):

> dydxb:=1/2/h*(u[m-2](zeta)+3*u[m](zeta)-4*u[m-1](zeta)):

> dydx:=1/2/h*(u[m+1](zeta)-u[m-1](zeta)):

> d2ydx2:=1/h^2*(u[m-1](zeta)-2*u[m](zeta)+u[m+1](zeta)):

> bc1:=subs(diff(u(x,y),x)=subs(m=0,dydxf),u(x,y)=u[0](zeta),bc1);

$$bc1 := u_0(\zeta)$$

6.1 Semianalytical and Numerical Method of Lines for Elliptic PDEs 549

> bc2:=subs(diff(u(x,y),x)=subs(m=N+1,dydxb),u(x,y)=u[N+1](zeta),bc2);

$$bc2 := \frac{1}{2}\,\frac{u_{N-1}(\zeta) + 3\,u_{N+1}(\zeta) - 4\,u_N(\zeta)}{h}$$

> N:=10;

$$N := 10$$

> eq[0]:=bc1;

$$eq_0 := u_0(\zeta)$$

> eq[N+1]:=bc2;

$$eq_{11} := \frac{1}{2}\,\frac{u_9(\zeta) + 3\,u_{11}(\zeta) - 4\,u_{10}(\zeta)}{h}$$

> for i from 1 to N do eq[N+1+i]:=diff(u[N+1+i](zeta),zeta)=
subs(diff(u(x,y),x$2) = subs(m=i,d2ydx2),diff(u(x,y),x) =
subs(m=i,dydx),diff(u(x,y),y)=epsilon/h*u[N+1+i](zeta),u(x,y)=u[i](zeta),
x=i*h,rhs(h^2/epsilon^2*ge));od:

> u[0](zeta):=(solve(eq[0],u[0](zeta)));

$$u_0(\zeta) := 0$$

> u[N+1](zeta):=solve(eq[N+1],u[N+1](zeta));

$$u_{11}(\zeta) := -\frac{1}{3}\,u_9(\zeta) + \frac{4}{3}\,u_{10}(\zeta)$$

> for i from 1 to N do eq[i]:=diff(u[i](zeta),zeta)= u[N+1+i](zeta);od:
> for i from 1 to N do eq[i]:=eval(eq[i]);od:for i from 1 to N do eq[N+1+i]:=
eval(eq[N+1+i]);od:
> eqns:=[seq(rhs(eq[j]),j=1..N),seq(rhs(eq[N+1+j]),j=1..N)]:
> Y:=[seq(u[i](zeta),i=1..N),seq(u[N+1+i](zeta),i=1..N)]:
> A:=genmatrix(eqns,Y,'b1'):
> if N>2 then A:=map(evalf,A):end:
> evalm(A):
> b:=matrix(2*N,1):for i from 1 to 2*N do b[i,1]:=-b1[i];od:evalm(b):
> h:=eval(1/(N+1));

$$h := \frac{1}{11}$$

```
> J:=jordan(A,S):
> mat:=evalm(S&*exponential(J,zeta)&*inverse(S)):
> mat1:=evalm(subs(zeta=zeta-zeta1,evalm(mat))):
> b2:=evalm(subs(zeta=zeta1,evalm(b))):
> mat2:=evalm(mat1&*b2):
> mat2:=map(expand,mat2):
> mat3:=map(int,mat2,zeta1=0..zeta):
> Y0:=matrix(2*N,1);
```

$$Y0 := array\,(1..20, 1..1, [\,])$$

```
> for i to N do Y0[i,1]:=p[i];od:
> for i to N do Y0[N+i,1]:=c[i]:od:
> evalm(Y0):
> Y:=evalm(mat&*Y0+mat3):
```

For calculating the constants, the solution is evaluated at y = 0 and y = H.

```
> sol0:=map(eval,evalm(subs(zeta=0,evalm(Y)))):
> sol1:=map(eval,evalm(subs(zeta=epsilon*H/h,evalm(Y)))):
```

The boundary condition at y = 0 (bc3) yields N nonlinear algebraic equations.

```
> for i to N do
Eq[i]:=subs(diff(u(x,y),y)=epsilon/h*c[i],u(x,y)=p[i],x=i*h,bc3);od;
```

$$Eq_1 := 11\,c_1 - p_1^4$$

$$Eq_2 := 11\,c_2 - p_2^4$$

$$Eq_3 := 11\,c_3 - p_3^4$$

$$Eq_4 := 11\,c_4 - p_4^4$$

$$Eq_5 := 11\,c_5 - p_5^4$$

$$Eq_6 := 11\,c_6 - p_6^4$$

$$Eq_7 := 11\,c_7 - p_7^4$$

$$Eq_8 := 11\,c_8 - p_8^4$$

6.1 Semianalytical and Numerical Method of Lines for Elliptic PDEs 551

$$Eq_9 := 11\, c_9 - p_9^4$$

$$Eq_{10} := 11\, c_{10} - p_{10}^4$$

The boundary condition at y = H yields N linear algebraic equations.

> for i to N do Eq[N+i]:=evalf(subs(diff(u(x,y),y)=epsilon/h*sol1[N+i,1], u(x,y)=sol1[i,1],bc4));od;

$Eq_{11} := 3858.514108\, p_1 - 4653.164388\, p_2 + 3033.298877\, p_3 - 1287.203492\, p_4 + 382.331788\, p_5 - 83.702002\, p_6$

$+ 13.125612\, p_7 - 2.6034360\, p_8 + 0.8163269\, p_9 - 2.1223312\, p_{10} + 2191.136869\, c_1 - 2493.624934\, c_2$

$+ 1498.253557\, c_3 - 574.6737795\, c_4 + 156.3783886\, c_5 - 29.2802035\, c_6 + 5.0155528\, c_7 - 1.2229915\, c_8$

$- 1.2044574\, c_9 - 5.9395771\, c_{10} - 1.$

$Eq_{12} := -4652.719290\, p_1 + 6892.416885\, p_2 - 5939.614617\, p_3 + 3416.111850\, p_4 - 1370.173173\, p_5$

$+ 396.812172\, p_6 - 83.997310\, p_7 + 15.273865\, p_8 - 2.875207\, p_9 + 3.2921156\, p_{10} - 2494.022170\, c_1$

$+ 3687.852500\, c_2 - 3071.147453\, c_3 + 1651.118777\, c_4 - 606.6428527\, c_5 + 160.702335\, c_6 - 28.066335\, c_7$

$+ 9.5454631\, c_8 + 5.4966905\, c_9 + 15.8502263\, c_{10} - 1.$

$Eq_{13} := 3033.288274\, p_1 - 5940.242143\, p_2 + 7274.187535\, p_3 - 6023.523132\, p_4 + 3429.538896\, p_5$

$- 1372.844114\, p_6 + 395.243572\, p_7 - 86.420787\, p_8 + 15.067269\, p_9 - 4.3477534\, p_{10} + 1495.359422\, c_1$

$- 3073.051677\, c_2 + 3839.738649\, c_3 - 3104.770099\, c_4 + 1651.150226\, c_5 - 614.375361\, c_6 + 150.269750\, c_7$

$- 42.654478\, c_8 - 8.8968508\, c_9 - 28.5534349\, c_{10} - 1.$

$Eq_{14} := -1286.902661\, p_1 + 3416.016222\, p_2 - 6022.990988\, p_3 + 7288.502688\, p_4 - 6024.985893\, p_5$

$+ 3430.417815\, p_6 - 1371.187253\, p_7 + 397.241891\, p_8 - 85.257533\, p_9 + 13.3418246\, p_{10} - 573.7732629\, c_1$

$+ 1656.218410\, c_2 - 3094.667384\, c_3 + 3855.746569\, c_4 - 3090.413999\, c_5 + 1670.648262\, c_6 - 589.820169\, c_7$

$+ 180.261203\, c_8 - 6.7962422\, c_9 + 47.9784160\, c_{10} - 1.$

$Eq_{15} := 382.327953\, p_1 - 1370.354689\, p_2 + 3429.497343\, p_3 - 6025.177349\, p_4 + 7288.463045\, p_5 - 6025.395860\, p_6$

$+ 3429.189310\, p_7 - 1371.968362\, p_8 + 390.770425\, p_9 - 67.9782262\, p_{10} + 150.5947305\, c_1 - 614.5838522\, c_2$

$+ 1644.758788\, c_3 - 3116.932738\, c_4 + 3827.778995\, c_5 - 3125.518022\, c_6 + 1628.056077\, c_7 - 639.716761\, c_8$

$+ 120.5422424\, c_9 - 88.5706356\, c_{10} - 1.$

$Eq_{16} := -83.259147\,p_1 + 396.141711\,p_2 - 1372.017216\,p_3 + 3429.889887\,p_4 - 6025.047339\,p_5 + 7288.569251\,p_6$

$- 6024.020787\,p_7 + 3423.954326\,p_8 - 1332.143362\,p_9 + 289.1913209\,p_{10} - 25.2914531\,c_1 + 169.234295\,c_2$

$- 592.494306\,c_3 + 1676.070554\,c_4 - 3076.609931\,c_5 + 3877.607220\,c_6 - 3065.709434\,c_7 + 1695.340411\,c_8$

$- 546.7321088\,c_9 + 198.0324038\,c_{10} - 1.$

$Eq_{17} := 13.760632\,p_1 - 85.040984\,p_2 + 396.297985\,p_3 - 1372.110554\,p_4 + 3429.753632\,p_5 - 6024.358765\,p_6$

$+ 7282.176824\,p_7 - 5984.943123\,p_8 + 3226.094309\,p_9 - 879.9651817\,p_{10} - 0.7480177\,c_1 - 41.454855\,c_2$

$+ 143.547581\,c_3 - 628.420370\,c_4 + 1629.690840\,c_5 - 3133.413202\,c_6 + 3807.845251\,c_7 - 3130.118407\,c_8$

$+ 1528.966175\,c_9 - 487.6494719\,c_{10} - 1.$

$Eq_{18} := -1.7763726\,p_1 + 13.936468\,p_2 - 85.048253\,p_3 + 396.252442\,p_4 - 1371.340087\,p_5 + 3423.655191\,p_6$

$- 5984.766884\,p_7 + 7084.757196\,p_8 - 5239.180737\,p_9 + 1765.252067\,p_{10} + 3.9038380\,c_1 + 13.4411524\,c_2$

$- 17.711601\,c_3 + 176.918170\,c_4 - 585.050152\,c_5 + 1680.713378\,c_6 - 3056.494542\,c_7 + 3803.238594\,c_8$

$- 2726.604513\,c_9 + 978.6155416\,c_{10} - 1.$

$Eq_{19} := 0.1835672\,p_1 - 1.781117\,p_2 + 13.865369\,p_3 - 84.306946\,p_4 + 390.136308\,p_5 - 1331.859755\,p_6$

$+ 3226.042819\,p_7 - 5239.289787\,p_8 + 5034.923131\,p_9 - 2007.630263\,p_{10} - 2.6973519\,c_1 - 6.0196810\,c_2$

$- 3.6378330\,c_3 - 41.7019359\,c_4 + 143.5340766\,c_5 - 609.6475231\,c_6 + 1557.556895\,c_7 - 2788.497173\,c_8$

$+ 2774.636147\,c_9 - 1182.285621\,c_{10} - 1.$

$Eq_{20} := -0.0095035\,p_1 + 0.1088333\,p_2 - 1.0335078\,p_3 + 7.7574718\,p_4 - 44.8267739\,p_5 + 192.5636291\,p_6$

$- 586.3217946\,p_7 + 1176.377570\,p_8 - 1337.998486\,p_9 + 594.5592720\,p_{10} + 0.57609030\,c_1 + 1.1856324\,c_2$

$+ 1.4177409\,c_3 + 4.9166688\,c_4 - 13.8086660\,c_5 + 83.4388374\,c_6 - 268.0681652\,c_7 + 613.0259280\,c_8$

$- 755.8352129\,c_9 + 373.2264554\,c_{10} - 1.$

Since the equations are nonlinear Maple's 'fsolve' is used to solve for the constants:

> csol:=fsolve({seq(Eq[i],i=1..2*N)},{seq(c[i],i=1..N),seq(p[i],i=1..N)});

$csol := \{c_9 = 0.08277548289, c_8 = 1.208912783, c_7 = 1.103795551, c_5 = 0.03605297735, c_6 = 0.1259009972, c_{10}$

$= 0.01460469850, p_1 = -.6902338498, p_3 = -1.096798643, p_2 = -2.191015105, c_4 = 0.03650493307, c_3$

6.1 Semianalytical and Numerical Method of Lines for Elliptic PDEs

$= 0.1315572943, c_1 = 0.02063442303, c_2 = 2.095023029, p_{10} = -.6330985520, p_9 = -.9768403758, p_8$

$= -1.909619821, p_6 = -1.084814456, p_7 = -1.866681963, p_5 = -.7935667361, p_4 = -.7960421434\}$

When 'fsolve' is used Maple returns values that do not make physical sense. The dependent variable at y = 0 (p_i, i=1..N) cannot be negative. Since the equations are nonlinear there can be more than one solution. To get the correct solution that makes physical sense the range should be provided for the constants (see chapter 1).

> csol:=fsolve({seq(Eq[i],i=1..2*N)},{seq(c[i]=0..1,i=1..N),seq(p[i]=0..1,i=1..N)});

$csol := \{p_2 = 0.3210078119, p_{10} = 0.7561371250, c_2 = 0.0009653164793, c_1 = 0.00007314833686, c_3$

$= 0.003700447301, c_4 = 0.008191872861, p_1 = 0.1684221804, p_6 = 0.6724477879, p_8 = 0.7322608786, c_6$

$= 0.01858838213, c_5 = 0.01352904201, c_7 = 0.02295104492, p_3 = 0.4491710189, c_8 = 0.02613789713, c_{10}$

$= 0.02971731457, c_9 = 0.02844614901, p_4 = 0.5478907548, p_5 = 0.6211048673, p_7 = 0.7088409305, p_9$

$= 0.7479180965\}$

> assign(csol);

> YY:=map(eval,Y):

> for i from 1 to N do u[i](zeta):=eval((YY[i,1]));od:

> for i from 0 to N+1 do u[i](zeta):=eval(u[i](zeta));od:

> for i from 0 to N+1 do u[i](y):=eval(subs(zeta=epsilon*y/h,u[i](zeta)));od:

> for i from 0 to N+1 do p[i]:=plot(u[i](y),y=0..H,thickness=3);od:

> pp:=plot([seq(u[i](y),i=0..N+1,2)],y=0..H,thickness=3,
legend=[seq(typeset(u[i],"(y)"),i=0..N+1,2)]);

$$pp := PLOT(...)$$

> display({pp},axes=boxed,title="Figure 6.10",labels=[y,"u"]);

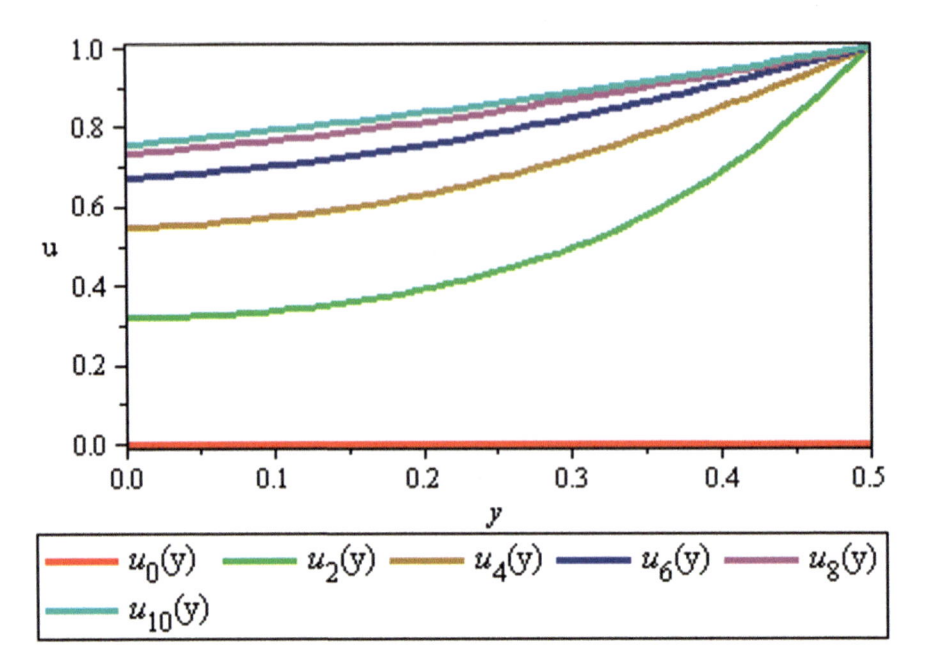

554   6 Method of Lines for Elliptic Partial Differential Equations

Fig. 6.10

> M:=5;

$$M := 5$$

> T1:=[seq(evalf(i*H/M),i=0..M)];

$$T1 := [0., 0.1000000000, 0.2000000000, 0.3000000000, 0.4000000000, 0.5000000000]$$

> for j from 1 to M do
P[j]:=plot([seq([h*i,evalf(subs(y=T1[j],evalf(u[i](y))))],i=0..N+1)],

style=line,thickness=3,axes=boxed,view=[0..1,0..1.1]):od:

> P[M+1]:=plot([seq([h*i,evalf(subs(x=i*h,1))],i=0..N+1)],style= line, thickness=3,axes=boxed):

> for j from 1 to M+1 do pt[j]:=textplot([0.5,evalf(subs(y=T1[j],u[5](y))), typeset(y,sprintf("=%4.2f",T1[j]))],align={above});od:

> display({seq(P[i],i=1..M+1),seq(pt[i],i=1..M+1)},title="Figure 6.11", labels=[x,u]);

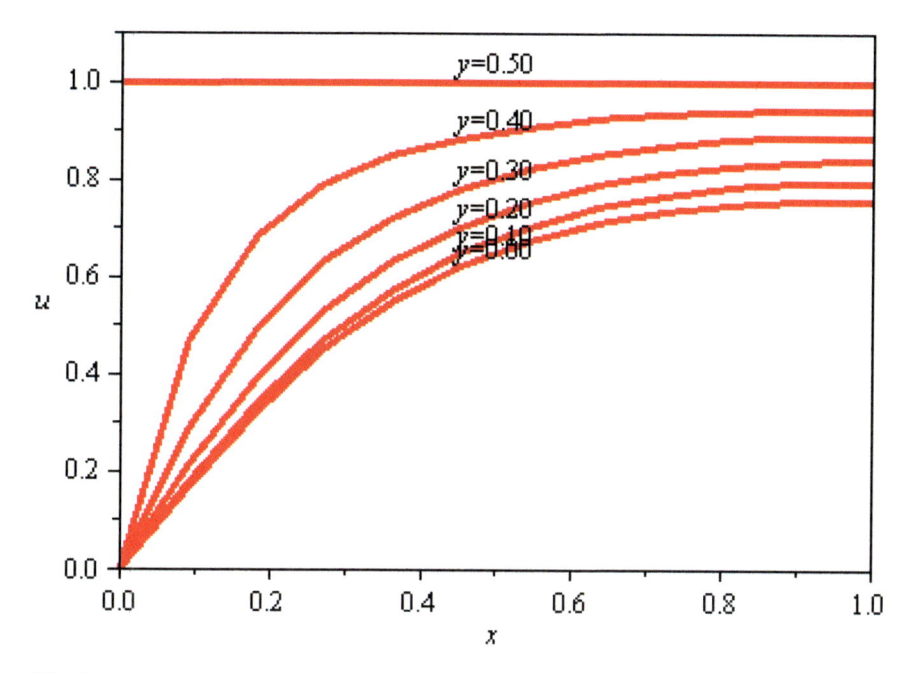

6.1 Semianalytical and Numerical Method of Lines for Elliptic PDEs   555

Fig. 6.11

> Ny:=30;

$$Ny := 30$$

> PP:=matrix(N+2,Ny);

$$PP := array\,(1..12,\,1..30,\,[\;])$$

First, the boundaries x = 0 and y = 1 are defined.

> for i to Ny do PP[1,i]:=0;od:

> for i to N+2 do PP[i,Ny]:=1;od:

The temperature inside the rectangle is obtained using the semianalytical solution:

> for i from 2 to N+2 do for j from 1 to Ny-1 do
PP[i,j]:=evalf(subs(y=(j-1)*H/(Ny-1),u[i-1](y)));od;od:

> plotdata := [seq([seq([(i-1)/(N+1),(j-1)*H/(Ny-1),PP[i,j]], i=1..N+2)],
j=1..Ny)]:

> surfdata(plotdata,axes=boxed,title="Figure 6.12",
labels=[x,y,u],orientation=[-120,60]);

556   6 Method of Lines for Elliptic Partial Differential Equations

Fig. 6.12

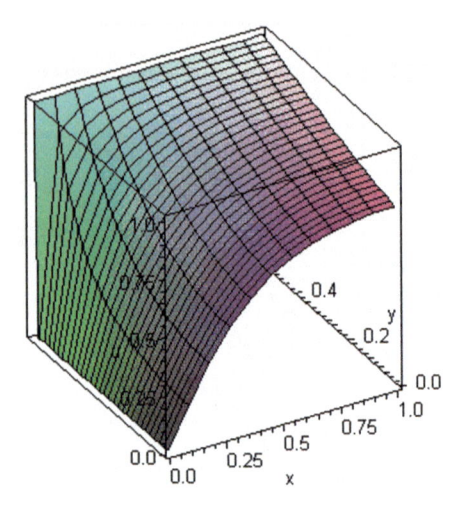

>

6.1.5 Semianalytical Method for Elliptic PDEs with Irregular Shapes

Example 6.5. Potential Distribution in a Hull Cell

Current density distributions in electrochemical systems are governed by Laplace equation (Newman, 1991) with linear/nonlinear boundary condition at the boundaries (electrodes). Consider a Hull cell (see Fig. 6.13) in which a metal is deposited at the cathode (Subramanian and White, 1999).

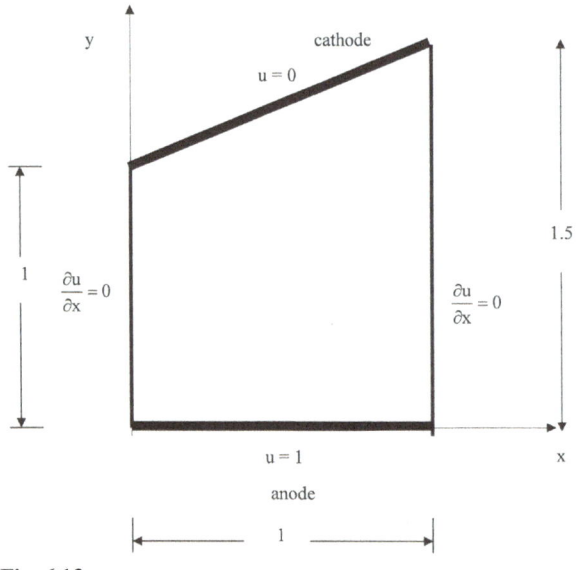

Fig. 6.13

6.1 Semianalytical and Numerical Method of Lines for Elliptic PDEs

The governing equation can be written as

$$\frac{\partial^2 u}{\partial x^2} + \frac{\partial^2 u}{\partial y^2} = 0$$

$$\frac{\partial u}{\partial x}(0, y) = 0 \quad 0 \le y < 1$$

$$\frac{\partial u}{\partial x}(1, y) = 0 \quad 0 \le y \le 1.5 \quad (6.24)$$

$$u(x,0) = 1 \quad 0 \le x \le 1$$

$$u(x, 1+0.5x) = 0 \quad 0 \le x \le 1$$

The cathode surface is defined by the equation y = 1+0.5x. The semianalytical technique developed earlier can be used for equation (6.24). The only change is the calculation of constants because of the fourth boundary condition (cathode). This is taken care of by using the equation for the cathode surface for calculating the constants.

Example 6.5 is solved in Maple below as:

> restart;with(plottools):with(linalg):with(plots):
> ge:=diff(u(x,y),y$2)=-diff(u(x,y),x$2);

$$ge := \frac{\partial^2}{\partial y^2} u(x, y) = -\left(\frac{\partial^2}{\partial x^2} u(x, y)\right)$$

> Digits:=20;

$$Digits := 20$$

> bc1:=diff(u(x,y),x);

$$bc1 := \frac{\partial}{\partial x} u(x, y)$$

> bc2:=diff(u(x,y),x);

$$bc2 := \frac{\partial}{\partial x} u(x, y)$$

> bc3:=u(x,y)-1;

$$bc3 := u(x, y) - 1$$

> bc4:=u(x,y);

$$bc4 := u(x, y)$$

558   6 Method of Lines for Elliptic Partial Differential Equations

Enter the equation for the cathode surface:

> eq_cathode:=y=1+0.5*x;

$$eq\_cathode := y = 1 + 0.5\,x$$

> epsilon:=1;

$$\varepsilon := 1$$

> dydxf:=1/2/h*(-u[m+2](zeta)-3*u[m](zeta)+4*u[m+1](zeta)):
> dydxb:=1/2/h*(u[m-2](zeta)+3*u[m](zeta)-4*u[m-1](zeta)):
> dydx:=1/2/h*(u[m+1](zeta)-u[m-1](zeta)):
> d2ydx2:=1/h^2*(u[m-1](zeta)-2*u[m](zeta)+u[m+1](zeta)):
> bc1:=subs(diff(u(x,y),x)=subs(m=0,dydxf),u(x,y)=u[0](zeta),bc1);

$$bc1 := \frac{1}{2}\,\frac{-u_2(\zeta) - 3\,u_0(\zeta) + 4\,u_1(\zeta)}{h}$$

> bc2:=subs(diff(u(x,y),x)=subs(m=N+1,dydxb),u(x,y)=u[N+1](zeta),bc2);

$$bc2 := \frac{1}{2}\,\frac{u_{N-1}(\zeta) + 3\,u_{N+1}(\zeta) - 4\,u_N(\zeta)}{h}$$

> N:=10;

$$N := 10$$

> eq[0]:=bc1;

$$eq_0 := \frac{1}{2}\,\frac{-u_2(\zeta) - 3\,u_0(\zeta) + 4\,u_1(\zeta)}{h}$$

> eq[N+1]:=bc2;

$$eq_{11} := \frac{1}{2}\,\frac{u_9(\zeta) + 3\,u_{11}(\zeta) - 4\,u_{10}(\zeta)}{h}$$

> for i from 1 to N do eq[N+1+i]:=diff(u[N+1+i](zeta),zeta)=
subs(diff(u(x,y),x$2) = subs(m=i,d2ydx2),diff(u(x,y),x) =
subs(m=i,dydx),diff(u(x,y),y)=epsilon/h*u[N+1+i](zeta),u(x,y)=u[i](zeta),
x=i*h,rhs(h^2/epsilon^2*ge));od:

> u[0](zeta):=(solve(eq[0],u[0](zeta)));

$$u_0(\zeta) := -\frac{1}{3}\,u_2(\zeta) + \frac{4}{3}\,u_1(\zeta)$$

6.1 Semianalytical and Numerical Method of Lines for Elliptic PDEs

```
> u[N+1](zeta):=solve(eq[N+1],u[N+1](zeta));
```

$$u_{11}(\zeta) := -\frac{1}{3} u_9(\zeta) + \frac{4}{3} u_{10}(\zeta)$$

```
> for i from 1 to N do eq[i]:=diff(u[i](zeta),zeta)= u[N+1+i](zeta);od:
> for i from 1 to N do eq[i]:=eval(eq[i]);od:for i from 1 to N do
eq[N+1+i]:=eval(eq[N+1+i]);od:
> eqns:=[seq(rhs(eq[j]),j=1..N),seq(rhs(eq[N+1+j]),j=1..N)]:
> Y:=[seq(u[i](zeta),i=1..N),seq(u[N+1+i](zeta),i=1..N)]:
> A:=genmatrix(eqns,Y,'b1'):
> if N>2 then A:=map(evalf,A):end:
> evalm(A):
> b:=matrix(2*N,1):for i from 1 to 2*N do b[i,1]:=-b1[i];od:evalm(b):
> h:=eval(1/(N+1));
```

$$h := \frac{1}{11}$$

```
> J:=jordan(A,S):
> mat:=evalm(S&*exponential(J,zeta)&*inverse(S)):
> mat1:=evalm(subs(zeta=zeta-zeta1,evalm(mat))):
> b2:=evalm(subs(zeta=zeta1,evalm(b))):
> mat2:=evalm(mat1&*b2):
> mat2:=map(expand,mat2):
> mat3:=map(int,mat2,zeta1=0..zeta):
> Y0:=matrix(2*N,1);
```

$$Y0 := array(1..20, 1..1, [\ ])$$

```
> for i to N do Y0[i,1]:=p[i];od:
> for i to N do Y0[N+i,1]:=c[i];od:
> evalm(Y0):
> Y:=evalm(mat&*Y0+mat3):
```

The boundary condition at y = 0 is applied as:

```
> sol0:=map(eval,evalm(subs(zeta=0,evalm(Y)))):
> for i to N do
Eq[i]:=subs(diff(u(x,y),y)=epsilon/h*c[i],u(x,y)=p[i],x=i*h,bc3);od;
```

$$Eq_1 := p_1 - 1$$
$$Eq_2 := p_2 - 1$$
$$Eq_3 := p_3 - 1$$
$$Eq_4 := p_4 - 1$$
$$Eq_5 := p_5 - 1$$
$$Eq_6 := p_6 - 1$$
$$Eq_7 := p_7 - 1$$
$$Eq_8 := p_8 - 1$$
$$Eq_9 := p_9 - 1$$
$$Eq_{10} := p_{10} - 1$$

The boundary condition at the cathode surface is defined as:

```
> for i to N do
Eq[N+i]:=evalf(subs(diff(u(x,y),y)=epsilon/h*Y[N+i,1],u(x,y)=Y[i,1],bc4));od:
> for i to N do Eq[N+i]:=evalf(subs(zeta=epsilon/h*(1+0.5*i*h),Eq[N+i]));od:
```

The constants are solved as:

```
> csol:=solve({seq(Eq[i],i=1..2*N)},{seq(c[i],i=1..N),seq(p[i],i=1..N)});
```

$csol := \{p_{10} = 1., c_8 = -0.074900981335000207319, c_9 = -0.074543758365780527131, c_{10}$

$= -0.074320440072131068373, c_4 = -0.077051148230933039534, c_5 = -0.076486416946393308610, c_6$

$= -0.075909686394537275958, c_7 = -0.075367481464442807825, c_2 = -0.077953408954862761189, c_3$

$= -0.077555422871608811953, c_1 = -0.078207174783026263289, p_1 = 1., p_2 = 1., p_3 = 1., p_5 = 1., p_6 = 1., p_4$

$= 1., p_7 = 1., p_8 = 1., p_9 = 1.\}$

```
> assign(csol);
> YY:=map(eval,Y):
> for i from 1 to N do u[i](zeta):=eval((YY[i,1]));od:
> for i from 0 to N+1 do u[i](zeta):=eval(u[i](zeta));od:
> for i from 0 to N+1 do u[i](y):=eval(subs(zeta=epsilon*y/h,u[i](zeta)));od:
```

6.1 Semianalytical and Numerical Method of Lines for Elliptic PDEs 561

```
> for i from 0 to N+1 do p[i]:=plot(u[i](y),y=0..(1+0.5*i*h),thickness=2);od:
> for i from 0 to N+1 do
  pl[i]:=line([1.2,0.3+i*0.05],[0.9,evalf(subs(y=0.9,u[i](y)))],thickness=1,
linestyle=solid);
  pt[i]:=textplot([1.2,0.3+i*0.05,typeset(u[i],"(y)")],align=right):
end do:
> display([seq(p[i],i=0..N+1,2),seq(pl[i],i=0..N+1,2),seq(pt[i],i=0..N+1,2)],
axes=boxed,title="Fig. 6.14",labels=[y,"u"]);
```

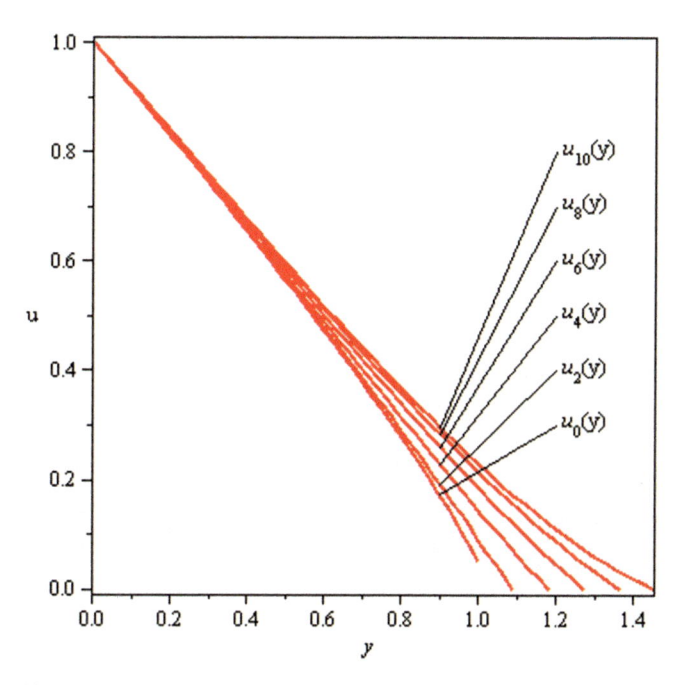

Fig. 6.14

```
> M:=10;
```

$$M := 10$$

```
> T1:=[seq(evalf(i/M),i=0..M)];
```

$T1 := [0., 0.1000000000000000000, 0.2000000000000000000, 0.3000000000000000000,$

$0.4000000000000000000, 0.5000000000000000000, 0.6000000000000000000, 0.7000000000000000000,$

$0.8000000000000000000, 0.9000000000000000000, 1.]$

```
> for j from 1 to M+1 do
P[j]:=plot([seq([h*i,evalf(subs(y=T1[j]*(1+0.5*i*h),evalf(u[i](y))))],i=0..N+1)],
style=line,thickness=3,axes=boxed,view=[0..1,0..1.1]):od:
```

```
> P[M+1]:=plot([seq([h*i,evalf(subs(x=i*h,0))],i=0..N+1)],style=line,
thickness=3,axes=boxed):
```

```
> for j from 1 to M+1 do  pt[j]:=textplot([0.5,evalf(subs(y=T1[j]*(1+0.5*5*h),
u[5](y))),typeset(y,sprintf("=%4.2f",T1[j]))],align=above);od:
```

```
> display({seq(P[i],i=1..M+1),seq(pt[i],i=1..M+1)},title="Fig. 6.15",
labels=[x,u]);
```

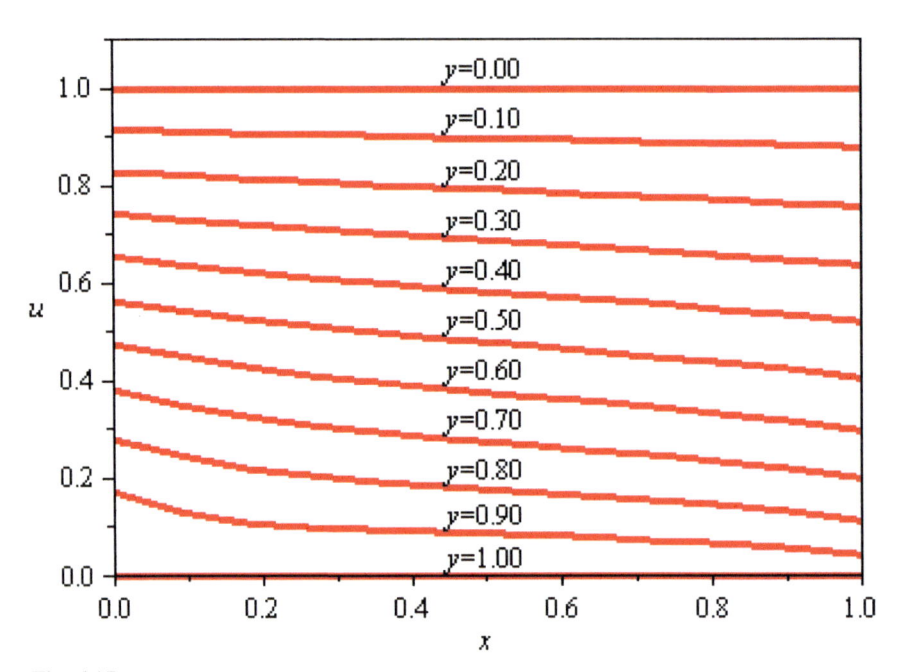

Fig. 6.15

```
> Ny:=30;
```

$$Ny := 30$$

```
> PP:=matrix(N+2,Ny);
```

$$PP := array(1..12, 1..30, [\,])$$

```
> for i to N+2 do PP[i,1]:=1;PP[i,Ny]:=0;od:
```

```
> for i from 1 to N+2 do for j from 2 to Ny-1 do
PP[i,j]:=evalf(subs(y=(j-1)*(1+0.5*(i-1)*h)/(Ny-1),u[i-1](y)));od;od:
```

6.1 Semianalytical and Numerical Method of Lines for Elliptic PDEs   563

> plotdata := [seq([seq([(i-1)/(N+1),(j-1)*(1+0.5*(i-1)*h)/(Ny-1),PP[i,j]], i=1..N+2)], j=1..Ny)]:

> surfdata(plotdata,axes=boxed,title="Fig. 6.16", labels=[x,y,u],orientation=[45,45]);

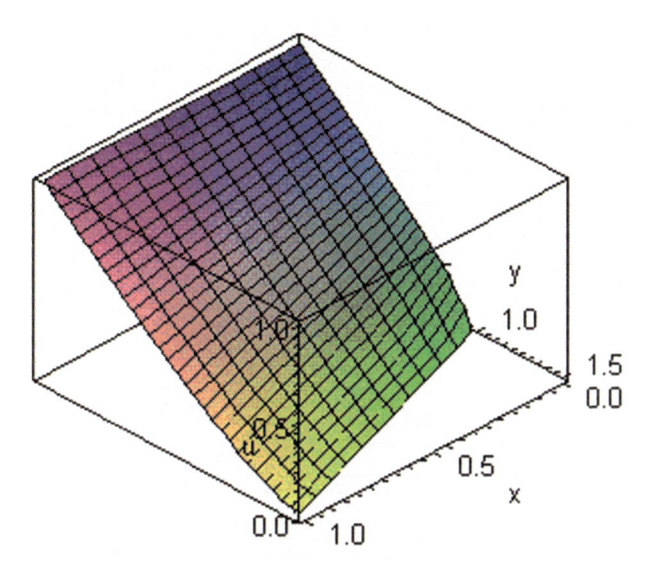

Fig. 6.16

> surfdata(plotdata,axes=boxed,title="Fig. 6.17", labels=[x,y,u],orientation=[120,0] ,style=patchnogrid);

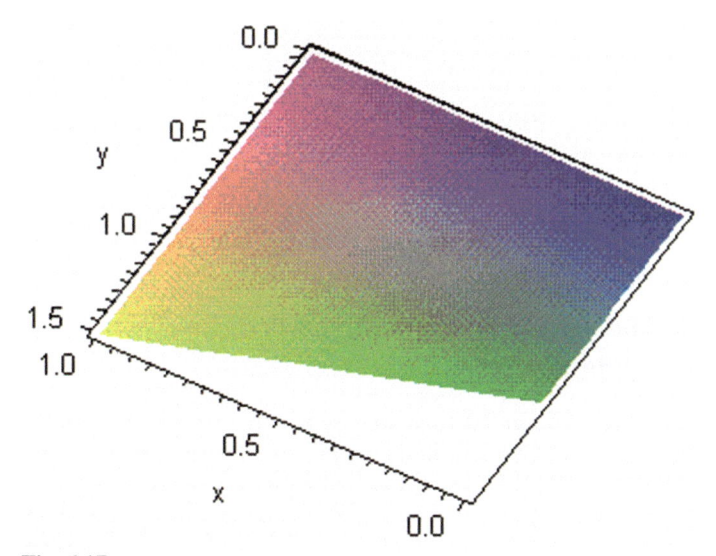

Fig. 6.17

> for i from 1 to N do curr[i]:=evalf(subs(y=1+0.5*i*h,diff(u[i](y),y)));od:
> curr[0]:=4/3*curr[1]-1/3*curr[2];

$$curr_0 := -1.5643261412418003435$$

> curr[N+1]:=4/3*curr[N]-1/3*curr[N-1];

$$curr_{11} := -.25598598997170694673$$

> avecurr:=sum(curr[k],k=0..N+1)/(N+2);

$$avecurr := -.72384858911456061823$$

> plot([seq([i*h,curr[i]/avecurr],i=0..N+1)],thickness=4,axes=boxed,
title="Fig. 6.18",style=point,labels=[x,'i/iavg']);

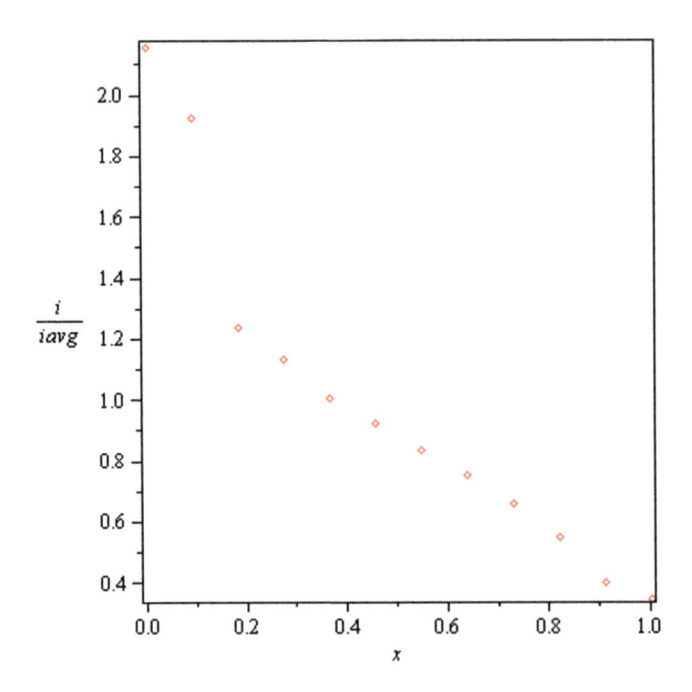

Fig. 6.18

6.1.6 Numerical Method of Lines for Elliptic PDEs in Rectangular Coordinates

For nonlinear elliptic partial differential equations, successive relaxation or finite difference approximations can be used in both the coordinates.[7] [12] [13] (Constantinides & Mostoufi, 1999; Davis, 1984, Finlayson, 1980) As illustrated by Schiesser (1991),[2] a method of lines was used for 2D and 3D steady state problems by adding a pseudo time derivative, applying finite differences in all the

6.1 Semianalytical and Numerical Method of Lines for Elliptic PDEs

spatial coordinates and integrating numerically in time. In this chapter, we apply finite differences in one of the directions (x), convert the governing equation and boundary conditions in x to finite difference form. The resulting system of coupled nonlinear boundary values problems (second order ordinary differential equations in y) are then solved using Maple's 'dsolve' numeric command for boundary value problems (see chapter 3.2.8).

Example 6.6. Numerical Solution for Heat Transfer in a Rectangle

Example 6.1 (heat transfer in a rectangle) is solved again using the numerical method of lines. The procedure involved in solving a linear or nonlinear steady state elliptic PDE numerically is summarized as follows:

1. Start the Maple worksheet with a 'restart' command to clear all variables.
2. Call 'with(linalg)' and 'with(plots)' commands.
3. Enter the governing equation.
4. Store the 'x' boundary conditions in bc1 (x = 0) and bc2 (x = 1).
5. Store the 'y' boundary conditions in bc3 (y = 0) and bc4 (y = 1). Note that a right hand side should be included so that Maple's 'dsolve' command can be used.
6. Enter the number of interior node points, N.
7. Enter the length of the domain, L and height of the domain.
8. Convert the boundary conditions in x (bc1, and bc2) to finite difference form by using 3-point forward and backward differences (accurate to the order h^2), respectively, for bc1 and bc2.
9. Convert the governing equation to finite difference form by using central difference expression accurate to the order h^2 for the first and second derivatives in the spatial variable, x. This gives raise to N second order ODEs in y.
10. The variable $u_i(y)$, i = 0..N+1 corresponds to the dependent variable, u_i at node point i.
11. Eliminate the boundary values ($u_0(y)$ and $u_{N+1}(y)$) using the boundary conditions. Note that this can be done only for linear boundary conditions at x = 0 and x = 1. If boundary conditions are nonlinear, differentiate boundary conditions to obtain differential equations in y (see chapter 5.2 and example 5.2.3).
12. Find the numerical solution using the 'dsolve' numeric command and boundary conditions bc3 and bc4.

Example 6.6 is solved below in Maple using this procedure.

> restart: with(linalg): with(plots):

The governing equation and boundary conditions are entered. Boundary conditions in y are entered with a right hand side.

> ge:=diff(u(x,y),y$2)=-diff(u(x,y),x$2);

$$ge := \frac{\partial^2}{\partial y^2} u(x,y) = -\left(\frac{\partial^2}{\partial x^2} u(x,y) \right)$$

> bc1:=u(x,y);
$$bc1 := u(x,y)$$

> bc2:=u(x,y);
$$bc2 := u(x,y)$$

> bc3:=u(x,y)=0;
$$bc3 := u(x,y) = 0$$

> bc4:=u(x,y)-1=0;
$$bc4 := u(x,y) - 1 = 0$$

The dimensions of the domain and the number of interior node points are entered:

> L:=1;H:=1;
$$L := 1$$
$$H := 1$$

> N:=10;
$$N := 10$$

Boundary conditions in x and the governing equation are converted to finite difference form:

> fd1:=(1/2)*(-u[m+2](y)-3*u[m](y)+4*u[m+1](y))/h;
$$fd1 := \frac{1}{2} \frac{-u_{m+2}(y) - 3 u_m(y) + 4 u_{m+1}(y)}{h}$$

> bd1:=(1/2)*(u[m-2](y)+3*u[m](y)-4*u[m-1](y))/h;
$$bd1 := \frac{1}{2} \frac{u_{m-2}(y) + 3 u_m(y) - 4 u_{m-1}(y)}{h}$$

> cd1:=(u[m+1](y)-u[m-1](y))/h^2;
$$cd1 := \frac{u_{m+1}(y) - u_{m-1}(y)}{h^2}$$

> cd2:=(u[m-1](y)-2*u[m](y)+u[m+1](y))/h^2;
$$cd2 := \frac{u_{m-1}(y) - 2 u_m(y) + u_{m+1}(y)}{h^2}$$

> bc1:=subs(diff(u(x,y),x)=subs(m=0,fd1),u(x,y)=u[0](y),x=0,bc1);
$$bc1 := u_0(y)$$

6.1 Semianalytical and Numerical Method of Lines for Elliptic PDEs

```
> bc2:=subs(diff(u(x,y),x)=subs(m=N+1,bd1),u(x,y)=u[N+1](y),x=1,bc2);
```

$$bc2 := u_{11}(y)$$

```
> eq[0]:=bc1;
```

$$eq_0 := u_0(y)$$

```
> eq[N+1]:=bc2;
```

$$eq_{11} := u_{11}(y)$$

```
> for i from 1 to N do
eq[i]:=diff(u[i](y),y$2)=expand(subs(diff(u(x,y),x$2)=subs(m=i,cd2),
diff(u(x,y),x)=subs(m=i,cd1),diff(u(x,y),y)=u[N+1+i](y),u(x,y)=u[i](y),
x=i*h,rhs(ge)));
end do;
```

$$eq_1 := \frac{d^2}{dy^2} u_1(y) = -\frac{u_0(y)}{h^2} + \frac{2u_1(y)}{h^2} - \frac{u_2(y)}{h^2}$$

$$eq_2 := \frac{d^2}{dy^2} u_2(y) = -\frac{u_1(y)}{h^2} + \frac{2u_2(y)}{h^2} - \frac{u_3(y)}{h^2}$$

$$eq_3 := \frac{d^2}{dy^2} u_3(y) = -\frac{u_2(y)}{h^2} + \frac{2u_3(y)}{h^2} - \frac{u_4(y)}{h^2}$$

$$eq_4 := \frac{d^2}{dy^2} u_4(y) = -\frac{u_3(y)}{h^2} + \frac{2u_4(y)}{h^2} - \frac{u_5(y)}{h^2}$$

$$eq_5 := \frac{d^2}{dy^2} u_5(y) = -\frac{u_4(y)}{h^2} + \frac{2u_5(y)}{h^2} - \frac{u_6(y)}{h^2}$$

$$eq_6 := \frac{d^2}{dy^2} u_6(y) = -\frac{u_5(y)}{h^2} + \frac{2u_6(y)}{h^2} - \frac{u_7(y)}{h^2}$$

$$eq_7 := \frac{d^2}{dy^2} u_7(y) = -\frac{u_6(y)}{h^2} + \frac{2u_7(y)}{h^2} - \frac{u_8(y)}{h^2}$$

$$eq_8 := \frac{d^2}{dy^2} u_8(y) = -\frac{u_7(y)}{h^2} + \frac{2u_8(y)}{h^2} - \frac{u_9(y)}{h^2}$$

$$eq_9 := \frac{d^2}{dy^2} u_9(y) = -\frac{u_8(y)}{h^2} + \frac{2u_9(y)}{h^2} - \frac{u_{10}(y)}{h^2}$$

$$eq_{10} := \frac{d^2}{dy^2} u_{10}(y) = -\frac{u_9(y)}{h^2} + \frac{2u_{10}(y)}{h^2} - \frac{u_{11}(y)}{h^2}$$

> u[0](y):=(solve(eq[0],u[0](y)));

$$u_0(y) := 0$$

> u[N+1](y):=solve(eq[N+1],u[N+1](y));

$$u_{11}(y) := 0$$

> h:=1/(N+1);

$$h := \frac{1}{11}$$

> for i from 1 to N do

  eq[i]:=eval(eq[i]);

end do;

$$eq_1 := \frac{d^2}{dy^2} u_1(y) = 242\, u_1(y) - 121\, u_2(y)$$

$$eq_2 := \frac{d^2}{dy^2} u_2(y) = -121\, u_1(y) + 242\, u_2(y) - 121\, u_3(y)$$

$$eq_3 := \frac{d^2}{dy^2} u_3(y) = -121\, u_2(y) + 242\, u_3(y) - 121\, u_4(y)$$

$$eq_4 := \frac{d^2}{dy^2} u_4(y) = -121\, u_3(y) + 242\, u_4(y) - 121\, u_5(y)$$

$$eq_5 := \frac{d^2}{dy^2} u_5(y) = -121\, u_4(y) + 242\, u_5(y) - 121\, u_6(y)$$

$$eq_6 := \frac{d^2}{dy^2} u_6(y) = -121\, u_5(y) + 242\, u_6(y) - 121\, u_7(y)$$

$$eq_7 := \frac{d^2}{dy^2} u_7(y) = -121\, u_6(y) + 242\, u_7(y) - 121\, u_8(y)$$

$$eq_8 := \frac{d^2}{dy^2} u_8(y) = -121\, u_7(y) + 242\, u_8(y) - 121\, u_9(y)$$

$$eq_9 := \frac{d^2}{dy^2} u_9(y) = -121\, u_8(y) + 242\, u_9(y) - 121\, u_{10}(y)$$

$$eq_{10} := \frac{d^2}{dy^2} u_{10}(y) = -121\, u_9(y) + 242\, u_{10}(y)$$

6.1 Semianalytical and Numerical Method of Lines for Elliptic PDEs

Boundary conditions in y are converted to index form and used with the governing equation in the interior node points to obtain a numerical solution.

```
> BC3:=seq(subs(diff(u(x,y),y)=(D(u[i]))(0),u(x,y)=u[i](0),x=i*h,bc3),i=1..N);
```

$BC3 := u_1(0) = 0, u_2(0) = 0, u_3(0) = 0, u_4(0) = 0, u_5(0) = 0, u_6(0)$
$= 0, u_7(0) = 0, u_8(0) = 0, u_9(0) = 0, u_{10}(0) = 0$

```
> BC4:=seq(subs(diff(u(x,y),y)=(D(u[i]))(H),u(x,y)=u[i](H),x = i*h,bc4), i=1..N);
```

$BC4 := u_1(1) - 1 = 0, u_2(1) - 1 = 0, u_3(1) - 1 = 0, u_4(1) - 1 = 0,$
$u_5(1) - 1 = 0, u_6(1) - 1 = 0, u_7(1) - 1 = 0, u_8(1) - 1 = 0,$
$u_9(1) - 1 = 0, u_{10}(1) - 1 = 0$

```
> BCS:=BC3,BC4;
```

$BCS := u_1(0) = 0, u_2(0) = 0, u_3(0) = 0, u_4(0) = 0, u_5(0) = 0, u_6(0)$
$= 0, u_7(0) = 0, u_8(0) = 0, u_9(0) = 0, u_{10}(0) = 0, u_1(1) - 1 = 0,$
$u_2(1) - 1 = 0, u_3(1) - 1 = 0, u_4(1) - 1 = 0, u_5(1) - 1 = 0,$
$u_6(1) - 1 = 0, u_7(1) - 1 = 0, u_8(1) - 1 = 0, u_9(1) - 1 = 0,$
$u_{10}(1) - 1 = 0$

```
> sol:=dsolve({BCS,seq(eq[i],i=1..N)},type=numeric,output=listprocedure, abserr = 0.1e-8);
```

$sol := \left[y = \text{proc}(y) \ldots \text{end proc}, u_1(y) = \text{proc}(y) \ldots \text{end proc}, \frac{d}{dy} u_1(y) = \text{proc}(y) \ldots \text{end proc}, u_2(y) = \text{proc}(y) \right.$

$\ldots$

$\text{end proc}, \frac{d}{dy} u_2(y) = \text{proc}(y) \ldots \text{end proc}, u_3(y) = \text{proc}(y) \ldots \text{end proc}, \frac{d}{dy} u_3(y) = \text{proc}(y) \ldots \text{end proc}, u_4(y)$

$= \text{proc}(y) \ldots \text{end proc}, \frac{d}{dy} u_4(y) = \text{proc}(y) \ldots \text{end proc}, u_5(y) = \text{proc}(y) \ldots \text{end proc}, \frac{d}{dy} u_5(y) =$

$\text{proc}(y)$

$\ldots$

$\text{end proc}, u_6(y) = \text{proc}(y) \ldots \text{end proc}, \frac{d}{dy} u_6(y) = \text{proc}(y) \ldots \text{end proc}, u_7(y) = \text{proc}(y) \ldots \text{end proc}, \frac{d}{dy} u_7(y)$

$= \text{proc}(y) \ldots \text{end proc}, u_8(y) = \text{proc}(y) \ldots \text{end proc}, \frac{d}{dy} u_8(y) = \text{proc}(y) \ldots \text{end proc}, u_9(y) = \text{proc}(y)$

$\ldots$

$\text{end proc}, \frac{d}{dy} u_9(y) = \text{proc}(y) \ldots \text{end proc}, u_{10}(y) = \text{proc}(y) \ldots \text{end proc}, \frac{d}{dy} u_{10}(y) = \text{proc}(y) \ldots \text{end proc} \left.\right]$

570   6 Method of Lines for Elliptic Partial Differential Equations

The solution obtained is used to make the following plots.

> for i from 1 to N do

  U[i]:=subs(sol,u[i](y));

  od:

> U[0]:=subs(u[1](y)=U[1],u[2](y)=U[2],u[0](y));

$$U_0 := 0$$

> U[N+1]:=subs(u[N](y)=U[N],u[N-1](y)=U[N-1],u[N+1](y));

$$U_{11} := 0$$

> for i from 0 to N+1 do

  py[i]:=plot(U[i](y),y=0..H,thickness=4):

  end do:

> display({seq(py[i],i=0..N+1)},labels=[y,"u"],axes=boxed,title="Fig. 6.19");

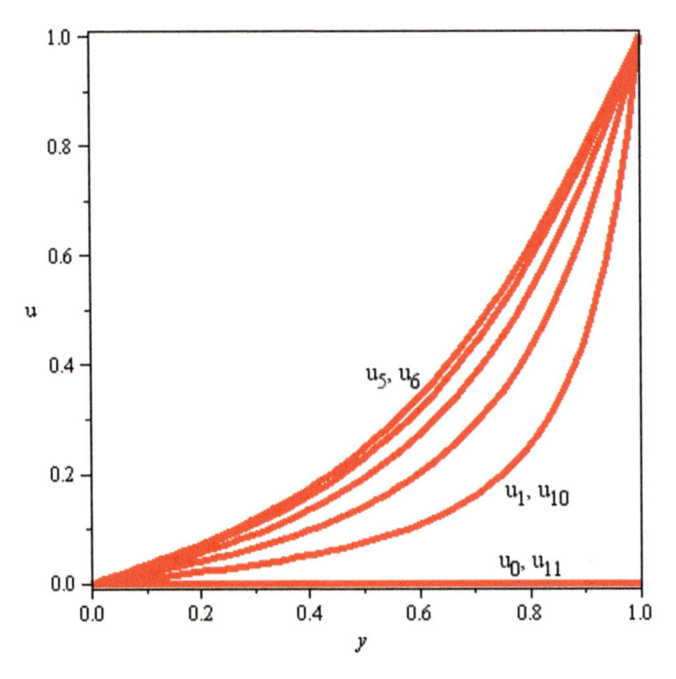

Fig. 6.19

> M:=20;

$$M := 20$$

6.1 Semianalytical and Numerical Method of Lines for Elliptic PDEs

> for i from 1 to N do

fy[i]:=subs(sol,u[i](y)):

end do:

> T1:=[seq(evalf(i*H/M),i=0..M)];

$T1 := [0., 0.05000000000, 0.1000000000, 0.1500000000, 0.2000000000, 0.2500000000, 0.3000000000,$
$0.3500000000, 0.4000000000, 0.4500000000, 0.5000000000, 0.5500000000, 0.6000000000, 0.6500000000,$
$0.7000000000, 0.7500000000, 0.8000000000, 0.8500000000, 0.9000000000, 0.9500000000, 1.]$

> for j from 1 to M+1 do

P[j]:=plot([seq([h*i,U[i](T1[j])],i=0..N+1)],style=line,thickness=3, axes=boxed,title="Fig. 6.20"):

end do:

> display({seq(P[i],i=1..M+1)},labels=[x,u]);

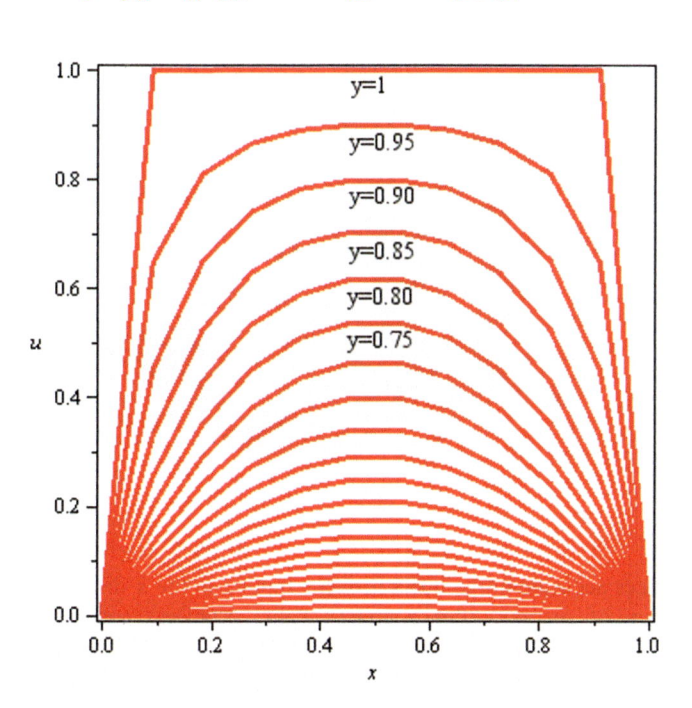

Fig. 6.20

> Ny:=30;

$$Ny := 30$$

572   6 Method of Lines for Elliptic Partial Differential Equations

```
> PP:=matrix(N+2,Ny);
```

$$PP := array(1..12, 1..30, [\ ])$$

```
> for i from 1 to N+2 do
  for j from 1 to M+1 do
    PP[i,j]:=evalf(subs(t=T1[j],U[i-1](t))):
  end do:
end do:
> plotdata:=[seq([seq([(i-1)*h,T1[j],PP[i,j]],i=1..N+2)],j=1..M+1)]:
> surfdata(plotdata,axes=boxed,title="Fig. 6.21",
labels=[x,y,u],orientation=[-145,60]);
```

Fig. 6.21

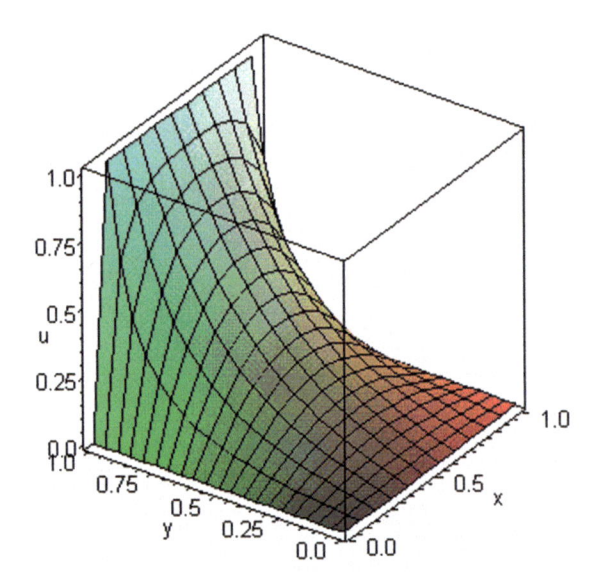

Note that finite differences accurate to the order h^2 were applied in the x-direction and the resulting system of second order boundary value problems (ordinary differential equations) was solved numerically using Maple's 'dsolve' numeric command in the y-direction. This is equivalent to applying finite differences in both x and y directions. The only difference is that in the y-direction Maple's interpolation technique is used to expedite convergence. In addition, one gets both the dependent variable and its derivative in y directly. The method described may not work for stiff problems and Maple's 'dsolve' command might take a long time to predict a numerical solution. In addition, one might have to use advanced commands to get a converged solution (see chapter 3.2).

6.1 Semianalytical and Numerical Method of Lines for Elliptic PDEs

Example 6.7. Numerical Solution for Heat Transfer for Nonlinear Elliptic PDEs

The numerical method of lines described in the previous example can be used for nonlinear elliptic partial differential equations, also. For example, consider the following nonlinear boundary value problem (diffusion with a second order reaction):

$$\frac{\partial^2 u}{\partial x^2} + \frac{\partial^2 u}{\partial y^2} = u^2$$

$$\frac{\partial u}{\partial x}(0, y) = 0 \quad 0 \le y \le 1$$

$$u(1, y) = 1 \quad 0 \le y \le 1 \quad (6.25)$$

$$u(x, 0) = 0 \quad 0 \le x \le 1$$

$$\frac{\partial u}{\partial y}(x, 1) = 0 \quad 0 < x < 1$$

The Maple program used for example 6.6 is used to solve this boundary value problem as shown in the worksheet that follows.

> restart:with(linalg):with(plots):
> ge:=diff(u(x,y),y$2)=-diff(u(x,y),x$2)+u(x,y)^2;

$$ge := \frac{\partial^2}{\partial y^2} u(x, y) = -\left(\frac{\partial^2}{\partial x^2} u(x, y)\right) + u(x, y)^2$$

> Digits:=50;

$$Digits := 50$$

> bc1:=diff(u(x,y),x);

$$bc1 := \frac{\partial}{\partial x} u(x, y)$$

> bc2:=u(x,y)-1;

$$bc2 := u(x, y) - 1$$

> bc3:=u(x,y)-0;

$$bc3 := u(x, y)$$

> bc4:=diff(u(x,y),y);

$$bc4 := \frac{\partial}{\partial y} u(x, y)$$

> L:=1;
$$L := 1$$

> H:=1;
$$H := 1$$

> N:=10;
$$N := 10$$

> fd1:=(1/2)*(-u[m+2](y)-3*u[m](y)+4*u[m+1](y))/h;
$$fd1 := \frac{1}{2} \frac{-u_{m+2}(y) - 3 u_m(y) + 4 u_{m+1}(y)}{h}$$

> bd1:=(1/2)*(u[m-2](y)+3*u[m](y)-4*u[m-1](y))/h;
$$bd1 := \frac{1}{2} \frac{u_{m-2}(y) + 3 u_m(y) - 4 u_{m-1}(y)}{h}$$

> cd1:=(u[m+1](y)-u[m-1](y))/h^2;
$$cd1 := \frac{u_{m+1}(y) - u_{m-1}(y)}{h^2}$$

> cd2:=(u[m-1](y)-2*u[m](y)+u[m+1](y))/h^2;
$$cd2 := \frac{u_{m-1}(y) - 2 u_m(y) + u_{m+1}(y)}{h^2}$$

> bc1:=subs(diff(u(x,y),x)=subs(m =0,fd1),u(x,y)=u[0](y),x=0,bc1);
$$bc1 := \frac{1}{2} \frac{-u_2(y) - 3 u_0(y) + 4 u_1(y)}{h}$$

> bc2:=subs(diff(u(x,y),x)=subs(m=N+1,bd1),u(x,y)=u[N+1](y),x=1,bc2);
$$bc2 := u_{11}(y) - 1$$

> eq[0]:=bc1;
$$eq_0 := \frac{1}{2} \frac{-u_2(y) - 3 u_0(y) + 4 u_1(y)}{h}$$

> eq[N+1]:=bc2;
$$eq_{11} := u_{11}(y) - 1$$

6.1 Semianalytical and Numerical Method of Lines for Elliptic PDEs

```
> for i from 1 to N do
eq[i]:=diff(u[i](y),y$2)=expand(subs(diff(u(x,y),x$2)=subs(m=i,cd2),
diff(u(x,y),x)=subs(m=i,cd1),diff(u(x,y),y)=u[N+1+i](y),u(x,y)=u[i](y),
x=i*h,rhs(ge)));
end do;
```

$$eq_1 := \frac{d^2}{dy^2} u_1(y) = -\frac{u_0(y)}{h^2} + \frac{2\,u_1(y)}{h^2} - \frac{u_2(y)}{h^2} + u_1(y)^2$$

$$eq_2 := \frac{d^2}{dy^2} u_2(y) = -\frac{u_1(y)}{h^2} + \frac{2\,u_2(y)}{h^2} - \frac{u_3(y)}{h^2} + u_2(y)^2$$

$$eq_3 := \frac{d^2}{dy^2} u_3(y) = -\frac{u_2(y)}{h^2} + \frac{2\,u_3(y)}{h^2} - \frac{u_4(y)}{h^2} + u_3(y)^2$$

$$eq_4 := \frac{d^2}{dy^2} u_4(y) = -\frac{u_3(y)}{h^2} + \frac{2\,u_4(y)}{h^2} - \frac{u_5(y)}{h^2} + u_4(y)^2$$

$$eq_5 := \frac{d^2}{dy^2} u_5(y) = -\frac{u_4(y)}{h^2} + \frac{2\,u_5(y)}{h^2} - \frac{u_6(y)}{h^2} + u_5(y)^2$$

$$eq_6 := \frac{d^2}{dy^2} u_6(y) = -\frac{u_5(y)}{h^2} + \frac{2\,u_6(y)}{h^2} - \frac{u_7(y)}{h^2} + u_6(y)^2$$

$$eq_7 := \frac{d^2}{dy^2} u_7(y) = -\frac{u_6(y)}{h^2} + \frac{2\,u_7(y)}{h^2} - \frac{u_8(y)}{h^2} + u_7(y)^2$$

$$eq_8 := \frac{d^2}{dy^2} u_8(y) = -\frac{u_7(y)}{h^2} + \frac{2\,u_8(y)}{h^2} - \frac{u_9(y)}{h^2} + u_8(y)^2$$

$$eq_9 := \frac{d^2}{dy^2} u_9(y) = -\frac{u_8(y)}{h^2} + \frac{2\,u_9(y)}{h^2} - \frac{u_{10}(y)}{h^2} + u_9(y)^2$$

$$eq_{10} := \frac{d^2}{dy^2} u_{10}(y) = -\frac{u_9(y)}{h^2} + \frac{2\,u_{10}(y)}{h^2} - \frac{u_{11}(y)}{h^2} + u_{10}(y)^2$$

```
> u[0](y):=(solve(eq[0],u[0](y)));
```

$$u_0(y) := -\frac{1}{3}\,u_2(y) + \frac{4}{3}\,u_1(y)$$

```
> u[N+1](y):=solve(eq[N+1],u[N+1](y));
```

$$u_{11}(y) := 1$$

```
> h:=1/(N+1);
```

$$h := \frac{1}{11}$$

```
> for i from 1 to N do
  eq[i]:=eval(eq[i]);
  end do;
```

$$eq_1 := \frac{d^2}{dy^2} u_1(y) = -\frac{242}{3} u_2(y) + \frac{242}{3} u_1(y) + u_1(y)^2$$

$$eq_2 := \frac{d^2}{dy^2} u_2(y) = -121 u_1(y) + 242 u_2(y) - 121 u_3(y) + u_2(y)^2$$

$$eq_3 := \frac{d^2}{dy^2} u_3(y) = -121 u_2(y) + 242 u_3(y) - 121 u_4(y) + u_3(y)^2$$

$$eq_4 := \frac{d^2}{dy^2} u_4(y) = -121 u_3(y) + 242 u_4(y) - 121 u_5(y) + u_4(y)^2$$

$$eq_5 := \frac{d^2}{dy^2} u_5(y) = -121 u_4(y) + 242 u_5(y) - 121 u_6(y) + u_5(y)^2$$

$$eq_6 := \frac{d^2}{dy^2} u_6(y) = -121 u_5(y) + 242 u_6(y) - 121 u_7(y) + u_6(y)^2$$

$$eq_7 := \frac{d^2}{dy^2} u_7(y) = -121 u_6(y) + 242 u_7(y) - 121 u_8(y) + u_7(y)^2$$

$$eq_8 := \frac{d^2}{dy^2} u_8(y) = -121 u_7(y) + 242 u_8(y) - 121 u_9(y) + u_8(y)^2$$

$$eq_9 := \frac{d^2}{dy^2} u_9(y) = -121 u_8(y) + 242 u_9(y) - 121 u_{10}(y) + u_9(y)^2$$

$$eq_{10} := \frac{d^2}{dy^2} u_{10}(y) = -121 u_9(y) + 242 u_{10}(y) - 121 + u_{10}(y)^2$$

```
> BC3:=seq(subs(diff(u(x,y),y)=(D(u[i]))(0),u(x,y)=u[i](0),x=i*h,bc3),i=1..N);
```

$$BC3 := u_1(0), u_2(0), u_3(0), u_4(0), u_5(0), u_6(0), u_7(0), u_8(0), u_9(0),$$
$$u_{10}(0)$$

6.1 Semianalytical and Numerical Method of Lines for Elliptic PDEs

```
> BC4:=seq(subs(diff(u(x,y),y)=(D(u[i]))(H),u(x,y)=u[i](H),
x = i*h,bc4),i=1..N);
```

$$BC4 := D(u_1)(1), D(u_2)(1), D(u_3)(1), D(u_4)(1), D(u_5)(1),$$
$$D(u_6)(1), D(u_7)(1), D(u_8)(1), D(u_9)(1), D(u_{10})(1)$$

```
> BCS:=BC3,BC4;
```

$BCS := u_1(0), u_2(0), u_3(0), u_4(0), u_5(0), u_6(0), u_7(0), u_8(0), u_9(0), u_{10}(0), D(u_1)(1), D(u_2)(1), D(u_3)(1),$

$D(u_4)(1), D(u_5)(1), D(u_6)(1), D(u_7)(1), D(u_8)(1), D(u_9)(1), D(u_{10})(1)$

```
> sol:=dsolve({BCS,seq(eq[i],i=1..N)},type=numeric,output=listprocedure,
abserr = 0.1e-8);
```

$sol := \left[y = \mathbf{proc}(y) \ldots \mathbf{end\ proc}, u_1(y) = \mathbf{proc}(y) \ldots \mathbf{end\ proc}, \frac{d}{dy} u_1(y) = \mathbf{proc}(y) \ldots \mathbf{end\ proc}, u_2(y) = \mathbf{proc}(y) \right.$

...

$\mathbf{end\ proc}, \frac{d}{dy} u_2(y) = \mathbf{proc}(y) \ldots \mathbf{end\ proc}, u_3(y) = \mathbf{proc}(y) \ldots \mathbf{end\ proc}, \frac{d}{dy} u_3(y) = \mathbf{proc}(y) \ldots \mathbf{end\ proc}, u_4(y)$

$= \mathbf{proc}(y) \ldots \mathbf{end\ proc}, \frac{d}{dy} u_4(y) = \mathbf{proc}(y) \ldots \mathbf{end\ proc}, u_5(y) = \mathbf{proc}(y) \ldots \mathbf{end\ proc}, \frac{d}{dy} u_5(y) =$

$\mathbf{proc}(y)$

...

$\mathbf{end\ proc}, u_6(y) = \mathbf{proc}(y) \ldots \mathbf{end\ proc}, \frac{d}{dy} u_6(y) = \mathbf{proc}(y) \ldots \mathbf{end\ proc}, u_7(y) = \mathbf{proc}(y) \ldots \mathbf{end\ proc}, \frac{d}{dy} u_7(y)$

$= \mathbf{proc}(y) \ldots \mathbf{end\ proc}, u_8(y) = \mathbf{proc}(y) \ldots \mathbf{end\ proc}, \frac{d}{dy} u_8(y) = \mathbf{proc}(y) \ldots \mathbf{end\ proc}, u_9(y) = \mathbf{proc}(y)$

...

$\mathbf{end\ proc}, \frac{d}{dy} u_9(y) = \mathbf{proc}(y) \ldots \mathbf{end\ proc}, u_{10}(y) = \mathbf{proc}(y) \ldots \mathbf{end\ proc}, \frac{d}{dy} u_{10}(y) = \mathbf{proc}(y) \ldots \mathbf{end\ proc} \Big]$

```
> for i from 1 to N do
  U[i]:=subs(sol,u[i](y));
od:
> U[0]:=subs(u[1](y)=U[1],u[2](y)=U[2],u[0](y));
```

$$U_0 := -\frac{1}{3} U_2 + \frac{4}{3} U_1$$

```
> U[N+1]:=subs(u[N](y)=U[N],u[N-1](y)=U[N-1],u[N+1](y));
```

$$U_{11} := 1$$

578 6 Method of Lines for Elliptic Partial Differential Equations

```
> for i from 0 to N+1 do

  py[i]:=plot(U[i](y),y=0..H,thickness=4):

  end do:

> display({seq(py[i],i=0..N+1)},labels=[y,"u"],axes=boxed,title="Fig. 6.22");
```

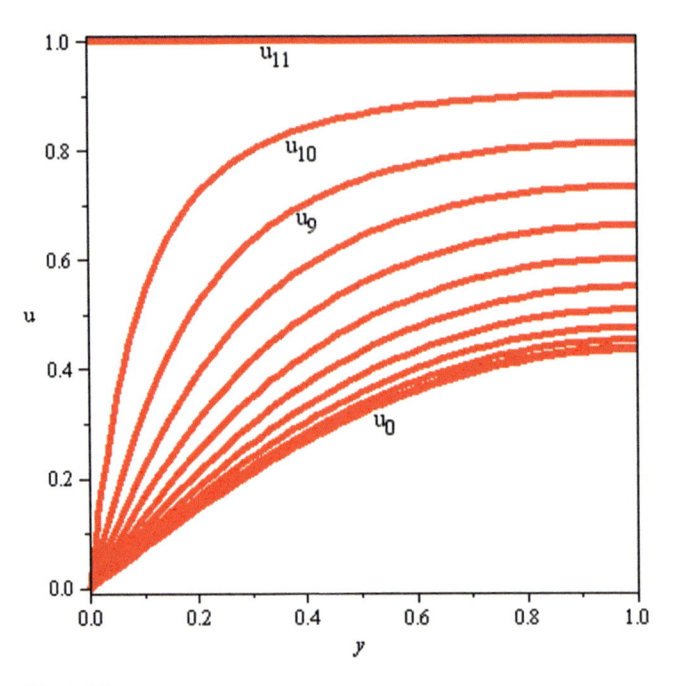

Fig. 6.22

```
> M:=20;
```

$$M := 20$$

```
> for i from 1 to N do

  fy[i]:=subs(sol,u[i](y)):

  end do:

> T1:=[seq(evalf(i*H/M),i=0..M)];
```

6.1 Semianalytical and Numerical Method of Lines for Elliptic PDEs

$T1 :=$ [0., 0.05000000000000000000000000000000000000000000000000,

0.1000000000000000000000000000000000000000000000000,

0.1500000000000000000000000000000000000000000000000,

0.2000000000000000000000000000000000000000000000000,

0.2500000000000000000000000000000000000000000000000,

0.3000000000000000000000000000000000000000000000000,

0.3500000000000000000000000000000000000000000000000,

0.4000000000000000000000000000000000000000000000000,

0.4500000000000000000000000000000000000000000000000,

0.5000000000000000000000000000000000000000000000000,

0.5500000000000000000000000000000000000000000000000,

0.6000000000000000000000000000000000000000000000000,

0.6500000000000000000000000000000000000000000000000,

0.7000000000000000000000000000000000000000000000000,

0.7500000000000000000000000000000000000000000000000,

0.8000000000000000000000000000000000000000000000000,

0.8500000000000000000000000000000000000000000000000,

0.9000000000000000000000000000000000000000000000000,

0.9500000000000000000000000000000000000000000000000, 1.]

> for j from 1 to M+1 do

P[j]:=plot([seq([h*i,U[i](T1[j])],i=0..N+1)],style=line,thickness=3, axes=boxed,title="Fig. 6.23"):

end do:

> display({seq(P[i],i=1..M+1)},labels=[x,u]);

580   6 Method of Lines for Elliptic Partial Differential Equations

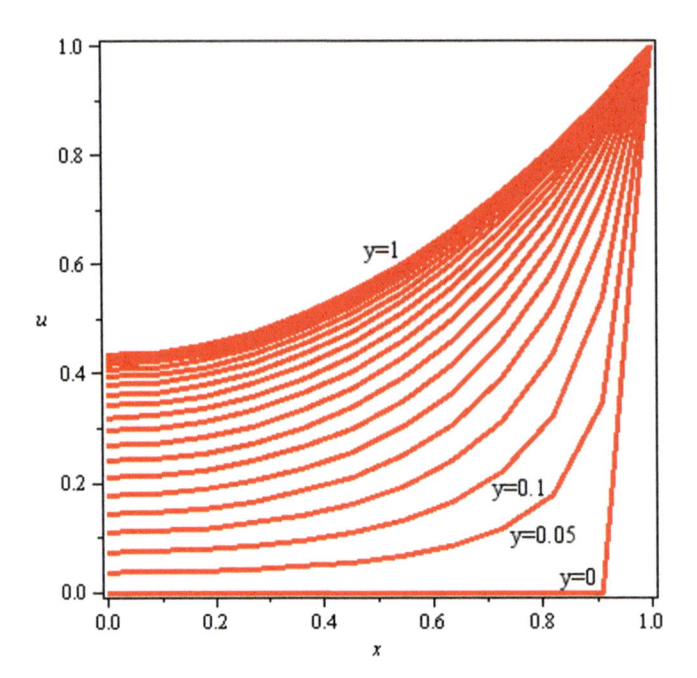

[Figure: plot of u vs x showing curves labeled y=1, y=0.1, y=0.05, y=0]

Fig. 6.23

> Ny:=30;

$$Ny := 30$$

> PP:=matrix(N+2,Ny);

$$PP := array(1..12, 1..30, [\ ])$$

> for i from 1 to N+2 do

for j from 1 to M+1 do

PP[i,j]:=evalf(subs(t=T1[j],U[i-1](t))):

end do:

end do:

> plotdata:=[seq([seq([(i-1)*h,T1[j],PP[i,j]],i=1..N+2)],j=1..M+1)]:

> surfdata(plotdata,axes=boxed,title="Fig. 6.24",labels=[x,y,u], orientation=[-145,60]);

Fig. 6.24

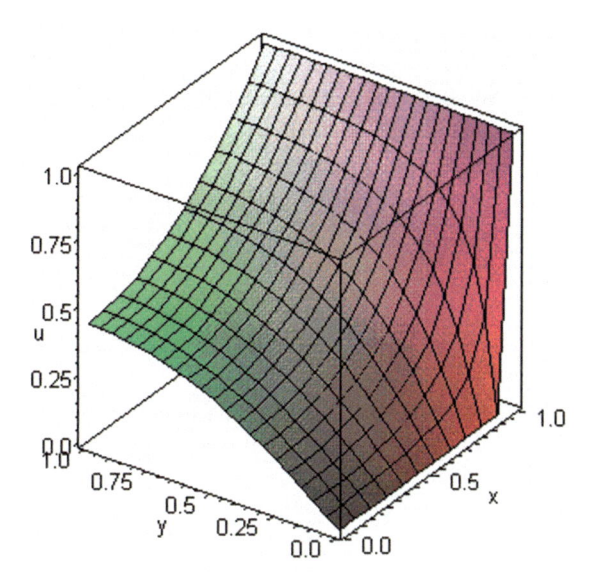

6.1.7 Summary

In this chapter, linear elliptic partial differential equations were solved using the analytical method of lines. This method involves applying finite differences in the x-direction and integrating the resulting system of coupled ordinary differential equations in the y-direction using the exponential method described in chapters 2, 3 and 5. In section 6.1.2, the given linear elliptic partial differential equation in rectangular boundary conditions was solved using a semianalytical method. In section 6.1.3, a semianalytical method was extended for problems in cylindrical coordinates. In section 6.1.4, this method was extended to elliptic partial differential equations with nonlinear boundary conditions. In section 6.1.5, a Hull cell (irregular shape in y) was solved using the semianalytical method. Note that the semianalytical method can be used for certain nonlinear elliptic partial differential equations (Subramanian and White, 2004).[14]

The numerical method of lines was used to solve linear and nonlinear elliptic partial differential equations in section 6.1.7. This method involves using finite differences in one direction and solving the resulting system of boundary value problems in y using Maple's 'dsolve' numeric command. This method provides a numerical solution for both the dependent variables and its derivative in the y-direction.

Both analytical and numerical methods of lines are presented in this chapter for elliptic partial differential equations. Semianalytical method, presented in this chapter is very powerful technique, and is valid for elliptic Partial differential equations with mixed boundaries also (Subramanian and White, 1999). Numerical method of lines presented in this chapter should be used with precaution, as it may not work for stiff problems. A total of seven examples were presented in this chapter.

Exercise Problems

1. Solve the following Poison's equation using analytical method of lines (semianalytical method):

    $$\frac{\partial^2 u}{\partial x^2} + \frac{\partial^2 u}{\partial y^2} = -1$$

    $u(0,y) = 0; u(1,y) = 0$
    $u(x,0) = 0; u(x,1) = 0$

2. Solve the following Laplace equation with non-homogeneous two-flux boundary conditions and plot the profiles:

    $$\frac{\partial^2 u}{\partial y^2} + \frac{\partial^2 u}{\partial x^2} = 0$$

    $\frac{\partial u}{\partial x}(0,y) = 0$ and $\frac{\partial u}{\partial x}(0,y) = 1$

    $u(x,0) = 0$ and $\frac{\partial u}{\partial y}(x,1) = 0$

3. Redo problem 2 after switching x and y.
4. Consider the steady state diffusion in a cylinder

    $$\frac{\partial^2 u}{\partial y^2} + \frac{\partial^2 u}{\partial x^2} + \frac{1}{x}\frac{\partial u}{\partial x} = 0$$

    $\frac{\partial u}{\partial x}(0,y) = 0$ and $u(1,y) = 0$

    $u(x,0) = 1$ and $\frac{\partial u}{\partial y}(x,1) = 0$

    Solve this problem using semianalytical method.

5. Consider a rectangle catalyst of two dimensions in which a first order chemical reaction is taking place. The governing equations and boundary conditions in dimensionless form are (Rice and Do, 1995):[10]

6.1 Semianalytical and Numerical Method of Lines for Elliptic PDEs

$$\frac{\partial^2 u}{\partial y^2} + \frac{\partial^2 u}{\partial x^2} - \Phi^2 u = 0$$

$$\frac{\partial u}{\partial x}(0,y) = 0 \text{ and } u(1,y) = 1$$

$$\frac{\partial u}{\partial y}(x,0) = 0 \text{ and } u(x,1) = 1$$

Solve this problem using the semianalytical method.

6. Consider current-potential distribution in a curvilinear Hull cell (Subramanian and White, 1999; Chapman, 1997 ?). The governing equations and boundary conditions are:

$$\frac{\partial^2 u}{\partial \theta^2} + x^2 \frac{\partial^2 u}{\partial x^2} + x \frac{\partial u}{\partial x} = 0$$

$$\frac{\partial u}{\partial x}(1,\theta) = 0 \text{ and } \frac{\partial u}{\partial x}(2,\theta) = 0$$

$$u(x,0) = 1 \text{ and } u(x,\frac{\pi}{2}) = 0$$

Solve this problem using semianalytical method. Hint: note that x varies from 1 to 2 and finite difference form of the governing equation should be programmed to take care of this.

7. Consider diffusion with reaction in a non-isothermal cylindrical pellet (Finlayson, 1980). The governing equations and boundary conditions are:

$$\frac{\partial^2 u}{\partial y^2} + \frac{\partial^2 u}{\partial x^2} + \frac{1}{x}\frac{\partial u}{\partial x} = \Phi^2 u \exp\left(\frac{\gamma \beta (1-u)}{1+\beta(1-u)}\right)$$

$$\frac{\partial u}{\partial x}(0,y) = 0 \text{ and } u(1,y) = 1$$

$$\frac{\partial u}{\partial y}(x,0) = 0 \text{ and } u(x,1) = 1$$

Solve this problem using numerical method of lines for $\Phi = 2$, $\gamma = 30$ and $\beta = 0.1$. Which method is more efficient - using the 'dsolve' command in x

(with finite differences in x) or 'dsolve' command in y (with finite differences in x) direction?
8. Complete the details missing in example 6.7.
9. Redo example 6.3 using numerical method of lines described in example 6.6. (Apply finite differences in x and integrate using dsolve in y).
10. Redo problem 2 by applying finite differences in y and using dsolve in x. Hint: for programming purposes rewrite equation (6.22) by switching x and y and use program given in example 6.6. Compare problem 2 and problem 3. Which method do you recommend and why? Does the value of Peclet number make have an impact in the decision?
11. Redo example 6.5 using numerical method of lines.
12. Redo example 6.5 if cathode shape is given by $y = 1 + 0.5x^2$.
13. (Note that this is an advanced problem involving an advanced matrix method.) Consider example 6.1 again (with $\varepsilon = 1$). If finite differences are applied only in the x-direction, the governing equation can be written in finite difference form as:

$$\frac{d^2 u_i}{dy^2} = \frac{-u_{i+1} + 2u_i - u_{i-1}}{h^2}, \quad i = 1..N \quad (6.26)$$

After eliminating the boundary values (u_0 and u_{N+1}), the system of equations can be written in matrix form as:

$$\frac{d^2 Y}{dy^2} = aY \quad (6.27)$$

How is this **a** matrix related to the **a** matrix defined in equation (6.19)? Find the eigenvalues of **a** matrix. Since all the eigenvalues are distinct, **a** matrix can be diagonalized ($a = PDP^{-1}$) and equation (6.27) is modified as:

$$\frac{d^2 Y}{dy^2} = PDP^{-1}Y \quad (6.28)$$

Pre-multiplying equation (6.28) by P^{-1} and defining $Z = P^{-1}Y$ we get:

$$\frac{d^2 Z}{dy^2} = DZ \quad (6.29)$$

Equation (6.29) is an easy equation to solve because D is a diagonal matrix and equation (6.29) yields the following decoupled equation for z_i, the ith row of **Z** matrix (d_{ii} is the diagonal element of **D** matrix)

$$\frac{d^2 z_i}{dy^2} = d_{i,i} z_i, \quad i = 1..N \quad (6.30)$$

Equation (6.30) can be easily solved using the methods described in chapter 3.1 or using the 'dsolve' command to get a closed form solution for z_i. What are the boundary conditions for z_i? How is the final solution obtained for u? What are the advantages and disadvantages of this method compared to the method described in example 6.1?

14. Apply the methodology described in problem 13 for the Graetz problem (example 6.3).
15. Apply the procedure described in problem 13 for problem 4. Hint: apply finite differences in x and integrate analytically in y.
16. Redo problem 15 by applying finite differences in y and integrating analytically in x. Once the equations are decoupled get analytical solutions in y using Maple's 'dsolve' command.

References

1. Schiesser, W.E., Silebi, C.A.: Dynamic Modeling of Transport Process Systems (1997)
2. Schiesser, W.E.: The Numerical Methods of Lines. Academic Press, Inc., New York (1991)
3. Varma, A., Morbidelli, M.: Mathematical Methods in Chemical Engineering. Oxford University Press, Inc., Oxford (1997)
4. Taylor, R., Krishna, R.: Multicomponent Mass Transfer. Wiley & Sons, Chichester (1993)
5. Amundson, N.R.: Mathematical Methods in Chemical Engineering: Matrices and Their Applications. Prentice Hall, Inc., Englewood Cliffs (1966)
6. Subramanian, V.R., White, R.E.: Solving Differential Equations with Maple. Chemical Engineering Education, 328–336 (Fall 2000)
7. Constantinides, A., Mostoufi, N.: Numerical Methods for Chemical Engineers with MATLAB Applications. Prentice-Hall PTR, Englewood Cliffs (1999)
8. Carslaw, H.S., Jaeger, J.C.: Conduction of Heat in Solids. Oxford University Press, Oxford (1972)
9. Subramanian, V.R., White, R.E.: A Semianlytical Method for Predicting Primary and Secondary Curent Density Distributions: Linear and Nonlinear Boundary Conditions. Journal of the Electrochemical Society 147(5), 1636–1644 (2000)
10. Rice, R.G., Do, D.D.: Applied Mathematics and Modeling for Chemical Engineers. John Wiley & Sons, Inc., Chichester (1995)
11. Varma, A., Morbidelli, M.: Mathematical Methods in Chemical Engineering. Oxford University Press, New York (1997)
12. Davis, M.E.: Numerical Methods and Modeling for Chemical Engineers. John Wiley & Sons, Chichester (1984)
13. Finlayson, B.A.: Nonlinear Analysis in Chemical Engineering. McGraw-Hill, New York (1980)
14. Subramanian, V.R., White, R.E.: Semianalytical method of lines for solving elliptic partial differential equations. Chemical Engineering Science 59(4), 781–788 (2004)

Chapter 7
Partial Differential Equations in Finite Domains

7.1 Separation of Variables Method for Partial Differential Equations (PDEs) in Finite Domains

7.1.1 Introduction

Transient heat conduction or mass transfer in solids with constant physical properties (diffusion coefficient, thermal diffusivity, thermal conductivity, etc.) is usually represented by a linear parabolic partial differential equation. Steady state heat or mass transfer in solids, potential distribution in electrochemical cells is usually represented by elliptic partial differential equations. In this chapter, we describe how one can arrive at the analytical solutions for linear parabolic partial differential equations and elliptic partial differential equations in finite domains using a separation of variables method. The methodology is illustrated using a transient one dimensional heat conduction in a rectangle.

7.1.2 Separation of Variables for Parabolic PDEs with Homogeneous Boundary Conditions

Parabolic partial differential equations with homogenous boundary conditions are solved in this section. The dependent variable u is assumed to take the form u = XT, where X is a function of x alone and T is a function of t alone. This leads to separate differential equations for X and T. This methodology is best illustrated using an example.

Example 7.1. Heat Conduction in a rectangle

Consider heat transfer in a finite slab.[1] The dimensionless temperature profile is governed by:

$$\frac{\partial u}{\partial t} = \frac{\partial^2 u}{\partial x^2} \qquad (7.1)$$

with the initial condition

$$u(x,0) = 1 \qquad (7.2)$$

The boundary conditions are

$$u(0,t) = 0 \qquad (7.3)$$

and

$$u(1,t) = 0 \qquad (7.4)$$

Next, the following transformation is used

$$u = XT \qquad (7.5)$$

where X is a function of x alone and T is a function of t alone. Equation (7.5) is substituted into equation (7.1) to obtain:

$$X\frac{dT}{dt} = T\frac{d^2X}{dx^2} \qquad (7.6)$$

Dividing both sides of equation (7.6) by XT we obtain:

$$\frac{1}{T}\frac{dT}{dt} = \frac{1}{X}\frac{d^2X}{dx^2} \qquad (7.7)$$

We observe that in equation (7.7) the left hand side is a function of t only and the right hand side is a function of x only. Hence, both sides of equation (7.7) should be equal to a constant:

$$\frac{1}{T}\frac{dT}{dt} = \frac{1}{X}\frac{d^2X}{dx^2} = -\lambda^2 \qquad (7.8)$$

We have used, $-\lambda^2$, a negative constant in equation (7.8). It can be shown that if the constant is positive the solution for T becomes infinite as t approaches infinity (steady state profile). Equation (7.8) can be separated into two equations for T and X as:

$$\frac{dT}{dt} + \lambda^2 T = 0 \qquad (7.9)$$

and

$$\frac{d^2X}{dx^2} + \lambda^2 X = 0 \qquad (7.10)$$

Equation (7.9) can be solved as:

$$T = T0 \exp(-2t) \qquad (7.11)$$

7.1 Separation of Variables Method for PDEs in Finite Domains

where T_0 is an unknown constant. Equation (7.10) can be solved as:

$$X = c_1 \sin(x) + c_2 \cos(x) \tag{7.12}$$

where c_1 and c_2 are unknown constants. The function X should satisfy the boundary conditions in x (equations (7.13) and (7.14)):

$$X(0) = 0 \tag{7.13}$$

and

$$X(1) = 0 \tag{7.14}$$

Applying the boundary conditions for X we get

$$c_2 = 0 \tag{7.15}$$

and

$$c_1 \sin(\lambda) = 0 \tag{7.16}$$

Since $c_2 = 0$ in equation (7.15), c_1 in equation (7.16) cannot be zero. Equation (7.16) can be simplified as

$$\sin(\lambda) = 0 \tag{7.17}$$

Equation (7.17) has an infinite number of solutions as given by:

$$\lambda = n\pi, \, n = 1, 2, 3, \ldots \infty \tag{7.18}$$

Note that n = 0 corresponds to a trivial solution X = 0 and is hence neglected. The solution for u is obtained by combining X and T (equation (7.5))

$$u = c_1 T_0 \sin(n\pi x)\exp(-n_2 \pi_2 t) \tag{7.19}$$

The constants c_1 and T_0 can be combined as a single constant A_n. Since there is an infinite number of eigenvalues, there are an infinite number of fundamental solutions that satisfy the given partial differential equations. The total solution can be expressed as the superposition of the individual solutions as:

$$u = \sum_{n=1}^{\infty} A_n \sin(n\pi x)\exp\left(-n^2\pi^2 t\right) \tag{7.20}$$

We have not used the initial condition from equation (7.2) until now. The initial condition is applied to equation (7.20) as

$$1 = \sum_{n=1}^{\infty} A_n \sin(n\pi x) \qquad (7.21)$$

Equation (7.21) is multiplied on both sides by $\sin(m\pi x)$ and integrated from 0 to 1:

$$\int_0^1 \sin(m\pi x)\, dx = \sum_{n=1}^{\infty} A_n \int_0^1 \sin(m\pi x)\sin(n\pi x)\, dx \qquad (7.22)$$

The integral in the right hand side simplifies as:

$$\int_0^1 \sin(m\pi x)\sin(n\pi x)\, dx = 0 \text{ if } m \neq n$$
$$= \frac{1}{2} \text{ if } m = n \qquad (7.23)$$

Hence, all the terms in the infinite series in equation (7.22) vanish except when $n = m$:

$$\int_0^1 \sin(m\pi x)\, dx = A_m \int_0^1 \sin^2(m\pi x)\, dx$$
$$\Rightarrow \qquad (7.24)$$
$$A_m = 2\int_0^1 \sin(m\pi x)\, dx$$

The dummy variable m can be changed to n in equation (7.24) to give

$$A_n = 2\int_0^1 \sin(n\pi x)\, dx$$
$$= 2\left(\frac{1-\cos(n\pi)}{n\pi}\right) \qquad (7.25)$$

Once the coefficient A_n is obtained, the solution is completed as:

$$u = \frac{2}{\pi}\sum_{n=1}^{\infty}\left(\frac{1-\cos(n\pi)}{n}\right)\sin(n\pi x)\exp(-n^2\pi^2 t) \qquad (7.26)$$

7.1 Separation of Variables Method for PDEs in Finite Domains

The Maple program used to solve Example 7.1 is given below:

> restart:

> with(plots):

The governing equation is entered here:

> eq:=diff(u(x,t),t)=diff(u(x,t),x$2);

$$eq := \frac{\partial}{\partial t} u(x, t) = \frac{\partial^2}{\partial x^2} u(x, t)$$

The initial condition is entered here:

> IC:=u(x,0)=1;

$$IC := u(x, 0) = 1$$

The boundary conditions are entered here:

> bc1:=u(x,t)=0;

$$bc1 := u(x, t) = 0$$

> bc2:=u(x,t)=0;

$$bc2 := u(x, t) = 0$$

Next, u is separated as u = XT (equation (7.5)):

> Eq:=subs(u(x,t)=X(x)*T(t),eq);

$$Eq := \frac{\partial}{\partial t} (X(x)\, T(t)) = \frac{\partial^2}{\partial x^2} (X(x)\, T(t))$$

> Eq:=Eq/X(x)/T(t);

$$Eq := \frac{\frac{d}{dt} T(t)}{T(t)} = \frac{\frac{d^2}{dx^2} X(x)}{X(x)}$$

The governing equation for T is:

> Eq_T:=lhs(Eq)=-lambda^2;

$$Eq\_T := \frac{\frac{d}{dt} T(t)}{T(t)} = -\lambda^2$$

T can be solved as;

> T(t):=rhs(dsolve({Eq_T,T(0)=T0},T(t)));

$$T(t) := T0\, e^{-\lambda^2 t}$$

The governing equation for X is:

> Eq_X:=rhs(Eq)=-lambda^2;

$$Eq\_X := \frac{\frac{d^2}{dx^2} X(x)}{X(x)} = -\lambda^2$$

X can be solved as:

> dsolve({Eq_X},X(x));

$$\{X(x) = \_C1 \sin(\lambda x) + \_C2 \cos(\lambda x)\}$$

Maple sometimes attaches _C1 with the sine function and sometimes with the cosine function. To avoid this, we can specify the initial conditions for X and its derivative:

> X(x):=rhs(dsolve({Eq_X,D(X)(0)=c[1]*lambda,X(0)=c[2]},X(x)));

$$X(x) := c_1 \sin(\lambda x) + c_2 \cos(\lambda x)$$

The boundary conditions for X are:

> Bc1:=X(x)=0;

$$Bc1 := c_1 \sin(\lambda x) + c_2 \cos(\lambda x) = 0$$

> Bc2:=X(x)=0;

$$Bc2 := c_1 \sin(\lambda x) + c_2 \cos(\lambda x) = 0$$

The constant c2 is solved using the boundary condition at X = 0:

> Eq_Bc1:=eval(subs(x=0,Bc1));

$$Eq\_Bc1 := c_2 = 0$$

> c[2]:=solve(Eq_Bc1,c[2]);

$$c_2 := 0$$

The second boundary condition at x = 1 gives:

> Eq_Bc2:=eval(subs(x=1,Bc2));

$$Eq\_Bc2 := c_1 \sin(\lambda) = 0$$

If c1 is zero, then we get the trivial solution X = 0. Hence, sin(λ) should be zero:

> Eq_Eig:=sin(lambda)=0;

$$Eq\_Eig := \sin(\lambda) = 0$$

7.1 Separation of Variables Method for PDEs in Finite Domains

The eigenvalue equation can be solved to get the eigenvalue, λ:

> solve(Eq_Eig,lambda);

$$0$$

By default, Maple picks only one eigenvalue. We can ask Maple to find all the eigenvalues using the following command:

> _EnvAllSolutions := true:

> solve(Eq_Eig,lambda);

$$\pi \_Z1\sim$$

_Z1 means the integer plane. Hence, the eigenvalues can be taken as $\lambda = n\pi$, $n = 1, 2, 3,...\infty$. The solution for u can be written as:

> U:=eval(X(x)*T(t));

$$U := c_1 \sin(\lambda x) \, T0 \, e^{-\lambda^2 t}$$

The constants c1 and T0 can be combined as a single constant A_n:

> Un:=subs(c[1]=A[n]/T0,lambda=lambda[n],U);

$$Un := A_n \sin(\lambda_n x) \, e^{-\lambda_n^2 t}$$

Hence, the solution can be written as an infinite series:

> u(x,t):=Sum(Un,n=1..infinity);

$$u(x,t) := \sum_{n=1}^{\infty} A_n \sin(\lambda_n x) \, e^{-\lambda_n^2 t}$$

The values of the eigenvalues can be substituted as:

> u(x,t):=subs(lambda[n]=n*Pi,u(x,t));

$$u(x,t) := \sum_{n=1}^{\infty} A_n \sin(n \pi x) \, e^{-n^2 \pi^2 t}$$

The constant A_n is obtained by applying the initial condition and by using the orthogonal property of the eigenfunction:

> eq_An:=eval(subs(t=0,u(x,t)))=rhs(IC);

$$eq\_An := \sum_{n=1}^{\infty} A_n \sin(n \pi x) = 1$$

> I1:=int((sin(n*Pi*x))^2,x=0..1);

$$I1 := \begin{cases} 0 & n = 0 \\ \dfrac{1}{2}\,\dfrac{-\cos(n\pi)\sin(n\pi) + n\pi}{n\pi} & \text{otherwise} \end{cases}$$

> I2:=int(sin(n*Pi*x),x=0..1);

$$I2 := \begin{cases} 0 & n = 0 \\ -\dfrac{\cos(n\pi) - 1}{n\pi} & \text{otherwise} \end{cases}$$

The integrals can be simplified by substituting $\sin(n\pi) = 0$:

> vars:={sin(n*Pi)=0};

$$vars := \{\sin(n\pi) = 0\}$$

> I1:=subs(vars,I1);

$$I1 := \begin{cases} 0 & n = 0 \\ \dfrac{1}{2} & \text{otherwise} \end{cases}$$

> A[n]:=I2/I1;

$$A_n := \dfrac{\begin{cases} 0 & n = 0 \\ -\dfrac{\cos(n\pi) - 1}{n\pi} & \text{otherwise} \end{cases}}{\begin{cases} 0 & n = 0 \\ \dfrac{1}{2} & \text{otherwise} \end{cases}}$$

The dimensionless temperature profile is given by:

> u(x,t):=eval(u(x,t));

7.1 Separation of Variables Method for PDEs in Finite Domains

$$u(x,t) := \sum_{n=1}^{\infty} \frac{\left(\begin{cases} 0 & n=0 \\ -\dfrac{\cos(n\pi)-1}{n\pi} & \text{otherwise} \end{cases}\right)\sin(n\pi x)\, e^{-n^2\pi^2 t}}{\begin{cases} 0 & n=0 \\ \dfrac{1}{2} & \text{otherwise} \end{cases}}$$

For plotting purposes, we replace ∞ by, N, an integer:

> u(x,t):=subs(infinity=N,u(x,t));

$$u(x,t) := \sum_{n=1}^{N} \frac{\left(\begin{cases} 0 & n=0 \\ -\dfrac{\cos(n\pi)-1}{n\pi} & \text{otherwise} \end{cases}\right)\sin(n\pi x)\, e^{-n^2\pi^2 t}}{\begin{cases} 0 & n=0 \\ \dfrac{1}{2} & \text{otherwise} \end{cases}}$$

N = 20 is enough for this problem:

> ua:=subs(N=20,u(x,t));

$$ua := \sum_{n=1}^{20} \frac{\left(\begin{cases} 0 & n=0 \\ -\dfrac{\cos(n\pi)-1}{n\pi} & \text{otherwise} \end{cases}\right)\sin(n\pi x)\, e^{-n^2\pi^2 t}}{\begin{cases} 0 & n=0 \\ \dfrac{1}{2} & \text{otherwise} \end{cases}}$$

The solution obtained is plotted at t = 0. We observe oscillations about initial condition 1. This is called Gibb's phenomenon. Theoretically it will take N = ∞ for this profile to become u = 1 at t = 0.

596   7 Partial Differential Equations in Finite Domains

```
> plot(eval(subs(t=0,ua)),x=0..1,axes=boxed,title="Figure 7.1",
thickness=3,labels=[x,"u(x,0)"]);
```

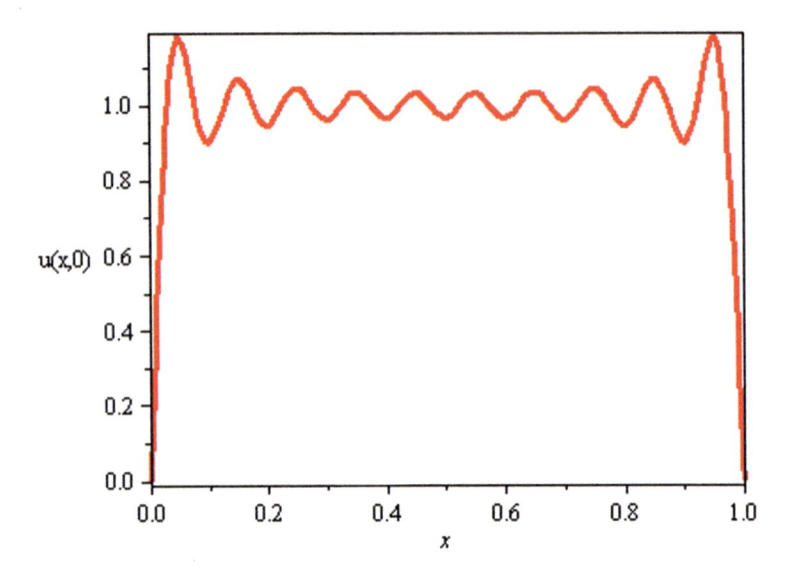

Fig. 7.1

Next, the initial condition is predicted using N = 100 terms in the series:

```
> ua2:=subs(N=100,u(x,t));
```

$$ua2 := \sum_{n=1}^{100} \frac{\left(\begin{cases} 0 & n=0 \\ -\dfrac{\cos(n\pi)-1}{n\pi} & \text{otherwise} \end{cases}\right) \sin(n\pi x)\, e^{-n^2\pi^2 t}}{\begin{cases} 0 & n=0 \\ \dfrac{1}{2} & \text{otherwise} \end{cases}}$$

```
> plot(eval(subs(t=0,ua2)),x=0..1,axes=boxed,title="Figure 7.2",
thickness=3,labels=[x,"u(x,0)"]);
```

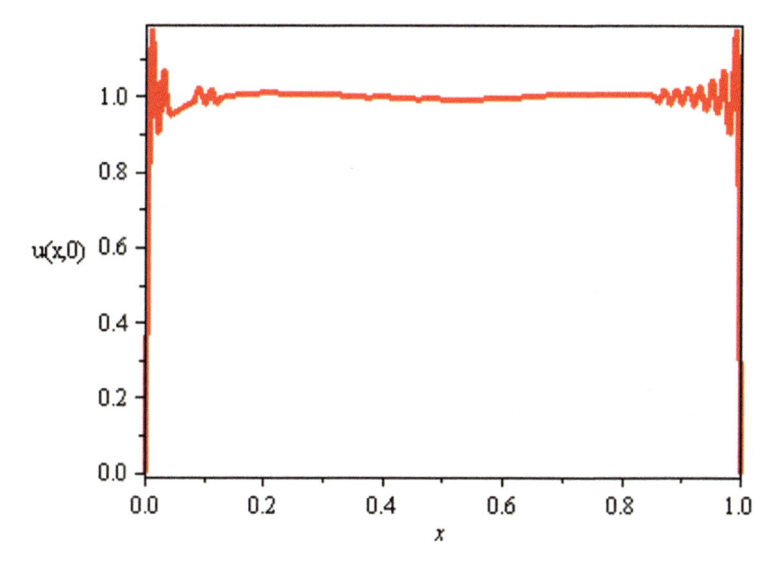

7.1 Separation of Variables Method for PDEs in Finite Domains

Fig. 7.2

We observe better predictions as the number of terms increase. However, we need more terms at t = 0. For values of t > 0, N = 20 is enough. Hence, a piecewise polynomial can be used to define the initial condition at t = 0 and the separation of variables solution obtained can be used for values of t > 0.

> uu:=piecewise(t=0,rhs(IC),t>0,ua);

$$uu := \begin{cases} \begin{cases} 1 & t = 0 \\ \sum_{n=1}^{20} \begin{cases} 0 & n = 0 \\ \dfrac{1}{2} & \text{otherwise} \end{cases} \begin{cases} 0 & n = 0 \\ -\dfrac{\cos(n\pi) - 1}{n\pi} & \text{otherwise} \end{cases} \sin(n\pi x) \, e^{-n^2 \pi^2 t} & 0 < t \end{cases} \end{cases}$$

598					7 Partial Differential Equations in Finite Domains

A three dimensional plot can be made as:

> plot3d(uu,x=1..0,t=0.3..0,axes=boxed,title="Figure 7.3",
labels=[x,t,"u"],orientation=[60,60]);

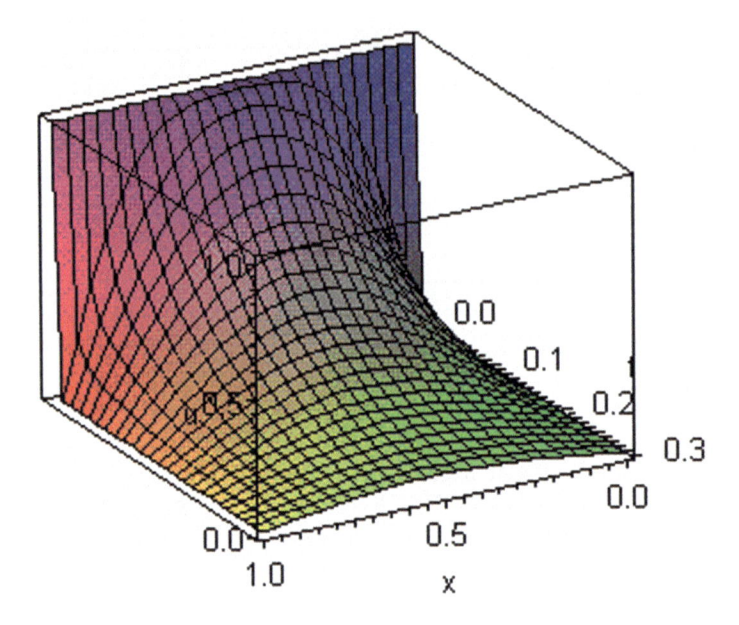

Fig. 7.3

The profiles across the slab at different times can be plotted as:

> plot([subs(t=0,uu),subs(t=0.01,uu),subs(t=0.05,uu),subs(t=0.1,uu),

subs(t=0.2,uu)],x=0..1,axes=boxed,title="Figure 7.4",

thickness=5,labels=[x,"u"],legend=["t=0","t=0.01","t=0.05","t=0.1","t=0.2"]);

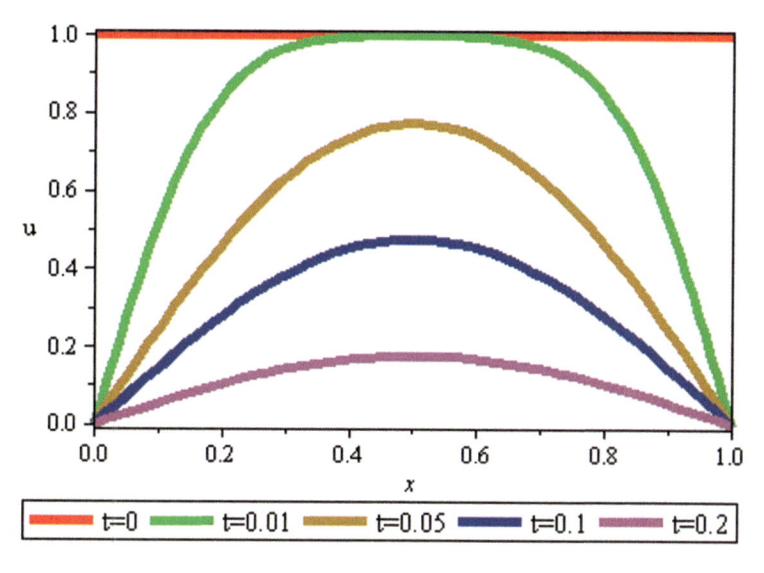

7.1 Separation of Variables Method for PDEs in Finite Domains

Fig. 7.4

Example 7.2. Heat Conduction with an Insulator Boundary Condition

In the previous example, temperature was specified at both boundaries. In this example, the temperature flux is taken to be zero because of symmetry.[1] The governing equation is

$$\frac{\partial u}{\partial t} = \frac{\partial^2 u}{\partial x^2}$$

$$u(x,0) = 1 \qquad (7.27)$$

$$\frac{\partial u}{\partial x}(0,t) = 0 \text{ and } u(1,t) = 0$$

Equation (7.27) is solved in Maple below by making few modifications in the program used for Example 7.1.

> restart:

> with(plots):

> eq:=diff(u(x,t),t)=diff(u(x,t),x$2);

$$eq := \frac{\partial}{\partial t} u(x,t) = \frac{\partial^2}{\partial x^2} u(x,t)$$

> IC:=u(x,0)=1;
$$IC := u(x, 0) = 1$$

> bc1:=diff(u(x,t),x)=0;
$$bc1 := \frac{\partial}{\partial x} u(x, t) = 0$$

> bc2:=u(x,t)=0;
$$bc2 := u(x, t) = 0$$

> Eq:=subs(u(x,t)=X(x)*T(t),eq):
> Eq:=Eq/X(x)/T(t):
> Eq_T:=lhs(Eq)=-lambda^2:
> T(t):=rhs(dsolve({Eq_T,T(0)=T0},T(t)));
$$T(t) := T0 \, e^{-\lambda^2 t}$$

> Eq_X:=rhs(Eq)=-lambda^2:
> X(x):=rhs(dsolve({Eq_X,D(X)(0)=c[1]*lambda,X(0)=c[2]},X(x))):
> Bc1:=diff(X(x),x)=0:
> Bc2:=X(x)=0:
> Eq_Bc1:=eval(subs(x=0,Bc1)):
> c[1]:=solve(Eq_Bc1,c[1]):
> Eq_Bc2:=eval(subs(x=1,Bc2)):
> Eq_Eig:=cos(lambda)=0;
$$Eq\_Eig := \cos(\lambda) = 0$$

> _EnvAllSolutions := true:
> solve(Eq_Eig,lambda);
$$\frac{1}{2} \pi + \pi \, \_Z1\sim$$

> U:=eval(X(x)*T(t)):
> Un:=subs(c[2]=A[n]/T0,lambda=lambda[n],U):
> u(x,t):=Sum(Un,n=0..infinity):
> u(x,t):=subs(lambda[n]=(2*n+1)/2*Pi,u(x,t));

7.1 Separation of Variables Method for PDEs in Finite Domains

$$u(x,t) := \sum_{n=0}^{\infty} A_n \cos\left(\frac{1}{2}(2n+1)\pi x\right) e^{-\frac{1}{4}(2n+1)^2 \pi^2 t}$$

```
> eq_An:=eval(subs(t=0,u(x,t)))=rhs(IC);
```

$$eq\_An := \sum_{n=0}^{\infty} A_n \cos\left(\frac{1}{2}(2n+1)\pi x\right) = 1$$

```
> I1:=int((cos(1/2*(2*n+1)*Pi*x))^2,x=0..1):
> I2:=int(cos(1/2*(2*n+1)*Pi*x),x=0..1):
> vars:={sin(n*Pi)=0}:
> I1:=subs(vars,I1):
> A[n]:=I2/I1;
```

$$A_n := \frac{\begin{cases} 1 & n = -\frac{1}{2} \\ \dfrac{2\cos(\pi n)}{(2n+1)\pi} & \text{otherwise} \end{cases}}{\begin{cases} 1 & n = -\frac{1}{2} \\ \dfrac{1}{2}\dfrac{2\pi n + \pi}{(2n+1)\pi} & \text{otherwise} \end{cases}}$$

```
> u(x,t):=eval(u(x,t));
```

$$u(x,t) := \sum_{n=0}^{\infty} \frac{\begin{cases} 1 & n = -\frac{1}{2} \\ \dfrac{2\cos(\pi n)}{(2n+1)\pi} & \text{otherwise} \end{cases}}{\begin{cases} 1 & n = -\frac{1}{2} \\ \dfrac{1}{2}\dfrac{2\pi n + \pi}{(2n+1)\pi} & \text{otherwise} \end{cases}} \cos\left(\frac{1}{2}(2n+1)\pi x\right) e^{-\frac{1}{4}(2n+1)^2 \pi^2 t}$$

```
> u(x,t):=subs(infinity=N,u(x,t)):
> ua:=subs(N=20,u(x,t)):
> uu:=piecewise(t=0,rhs(IC),t>0,ua);
```

602   7 Partial Differential Equations in Finite Domains

$$uu := \begin{cases} 1 & t=0 \\ \sum_{n=0}^{20} \dfrac{\left(\begin{cases} 1 & n=-\frac{1}{2} \\ \dfrac{2\cos(\pi n)}{(2n+1)\pi} & \text{otherwise} \end{cases}\right) \cos\left(\frac{1}{2}(2n+1)\pi x\right) e^{-\frac{1}{4}(2n+1)^2 \pi^2 t}}{\begin{cases} 1 & n=-\frac{1}{2} \\ \dfrac{1}{2}\dfrac{2\pi n+\pi}{(2n+1)\pi} & \text{otherwise} \end{cases}} & 0<t \end{cases}$$

> plot3d(uu,x=1..0,t=0.3..0,axes=boxed,title="Figure 7.5",

labels=[x,t,"u"],orientation=[60,60]);

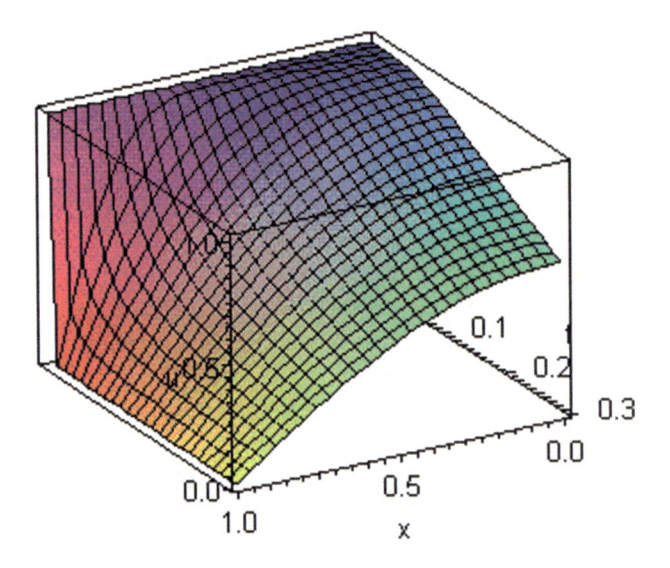

Fig. 7.5

> plot([subs(t=0,uu),subs(t=0.01,uu),subs(t=0.05,uu),subs(t=0.1,uu),

subs(t=0.2,uu)],x=0..1,axes=boxed,title="Figure 7.6",

thickness=5,labels=[x,"u"],legend=["t=0","t=0.01","t=0.05","t=0.1","t=0.2"]);

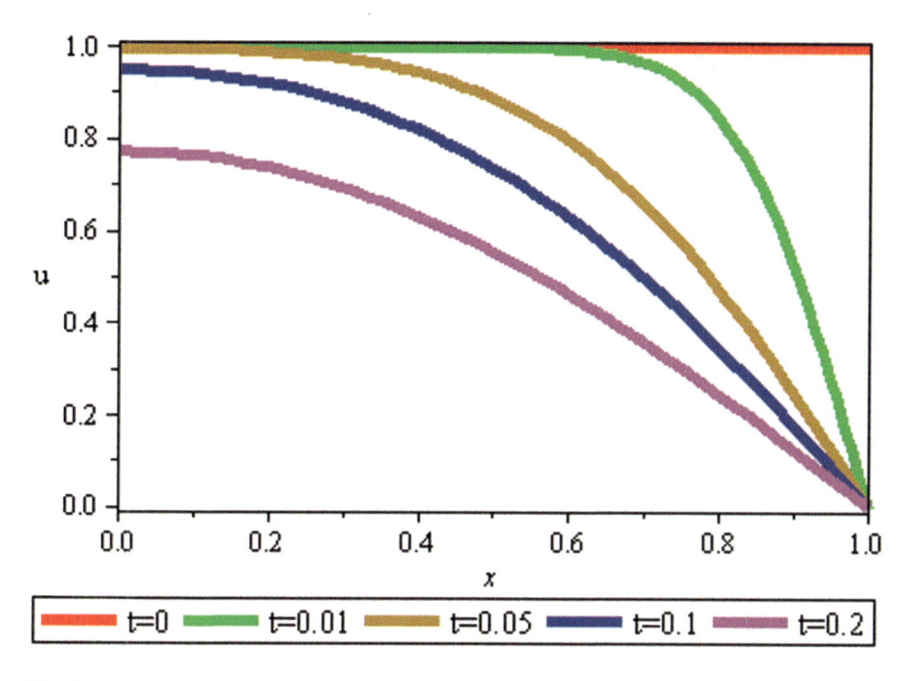

7.1 Separation of Variables Method for PDEs in Finite Domains

Fig. 7.6

In the previous two examples, we obtained the coefficient A_n in the infinite series by using the orthogonal property of the eigenfunction. For certain cases it is not trivial to choose the eigenfunctions for obtaining the integrals. A general criterion for obtaining the coefficient A_n is given by the Sturm-Liouville equation.[2] If the governing equation for X is given by:

$$\frac{d}{dx}\left(p(x)\frac{dX}{dx}\right)+(q(x)+\beta r(x))X=0 \qquad (7.28)$$

where, β is a constant. The coefficient A_n is given by:

$$A_n = \frac{\int_0^1 IC\, \phi_n r(x)\, dx}{\int_0^1 \phi_n^2 r(x)\, dx} \qquad (7.29)$$

where IC is the initial condition, ϕ_n is the eigenfunction that satisfies equation (7.28). In Example 7.1, $\sin(\lambda x)$ was the eigenfunction and in Example 7.2 $\cos(\lambda x)$ was the eigenfunction. In both the examples, the weighting function r(x) was 1. For problems in cylindrical coordinates and spherical coordinates, r(x) is x and x^2, respectively.

Example 7.3. Mass Transfer in a Spherical Pellet

Consider mass transfer in a spherical pellet.[1] The governing equation in dimensionless form is

$$\frac{\partial u}{\partial t} = \frac{\partial^2 u}{\partial x^2} + \frac{2}{x}\frac{\partial u}{\partial x}$$

$$u(x,0) = 1 \qquad (7.30)$$

$$\frac{\partial u}{\partial x}(0,t) = 0 \text{ and } u(1,t) = 0$$

Equation (7.30) is solved in Maple below:

> restart:

> with(plots):

> eq:=diff(u(x,t),t)=diff(u(x,t),x$2)+2/x*diff(u(x,t),x);

$$eq := \frac{\partial}{\partial t}u(x,t) = \frac{\partial^2}{\partial x^2}u(x,t) + \frac{2\left(\frac{\partial}{\partial x}u(x,t)\right)}{x}$$

> IC:=u(x,0)=1;

$$IC := u(x,0) = 1$$

> bc1:=diff(u(x,t),x)=0;

$$bc1 := \frac{\partial}{\partial x}u(x,t) = 0$$

> bc2:=u(x,t)=0;

$$bc2 := u(x,t) = 0$$

> Eq:=subs(u(x,t)=X(x)*T(t),eq):

> Eq:=expand(Eq/X(x)/T(t)):

> Eq_T:=lhs(Eq)=-lambda^2;

7.1 Separation of Variables Method for PDEs in Finite Domains

$$Eq\_T := \frac{\frac{d}{dt} T(t)}{T(t)} = -\lambda^2$$

> T(t):=rhs(dsolve({Eq_T,T(0)=T0},T(t)));

$$T(t) := T0\, e^{-\lambda^2 t}$$

> Eq_X:=rhs(Eq)=-lambda^2:

> Eq_X:=expand(Eq_X*X(x));

$$Eq\_X := \frac{d^2}{dx^2} X(x) + \frac{2\left(\frac{d}{dx} X(x)\right)}{x} = -X(x)\, \lambda^2$$

> dsolve({Eq_X},X(x));

$$\left\{ X(x) = \frac{\_C1\, \sin(\lambda x)}{x} + \frac{\_C2\, \cos(\lambda x)}{x} \right\}$$

The solution for X(x) can be taken as:

> X(x):=c[1]*sin(lambda*x)/x+c[2]*cos(lambda*x)/x;

$$X(x) := \frac{c_1 \sin(\lambda x)}{x} + \frac{c_2 \cos(\lambda x)}{x}$$

>Bc1:=diff(X(x),x)=0;

$$Bc1 := \frac{c_1 \cos(\lambda x)\, \lambda}{x} - \frac{c_1 \sin(\lambda x)}{x^2} - \frac{c_2 \sin(\lambda x)\, \lambda}{x} - \frac{c_2 \cos(\lambda x)}{x^2} = 0$$

> Bc2:=X(x)=0;

$$Bc2 := \frac{c_1 \sin(\lambda x)}{x} + \frac{c_2 \cos(\lambda x)}{x} = 0$$

x = 0 cannot be substituted. Hence, the limit at x = 0 is obtained:

> Eq_Bc1:=eval(subs(x=0,Bc1));

Error, numeric exception: division by zero

> limit(Bc1,x=0);

$$-\operatorname{signum}(c_2)\, \infty = 0$$

So the constant c2 should be zero.

> c[2]:=0;

$$c_2 := 0$$

> Eq_Bc2:=eval(subs(x=1,Bc2)):
> Eq_Eig:=sin(lambda)=0;

$$Eq\_Eig := \sin(\lambda) = 0$$

> _EnvAllSolutions := true:
> solve(Eq_Eig,lambda);

$$\pi\_Z1\sim$$

> U:=eval(X(x)*T(t)):
> Un:=subs(c[1]=A[n]/T0,lambda=lambda[n],U):
> u(x,t):=Sum(Un,n=1..infinity):
> u(x,t):=subs(lambda[n]=n*Pi,u(x,t));

$$u(x,t) := \sum_{n=1}^{\infty} \frac{A_n \sin(n \pi x) e^{-n^2 \pi^2 t}}{x}$$

> eq_An:=eval(subs(t=0,u(x,t)))=rhs(IC);

$$eq\_An := \sum_{n=1}^{\infty} \frac{A_n \sin(n \pi x)}{x} = 1$$

The eigenfunction is:

> phi[n]:=sin(n*Pi*x)/x;

$$\phi_n := \frac{\sin(n \pi x)}{x}$$

The weighting function is:

> r(x):=x^2;

$$r(x) := x^2$$

Next, the integrals are calculated to obtain A_n:

> I1:=int(phi[n]^2*r(x),x=0..1):
> IC;

$$u(x, 0) = 1$$

7.1 Separation of Variables Method for PDEs in Finite Domains

```
> I2:=int(rhs(IC)*phi[n]*r(x),x=0..1):
> vars:={sin(n*Pi)=0}:
> I1:=subs(vars,I1):
> I2:=subs(vars,I2):
> A[n]:=I2/I1;
```

$$A_n := \frac{\begin{cases} 0 & n = 0 \\ -\dfrac{\cos(n\pi)}{n\pi} & \text{otherwise} \end{cases}}{\begin{cases} 0 & n = 0 \\ \dfrac{1}{2} & \text{otherwise} \end{cases}}$$

The dimensionless concentration profile is given by:

```
> u(x,t):=eval(u(x,t)):
> u(x,t):=subs(infinity=N,u(x,t)):
> ua:=subs(N=20,u(x,t)):
> uu:=piecewise(t=0,rhs(IC),t>0,ua);
```

$$uu := \begin{cases} 1 & t = 0 \\ \displaystyle\sum_{n=1}^{20} \frac{\left(\begin{cases} 0 & n=0 \\ -\dfrac{\cos(n\pi)}{n\pi} & \text{otherwise}\end{cases}\right) \sin(n\pi x)\, e^{-n^2 \pi^2 t}}{\left(\begin{cases} 0 & n=0 \\ \dfrac{1}{2} & \text{otherwise}\end{cases}\right)} x & 0 < t \end{cases}$$

```
> plot3d(uu,x=1..0,t=0.3..0,axes=boxed,title="Figure 7.7",
labels=[x,t,"u"],orientation=[60,60]);
```

608   7 Partial Differential Equations in Finite Domains

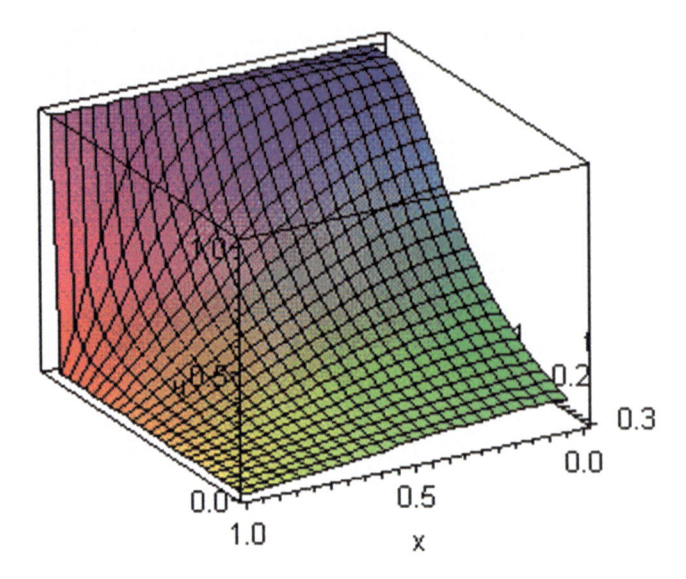

Fig. 7.7

> plot([subs(t=0,uu),subs(t=0.01,uu),subs(t=0.05,uu),subs(t=0.1,uu),
subs(t=0.2,uu)],x=0..1,axes=boxed,title="Figure 7.8",
thickness=5,labels=[x,"u"],legend=["t=0","t=0.01","t=0.05","t=0.1","t=0.2"]);

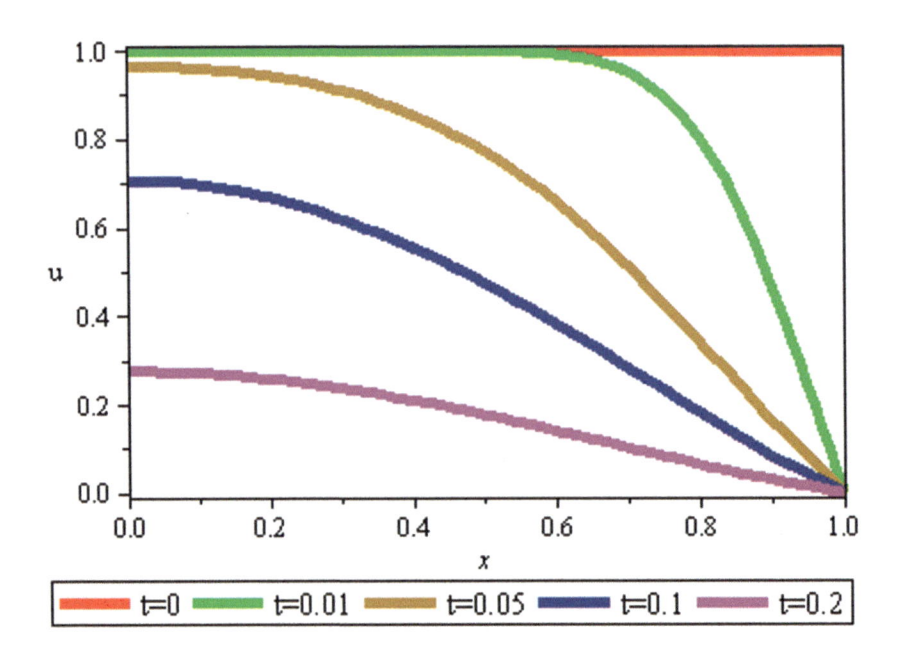

Fig. 7.8

7.1 Separation of Variables Method for PDEs in Finite Domains

7.1.3 Separation of Variables for Parabolic PDEs with an Initial Profile

In the previous three examples, the initial condition was uniform (= 1). In this section, partial differential equations with an initial profile in x are considered.

Example 7.4. Heat Conduction in a rectangle with an Initial Profile

Example 7.1 is solved again with a sinusoidal profile in x as the initial condition.[1]

The dimensionless temperature profile is governed by:

$$\frac{\partial u}{\partial t} = \frac{\partial^2 u}{\partial x^2}$$

$$u(x,0) = \sin(\pi x) \tag{7.31}$$

$$u(0,t) = 0 \text{ and } u(1,t) = 0$$

Equation (7.31) is solved in Maple below:

> restart:

> with(plots):

The governing equation is entered here:

> eq:=diff(u(x,t),t)=diff(u(x,t),x$2);

$$eq := \frac{\partial}{\partial t} u(x, t) = \frac{\partial^2}{\partial x^2} u(x, t)$$

The initial condition is entered here:

> IC:=u(x,0)=sin(Pi*x);

$$IC := u(x, 0) = \sin(\pi x)$$

The boundary conditions are entered here:

> bc1:=u(x,t)=0;

$$bc1 := u(x, t) = 0$$

> bc2:=u(x,t)=0;

$$bc2 := u(x, t) = 0$$

```
> Eq:=subs(u(x,t)=X(x)*T(t),eq):
> Eq:=expand(Eq/X(x)/T(t)):
> Eq_T:=lhs(Eq)=-lambda^2:
> T(t):=rhs(dsolve({Eq_T,T(0)=T0},T(t))):
> Eq_X:=rhs(Eq)=-lambda^2:
> Eq_X:=expand(Eq_X*X(x)):
> dsolve({Eq_X},X(x)):
> X(x):=c[1]*sin(lambda*x)+c[2]*cos(lambda*x):
> Bc1:=X(x)=0:
> Bc2:=X(x)=0:
> Eq_Bc1:=eval(subs(x=0,Bc1)):
> c[2]:=solve(Eq_Bc1,c[2]):
> Eq_Bc2:=eval(subs(x=1,Bc2)):
> Eq_Eig:=sin(lambda)=0;
```

$$Eq\_Eig := \sin(\lambda) = 0$$

```
> _EnvAllSolutions := true:
> solve(Eq_Eig,lambda):
> U:=eval(X(x)*T(t)):
> Un:=subs(c[1]=A[n]/T0,lambda=lambda[n],U):
> u(x,t):=Sum(Un,n=1..infinity):
> u(x,t):=subs(lambda[n]=n*Pi,u(x,t));
```

$$u(x,t) := \sum_{n=1}^{\infty} A_n \sin(n\pi x)\, e^{-n^2\pi^2 t}$$

```
> eq_An:=eval(subs(t=0,u(x,t)))=rhs(IC);
```

$$eq\_An := \sum_{n=1}^{\infty} A_n \sin(n\pi x) = \sin(\pi x)$$

```
> phi[n]:=sin(n*Pi*x):
> r(x):=1:
> I1:=int(phi[n]^2*r(x),x=0..1):
> IC;
```

7.1 Separation of Variables Method for PDEs in Finite Domains 611

$$u(x, 0) = \sin(\pi x)$$

> I2:=int(rhs(IC)*phi[n]*r(x),x=0..1);

$$I2 := \begin{cases} -\dfrac{1}{2} & n = -1 \\ \dfrac{1}{2} & n = 1 \\ -\dfrac{\sin(n\pi)}{\pi(-1+n^2)} & otherwise \end{cases}$$

> vars:={sin(n*Pi)=0}:
> I1:=subs(vars,I1);

$$I1 := \begin{cases} 0 & n = 0 \\ \dfrac{1}{2} & otherwise \end{cases}$$

The coefficient of A_n is obtained as:

> A[n]:=I2/I1;

$$A_n := \dfrac{\begin{cases} -\dfrac{1}{2} & n = -1 \\ \dfrac{1}{2} & n = 1 \\ -\dfrac{\sin(n\pi)}{\pi(-1+n^2)} & otherwise \end{cases}}{\begin{cases} 0 & n = 0 \\ \dfrac{1}{2} & otherwise \end{cases}}$$

For n = any integer value other than 1, A_n is zero:

> eval(subs(n=2,A[n]));

$$0$$

> eval(subs(n=3,A[n]));

$$0$$

> eval(subs(n=100,A[n]));

$$0$$

612   7 Partial Differential Equations in Finite Domains

> eval(subs(n=1,A[n]));

Error, numeric exception: division by zero

The limit command is applied to n = 1:

> A[1]:=limit(A[n],n=1);

$$A_1 := 1$$

There is only one nonzero term in the infinite series:

> u(x,t):=A[1]*sin(1*Pi*x)*exp(-Pi^2*t);

$$u(x,t) := \sin(\pi x) \, e^{-\pi^2 t}$$

> plot3d(u(x,t),x=1..0,t=0.3..0,axes=boxed,title="Figure 7.9",
labels=[x,t,"u"],orientation=[60,60]);

Fig. 7.9

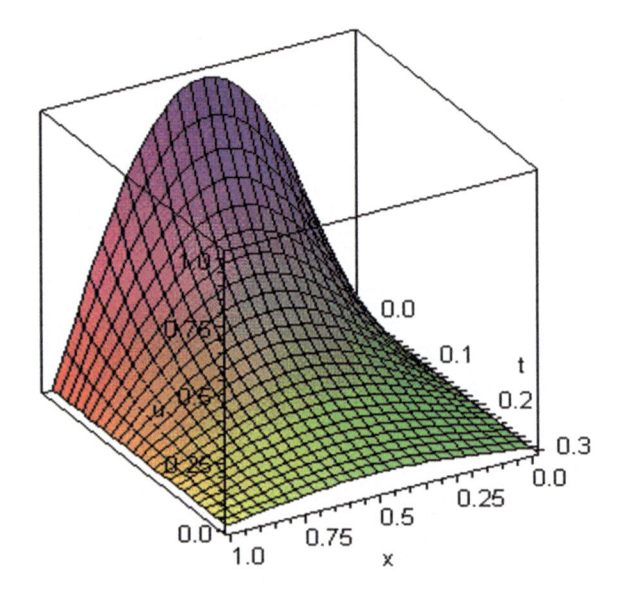

> plot([subs(t=0,u(x,t)),subs(t=0.05,u(x,t)),subs(t=0.1,u(x,t)),
subs(t=0.2,u(x,t)),subs(t=.3,u(x,t))],x=0..1,title="Figure 7.10",
axes=boxed,thickness=5,labels=[x,"u"],legend=["t=0","t=0.05","t=0.1",
"t=0.2","t=0.3"]);

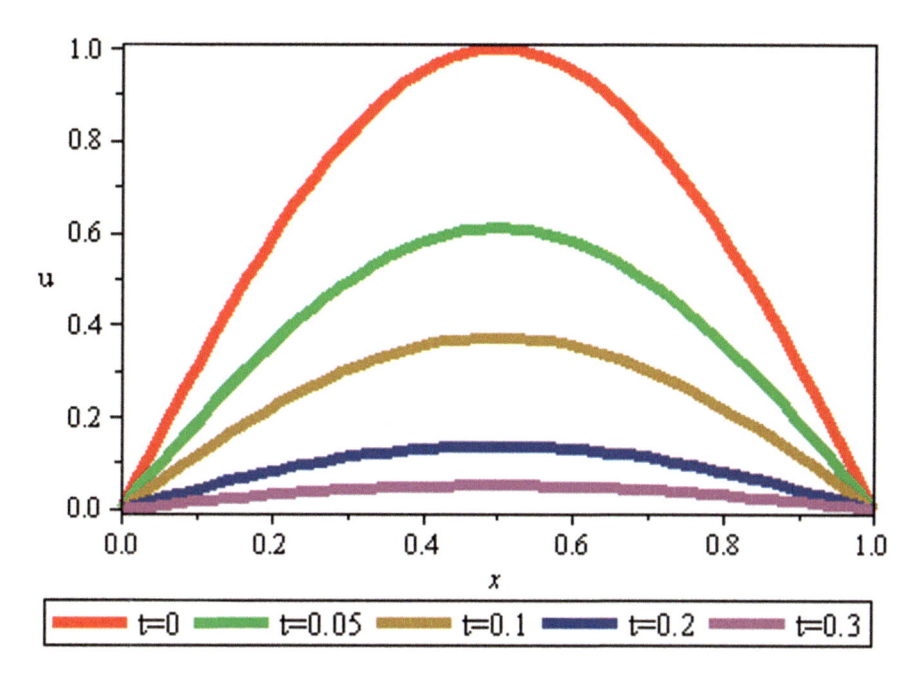

7.1 Separation of Variables Method for PDEs in Finite Domains

Fig. 7.10

Example 7.5. Heat Conduction in a Slab with a Linear Initial Profile

Example 7.2 is solved again with a linear profile in x as the initial condition.[1] The dimensionless temperature profile is governed by:

$$\frac{\partial u}{\partial t} = \frac{\partial^2 u}{\partial x^2}$$

$$u(x,0) = 1 - x \qquad (7.32)$$

$$\frac{\partial u}{\partial x}(0,t) = 0 \text{ and } u(1,t) = 0$$

Equation (7.32) is solved in Maple below by changing a few commands as:

> restart:

> with(plots):

> eq:=diff(u(x,t),t)=diff(u(x,t),x$2);

$$eq := \frac{\partial}{\partial t} u(x,t) = \frac{\partial^2}{\partial x^2} u(x,t)$$

> IC:=u(x,0)=1-x;

$$IC := u(x, 0) = 1 - x$$

> bc1:=diff(u(x,t),x)=0;

$$bc1 := \frac{\partial}{\partial x} u(x, t) = 0$$

> bc2:=u(x,t)=0;

$$bc2 := u(x, t) = 0$$

> Eq:=subs(u(x,t)=X(x)*T(t),eq):
> Eq:=expand(Eq/X(x)/T(t)):
> Eq_T:=lhs(Eq)=-lambda^2:
> T(t):=rhs(dsolve({Eq_T,T(0)=T0},T(t)));

$$T(t) := T0\, e^{-\lambda^2 t}$$

> Eq_X:=rhs(Eq)=-lambda^2:
> Eq_X:=expand(Eq_X*X(x)):
> dsolve({Eq_X},X(x)):
> X(x):=c[1]*sin(lambda*x)+c[2]*cos(lambda*x):
> Bc1:=diff(X(x),x)=0:
> Bc2:=X(x)=0:
> Eq_Bc1:=eval(subs(x=0,Bc1)):
> c[1]:=solve(Eq_Bc1,c[1]):
> Eq_Bc2:=eval(subs(x=1,Bc2)):
> Eq_Eig:=cos(lambda)=0:
> _EnvAllSolutions := true:
> solve(Eq_Eig,lambda);

$$\frac{1}{2}\pi + \pi\, \_Z1\!\sim$$

> U:=eval(X(x)*T(t)):
> Un:=subs(c[2]=A[n]/T0,lambda=lambda[n],U):
> u(x,t):=Sum(Un,n=0..infinity):
> u(x,t):=subs(lambda[n]=(2*n+1)/2*Pi,u(x,t));

7.1 Separation of Variables Method for PDEs in Finite Domains 615

$$u(x,t) := \sum_{n=0}^{\infty} A_n \cos\left(\frac{1}{2}(2n+1)\pi x\right) e^{-\frac{1}{4}(2n+1)^2 \pi^2 t}$$

> eq_An:=eval(subs(t=0,u(x,t)))=rhs(IC);

$$eq\_An := \sum_{n=0}^{\infty} A_n \cos\left(\frac{1}{2}(2n+1)\pi x\right) = 1 - x$$

> phi[n]:=cos((2*n+1)/2*Pi*x):

> r(x):=1:

> I1:=int(phi[n]^2*r(x),x=0..1):

> IC;

$$u(x,0) = 1 - x$$

> I2:=int(rhs(IC)*phi[n]*r(x),x=0..1);

$$I2 := \begin{cases} \dfrac{1}{2} & n = -\dfrac{1}{2} \\ \dfrac{4(1+\sin(\pi n))}{\pi^2(4n^2+4n+1)} & \text{otherwise} \end{cases}$$

> vars:={sin(n*Pi)=0}:

> I1:=subs(vars,I1):

> I2:=subs(vars,I2):

> A[n]:=I2/I1:

> A[n]:=simplify(A[n]);

$$A_n := \begin{cases} \dfrac{1}{2} & n = -\dfrac{1}{2} \\ \dfrac{8}{\pi^2(4n^2+4n+1)} & \text{otherwise} \end{cases}$$

> u(x,t):=eval(u(x,t));

$$u(x,t) := \sum_{n=0}^{\infty} \left(\begin{cases} \dfrac{1}{2} & n = -\dfrac{1}{2} \\ \dfrac{8}{\pi^2(4n^2+4n+1)} & \text{otherwise} \end{cases} \right) \cos\left(\frac{1}{2}(2n+1)\pi x\right) e^{-\frac{1}{4}(2n+1)^2 \pi^2 t}$$

> u(x,t):=subs(infinity=N,u(x,t)):

> ua:=subs(N=20,u(x,t)):

The dimensionless temperature distribution is given by:

> uu:=piecewise(t=0,rhs(IC),t>0,ua);

$$uu := \begin{cases} 1-x & t=0 \\ \sum_{n=0}^{20} \left(\begin{cases} \frac{1}{2} & n=-\frac{1}{2} \\ \frac{8}{\pi^2(4n^2+4n+1)} & \text{otherwise} \end{cases} \right) \cos\left(\frac{1}{2}(2n+1)\pi x\right) e^{-\frac{1}{4}(2n+1)^2 \pi^2 t} & 0 < t \end{cases}$$

> plot3d(uu,x=1..0,t=0.5..0,axes=boxed,title="Figure 7.11",

labels=[x,t,"u"],orientation=[45,60]);

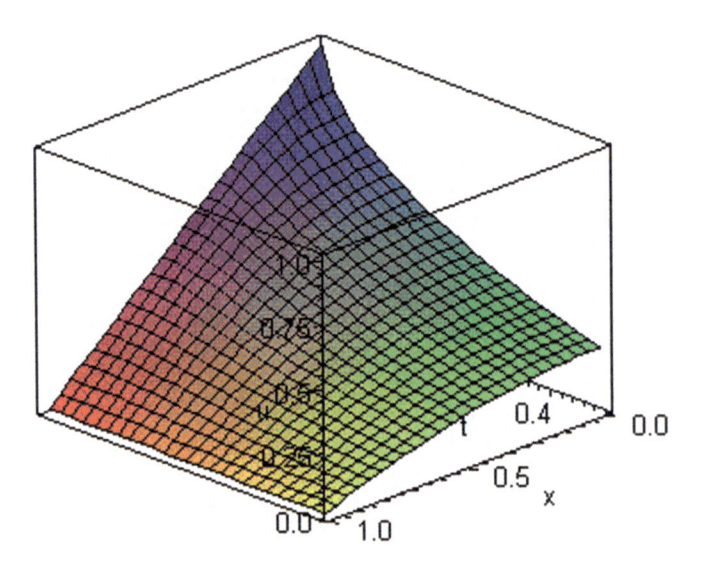

Fig. 7.11

> plot([subs(t=0,uu),subs(t=0.05,uu),subs(t=0.1,uu),subs(t=0.2,uu),

subs(t=0.5,uu)],x=0..1,axes=boxed,title="Figure 7.12",

thickness=5,labels=[x,"u"],legend=["t=0","t=0.05","t=0.1","t=0.2","t=0.5"]);

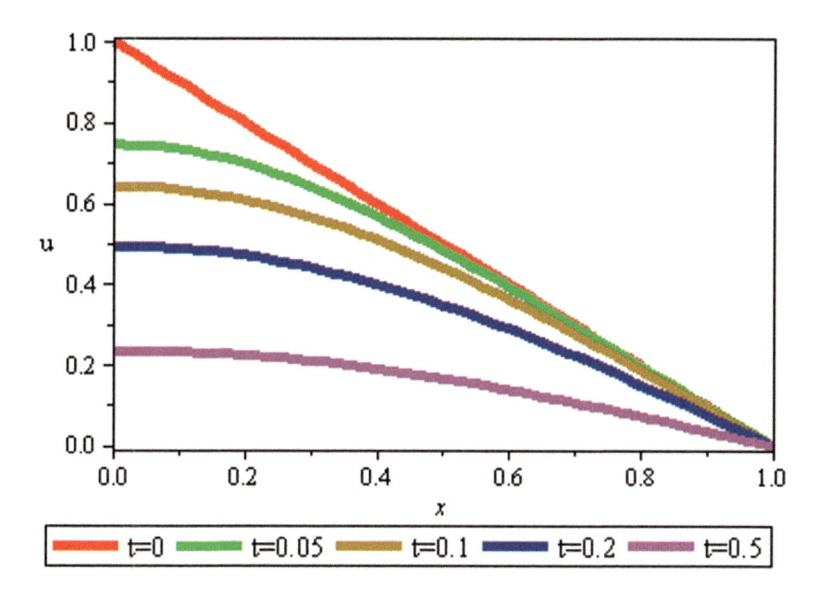

7.1 Separation of Variables Method for PDEs in Finite Domains   617

Fig. 7.12

Temperature at the surface, x = 0, can be found and plotted:

> us:=eval(subs(x=0,uu));

$$us := \begin{cases} 1 & t = 0 \\ \sum_{n=0}^{20} \left(\begin{cases} \frac{1}{2} & n = -\frac{1}{2} \\ \frac{8}{\pi^2 (4n^2 + 4n + 1)} & \text{otherwise} \end{cases} \right) e^{-\frac{1}{4}(2n+1)^2 \pi^2 t} & 0 < t \end{cases}$$

> plot(us,t=0..1,axes=boxed,title="Figure 7.13",thickness=3,labels=[t,"u"]);

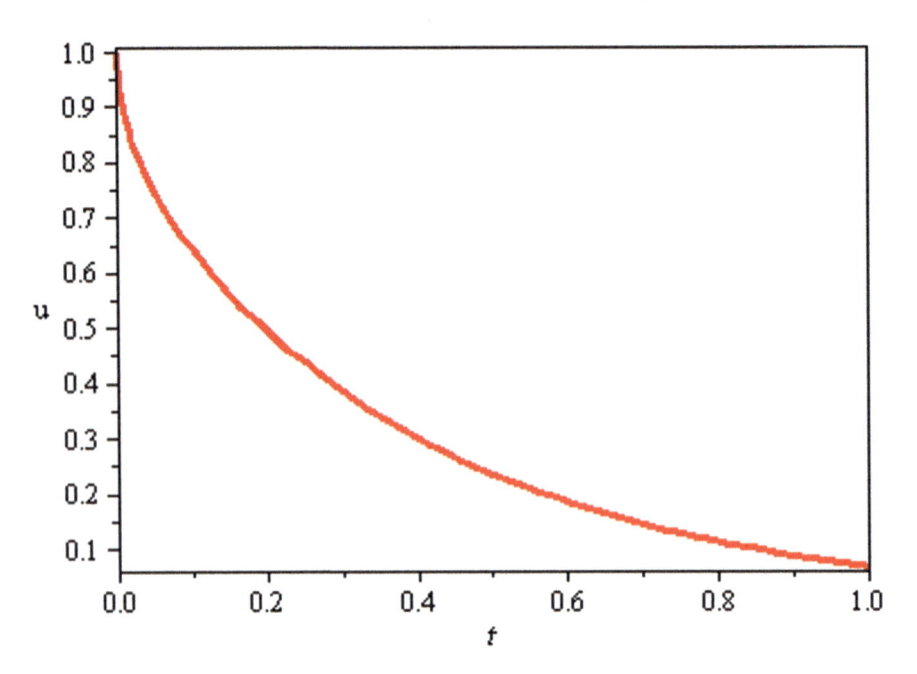

Fig. 7.13

7.1.4 Separation of Variables for Parabolic PDEs with Eigenvalues Governed by Transcendental Equations

In the previous five examples, eigenvalues were explicitly solvable ($n\pi$ or $\dfrac{2n+1}{2}\pi$). For some problems, the eigenvalues cannot be solved explicitly. In this section, we solve problems in which eigenvalues are obtained numerically.

Example 7.6. Heat Conduction in a Slab with Radiation Boundary Conditions

Consider heat conduction in a rectangle with radiation at the surface.[1] The dimensionless temperature profile is governed by:

$$\frac{\partial u}{\partial t} = \frac{\partial^2 u}{\partial x^2}$$

$$u(x,0) = 1 \tag{7.33}$$

$$\frac{\partial u}{\partial x}(0,t) = 0 \text{ and } \frac{\partial u}{\partial x}(1,t) + u(1,t) = 0$$

7.1 Separation of Variables Method for PDEs in Finite Domains

Equation (7.33) is solved in Maple below:

> restart:

> with(plots):

The governing equation, initial and boundary conditions are entered here:

> eq:=diff(u(x,t),t)=diff(u(x,t),x$2);

$$eq := \frac{\partial}{\partial t} u(x,t) = \frac{\partial^2}{\partial x^2} u(x,t)$$

> IC:=u(x,0)=1;

$$IC := u(x,0) = 1$$

> bc1:=diff(u(x,t),x)=0;

$$bc1 := \frac{\partial}{\partial x} u(x,t) = 0$$

> bc2:=diff(u(x,t),x)+u(x,t)=0;

$$bc2 := \frac{\partial}{\partial x} u(x,t) + u(x,t) = 0$$

> Eq:=subs(u(x,t)=X(x)*T(t),eq):
> Eq:=expand(Eq/X(x)/T(t)):
> Eq_T:=lhs(Eq)=-lambda^2:
> T(t):=rhs(dsolve({Eq_T,T(0)=T0},T(t))):
> Eq_X:=rhs(Eq)=-lambda^2:
> Eq_X:=expand(Eq_X*X(x)):
> dsolve({Eq_X},X(x)):
> X(x):=c[1]*sin(lambda*x)+c[2]*cos(lambda*x):
> Bc1:=diff(X(x),x)=0:
> Bc2:=diff(X(x),x)+X(x)=0:
> Eq_Bc1:=eval(subs(x=0,Bc1)):
> c[1]:=solve(Eq_Bc1,c[1]):
> Eq_Bc2:=eval(subs(x=1,Bc2)):

The eigenvalue equation is:

> Eq_Eig:=cos(lambda)-lambda*sin(lambda);

$$Eq\_Eig := \cos(\lambda) - \lambda \sin(\lambda)$$

> solve(Eq_Eig,lambda);

$$\text{RootOf}(\_Z \tan(\_Z) - 1)$$

Maple predicts negative eigenvalues:

> fsolve(Eq_Eig,lambda);

$$-.0.8603335890$$

Positive eigenvalues can be obtained by specifying a positive range in the 'fsolve' command.

> fsolve(Eq_Eig,lambda=0..3);

$$0.8603335890$$

The eigenvalue equation can be plotted as a function of the eigenvalue as:

> plot(Eq_Eig,lambda=0..20,thickness=3,title="Figure 7.14");

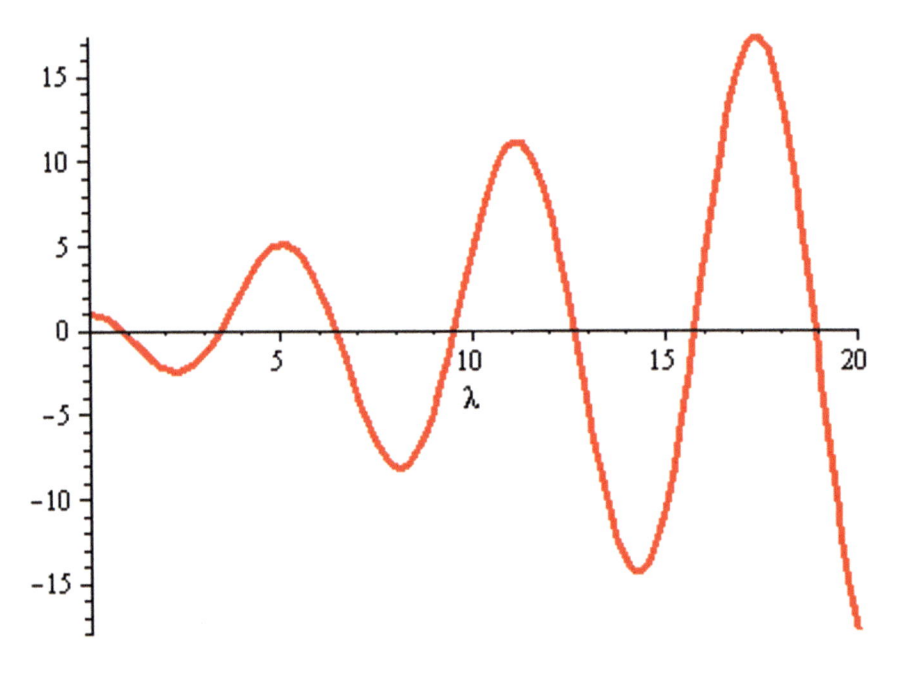

Fig. 7.14

The first 20 eigenvalues are obtained as:

> N:=20;

$$N := 20$$

7.1 Separation of Variables Method for PDEs in Finite Domains

> l[1]:=fsolve(Eq_Eig,lambda=0..3);

$$l_1 := 0.8603335890$$

> for i from 2 to N do l[i]:=fsolve(Eq_Eig,lambda=l[i-1]..l[i-1]+4);od:

> seq(l[i],i=1..N);

0.8603335890, 3.425618459, 6.437298179, 9.529334405, 12.64528722, 15.77128487, 18.90240996, 22.03649673,

25.17244633, 28.30964285, 31.44771464, 34.58642422, 37.72561283, 40.86517033, 44.00501792, 47.14509774,

50.28536634, 53.42579048, 56.56634428, 59.70700731

> U:=eval(X(x)*T(t)):

> Un:=subs(c[2]=A[n]/T0,lambda=lambda[n],U):

> u(x,t):=Sum(Un,n=1..infinity);

$$u(x,t) := \sum_{n=1}^{\infty} A_n \cos(\lambda_n x) \, e^{-\lambda_n^2 t}$$

> eq_An:=eval(subs(t=0,u(x,t)))=rhs(IC);

$$eq\_An := \sum_{n=1}^{\infty} A_n \cos(\lambda_n x) = 1$$

> phi[n]:=cos(lambda[n]*x):

> r(x):=1:

> l1:=int(phi[n]^2*r(x),x=0..1):

> IC;

$$u(x,0) = 1$$

> l2:=int(rhs(IC)*phi[n]*r(x),x=0..1):

> A[n]:=l2/l1;

$$A_n := \frac{2 \sin(\lambda_n)}{\cos(\lambda_n) \sin(\lambda_n) + \lambda_n}$$

> u(x,t):=eval(u(x,t));

$$u(x,t) := \sum_{n=1}^{\infty} \frac{2 \sin(\lambda_n) \cos(\lambda_n x) \, e^{-\lambda_n^2 t}}{\cos(\lambda_n) \sin(\lambda_n) + \lambda_n}$$

```
> u(x,t):=subs(infinity=N,u(x,t)):
> for i to N do lambda[i]:=l[i];od:
```

The transient analytical solution is:

```
> Digits:=5:u(x,t):=evalf(u(x,t));
```

$u(x, t) := 1.1191 \cos(0.86033\, x)\, e^{-0.74017\, t} - 0.15169 \cos(3.4256\, x)\, e^{-11.735\, t} + 0.046594 \cos(6.4373\, x)\, e^{-41.439\, t}$

$- 0.021668 \cos(9.5293\, x)\, e^{-90.808\, t} + 0.012392 \cos(12.645\, x)\, e^{-159.90\, t} - 0.0079920 \cos(15.771\, x)\, e^{-248.73\, t}$

$+ 0.0055741 \cos(18.902\, x)\, e^{-357.30\, t} - 0.0041062 \cos(22.036\, x)\, e^{-485.61\, t} + 0.0031491 \cos(25.172\, x)\, e^{-633.65\, t}$

$- 0.0024907 \cos(28.310\, x)\, e^{-801.44\, t} + 0.0020190 \cos(31.448\, x)\, e^{-988.96\, t} - 0.0016696 \cos(34.586\, x)\, e^{-1196.2\, t}$

$+ 0.0014036 \cos(37.726\, x)\, e^{-1423.2\, t} - 0.0011965 \cos(40.865\, x)\, e^{-1670.0\, t} + 0.0010321 \cos(44.005\, x)\, e^{-1936.4\, t}$

$- 0.00089931 \cos(47.145\, x)\, e^{-2222.7\, t} + 0.00079062 \cos(50.285\, x)\, e^{-2528.6\, t}$

$- 0.00070031 \cos(53.426\, x)\, e^{-2854.3\, t} + 0.00062460 \cos(56.566\, x)\, e^{-3199.8\, t}$

$- 0.00056088 \cos(59.707\, x)\, e^{-3564.9\, t}$

```
> uu:=piecewise(t=0,rhs(IC),t>0,u(x,t)):
```

The plots are obtained as:

```
> plot3d(uu,x=1..0,t=0.5..0,axes=boxed,title="Figure 7.15",
labels=[x,t,"u"],orientation=[45,60]);
```

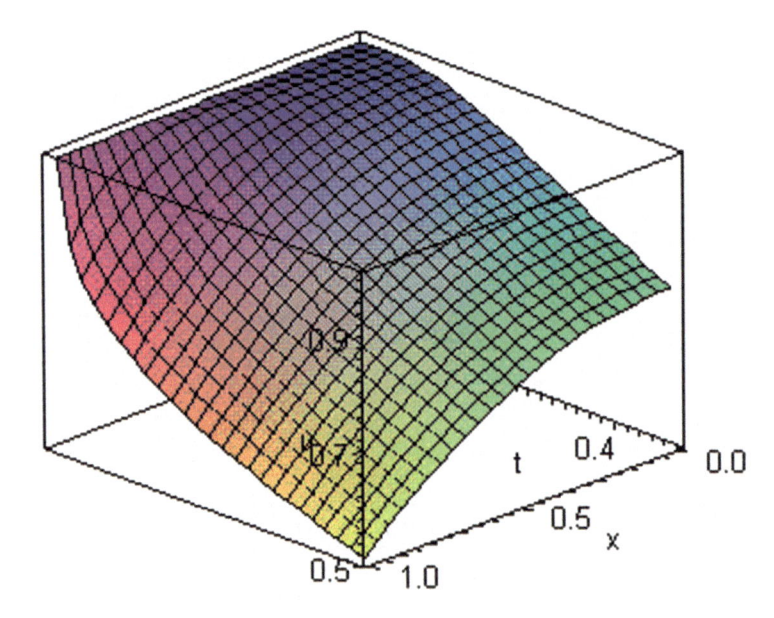

Fig. 7.15

7.1 Separation of Variables Method for PDEs in Finite Domains

```
> plot([subs(t=0,uu),subs(t=0.05,uu),subs(t=0.1,uu),subs(t=0.2,uu),
subs(t=0.5,uu)],x=0..1,axes=boxed,title="Figure 7.16",
thickness=5,labels=[x,"u"],legend=["t=0","t=0.05","t=0.1","t=0.2","t=0.5"]);
```

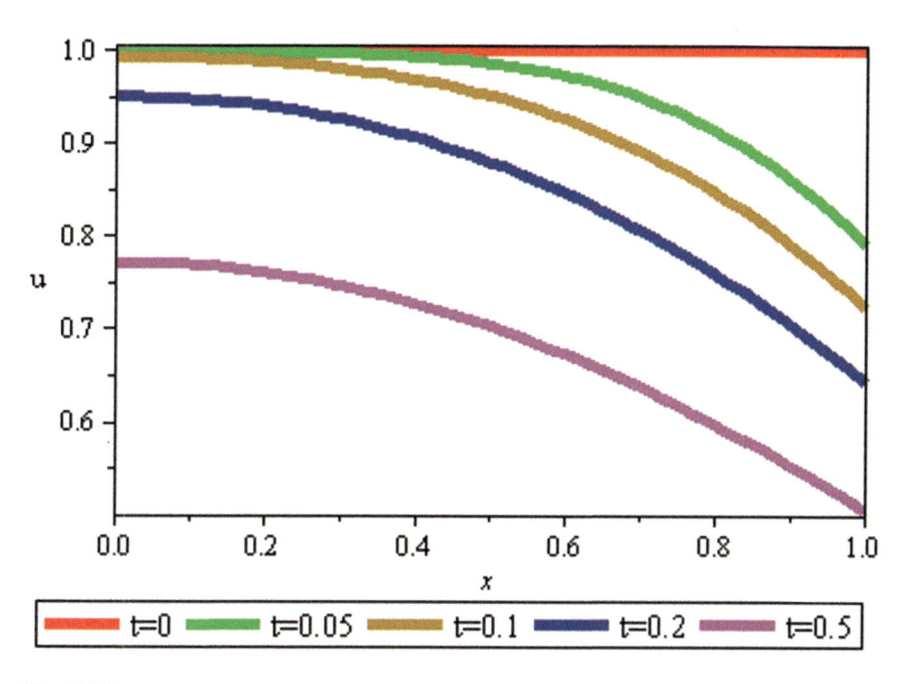

Fig. 7.16

7.1.5 Separation of Variables for Parabolic PDEs with Nonhomogeneous Boundary Conditions

In the previous examples, both of the boundary conditions were homogeneous. For non-homogeneous boundary conditions, the separation of variables method cannot be applied directly. Alternatively, functionality in x (w(x)) is introduced to take care of the non-homogeneity of the boundary conditions and separation of variables method is applied for the original partial differential equation with the homogeneous boundary conditions.

Example 7.7. Heat Conduction in a slab with Nonhomogeneous Boundary Conditions

Consider heat conduction in a rectangle with a nonhomogeneous boundary condition at x = 1.[1]

$$\frac{\partial u}{\partial t} = \frac{\partial^2 u}{\partial x^2}$$

$$u(x,0) = 0 \tag{7.34}$$

$$u(0,t) = 0 \text{ and } u(1,t) = 1$$

The separation of variables method cannot be directly applied because of the nonhomogeneous boundary condition at $x = 1$. To take care of this nonhomogeneity the following transformation is introduced

$$u(x,t) = g(x,t) + w(x) \tag{7.35}$$

where w(x) satisfies the nonhomogeneous boundary conditions

$$w(0) = 0 \text{ and } w(1) = 1 \tag{7.36}$$

and g(x,t) satisfies the homogeneous boundary conditions

$$g(0,t) = 0 \text{ and } g(1,t) = 0 \tag{7.37}$$

Substituting the transformation equation (7.35) into the governing equation (7.34) we get

$$\frac{\partial g}{\partial t} = \frac{\partial^2 g}{\partial x^2} + \frac{d^2 w}{dx^2} \tag{7.38}$$

Equation (7.38) can be separated for g and w as:

$$\frac{\partial g}{\partial t} = \frac{\partial^2 g}{\partial x^2} \tag{7.39}$$

$$\frac{d^2 w}{dx^2} = 0 \tag{7.40}$$

Equation (7.40) can be solved for the boundary conditions (equation (7.36)) as:

$$w = x \tag{7.41}$$

Next, equation (7.39) can be solved using separation of variables method with the homogeneous boundary conditions (equation (7.37)) as in Example 7.1 as:

$$g = \sum_{n=1}^{\infty} A_n \sin(n\pi x) \exp(-n^2 \pi^2 t) \tag{7.42}$$

7.1 Separation of Variables Method for PDEs in Finite Domains

The final solution is:

$$u = g + w = \sum_{n=1}^{\infty} A_n \sin(n\pi x) \exp(-n^2\pi^2 t) + x \quad (7.43)$$

Initial condition is applied to equation (7.42) as

$$0 = \sum_{n=1}^{\infty} A_n \sin(n\pi x) + x \quad (7.44)$$

By applying the Sturm-Liouville theorem, the coefficient A_n for partial differential equations with nonhomogeneous boundary conditions is obtained as:

$$A_n = \frac{\int_0^1 (IC - w)\, \phi_n r(x)\, dx}{\int_0^1 \phi_n^2 r(x)\, dx} \quad (7.45)$$

For this example, $r(x) = 1$, $IC = 0$, $w = x$, and $f_n = \sin(npx)$.

The Maple program used to solve Example 7.7 is given below:

```
> restart:
> with(plots):
> eq:=diff(u(x,t),t)=diff(u(x,t),x$2);
```

$$eq := \frac{\partial}{\partial t} u(x,t) = \frac{\partial^2}{\partial x^2} u(x,t)$$

```
> IC:=u(x,0)=0;
```

$$IC := u(x,0) = 0$$

```
> bc1:=u(x,t)=0;
```

$$bc1 := u(x,t) = 0$$

```
> bc2:=u(x,t)=1;
```

$$bc2 := u(x,t) = 1$$

```
> eq1:=eval(subs(u(x,t)=g(x,t)+w(x),eq));
```

$$eq1 := \frac{\partial}{\partial t} g(x,t) = \frac{\partial^2}{\partial x^2} g(x,t) + \frac{d^2}{dx^2} w(x)$$

> eqw:=diff(w(x),x$2);

$$eqw := \frac{d^2}{dx^2} w(x)$$

> eqg:=diff(g(x,t),t)=diff(g(x,t),x$2);

$$eqg := \frac{\partial}{\partial t} g(x,t) = \frac{\partial^2}{\partial x^2} g(x,t)$$

> bc1w:=w(x)=0;

$$bc1w := w(x) = 0$$

> bc2w:=w(x)=1;

$$bc2w := w(x) = 1$$

> w(x):=rhs(dsolve({eqw,w(0)=0,w(1)=1},w(x)));

$$w(x) := x$$

> Eq:=subs(g(x,t)=X(x)*T(t),eqg):

> Eq:=expand(Eq/X(x)/T(t)):

> Eq_T:=lhs(Eq)=-lambda^2:

> T(t):=rhs(dsolve({Eq_T,T(0)=T0},T(t))):

> Eq_X:=rhs(Eq)=-lambda^2:

> Eq_X:=expand(Eq_X*X(x)):

> dsolve({Eq_X},X(x)):

> X(x):=c[1]*sin(lambda*x)+c[2]*cos(lambda*x);

$$X(x) := c_1 \sin(\lambda x) + c_2 \cos(\lambda x)$$

> Bc1:=X(x)=0;

$$Bc1 := c_1 \sin(\lambda x) + c_2 \cos(\lambda x) = 0$$

> Bc2:=X(x)=0;

$$Bc2 := c_1 \sin(\lambda x) + c_2 \cos(\lambda x) = 0$$

> Eq_Bc1:=eval(subs(x=0,Bc1)):

> c[2]:=solve(Eq_Bc1,c[2]):

> Eq_Bc2:=eval(subs(x=1,Bc2)):

7.1 Separation of Variables Method for PDEs in Finite Domains

> Eq_Eig:=sin(lambda)=0:

> solve(Eq_Eig,lambda):

> _EnvAllSolutions := true:

> solve(Eq_Eig,lambda):

> G1:=eval(X(x)*T(t)):

> Gn:=subs(c[1]=A[n]/T0,lambda=lambda[n],G1):

> g(x,t):=Sum(Gn,n=1..infinity):

> g(x,t):=subs(lambda[n]=n*Pi,g(x,t));

$$g(x,t) := \sum_{n=1}^{\infty} A_n \sin(n \pi x) \, e^{-n^2 \pi^2 t}$$

> u(x,t):=g(x,t)+w(x);

$$u(x,t) := \sum_{n=1}^{\infty} A_n \sin(n \pi x) \, e^{-n^2 \pi^2 t} + x$$

> eq_An:=eval(subs(t=0,u(x,t)))=rhs(IC);

$$eq\_An := \sum_{n=1}^{\infty} A_n \sin(n \pi x) + x = 0$$

> phi[n]:=sin(n*Pi*x):

> r(x):=1:

> I1:=int(phi[n]^2*r(x),x=0..1):

> I2:=int((rhs(IC)-w(x))*phi[n]*r(x),x=0..1):

> vars:={sin(n*Pi)=0}:

> I1:=subs(vars,I1):

> I2:=subs(vars,I2):

> A[n]:=I2/I1:

> A[n]:=simplify(A[n]):

```
> u(x,t):=eval(u(x,t)):
```

```
> u(x,t):=subs(infinity=N,u(x,t)):
```

```
> ua:=subs(N=20,u(x,t)):
```

```
> uu:=piecewise(t=0,rhs(IC),t>0,ua);
```

$$uu := \begin{cases} 0 & t = 0 \\ \sum_{n=1}^{20} \left(\begin{cases} \text{undefined} & n = 0 \\ \dfrac{2\cos(n\pi)}{n\pi} & \text{otherwise} \end{cases} \sin(n\pi x)\, e^{-n^2 \pi^2 t} + x \right) & 0 < t \end{cases}$$

```
> plot3d(uu,x=1..0,t=0.5..0,axes=boxed,title="Figure 7.17",
labels=[x,t,"u"],orientation=[-135,60]);
```

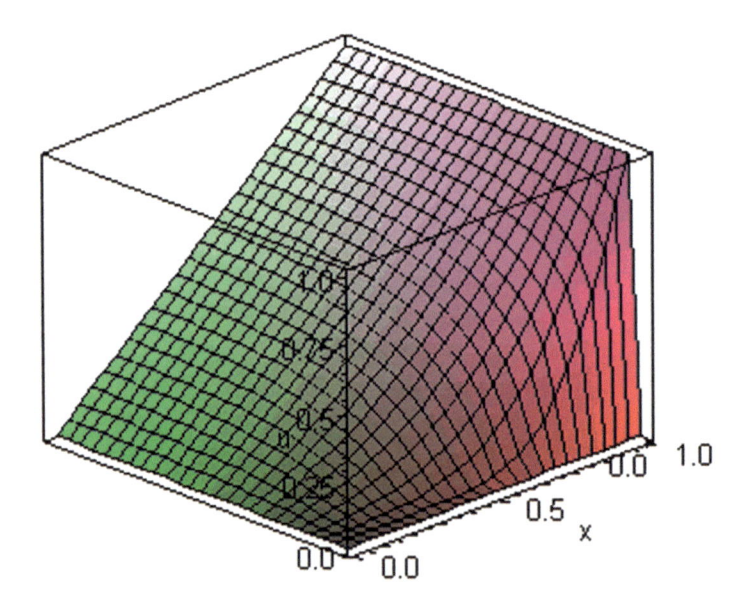

Fig. 7.17

```
> plot([subs(t=0,uu),subs(t=0.05,uu),subs(t=0.1,uu),subs(t=0.2,uu),
subs(t=0.5,uu)],x=0..1,axes=boxed,title="Figure 7.18",
thickness=5,labels=[x,"u"],legend=["t=0","t=0.05","t=0.1","t=0.2","t=0.5"]);
```

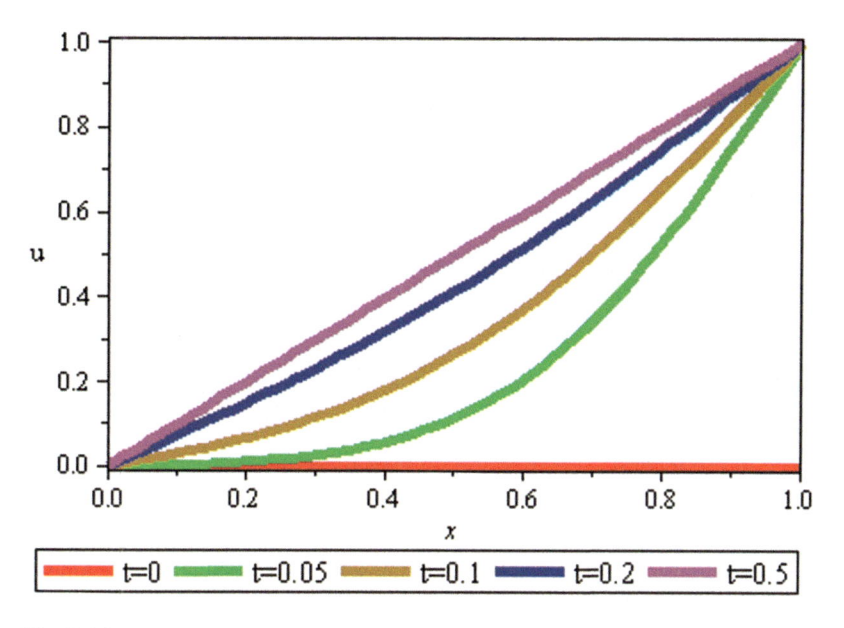

7.1 Separation of Variables Method for PDEs in Finite Domains

Fig. 7.18

In this example, w is just a linear function of x. However, for some complicated problems, w can be complicated functions of x as shown in the next example.

Example 7.8. Diffusion with Reaction

Consider the diffusion of a gas (A) through a stagnant liquid (B) in a container.[3] A reacts with B according to the irreversible reaction $A + B \xrightarrow{k} C$ (see Example 5.5). The governing equation for this problem in dimensionless form is,

$$\frac{\partial u}{\partial t} = \frac{\partial^2 u}{\partial x^2} - \Phi^2 u$$

$$u(x, 0) = 0 \tag{7.46}$$

$$u(0, t) = 1 \text{ and } \frac{\partial u}{\partial x}(1, t) = 0$$

where $\Phi = \sqrt{\dfrac{kL^2}{D_{AB}}}$ is the Thiele modulus. This problem is solved in Maple below.

630   7 Partial Differential Equations in Finite Domains

> restart:
> with(plots):
> eq:=diff(u(x,t),t)=diff(u(x,t),x$2)-Phi^2*u(x,t);

$$eq := \frac{\partial}{\partial t} u(x,t) = \frac{\partial^2}{\partial x^2} u(x,t) - \Phi^2 u(x,t)$$

> IC:=u(x,0)=0;

$$IC := u(x,0) = 0$$

> bc1:=u(x,t)=1;

$$bc1 := u(x,t) = 1$$

> bc2:=diff(u(x,t),x)=0;

$$bc2 := \frac{\partial}{\partial x} u(x,t) = 0$$

> eq1:=expand(eval(subs(u(x,t)=g(x,t)+w(x),eq)));

$$eq1 := \frac{\partial}{\partial t} g(x,t) = \frac{\partial^2}{\partial x^2} g(x,t) + \frac{d^2}{dx^2} w(x) - \Phi^2 g(x,t) - \Phi^2 w(x)$$

> eqw:=diff(w(x),x$2)-Phi^2*w(x);

$$eqw := \frac{d^2}{dx^2} w(x) - \Phi^2 w(x)$$

> eqg:=diff(g(x,t),t)=diff(g(x,t),x$2)-Phi^2*g(x,t);

$$eqg := \frac{\partial}{\partial t} g(x,t) = \frac{\partial^2}{\partial x^2} g(x,t) - \Phi^2 g(x,t)$$

> bc1w:=w(x)=1;

$$bc1w := w(x) = 1$$

> bc2w:=diff(w(x),x)=0;

$$bc2w := \frac{d}{dx} w(x) = 0$$

> dsolve({eqw},w(x));

$$\{w(x) = \_C1\, e^{-\Phi x} + \_C2\, e^{\Phi x}\}$$

w(x) can be written as:

> w(x):=w1*cosh(Phi*x)+w2*sinh(Phi*x);

$$w(x) := w1 \cosh(\Phi x) + w2 \sinh(\Phi x)$$

7.1 Separation of Variables Method for PDEs in Finite Domains

```
> eq_bc1w:=eval(subs(x=0,bc1w)):
> eq_bc2w:=eval(subs(x=1,bc2w)):
> w1:=solve(eq_bc1w,w1):
> w2:=solve(eq_bc2w,w2):
> w(x):=eval(w(x)):
> w(x):=combine(w(x)):
> w(x):=cosh(Phi*(1-x))/cosh(Phi);
```

$$w(x) := \frac{\cosh(\Phi\,(1-x))}{\cosh(\Phi)}$$

```
> Eq:=subs(g(x,t)=X(x)*T(t),eqg);
```

$$Eq := \frac{\partial}{\partial t}\,(X(x)\,T(t)) = \frac{\partial^2}{\partial x^2}\,(X(x)\,T(t)) - \Phi^2\,X(x)\,T(t)$$

```
> Eq:=expand(Eq/X(x)/T(t));
```

$$Eq := \frac{\frac{d}{dt}\,T(t)}{T(t)} = \frac{\frac{d^2}{dx^2}\,X(x)}{X(x)} - \Phi^2$$

For convenience, Φ^2 is written to the left hand side.

```
> Eq:=lhs(Eq)+Phi^2=rhs(Eq)+Phi^2;;
```

$$Eq := \frac{\frac{d}{dt}\,T(t)}{T(t)} + \Phi^2 = \frac{\frac{d^2}{dx^2}\,X(x)}{X(x)}$$

```
> Eq_T:=lhs(Eq)=-lambda^2;
```

$$Eq\_T := \frac{\frac{d}{dt}\,T(t)}{T(t)} + \Phi^2 = -\lambda^2$$

```
> T(t):=rhs(dsolve({Eq_T,T(0)=T0},T(t)));
```

$$T(t) := T0\,e^{(-\Phi^2 - \lambda^2)\,t}$$

T(t) can be written as:

```
> T(t):=T0*exp(-lambda^2*t)*exp(-Phi^2*t);
```

$$T(t) := T0\,e^{-t\lambda^2}\,e^{-t\Phi^2}$$

```
> Eq_X:=rhs(Eq)=-lambda^2:
> Eq_X:=expand(Eq_X*X(x)):
> dsolve({Eq_X},X(x)):
> X(x):=c[1]*sin(lambda*x)+c[2]*cos(lambda*x):
> Bc1:=X(x)=0:
> Bc2:=diff(X(x),x)=0:
> Eq_Bc1:=eval(subs(x=0,Bc1)):
> c[2]:=solve(Eq_Bc1,c[2]):
> Eq_Bc2:=eval(subs(x=1,Bc2)):
> Eq_Eig:=cos(lambda)=0:
> solve(Eq_Eig,lambda):
> _EnvAllSolutions := true:
> solve(Eq_Eig,lambda):
> G1:=eval(X(x)*T(t)):
> Gn:=subs(c[1]=A[n]/T0,lambda=lambda[n],G1):
> g(x,t):=Sum(Gn,n=1..infinity):
> g(x,t):=subs(lambda[n]=(2*n-1)/2*Pi,g(x,t));
```

$$g(x,t) := \sum_{n=1}^{\infty} A_n \sin\left(\frac{1}{2}(2n-1)\pi x\right) e^{-\frac{1}{4}t(2n-1)^2\pi^2} e^{-t\Phi^2}$$

```
> u(x,t):=g(x,t)+w(x);
```

$$u(x,t) := \sum_{n=1}^{\infty} A_n \sin\left(\frac{1}{2}(2n-1)\pi x\right) e^{-\frac{1}{4}t(2n-1)^2\pi^2} e^{-t\Phi^2} + \frac{\cosh(\Phi(1-x))}{\cosh(\Phi)}$$

```
> eq_An:=eval(subs(t=0,u(x,t)))=rhs(IC);
```

$$eq\_An := \sum_{n=1}^{\infty} A_n \sin\left(\frac{1}{2}(2n-1)\pi x\right) + \frac{\cosh(\Phi(1-x))}{\cosh(\Phi)} = 0$$

```
> phi[n]:=sin((2*n-1)/2*Pi*x):
> r(x):=1:
> I1:=int(phi[n]^2*r(x),x=0..1):
> I2:=int((rhs(IC)-w(x))*phi[n]*r(x),x=0..1):
> vars:={sin(n*Pi)=0}:
```

7.1 Separation of Variables Method for PDEs in Finite Domains

```
> I1:=subs(vars,I1):
> I1:=simplify(I1);
```

$$I1 := \begin{cases} 0 & n = \dfrac{1}{2} \\ \dfrac{1}{2} & otherwise \end{cases}$$

```
> I2:=subs(vars,I2):
> I2:=simplify(I2);
```

$$I2 := -\frac{2\,(2n-1)\,\pi}{4\,\Phi^2 + 4\,\pi^2\,n^2 - 4\,\pi^2\,n + \pi^2}$$

I2 is simplified further:

```
> I2n:=numer(I2):
> I2d:=denom(I2):
> I2d:=expand(I2d):
> I2:=I2n/I2d;
```

$$I2 := -\frac{2\,(2n-1)\,\pi}{4\,\Phi^2 + 4\,\pi^2\,n^2 - 4\,\pi^2\,n + \pi^2}$$

```
> A[n]:=I2/I1:
> A[n]:=simplify(A[n]):
> u(x,t):=eval(u(x,t));
```

$$u(x,t) := \left(\sum_{n=1}^{\infty} \left(\begin{cases} \dfrac{undefined}{\Phi^2} & n = \dfrac{1}{2} \\ -\dfrac{4\,(2n-1)\,\pi}{4\,\Phi^2 + 4\,\pi^2\,n^2 - 4\,\pi^2\,n + \pi^2} & otherwise \end{cases} \right) \sin\!\left(\dfrac{1}{2}(2n-1)\pi x\right) e^{-\tfrac{1}{4}t(2n-1)^2\pi^2} \right) e^{-t\Phi^2}$$

$$+\;\frac{\cosh(\Phi\,(1-x))}{\cosh(\Phi)}$$

```
> u(x,t):=subs(infinity=N,u(x,t)):
> ua:=subs(N=20,u(x,t)):
> uu:=piecewise(t=0,rhs(IC),t>0,ua):
> plot3d(subs(Phi=1,uu),x=1..0,t=0.5..0,axes=boxed,title="Figure 7.19",
labels=[x,t,"u"],orientation=[-45,60]);
```

634   7 Partial Differential Equations in Finite Domains

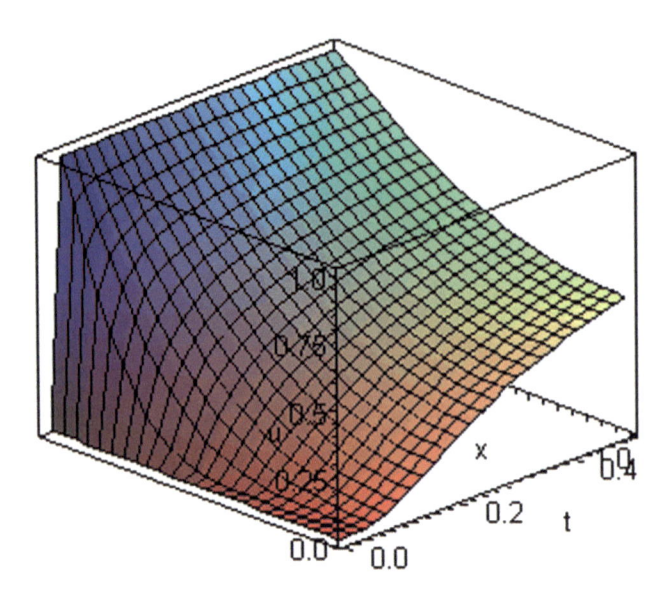

Fig. 7.19

> plot([subs(t=0,Phi=1,uu),subs(t=0.05,Phi=1,uu),subs(t=0.1,Phi=1,uu),
subs(t=0.2,Phi=1,uu),subs(t=0.5,Phi=1,uu)],x=0..1,axes=boxed,
title="Figure 7.20.",thickness=5,labels=[x,"u"],
legend=["t=0","t=0.05","t=0.1","t=0.2","t=0.5"]);

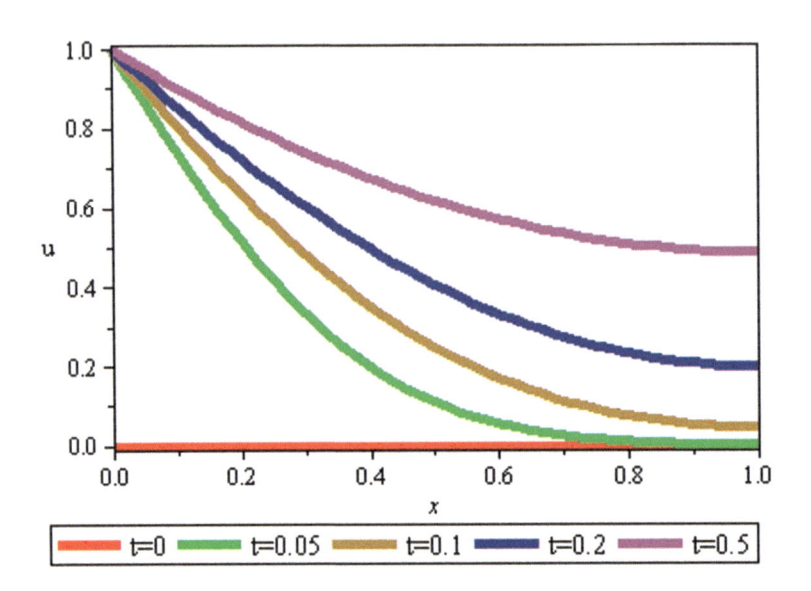

Fig. 7.20

7.1.6 Separation of Variables for Parabolic PDEs with Two Flux Boundary Conditions

In section 7.1.5, non-homogeneity in the boundary conditions was removed by adding w(x) to the solution. However, for partial differential equations with two nonhomogeneous flux boundary conditions, this method does not work.[4] For this case two separate functions (w(x) v(t)) are introduced to take care of the non-homogeneity of the boundary conditions and the separation of variables method is applied for the partial differential equation with the homogeneous boundary conditions.

Example 7.9. Diffusion in a Slab with Nonhomogeneous Flux Boundary Conditions

Consider diffusion in a rectangle with a nonhomogeneous boundary condition at x = 1 x[4]

$$\frac{\partial u}{\partial t} = \frac{\partial^2 u}{\partial x^2}$$
$$u(x,0) = 0 \qquad (7.47)$$
$$\frac{\partial u}{\partial x}(0,t) = 0 \text{ and } \frac{\partial u}{\partial x}(1,t) = \delta$$

The technique illustrated in section 7.1.5 cannot be directly applied for this problem because w(x) cannot be solved to satisfy the two flux boundary conditions at x = 0 and x = 1. To take care of the non-homogeneity the following transformation is introduced

$$u(x,t) = g(x,t) + w(x) + v(t) \qquad (7.48)$$

where w(x) satisfies the boundary conditions

$$\frac{dw}{dx}(0) = 0 \text{ and } \frac{dw}{dx}(1) = \delta \qquad (7.49)$$

and g(x,t) satisfies the homogeneous boundary conditions

$$\frac{\partial g}{\partial x}(0,t) = 0 \text{ and } \frac{\partial g}{\partial x}(1,t) = 0 \qquad (7.50)$$

The variable v(t) satisfies the initial condition:

$$v(0) = 0 \qquad (7.51)$$

Substituting the transformation (equation (7.48)) into the governing equation (7.47) we get

636   7 Partial Differential Equations in Finite Domains

$$\frac{\partial g}{\partial t} + \frac{dv}{dt} = \frac{\partial^2 g}{\partial x^2} + \frac{d^2 w}{dx^2} \qquad (7.52)$$

Equation (7.52) can be separated as:

$$\frac{\partial g}{\partial t} = \frac{\partial^2 g}{\partial x^2} \qquad (7.53)$$

$$\frac{dv}{dt} = \frac{d^2 w}{dx^2} \qquad (7.54)$$

The left hand side of equation (7.54) is a function of time alone and the right hand side is a function of x alone. Hence, both sides should be equal to a constant k,

$$\frac{dv}{dt} = \frac{d^2 w}{dx^2} = k \qquad (7.55)$$

The second half of equation (7.55) can be solved with the boundary conditions to give,

$$k = d \qquad (7.56)$$

and

$$w(x) = \frac{\delta}{2} x^2 + B \qquad (7.57)$$

where B is an arbitrary constant. The left hand side of equation (7.55) can be solved with the initial condition for v(t) (equation (7.51)) to give

$$v(t) = dt \qquad (7.58)$$

Hence, the solution is given by

$$\begin{aligned} u(x,t) &= g(x,t) + w(x) + v(t) \\ &= g(x,t) + \frac{\delta}{2} x^2 + \delta t + B \end{aligned} \qquad (7.59)$$

Now g(x,t) is obtained by solving equation (7.53) with the homogeneous boundary conditions (equation (7.50)) to give,

$$g(x,t) = \sum_{n=1}^{\infty} A_n \exp(-n^2 \pi^2 t) \cos(n\pi x) \qquad (7.60)$$

where A_n, n = 1,2,..., are constants. Hence, the final solution is given by,

$$u(x,t) = \frac{\delta}{2} x^2 + \delta t + B + \sum_{n=1}^{\infty} A_n \exp(-n^2 \pi^2 t) \cos(n\pi x) \qquad (7.61)$$

7.1 Separation of Variables Method for PDEs in Finite Domains

The constants A_n and B are obtained by imposing the initial condition,

$$u(x,0) = 0 = \frac{\delta}{2}x^2 + B + \sum_{n=1}^{\infty} A_n \cos(n\pi x) \tag{7.62}$$

Next, A_n is obtained as in section 7.1.5 as:

$$A_n = \frac{\int_0^1 (0-w)\phi_n r(x)dx}{\int_0^1 \phi_n^2 r(x)dx} = -\delta\frac{2(-1)^n}{n^2\pi^2} \tag{7.63}$$

For this example $r(x) = 1$, IC = 0, $w = \frac{\delta}{2}x^2 + Bx$ and $\phi_n = \cos(n\pi x)$. Next B is obtained by multiplying both sides of equation (7.63) by r(x) and integrating from 0 to 1:

$$\int_0^1 IC r(x)dx = \frac{\delta}{2}\int_0^1 r(x)x^2 dx + B\int_0^1 r(x)dx + \sum_{n=1}^{\infty} A_n \int_0^1 r(x)\phi_n dx \tag{7.64}$$

Note that the last integral in equation (7.64) goes to zero. For this example IC = 0, r(x) = 1, and $\phi_n = \cos(n\pi x)$. Substituting these values in equation (7.64) we get,

$$B = \frac{-\delta}{6} \tag{7.65}$$

Substituting equations (7.63) and (7.65) into equation (7.62) we get the complete solution:

$$u = \delta\left[t + \frac{1}{6}(3x^2 - 1) - \frac{2}{\pi^2}\sum_{n=1}^{\infty}\frac{(-1)^n}{n^2}\exp(-n^2\pi^2 t)\cos(n\pi x)\right] \tag{7.66}$$

The Maple program used to solve Example 7.9 is given below:

> restart:

> with(plots):

> eq:=diff(u(x,t),t)=diff(u(x,t),x$2);

$$eq := \frac{\partial}{\partial t}u(x,t) = \frac{\partial^2}{\partial x^2}u(x,t)$$

> IC:=u(x,0)=0;

$$IC := u(x, 0) = 0$$

> bc1:=diff(u(x,t),x)=0;

$$bc1 := \frac{\partial}{\partial x} u(x, t) = 0$$

> bc2:=diff(u(x,t),x)=delta;

$$bc2 := \frac{\partial}{\partial x} u(x, t) = \delta$$

> eq1:=eval(subs(u(x,t)=g(x,t)+w(x)+v(t),eq));

$$eq1 := \frac{\partial}{\partial t} g(x, t) + \frac{d}{dt} v(t) = \frac{\partial^2}{\partial x^2} g(x, t) + \frac{d^2}{dx^2} w(x)$$

The governing equations for v(t), w(x) and g(x, t) are:

> eqv:=diff(v(t),t)=k;

$$eqv := \frac{d}{dt} v(t) = k$$

> eqw:=diff(w(x),x$2)=k;

$$eqw := \frac{d^2}{dx^2} w(x) = k$$

> eqg:=diff(g(x,t),t)=diff(g(x,t),x$2);

$$eqg := \frac{\partial}{\partial t} g(x, t) = \frac{\partial^2}{\partial x^2} g(x, t)$$

The boundary conditions for w(x) are:

> bc1w:=diff(w(x),x)=0;

$$bc1w := \frac{d}{dx} w(x) = 0$$

> bc2w:=diff(w(x),x)=delta;

$$bc2w := \frac{d}{dx} w(x) = \delta$$

7.1 Separation of Variables Method for PDEs in Finite Domains

w(x) can be solved with the boundary condition at x = 0 as:

> w(x):=rhs(dsolve({eqw,D(w)(0)=0},w(x)));

$$w(x) := \frac{1}{2} k x^2 + \_C2$$

The value of constant k is found using the boundary condtion at x = 1:

> bc2w:=subs(x=1,diff(w(x),x))=delta;

$$bc2w := k = \delta$$

> k:=solve(bc2w,k);

$$k := \delta$$

> w(x):=1/2*delta*x^2+B;

$$w(x) := \frac{1}{2} \delta x^2 + B$$

v(t) is solved as:

> v(t):=rhs(dsolve({eqv,v(0)=0},v(t)));

$$v(t) := \delta t$$

> Eq:=subs(g(x,t)=X(x)*T(t),eqg):

g(x, t) is obtained as:

> Eq:=expand(Eq/X(x)/T(t)):
> Eq_T:=lhs(Eq)=-lambda^2:
> T(t):=rhs(dsolve({Eq_T,T(0)=T0},T(t))):
> Eq_X:=rhs(Eq)=-lambda^2:
> Eq_X:=expand(Eq_X*X(x)):
> dsolve({Eq_X},X(x)):
> X(x):=c[1]*sin(lambda*x)+c[2]*cos(lambda*x):
> Bc1:=diff(X(x),x)=0:
> Bc2:=diff(X(x),x)=0:
> Eq_Bc1:=eval(subs(x=0,Bc1)):
> c[1]:=solve(Eq_Bc1,c[1]):
> Eq_Bc2:=eval(subs(x=1,Bc2)):
> Eq_Eig:=sin(lambda)=0:
> solve(Eq_Eig,lambda):

> _EnvAllSolutions := true:
> solve(Eq_Eig,lambda):
> G1:=eval(X(x)*T(t)):
> Gn:=subs(c[2]=A[n]/T0,lambda=lambda[n],G1):
> g(x,t):=Sum(Gn,n=1..infinity):
> g(x,t):=subs(lambda[n]=n*Pi,g(x,t));

$$g(x,t) := \sum_{n=1}^{\infty} A_n \cos(n\pi x) \, e^{-n^2 \pi^2 t}$$

> u(x,t):=g(x,t)+w(x)+v(t);

$$u(x,t) := \sum_{n=1}^{\infty} A_n \cos(n\pi x) \, e^{-n^2 \pi^2 t} + \frac{1}{2}\delta x^2 + B + \delta t$$

> eq_An:=eval(subs(t=0,u(x,t)))=rhs(IC);

$$eq\_An := \sum_{n=1}^{\infty} A_n \cos(n\pi x) + \frac{1}{2}\delta x^2 + B = 0$$

The constant A_n is found as:

> phi[n]:=cos(n*Pi*x):
> r(x):=1:
> I1:=int(phi[n]^2*r(x),x=0..1):
> I2:=int((rhs(IC)-w(x))*phi[n]*r(x),x=0..1):
> vars:={sin(n*Pi)=0}:
> I1:=subs(vars,I1):
> I2:=subs(vars,I2):
> A[n]:=I2/I1;

$$A_n := \frac{\begin{cases} -B - \frac{1}{6}\delta & n = 0 \\ -\dfrac{\delta \cos(n\pi)}{n^2 \pi^2} & \text{otherwise} \end{cases}}{\begin{cases} 1 & n = 0 \\ \dfrac{1}{2} & \text{otherwise} \end{cases}}$$

7.1 Separation of Variables Method for PDEs in Finite Domains

Next, constant B is found as:

```
> eqB:=int(lhs(eq_An*r(x)),x=0..1)=int(rhs(eq_An*r(x)),x=0..1);
```

  Error, (in simpl/Im) too many levels of recursion

```
> eqB:=subs(vars,eqB);
```

$$eqB := eqB$$

```
> B:=-delta/6;
```

$$B := -\frac{1}{6}\delta$$

```
> u(x,t):=eval(u(x,t));
```

$$u(x,t) := \sum_{n=1}^{\infty} \frac{\left(\begin{cases} 0 & n=0 \\ -\dfrac{\delta \cos(n\pi)}{n^2 \pi^2} & \text{otherwise} \end{cases}\right) \cos(n\pi x)\, e^{-n^2 \pi^2 t}}{\begin{cases} 1 & n=0 \\ \dfrac{1}{2} & \text{otherwise} \end{cases}} + \frac{1}{2}\delta x^2 - \frac{1}{6}\delta + \delta t$$

```
> u(x,t):=subs(infinity=N,u(x,t)):
```

```
> ua:=subs(N=20,u(x,t)):
```

```
> uu:=piecewise(t=0,rhs(IC),t>0,ua);
```

$$uu := \begin{cases} 0 & t=0 \\ \displaystyle\sum_{n=1}^{20} \dfrac{\left(\begin{cases} 0 & n=0 \\ -\dfrac{\delta \cos(n\pi)}{n^2 \pi^2} & \text{otherwise} \end{cases}\right)\cos(n\pi x)\, e^{-n^2 \pi^2 t}}{\begin{cases} 1 & n=0 \\ \dfrac{1}{2} & \text{otherwise} \end{cases}} + \dfrac{1}{2}\delta x^2 - \dfrac{1}{6}\delta + \delta t & 0 < t \end{cases}$$

```
> plot3d(subs(delta=1,uu),x=1..0,t=0.5..0,axes=boxed,title="Figure 7.21",
labels=[x,t,"u"],orientation=[-135,60]);
```

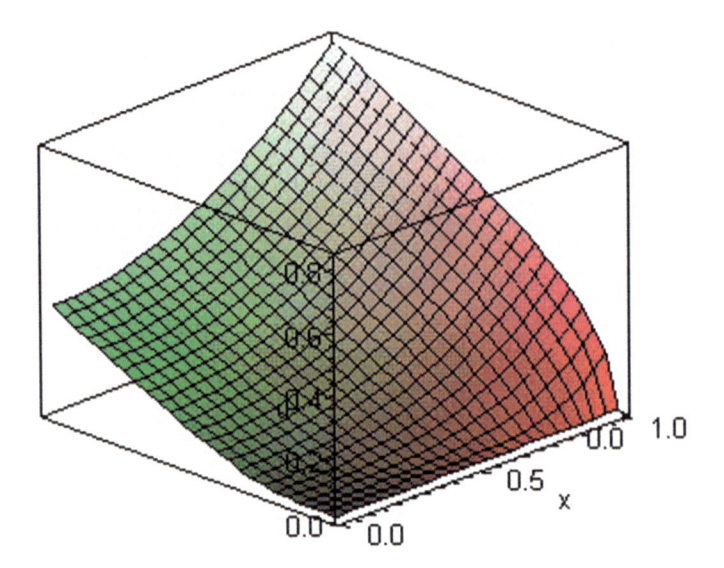

642   7 Partial Differential Equations in Finite Domains

Fig. 7.21

> plot([subs(t=0,delta=1,uu),subs(t=0.2,delta=1,uu),subs(t=0.5,delta=1,uu),
subs(t=0.7,delta=1,uu),subs(t=1,delta=1,uu)],x=0..1,title="Figure 7.22",
axes=boxed,thickness=5,labels=[x,"u"],legend=["t=0","t=0.2","t=0.5",
"t=0.7","t=1.0"]);

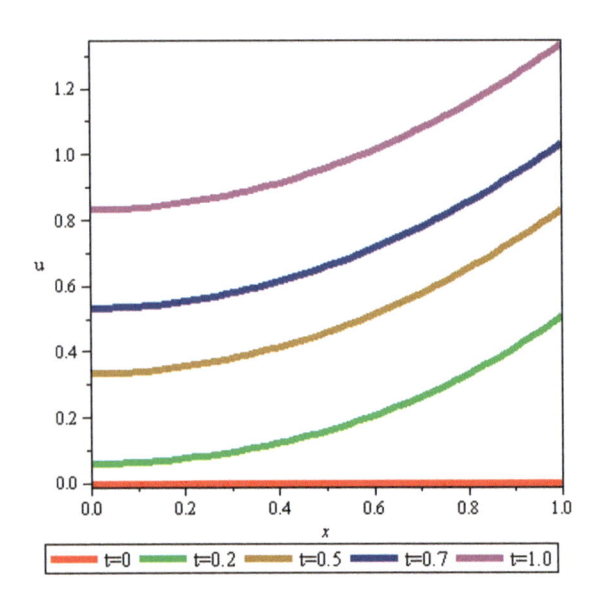

Fig. 7.22

7.1 Separation of Variables Method for PDEs in Finite Domains

7.1.7 Numerical Separation of Variables for Parabolic PDEs

In the previous sections, analytical expressions were derived for the eigenfunction (X(x)) and eigenvalues were obtained analytically or numerically from a transcendental equation. Alternatively, one can numerically obtain the eigenfunctions and eigenvalues. The advantage with the numerical approach is that the method is very general and one does not need to use Bessel or other special functions for the eigenfunctions. Also, there is no need to solve the transcendental equation. The methodology is illustrated by solving Example 7.1 numerically.

Example 7.10. Heat Transfer in a Rectangle

Consider the heat transfer problem

$$\frac{\partial u}{\partial t} = \frac{\partial^2 u}{\partial x^2}$$

$$u(x,0) = 1 \qquad (7.67)$$

$$u(0,t) = 0 \text{ and } u(1,t) = 0$$

The steps involved are the same as those in Example 7.1. The only difference is that the eigenfunction X(x) and the eigenvalues λ's are obtained numerically. An additional boundary condition at x = 0 $\left(\frac{dX}{dx}(0) = 0\right)$ is taken to obtain the eigenvalue. Read Example 3.2.15 for additional details. The Maple program used to solve this example using numerical separation of variables is given below:

> restart:

> with(plots):

> eq:=diff(u(x,t),t)=diff(u(x,t),x$2);

$$eq := \frac{\partial}{\partial t} u(x, t) = \frac{\partial^2}{\partial x^2} u(x, t)$$

> IC:=u(x,0)=1;

$$IC := u(x, 0) = 1$$

> bc1:=u(x,t)=0;

$$bc1 := u(x, t) = 0$$

> bc2:=u(x,t)=0;

$$bc2 := u(x, t) = 0$$

```
> Eq:=subs(u(x,t)=X(x)*T(t),eq):
> Eq:=Eq/X(x)/T(t):
> Eq_T:=lhs(Eq)=-lambda^2:
> T(t):=rhs(dsolve({Eq_T,T(0)=T0},T(t)));
```

$$T(t) := T0\, e^{-\lambda^2 t}$$

```
> Eq_X:=rhs(Eq)=-lambda^2:
> Eq_X:=expand(Eq_X*X(x));
```

$$Eq\_X := \frac{d^2}{dx^2} X(x) = -X(x)\, \lambda^2$$

The eigenvalue λ is obtained using the shooting technique. The third condition at $x = 0$ is assumed for $\frac{dX}{d\lambda}$. The sensitivity equation is developed for the sensitivity variable $X2 = \frac{dX}{d\lambda}$.

```
> eqlambda:=subs(X(x)=X(x,lambda),Eq_X):
> eqlambda:=diff(eqlambda,lambda):
> eqlambda:=subs(diff(X(x,lambda),lambda)=X2(x),eqlambda):
> eqlambda:=subs(X(x,lambda)=X(x),eqlambda);
```

$$eqlambda := \frac{d^2}{dx^2} X2(x) = -X2(x)\, \lambda^2 - 2\, X(x)\, \lambda$$

The dependent variables are:

```
> vars:=(X(x),X2(x));
```

$$vars := X(x), X2(x)$$

The initial conditions for X and X2 are:

```
> ICs:=(X(0)=0,D(X)(0)=1,X2(0)=0,D(X2)(0)=0);
```

$$ICs := X(0) = 0,\, D(X)(0) = 1,\, X2(0) = 0,\, D(X2)(0) = 0$$

A tolerance 1e - 9 is set. A scaling factor of 1/2 is used for obtaining eigenvalues.

```
> tol:=1e-9;rho:=1/2;
```

$$tol := 1.\, 10^{-9}$$

$$\rho := \frac{1}{2}$$

7.1 Separation of Variables Method for PDEs in Finite Domains

Different initial guesses are used to predict the first ten eigenvalues:

> lambdaguess:=[3,6,9,12,15,18,21,24,27,30];

$$lambdaguess := [3, 6, 9, 12, 15, 18, 21, 24, 27, 30]$$

> MM:=nops(lambdaguess);

$$MM := 10$$

> Xexp:=0:

> for i from 1 to MM do

> lambda0:=lambdaguess[i];

> k:=1;err:=1;

> while err>tol do

> eqs:=subs(lambda=lambda0,Eq_X),subs(lambda=lambda0,eqlambda);

> Sol[i]:=dsolve({eqs,ICs},{vars},type=numeric,output=listprocedure);

> Xpred:=rhs(Sol[i](1)[2]);

> X2pred:=rhs(Sol[i](1)[4]);

> lambda1:=lambda0+rho*(Xexp-Xpred)/X2pred;

> err:=abs(lambda1-lambda0);

> lambda0:=lambda1;k:=k+1;

> end;

> l[i]:=lambda0;

> kk[i]:=k;

> Err[i]:=err;

> od:

The first ten eigenvalues are:

> seq(l[i],i=1..MM);

3.141592666, 6.283185394, 9.424778160, 12.56637090, 15.70796362, 18.84955636, 21.99114908, 25.13274183, 28.27433459, 31.41592732

The number of iterations required to get the eigenvalues are:

> seq(kk[i],i=1..MM);

$$28, 29, 30, 28, 28, 28, 28, 27, 28, 29$$

The error associated with the eigenvalues is:

> seq(Err[i],i=1..MM);

$$1.\,10^{-9},\,1.\,10^{-9},\,1.\,10^{-9},\,0.,\,0.,\,0.,\,0.,\,0.,\,0.,\,0.$$

The first ten eigenfunctions are plotted as:

> for i to MM do XX[i]:=subs(Sol[i],X(x)):od:

> for i to MM do p[i]:=plot(XX[i](x),x=0..1,thickness=3,title="Figure 7.23", axes=boxed):od:

> display({seq(p[i],i=1..MM)},labels=[x,"X"]);

Fig. 7.23

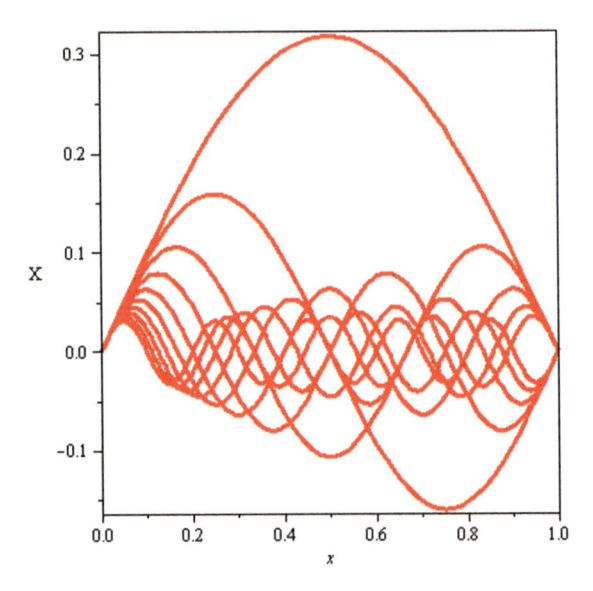

The first eigenfunction becomes negative. The second eigenfunction becomes negative one then returns to zero. Similarly, the nth eigenfunction crosses the x axis n - 1 times.

> U:=eval(X(x)*T(t)):

The solution can be taken as:

> Un:=A[n]*XX[n](x)*exp(-lambda[n]^2*t);

7.1 Separation of Variables Method for PDEs in Finite Domains

$$Un := A_n \, XX_n(x) \, e^{-\lambda_n^2 t}$$

> for i to MM do lambda[i]:=l[i];od:

> u(x,t):=Sum(Un,n=1..MM);

$$u(x,t) := \sum_{n=1}^{10} A_n \, XX_n(x) \, e^{-\lambda_n^2 t}$$

The constant A_n is obtained as:

> eq_An:=eval(subs(t=0,u(x,t)))=rhs(IC);

$$eq\_An := \sum_{n=1}^{10} A_n \, XX_n(x) = 1$$

> I1:=int(X(x)^2,x=0..1):

> I2:=int(X(x),x=0..1):

> An:=I2/I1;

$$An := \frac{\int_0^1 X(x) \, dx}{\int_0^1 X(x)^2 \, dx}$$

The coefficient A_n, n = 1...10 are numerically obtained as:

> for j to MM do

> N:=200:I1:=0:I2:=0:

> for i from 1 to N-1 do I1:=I1+XX[j](i/N)^2*2/(2*N);I2:=I2+XX[j](i/N)*2/(2*N)od:

> I1:=I1+(XX[j](0)^2+XX[j](1)^2)/(2*N):

> I2:=I2+(XX[j](0)+XX[j](1))/(2*N):

> A[j]:=I2/I1;od:

> seq(A[j],j=1..MM);

3.999917395, −0.000001708160921, 3.999256880, −0.000003978565951, 3.997937659, −0.000006108475765,

3.995960133, −0.000008302711514, 3.993323724, −0.00001021964161

> u(x,t):=eval(u(x,t)):

> ua:=evalf(u(x,t));

$ua := 3.999917395 \, XX_1(x) \, e^{-9.869604479 \, t} - 0.000001708160921 \, XX_2(x) \, e^{-39.47841870 \, t}$

$+ 3.999256880 \, XX_3(x) \, e^{-88.82644336 \, t} - 0.000003978565951 \, XX_4(x) \, e^{-157.9136776 \, t}$

$+ 3.997937659 \, XX_5(x) \, e^{-246.7401211 \, t} - 0.000006108475765 \, XX_6(x) \, e^{-355.3057750 \, t}$

$+ 3.995960133 \, XX_7(x) \, e^{-483.6106379 \, t} - 0.000008302711514 \, XX_8(x) \, e^{-631.6547119 \, t}$

$+ 3.993323724 \, XX_9(x) \, e^{-799.4379965 \, t} - 0.00001021964161 \, XX_{10}(x) \, e^{-986.9604894 \, t}$

```
> uu:=piecewise(t=0,rhs(IC),t>0,ua):
> plot3d(uu,x=1..0,t=0.3..0,axes=boxed,title="Figure 7.24",
labels=[x,t,"u"],orientation=[60,60]);
```

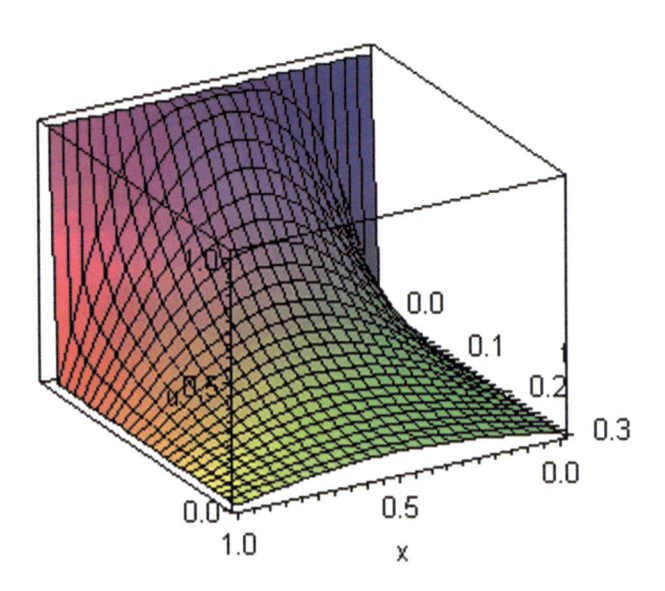

Fig. 7.24

```
> plot([subs(t=0,uu),subs(t=0.01,uu),subs(t=0.05,uu),subs(t=0.1,uu),
subs(t=0.2,uu)],x=0..1,axes=boxed,title="Figure 7.25",
thickness=5,labels=[x,"u"],legend=["t=0","t=0.01","t=0.05","t=0.1","t=0.2"]);
```

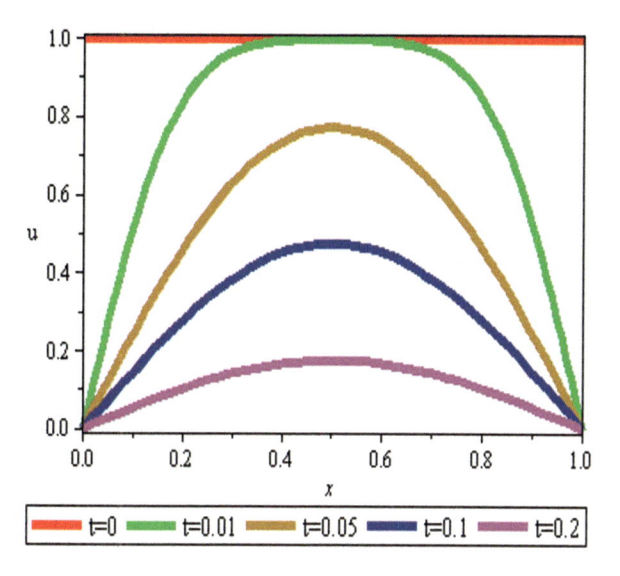

Fig. 7.25

7.1.8 Separation of Variables for Elliptic PDEs

The separation of variables method can be used for steady state elliptic PDEs also. In this case, the dependent variable is separated in the x and y coordinates. In order to apply separation of variables method, only one boundary condition can be nonhomogeneous. If more than one boundary condition is nonhomogeneous, then the problem has to be reduced to the case where only one boundary condition is nonhomogeneous. One has to separate the original boundary value problem into two boundary value problems, each of which will have only one nonhomogeneous boundary condition. The methodology is illustrated using steady state heat transfer in a rectangle.

Example 7.11. Heat Transfer in a Rectangle

Consider the steady state heat transfer problem.[1] The boundary condition at y = 1 (the nonhomogeneous boundary condition) is used to find the coefficient A_n in the infinite series after separating the variables.

$$\frac{\partial^2 u}{\partial y^2} + \frac{\partial^2 u}{\partial x^2} = 0$$

$$u(0,y) = 0 \text{ and } u(1,y) = 0 \tag{7.68}$$

$$u(x,0) = 0 \text{ and } u(x,1) = 1$$

The Maple program used to solve this example is given below:

> restart:

> with(plots):

Derivatives in y are entered on the left hand side and derivatives in x are entered on the right hand side.

> eq:=diff(u(x,y),y$2)=-diff(u(x,y),x$2);

$$eq := \frac{\partial^2}{\partial y^2} u(x,y) = -\left(\frac{\partial^2}{\partial x^2} u(x,y)\right)$$

The boundary conditions at x = 0 and x = 1 are entered:

> bc1:=u(x,y)=0;

$$bc1 := u(x,y) = 0$$

> bc2:=u(x,y)=0;

$$bc2 := u(x,y) = 0$$

The boundary conditions at y = 0 and y = 1 are entered:

> bc3:=u(x,y)=0;

$$bc3 := u(x,y) = 0$$

> bc4:=u(x,y)=1;

$$bc4 := u(x,y) = 1$$

The dependent variable is separated as:

> Eq:=subs(u(x,y)=X(x)*Y(y),eq);

$$Eq := \frac{\partial^2}{\partial y^2}(X(x)\,Y(y)) = -\left(\frac{\partial^2}{\partial x^2}(X(x)\,Y(y))\right)$$

> Eq:=expand(Eq/X(x)/Y(y));

$$Eq := \frac{\frac{d^2}{dy^2} Y(y)}{Y(y)} = -\frac{\frac{d^2}{dx^2} X(x)}{X(x)}$$

> Eq_Y:=lhs(Eq)=lambda^2:
> Eq_Y:=eval(Eq_Y*Y(y));

$$Eq\_Y := \frac{d^2}{dy^2} Y(y) = Y(y)\,\lambda^2$$

7.1 Separation of Variables Method for PDEs in Finite Domains

> dsolve(Eq_Y,Y(y));

$$Y(y) = \_C1\ e^{\lambda y} + \_C2\ e^{-\lambda y}$$

Y(y) can be written as:

> Y(y):=C[1]*sinh(lambda*y)+C[2]*cosh(lambda*y);

$$Y(y) := C_1 \sinh(\lambda y) + C_2 \cosh(\lambda y)$$

The homogeneous boundary condition in y (at y = 0) is used to eliminate one of the constants.

> Bc3:=Y(y)=0;

$$Bc3 := C_1 \sinh(\lambda y) + C_2 \cosh(\lambda y) = 0$$

> Eq_Bc3:=eval(subs(y=0,Bc3));

$$Eq\_Bc3 := C_2 = 0$$

> C[2]:=solve(Eq_Bc3);

$$C_2 := 0$$

> Y(y):=eval(Y(y));

$$Y(y) := C_1 \sinh(\lambda y)$$

X(x) is obtained as in Example 7.1:

> Eq_X:=rhs(Eq)=lambda^2:
> Eq_X:=expand(Eq_X*X(x));

$$Eq\_X := -\left(\frac{d^2}{dx^2} X(x)\right) = X(x)\, \lambda^2$$

> dsolve({Eq_X},X(x));

$$\{X(x) = \_C1\ \sin(\lambda x) + \_C2 \cos(\lambda x)\}$$

> X(x):=c[1]*sin(lambda*x)+c[2]*cos(lambda*x);

$$X(x) := c_1 \sin(\lambda x) + c_2 \cos(\lambda x)$$

> Bc1:=X(x)=0;

$$Bc1 := c_1 \sin(\lambda x) + c_2 \cos(\lambda x) = 0$$

> Bc2:=X(x)=0;

$$Bc2 := c_1 \sin(\lambda x) + c_2 \cos(\lambda x) = 0$$

> Eq_Bc1:=eval(subs(x=0,Bc1));

$$Eq\_Bc1 := c_2 = 0$$

> c[2]:=solve(Eq_Bc1,c[2]);

$$c_2 := 0$$

> Eq_Bc2:=eval(subs(x=1,Bc2));

$$Eq\_Bc2 := c_1 \sin(\lambda) = 0$$

> Eq_Eig:=sin(lambda)=0;

$$Eq\_Eig := \sin(\lambda) = 0$$

> _EnvAllSolutions := true:
> solve(Eq_Eig,lambda);

$$\pi\_Z1\sim$$

> U:=eval(X(x)*Y(y));

$$U := c_1 \sin(\lambda x) \, C_1 \sinh(\lambda y)$$

> Un:=subs(c[1]=A[n]/C[1],lambda=lambda[n],U):
> u(x,y):=Sum(Un,n=1..infinity):

The dimensionless temperature is:

> u(x,y):=subs(lambda[n]=n*Pi,u(x,y));

$$u(x,y) := \sum_{n=1}^{\infty} A_n \sin(n \pi x) \sinh(n \pi y)$$

The constant A_n is obtained using the nonhomogeneous boundary condition at y = 1:

> eq_An:=eval(subs(y=1,u(x,y)))=rhs(bc4);

$$eq\_An := \sum_{n=1}^{\infty} A_n \sin(n \pi x) \sinh(n \pi) = 1$$

> phi[n]:=sin(n*Pi*x);

$$\phi_n := \sin(n \pi x)$$

> r(x):=1;

$$r(x) := 1$$

7.1 Separation of Variables Method for PDEs in Finite Domains

```
> l1:=int(phi[n]^2*r(x),x=0..1):
> l2:=int(1*phi[n]*r(x),x=0..1):
> vars:={sin(n*Pi)=0}:
> l1:=subs(vars,l1):
> l2:=subs(vars,l2):
> A[n]:=l2/l1/sinh(n*Pi);
```

$$A_n := \frac{\begin{cases} 0 & n=0 \\ -\dfrac{\cos(n\pi)-1}{n\pi} & otherwise \end{cases}}{\left(\begin{cases} 0 & n=0 \\ \dfrac{1}{2} & otherwise \end{cases}\right) \sinh(n\pi)}$$

```
> u(x,y):=eval(u(x,y));
```

$$u(x,y) := \sum_{n=1}^{\infty} \frac{\begin{cases} 0 & n=0 \\ -\dfrac{\cos(n\pi)-1}{n\pi} & otherwise \end{cases} \sin(n\pi x) \sinh(n\pi y)}{\left(\begin{cases} 0 & n=0 \\ \dfrac{1}{2} & otherwise \end{cases}\right) \sinh(n\pi)}$$

```
> u(x,y):=subs(infinity=N,u(x,y)):
> ua:=subs(N=20,u(x,y)):
> uu:=piecewise(y<0.99,ua,y>0.99,1);
```

$$uu := \begin{cases} \displaystyle\sum_{n=1}^{20} \frac{\begin{cases} 0 & n=0 \\ -\dfrac{\cos(n\pi)-1}{n\pi} & otherwise \end{cases} \sin(n\pi x) \sinh(n\pi y)}{\left(\begin{cases} 0 & n=0 \\ \dfrac{1}{2} & otherwise \end{cases}\right) \sinh(n\pi)} & y<0.99 \\ 1 & 0.99<y \end{cases}$$

```
> plot3d(evalf(uu),x=0..1,y=1..0,axes=boxed,title="Figure 7.26",
labels=[x,y,"u"],orientation=[-120,60]);
```

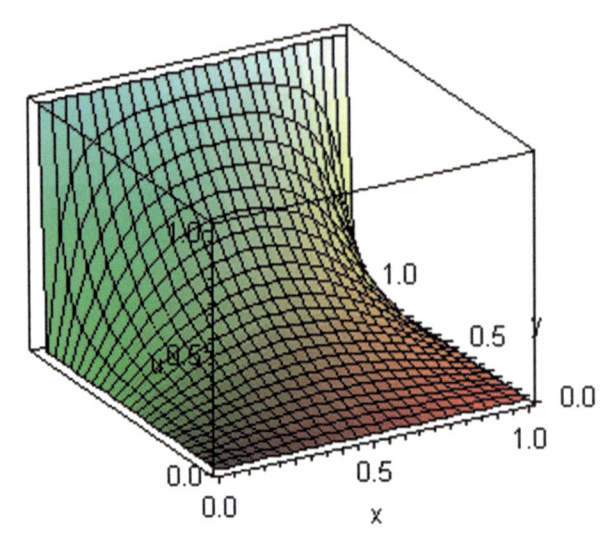

654   7 Partial Differential Equations in Finite Domains

Fig. 7.26

```
> plot([subs(y=0,uu),subs(y=0.4,uu),subs(y=0.6,uu),subs(y=0.8,uu),
subs(y=0.9,uu),subs(y=1,uu)],x=0..1,title="Figure 7.27",
axes=boxed,thickness=5,labels=[x,"u"],legend=["y=0","y=0.4","y=0.6",
"y=0.8","y=0.9","y=1.0"]);
```

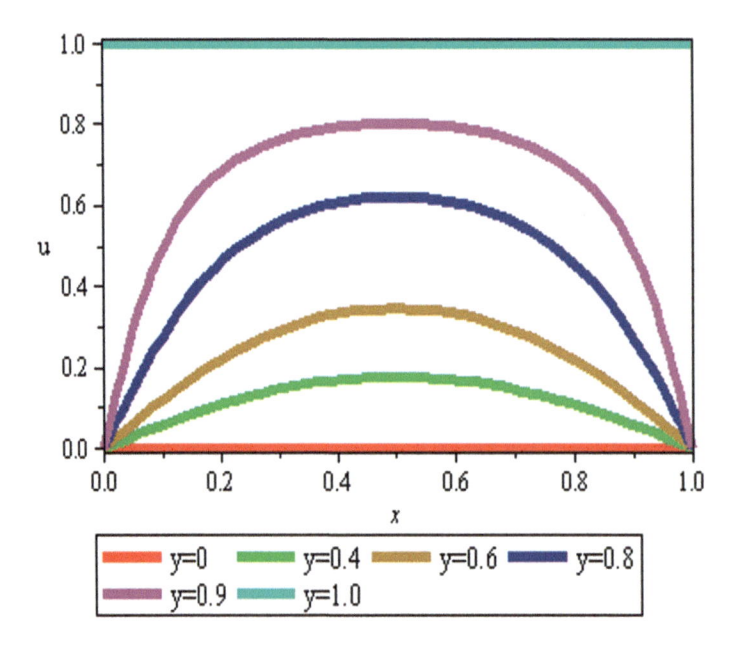

Fig. 7.27

7.1 Separation of Variables Method for PDEs in Finite Domains

Example 7.12. Diffusion in a Cylinder

Consider the steady state diffusion in a cylinder

$$\frac{\partial^2 u}{\partial y^2} + \frac{\partial^2 u}{\partial x^2} + \frac{1}{x}\frac{\partial u}{\partial x} = 0$$

$$\frac{\partial u}{\partial x}(0,y) = 0 \text{ and } u(1,y) = 0 \qquad (7.69)$$

$$u(x,0) = 1 \text{ and } \frac{\partial u}{\partial x}(x,1) = 0$$

The boundary condition at y = 0 (the nonhomogeneous boundary condition) is used to find the coefficient A_n in for this example. Also, this example involves calculating the eigenvalues from the transcendental equation as illustrated in Example 7.14. The Maple output for this problem is given below:

> restart:

> with(plots):

> eq:=diff(u(x,y),y$2)=-diff(u(x,y),x$2)-1/x*diff(u(x,y),x);

$$eq := \frac{\partial^2}{\partial y^2} u(x,y) = -\left(\frac{\partial^2}{\partial x^2} u(x,y)\right) - \frac{\frac{\partial}{\partial x} u(x,y)}{x}$$

> bc1:=diff(u(x,y),x)=0;

$$bc1 := \frac{\partial}{\partial x} u(x,y) = 0$$

> bc2:=u(x,y)=0;

$$bc2 := u(x,y) = 0$$

> bc3:=u(x,y)=1;

$$bc3 := u(x,y) = 1$$

> bc4:=diff(u(x,y),x)=0;

$$bc4 := \frac{\partial}{\partial x} u(x,y) = 0$$

> Eq:=subs(u(x,y)=X(x)*Y(y),eq);

$$Eq := \frac{\partial^2}{\partial y^2}(X(x)\,Y(y)) = -\left(\frac{\partial^2}{\partial x^2}(X(x)\,Y(y))\right) - \frac{\frac{\partial}{\partial x}(X(x)\,Y(y))}{x}$$

> Eq:=expand(Eq/X(x)/Y(y));

$$Eq := \frac{\frac{d^2}{dy^2}Y(y)}{Y(y)} = -\frac{\frac{d^2}{dx^2}X(x)}{X(x)} - \frac{\frac{d}{dx}X(x)}{X(x)\,x}$$

> Eq_Y:=lhs(Eq)=lambda^2:
> Eq_Y:=eval(Eq_Y*Y(y));

$$Eq\_Y := \frac{d^2}{dy^2}Y(y) = Y(y)\,\lambda^2$$

> dsolve(Eq_Y,Y(y));

$$Y(y) = \_C1\,e^{-\lambda y} + \_C2\,e^{\lambda y}$$

> Y(y):=C[1]*sinh(lambda*y)+C[2]*cosh(lambda*y);

$$Y(y) := C_1 \sinh(\lambda y) + C_2 \cosh(\lambda y)$$

> Bc4:=diff(Y(y),y)=0;

$$Bc4 := C_1 \cosh(\lambda y)\,\lambda + C_2 \sinh(\lambda y)\,\lambda = 0$$

> Eq_Bc4:=eval(subs(y=1,Bc4));

$$Eq\_Bc4 := C_1 \cosh(\lambda)\,\lambda + C_2 \sinh(\lambda)\,\lambda = 0$$

> C[2]:=solve(Eq_Bc4,C[2]);

$$C_2 := -\frac{C_1 \cosh(\lambda)}{\sinh(\lambda)}$$

> Y(y):=eval(Y(y));

$$Y(y) := C_1 \sinh(\lambda y) - \frac{C_1 \cosh(\lambda)\cosh(\lambda y)}{\sinh(\lambda)}$$

> combine(Y(y));

$$-\frac{C_1 \cosh(-\lambda + \lambda y)}{\sinh(\lambda)}$$

7.1 Separation of Variables Method for PDEs in Finite Domains 657

```
> Y(y):=C[1]*cosh(lambda*(1-y))/sinh(lambda);
```

$$Y(y) := \frac{C_1 \cosh(\lambda\,(1-y))}{\sinh(\lambda)}$$

```
> Eq_X:=rhs(Eq)=lambda^2:
> Eq_X:=expand(Eq_X*X(x));
```

$$Eq\_X := -\left(\frac{d^2}{dx^2} X(x)\right) - \frac{\frac{d}{dx} X(x)}{x} = X(x)\,\lambda^2$$

```
> dsolve({Eq_X},X(x));
```

$$\{X(x) = \_C1\ \text{BesselJ}(0,\lambda\,x) + \_C2\ \text{BesselY}(0,\lambda\,x)\}$$

```
> X(x):=c[1]*BesselJ(0,lambda*x)+c[2]*BesselY(0,lambda*x);
```

$$X(x) := c_1\ \text{BesselJ}(0,\lambda\,x) + c_2\ \text{BesselY}(0,\lambda\,x)$$

```
> Bc1:=diff(X(x),x)=0;
```

$$Bc1 := -c_1\ \text{BesselJ}(1,\lambda\,x)\,\lambda - c_2\ \text{BesselY}(1,\lambda\,x)\,\lambda = 0$$

```
> Bc2:=X(x)=0;
```

$$Bc2 := c_1\ \text{BesselJ}(0,\lambda\,x) + c_2\ \text{BesselY}(0,\lambda\,x) = 0$$

```
> BesselJ(0,0);
```

$$1$$

```
> BesselY(0,0);
```

`Error, (in BesselY) numeric exception: division by zero`

```
> c[2]:=0;
```

$$c_2 := 0$$

```
> Eq_Bc2:=eval(subs(x=1,Bc2));
```

$$Eq\_Bc2 := c_1\ \text{BesselJ}(0,\lambda) = 0$$

```
> Eq_Eig:=BesselJ(0,lambda)=0;
```

$$Eq\_Eig := \text{BesselJ}(0,\lambda) = 0$$

658   7 Partial Differential Equations in Finite Domains

> l[1]:=fsolve(Eq_Eig,lambda=0..3);

$$l_1 := 2.404825558$$

> N:=20;

$$N := 20$$

> for i from 2 to N do l[i]:=fsolve(Eq_Eig,lambda=l[i-1]..l[i-1]+4);od:
> seq(l[i],i=1..N);

2.404825558, 5.520078110, 8.653727913, 11.79153444, 14.93091771, 18.07106397, 21.21163663, 24.35247153,

27.49347913, 30.63460647, 33.77582021, 36.91709835, 40.05842576, 43.19979171, 46.34118837, 49.48260990,

52.62405184, 55.76551076, 58.90698393, 62.04846919

> U:=eval(X(x)*Y(y)):
> Un:=subs(c[1]=A[n]/C[1],lambda=lambda[n],U);

$$Un := \frac{A_n \, \text{BesselJ}(0, \lambda_n \, x) \, \cosh(\lambda_n \, (1-y))}{\sinh(\lambda_n)}$$

> u(x,y):=Sum(Un,n=1..N):
> eq_An:=eval(subs(y=0,u(x,y)))=rhs(bc3);

$$eq\_An := \sum_{n=1}^{20} \frac{A_n \, \text{BesselJ}(0, \lambda_n \, x) \, \cosh(\lambda_n)}{\sinh(\lambda_n)} = 1$$

> phi[n]:=BesselJ(0,lambda[n]*x);

$$\phi_n := \text{BesselJ}(0, \lambda_n \, x)$$

> r(x):=x;

$$r(x) := x$$

> I1:=int(phi[n]^2*r(x),x=0..1):
> I2:=int(1*phi[n]*r(x),x=0..1):
> A[n]:=I2/I1/cosh(lambda[n])*sinh(lambda[n]);

$$A_n := \frac{2 \, \lambda_n \, \text{BesselJ}(1, \lambda_n) \, \sqrt{\pi} \, \sinh(\lambda_n)}{\left(\sqrt{\pi} \, \lambda_n^2 \, \text{BesselJ}(0, \lambda_n)^2 + \sqrt{\pi} \, \lambda_n^2 \, \text{BesselJ}(1, \lambda_n)^2\right) \cosh(\lambda_n)}$$

7.1 Separation of Variables Method for PDEs in Finite Domains

> u(x,y):=eval(u(x,y));

$$u(x,y) := \sum_{n=1}^{20} \frac{2\lambda_n \, \text{BesselJ}(1,\lambda_n) \sqrt{\pi} \, \text{BesselJ}(0,\lambda_n x) \cosh(\lambda_n (1-y))}{\left(\sqrt{\pi} \, \lambda_n^2 \, \text{BesselJ}(0,\lambda_n)^2 + \sqrt{\pi} \, \lambda_n^2 \, \text{BesselJ}(1,\lambda_n)^2\right) \cosh(\lambda_n)}$$

> for i to N do lambda[i]:=l[i];od:
> ua:=evalf(u(x,y)):
> uu:=piecewise(y=0,1,y>0,ua):
> plot3d(evalf(uu),x=0..1,y=1..0,axes=boxed,title="Figure 7.28",
labels=[x,y,"u"],orientation=[30,60]);

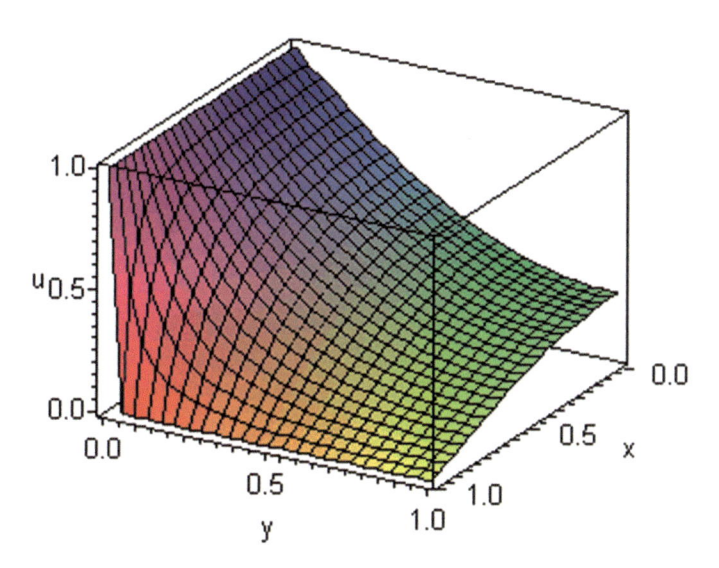

Fig. 7.28

> plot([subs(y=0,uu),subs(y=0.1,uu),subs(y=0.2,uu),subs(y=0.3,uu),
subs(y=0.4,uu),subs(y=1,uu)],x=0..1,axes=boxed,title="Figure 7.29",
thickness=5,labels=[x,"u"],legend=["y=0","y=0.1","y=0.2","y=0.3","y=0.4",
"y=1.0"]);

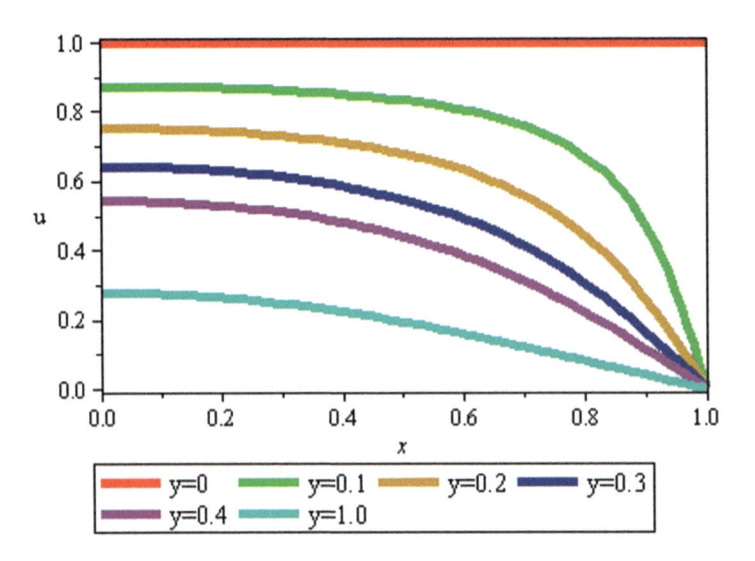

Fig. 7.29

Example 7.13. Heat Transfer with Nonhomogeneous Boundary Conditions

Consider the steady state heat transfer problem with nonhomogeneous boundary conditions in both x and y

$$\frac{\partial^2 u}{\partial y^2} + \frac{\partial^2 u}{\partial x^2} = 0$$

$$u(0,y) = 0 \text{ and } u(1,y) = 1 \quad (7.70)$$

$$u(x,0) = 0 \text{ and } u(x,1) = 1$$

If we define $u = v(x,y) + w(x,y)$ then both v and w satisfy the Laplace equation

$$\frac{\partial^2 v}{\partial y^2} + \frac{\partial^2 v}{\partial x^2} = 0 \text{ and } \frac{\partial^2 w}{\partial y^2} + \frac{\partial^2 w}{\partial x^2} = 0 \quad (7.71)$$

where v(x,y) has to satisfy only one nonhomogeneous boundary condition (at y = 1):

$$v(0,y) = 0 \text{ and } v(1,y) = 1$$

$$(7.72)$$

$$v(x,0) = 0 \text{ and } v(x,1) = 0$$

7.1 Separation of Variables Method for PDEs in Finite Domains

and w(x,y) has to satisfy only one nonhomogeneous boundary condition (at x = 1):

$$w(0,y) = 0 \text{ and } w(1,y) = 0$$
(7.73)
$$w(x,0) = 0 \text{ and } w(x,1) = 1$$

The final solution u = v + w satisfies nonhomogeneous boundary conditions at both y = 1 and x = 1. The solution for v was derived in Example 7.11. The solution for w can be obtained by interchanging x and y in the solution for w. The Maple output for this example is given below:

> restart:

> with(plots):

> eq:=diff(v(x,y),y$2)=-diff(v(x,y),x$2);

$$eq := \frac{\partial^2}{\partial y^2} v(x,y) = -\left(\frac{\partial^2}{\partial x^2} v(x,y)\right)$$

> bc1:=v(x,y)=0;

$$bc1 := v(x,y) = 0$$

> bc2:=v(x,y)=0;

$$bc2 := v(x,y) = 0$$

> bc3:=v(x,y)=0;

$$bc3 := v(x,y) = 0$$

> bc4:=u(x,y)=1;

$$bc4 := u(x,y) = 1$$

> Eq:=subs(v(x,y)=X(x)*Y(y),eq);

$$Eq := \frac{\partial^2}{\partial y^2}(X(x)Y(y)) = -\left(\frac{\partial^2}{\partial x^2}(X(x)Y(y))\right)$$

> Eq:=expand(Eq/X(x)/Y(y));

$$Eq := \frac{\frac{d^2}{dy^2}Y(y)}{Y(y)} = -\frac{\frac{d^2}{dx^2}X(x)}{X(x)}$$

> Eq_Y:=lhs(Eq)=lambda^2:

> Eq_Y:=eval(Eq_Y*Y(y));

$$Eq\_Y := \frac{d^2}{dy^2} Y(y) = Y(y) \lambda^2$$

> dsolve(Eq_Y,Y(y));

$$Y(y) = C1\, e^{-\lambda y} + C2\, e^{\lambda y}$$

> Y(y):=C[1]*sinh(lambda*y)+C[2]*cosh(lambda*y);

$$Y(y) := C_1 \sinh(\lambda y) + C_2 \cosh(\lambda y)$$

> Bc3:=Y(y)=0;

$$Bc3 := C_1 \sinh(\lambda y) + C_2 \cosh(\lambda y) = 0$$

> Eq_Bc3:=eval(subs(y=0,Bc3));

$$Eq\_Bc3 := C_2 = 0$$

> C[2]:=solve(Eq_Bc3);

$$C_2 := 0$$

> Y(y):=eval(Y(y));

$$Y(y) := C_1 \sinh(\lambda y)$$

> Eq_X:=rhs(Eq)=lambda^2:
> Eq_X:=expand(Eq_X*X(x));

$$Eq\_X := -\left(\frac{d^2}{dx^2} X(x)\right) = X(x) \lambda^2$$

> dsolve({Eq_X},X(x));

$$\{X(x) = \_C1 \sin(\lambda x) + \_C2 \cos(\lambda x)\}$$

> X(x):=c[1]*sin(lambda*x)+c[2]*cos(lambda*x);

$$X(x) := c_1 \sin(\lambda x) + c_2 \cos(\lambda x)$$

> Bc1:=X(x)=0;

$$Bc1 := c_1 \sin(\lambda x) + c_2 \cos(\lambda x) = 0$$

> Bc2:=X(x)=0;

$$Bc2 := c_1 \sin(\lambda x) + c_2 \cos(\lambda x) = 0$$

> Eq_Bc1:=eval(subs(x=0,Bc1));

$$Eq\_Bc1 := c_2 = 0$$

7.1 Separation of Variables Method for PDEs in Finite Domains 663

> c[2]:=solve(Eq_Bc1,c[2]);

$$c_2 := 0$$

> Eq_Bc2:=eval(subs(x=1,Bc2));

$$Eq\_Bc2 := c_1 \sin(\lambda) = 0$$

> Eq_Eig:=sin(lambda)=0;

$$Eq\_Eig := \sin(\lambda) = 0$$

> _EnvAllSolutions := true:

> solve(Eq_Eig,lambda);

$$\pi \_Z1\~$$

> V:=eval(X(x)*Y(y));

$$V := c_1 \sin(\lambda x) \, C_1 \sinh(\lambda y)$$

> Vn:=subs(c[1]=A[n]/C[1],lambda=lambda[n],V):

> v(x,y):=Sum(Vn,n=1..infinity):

> v(x,y):=subs(lambda[n]=n*Pi,v(x,y));

$$v(x,y) := \sum_{n=1}^{\infty} A_n \sin(n \pi x) \sinh(n \pi y)$$

> eq_An:=eval(subs(y=1,v(x,y)))=rhs(bc4);

$$eq\_An := \sum_{n=1}^{\infty} A_n \sin(n \pi x) \sinh(n \pi) = 1$$

> phi[n]:=sin(n*Pi*x);

$$\phi_n := \sin(n \pi x)$$

> r(x):=1;

$$r(x) := 1$$

> I1:=int(phi[n]^2*r(x),x=0..1):

> I2:=int(1*phi[n]*r(x),x=0..1):

> vars:={sin(n*Pi)=0}:

> I1:=subs(vars,I1):

> l2:=subs(vars,l2):

> A[n]:=l2/l1/sinh(n*Pi);

$$A_n := \frac{\begin{cases} 0 & n=0 \\ -\dfrac{\cos(n\pi)-1}{n\pi} & \text{otherwise} \end{cases}}{\left(\begin{cases} \dfrac{1}{2} & n<0 \\ 0 & n=0 \\ \dfrac{1}{2} & 0<n \end{cases}\right)\sinh(n\pi)}$$

> v(x,y):=eval(v(x,y));

$$v(x,y) := \sum_{n=1}^{\infty} \frac{\begin{cases} 0 & n=0 \\ -\dfrac{\cos(n\pi)-1}{n\pi} & \text{otherwise} \end{cases} \sin(n\pi x)\sinh(n\pi y)}{\left(\begin{cases} 0 & n=0 \\ \dfrac{1}{2} & \text{otherwise} \end{cases}\right)\sinh(n\pi)}$$

> v(x,y):=subs(infinity=N,v(x,y)):

> va:=subs(N=20,v(x,y));

$$va := \sum_{n=1}^{20} \frac{\begin{cases} 0 & n=0 \\ -\dfrac{\cos(n\pi)-1}{n\pi} & \text{otherwise} \end{cases} \sin(n\pi x)\sinh(n\pi y)}{\left(\begin{cases} 0 & n=0 \\ \dfrac{1}{2} & \text{otherwise} \end{cases}\right)\sinh(n\pi)}$$

> wa:=subs(x=Y,y=X,va):wa:=subs(X=x,Y=y,wa);

$$wa := \sum_{n=1}^{20} \frac{\begin{cases} 0 & n=0 \\ -\dfrac{\cos(n\pi)-1}{n\pi} & \text{otherwise} \end{cases} \sin(n\pi y)\sinh(n\pi x)}{\left(\begin{cases} 0 & n=0 \\ \dfrac{1}{2} & \text{otherwise} \end{cases}\right)\sinh(n\pi)}$$

7.1 Separation of Variables Method for PDEs in Finite Domains 665

> ua:=va+wa;

$$ua := \sum_{n=1}^{20} \frac{\left(\left\{\begin{array}{ll} 0 & n=0 \\ -\frac{\cos(n\pi)-1}{n\pi} & \text{otherwise} \end{array}\right| \sin(n\pi x)\sinh(n\pi y)\right)}{\left(\left\{\begin{array}{ll} 0 & n=0 \\ \frac{1}{2} & \text{otherwise} \end{array}\right| \sinh(n\pi)\right)} + \sum_{n=1}^{20} \frac{\left(\left\{\begin{array}{ll} 0 & n=0 \\ -\frac{\cos(n\pi)-1}{n\pi} & \text{otherwise} \end{array}\right| \sin(n\pi y)\sinh(n\pi x)\right)}{\left(\left\{\begin{array}{ll} \frac{1}{2} & n<0 \\ 0 & n=0 \\ \frac{1}{2} & 0<n \end{array}\right| \sinh(n\pi)\right)}$$

> plot(subs(x=1,ua),y=0..1,thickness=3,title="Figure 7.30",axes=boxed);

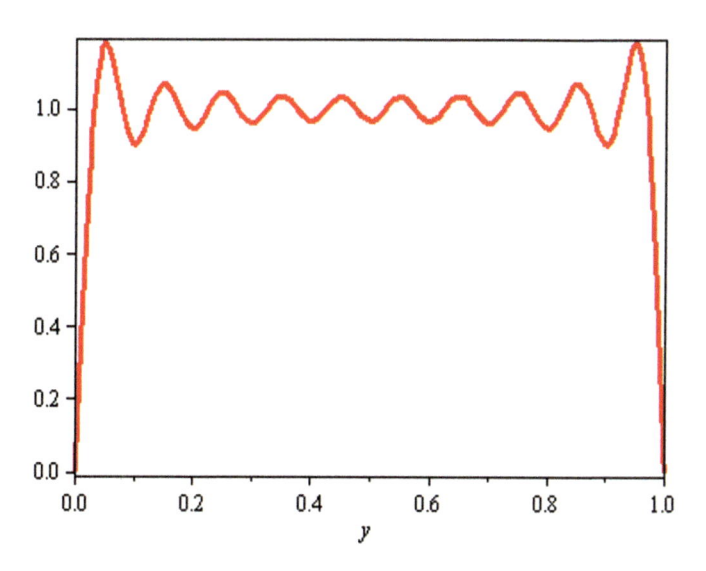

Fig. 7.30

666   7 Partial Differential Equations in Finite Domains

> uu:=piecewise(x=1,1,y=1,1,ua);

$$uu := \begin{cases} \sum_{n=1}^{20} \dfrac{\begin{cases} 0 & n=0 \\ -\dfrac{\cos(n\pi)-1}{n\pi} & \text{otherwise} \end{cases} \sin(n\pi x)\sinh(n\pi y)}{\begin{cases} 0 & n=0 \\ \dfrac{1}{2} & \text{otherwise} \end{cases} \sinh(n\pi)} + \sum_{n=1}^{20} \dfrac{\begin{cases} 0 & n=0 \\ -\dfrac{\cos(n\pi)-1}{n\pi} & \text{otherwise} \end{cases} \sin(n\pi y)\sinh(n\pi x)}{\begin{cases} 0 & n=0 \\ \dfrac{1}{2} & \text{otherwise} \end{cases} \sinh(n\pi)} & \text{otherwise} \\ 1 & x=1 \\ 1 & y=1 \end{cases}$$

> plot3d(evalf(uu),x=0..1,y=1..0,axes=boxed,title="Figure 7.31",
labels=[x,y,"u"],orientation=[-120,60],view=[0..1,0..1,0..1]);

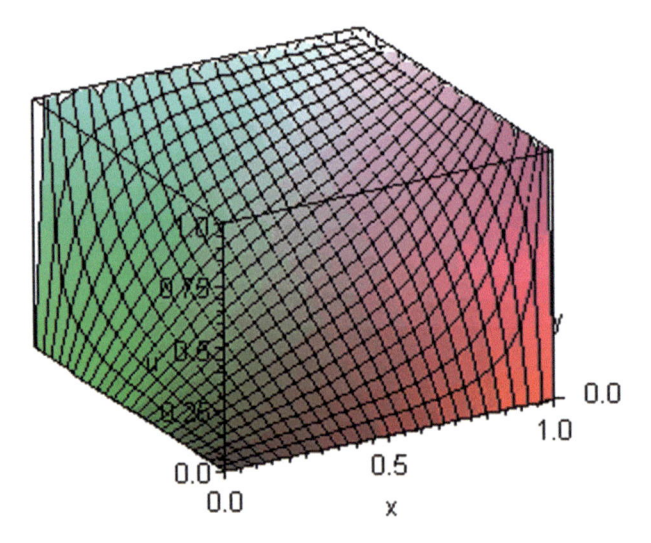

Fig. 7.31

> plot([subs(y=0,uu),subs(y=0.4,uu),subs(y=0.6,uu),subs(y=0.8,uu),subs
(y=0.9,uu),subs(y=1,uu)],x=0..1,axes=boxed,title="Figure 32",
thickness=5,labels=[x,"u"],legend=["y=0","y=0.4","y=0.6","y=0.8","y=0.9",
"y=1.0"]);

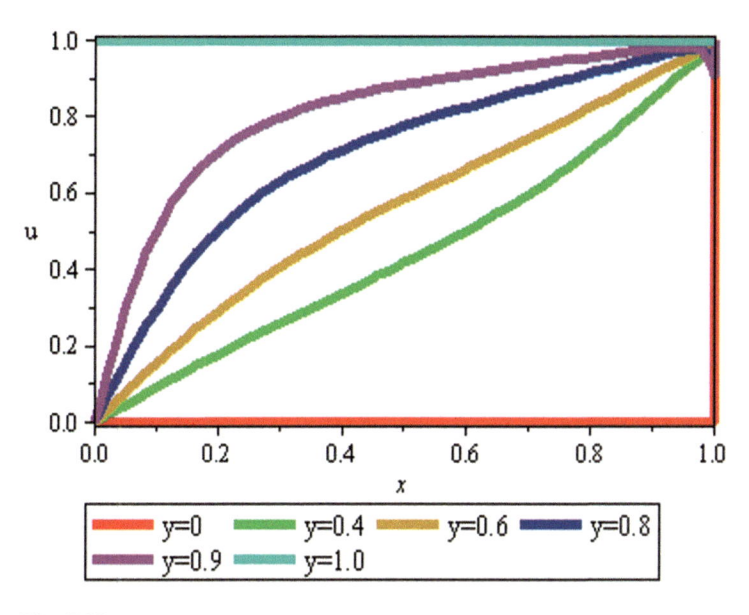

7.1 Separation of Variables Method for PDEs in Finite Domains

Fig. 7.32

Example 7.14. Heat Transfer with a Nonhomogeneous Governing Equation

> restart:

> with(plots):

> eq:=diff(u(x,y),y$2)=-diff(u(x,y),x$2)-1;

$$eq := \frac{\partial^2}{\partial y^2} u(x,y) = -\left(\frac{\partial^2}{\partial x^2} u(x,y)\right) - 1$$

> bc1:=u(x,y)=0;

$$bc1 := u(x,y) = 0$$

> bc2:=u(x,y)=0;

$$bc2 := u(x,y) = 0$$

> bc3:=u(x,y)=0;

$$bc3 := u(x,y) = 0$$

> bc4:=u(x,y)=0;

$$bc4 := u(x,y) = 0$$

```
> eq:=expand(subs(u(x,y)=g(x,y)+w(x),eq));
```

$$eq := \frac{\partial^2}{\partial y^2} g(x,y) = -\left(\frac{\partial^2}{\partial x^2} g(x,y)\right) - \left(\frac{d^2}{dx^2} w(x)\right) - 1$$

```
> eq_w:=diff(w(x),`$`(x,2))+1=0;
```

$$eq\_w := \frac{d^2}{dx^2} w(x) + 1 = 0$$

```
> w(x):=rhs(dsolve({eq_w,w(0)=0,w(1)=0},w(x)));
```

$$w(x) := -\frac{1}{2} x^2 + \frac{1}{2} x$$

```
> eqg:=diff(g(x,y),`$`(y,2)) = -diff(g(x,y),`$`(x,2));
```

$$eqg := \frac{\partial^2}{\partial y^2} g(x,y) = -\left(\frac{\partial^2}{\partial x^2} g(x,y)\right)$$

```
> Eq:=subs(g(x,y)=X(x)*Y(y),eqg):
> Eq:=expand(Eq/X(x)/Y(y)):
> Eq_Y:=lhs(Eq)=lambda^2:
> Eq_Y:=eval(Eq_Y*Y(y)):
> dsolve(Eq_Y,Y(y)):
> Y(y):=C[1]*sinh(lambda*y)+C[2]*cosh(lambda*y):
> Bc3:=Y(y)=0;
```

$$Bc3 := C_1 \sinh(\lambda y) + C_2 \cosh(\lambda y) = 0$$

```
> Eq_Bc3:=eval(subs(y=0,Bc3)):
> C[2]:=solve(Eq_Bc3):
> Y(y):=eval(Y(y));
```

$$Y(y) := C_1 \sinh(\lambda y)$$

```
> Eq_X:=rhs(Eq)=lambda^2:
> Eq_X:=expand(Eq_X*X(x)):
> dsolve({Eq_X},X(x)):
> X(x):=c[1]*sin(lambda*x)+c[2]*cos(lambda*x):
> Bc1:=X(x)=0;
```

$$Bc1 := c_1 \sin(\lambda x) + c_2 \cos(\lambda x) = 0$$

7.1 Separation of Variables Method for PDEs in Finite Domains 669

> Bc2:=X(x)=0;

$$Bc2 := c_1 \sin(\lambda x) + c_2 \cos(\lambda x) = 0$$

> Eq_Bc1:=eval(subs(x=0,Bc1)):

> c[2]:=solve(Eq_Bc1,c[2]);

$$c_2 := 0$$

> Eq_Bc2:=eval(subs(x=1,Bc2));

$$Eq\_Bc2 := c_1 \sin(\lambda) = 0$$

> Eq_Eig:=sin(lambda)=0;

$$Eq\_Eig := \sin(\lambda) = 0$$

> _EnvAllSolutions := true:

> solve(Eq_Eig,lambda);

$$\pi\_Z1\sim$$

> G:=eval(X(x)*Y(y));

$$G := c_1 \sin(\lambda x) \, C_1 \sinh(\lambda y)$$

> Gn:=subs(c[1]=A[n]/C[1],lambda=lambda[n],G):

> u(x,y):=Sum(Gn,n=1..infinity)+w(x):

> u(x,y):=subs(lambda[n]=n*Pi,u(x,y));

$$u(x,y) := \sum_{n=1}^{\infty} A_n \sin(n\pi x) \sinh(n\pi y) - \frac{1}{2}x^2 + \frac{1}{2}x$$

> eq_An:=eval(subs(y=1,u(x,y)))=rhs(bc4);

$$eq\_An := \sum_{n=1}^{\infty} A_n \sin(n\pi x) \sinh(n\pi) - \frac{1}{2}x^2 + \frac{1}{2}x = 0$$

> phi[n]:=sin(n*Pi*x);

$$\phi_n := \sin(n\pi x)$$

> r(x):=1;

$$r(x) := 1$$

> l1:=int(phi[n]^2*r(x),x=0..1):
> l2:=int((0-w(x))*phi[n]*r(x),x=0..1):
> vars:={sin(n*Pi)=0}:
> l1:=subs(vars,l1):
> l2:=subs(vars,l2):
> A[n]:=l2/l1/sinh(n*Pi);

$$A_n := \frac{\begin{cases} 0 & n=0 \\ \dfrac{1}{2}\dfrac{-2+2\cos(n\pi)}{n^3\pi^3} & \text{otherwise} \end{cases}}{\left(\begin{cases} 0 & n=0 \\ \dfrac{1}{2} & \text{otherwise} \end{cases}\right)\sinh(n\pi)}$$

> u(x,y):=eval(u(x,y));

$$u(x,y) := \sum_{n=1}^{\infty} \frac{\begin{cases} 0 & n=0 \\ \dfrac{1}{2}\dfrac{-2+2\cos(n\pi)}{n^3\pi^3} & \text{otherwise} \end{cases}\sin(n\pi x)\sinh(n\pi y)}{\left(\begin{cases} 0 & n=0 \\ \dfrac{1}{2} & \text{otherwise} \end{cases}\right)\sinh(n\pi)} - \frac{1}{2}x^2 + \frac{1}{2}x$$

> u(x,y):=subs(infinity=N,u(x,y)):
> ua:=subs(N=20,u(x,y)):
> uu:=piecewise(y<0.9999,ua,y>0.9999,0);

$$uu := \begin{cases} \displaystyle\sum_{n=1}^{20} \frac{\begin{cases} 0 & n=0 \\ \dfrac{1}{2}\dfrac{-2+2\cos(n\pi)}{n^3\pi^3} & \text{otherwise} \end{cases}\sin(n\pi x)\sinh(n\pi y)}{\left(\begin{cases} 0 & n=0 \\ \dfrac{1}{2} & \text{otherwise} \end{cases}\right)\sinh(n\pi)} - \dfrac{1}{2}x^2 + \dfrac{1}{2}x & y<0.9999 \\ 0 & 0.9999<y \end{cases}$$

> plot3d(evalf(uu),x=0..1,y=1..0,axes=boxed,title="Figure 7.33",
labels=[x,y,"u"],orientation=[-120,60]);

7.1 Separation of Variables Method for PDEs in Finite Domains 671

Fig. 7.33

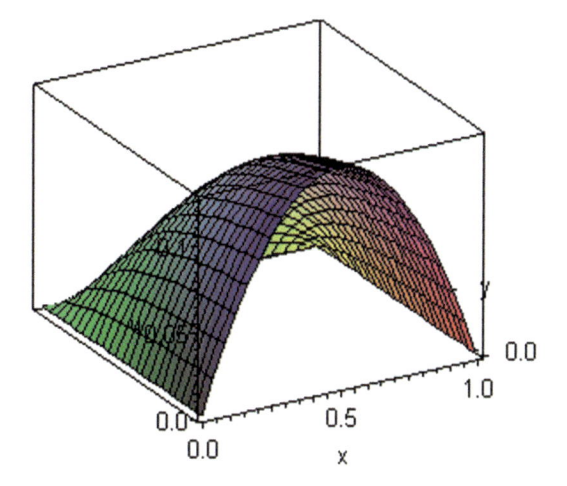

```
> plot([subs(y=0,uu),subs(y=0.6,uu),subs(y=0.8,uu),subs(y=0.9,uu),
subs(y=0.95,uu),subs(y=1,uu)],x=0..1,axes=boxed,title="Figure 7.34",
thickness=5,labels=[x,"u"],legend=["y=0","y=0.6","y=0.8","y=0.9","y=0.95",
"y=1.0"]);
```

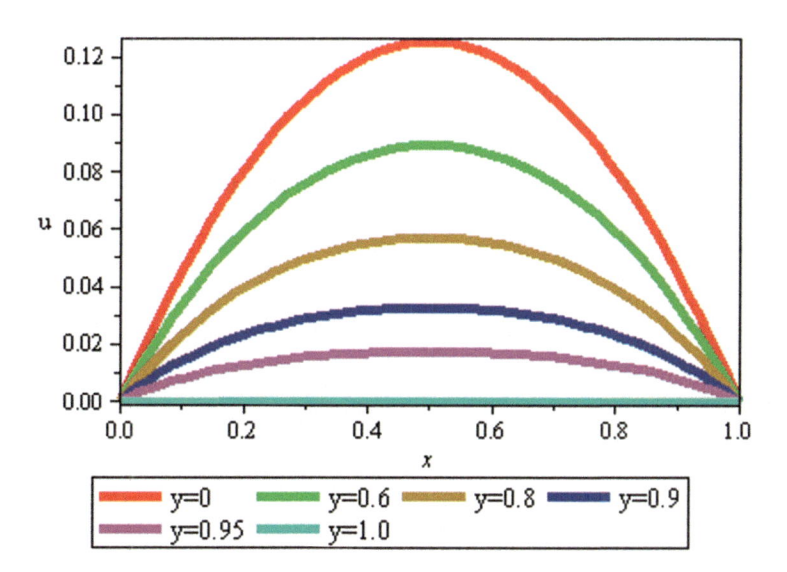

Fig. 7.34

7.1.9 Summary

In this chapter, analytical solutions were obtained for parabolic and elliptic partial differential equations in finite domains using separation of variables method. In section 7.1.2, a linear parabolic partial differential equation with homogeneous boundary conditions was solved using the separation of variables method. The dependent variable was assumed to be a product of two separate functions of x and t. These functions were then solved using the corresponding boundary and initial conditions. The final solution was obtained by superposition of individual solutions. In section 7.1.3 this method was then extended to problems with an initial profile in x. In sections 7.1.2 and 7.1.3 analytical explicit relations were obtained for the eigenvalues. In section 7.1.4 eigenvalues were obtained from a nonlinear implicit transcendental equation. In section 7.1.5, the method was extended to parabolic partial differential equations with nonhomogeneous boundary conditions. In section 7.1.6, the method was extended to parabolic partial differential equations with nonhomogeneous flux boundary conditions (at both of the boundaries).

In section 7.1.7, eigenfunctions and eigenvalues were obtained numerically. This method is very general and can be used to avoid the use of complicated special function solutions. In section 7.1.8, the separation of variables method which was illustrated earlier for parabolic partial differential equations was extended to elliptic partial differential equations. A total of fourteen examples were presented in this chapter.

7.1.10 Exercise Problems

1. Complete the details missing in Example 7.5.
2. Obtain an analytical solution for Example 7.5 if the initial condition is replaced by $u(x,0) = 1-x^m$, where m is an integer.
3. Redo Example 7.4 if the initial condition is given by the piecewise function

$$u(x,0) = x \quad 0 \le x \le 0.5$$
$$= 1 - x \quad 0.5 < x \le 1$$

4. Solve the following parabolic PDE using the separation of variables method:

$$\frac{\partial u}{\partial t} = \frac{\partial^2 u}{\partial x^2}$$

$$u(x,0) = \cos\left(\frac{\pi x}{2}\right)$$

$$\frac{\partial u}{\partial x}(0,t) = 0 \text{ and } u(1,t) = 0$$

7.1 Separation of Variables Method for PDEs in Finite Domains

5. Obtain an analytical solution for Example 5.2 using the separation of variables method and plot the profiles.
6. Obtain an analytical solution for the Graetz problem described in Example 5.6. If you are not able to find the eigenfunction and eigenvalues analytically, find them numerically.
7. Consider diffusion with convection in a coated wall reactor where the reaction takes place at the wall (Rice and Do, 1995;[2] chapter 5.1, exercise problem 1). The governing equation and boundary conditions for concentration in dimensionless form are:

$$\frac{\partial u}{\partial Z} = \frac{\partial^2 u}{\partial x^2} + \frac{1}{x}\frac{\partial u}{\partial x}$$

$$\frac{\partial u}{\partial x}(0,Z) = 0 \text{ and } \frac{\partial u}{\partial x}(1,Z) + \text{Ha } u(1,Z) = 0$$

$$u(x,0) = 1$$

where Ha is the Hatta number. Obtain an analytical solution for this problem using the separation of variables method.

8. Consider the cooling of spherical nuclear pellets (Rice and Do, 1995;[2] chapter 5.1, exercise problem 7). The dimensionless temperature distribution is governed by:

$$\frac{\partial u}{\partial t} = \frac{\partial^2 u}{\partial x^2} + \frac{2}{x}\frac{\partial u}{\partial x} + Q$$

$$\frac{\partial u}{\partial x}(0,t) = 0 \text{ and } \frac{\partial u}{\partial x}(1,t) + \text{Bi } u(1,t) = 0$$

$$u(x,0) = 1$$

where Q is the ratio of heat generation to heat conduction and Bi is the Biot number. Obtain an analytical solution for this problem using the separation of variables method. Plot the profiles for Q = 1, Bi = 0.2 and Q = 1, Bi = 10.

9. Consider dispersion of a linear kinematic wave in dimensionless form (Aris, 1999;[5] chapter 5.1, exercise problem 9). The governing equation and boundary/initial conditions are:

$$\frac{\partial u}{\partial t} = \frac{\partial^2 u}{\partial x^2} - \text{Pe}\frac{\partial u}{\partial x}$$

$$u(0,t) = 1; u(1,t) = 0$$

$$u(x,0) = 0$$

Obtain an analytical solution for this problem using the separation of variables method. Plot the profiles for Pe = 1, 10 and 50.

10. Consider the fluid-flow problem (Davis, 1984;[6] chapter 5.1, exercise problem 10):

$$\frac{\partial u}{\partial t} = \frac{\partial^2 u}{\partial x^2} + \frac{1}{x}\frac{\partial u}{\partial x} + 4$$

$$\frac{\partial u}{\partial x}(0,t) = 0 \text{ and } u(1,t) = 0$$

$$u(x,0) = 0$$

Obtain an analytical solution using separation of variables method and plot the dimensionless velocity profiles.

11. Consider the Graetz problem in planar geometry (chapter 5.1, exercise problem 11). The governing equations and boundary/initial conditions are:

$$2Pe(1-x^2)\frac{\partial u}{\partial z} = \frac{\partial^2 u}{\partial x^2}$$

$$u(x,0) = 0$$

$$\frac{\partial u}{\partial x}(0,z) = 0 \text{ and } u(1,z) = 1$$

Solve this problem analytically and plot the profiles for different values of Peclet number. Hint: if you can't find the eigenfunction and eigenvalues analytically, find them numerically.

12. Consider heat conduction in a slab with radiation at both ends (Carslaw and Jaeger, 1959[1]; chapter 5.1, exercise problem 12). The dimensionless governing equations and boundary/initial conditions are:

$$\frac{\partial u}{\partial t} = \frac{\partial^2 u}{\partial x^2}$$

$$-\frac{\partial u}{\partial x}(0,t) + Hu(0,t) = 0 \text{ and } \frac{\partial u}{\partial x}(1,t) + Hu(1,t) = 0$$

$$u(x,0) = 1$$

where H is the dimensionless heat transfer coefficient. Obtain an analytical solution and plot the profiles for H = 1, 10.

13. Consider the particle electrode problem discussed in Example 7.6. Example 7.6 describes the charging of a particle. The governing equations for electrochemical discharge of a particle electrode are:

7.1 Separation of Variables Method for PDEs in Finite Domains   675

$$\frac{\partial u}{\partial t} = \frac{\partial^2 u}{\partial x^2}$$

$$u(x,0) = 1$$

$$\frac{\partial u}{\partial x}(0,t) = 0 \text{ and } \frac{\partial u}{\partial x}(1,t) = -\delta$$

Explain how one can find an analytical solution for this problem using the solution obtained in Example 7.9. Plot the profiles for $\delta = 0.1, 1, 2$ and 5.

14. The electrochemical discharge of spherical particles is very similar to problem 13 (Subramanian and White, 2001[4]). Governing equations and boundary/initial conditions for this problem are:

$$\frac{\partial u}{\partial t} = \frac{\partial^2 u}{\partial x^2} + \frac{2}{x}\frac{\partial u}{\partial x}$$

$$u(x,0) = 1$$

$$\frac{\partial u}{\partial x}(0,t) = 0 \text{ and } \frac{\partial u}{\partial x}(1,t) = -\delta$$

Obtain an analytical solution for this problem using the separation of variables method. Plot the profiles for $\delta = 0.1, 1, 2$ and 5.

15. Consider problem 13 and 14. How does the governing equation change for cylindrical coordinates? Obtain an analytical solution for the cylindrical geometry with the same boundary/initial conditions in problem 14 and 15 using the separation of variables method and plot the profiles for $\delta = 0.1, 1, 2$ and 5.

16. Consider the diffusion reaction discussed in Example 7.8. A similar equation governs the overpotential during galvanostatic charge/discharge of porous electrodes in the absence of concentration gradients:[7]

$$\frac{\partial u}{\partial t} = \frac{\partial^2 u}{\partial x^2} - v^2 u$$

$$u(x,0) = 0$$

$$\frac{\partial u}{\partial x}(0,t) = \delta \text{ and } \frac{\partial u}{\partial x}(1,t) = -\delta\beta$$

where v is the modified exchange current density, δ is the applied current density and β is the ratio of solution phase conductivity to solid phase conductivity (usually < 1). Obtain an analytical solution for this problem using the separation of variables method and plot the profiles.

17. Solve the following Poison's equation using the separation of variables method and plot the profiles:

$$\frac{\partial^2 u}{\partial x^2} + \frac{\partial^2 u}{\partial y^2} = -1$$

$$u(0,y) = 0;\ u(1,y) = 0$$
$$u(x,0) = 0;\ u(x,1) = 0$$

18. Solve the following Laplace equation with nonhomogeneous two flux boundary conditions and plot the profiles:

$$\frac{\partial^2 u}{\partial y^2} + \frac{\partial^2 u}{\partial x^2} = 0$$

$$\frac{\partial u}{\partial x}(0,y) = 0 \text{ and } \frac{\partial u}{\partial x}(0,y) = 1$$

$$u(x,0) = 0 \text{ and } \frac{\partial u}{\partial x}(x,1) = 0$$

19. Consider the Graetz problem with axial conduction.[8] [7] The governing equation is:

$$2\mathrm{Pe}(1-r^2)\frac{\partial T}{\partial z} = \frac{\partial^2 T}{\partial r^2} + \frac{1}{r}\frac{\partial T}{\partial r} + \frac{\partial^2 T}{\partial z^2}$$

with the following boundary conditions

$$\frac{\partial T}{\partial r}(0,z) = 0 \text{ for } 0 \leq z \leq z_L$$

$$T(1,z) = 1 \text{ for } 0 \leq z \leq z_L$$

$$T(r,0) = 0 \text{ for } 0 \leq r < 1$$

and

$$\frac{\partial T}{\partial z}(r, z_L) = 0 \quad \text{for } 0 \leq r \leq 1$$

Obtain an analytical solution for this problem and plot the temperature distribution for Pe = 10 and z_L = 2. Hint: if you can't find the eigenfunction and eigenvalues analytically, find them numerically.

20. Consider the transient diffusion problem in a composite plate consisting of two regions of different conductivities. The governing equations for

7.1 Separation of Variables Method for PDEs in Finite Domains

dimensional concentration in region I ($0 \leq x \leq \alpha$), u_1 and region II ($\alpha \leq x \leq 1$), u_2 are:

$$\frac{\partial u_1}{\partial t} = \frac{1}{\beta^2} \frac{\partial^2 u_1}{\partial x^2} \quad 0 \leq x \leq \alpha$$

$$\frac{\partial u_2}{\partial t} = \frac{\partial^2 u_2}{\partial x^2} \quad \alpha \leq x \leq 1$$

$$u_1(x,0) = u_2(x,0) = 1$$

$$\frac{\partial u_1}{\partial x}(0,t) = 0$$

$$u_2(1,t) = 0$$

$$u_1(\alpha,t) = u_2(\alpha,t)$$

$$\frac{1}{\beta^2} \frac{\partial u_1}{\partial x}(\alpha,t) = \frac{\partial u_2}{\partial x}(\alpha,t)$$

where, β^2 is the ratio of diffusion coefficients in region II to region I. Using the governing equations and boundary conditions at $x = 0$ and 1, show that eigenfunctions for regions I and II are:

$$X_1 = A_n \cos(\beta \lambda_n x)$$
$$X_2 = B_n \sin(\lambda_n [1-x])$$

where, λ_n is the eigenvalue. Using the third boundary condition (concentrations are continuous at $x = \alpha$), show that the eigenfunctions can be rewritten as:

$$X_1 = C_n \cos(\beta \lambda_n x) \sin(\lambda_n [1-\alpha])$$
$$X_2 = C_n \cos(\beta \lambda_n \alpha) \sin(\lambda_n [1-x])$$

Next, use the fourth boundary condition to obtain the eigenvalue λ_n. Thus, the transient solutions can be obtained as:

$$u_1 = \sum_{n=1}^{\infty} C_n \cos(\beta \lambda_n x) \sin(\lambda_n [1-\alpha]) \exp(-\lambda_n^2 t)$$

$$u_2 = \sum_{n=1}^{\infty} C_n \cos(\beta \lambda_n \alpha) \sin(\lambda_n [1-x]) \exp(-\lambda_n^2 t)$$

Next, use the initial condition to find the coefficient C_n:

$$1 = \sum_{n=1}^{\infty} C_n \cos(\beta\lambda_n x) \sin(\lambda_n [1-\alpha])$$

$$1 = \sum_{n=1}^{\infty} C_n \cos(\beta\lambda_n \alpha) \sin(\lambda_n [1-x])$$

Multiply both sides of each equation by the corresponding eigenfunction and integrate over the domain of interest:

$$\int_0^\alpha \sin(\lambda_n [1-\alpha]) \cos(\beta\lambda_m x) dx = \sum_{n=1}^{\infty} C_n \sin(\lambda_n [1-\alpha]) \sin(\lambda_m [1-\alpha]) \int_0^\alpha \cos(\beta\lambda_n x) \cos(\beta\lambda_m x) dx$$

$$\int_\alpha^1 \sin(\lambda_m [1-x]) \cos(\beta\lambda_m \alpha) dx = \sum_{n=1}^{\infty} C_n \cos(\beta\lambda_n \alpha) \cos(\beta\lambda_m \alpha) \int_\alpha^1 \sin(\lambda_n [1-x]) \sin(\lambda_m [1-x]) dx$$

Simplify the integrals in both the equations and add both the equations to obtain the constant C_n (you might have to use the eigenvalue equation to simplify the integrals). Once an analytical solution is obtained, plot the profiles for $\alpha = 0.4$ and $\beta = 0.5$.

21. Consider electrochemical discharge composite planar electrodes.[4] The governing equations are same as problem 20 with the only difference being the boundary condition at $x = 1$:

$$\frac{\partial u_2}{\partial x}(1,t) = -\delta$$

where, δ is the dimensionless applied current density. Obtain an analytical solution for this problem and plot the profiles for $\alpha = 0.4$, $\beta = 0.5$ and $\delta = -1$.

References

1. Carslaw, H.S., Jaeger, J.C.: Conduction of Heat in Solids. Oxford University Press, Oxford (1972)
2. Rice, R.G., Do, D.D.: Applied Mathematics and Modeling for Chemical Engineers. John Wiley & Sons, Inc., Chichester (1995)
3. Constantinides, A., Mostoufi, N.: Numerical Methods for Chemical Engineers with MATLAB Applications. Prentice-Hall PTR, Englewood Cliffs (1999)
4. Subramanian, V.R., White, R.E.: New separation of variables method for composite electrodes with galvanostatic boundary conditions. Journal of Power Sources 96(2), 385–395 (2001)
5. Aris, R.: Mathematical Modeling: A Chemical Engineer's Perspective. Academic Press, London (1999)
6. Davis, M.E.: Numerical Methods and Modeling for Chemical Engineers. John Wiley & Sons, Chichester (1984)
7. Subramanian, V.R., Devan, S., White, R.E.: An approximate solution for a pseudocapacitor. Journal of Power Sources 135(1-2), 361–367 (2004)
8. Schiesser, W.E., Silebi, C.A.: Dynamic Modeling of Transport Process Systems (1997)

Chapter 8
Laplace Transform Technique for Partial Differential Equations

8.1 Laplace Transform Technique for Partial Differential Equations (PDEs) in Finite Domains

8.1.1 Introduction

Transient heat conduction or mass transfer in solids with constant physical properties (diffusion coefficient, thermal diffusivity, thermal conductivity, etc.) is usually represented by a linear parabolic partial differential equation. In this chapter, we describe how one can arrive at the analytical solutions for linear first order hyperbolic partial differential equations and parabolic partial differential equations in finite domains using the Laplace transform technique.

8.1.2 Laplace Transform Technique for Hyperbolic PDEs

Linear first order hyperbolic partial differential equations are solved using Laplace transform techniques in this section. Hyperbolic partial differential equations are first order in the time variable and first order in the spatial variable. The method involves applying Laplace transform in the time variable to convert the partial differential equation to an ordinary differential equation in the Laplace domain. This becomes an initial value problem (IVP) in the spatial direction with s, the Laplace variable, as a parameter. The boundary conditions in x are converted to the Laplace domain and the differential equation in the Laplace domain is solved by using the techniques illustrated in chapter 2.1 for solving linear initial value problems. Once an analytical solution is obtained in the Laplace domain, the solution is inverted to the time domain to obtain the final analytical solution (in time and spatial coordinates). This is best illustrated with the following example.

Example 8.1. Wave Propagation in a Rectangle

Consider the propagation of a wave in a rectangle.[1] The dimensionless concentration profile is governed by:

$$\frac{\partial u}{\partial t} + v\frac{\partial u}{\partial x} = 0$$
$$u(x,0) = 1 \quad (8.1)$$
$$u(0,t) = 0$$

Equation (8.1) is solved in Maple below:

> restart:with(inttrans):with(plots):

The governing equation is entered here:

> eq:=diff(u(x,t),t)+v*diff(u(x,t),x);

$$eq := \frac{\partial}{\partial t} u(x,t) + v\left(\frac{\partial}{\partial x} u(x,t)\right)$$

The initial and boundary conditions are entered here.

> u(x,0):=1;

$$u(x,0) := 1$$

> bc1:=u(0,t)=0;

$$bc1 := u(0,t) = 0$$

The governing equation and the boundary condition are converted to the Laplace domain:

> eqs:=laplace(eq,t,s);

$$eqs := s\, laplace(u(x,t),t,s) - 1 + v\left(\frac{\partial}{\partial x} laplace(u(x,t),t,s)\right)$$

> eqs:=subs(laplace(u(x,t),t,s)=U(x),eqs);

$$eqs := s\, U(x) - 1 + v\left(\frac{d}{dx} U(x)\right)$$

> bc1:=laplace(bc1,t,s);

$$bc1 := laplace(u(0,t),t,s) = 0$$

> bc1:=subs(laplace(u(0,t),t,s)=U(0),bc1);

$$bc1 := U(0) = 0$$

8.1 Laplace Transform Technique for PDEs in Finite Domains   681

> U(x):=rhs(dsolve({eqs,bc1},U(x)));

$$U(x) := \frac{1}{s} - \frac{e^{-\frac{sx}{v}}}{s}$$

The solution obtained in the Laplace domain is converted to the time domain here:

> u:=invlaplace(U(x),s,t);

$$u := 1 - invlaplace\left(\frac{e^{-\frac{sx}{v}}}{s}, s, t\right)$$

The following plots can be obtained:

> plot3d(subs(v=1.,u),x=0..1,t=1e-6..1,axes=boxed,title="Figure 8.1",
labels=[x,t,"u"],orientation=[-137,50]);

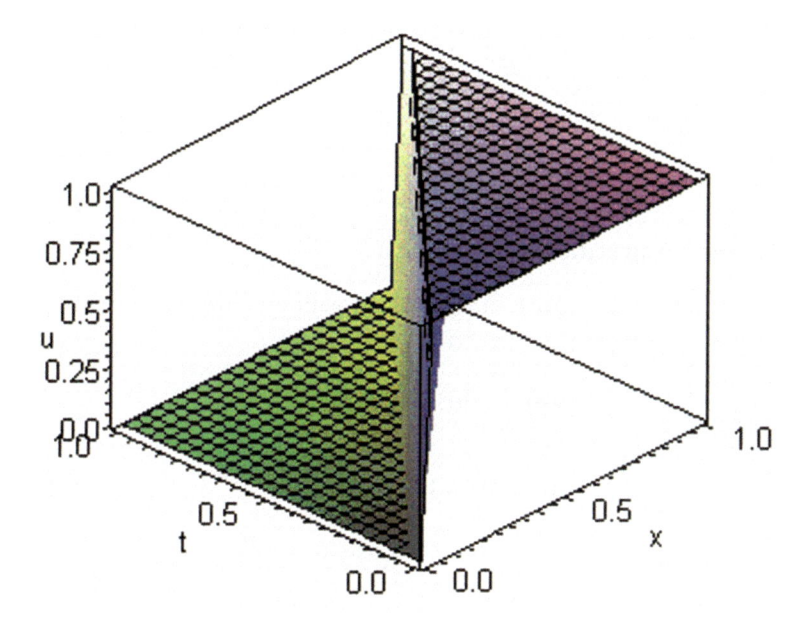

Fig. 8.1

> plot([subs(v=1,t=0.1,u),subs(v=1,t=0.25,u),subs(v=1,t=0.5,u),
subs(v=1,t=1,u)],x=0..1,axes=boxed,title="Figure 8.2",
thickness=5,labels=[x,"u"]);

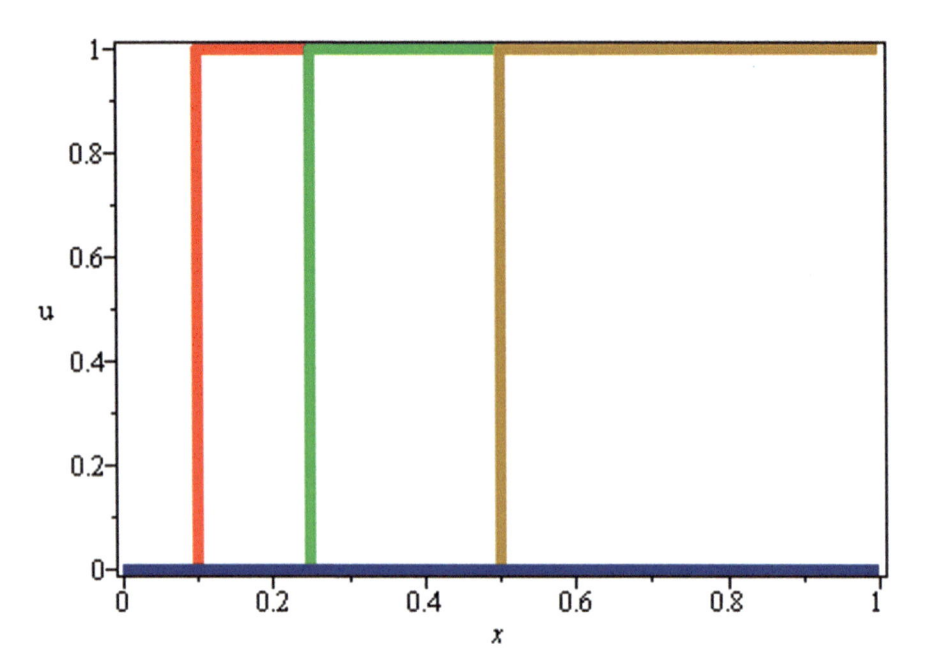

Fig. 8.2

Example 8.2. Wave Propagation in a Rectangle

Consider the propagation of a wave in a rectangle with a known initial profile. The dimensionless concentration profile is governed by:

$$\frac{\partial u}{\partial t} + \frac{\partial u}{\partial x} = 0$$
$$u(x,0) = 1 - \exp(-x) \qquad (8.2)$$
$$u(0,t) = 0$$

Equation (8.2) is solved in Maple below:

> restart:with(inttrans):with(plots):

> eq:=diff(u(x,t),t)+diff(u(x,t),x);

$$eq := \frac{\partial}{\partial t} u(x,t) + \frac{\partial}{\partial x} u(x,t)$$

8.1 Laplace Transform Technique for PDEs in Finite Domains

```
> u(x,0):=1-exp(-x);
```

$$u(x, 0) := 1 - e^{-x}$$

```
> bc1:=u(0,t)=0;
```

$$bc1 := u(0, t) = 0$$

The solution obtained in the Laplace domain is:

```
> eqs:=laplace(eq,t,s):
> eqs:=subs(laplace(u(x,t),t,s)=U(x),eqs);
```

$$eqs := s\,U(x) - 1 + e^{-x} + \frac{d}{dx}U(x)$$

```
> bc1:=laplace(bc1,t,s):
> bc1:=subs(laplace(u(0,t),t,s)=U(0),bc1);
```

$$bc1 := U(0) = 0$$

```
> U(x):=rhs(dsolve({eqs,bc1},U(x)));
```

$$U(x) := \frac{e^{-sx}}{s\,(s-1)} - \frac{-s + 1 + e^{-x}\,s}{s\,(s-1)}$$

The solution obtained in the time domain is obtained as:

```
> u:=invlaplace(U(x),s,t);
```

$$u := -invlaplace\left(\frac{e^{-sx}}{s}, s, t\right) + invlaplace\left(\frac{e^{-sx}}{s-1}, s, t\right) + 1 - e^{-x+t}$$

```
> plot3d(u,x=1e-6..1,t=0..1,axes=boxed,title="Figure 8.3",
labels=[x,t,"u"],orientation=[120,60]);
```

684   8 Laplace Transform Technique for Partial Differential Equations

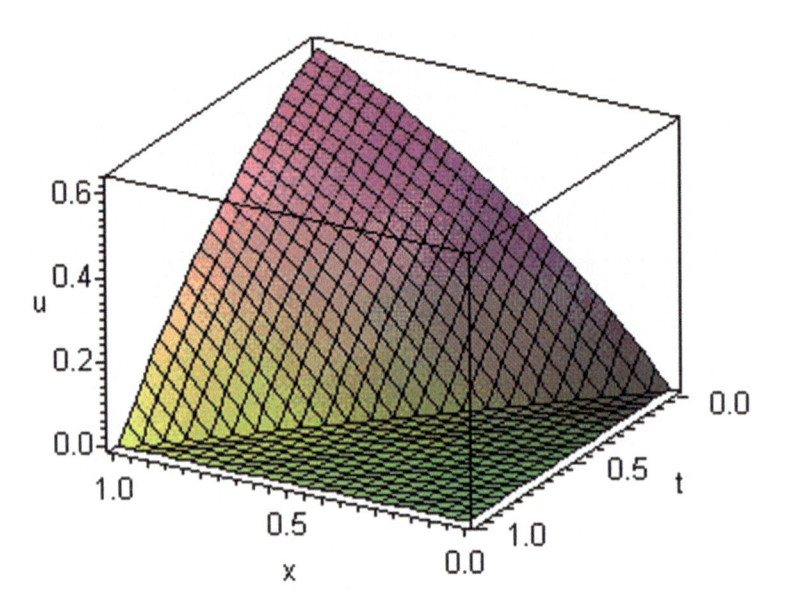

Fig. 8.3

> plot([subs(t=0.1,u),subs(t=0.25,u),subs(t=0.5,u),subs(t=1,u)],x=0..1,
axes=boxed,title="Figure 8.4",thickness=5,labels=[x,"u"]);

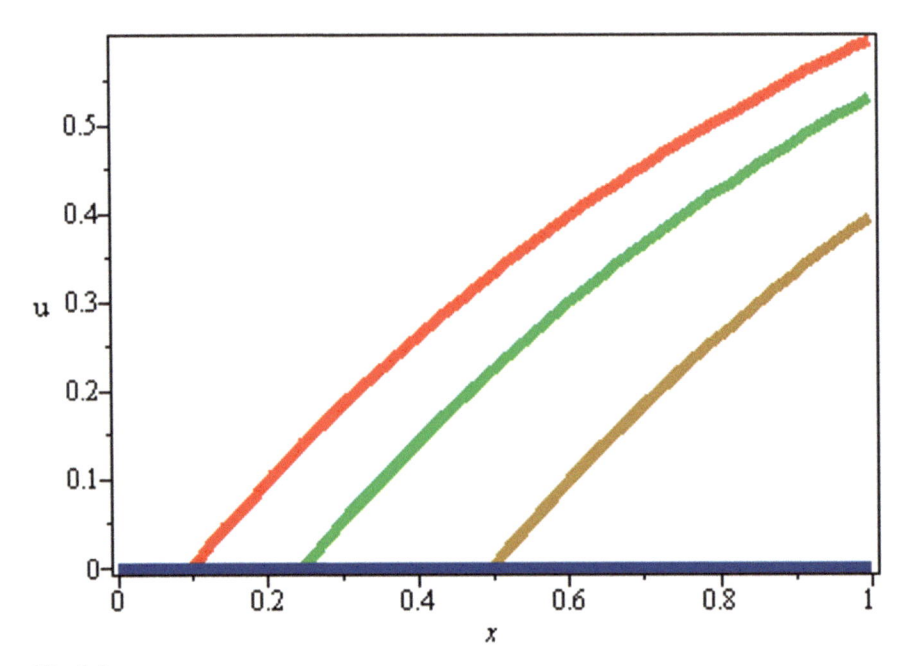

Fig. 8.4

8.1 Laplace Transform Technique for PDEs in Finite Domains 685

8.1.3 *Laplace Transform Technique for Parabolic Partial Differential Equations – Simple Solutions*

Linear first order parabolic partial differential equations in finite domains are solved using the Laplace transform technique in this section. Parabolic PDEs are first order in the time variable and second order in the spatial variable. The method involves applying the Laplace transform in the time variable to convert the partial differential equation to an ordinary differential equation in the Laplace domain. This becomes a boundary value problem (BVP) in the spatial direction with s, the Laplace variable as a parameter. The boundary conditions in x are converted to the Laplace domain and the differential equation in the Laplace domain is solved by using the techniques illustrated in chapter 3.1 for solving linear boundary value problems. Once an analytical solution is obtained in the Laplace domain, the solution is inverted to the time domain to obtain the final analytical solution (in time and spatial coordinates). Certain simple problems can be inverted to the time domain using Maple. This is best illustrated with the following examples.

Example 8.3. Heat Transfer in a Rectangle

Example 7.4, heat transfer in a rectangle with a sinusoidal initial profile,[2] is solved here again using the Laplace transform technique. The dimensionless temperature profile is governed by:

$$\frac{\partial u}{\partial t} = \frac{\partial^2 u}{\partial x^2}$$
$$u(x,0) = \sin(\pi x) \tag{8.3}$$
$$u(0,t) = 0 \text{ and } u(1,t) = 0$$

Equation (8.3) is solved in Maple below:

> restart:with(inttrans):with(plots):
> eq:=diff(u(x,t),t)=diff(u(x,t),x$2);

$$eq := \frac{\partial}{\partial t} u(x,t) = \frac{\partial^2}{\partial x^2} u(x,t)$$

> u(x,0):=sin(Pi*x);

$$u(x,0) := \sin(\pi x)$$

> bc1:=u(x,t)=0;

$$bc1 := u(x,t) = 0$$

> bc2:=u(x,t)=0;

$$bc2 := u(x, t) = 0$$

The governing equation and the boundary conditions are converted to the Laplace domain:

> eqs:=laplace(eq,t,s):
> eqs:=subs(laplace(u(x,t),t,s)=U(x),eqs);

$$eqs := s\,U(x) - \sin(\pi x) = \frac{d^2}{dx^2} U(x)$$

> bc1:=laplace(bc1,t,s):
> bc1:=subs(laplace(u(x,t),t,s)=U(0),bc1);

$$bc1 := U(0) = 0$$

> bc2:=laplace(bc2,t,s):
> bc2:=subs(laplace(u(x,t),t,s)=U(1),bc2);

$$bc2 := U(1) = 0$$

The governing equation in the Laplace domain is solved as:

> dsolve(eqs,U(x));

$$U(x) = e^{\sqrt{s}\,x}\_C2 + e^{-\sqrt{s}\,x}\_C1 + \frac{\sin(\pi x)}{s + \pi^2}$$

The governing equation is solved with the boundary conditions as:

> U(x):=rhs(dsolve({eqs,bc1,bc2},U(x)));

$$U(x) := \frac{\sin(\pi x)}{s + \pi^2}$$

The solution obtained is converted to the time domain as:

> u:=invlaplace(U(x),s,t);

$$u := \sin(\pi x)\,e^{-\pi^2 t}$$

> plot3d(u,x=1..0,t=0..0.2,axes=boxed,title="Figure 8.5", labels=[x,t,"u"],orientation=[120,60]);

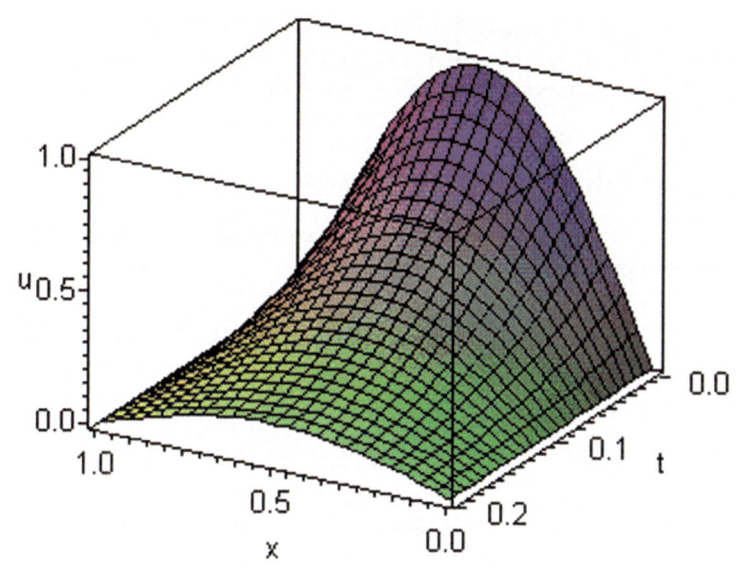

8.1 Laplace Transform Technique for PDEs in Finite Domains 687

Fig. 8.5

> plot([subs(t=0,u),subs(t=0.05,u),subs(t=0.1,u),subs(t=0.2,u)],x=0..1,
title="Figure 8.6",axes=boxed,thickness=5,labels=[x,"u"]);

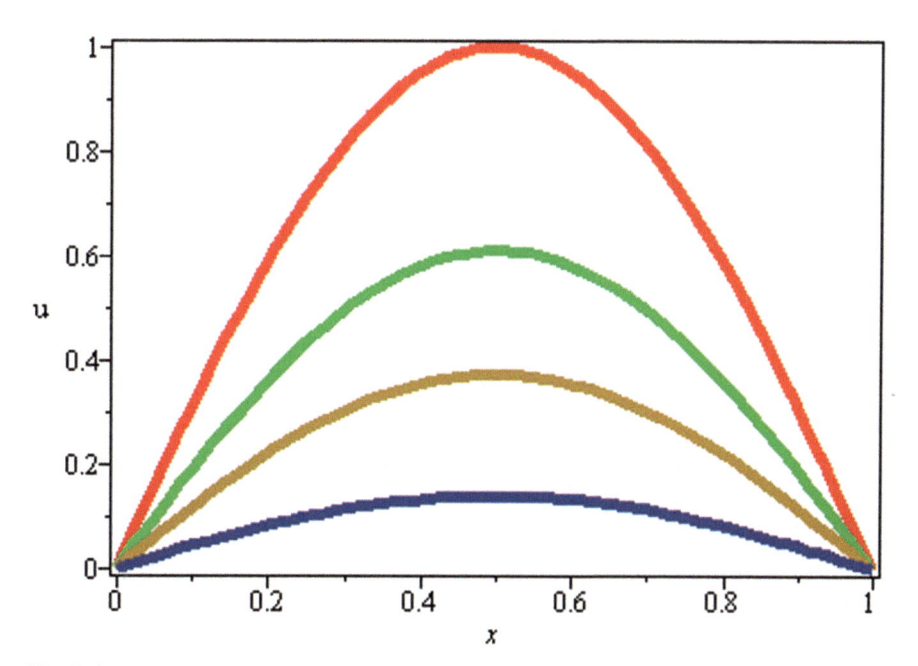

Fig. 8.6

In all the examples previously discussed, the boundary conditions did not involve derivatives until now. When there is a derivative in the boundary condition it has to be taken care of while applying the Laplace transform as shown in the next example.

Example 8.4. Transient Heat Transfer in a Rectangle

Consider heat transfer in a rectangle with a derivative boundary condition. The dimensionless temperature profile is governed by:

$$\frac{\partial u}{\partial t} = \frac{\partial^2 u}{\partial x^2}$$

$$u(x,0) = \sin\left(\frac{\pi x}{2}\right) \qquad (8.4)$$

$$u(0,t) = 0 \text{ and } \frac{\partial u}{\partial x}(1,t) = 0$$

Equation (8.4) is solved in Maple below:

> restart:with(inttrans):with(plots):
> eq:=diff(u(x,t),t)=diff(u(x,t),x$2);

$$eq := \frac{\partial}{\partial t} u(x,t) = \frac{\partial^2}{\partial x^2} u(x,t)$$

> u(x,0):=sin(Pi*x/2);

$$u(x,0) := \sin\left(\frac{1}{2} \pi x\right)$$

> bc1:=u(x,t)=0;

$$bc1 := u(x,t) = 0$$

> bc2:=diff(u(x,t),x)=0;

$$bc2 := \frac{\partial}{\partial x} u(x,t) = 0$$

> eqs:=laplace(eq,t,s):
> eqs:=subs(laplace(u(x,t),t,s)=U(x),eqs);

$$eqs := s\, U(x) - \sin\left(\frac{1}{2} \pi x\right) = \frac{d^2}{dx^2} U(x)$$

8.1 Laplace Transform Technique for PDEs in Finite Domains

```
> bc1:=laplace(bc1,t,s):
> bc1:=subs(diff(laplace(u(x,t),t,s),x)=D(U)(0),laplace(u(x,t),t,s)=U(0),bc1);
```

$$bc1 := U(0) = 0$$

```
> bc2:=laplace(bc2,t,s):
> bc2:=subs(diff(laplace(u(x,t),t,s),x)=D(U)(1),laplace(u(x,t),t,s)=U(1),bc2);
```

$$bc2 := D(U)(1) = 0$$

```
> U(x):=rhs(dsolve({eqs,bc1,bc2},U(x)));
```

$$U(x) := \frac{4 \sin\left(\frac{1}{2} \pi x\right)}{4s + \pi^2}$$

```
> u:=invlaplace(U(x),s,t);
```

$$u := \sin\left(\frac{1}{2} \pi x\right) e^{-\frac{1}{4} \pi^2 t}$$

```
> plot3d(u,x=1..0,t=0..0.2,axes=boxed,title="Figure 8.7",
labels=[x,t,"u"],orientation=[120,60]);
```

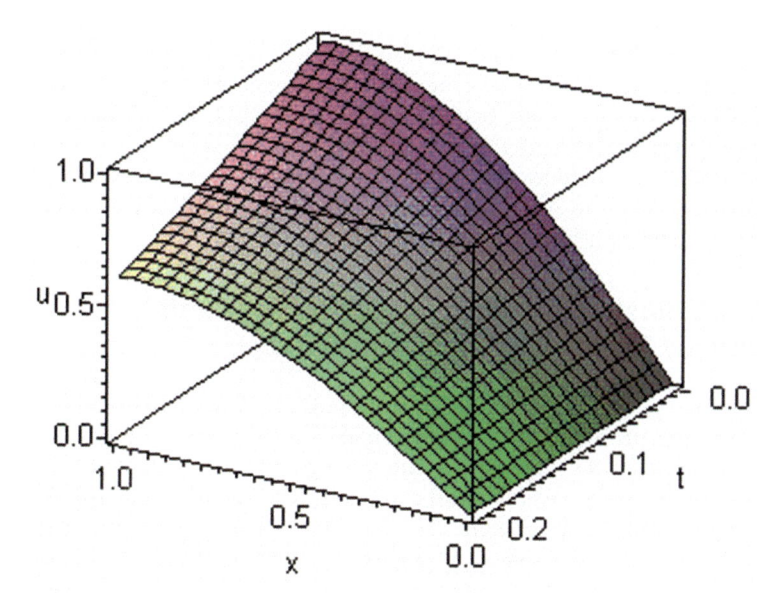

Fig. 8.7

```
> plot([subs(t=0,u),subs(t=0.1,u),subs(t=0.2,u),subs(t=0.5,u)],x=0..1,
axes=boxed,title="Figure 8.8.",thickness=5,labels=[x,"u"]);
```

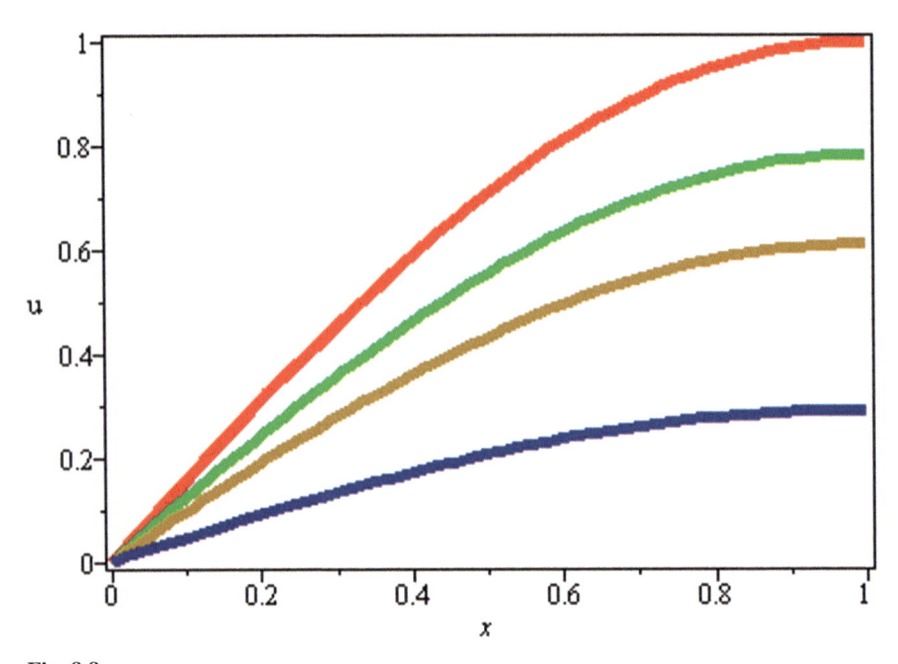

Fig. 8.8

In examples 8.3 and 8.4 Maple was used to invert from the Laplace domain to the time domain. Unfortunately, these two examples are very simple and, hence, we could invert to the time domain using Maple. For practical problems, inversion is not straightforward. The inversion to the time domain can be done in two different ways. In section 8.1.4, short time solutions will be obtained by converting the solution in Laplace domain to an infinite series. In section 8.1.5, a long time solution will be obtained by using the Heaviside expansion theorem.

8.1.4 Laplace Transform Technique for Parabolic Partial Differential Equations – Short Time Solution

The methodology is the same as that used in section 8.1.3. When Maple fails to invert the Laplace domain solution to the time domain, a short time solution can be obtained by converting the Laplace domain solution to an infinite series in which each term can be easily inverted to time domain. The solution obtained for heat transfer in a rectangle in example 7.1 using the separation of variables method cannot be used at short times. At time $t = 0$, one would need infinite number of terms in the separation of variables solution. Fortunately, the Laplace transform technique helps us obtain a solution, which can be used efficiently at short times also. This is best illustrated with the following examples.

8.1 Laplace Transform Technique for PDEs in Finite Domains

Example 8.5. Heat Transfer in a Rectangle

Consider Example 7.1 heat transfer in a rectangle[2] which is solved here again using the Laplace transform technique. The dimensionless temperature profile is governed by:

$$\frac{\partial u}{\partial t} = \frac{\partial^2 u}{\partial x^2}$$
$$u(x,0) = 1 \qquad (8.5)$$
$$u(0,t) = 0 \text{ and } u(1,t) = 0$$

Equation (8.5) is solved in Maple below:

> restart:with(inttrans):with(plots):

The governing equation and boundary conditions are entered and converted to the Laplace domain.

> eq:=diff(u(x,t),t)=diff(u(x,t),x$2);

$$eq := \frac{\partial}{\partial t} u(x, t) = \frac{\partial^2}{\partial x^2} u(x, t)$$

> u(x,0):=1;

$$u(x, 0) := 1$$

> bc1:=u(x,t)=0;

$$bc1 := u(x, t) = 0$$

> bc2:=u(x,t)=0;

$$bc2 := u(x, t) = 0$$

> eqs:=laplace(eq,t,s):
> eqs:=subs(laplace(u(x,t),t,s)=U(x),eqs);

$$eqs := s\, U(x) - 1 = \frac{d^2}{dx^2} U(x)$$

> bc1:=laplace(bc1,t,s):
> bc1:=subs(diff(laplace(u(x,t),t,s),x)=D(U)(0),laplace(u(x,t),t,s)=U(0),bc1);

$$bc1 := U(0) = 0$$

```
> bc2:=laplace(bc2,t,s):
> bc2:=subs(diff(laplace(u(x,t),t,s),x)=D(U)(1),laplace(u(x,t),t,s)=U(1),bc2);
```

$$bc2 := U(1) = 0$$

The solution obtained in the Laplace domain is:

```
> U(x):=rhs(dsolve({eqs,bc1,bc2},U(x)));
```

$$U(x) := -\frac{e^{\sqrt{s}\,x}\left(-1+e^{-\sqrt{s}}\right)}{s\left(-e^{\sqrt{s}}+e^{-\sqrt{s}}\right)} + \frac{e^{-\sqrt{s}\,x}\left(-1+e^{\sqrt{s}}\right)}{s\left(-e^{\sqrt{s}}+e^{-\sqrt{s}}\right)} + \frac{1}{s}$$

Maple fails to invert the solution obtained:

```
> invlaplace(U(x),s,t);
```

$$-\frac{1}{2}\,invlaplace\left(\frac{e^{\sqrt{s}\,x}}{s\sinh(\sqrt{s})},s,t\right) - invlaplace\left(\frac{e^{\sqrt{s}\,x-\sqrt{s}}}{s\left(-e^{\sqrt{s}}+e^{-\sqrt{s}}\right)},s,t\right) + \frac{1}{2}\,invlaplace\left(\frac{e^{-\sqrt{s}\,x}}{s\sinh(\sqrt{s})},s,t\right)$$
$$+ invlaplace\left(\frac{e^{-\sqrt{s}\,x+\sqrt{s}}}{s\left(-e^{\sqrt{s}}+e^{-\sqrt{s}}\right)},s,t\right) + 1$$

The first two terms of U(x) are expressed as an infinite series below:

```
> U1s:=-exp(s^(1/2)*x)/s/(exp(s^(1/2))+1);
```

$$U1s := -\frac{e^{\sqrt{s}\,x}}{s\left(e^{\sqrt{s}}+1\right)}$$

```
> U2s:=-exp(-s^(1/2)*x)*exp(s^(1/2))/s/(exp(s^(1/2))+1);
```

$$U2s := -\frac{e^{-\sqrt{s}\,x}e^{\sqrt{s}}}{s\left(e^{\sqrt{s}}+1\right)}$$

We want to write a series in terms of S=exp(-s^(1/2)) so that the series will converge:

```
> U1S:=series(subs(exp(s^(1/2))=1/S,U1s),S);
```

$$U1S := -\frac{e^{\sqrt{s}\,x}}{s}S + \frac{e^{\sqrt{s}\,x}}{s}S^2 - \frac{e^{\sqrt{s}\,x}}{s}S^3 + \frac{e^{\sqrt{s}\,x}}{s}S^4 - \frac{e^{\sqrt{s}\,x}}{s}S^5 + O(S^6)$$

```
> U1S:=subs(S=exp(-s^(1/2)),U1S);
```

$$U1S := -\frac{e^{\sqrt{s}\,x}e^{-\sqrt{s}}}{s} + \frac{e^{\sqrt{s}\,x}\left(e^{-\sqrt{s}}\right)^2}{s} - \frac{e^{\sqrt{s}\,x}\left(e^{-\sqrt{s}}\right)^3}{s} + \frac{e^{\sqrt{s}\,x}\left(e^{-\sqrt{s}}\right)^4}{s} - \frac{e^{\sqrt{s}\,x}\left(e^{-\sqrt{s}}\right)^5}{s} + O\left(\left(e^{-\sqrt{s}}\right)^6\right)$$

```
> simplify(U1S);
```

$$\frac{-e^{\sqrt{s}\,(x-1)} + e^{\sqrt{s}\,(x-2)} - e^{\sqrt{s}\,(x-3)} + e^{\sqrt{s}\,(x-4)} - e^{\sqrt{s}\,(x-5)} + O\left(e^{-6\sqrt{s}}\right)}{s}$$

8.1 Laplace Transform Technique for PDEs in Finite Domains

Hence, U1S can be written as the infinite series:

> U1S:=Sum((-1)^n*exp(s^(1/2)*(x-n))/s,n=1..infinity);

$$U1S := \sum_{n=1}^{\infty} \frac{(-1)^n e^{\sqrt{s}\,(x-n)}}{s}$$

The general term in the above series is:

> u1s:=(-1)^n*exp(s^(1/2)*(x-n))/s;

$$u1s := \frac{(-1)^n e^{\sqrt{s}\,(x-n)}}{s}$$

The time domain solution for this expression is:

> u1t:=invlaplace(u1s,s,t);

$$u1t := (-1)^n \, invlaplace\left(\frac{e^{\sqrt{s}\,(x-n)}}{s}, s, t\right)$$

Hence, the inverse of U1S is the infinite series given by:

> U1t:=Sum(u1t,n=1..infinity);

$$U1t := \sum_{n=1}^{\infty} (-1)^n \, invlaplace\left(\frac{e^{\sqrt{s}\,(x-n)}}{s}, s, t\right)$$

Similarly, U2S is inverted below:

> U2S:=series(subs(exp(s^(1/2))=1/S,U2s),S);

$$U2S := -\frac{e^{-\sqrt{s}\,x}}{s} + \frac{e^{-\sqrt{s}\,x}}{s} S - \frac{e^{-\sqrt{s}\,x}}{s} S^2 + \frac{e^{-\sqrt{s}\,x}}{s} S^3$$
$$- \frac{e^{-\sqrt{s}\,x}}{s} S^4 + \frac{e^{-\sqrt{s}\,x}}{s} S^5 + O(S^6)$$

> U2S:=subs(S=exp(-s^(1/2)),U2S);

$$U2S := -\frac{e^{-\sqrt{s}\,x}}{s} + \frac{e^{-\sqrt{s}\,x} e^{-\sqrt{s}}}{s} - \frac{e^{-\sqrt{s}\,x}\left(e^{-\sqrt{s}}\right)^2}{s}$$
$$+ \frac{e^{-\sqrt{s}\,x}\left(e^{-\sqrt{s}}\right)^3}{s} - \frac{e^{-\sqrt{s}\,x}\left(e^{-\sqrt{s}}\right)^4}{s} + \frac{e^{-\sqrt{s}\,x}\left(e^{-\sqrt{s}}\right)^5}{s}$$
$$+ O\left(\left(e^{-\sqrt{s}}\right)^6\right)$$

> simplify(U2S);

$$\frac{1}{s}\left(-e^{-\sqrt{s}\,x} + e^{-\sqrt{s}\,(x+1)} - e^{-\sqrt{s}\,(x+2)} + e^{-\sqrt{s}\,(x+3)}\right.$$
$$\left. - e^{-\sqrt{s}\,(x+4)} + e^{-\sqrt{s}\,(x+5)} + O\!\left(e^{-6\sqrt{s}}\right)s\right)$$

> U2S:=Sum((-1)^n*exp(-s^(1/2)*(x+n-1))/s,n=1..infinity);

$$U2S := \sum_{n=1}^{\infty} \frac{(-1)^n\, e^{-\sqrt{s}\,(x+n-1)}}{s}$$

> u2s:=(-1)^n*exp(-s^(1/2)*(x+n-1))/s;

$$u2s := \frac{(-1)^n\, e^{-\sqrt{s}\,(x+n-1)}}{s}$$

> u2t:=invlaplace(u2s,s,t);

$$u2t := (-1)^n\, \mathit{invlaplace}\!\left(\frac{e^{(-x-n+1)\sqrt{s}}}{s}, s, t\right)$$

> U2t:=Sum(u2t,n=1..infinity);

$$U2t := \sum_{n=1}^{\infty} (-1)^n\, \mathit{invlaplace}\!\left(\frac{e^{(-x-n+1)\sqrt{s}}}{s}, s, t\right)$$

The final solution for u in the time domain is:

> Ut:=U1t+U2t+1;

$$Ut := \sum_{n=1}^{\infty} (-1)^n\, \mathit{invlaplace}\!\left(\frac{e^{\sqrt{s}\,(x-n)}}{s}, s, t\right) + \sum_{n=1}^{\infty} (-1)^n\, \mathit{invlaplace}\!\left(\frac{e^{(-x-n+1)\sqrt{s}}}{s}, s, t\right) + 1$$

For plotting purposes, infinity is replaced by N = 20:

> u:=subs(infinity=N,Ut);

8.1 Laplace Transform Technique for PDEs in Finite Domains

$$u := \sum_{n=1}^{N} (-1)^n \, invlaplace\left(\frac{e^{\sqrt{s}\,(x-n)}}{s}, s, t\right) + \sum_{n=1}^{N} (-1)^n \, invlaplace\left(\frac{e^{(-x-n+1)\sqrt{s}}}{s}, s, t\right) + 1$$

> u:=subs(N=20,u);

$$u := \sum_{n=1}^{20} (-1)^n \, invlaplace\left(\frac{e^{\sqrt{s}\,(x-n)}}{s}, s, t\right) + \sum_{n=1}^{20} (-1)^n \, invlaplace\left(\frac{e^{(-x-n+1)\sqrt{s}}}{s}, s, t\right) + 1$$

The following plots can be obtained:

> plot3d(u,x=0..1,t=1e-6..0.1,axes=boxed,title="Figure 8.9",
labels=[x,t,"u"],orientation=[60,60]);

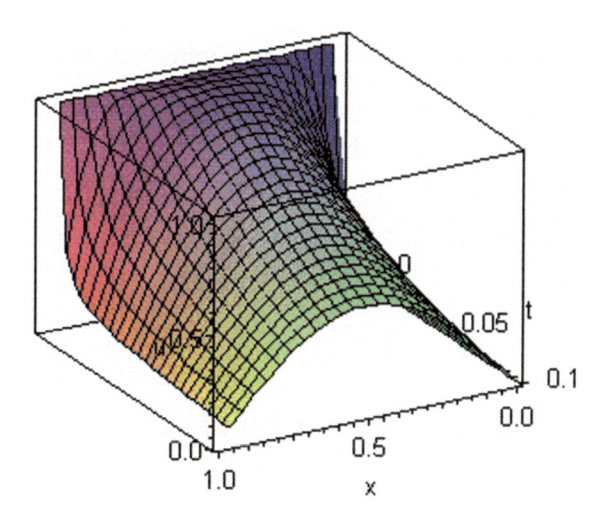

Fig. 8.9

> plot([subs(t=1e-6,u),subs(t=1e-3,u),subs(t=0.01,u),subs(t=0.05,u)],
x=0..1,axes=boxed,title="Figure 8.10",thickness=5,labels=[x,"u"]);

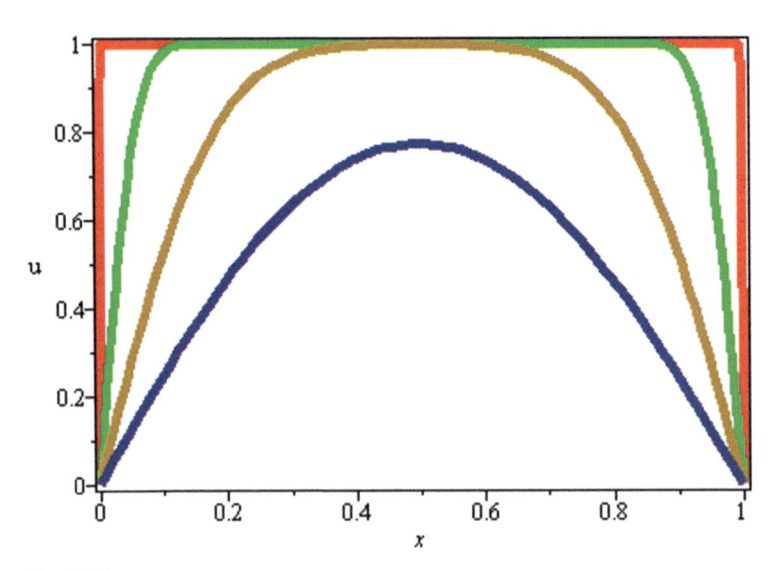

Fig. 8.10

Note that for plotting purposes t = 0 is replaced by t = 10^{-6} to avoid singularity at t = 0.

Example 8.6. Mass Transfer in a Spherical Pellet

Consider Example 7.3, mass transfer in a spherical pellet,[2] which is solved again here. The governing equation in dimensionless form is

$$\frac{\partial u}{\partial t} = \frac{\partial^2 u}{\partial x^2} + \frac{2}{x}\frac{\partial u}{\partial x}$$
$$u(x,0) = 1 \qquad (8.6)$$
$$\frac{\partial u}{\partial x}(0,t) = 0 \text{ and } u(1,t) = 0$$

Equation (8.6) is solved in Maple and the results obtained are given below:

```
> restart:with(inttrans):with(plots):
> eq:=diff(u(x,t),t)=diff(u(x,t),x$2)+2/x*diff(u(x,t),x);
```

$$eq := \frac{\partial}{\partial t} u(x,t) = \frac{\partial^2}{\partial x^2} u(x,t) + \frac{2\left(\frac{\partial}{\partial x} u(x,t)\right)}{x}$$

```
> u(x,0):=0;
```

$$u(x,0) := 0$$

8.1 Laplace Transform Technique for PDEs in Finite Domains

> bc1:=diff(u(x,t),x)=0;

$$bc1 := \frac{\partial}{\partial x} u(x,t) = 0$$

> bc2:=u(x,t)=1;

$$bc2 := u(x,t) = 1$$

> eqs:=laplace(eq,t,s):
> eqs:=subs(laplace(u(x,t),t,s)=U(x),eqs);

$$eqs := s\, U(x) = \frac{d^2}{dx^2} U(x) + \frac{2\left(\frac{d}{dx} U(x)\right)}{x}$$

> bc1:=laplace(bc1,t,s):
> bc1:=subs(diff(laplace(u(x,t),t,s),x)=D(U)(0),laplace(u(x,t),t,s)=U(0),bc1);

$$bc1 := D(U)(0) = 0$$

> bc2:=laplace(bc2,t,s):
> bc2:=subs(diff(laplace(u(x,t),t,s),x)=D(U)(1),laplace(u(x,t),t,s)=U(1),bc2);

$$bc2 := U(1) = \frac{1}{s}$$

> U(x):=rhs(dsolve({eqs,bc2},U(x)));

$$U(x) := \frac{\_C1\, \sinh(\sqrt{s}\, x)}{x} - \frac{\left(s\, \_C1\, e^{8\sqrt{s}} - s\, \_C1\, e^{6\sqrt{s}} - 2 e^{7\sqrt{s}}\right) \cosh(\sqrt{s}\, x)}{s\left(e^{8\sqrt{s}} + e^{6\sqrt{s}}\right) x}$$

> U(x):=subs(_C2=0,U(x));

$$U(x) := \frac{\_C1\, \sinh(\sqrt{s}\, x)}{x} - \frac{\left(s\, \_C1\, e^{8\sqrt{s}} - s\, \_C1\, e^{6\sqrt{s}} - 2 e^{7\sqrt{s}}\right) \cosh(\sqrt{s}\, x)}{s\left(e^{8\sqrt{s}} + e^{6\sqrt{s}}\right) x}$$

> convert(U(x),exp);

$$\frac{\_C1\left(\frac{1}{2}e^{\sqrt{s}\,x} - \frac{1}{2}e^{-\sqrt{s}\,x}\right)}{x}$$

$$-\frac{1}{s\left(e^{8\sqrt{s}} + e^{6\sqrt{s}}\right)x}\left(\left(s\,\_C1\,e^{8\sqrt{s}} - s\,\_C1\,e^{6\sqrt{s}}\right.\right.$$

$$\left.\left. - 2e^{7\sqrt{s}}\right)\left(\frac{1}{2}e^{\sqrt{s}\,x} + \frac{1}{2}e^{-\sqrt{s}\,x}\right)\right)$$

> U1s:=exp(s^(1/2))/s/(exp(s^(1/2))^2-1)*exp(s^(1/2)*x)/x;

$$U1s := \frac{e^{\sqrt{s}}\,e^{\sqrt{s}\,x}}{s\left(\left(e^{\sqrt{s}}\right)^2 - 1\right)x}$$

> U2s:=-exp(s^(1/2))/s/(exp(s^(1/2))^2-1)*exp(-s^(1/2)*x)/x;

$$U2s := -\frac{e^{\sqrt{s}}\,e^{-\sqrt{s}\,x}}{s\left(\left(e^{\sqrt{s}}\right)^2 - 1\right)x}$$

> U1S:=series(subs(exp(s^(1/2))=1/S,U1s),S):
> U1S:=subs(S=exp(-s^(1/2)),U1S):
> simplify(U1S);

$$\frac{e^{\sqrt{s}\,(x-1)} + e^{\sqrt{s}\,(x-3)} + e^{\sqrt{s}\,(x-5)} + O\!\left(e^{-7\sqrt{s}}\right)s\,x}{s\,x}$$

> U1S:=1/x*Sum(exp(s^(1/2)*(x-2*n+1))/s,n=1..infinity);

$$U1S := \frac{\sum_{n=1}^{\infty} \frac{e^{\sqrt{s}\,(x-2n+1)}}{s}}{x}$$

> u1s:=exp(s^(1/2)*(x-2*n+1))/s;

$$u1s := \frac{e^{\sqrt{s}\,(x-2n+1)}}{s}$$

> u1t:=invlaplace(u1s,s,t);

$$u1t := \mathrm{invlaplace}\!\left(\frac{e^{\sqrt{s}\,(x-2n+1)}}{s},\,s,\,t\right)$$

8.1 Laplace Transform Technique for PDEs in Finite Domains 699

> U1t:=1/x*Sum(u1t,n=1..infinity);

$$U1t := \frac{\sum_{n=1}^{\infty} invlaplace\left(\frac{e^{\sqrt{s}\,(x-2n+1)}}{s}, s, t\right)}{x}$$

> U2S:=series(subs(exp(s^(1/2))=1/S,U2s),S):
> U2S:=subs(S=exp(-s^(1/2)),U2S):
> simplify(U2S);

$$\frac{-e^{-\sqrt{s}\,(x+1)} - e^{-\sqrt{s}\,(x+3)} - e^{-\sqrt{s}\,(x+5)} + O\!\left(e^{-7\sqrt{s}}\right) s\,x}{s\,x}$$

> U2S:=-1/x*Sum(exp(-s^(1/2)*(x+2*n-1))/s,n=1..infinity);

$$U2S := -\frac{\sum_{n=1}^{\infty} \frac{e^{-\sqrt{s}\,(x+2n-1)}}{s}}{x}$$

> u2s:=exp(-s^(1/2)*(x+2*n-1))/s:
> u2t:=invlaplace(u2s,s,t);

$$u2t := invlaplace\left(\frac{e^{(-x-2n+1)\sqrt{s}}}{s}, s, t\right)$$

> U2t:=-1/x*Sum(u2t,n=1..infinity);

$$U2t := -\frac{\sum_{n=1}^{\infty} invlaplace\left(\frac{e^{(-x-2n+1)\sqrt{s}}}{s}, s, t\right)}{x}$$

> Ut:=U1t+U2t;

$$Ut := \frac{\sum_{n=1}^{\infty} invlaplace\left(\frac{e^{\sqrt{s}\,(x-2n+1)}}{s}, s, t\right)}{x}$$
$$- \frac{\sum_{n=1}^{\infty} invlaplace\left(\frac{e^{(-x-2n+1)\sqrt{s}}}{s}, s, t\right)}{x}$$

```
> u:=subs(infinity=N,Ut):
> u:=subs(N=20,u);
```

$$u := \frac{\displaystyle\sum_{n=1}^{20} invlaplace\left(\frac{e^{\sqrt{s}\,(x-2n+1)}}{s}, s, t\right)}{x}$$

$$- \frac{\displaystyle\sum_{n=1}^{20} invlaplace\left(\frac{e^{(-x-2n+1)\sqrt{s}}}{s}, s, t\right)}{x}$$

```
> plot3d(u,x=1e-6..1,t=1e-6..0.1,axes=boxed,title="Figure 8.11",
labels=[x,t,"u"],orientation=[-150,60]);
```

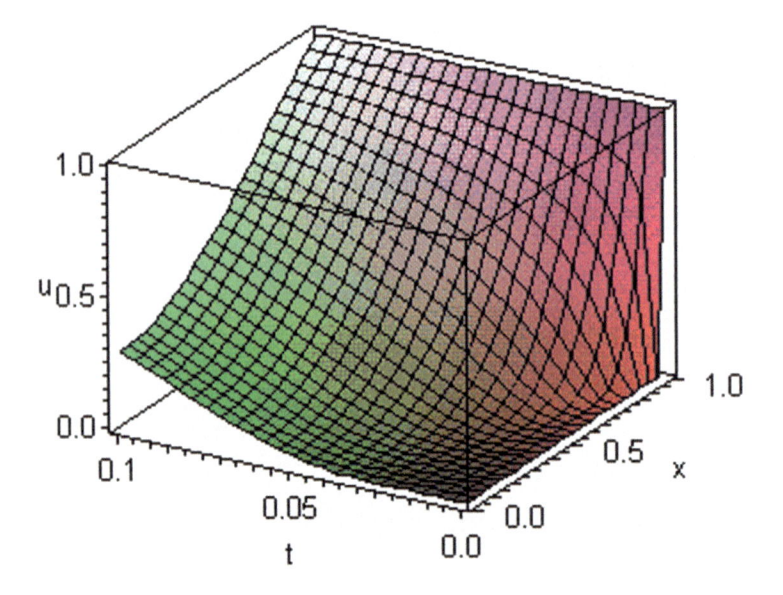

Fig. 8.11

```
> plot([subs(t=1e-6,u),subs(t=1e-2,u),subs(t=0.05,u),subs(t=0.1,u)],x=0..1,
axes=boxed,title="Figure 8.12",thickness=5,labels=[x,"u"]);
```

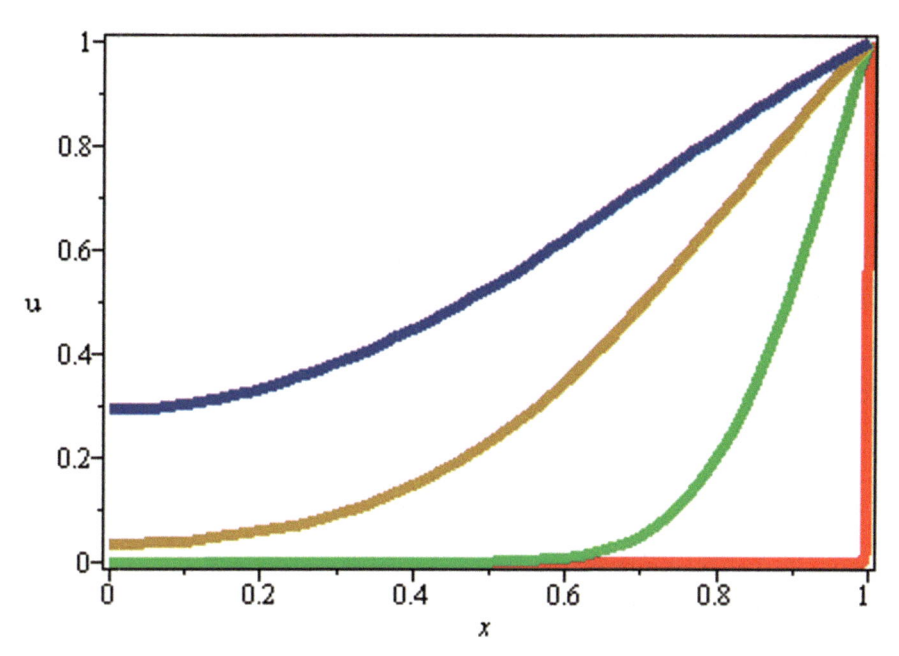

Fig. 8.12

8.1.5 Laplace Transform Technique for Parabolic Partial Differential Equations – Long Time Solution

The short time solutions obtained in section 8.1.4 (examples 8.1.5 and 8.1.6) require only a few terms in the infinite series at short times to converge. However, at long times the series requires a large number of terms and cannot be used efficiently. The long time solution can be obtained using Heaviside expansion theorem.[1] If we denote the solution obtained in the Laplace domain as F(s):

$$F(s) = \frac{p(s)}{q(s)} \qquad (8.7)$$

Typically when linear partial differential equations are solved using the Laplace transform method the solution obtained in the Laplace domain can be represented as in equation (8.7) and q(s) usually has an infinite number of roots. If $s = \mu_n$, $n = 1..\infty$ are the distinct roots of q(s), q(s) can be factorized as

$$q(s) = (s - \mu_1)(s - \mu_2)...(s - \mu_n)...(s - \mu_\infty) \qquad (8.8)$$

Using equation (8.8), equation (8.7) (if the order of q(s) is greater than the order of p(s)) can be converted to partial fractions as:

$$F(s) = \frac{p(s)}{q(s)} = \frac{A_1}{s-\mu_1} + \frac{A_2}{s-\mu_2} + \ldots \frac{A_n}{s-\mu_n} + \frac{A_{n+1}}{s-\mu_{n+1}} + \ldots \quad (8.9)$$

From equation (8.9) the coefficient A_n can be obtained by multiplying both sides by $s - \mu_n$.

$$\frac{p(s)(s-\mu_n)}{q(s)} = \frac{A_1(s-\mu_n)}{s-\mu_1} + \frac{A_2(s-\mu_n)}{s-\mu_2} + \ldots A_n + \frac{A_{n+1}(s-\mu_n)}{s-\mu_{n+1}} + \ldots$$

$$(8.10)$$

Next, the limit $s \to \mu_n$ is obtained from equation (8.10):

$$\lim_{s \to \mu_n}\left(\frac{p(s)(s-\mu_n)}{q(s)}\right) = A_n \quad (8.11)$$

Since $s = \mu_n$ is a root of $q(s)$ both the numerator and the denominator become zero when the limit is applied. Consequently, L'Hopital's rule is applied to find the limit of equation (8.11).

$$A_n = \lim_{s \to \mu_n}\left(\frac{p(s)(s-\mu_n)}{q(s)}\right) = \lim_{s \to \mu_n}\left(\frac{p'(s)(s-\mu_n)+p(s)}{q'(s)}\right) = \frac{p(\mu_n)}{q'(\mu_n)}$$

$$(8.12)$$

Once the coefficients are obtained, the inverse Laplace transform can be obtained from equation (8.9) as:

$$F(s) = \sum_{n=1}^{\infty} \frac{A_n}{s-\mu_n}$$

$$\Rightarrow \quad (8.13)$$

$$f(t) = L^{-1}(F(s)) = \sum_{n=1}^{\infty} L^{-1}\frac{A_n}{s-\mu_n} = \sum_{n=1}^{\infty} A_n \exp(\mu_n t)$$

Equation (8.13) gives the solution in the time domain. Often $s = 0$ happens to be an additional root of $q(s)$. In this case $F(s)$ can be written as:

$$F(s) = \frac{p(s)}{q(s)} = \frac{A_0}{s} + \sum_{n=1}^{\infty} \frac{A_n}{s-\mu_n} \quad (8.14)$$

A_0 is obtained by multiplying both sides of equation (8.14) and applying the limit $s \to 0$. A_0 is obtained by applying L'Hopital's rule as before as:

8.1 Laplace Transform Technique for PDEs in Finite Domains

$$A_0 = \lim_{s \to 0} \left(\frac{p(s)(s-\mu_n)}{q(s)} \right) = \lim_{s \to 0} \left(\frac{p'(s)(s-\mu_n) + p(s)}{q'(s)} \right) = \frac{p(0)}{q'(0)}$$

(8.15)

The solution in time domain is obtained as:

$$f(t) = L^{-1}(F(s)) = L^{-1}\left(\frac{A_0}{s}\right) + \sum_{n=1}^{\infty} L^{-1} \frac{A_n}{s - \mu_n} = A_0 + \sum_{n=1}^{\infty} A_n \exp(\mu_n t)$$

(8.16)

Equation (8.16) can be used to invert the solution obtained in the Laplace domain as long as all the roots of q(s) are distinct and the order of q(s) is greater than p(s).

Example 8.7. Heat Conduction with an Insulator Boundary Condition

Consider heat transfer in a rectangle with an insulator boundary condition at one end.[2] The dimensionless temperature profile is governed by:

$$\frac{\partial u}{\partial t} = \frac{\partial^2 u}{\partial x^2}$$

$$u(x,0) = 0$$

(8.17)

$$\frac{\partial u}{\partial x}(0,t) = 0 \text{ and } u(1,t) = 1$$

Equation (8.17) is solved in Maple below:

> restart:with(inttrans):with(plots):

First, the governing equations and boundary conditions are converted to the Laplace domain and solved in the Laplace domain:

> eq:=diff(u(x,t),t)=diff(u(x,t),x$2);

$$eq := \frac{\partial}{\partial t} u(x, t) = \frac{\partial^2}{\partial x^2} u(x, t)$$

> u(x,0):=0;

$$u(x, 0) := 0$$

> bc1:=diff(u(x,t),x)=0;

$$bc1 := \frac{\partial}{\partial x} u(x,t) = 0$$

> bc2:=u(x,t)=1;

$$bc2 := u(x,t) = 1$$

> eqs:=laplace(eq,t,s):
> eqs:=subs(laplace(u(x,t),t,s)=U(x),eqs);

$$eqs := s\,U(x) = \frac{d^2}{dx^2} U(x)$$

> bc1:=laplace(bc1,t,s):
> bc1:=subs(laplace(u(x,t),t,s)=U(x),bc1);

$$bc1 := \frac{d}{dx} U(x) = 0$$

> bc2:=laplace(bc2,t,s):
> bc2:=subs(laplace(u(x,t),t,s)=U(x),bc2);

$$bc2 := U(x) = \frac{1}{s}$$

> dsolve(eqs,U(x));

$$U(x) = \_C1\, e^{\sqrt{s}\, x} + \_C2\, e^{-\sqrt{s}\, x}$$

> U(x):=c[1]*cosh(s^(1/2)*x)+c[2]*sinh(s^(1/2)*x);

$$U(x) := c_1 \cosh(\sqrt{s}\, x) + c_2 \sinh(\sqrt{s}\, x)$$

> eq0:=eval(subs(x=0,bc1));

$$eq0 := c_2 \sqrt{s} = 0$$

> eq1:=eval(subs(x=1,bc2));

$$eq1 := c_1 \cosh(\sqrt{s}) + c_2 \sinh(\sqrt{s}) = \frac{1}{s}$$

> con:=solve({eq0,eq1},{c[1],c[2]});

$$con := \left\{ c_1 = \frac{1}{\cosh(\sqrt{s})\, s},\, c_2 = 0 \right\}$$

8.1 Laplace Transform Technique for PDEs in Finite Domains

The solution obtained in the Laplace domain is:

> U(x):=subs(con,U(x));

$$U(x) := \frac{\cosh(\sqrt{s}\ x)}{\cosh(\sqrt{s})\ s}$$

The polynomials are:

> P(s):=numer(U(x));

$$P(s) := \cosh(\sqrt{s}\ x)$$

> Q(s):=denom(U(x));

$$Q(s) := \cosh(\sqrt{s})\ s$$

Note that the order of q(s) is greater than the order of p(s).

> A(s):=P(s)/diff(Q(s),s);

$$A(s) := \frac{\cosh(\sqrt{s}\ x)}{\frac{1}{2}\sinh(\sqrt{s})\sqrt{s} + \cosh(\sqrt{s})}$$

The roots of Q(s) are found as:

> solve(Q(s),s);

$$-\frac{1}{4}\pi^2, 0$$

> _EnvAllSolutions := true;

$$\_EnvAllSolutions := true$$

> solve(Q(s),s);

$$-\frac{1}{4}\pi^2(1+2\_Z1\sim)^2, 0$$

The roots can be taken as:

> 0,-((2*n-1)*Pi/2)^2;

$$0, -\frac{1}{4}(2n-1)^2\pi^2$$

Next, the coefficients are found:

> A[n]:=simplify(subs(s=mu,A(s)));

$$A_n := \frac{2\cosh(\sqrt{\mu}\, x)}{\sinh(\sqrt{\mu})\sqrt{\mu} + 2\cosh(\sqrt{\mu})}$$

First A0 is found:

> A[0]:=subs(mu=0,A[n]);

$$A_0 := 1$$

The coefficient An for values n = 1..∞ can be found as:

> A[n]:=simplify(subs(mu^(1/2)=I*(2*n-1)/2*Pi,A[n]));

$$A_n := -\left(4\cos\left(\frac{1}{2}(2n-1)\pi x\right)\right) \Big/ \left(2\sin\left(\frac{1}{2}(2n-1)\pi\right)\pi n - \sin\left(\frac{1}{2}(2n-1)\pi\right)\pi - 4\cos\left(\frac{1}{2}(2n-1)\pi\right)\right)$$

A_n is simplified as:

> vars:={cos(1/2*(2*n-1)*Pi)=0,sin(1/2*(2*n-1)*Pi)=(-1)^(n-1)};

$$vars := \left\{\cos\left(\frac{1}{2}(2n-1)\pi\right) = 0, \sin\left(\frac{1}{2}(2n-1)\pi\right) = (-1)^{n-1}\right\}$$

> A[n]:=simplify(subs(vars,A[n]));

$$A_n := \frac{4(-1)^{-n}\cos\left(\frac{1}{2}(2n-1)\pi x\right)}{\pi(2n-1)}$$

The general terms in the Laplace domain solution are (see equation (8.16)):

> u0s:=A[0]*1/s;

$$u0s := \frac{1}{s}$$

The inverse Laplace transform is:

> u0t:=invlaplace(u0s,s,t);

$$u0t := 1$$

The term in the infinite series is

> uns:=A[n]/(s-mu);

8.1 Laplace Transform Technique for PDEs in Finite Domains

$$uns := \frac{4(-1)^{-n} \cos\left(\frac{1}{2}(2n-1)\pi x\right)}{\pi(2n-1)(s-\mu)}$$

> unt:=invlaplace(uns,s,t);

$$unt := \frac{4(-1)^{-n} \cos\left(\frac{1}{2}(2n-1)\pi x\right) e^{\mu t}}{\pi(2n-1)}$$

> unt:=subs(mu=-((2*n-1)/2*Pi)^2,unt);

$$unt := \frac{4(-1)^{-n} \cos\left(\frac{1}{2}(2n-1)\pi x\right) e^{-\frac{1}{4}(2n-1)^2 \pi^2 t}}{\pi(2n-1)}$$

The final solution is obtained as:

> U:=u0t+Sum(unt,n=1..infinity);

$$U := 1 + \sum_{n=1}^{\infty} \frac{4(-1)^{-n} \cos\left(\frac{1}{2}(2n-1)\pi x\right) e^{-\frac{1}{4}(2n-1)^2 \pi^2 t}}{\pi(2n-1)}$$

As in chapter 7, the initial condition is used at time, t = 0, to avoid Gibb's phenomenon:

> u:=piecewise(t=0,0,t>0,subs(infinity=20,U));

$$u := \begin{cases} 0 & t=0 \\ 1 + \sum_{n=1}^{20} \frac{4(-1)^{-n} \cos\left(\frac{1}{2}(2n-1)\pi x\right) e^{-\frac{1}{4}(2n-1)^2 \pi^2 t}}{\pi(2n-1)} & 0<t \end{cases}$$

The following plots are obtained:

> plot3d(u,x=0..1,t=0..1,axes=boxed,title="Figure 8.13",
labels=[x,t,"u"],orientation=[-150,60]);

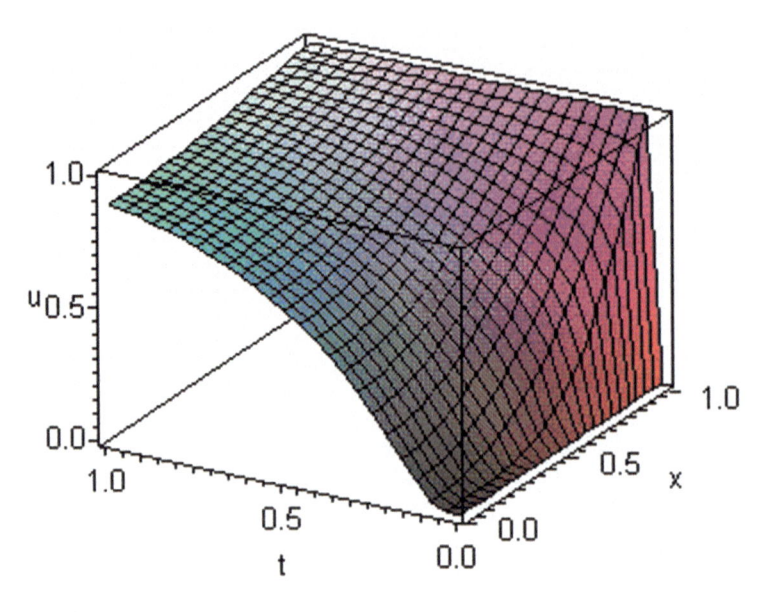

Fig. 8.13

> plot([subs(t=0,u),subs(t=0.1,u),subs(t=0.2,u),subs(t=0.3,u)],x=0..1,
axes=boxed,title="Figure 8.14",thickness=5,labels=[x,"u"]);

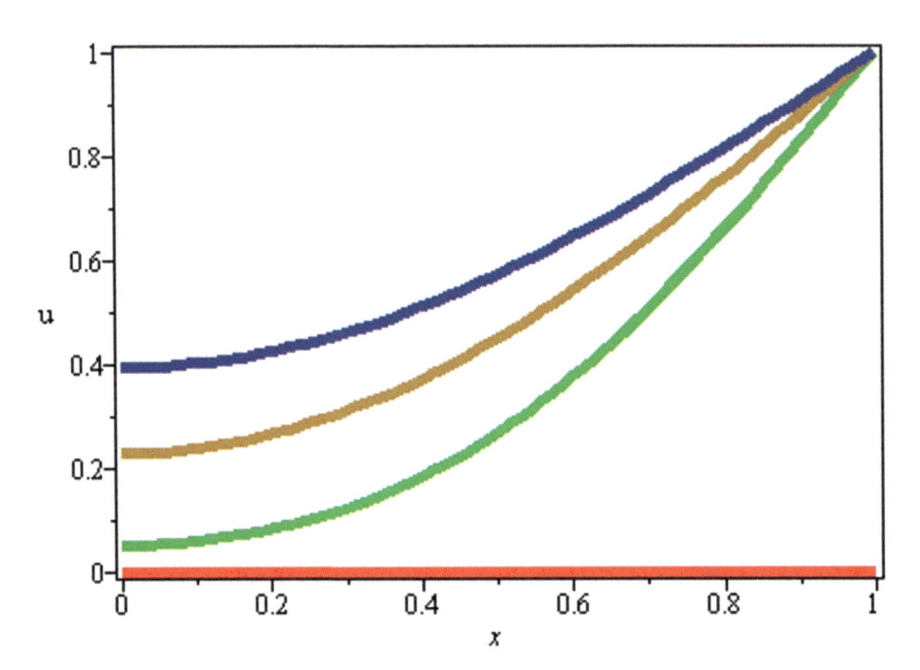

Fig. 8.14

8.1 Laplace Transform Technique for PDEs in Finite Domains

The short time solution for the same problem can be obtained using the methodology described in section 8.1.4 (examples 8.5 and 8.6) as:

> Ut:=U1t+U2t;

$$Ut := \sum_{n=1}^{\infty} (-1)^{1+n} \, invlaplace\left(\frac{e^{\sqrt{s}\,(x-2n+1)}}{s}, s, t\right) + \sum_{n=1}^{\infty} (-1)^1$$
$$+n \, invlaplace\left(\frac{e^{(-x-2n+1)\sqrt{s}}}{s}, s, t\right)$$

Example 8.8. Diffusion with Reaction

Consider Example 7.8, diffusion with reaction in a rectangle, which is solved here using the Laplace transform technique. The dimensionless concentration profile is governed by:

$$\frac{\partial u}{\partial t} = \frac{\partial^2 u}{\partial x^2} - \Phi^2 u$$

$$u(x,0) = 0 \qquad (8.18)$$

$$\frac{\partial u}{\partial x}(0,t) = 0 \text{ and } u(1,t) = 1$$

Equation (8.18) is solved in Maple below:

>restart:with(inttrans):with(plots):

> eq:=diff(u(x,t),t)=diff(u(x,t),x$2)-Phi^2*u(x,t);

$$eq := \frac{\partial}{\partial t} u(x,t) = \frac{\partial^2}{\partial x^2} u(x,t) - \Phi^2 u(x,t)$$

> u(x,0):=0;

$$u(x,0) := 0$$

> bc1:=u(x,t)=1;

$$bc1 := u(x,t) = 1$$

> bc2:=diff(u(x,t),x)=0;

$$bc2 := \frac{\partial}{\partial x} u(x,t) = 0$$

```
> eqs:=laplace(eq,t,s):
> eqs:=subs(laplace(u(x,t),t,s)=U(x),eqs);
```

$$eqs := s\, U(x) = \frac{d^2}{dx^2}\, U(x) - \Phi^2\, U(x)$$

```
> bc1:=laplace(bc1,t,s):
> bc1:=subs(laplace(u(x,t),t,s)=U(x),bc1);
```

$$bc1 := U(x) = \frac{1}{s}$$

```
> bc2:=laplace(bc2,t,s):
> bc2:=subs(laplace(u(x,t),t,s)=U(x),bc2);
```

$$bc2 := \frac{d}{dx}\, U(x) = 0$$

```
> dsolve(eqs,U(x));
```

$$U(x) = \_C1\, \sin\!\left(\sqrt{-s - \Phi^2}\, x\right) + \_C2\, \cos\!\left(\sqrt{-s - \Phi^2}\, x\right)$$

```
[1]> U(x):=c[1]*cosh((s+Phi^2)^(1/2)*x)+c[2]*sinh((s+Phi^2)^(1/2)*x);
```

$$U(x) := c_1\, \cosh\!\left(\sqrt{s + \Phi^2}\, x\right) + c_2\, \sinh\!\left(\sqrt{s + \Phi^2}\, x\right)$$

```
> eq0:=eval(subs(x=0,bc1)):
> eq1:=eval(subs(x=1,bc2)):
> con:=solve({eq0,eq1},{c[1],c[2]}):
> U(x):=subs(con,U(x));
```

$$U(x) := \frac{\cosh\!\left(\sqrt{s + \Phi^2}\, x\right)}{s}$$
$$- \frac{\sinh\!\left(\sqrt{s + \Phi^2}\right) \sinh\!\left(\sqrt{s + \Phi^2}\, x\right)}{\cosh\!\left(\sqrt{s + \Phi^2}\right) s}$$

```
> U(x):=combine(simplify(U(x))):
```

The shifting theorem is used to find the inverse Laplace transform[1] as: L-1F(s)=exp(-Φ2t)L=1F(s-Φ2).

8.1 Laplace Transform Technique for PDEs in Finite Domains 711

\> U(x):=factor(U(x));

$$U(x) := \frac{\cosh\left(\sqrt{s + \Phi^2}\ (x - 1)\right)}{\cosh\left(\sqrt{s + \Phi^2}\right) s}$$

\> U1(x):=subs(s=s-Phi^2,U(x));

$$U1(x) := \frac{\cosh\left(\sqrt{s}\ (x - 1)\right)}{\cosh\left(\sqrt{s}\right) \left(s - \Phi^2\right)}$$

\> P(s):=numer(U1(x));

$$P(s) := -\cosh\left(\sqrt{s}\ (x - 1)\right)$$

\> Q(s):=denom(U1(x));

$$Q(s) := \cosh\left(\sqrt{s}\right) \left(-s + \Phi^2\right)$$

\> A(s):=P(s)/diff(Q(s),s);

$$A(s) := -\frac{\cosh\left(\sqrt{s}\ (x - 1)\right)}{\frac{1}{2} \frac{\sinh\left(\sqrt{s}\right) \left(-s + \Phi^2\right)}{\sqrt{s}} - \cosh\left(\sqrt{s}\right)}$$

\> solve(Q(s),s);

$$-\frac{1}{4} \pi^2, \Phi^2$$

\> _EnvAllSolutions := true:

\> solve(Q(s),s):

The roots are:

\> Phi^2,-((2*n-1)*Pi/2)^2;

$$\Phi^2, -\frac{1}{4} (2n - 1)^2 \pi^2$$

\> A[n]:=simplify(subs(s=mu,A(s)));

$$A_n := -\frac{2 \cosh\left(\sqrt{\mu}\ (x - 1)\right) \sqrt{\mu}}{-\sinh\left(\sqrt{\mu}\right) \mu + \sinh\left(\sqrt{\mu}\right) \Phi^2 - 2 \cosh\left(\sqrt{\mu}\right) \sqrt{\mu}}$$

> A[0]:=subs(mu^(1/2)=Phi,mu=Phi^2,A[n]):
> A[0]:=simplify(A[0]);

$$A_0 := \frac{\cosh(\Phi(x-1))}{\cosh(\Phi)}$$

> A[n]:=simplify(subs(mu^(1/2)=I*(2*n-1)/2*Pi,mu=-((2*n-1)*Pi/2)^2,A[n])):
> vars:={cos(1/2*(2*n-1)*Pi)=0,sin(1/2*(2*n-1)*Pi)=(-1)^(n-1)}:
> A[n]:=simplify(subs(vars,A[n]));

$$A_n := \frac{4(-1)^{-n}(2n-1)\pi \cos\left(\frac{1}{2}(2n-1)\pi(x-1)\right)}{4\pi^2 n^2 - 4\pi^2 n + \pi^2 + 4\Phi^2}$$

> u0s:=A[0]*subs(mu=Phi^2,1/(s-mu));

$$u0s := \frac{\cosh(\Phi(x-1))}{\cosh(\Phi)(s-\Phi^2)}$$

> u0t:=invlaplace(u0s,s,t);

$$u0t := \frac{\cosh(\Phi(x-1))\, e^{\Phi^2 t}}{\cosh(\Phi)}$$

> uns:=A[n]/(s-mu);

$$uns := \frac{4(-1)^{-n}(2n-1)\pi \cos\left(\frac{1}{2}(2n-1)\pi(x-1)\right)}{\left(4\pi^2 n^2 - 4\pi^2 n + \pi^2 + 4\Phi^2\right)(s-\mu)}$$

> unt:=invlaplace(uns,s,t);

$$unt := \frac{4(-1)^{-n}(2n-1)\pi \cos\left(\frac{1}{2}(2n-1)\pi(x-1)\right) e^{\mu t}}{4\pi^2 n^2 - 4\pi^2 n + \pi^2 + 4\Phi^2}$$

> unt:=subs(mu=-((2*n-1)/2*Pi)^2,unt);

$$unt := \frac{1}{4\pi^2 n^2 - 4\pi^2 n + \pi^2 + 4\Phi^2}\left(4(-1)^{-n}(2n-1)\pi \cos\left(\frac{1}{2}(2n-1)\pi(x-1)\right) e^{-\frac{1}{4}(2n-1)^2 \pi^2 t}\right)$$

8.1 Laplace Transform Technique for PDEs in Finite Domains

The time domain solution is obtained by multiplying the inverse Laplace transform of U1(x) by exp(-Φ2t):

> U:=simplify(u0t*exp(-Phi^2*t))+Sum(unt,n=1..infinity)*exp(-Phi^2*t);

$$U := \frac{\cosh(\Phi(x-1))}{\cosh(\Phi)} + \left(\sum_{n=1}^{\infty} \frac{1}{4\pi^2 n^2 - 4\pi^2 n + \pi^2 + 4\Phi^2} \left(4 \right. \right.$$

$$\left. \left. (-1)^{-n} (2n-1) \pi \cos\left(\frac{1}{2} (2n-1) \pi (x-1)\right) e^{-\frac{1}{4}(2n-1)^2 \pi^2 t} \right) \right) e^{-\Phi^2 t}$$

> u:=piecewise(t=0,0,t>0,subs(infinity=20,U)):

The following plots are obtained:

> plot3d(subs(Phi=1,u),x=1..0,t=0.5..0,axes=boxed,title="Figure 8.15", labels=[x,t,"u"],orientation=[-45,60]);

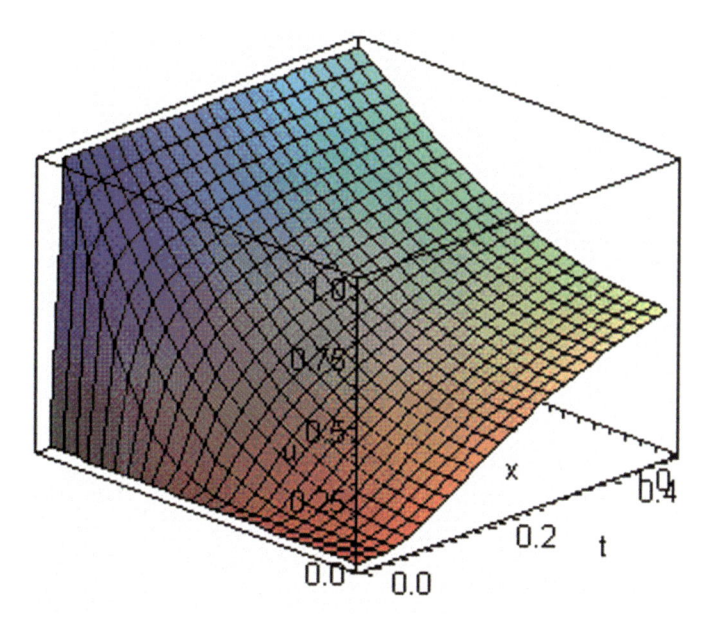

Fig. 8.15

The solution obtained matches the separation of variables solution obtained in example 7.8.

Example 8.9. Heat Conduction with Time Dependent Boundary Conditions

Consider that one of the main advantages of the Laplace transform technique is that it can be used for time dependent boundary conditions, also. The separation of variables technique cannot be directly used and one has to use Duhamel's superposition theorem[1] for this purpose. Consider the modification of example 8.7:

$$\frac{\partial u}{\partial t} = \frac{\partial^2 u}{\partial x^2}$$
$$u(x,0) = 0 \qquad (8.19)$$
$$\frac{\partial u}{\partial x}(0,t) = 0 \text{ and } u(1,t) = \exp(-t)$$

Equation (8.19) is solved by slightly modifying the Maple program used for example 8.7 as:

> restart:with(inttrans):with(plots):
> eq:=diff(u(x,t),t)=diff(u(x,t),x$2);

$$eq := \frac{\partial}{\partial t} u(x,t) = \frac{\partial^2}{\partial x^2} u(x,t)$$

> u(x,0):=0;

$$u(x,0) := 0$$

> bc1:=diff(u(x,t),x)=0;

$$bc1 := \frac{\partial}{\partial x} u(x,t) = 0$$

> bc2:=u(x,t)=exp(-t);

$$bc2 := u(x,t) = e^{-t}$$

> eqs:=laplace(eq,t,s):
> eqs:=subs(laplace(u(x,t),t,s)=U(x),eqs);

$$eqs := s\, U(x) = \frac{d^2}{dx^2} U(x)$$

8.1 Laplace Transform Technique for PDEs in Finite Domains

> bc1:=laplace(bc1,t,s):
> bc1:=subs(laplace(u(x,t),t,s)=U(x),bc1);

$$bc1 := \frac{d}{dx} U(x) = 0$$

> bc2:=laplace(bc2,t,s):
> bc2:=subs(laplace(u(x,t),t,s)=U(x),bc2);

$$bc2 := U(x) = \frac{1}{1+s}$$

> dsolve(eqs,U(x));

$$U(x) = \_C1 \, e^{\sqrt{s}\, x} + \_C2 \, e^{-\sqrt{s}\, x}$$

> U(x):=c[1]*cosh(s^(1/2)*x)+c[2]*sinh(s^(1/2)*x);

$$U(x) := c_1 \cosh(\sqrt{s}\, x) + c_2 \sinh(\sqrt{s}\, x)$$

> eq0:=eval(subs(x=0,bc1)):
> eq1:=eval(subs(x=1,bc2)):
> con:=solve({eq0,eq1},{c[1],c[2]}):
> U(x):=subs(con,U(x));

$$U(x) := \frac{\cosh(\sqrt{s}\, x)}{\cosh(\sqrt{s})\,(1+s)}$$

> P(s):=numer(U(x));

$$P(s) := \cosh(\sqrt{s}\, x)$$

> Q(s):=denom(U(x));

$$Q(s) := \cosh(\sqrt{s})\,(1+s)$$

> A(s):=P(s)/diff(Q(s),s);

$$A(s) := \frac{\cosh(\sqrt{s}\, x)}{\frac{1}{2} \frac{\sinh(\sqrt{s})\,(1+s)}{\sqrt{s}} + \cosh(\sqrt{s})}$$

> solve(Q(s),s);
$$-\frac{1}{4}\pi^2, -1$$

> _EnvAllSolutions := true:
> solve(Q(s),s):

The roots are:

> -1,-((2*n-1)*Pi/2)^2;
$$-1, -\frac{1}{4}(2n-1)^2\pi^2$$

> A[n]:=simplify(subs(s=mu,A(s)));
$$A_n := \frac{2\cosh(\sqrt{\mu}\,x)\sqrt{\mu}}{\sinh(\sqrt{\mu}) + \sinh(\sqrt{\mu})\mu + 2\cosh(\sqrt{\mu})\sqrt{\mu}}$$

> A[0]:=subs(mu^(1/2)=I,mu=-1,A[n]):
> A[0]:=simplify(A[0]);
$$A_0 := \frac{\cos(x)}{\cos(1)}$$

> A[n]:=simplify(subs(mu^(1/2)=I*(2*n-1)/2*Pi,mu=-((2*n-1)*Pi/2)^2,A[n])):
> vars:={cos(1/2*(2*n-1)*Pi)=0,sin(1/2*(2*n-1)*Pi)=(-1)^(n-1)}:
> A[n]:=simplify(subs(vars,A[n]));
$$A_n := \frac{4(-1)^{-n}(2n-1)\pi\cos\left(\frac{1}{2}(2n-1)\pi x\right)}{-4 + 4\pi^2 n^2 - 4\pi^2 n + \pi^2}$$

> u0s:=A[0]*subs(mu=-1,1/(s-mu));
$$u0s := \frac{\cos(x)}{\cos(1)(1+s)}$$

> u0t:=invlaplace(u0s,s,t);
$$u0t := \frac{\cos(x)\,e^{-t}}{\cos(1)}$$

8.1 Laplace Transform Technique for PDEs in Finite Domains

> uns:=A[n]/(s-mu);

$$uns := \frac{4\,(-1)^{-n}\,(2n-1)\,\pi\,\cos\!\left(\frac{1}{2}(2n-1)\,\pi\,x\right)}{\left(-4+4\pi^2 n^2 - 4\pi^2 n + \pi^2\right)(s-\mu)}$$

> unt:=invlaplace(uns,s,t);

$$unt := \frac{4\cos\!\left(\frac{1}{2}(2n-1)\,\pi\,x\right) e^{\mu t}\,\pi\,(-1)^{-n}\,(2n-1)}{(2\pi n - \pi)^2 - 4}$$

> unt:=subs(mu=-((2*n-1)/2*Pi)^2,unt);

$$unt := \frac{4\cos\!\left(\frac{1}{2}(2n-1)\,\pi\,x\right) e^{-\frac{1}{4}(2n-1)^2\pi^2 t}\,\pi\,(-1)^{-n}\,(2n-1)}{(2\pi n - \pi)^2 - 4}$$

> U:=u0t+Sum(unt,n=1..infinity);

$$U := \frac{\cos(x)\,e^{-t}}{\cos(1)} + \sum_{n=1}^{\infty} \frac{4\cos\!\left(\frac{1}{2}(2n-1)\,\pi\,x\right) e^{-\frac{1}{4}(2n-1)^2\pi^2 t}\,\pi\,(-1)^{-n}\,(2n-1)}{(2\pi n - \pi)^2 - 4}$$

> u:=piecewise(t=0,0,t>0,subs(infinity=20,U)):

The following plots are obtained:

> plot3d(u,x=0..1,t=0..1,axes=boxed,title="Figure 8.16",
labels=[x,t,"u"],orientation=[-150,50]);

718   8 Laplace Transform Technique for Partial Differential Equations

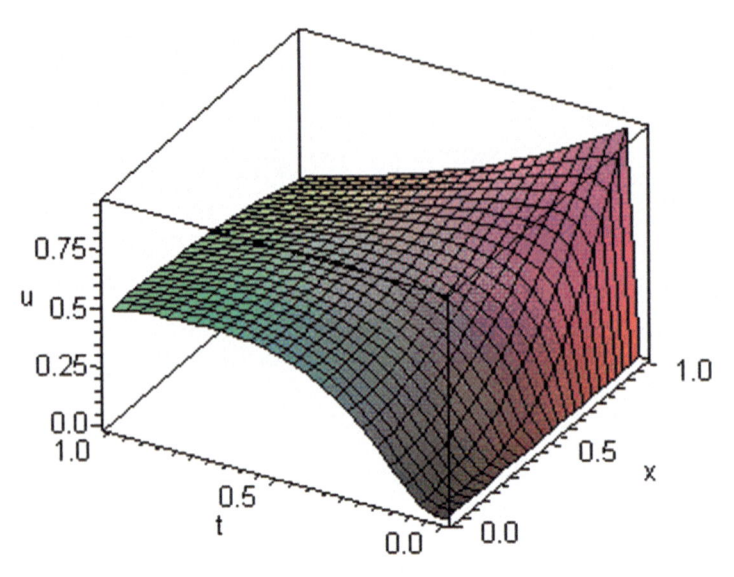

Fig. 8.16

The dimensionless temperature at the surface x =0 reaches a maximum and then decreases as a function of time:

> plot(subs(x=0,u),t=0..2,thickness=3,axes=boxed,title="Figure 8.17", labels=[t,"u(0,t)"]);

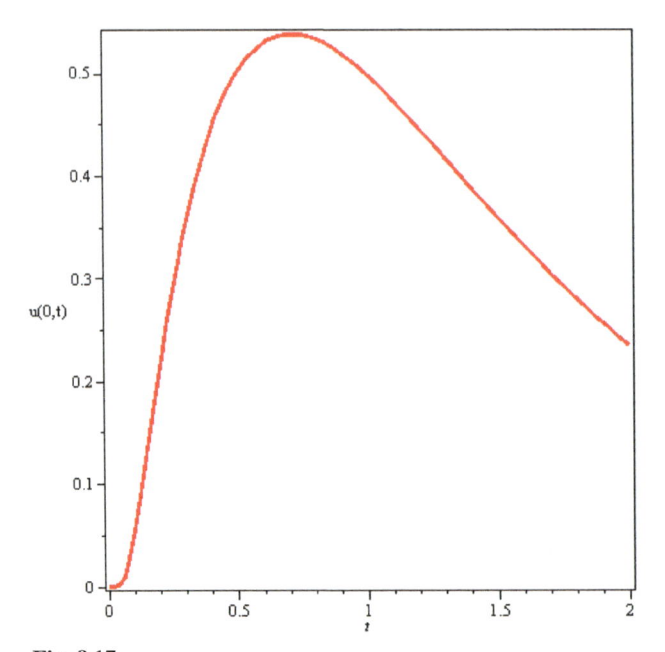

Fig. 8.17

8.1.6 Laplace Transform Technique for Parabolic Partial Differential Equations – Heaviside Expansion Theorem for Multiple Roots

In section 8.1.5, q(s) had only distinct roots. Some times when partial differential equations are solved using the Laplace transform technique, the polynomial q(s) has multiple roots in addition to the infinite number of distinct roots. In this section, we consider q(s) as having an infinite number of distinct roots and a different root s = µ0 repeated twice. Even though one can invert when q(s) has any number of roots repeated any number of times, for most of the practical problems we will encounter roots being repeated only twice. In this case, the solution obtained in the Laplace domain can be expressed as:

$$F(s) = \frac{p(s)}{q(s)} = \frac{B_1}{s-\mu_0} + \frac{B_2}{(s-\mu_0)^2} + \sum_{n=1}^{\infty} \frac{A_n}{s-\mu_n} \quad (8.20)$$

First, B_2 is obtained by multiplying both sides by $(s - \mu_0)^2$ and applying the limit $s \to \mu_0$:

$$\frac{p(s)(s-\mu_0)^2}{q(s)} = B_1(s-\mu_0) + B_2 + (s-\mu_0)^2 \sum_{n=1}^{\infty} \frac{A_n}{s-\mu_n} \quad (8.21)$$

B_2 is obtained by applying the limit $s \to \mu_0$:

$$B_2 = \lim_{s \to \mu_0} \left(\frac{p(s)(s-\mu_0)^2}{q(s)} \right) \quad (8.22)$$

Maple's 'limit' command can be used to find the limit. Next, B_1 is obtained by differentiating both sides of equation (8.21) with respect to s:

$$\frac{d}{ds}\left(\frac{p(s)(s-\mu_0)^2}{q(s)} \right) = B_1 + 2(s-\mu_0) \sum_{n=1}^{\infty} \frac{A_n}{s-\mu_n} \quad (8.23)$$

B_2 is obtained by applying the limit $s \to \mu_0$ in equation (8.23):

$$B_1 = \lim_{s \to \mu_0} \left(\frac{d}{ds}\left(\frac{p(s)(s-\mu_0)^2}{q(s)} \right) \right) \quad (8.24)$$

Once the constants B_1 and B_2 are obtained equation (8.20) can be inverted as:

$$f(t) = L^{-1}(F(s)) = L^{-1}\left(\frac{B_1}{s-\mu_n}\right) + L^{-1}\left(\frac{B_2}{(s-\mu_n)^2}\right) + \sum_{n=1}^{\infty} L^{-1}\frac{A_n}{s-\mu_n}$$

$$= B_1 \exp(\mu_0 t) + B_2 t \exp(\mu_0 t) + \sum_{n=1}^{\infty} A_n \exp(\mu_n t)$$

(8.25)

Example 8.10. Heat Transfer in a Rectangle

Consider Example 7.1 heat transfer in a rectangle[2] which is solved here again using the Laplace transform technique. The dimensionless temperature profile is governed by:

$$\frac{\partial u}{\partial t} = \frac{\partial^2 u}{\partial x^2}$$
$$u(x,0) = 1 \qquad (8.26)$$
$$u(0,t) = 0 \text{ and } u(1,t) = 0$$

Equation (8.26) is solved in Maple below:

> restart:with(inttrans):with(plots):
> eq:=diff(u(x,t),t)=diff(u(x,t),x$2);

$$eq := \frac{\partial}{\partial t} u(x,t) = \frac{\partial^2}{\partial x^2} u(x,t)$$

> u(x,0):=1;

$$u(x,0) := 1$$

> bc1:=u(x,t)=0;

$$bc1 := u(x,t) = 0$$

> bc2:=u(x,t)=0;

$$bc2 := u(x,t) = 0$$

> eqs:=laplace(eq,t,s):
> eqs:=subs(laplace(u(x,t),t,s)=U(x),eqs);

8.1 Laplace Transform Technique for PDEs in Finite Domains

$$eqs := s\,U(x) - 1 = \frac{d^2}{dx^2}U(x)$$

> bc1:=laplace(bc1,t,s):
> bc1:=subs(laplace(u(x,t),t,s)=U(x),bc1);

$$bc1 := U(x) = 0$$

> bc2:=laplace(bc2,t,s):
> bc2:=subs(laplace(u(x,t),t,s)=U(x),bc2);

$$bc2 := U(x) = 0$$

> dsolve(eqs,U(x));

$$U(x) = e^{\sqrt{s}\,x}\_C2 + e^{-\sqrt{s}\,x}\_C1 + \frac{1}{s}$$

> U(x):=c[1]*cosh(s^(1/2)*x)+c[2]*sinh(s^(1/2)*x)+1/s;

$$U(x) := c_1 \cosh(\sqrt{s}\,x) + c_2 \sinh(\sqrt{s}\,x) + \frac{1}{s}$$

> eq0:=eval(subs(x=0,bc1)):
> eq1:=eval(subs(x=1,bc2)):
> con:=solve({eq0,eq1},{c[1],c[2]}):
> U(x):=subs(con,U(x));

$$U(x) := -\frac{\cosh(\sqrt{s}\,x)}{s} + \frac{(\cosh(\sqrt{s})-1)\sinh(\sqrt{s}\,x)}{\sinh(\sqrt{s})\,s} + \frac{1}{s}$$

> U(x):=factor(combine(simplify(U(x))));

$$U(x) := \frac{\sinh(\sqrt{s}\,(x-1)) - \sinh(\sqrt{s}\,x) + \sinh(\sqrt{s})}{\sinh(\sqrt{s})\,s}$$

> P(s):=numer(U(x));

$$P(s) := \sinh(\sqrt{s}\,(x-1)) - \sinh(\sqrt{s}\,x) + \sinh(\sqrt{s})$$

> Q(s):=denom(U(x));

$$Q(s) := \sinh(\sqrt{s})\,s$$

> solve(Q(s),s);

$$0$$

> _EnvAllSolutions := true;

$$\_EnvAllSolutions := true$$

> solve(Q(s),s);

$$-\pi^2 \_Z1\sim^2, 0$$

s = 0 is repeated twice (one root coming from 2 and the other coming from the sine term in q(s)). The roots are (where n goes from 1 to infinity):

> 0,0,-n^2*Pi^2;

$$0, 0, -n^2 \pi^2$$

> mu0:=0;

$$\mu 0 := 0$$

The coefficients B2 and B1 are found as (equations (8.22) and (8.24)):

> b[2]:=(s-mu0)^2*P(s)/Q(s);

$$b_2 := \frac{s\left(\sinh(\sqrt{s}\ (x-1)) - \sinh(\sqrt{s}\ x) + \sinh(\sqrt{s})\right)}{\sinh(\sqrt{s})}$$

> B[2]:=limit(b[2],s=0);

$$B_2 := 0$$

> b[1]:=diff(b[2],s):
> B[1]:=limit(b[1],s=0);

$$B_1 := 0$$

For this problem the contribution from the repeated root s = 0 is zero. This is not always true as shown in the next example.

> A(s):=P(s)/diff(Q(s),s):
> A[n]:=simplify(subs(s=mu,A(s)));

$$A_n := \frac{2\left(\sinh(\sqrt{\mu}\ (x-1)) - \sinh(\sqrt{\mu}\ x) + \sinh(\sqrt{\mu})\right)}{\cosh(\sqrt{\mu})\sqrt{\mu} + 2\sinh(\sqrt{\mu})}$$

> A[n]:=simplify(subs(mu^(1/2)=I*n*Pi,mu=-n^2*Pi^2,A[n])):
> vars:={cos(n*Pi)=(-1)^n,sin(n*Pi)=0};

8.1 Laplace Transform Technique for PDEs in Finite Domains 723

$$vars := \{\cos(n\pi) = (-1)^n, \sin(n\pi) = 0\}$$

> A[n]:=simplify(subs(vars,A[n])):
> A[n]:=simplify(subs(vars,expand(A[n])));

$$A_n := -\frac{2\sin(n\pi x)\left(-1 + (-1)^{-n}\right)}{n\pi}$$

> b1s:=B[1]*subs(mu0=0,1/(s-mu0));

$$b1s := 0$$

> b1t:=invlaplace(b1s,s,t);

$$b1t := 0$$

> b2s:=B[2]*subs(mu0=0,1/(s-mu0)^2);

$$b2s := 0$$

> b2t:=invlaplace(b2s,s,t);

$$b2t := 0$$

> uns:=A[n]/(s-mu);

$$uns := -\frac{2\sin(n\pi x)\left(-1 + (-1)^{-n}\right)}{n\pi(s-\mu)}$$

> unt:=invlaplace(uns,s,t);

$$unt := \frac{2\sin(n\pi x)\,e^{\mu t}\left(1 + (-1)^{1-n}\right)}{n\pi}$$

> unt:=subs(mu=-n^2*Pi^2,unt);

$$unt := \frac{2\sin(n\pi x)\,e^{-n^2\pi^2 t}\left(1 + (-1)^{1-n}\right)}{n\pi}$$

The solution is obtained and plotted as:

> U:=b1t+b2t+Sum(unt,n=1..infinity);

$$U := \sum_{n=1}^{\infty} \frac{2\sin(n\pi x)\,e^{-n^2\pi^2 t}\left(1 + (-1)^{1-n}\right)}{n\pi}$$

724   8 Laplace Transform Technique for Partial Differential Equations

> u:=piecewise(t=0,1,t>0,subs(infinity=20,U)):

> plot3d(u,x=0..1,t=0..0.5,axes=boxed,title="Figure 8.18",
labels=[x,t,"u"],orientation=[45,60]);

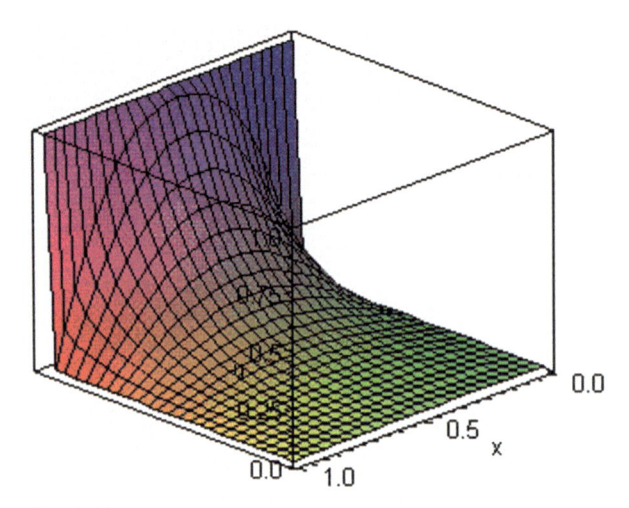

Fig. 8.18

> plot([subs(t=0,u),subs(t=0.01,u),subs(t=0.05,u),subs(t=0.1,u)],x=0..1,
axes=boxed,title="Figure 8.19",thickness=5,labels=[x,"u"]);

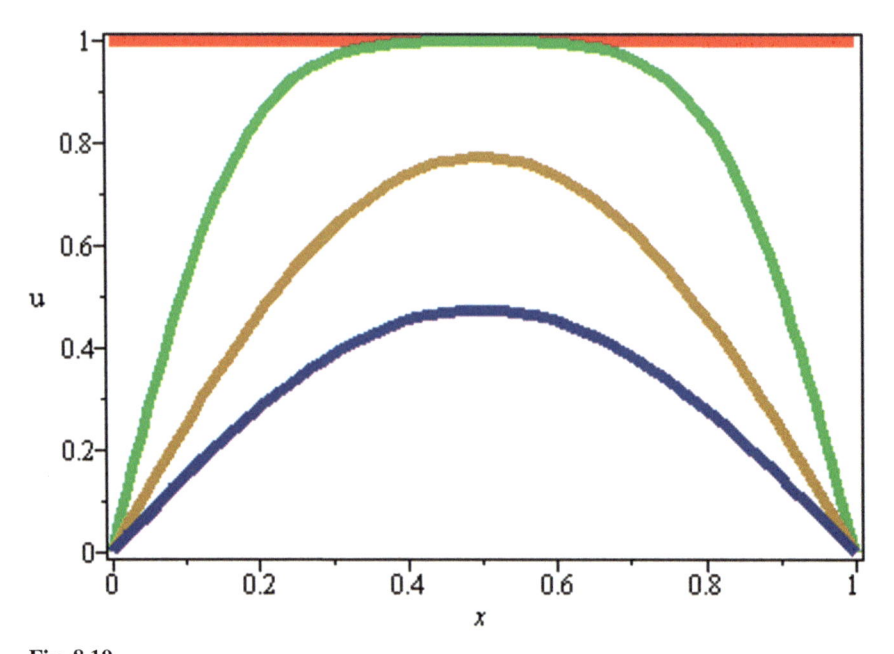

Fig. 8.19

8.1 Laplace Transform Technique for PDEs in Finite Domains

Example 8.11. Diffusion in a Slab with Nonhomogeneous Flux Boundary Conditions during Charging of a Battery

Consider Example 7.9, charging of a planar battery electrode, which is solved here using the Laplace transform technique. The dimensionless concentration profile is governed by:

$$\frac{\partial u}{\partial t} = \frac{\partial^2 u}{\partial x^2}$$

$$u(x,0) = 0 \quad (8.27)$$

$$\frac{\partial u}{\partial x}(0,t) = 0 \text{ and } \frac{\partial u}{\partial x}(1,t) = \delta$$

Equation (8.27) is solved in Maple and the results obtained are given below:

> restart:with(inttrans):with(plots):
> eq:=diff(u(x,t),t)=diff(u(x,t),x$2);

$$eq := \frac{\partial}{\partial t} u(x, t) = \frac{\partial^2}{\partial x^2} u(x, t)$$

> u(x,0):=0;

$$u(x, 0) := 0$$

> bc1:=diff(u(x,t),x)=0;

$$bc1 := \frac{\partial}{\partial x} u(x, t) = 0$$

> bc2:=diff(u(x,t),x)=delta;

$$bc2 := \frac{\partial}{\partial x} u(x, t) = \delta$$

> eqs:=laplace(eq,t,s):
> eqs:=subs(laplace(u(x,t),t,s)=U(x),eqs);

$$eqs := s\, U(x) = \frac{d^2}{dx^2} U(x)$$

726   8 Laplace Transform Technique for Partial Differential Equations

> bc1:=laplace(bc1,t,s):
> bc1:=subs(laplace(u(x,t),t,s)=U(x),bc1);

$$bc1 := \frac{d}{dx} U(x) = 0$$

> bc2:=laplace(bc2,t,s):
> bc2:=subs(laplace(u(x,t),t,s)=U(x),bc2);

$$bc2 := \frac{d}{dx} U(x) = \frac{\delta}{s}$$

> dsolve(eqs,U(x));

$$U(x) = \_C1\, e^{\sqrt{s}\, x} + \_C2\, e^{-\sqrt{s}\, x}$$

> U(x):=c[1]*cosh(s^(1/2)*x)+c[2]*sinh(s^(1/2)*x);

$$U(x) := c_1 \cosh(\sqrt{s}\, x) + c_2 \sinh(\sqrt{s}\, x)$$

> eq0:=eval(subs(x=0,bc1)):
> eq1:=eval(subs(x=1,bc2)):
> con:=solve({eq0,eq1},{c[1],c[2]}):
> U(x):=subs(con,U(x));

$$U(x) := \frac{\delta \cosh(\sqrt{s}\, x)}{\sinh(\sqrt{s})\, s^{3/2}}$$

> U(x):=factor(combine(simplify(U(x))));

$$U(x) := \frac{\delta \cosh(\sqrt{s}\, x)}{\sinh(\sqrt{s})\, s^{3/2}}$$

> P(s):=numer(U(x));

$$P(s) := \delta \cosh(\sqrt{s}\, x)$$

> Q(s):=denom(U(x));

$$Q(s) := \sinh(\sqrt{s})\, s^{3/2}$$

> solve(Q(s),s);

$$0$$

8.1 Laplace Transform Technique for PDEs in Finite Domains

> _EnvAllSolutions := true;

$$\_EnvAllSolutions := true$$

> solve(Q(s),s);

$$-\pi^2 \_Z1\!\sim^2, 0$$

> 0,0,-n^2*Pi^2;

$$0, 0, -n^2 \pi^2$$

> mu0:=0;

$$\mu 0 := 0$$

> b[2]:=(s-mu0)^2*P(s)/Q(s);

$$b_2 := \frac{\sqrt{s}\ \delta \cosh(\sqrt{s}\ x)}{\sinh(\sqrt{s})}$$

> B[2]:=limit(b[2],s=0);

$$B_2 := \delta$$

> b[1]:=diff(b[2],s):
> B[1]:=limit(b[1],s=0);

$$B_1 := \frac{1}{2}\delta x^2 - \frac{1}{6}\delta$$

> A(s):=P(s)/diff(Q(s),s):
> A[n]:=simplify(subs(s=mu,A(s)));

$$A_n := \frac{2\delta \cosh(\sqrt{\mu}\ x)}{\cosh(\sqrt{\mu})\ \mu + 3 \sinh(\sqrt{\mu})\ \sqrt{\mu}}$$

> A[n]:=simplify(subs(mu^(1/2)=I*n*Pi,mu=-n^2*Pi^2,A[n])):
> vars:={cos(n*Pi)=(-1)^n,sin(n*Pi)=0};

$$vars := \{\cos(n\ \pi) = (-1)^n, \sin(n\ \pi) = 0\}$$

> A[n]:=simplify(subs(vars,A[n])):
> A[n]:=simplify(subs(vars,expand(A[n])));

$$A_n := \frac{2\,(-1)^{1-n}\,\delta\,\cos(n\pi x)}{n^2\pi^2}$$

> b1s:=B[1]*subs(mu0=0,1/(s-mu0));

$$b1s := \frac{\frac{1}{2}\delta x^2 - \frac{1}{6}\delta}{s}$$

> b1t:=invlaplace(b1s,s,t);

$$b1t := \frac{1}{6}\delta\left(3x^2 - 1\right)$$

> b2s:=B[2]*subs(mu0=0,1/(s-mu0)^2);

$$b2s := \frac{\delta}{s^2}$$

> b2t:=invlaplace(b2s,s,t);

$$b2t := \delta\,t$$

> uns:=A[n]/(s-mu);

$$uns := \frac{2\,(-1)^{1-n}\,\delta\,\cos(n\pi x)}{n^2\pi^2\,(s-\mu)}$$

> unt:=invlaplace(uns,s,t);

$$unt := -\frac{2\,(-1)^{-n}\,\delta\,\cos(n\pi x)\,e^{\mu t}}{n^2\pi^2}$$

> unt:=subs(mu=-n^2*Pi^2,unt);

$$unt := -\frac{2\,(-1)^{-n}\,\delta\,\cos(n\pi x)\,e^{-n^2\pi^2 t}}{n^2\pi^2}$$

> U:=b1t+b2t+Sum(unt,n=1..infinity);

$$U := \frac{1}{6}\delta\left(3x^2 - 1\right) + \delta\,t + \sum_{n=1}^{\infty}\left(-\frac{2\,(-1)^{-n}\,\delta\,\cos(n\pi x)\,e^{-n^2\pi^2 t}}{n^2\pi^2}\right)$$

8.1 Laplace Transform Technique for PDEs in Finite Domains

```
> u:=piecewise(t=0,0,t>0,subs(infinity=20,U)):
> plot3d(subs(delta=1,u),x=0..1,t=0..0.5,axes=boxed,title="Figure 8.20",
labels=[x,t,"u"],orientation=[-135,60]);
```

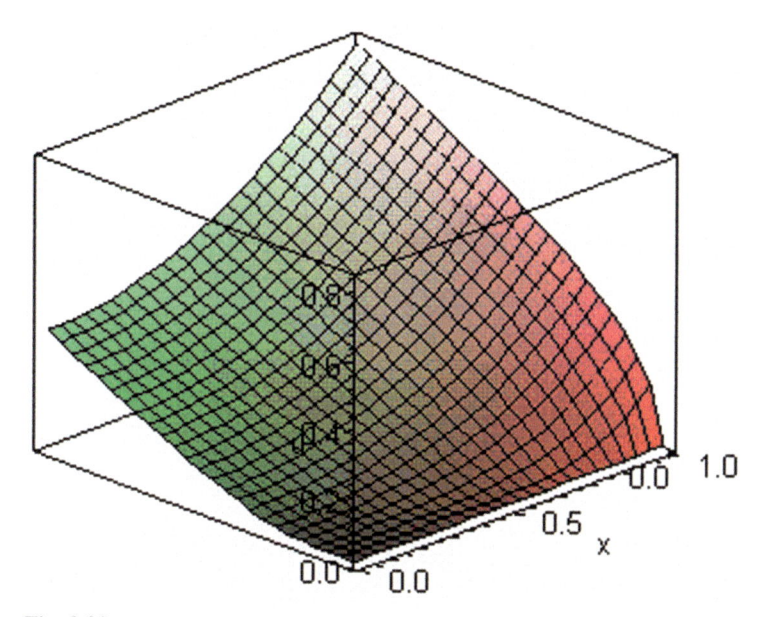

Fig. 8.20

Example 8.12. Distribution of Overpotential in a Porous Electrode

During the galvanostatic discharge of porous electrodes in the absence of concentration gradients, overpotential is governed by the following equation:

$$\frac{\partial u}{\partial t} = \frac{\partial^2 u}{\partial x^2} - v^2 u$$

$$u(x,0) = 0 \qquad (8.28)$$

$$\frac{\partial u}{\partial x}(0,t) = -\delta \text{ and } \frac{\partial u}{\partial x}(1,t) = \delta\beta$$

where v is the modified exchange current density, δ is the applied current density and β is the ratio of solution phase conductivity to solid phase conductivity (usually < 1). Equation (8.28) is solved in Maple and the results obtained are given below:

Example 8.12

```
> restart:with(inttrans):with(plots):
> eq:=diff(u(x,t),t)=diff(u(x,t),x$2)-nu^2*u(x,t);
```

$$eq := \frac{\partial}{\partial t} u(x,t) = \frac{\partial^2}{\partial x^2} u(x,t) - v^2 u(x,t)$$

```
> u(x,0):=0;
```

$$u(x,0) := 0$$

```
> bc1:=diff(u(x,t),x)=-delta;
```

$$bc1 := \frac{\partial}{\partial x} u(x,t) = -\delta$$

```
> bc2:=diff(u(x,t),x)=delta*beta;
```

$$bc2 := \frac{\partial}{\partial x} u(x,t) = \delta \beta$$

```
> eqs:=laplace(eq,t,s):
> eqs:=subs(laplace(u(x,t),t,s)=U(x),eqs);
```

$$eqs := s\, U(x) = \frac{d^2}{dx^2} U(x) - v^2 U(x)$$

```
> bc1:=laplace(bc1,t,s):
> bc1:=subs(laplace(u(x,t),t,s)=U(x),bc1);
```

$$bc1 := \frac{d}{dx} U(x) = -\frac{\delta}{s}$$

```
> bc2:=laplace(bc2,t,s):
> bc2:=subs(laplace(u(x,t),t,s)=U(x),bc2);
```

$$bc2 := \frac{d}{dx} U(x) = \frac{\delta \beta}{s}$$

```
> dsolve(eqs,U(x));
```

$$U(x) = \_C1 \sin\left(\sqrt{-s-v^2}\, x\right) + \_C2 \cos\left(\sqrt{-s-v^2}\, x\right)$$

8.1 Laplace Transform Technique for PDEs in Finite Domains 731

> U(x):=c[1]*cosh((s+nu^2)^(1/2)*x)+c[2]*sinh((s+nu^2)^(1/2)*x);

$$U(x) := c_1 \cosh\left(\sqrt{s+v^2}\, x\right) + c_2 \sinh\left(\sqrt{s+v^2}\, x\right)$$

> eq0:=eval(subs(x=0,bc1)):

> eq1:=eval(subs(x=1,bc2)):

> con:=solve({eq0,eq1},{c[1],c[2]}):

> U(x):=subs(con,U(x));

$$U(x) := \frac{\delta\left(\cosh\left(\sqrt{s+v^2}\right) + \beta\right)\cosh\left(\sqrt{s+v^2}\, x\right)}{\sinh\left(\sqrt{s+v^2}\right)\sqrt{s+v^2}\, s}$$
$$- \frac{\delta \sinh\left(\sqrt{s+v^2}\, x\right)}{\sqrt{s+v^2}\, s}$$

> U(x):=factor(combine(simplify(U(x))));

$$U(x) := \frac{\delta\left(\cosh\left(\sqrt{s+v^2}\,(x-1)\right) + \cosh\left(\sqrt{s+v^2}\, x\right)\beta\right)}{\sinh\left(\sqrt{s+v^2}\right)\sqrt{s+v^2}\, s}$$

> U1(x):=subs(s=s-nu^2,U(x));

$$U1(x) := \frac{\delta\left(\cosh\left(\sqrt{s}\,(x-1)\right) + \cosh\left(\sqrt{s}\, x\right)\beta\right)}{\sinh\left(\sqrt{s}\right)\sqrt{s}\,(s-v^2)}$$

> P(s):=numer(U1(x));

$$P(s) := -\delta\left(\cosh\left(\sqrt{s}\,(x-1)\right) + \cosh\left(\sqrt{s}\, x\right)\beta\right)$$

> Q(s):=denom(U1(x));

$$Q(s) := \sinh\left(\sqrt{s}\right)\sqrt{s}\,(-s+v^2)$$

> solve(Q(s),s);

$$0, v^2$$

> _EnvAllSolutions := true;

$$\_EnvAllSolutions := true$$

> solve(Q(s),s);

$$-\pi^2 \_Z1{\sim}^2, 0, v^2$$

The roots are:

> 0,0,nu^2,-n^2*Pi^2;

$$0, 0, v^2, -n^2 \pi^2$$

> mu0:=0;

$$\mu 0 := 0$$

> b[2]:=(s-mu0)^2*P(s)/Q(s);

$$b_2 := -\frac{s^{3/2} \delta \left(\cosh(\sqrt{s}\ (x-1)) + \cosh(\sqrt{s}\ x)\ \beta\right)}{\sinh(\sqrt{s})\ (-s+v^2)}$$

> B[2]:=limit(b[2],s=0);

$$B_2 := 0$$

> b[1]:=diff(b[2],s):

> B[1]:=limit(b[1],s=0);

$$B_1 := \frac{-\delta \beta - \delta}{v^2}$$

> A(s):=P(s)/diff(Q(s),s):
> A[n]:=simplify(subs(s=mu,A(s)));

$$A_n := \left(2\delta\left(\cosh(\sqrt{\mu}\ (x-1)) + \cosh(\sqrt{\mu}\ x)\ \beta\right)\sqrt{\mu}\right) \Big/$$
$$\left(\cosh(\sqrt{\mu})\ \mu^{3/2} - \cosh(\sqrt{\mu})\sqrt{\mu}\ v^2 + 3\sinh(\sqrt{\mu})\mu\right.$$
$$\left. - \sinh(\sqrt{\mu})\ v^2\right)$$

8.1 Laplace Transform Technique for PDEs in Finite Domains

> A[0]:=subs(mu^(1/2)=nu,mu^(3/2)=nu^3,mu=nu^2,A[n]):

> A[0]:=simplify(A[0]);

$$A_0 := \frac{\delta \left(\cosh(\nu \ (x-1)) + \cosh(\nu \ x)\ \beta\right)}{\nu \ \sinh(\nu)}$$

> A[n]:=simplify(subs(mu^(1/2)=I*n*Pi,mu^(3/2)=-I*n^3*Pi^3, mu=-n^2*Pi^2,A[n])):

> vars:={cos(n*Pi)=(-1)^n,sin(n*Pi)=0};

$$vars := \{\cos(n\ \pi) = (-1)^n,\ \sin(n\ \pi) = 0\}$$

> A[n]:=simplify(subs(vars,A[n])):

> A[n]:=simplify(subs(vars,expand(A[n])));

$$A_n := -\frac{2\ \delta \cos(n\ \pi\ x)\ \left(1 + (-1)^{-n}\ \beta\right)}{n^2\ \pi^2 + \nu^2}$$

> b1s:=B[1]*subs(mu0=0,1/(s-mu0));

$$b1s := \frac{-\delta\ \beta - \delta}{\nu^2\ s}$$

> b1t:=invlaplace(b1s,s,t);

$$b1t := -\frac{(1+\beta)\ \delta}{\nu^2}$$

> b2s:=B[2]*subs(mu0=0,1/(s-mu0)^2);

$$b2s := 0$$

> b2t:=invlaplace(b2s,s,t);

$$b2t := 0$$

> u0s:=subs(mu=nu^2,A[0]/(s-mu));

$$u0s := \frac{\delta \left(\cosh(\nu \ (x-1)) + \cosh(\nu \ x)\ \beta\right)}{\nu \ \sinh(\nu)\ (s - \nu^2)}$$

> u0t:=invlaplace(u0s,s,t);

$$u0t := \frac{\delta\left(\cosh(v\,(x-1)) + \cosh(v\,x)\,\beta\right)\,e^{v^2 t}}{v\,\sinh(v)}$$

> uns:=A[n]/(s-mu);

$$uns := -\frac{2\,\delta\,\cos(n\,\pi\,x)\,\left(1 + (-1)^{-n}\,\beta\right)}{\left(n^2\,\pi^2 + v^2\right)(s - \mu)}$$

> unt:=invlaplace(uns,s,t);

$$unt := -\frac{2\,\delta\,\cos(n\,\pi\,x)\,\left(1 + (-1)^{-n}\,\beta\right)\,e^{\mu t}}{n^2\,\pi^2 + v^2}$$

> unt:=subs(mu=-n^2*Pi^2,unt);

$$unt := -\frac{2\,\delta\,\cos(n\,\pi\,x)\,\left(1 + (-1)^{-n}\,\beta\right)\,e^{-n^2\,\pi^2\,t}}{n^2\,\pi^2 + v^2}$$

> U:=b1t*exp(-nu^2*t)+b2t*exp(-nu^2*t)+simplify(u0t*exp(-nu^2*t))+exp(-nu^2*t)*Sum(unt,n=1..infinity);

$$U := -\frac{(1+\beta)\,\delta\,e^{-v^2 t}}{v^2} + \frac{\delta\left(\cosh(v\,(x-1)) + \cosh(v\,x)\,\beta\right)}{v\,\sinh(v)}$$

$$+ e^{-v^2 t}\left(\sum_{n=1}^{\infty}\left(-\frac{2\,\delta\,\cos(n\,\pi\,x)\,\left(1 + (-1)^{-n}\,\beta\right)\,e^{-n^2\,\pi^2\,t}}{n^2\,\pi^2 + v^2}\right)\right)$$

> u:=piecewise(t=0,0,t>0,subs(infinity=20,U)):

> pars:={nu=1,delta=1,beta=0.1};

$$pars := \{v = 1,\,\beta = 0.1,\,\delta = 1\}$$

> plot3d(subs(pars,u),x=0..1,t=0..0.5,axes=boxed,title="Figure 8.21",labels=[x,t,"u"],orientation=[-60,60]);

8.1 Laplace Transform Technique for PDEs in Finite Domains 735

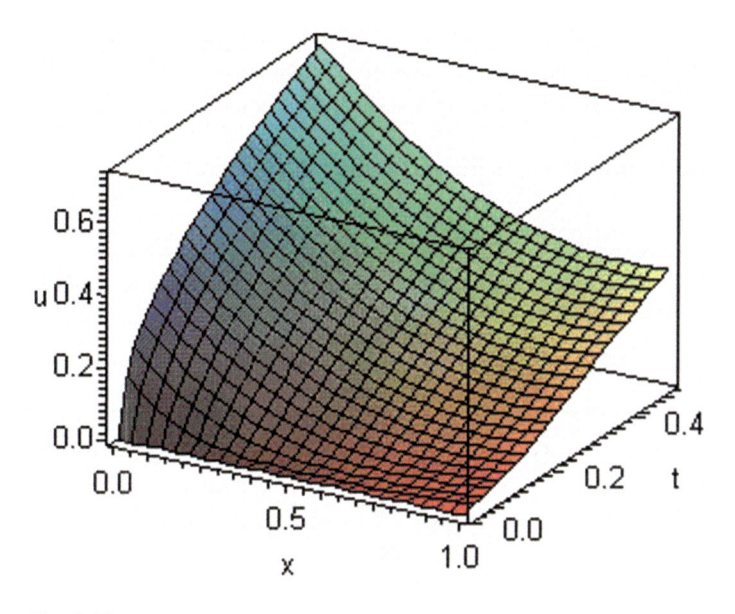

Fig. 8.21

> plot([subs(t=0,pars,u),subs(t=0.05,pars,u),subs(t=0.1,pars,u),
subs(t=0.2,pars,u)],x=0..1,axes=boxed,title="Figure 8.22",
thickness=5,labels=[x,"u"]);

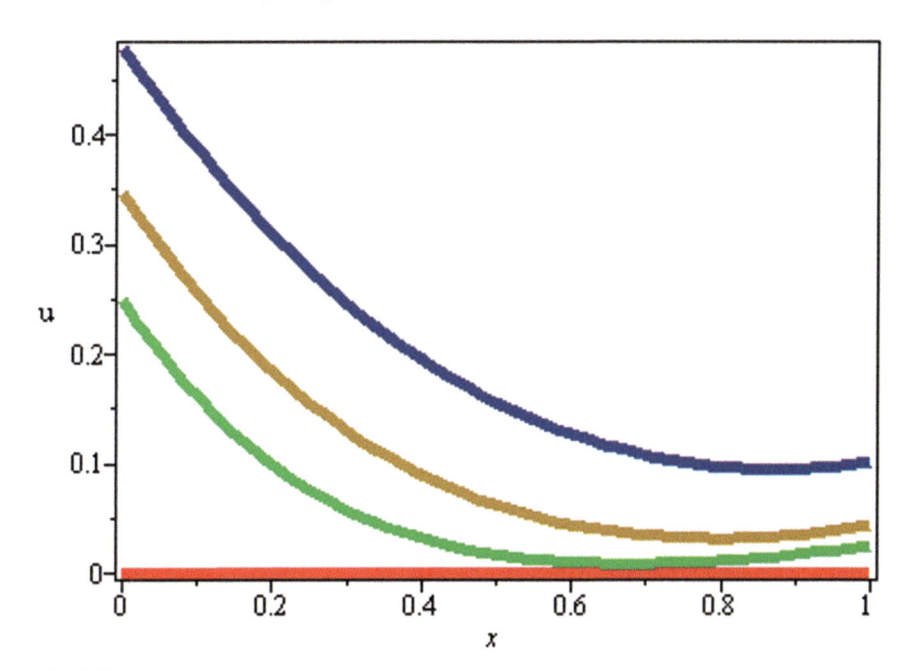

Fig. 8.22

In all the examples discussed in this chapter until now, the roots of q(s) were obtained analytically. This is not always possible. Often the roots should be obtained numerically as in section 7.1.4. This is illustrated in the next example.

Example 8.13. Heat Conduction in a Slab with Radiation Boundary Conditions

Consider Example 7.6, heat conduction in a rectangle with radiation at the surface,[2] which is solved here using the Laplace transform technique. The dimensionless temperature profile is governed by:

$$\frac{\partial u}{\partial t} = \frac{\partial^2 u}{\partial x^2}$$
$$u(x,0) = 1 \qquad (8.29)$$
$$\frac{\partial u}{\partial x}(0,t) = 0 \text{ and } \frac{\partial u}{\partial x}(1,t) + u(1,t) = 0$$

Equation (8.29) is solved in Maple below:

> restart:with(inttrans):with(plots):
> eq:=diff(u(x,t),t)=diff(u(x,t),x$2);

$$eq := \frac{\partial}{\partial t} u(x, t) = \frac{\partial^2}{\partial x^2} u(x, t)$$

> u(x,0):=1;

$$u(x, 0) := 1$$

> bc1:=diff(u(x,t),x)=0;

$$bc1 := \frac{\partial}{\partial x} u(x, t) = 0$$

> bc2:=diff(u(x,t),x)+u(x,t)=0;

$$bc2 := \frac{\partial}{\partial x} u(x, t) + u(x, t) = 0$$

> eqs:=laplace(eq,t,s):
> eqs:=subs(laplace(u(x,t),t,s)=U(x),eqs);

$$eqs := s\, U(x) - 1 = \frac{d^2}{dx^2} U(x)$$

8.1 Laplace Transform Technique for PDEs in Finite Domains 737

```
> bc1:=laplace(bc1,t,s):
> bc1:=subs(laplace(u(x,t),t,s)=U(x),bc1);
```

$$bc1 := \frac{d}{dx} U(x) = 0$$

```
> bc2:=laplace(bc2,t,s):
> bc2:=subs(laplace(u(x,t),t,s)=U(x),bc2);
```

$$bc2 := \frac{d}{dx} U(x) + U(x) = 0$$

```
> dsolve(eqs,U(x));
```

$$U(x) = e^{\sqrt{s}\, x}\, \_C2 + e^{-\sqrt{s}\, x}\, \_C1 + \frac{1}{s}$$

```
> U(x):=c[1]*cosh(s^(1/2)*x)+c[2]*sinh(s^(1/2)*x)+1/s;
```

$$U(x) := c_1 \cosh(\sqrt{s}\, x) + c_2 \sinh(\sqrt{s}\, x) + \frac{1}{s}$$

```
> eq0:=eval(subs(x=0,bc1)):
> eq1:=eval(subs(x=1,bc2)):
> con:=solve({eq0,eq1},{c[1],c[2]}):
> U(x):=subs(con,U(x)):
> U(x):=factor(simplify(U(x)));
```

$$U(x) := -\frac{\cosh(\sqrt{s}\, x) - \sinh(\sqrt{s}\,)\sqrt{s} - \cosh(\sqrt{s}\,)}{s\,(\sinh(\sqrt{s}\,)\sqrt{s} + \cosh(\sqrt{s}\,))}$$

```
> P(s):=numer(U(x));
```

$$P(s) := -\cosh(\sqrt{s}\, x) + \sinh(\sqrt{s}\,)\sqrt{s} + \cosh(\sqrt{s}\,)$$

```
> Q(s):=denom(U(x));
```

$$Q(s) := s\,(\sinh(\sqrt{s}\,)\sqrt{s} + \cosh(\sqrt{s}\,))$$

Maple cannot find the eigenvalues directly.

```
> solve(Q(s),s);
```

$$0, RootOf\left(\_Z\,(e^{\_Z})^2 - \_Z + (e^{\_Z})^2 + 1\right)^2$$

```
> eig:=sinh(s^(1/2))*s^(1/2)+cosh(s^(1/2));
```
$$eig := \sinh(\sqrt{s})\sqrt{s} + \cosh(\sqrt{s})$$

For convenience $s=-\lambda_2$ is substituted to find the eigenvalues.

```
> eiglambda:=simplify(subs(s^(1/2)=I*lambda,s=-lambda^2,eig));
```
$$eiglambda := -\sin(\lambda)\,\lambda + \cos(\lambda)$$

```
> plot(eiglambda,lambda=0..20,thickness=3,title="Figure 8.23",
axes=boxed);
```

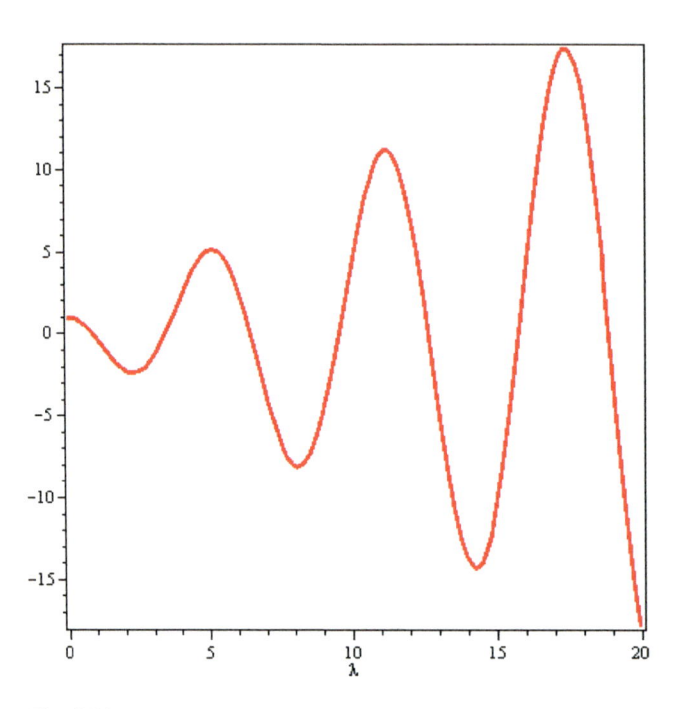

Fig. 8.23

The roots are:
```
> 0,0,-lambda^2;
```
$$0, 0, -\lambda^2$$

```
> fsolve(eiglambda,lambda=1);
```
$$0.8603335890$$

8.1 Laplace Transform Technique for PDEs in Finite Domains

The first 20 eigenvalues are obtained numerically.

> N:=20;

$$N := 20$$

> l[1]:=fsolve(eiglambda,lambda=0..2);

$$l_1 := 0.860333589$$

> for i from 2 to N do l[i]:=fsolve(eiglambda,lambda=l[i-1]..l[i-1]+4);od:
> seq(l[i],i=1..N);

$$0.8603335890, 3.425618459, 6.437298179, 9.529334405, 12.64528722,$$
$$15.77128487, 18.90240996, 22.03649673, 25.17244633, 28.30964285,$$
$$31.44771464, 34.58642422, 37.72561283, 40.86517033, 44.00501792,$$
$$47.14509774, 50.28536634, 53.42579048, 56.56634428, 59.70700731$$

> mu0:=0;

$$\mu 0 := 0$$

> b[2]:=(s-mu0)^2*P(s)/Q(s);

$$b_2 := \frac{s\left(-\cosh(\sqrt{s}\, x) + \sinh(\sqrt{s})\sqrt{s} + \cosh(\sqrt{s})\right)}{\sinh(\sqrt{s})\sqrt{s} + \cosh(\sqrt{s})}$$

> B[2]:=limit(b[2],s=0);

$$B_2 := 0$$

> b[1]:=diff(b[2],s):
> B[1]:=limit(b[1],s=0);

$$B_1 := 0$$

> A(s):=P(s)/diff(Q(s),s):
> A[n]:=simplify(subs(s=mu,A(s)));

$$A_n := -\frac{2\left(\cosh(\sqrt{\mu}\, x) - \sinh(\sqrt{\mu})\sqrt{\mu} - \cosh(\sqrt{\mu})\right)\sqrt{\mu}}{4\sinh(\sqrt{\mu})\mu + 2\cosh(\sqrt{\mu})\sqrt{\mu} + \mu^{3/2}\cosh(\sqrt{\mu})}$$

> A[n]:=simplify(subs(mu^(1/2)=I*lambda,mu^(3/2)=-I*lambda^3, mu=-lambda^2,A[n]));

$$A_n := \frac{2\left(\cos(\lambda x) + \sin(\lambda)\lambda - \cos(\lambda)\right)}{4\sin(\lambda)\lambda - 2\cos(\lambda) + \lambda^2 \cos(\lambda)}$$

> vars:={cos(lambda)=lambda*sin(lambda)};

$$vars := \{\cos(\lambda) = \sin(\lambda)\lambda\}$$

> A[n]:=simplify(subs(vars,expand(A[n])));

$$A_n := \frac{2\cos(\lambda x)}{\sin(\lambda)\lambda\left(2 + \lambda^2\right)}$$

> b1s:=B[1]*subs(mu0=0,1/(s-mu0));

$$b1s := 0$$

> b1t:=invlaplace(b1s,s,t);

$$b1t := 0$$

> b2s:=B[2]*subs(mu0=0,1/(s-mu0)^2);

$$b2s := 0$$

> b2t:=invlaplace(b2s,s,t);

$$b2t := 0$$

> uns:=A[n]/(s-mu);

$$uns := \frac{2\cos(\lambda x)}{\sin(\lambda)\lambda\left(2 + \lambda^2\right)(s - \mu)}$$

> unt:=invlaplace(uns,s,t);

$$unt := \frac{2\cos(\lambda x)\, e^{\mu t}}{\sin(\lambda)\lambda\left(2 + \lambda^2\right)}$$

8.1 Laplace Transform Technique for PDEs in Finite Domains 741

```
> unt:=subs(mu=-l[n]^2,lambda=l[n],unt);
```

$$unt := \frac{2\cos(l_n x)\, e^{-l_n^2 t}}{\sin(l_n)\, l_n \left(2 + l_n^2\right)}$$

The solution obtained can then be plotted.

```
> U:=b1t+b2t+Sum(unt,n=1..infinity);
```

$$U := \sum_{n=1}^{\infty} \frac{2\cos(l_n x)\, e^{-l_n^2 t}}{\sin(l_n)\, l_n \left(2 + l_n^2\right)}$$

```
> u:=piecewise(t=0,1,t>0,subs(infinity=20,U)):
> u:=evalf(u):
> plot3d(u,x=0..1,t=0..0.5,axes=boxed,title="Figure 8.24",
labels=[x,t,"u"],orientation=[45,60]);
```

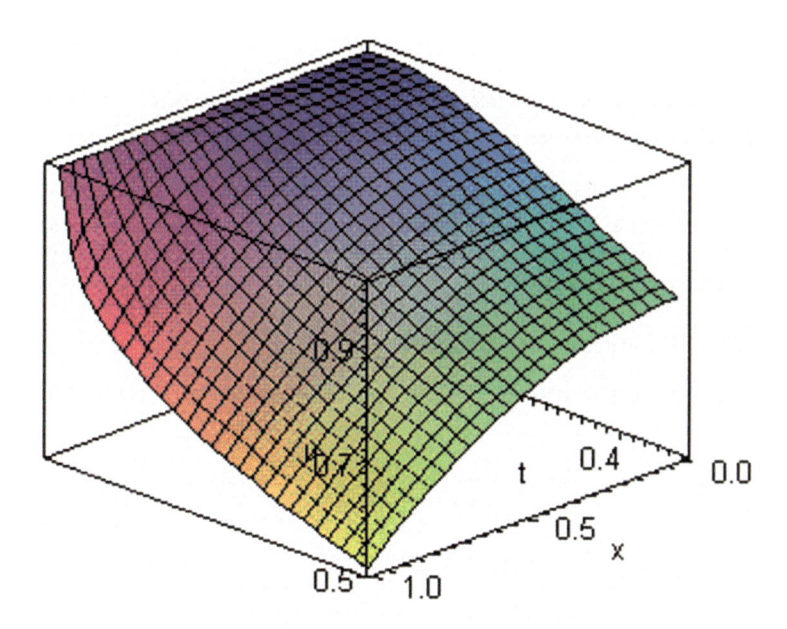

Fig. 8.24

```
> plot([subs(t=0,u),subs(t=0.1,u),subs(t=0.2,u),subs(t=0.5,u)],x=0..1,
title="Figure 8.25",axes=boxed,thickness=5,labels=[x,"u"]);
```

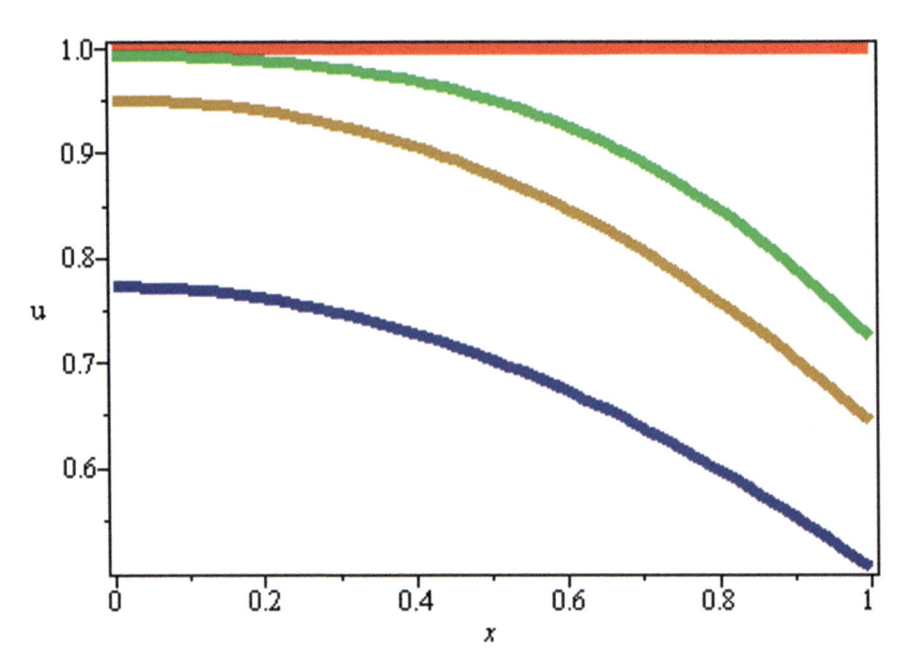

Fig. 8.25

8.1.7 Laplace Transform Technique for Parabolic Partial Differential Equations in Cylindrical Coordinates

The Laplace transform technique can be used for problems in cylindrical coordinates also. Problems in cylindrical coordinates involve Bessel functions. Maple's inbuilt Bessel functions can be used for modeling these problems. This is illustrated in the following example.

Example 8.14. Heat Conduction in a Cylinder

Consider heat conduction in a cylinder.[2] The dimensionless temperature profile is governed by

$$\frac{\partial u}{\partial t} = \frac{\partial^2 u}{\partial x^2} + \frac{1}{x}\frac{\partial u}{\partial x}$$
$$u(x,0) = 0 \tag{8.30}$$
$$\frac{\partial u}{\partial x}(0,t) = 0 \text{ and } u(1,t) = 1$$

Equation (8.30) is solved in Maple below:

```
> restart:with(inttrans):with(plots):
> eq:=diff(u(x,t),t)=diff(u(x,t),x$2)+1/x*diff(u(x,t),x);
```

8.1 Laplace Transform Technique for PDEs in Finite Domains

$$eq := \frac{\partial}{\partial t} u(x,t) = \frac{\partial^2}{\partial x^2} u(x,t) + \frac{\frac{\partial}{\partial x} u(x,t)}{x}$$

> u(x,0):=0;

$$u(x,0) := 0$$

> bc1:=diff(u(x,t),x)=0;

$$bc1 := \frac{\partial}{\partial x} u(x,t) = 0$$

> bc2:=u(x,t)=1;

$$bc2 := u(x,t) = 1$$

> eqs:=laplace(eq,t,s):
> eqs:=subs(laplace(u(x,t),t,s)=U(x),eqs);

$$eqs := s\, U(x) = \frac{d^2}{dx^2} U(x) + \frac{\frac{d}{dx} U(x)}{x}$$

> bc1:=laplace(bc1,t,s):
> bc1:=subs(laplace(u(x,t),t,s)=U(x),bc1);

$$bc1 := \frac{d}{dx} U(x) = 0$$

> bc2:=laplace(bc2,t,s):
> bc2:=subs(laplace(u(x,t),t,s)=U(x),bc2);

$$bc2 := U(x) = \frac{1}{s}$$

> dsolve(eqs,U(x));

$$U(x) = \_C1\, \text{BesselJ}\left(0, \sqrt{-s}\, x\right) + \_C2\, \text{BesselY}\left(0, \sqrt{-s}\, x\right)$$

Since BesselY becomes infinite x = 0_C2 should be zero and the solution is taken as:

> U(x):=c[1]*BesselJ(0,(-s)^(1/2)*x);

$$U(x) := c_1\, \text{BesselJ}\left(0, \sqrt{-s}\, x\right)$$

> eq0:=eval(subs(x=0,bc1));

$$eq0 := 0 = 0$$

> eq1:=eval(subs(x=1,bc2));

$$eq1 := c_1 \, \text{BesselJ}\left(0, \sqrt{-s}\right) = \frac{1}{s}$$

> con:=solve({eq1},{c[1]}):
> U(x):=subs(con,U(x)):
> U(x):=factor(simplify(U(x)));

$$U(x) := \frac{\text{BesselJ}\left(0, \sqrt{-s}\, x\right)}{\text{BesselJ}\left(0, \sqrt{-s}\right) s}$$

> P(s):=numer(U(x));

$$P(s) := \text{BesselJ}\left(0, \sqrt{-s}\, x\right)$$

> Q(s):=denom(U(x));

$$Q(s) := \text{BesselJ}\left(0, \sqrt{-s}\right) s$$

> solve(Q(s),s);

$$RootOf\left(\text{BesselJ}\left(0, \sqrt{-\_Z}\right)\right), 0$$

> eig:=BesselJ(0,(-s)^(1/2));

$$eig := \text{BesselJ}\left(0, \sqrt{-s}\right)$$

> convert(eig,BesselI);

$$\text{BesselJ}\left(0, \sqrt{-s}\right)$$

> eiglambda:=simplify(subs(s^(1/2)=I*lambda,(-s)^(1/2)=lambda, s=-lambda^2,eig));

$$eiglambda := \text{BesselJ}(0, \lambda)$$

> plot(eiglambda,lambda=0..20,thickness=3,axes=boxed):

The roots are:

> 0,-lambda^2;

$$0, -\lambda^2$$

8.1 Laplace Transform Technique for PDEs in Finite Domains

> N:=20;

$$N := 20$$

> l[1]:=fsolve(eiglambda,lambda=0..3);

$$l_1 := 2.404825558$$

> for i from 2 to N do l[i]:=fsolve(eiglambda,lambda=l[i-1]..l[i-1]+4);od:
> seq(l[i],i=1..N);

2.404825558, 5.520078110, 8.653727913, 11.79153444, 14.93091771,
18.07106397, 21.21163663, 24.35247153, 27.49347913, 30.63460647,
33.77582021, 36.91709835, 40.05842576, 43.19979171, 46.34118837,
49.48260990, 52.62405184, 55.76551076, 58.90698393, 62.04846919

> A(s):=P(s)/diff(Q(s),s):
> A[n]:=simplify(subs(s=mu,A(s)));

$$A_n := \frac{2\,\text{BesselJ}(0, \sqrt{-\mu}\, x) \sqrt{-\mu}}{\text{BesselJ}(1, \sqrt{-\mu})\, \mu + 2\,\text{BesselJ}(0, \sqrt{-\mu}) \sqrt{-\mu}}$$

> A[0]:=limit(A[n],mu=0);

$$A_0 := 1$$

> A[n]:=simplify(subs(mu^(1/2)=I*lambda,(-mu)^(1/2)=lambda, mu^(3/2)=-I*lambda^3,mu=-lambda^2,A[n]));

$$A_n := -\frac{2\,\text{BesselJ}(0, \lambda x)}{\text{BesselJ}(1, \lambda)\, \lambda - 2\,\text{BesselJ}(0, \lambda)}$$

> u0s:=A[0]*subs(mu=0,1/(s-mu));

$$u0s := \frac{1}{s}$$

> u0t:=invlaplace(u0s,s,t);

$$u0t := 1$$

> uns:=A[n]/(s-mu);

$$uns := -\frac{2\,\text{BesselJ}(0, \lambda x)}{(\text{BesselJ}(1, \lambda)\, \lambda - 2\,\text{BesselJ}(0, \lambda))\,(s - \mu)}$$

> unt:=invlaplace(uns,s,t);

$$unt := \frac{2\,\mathrm{BesselJ}(0,\lambda x)\,e^{\mu t}}{-\mathrm{BesselJ}(1,\lambda)\,\lambda + 2\,\mathrm{BesselJ}(0,\lambda)}$$

> unt:=subs(mu=-l[n]^2,lambda=l[n],unt);

$$unt := \frac{2\,\mathrm{BesselJ}(0,l_n x)\,e^{-l_n^2 t}}{-\mathrm{BesselJ}(1,l_n)\,l_n + 2\,\mathrm{BesselJ}(0,l_n)}$$

The following solution and plots are obtained:

> U:=u0t+Sum(unt,n=1..infinity);

$$U := 1 + \sum_{n=1}^{\infty} \frac{2\,\mathrm{BesselJ}(0,l_n x)\,e^{-l_n^2 t}}{-\mathrm{BesselJ}(1,l_n)\,l_n + 2\,\mathrm{BesselJ}(0,l_n)}$$

> u:=piecewise(t=0,0,t>0,subs(infinity=20,U)):

> u:=evalf(u):

> plot3d(u,x=0..1,t=0..0.5,axes=boxed,title="Figure 8.26", labels=[x,t,"u"],orientation=[-145,60]);

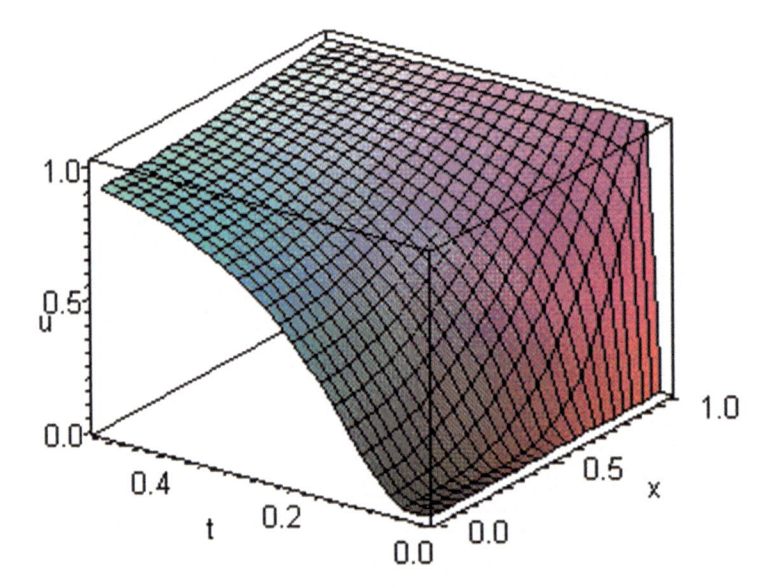

Fig. 8.26

8.1 Laplace Transform Technique for PDEs in Finite Domains   747

```
> plot([subs(t=0,u),subs(t=0.1,u),subs(t=0.2,u),subs(t=0.5,u)],x=0..1,
title="Figure 8.27",axes=boxed,thickness=5,labels=[x,"u"]);
```

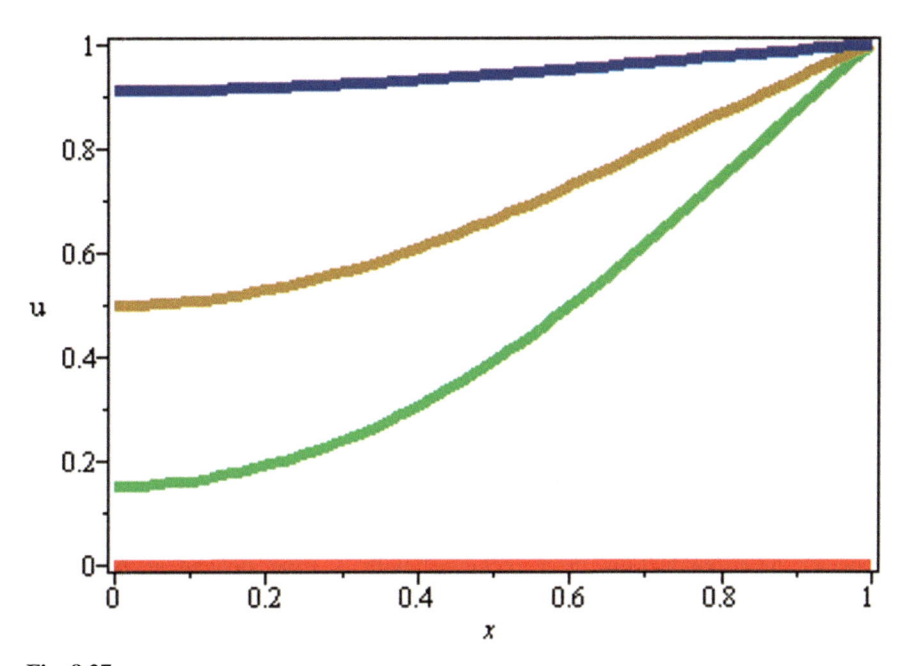

Fig. 8.27

8.1.8 Laplace Transform Technique for Parabolic Partial Differential Equations for Time Dependent Boundary Conditions – Use of Convolution Theorem

In example 8.9 the Laplace transform technique was used to solve a time dependent problem. Inversing the Laplace transform is not straightforward. For complicated time dependent boundary conditions the convolution theorem can be used to find the inverse Laplace transform efficiently. If H(s) is the solution obtained in the Laplace domain, H(s) is represented as a product of two functions:

$$H(s) = F(s)G(s) \qquad (8.31)$$

where F(s) is chosen such that it can be represented as:

$$F(s) = \frac{p(s)}{q(s)} \qquad (8.32)$$

F(s) is chosen so that the order of q(s) is greater than the order of p(s). F(s) is then inverted using the methodology illustrated in section 8.1.5 and 8.1.6. If F(s) and

G(s) are inverted to time domain individually, H(s) is then found using the convolution theorem[1] (Varma and Morbidelli, 1997):

$$h(t) = \int_0^t f(t-\tau)g(\tau)d\tau$$

where (8.33)

$$f(t) = L^{-1}(F(s)) \text{ and } g(t) = L^{-1}(G(s))$$

In equation (8.33), τ is a dummy variable. An alternate form of equation (8.34) can also be used:

$$h(t) = \int_0^t f(\tau)g(t-\tau)d\tau \qquad (8.34)$$

The methodology is illustrated using the following example.

Example 8.15. Heat Conduction in a Rectangle with a Time Dependent Boundary Condition

Consider heat conduction in a rectangle with a time dependent boundary condition.[2] The dimensionless temperature profile is governed by:

$$\frac{\partial u}{\partial t} = \frac{\partial^2 u}{\partial x^2}$$

$$u(x,0) = 0 \qquad (8.35)$$

$$u(0,t) = 0 \text{ and } u(1,t) = w(t)$$

Equation (8.35) is solved in Maple below for a general time dependent function, w(t), and plots are obtained for a particular step function.

> restart:with(inttrans):with(plots):
> eq:=diff(u(x,t),t)=diff(u(x,t),x$2);

$$eq := \frac{\partial}{\partial t}u(x,t) = \frac{\partial^2}{\partial x^2}u(x,t)$$

> u(x,0):=0;

$$u(x,0) := 0$$

8.1 Laplace Transform Technique for PDEs in Finite Domains

> bc1:=u(x,t)=0;

$$bc1 := u(x, t) = 0$$

> bc2:=u(x,t)=w(t);

$$bc2 := u(x, t) = w(t)$$

> eqs:=laplace(eq,t,s):
> eqs:=subs(laplace(u(x,t),t,s)=U(x),eqs);

$$eqs := s\, U(x) = \frac{d^2}{dx^2} U(x)$$

> bc1:=laplace(bc1,t,s):
> bc1:=subs(laplace(u(x,t),t,s)=U(x),laplace(w(t),t,s)=W(s),bc1);

$$bc1 := U(x) = 0$$

> bc2:=laplace(bc2,t,s):
> bc2:=subs(laplace(u(x,t),t,s)=U(x),laplace(w(t),t,s)=W(s),bc2);

$$bc2 := U(x) = W(s)$$

> dsolve(eqs,U(x));

$$U(x) = \_C1\, e^{\sqrt{s}\, x} + \_C2\, e^{-\sqrt{s}\, x}$$

> U(x):=c[1]*cosh(s^(1/2)*x)+c[2]*sinh(s^(1/2)*x);

$$U(x) := c_1 \cosh(\sqrt{s}\, x) + c_2 \sinh(\sqrt{s}\, x)$$

> eq0:=eval(subs(x=0,bc1)):
> eq1:=eval(subs(x=1,bc2)):
> con:=solve({eq0,eq1},{c[1],c[2]}):
> U(x):=subs(con,U(x));

$$U(x) := \frac{W(s) \sinh(\sqrt{s}\, x)}{\sinh(\sqrt{s})}$$

> U(x):=factor(combine(simplify(U(x))));

$$U(x) := \frac{W(s) \sinh(\sqrt{s}\, x)}{\sinh(\sqrt{s})}$$

Next U1(x) is written as a product of two functions (see equations (8.31) and (8.33)):

> U1(x):=simplify(U(x)/W(s)/s);

$$U1(x) := \frac{\sinh(\sqrt{s}\ x)}{\sinh(\sqrt{s})\ s}$$

U1(x) is chosen so that q(s) is a higher order than p(s) (see equation (8.32)):

> G(s):=W(s)*s;

$$G(s) := W(s)\ s$$

where W(s) is the Laplace transform of the time dependent boundary condition w(t) in equation (8.11.35.). Next, U1(X) is inverted to the time domain as illustrated in section 8.1.7 to obtain f(t) as:

> P(s):=numer(U1(x));

$$P(s) := \sinh(\sqrt{s}\ x)$$

> Q(s):=denom(U1(x));

$$Q(s) := \sinh(\sqrt{s})\ s$$

> solve(Q(s),s);

$$0$$

> _EnvAllSolutions := true;

$$\_EnvAllSolutions := true$$

> solve(Q(s),s);

$$-\pi^2\ \_Z1\text{\textasciitilde}^2,\ 0$$

> 0,0,-n^2*Pi^2;

$$0, 0, -n^2\ \pi^2$$

> mu0:=0;

$$\mu 0 := 0$$

> b[2]:=(s-mu0)^2*P(s)/Q(s);

$$b_2 := \frac{s\ \sinh(\sqrt{s}\ x)}{\sinh(\sqrt{s})}$$

8.1 Laplace Transform Technique for PDEs in Finite Domains

> B[2]:=limit(b[2],s=0);

$$B_2 := 0$$

> b[1]:=diff(b[2],s):
> B[1]:=limit(b[1],s=0);

$$B_1 := x$$

> A(s):=P(s)/diff(Q(s),s):
> A[n]:=simplify(subs(s=mu,A(s)));

$$A_n := \frac{2\sinh(\sqrt{\mu}\,x)}{\cosh(\sqrt{\mu})\sqrt{\mu} + 2\sinh(\sqrt{\mu})}$$

> A[n]:=simplify(subs(mu^(1/2)=I*n*Pi,mu^(3/2)=-I*n^3*Pi^3, mu=-n^2*Pi^2,A[n])):
> vars:={sin(n*Pi)=0};

$$vars := \{\sin(n\,\pi) = 0\}$$

> A[n]:=simplify(subs(vars,A[n])):
> A[n]:=simplify(subs(vars,expand(A[n])));

$$A_n := \frac{2\sin(n\,\pi\,x)}{\cos(n\,\pi)\,n\,\pi}$$

> b1s:=B[1]*subs(mu0=0,1/(s-mu0));

$$b1s := \frac{x}{s}$$

> b1t:=invlaplace(b1s,s,t);

$$b1t := x$$

> b2s:=B[2]*subs(mu0=0,1/(s-mu0)^2);

$$b2s := 0$$

> b2t:=invlaplace(b2s,s,t);

$$b2t := 0$$

> uns:=A[n]/(s-mu);

$$uns := \frac{2\sin(n\,\pi\,x)}{\cos(n\,\pi)\,n\,\pi\,(s-\mu)}$$

> unt:=invlaplace(uns,s,t);

$$unt := \frac{2\sin(n\pi x)\, e^{\mu t}}{\cos(n\pi)\, n\pi}$$

> unt:=subs(mu=-n^2*Pi^2,unt);

$$unt := \frac{2\sin(n\pi x)\, e^{-n^2\pi^2 t}}{\cos(n\pi)\, n\pi}$$

> f(t):=b1t+b2t+Sum(unt,n=1..infinity);

$$f(t) := x + \sum_{n=1}^{\infty} \frac{2\sin(n\pi x)\, e^{-n^2\pi^2 t}}{\cos(n\pi)\, n\pi}$$

Next, a step function is chosen for w(t) and plotted:

> w(t):=Heaviside(t-1)-1/2*Heaviside(t-2);

$$w(t) := \text{Heaviside}(t-1) - \frac{1}{2}\text{Heaviside}(t-2)$$

> plot(w(t),t=0..5,thickness=3,title="Figure 8.28",axes=boxed);

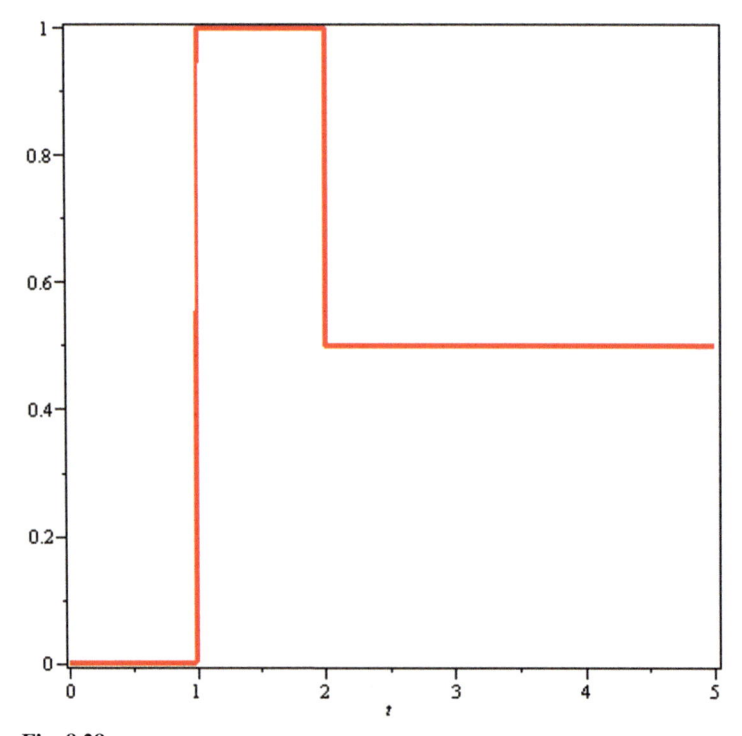

Fig. 8.28

8.1 Laplace Transform Technique for PDEs in Finite Domains

The Laplace transform of w(t) is:
> W(s):=laplace(w(t),t,s);

$$W(s) := \frac{1}{2} \frac{2e^{-s} - e^{-2s}}{s}$$

The function g(t) is obtained by inverting G(s):
> G(s):=s*W(s);

$$G(s) := e^{-s} - \frac{1}{2} e^{-2s}$$

> g(t):=invlaplace(G(s),s,t);

$$g(t) := \text{Dirac}(t-1) - \frac{1}{2} \text{Dirac}(t-2)$$

Next, the convolution integral is carried out to obtain the final time domain solution as:
> gtau:=subs(t=tau,g(t));

$$gtau := \text{Dirac}(\tau - 1) - \frac{1}{2} \text{Dirac}(\tau - 2)$$

> ftau:=subs(t=t-tau,f(t));

$$ftau := x + \sum_{n=1}^{\infty} \frac{2 \sin(n \pi x) e^{-n^2 \pi^2 (t-\tau)}}{\cos(n \pi) n \pi}$$

> U:=int(ftau*gtau,tau=0..t);

$$U := -\frac{1}{2} \text{Heaviside}(t-2) \left(x + \sum_{n=1}^{\infty} \frac{2 \sin(n \pi x) e^{-n^2 \pi^2 (t-2)}}{\cos(n \pi) n \pi} \right)$$
$$+ \text{Heaviside}(t-1) \left(x + \sum_{n=1}^{\infty} \frac{2 \sin(n \pi x) e^{-n^2 \pi^2 (t-1)}}{\cos(n \pi) n \pi} \right)$$

> u:=piecewise(t=0,0,t>0,subs(infinity=20,U)):
> plot3d(simplify(u),x=0..1,t=0..5,axes=boxed,title="Figure 8.29", labels=[x,t,"u"],orientation=[135,45]);

754   8 Laplace Transform Technique for Partial Differential Equations

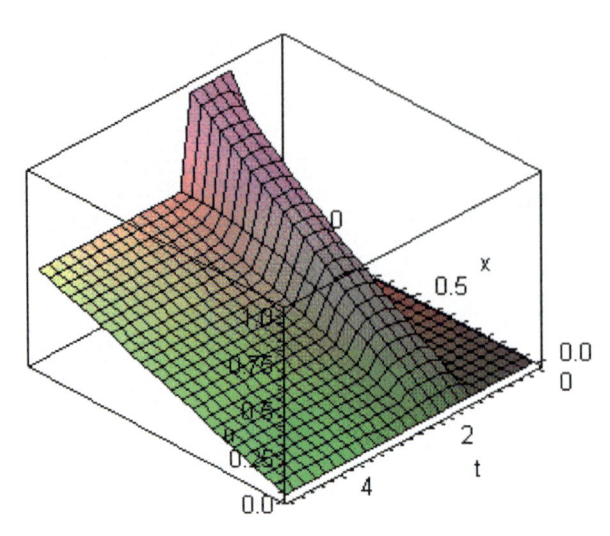

Fig. 8.29

The dimensionless temperature at different points inside the rectangle are plotted as:

> plot([simplify(subs(x=1,u)),simplify(subs(x=0.75,u)),simplify(subs(x=0.5,u)), simplify(subs(x=0.25,u)),simplify(subs(x=0.0,u))],t=0..5.,thickness=3, axes=boxed,title="Figure 8.30",labels=[t,"u"]);

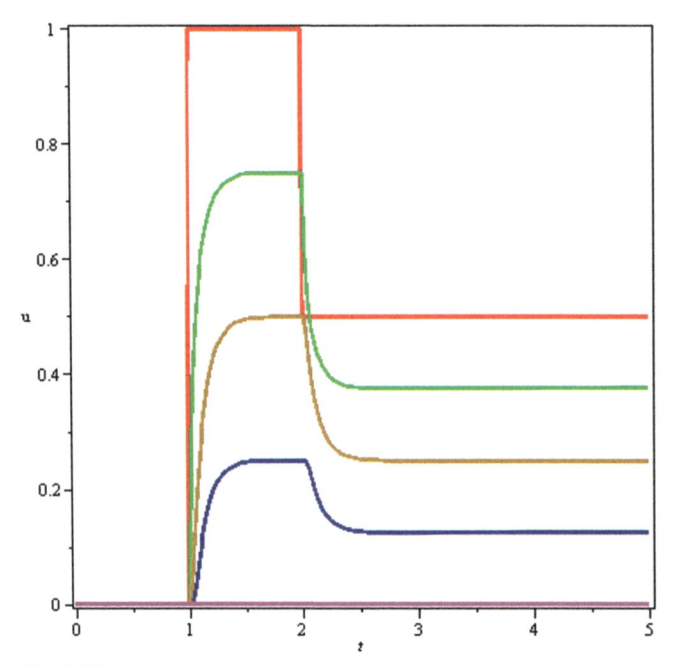

Fig. 8.30

8.1.9 Summary

In this chapter, analytical solutions were obtained for linear hyperbolic and parabolic partial differential equations in finite domains using Laplace transform technique. In section 8.1.2, a linear hyperbolic partial differential equations was solved using the Laplace transform technique. First, the partial differential equation was converted to an ordinary differential equation by converting the PDE from the time domain to the Laplace domain. For hyperbolic partial differential equations this results in an initial value problem (IVP), which is solved analytically in the Laplace domain as illustrated in chapter 2.1. The analytical solution obtained in the Laplace domain was converted easily to the time domain using Maple's inbuilt Laplace transform package. For parabolic partial differential equations, the governing equation in the Laplace domain is a boundary value problem (BVP), which is solved analytically as in chapter 3.1. For certain simple parabolic partial differential equations, the Laplace domain solution can be inverted to time domain easily using Maple as illustrated in section 8.1.3.

Often inversion to time domain solution is not trivial and the time domain involves an infinite series. In section 8.1.4 short time solution for parabolic partial differential equations was obtained by converting the solution obtained in the Laplace domain to an infinite series, in which each term can easily inverted to time domain. This short time solution is very useful for predicting the behavior at short time and medium times. For long times, a long term solution was obtained in section 8.1.5 using Heaviside expansion theorem. This solution is analogous to the separation of variables solution obtained in chapter 7. In section 8.1.6, the Heaviside expansion theorem was used for parabolic partial differential equations in which the solution obtained has multiple roots. In section 8.1.7, the Laplace transform technique was extended to parabolic partial differential equations in cylindrical coordinates. In section 8.1.8, the convolution theorem was used to solve the linear parabolic partial differential equations with complicated time dependent boundary conditions. For time dependent boundary conditions the Laplace transform technique was shown to be advantageous compared to the separation of variables technique. A total of fifteen examples were presented in this chapter.

8.1.10 Exercise Problems

1. Complete the details missing in example 8.2 (i.e., complete the Maple program). Can you obtain an analytical solution if the initial condition is replaced by $u(x,0) = x$?
2. Complete the details missing in example 8.6.
3. Obtain the short time solution reported in example 8.7.
4. Consider charging a battery as discussed in example 8.11. Complete the details missing in this example. Obtain the short time solution for the same problem.
5. Solve the following parabolic PDE using the Laplace transform technique (see examples 8.3 and 8.4):

$$\frac{\partial u}{\partial t} = \frac{\partial^2 u}{\partial x^2}$$

$$u(x,0) = \cos\left(\frac{\pi x}{2}\right)$$

$$\frac{\partial u}{\partial x}(0,t) = 0 \text{ and } u(1,t) = 0$$

6. Solve the following simple parabolic PDE using the Laplace transform technique:

$$\frac{\partial u}{\partial t} = \frac{\partial^2 u}{\partial x^2}$$

$$u(x,0) = \sin\left(\frac{\pi x}{2}\right)$$

$$u(0,t) = 0 \text{ and } u(1,t) = \exp\left(-\frac{\pi^2}{4}t\right)$$

7. Solve the following steady state heat transfer problem by applying the Laplace transformation in y coordinate:

$$\frac{\partial^2 u}{\partial x^2} + \frac{\partial^2 u}{\partial y^2} = 0$$

$$u(0,y) = 0;\ u(1,y) = 0$$

$$u(x,0) = \sin(\pi x);\ \frac{\partial u}{\partial y}(x,0) = -\pi \sin(\pi x)$$

8. Solve the following wave equation using the Laplace transformation technique:

$$\frac{\partial^2 u}{\partial t^2} - \frac{\partial^2 u}{\partial x^2} = 0$$

$$u(0,t) = 0;\ u((1,t)) = 0$$

$$u(x,0) = \sin(\pi x);\ \frac{\partial u}{\partial t}(x,0) = 0$$

9. Obtain an analytical solution for the Graetz problem described in example 5.6 and exercise problem 6 in chapter 7.

10. Consider diffusion with convection in a coated wall reactor, where the reaction takes place at the wall (Rice and Do, 1995;[3] chapter 5.1, exercise problem 1; chapter 7, exercise problem 7). The governing equation and boundary conditions for concentration in dimensionless form are:

$$\frac{\partial u}{\partial Z} = \frac{\partial^2 u}{\partial x^2} + \frac{1}{x}\frac{\partial u}{\partial x}$$

$$\frac{\partial u}{\partial x}(0,Z) = 0 \text{ and } \frac{\partial u}{\partial x}(1,Z) + \text{Ha } u(1,Z) = 0$$

$$u(x,0) = 1$$

where Ha is the Hatta number. Obtain an analytical solution for this problem using the Laplace transform technique.

11. Consider cooling of spherical nuclear pellets (Rice and Do, 1995;[3] chapter 5.1, exercise problem 7; chapter 7, exercise problem 8). The dimensionless temperature distribution is governed by:

$$\frac{\partial u}{\partial t} = \frac{\partial^2 u}{\partial x^2} + \frac{2}{x}\frac{\partial u}{\partial x} + Q$$

$$\frac{\partial u}{\partial x}(0,t) = 0 \text{ and } \frac{\partial u}{\partial x}(1,t) + \text{Bi } u(1,t) = 0$$

$$u(x,0) = 1$$

where Q is the ratio of heat generation to heat conduction and Bi is the Biot number. Obtain an analytical solution for this problem using the Laplace transform technique. Plot the profiles for Q = 1, Bi = 0.2 and Q = 1, Bi = 10.

12. Consider dispersion of a linear kinematic wave in a dimensionless form (Aris, 1999;[4] chapter 5.1, exercise problem 9; chapter 7, exercise problem 9). The governing equation and boundary/initial conditions are:

$$\frac{\partial u}{\partial t} = \frac{\partial^2 u}{\partial x^2} - \text{Pe}\frac{\partial u}{\partial x}$$

$$u(0,t) = 1;\ u(1,t) = 0$$

$$u(x,0) = 0$$

Obtain an analytical solution for this problem using the Laplace transform technique. Plot the profiles for Pe = 1, 10 and 50.

13. Consider the fluid flow problem (Davis, 1984;[5] chapter 5.1, exercise problem 10; chapter 7, exercise problem 10):

$$\frac{\partial u}{\partial t} = \frac{\partial^2 u}{\partial x^2} + \frac{1}{x}\frac{\partial u}{\partial x} + 4$$

$$\frac{\partial u}{\partial x}(0,t) = 0 \text{ and } u(1,t) = 0$$

$$u(x,0) = 0$$

Obtain an analytical solution using the Laplace transform technique and plot the dimensionless velocity profiles.

14. Consider the Graetz problem in planar geometry[5] (chapter 5.1, exercise problem 11, chapter 7, exercise problem 11). The governing equations and boundary/initial conditions are:

$$2Pe(1-x^2)\frac{\partial u}{\partial z} = \frac{\partial^2 u}{\partial x^2}$$

$$u(x,0) = 0$$

$$\frac{\partial u}{\partial x}(0,z) = 0 \text{ and } u(1,z) = 1$$

Solve this problem analytically and plot the profiles for different values of the Peclet number.

15. Consider heat conduction in a slab with radiation at both ends (Carslaw and Jaeger, 1959;[2] chapter 5.1, exercise problem 12; chapter 7, exercise problem 12). The dimensionless governing equations, and the boundary and initial conditions are:

$$\frac{\partial u}{\partial t} = \frac{\partial^2 u}{\partial x^2}$$

$$-\frac{\partial u}{\partial x}(0,t) + Hu(0,t) = 0 \text{ and } \frac{\partial u}{\partial x}(1,t) + Hu(1,t) = 0$$

$$u(x,0) = 1$$

where H is the dimensionless heat transfer coefficient. Obtain an analytical solution and plot the profiles for H = 1, 10.

16. Consider the electrochemical discharge of spherical particles[6] (Subramanian and White, 2001; chapter 7, exercise problem 14). Governing equations, and boundary and initial conditions for this problem are:

$$\frac{\partial u}{\partial t} = \frac{\partial^2 u}{\partial x^2} + \frac{2}{x}\frac{\partial u}{\partial x}$$

$$u(x,0) = 1$$

$$\frac{\partial u}{\partial x}(0,t) = 0 \text{ and } \frac{\partial u}{\partial x}(1,t) = -\delta$$

8.1 Laplace Transform Technique for PDEs in Finite Domains

Obtain the short time and the long time solution for this problem using the Laplace transform technique. Plot the profiles for $\delta = 0.1, 1, 2$ and 5.

17. Redo exercise problem 13, chapter 7 using the Laplace transform technique if the flux at the surface is a function of time.

$$\frac{\partial u}{\partial x}(1,t) = f(t)$$

Solve this problem for the following functions:

$$f(t) = -e^{-t}$$
$$f(t) = -\sin(t)$$
$$f(t) = -\text{Delta}(t) \text{ (delta function)}$$

18. Consider a stirred pot in which pure solvent is used to extract oil from spherically shaped seeds.[3] Dimensionless concentrations of oil in the seeds (u) and in the solvent (C) are governed by the following equations.

$$\frac{\partial u}{\partial t} = \frac{\partial^2 u}{\partial x^2} + \frac{2}{x}\frac{\partial u}{\partial x}$$

$$\frac{dC}{dt} = -3\alpha \frac{\partial u}{\partial x}\bigg|_{x=1}$$

$$\frac{\partial u}{\partial x}(0,t) = 0 \text{ and } u(1,t) = C(t)$$

$$u(x,0) = C(0) = 1$$

where α is the capacity ratio of seed to the solvent. Obtain an analytical solution for this problem using the Laplace transform technique and plot the profiles for $\alpha = 0.2$ and $\alpha = 1$.

19. Consider a porous electrode in contact with a separator (at $x = 0$) and a current collector (at $x = r$).[7] Dimensionless electrolyte concentrations in the porous electrode (u) and in the separator (C) are governed by the following equations:

$$\frac{\partial u}{\partial t} = \sqrt{\varepsilon}\frac{\partial^2 u}{\partial x^2} + J$$

$$\frac{dC}{dt} = -\frac{3rJ\varepsilon}{2} + 3u(0,t) - 3C$$

$$\varepsilon^{1.5}\frac{\partial u}{\partial x}(0,t) = -\frac{rJ\varepsilon}{2} + 3u(0,t) - 3C \text{ and } \frac{\partial u}{\partial x}(r,t) = 0$$

$$u(x,0) = C(0) = 1$$

where ε is the porosity and r is the ratio of electrode thickness to separator thickness. Note that separator concentration (C) cannot be eliminated from the governing equation for electrode concentration (u) in the time domain. However, the separator concentration C can be eliminated in the Laplace domain. Obtain an analytical solution for this problem using the Laplace transform technique and plot the profiles for $\varepsilon = 0.4$ and $r = 4$.

20. Does problem 18 have a steady state solution? If so, explain how you would obtain it.
21. Does problem 19 have a steady state solution? If so, explain how you would obtain it.

References

1. Varma, A., Morbidelli, M.: Mathematical Methods in Chemical Engineering. Oxford University Press, Inc., Oxford (1997)
2. Carslaw, H.S., Jaeger, J.C.: Conduction of Heat in Solids. Oxford University Press, Oxford (1972)
3. Rice, R.G., Do, D.D.: Applied Mathematics and Modeling for Chemical Engineers. John Wiley & Sons, Inc., Chichester (1995)
4. Aris, R.: Mathematical Modeling: A Chemical Engineer's Perspective. Academic Press, London (1999)
5. Davis, M.E.: Numerical Methods and Modeling for Chemical Engineers. John Wiley & Sons, Chichester (1984)
6. Subramanian, V.R., White, R.E.: Separation of Variables for Diffusion in Composite Electrodes with Flux Boundary Conditions. In: Electrochemical Society Proceedings, vol. 99-14 (1999)
7. Subramanian, V.R., Tapriyal, D., White, R.E.: A Boundary Condition for Porous Electrodes. Electrochemical and Solid-State Letters 7(9), A259-A263 (2004)

Chapter 9
Parameter Estimation

9.1 Introduction

Chemical engineers develop mathematical models of systems of interest that usually include parameters. This chapter describes methodology that can be used to determine these parameters, which usually appear in a nonlinear manner. For example, the material balance equation for the reaction

$$A \xrightarrow{k_1} B \qquad (9.1)$$

in a constant volume isothermal batch reactor is

$$\frac{dC_A}{dt} = -k_1 C_A \qquad (9.2)$$

Equation (9.2) can be integrated easily to obtain,

$$C_A(t) = C_A(0) e^{-k_1 t} \qquad (9.3)$$

Equation (9.3) shows that the model equation for $C_A(t)$ depends on k_1 in a nonlinear manner (i.e., k_1 appears in the exponential term in equation (9.3)). Bequette [1] (p. 458) shows how to use experimental data to determine both k_1 and $C_A(0)$ by using the least squares method for a linearized form of equation (9.3). That is, taking the natural logarithm of both sides of equation (9.3) yields

$$\ln C_A(t) = \ln C_A(0) - k_1 t \qquad (9.4)$$

In equation (9.4), the dependent variable $\ln C_A(t)$ depends linearly on the parameter k_1 and the parameter $\ln C_A(0)$. Bequette rewrites equation (9.4) as

$$y = \theta_1 t + \theta_2 \qquad (9.5)$$

where

$$y = \ln C_A(t) \tag{9.6}$$

$$\theta_1 = -k_1 \tag{9.7}$$

and

$$\theta_2 = \ln C_A(0) \tag{9.8}$$

Bequette then uses the least squares method to determine θ_1 and θ_2 (Bequette calls these (p_1 and p_2)) since equation (9.5) is in the form of a straight line

$$y = mx + b \tag{9.9}$$

Unfortunately, this method is of limited value because chemical engineers often want to determine simultaneously more than two parameter values. For example, we would like to develop a procedure for determining the rate contents k_1, k_2, k_3, and k_4 for the reversible reactions

$$A \underset{k_2}{\overset{k_1}{\rightleftarrows}} B \underset{k_4}{\overset{k_3}{\rightleftarrows}} C \tag{9.10}$$

In this case (see equation 2.10), for a constant volume, isothermal batch reactor

$$\frac{dC_A}{dt} = -k_1 C_A + k_2 C_B \tag{9.11}$$

$$\frac{dC_B}{dt} = k_1 C_A - k_2 C_B - k_3 C_B + k_4 C_C \tag{9.12}$$

$$\frac{dC_C}{dt} = k_3 C_B - k_4 C_C \tag{9.13}$$

Unfortunately, it is not possible to recast equations (9.10)-(9.13) into a simple equation like equation (9.9).

9.2 Least Squares Method

The determination of parameters such as rate constants in mathematical models of interest to chemical engineers is often done by using the least squares method. This method is based on some assumptions about the independent and the dependent variables. Typically we begin by assuming that the independent variables such as time are known exactly and the dependent variables are random or stochastic variables (i.e., they are not known precisely). These variables are measured and

9.2 Least Squares Method

have measurement noise associated with them. We define the true values of these variables that would be obtained when measured if there were no randomness associated with the measurement. This true value is an hypothetical value which is postulated to exist. Consequently, a measured value will differ from the true value by a measurement error. We call this error a random error because it represents the difference between a random variable and its true value.

To develop the linear least squares equations for a model that is linear in the parameters $\underline{\theta}$ let the objective function $\Phi(\underline{\theta})$ be defined as

$$\Phi(\underline{\theta}) = (\underline{Y}^* - \underline{Y})^T (\underline{Y}^* - \underline{Y}) \tag{9.14}$$

where $\underline{Y}^*$ is a vector of measured values and $\underline{Y}$ is a vector of predicted values, which are obtained by multiplying the independent variables matrix $\underline{\underline{X}}$ by the vector of parameters $\underline{\theta}$:

$$\underline{Y} = \underline{\underline{X}}\,\underline{\theta} \tag{9.15}$$

Equation (9.15) can be written for a straight line model (see equation (9.9) with x_1 as the independent variable (t, e.g., see equation (9.5)) specified for the first data point, x_2 specified as the second data point, etc.:

$$\underline{\underline{X}} = \begin{pmatrix} x_1 & 1 \\ x_2 & 1 \\ \vdots & \vdots \\ x_n & 1 \end{pmatrix} \tag{9.16}$$

where n is the number of data points. The column of 1's in equation (9.16) is for the intercept b; that is, $\underline{\theta}$ is the vector of parameters and for a straight line model and is given by

$$\underline{\theta} = \begin{pmatrix} \theta_1 \\ \theta_2 \end{pmatrix} \tag{9.17}$$

where for equation (9.9) the parameter vector is

$$\underline{\theta} = \begin{pmatrix} m \\ b \end{pmatrix} \tag{9.18}$$

It should be mentioned that $\underline{\underline{X}}$ could be written as

$$\underline{\underline{X}} = \begin{pmatrix} 1 & x_1 \\ 1 & x_2 \\ \vdots & \vdots \\ 1 & x_n \end{pmatrix} \quad (9.19)$$

where in this case

$$\underline{\theta} = \begin{pmatrix} b \\ m \end{pmatrix} \quad (9.20)$$

since the constant b would come first in the model equation

$$y = \theta_1 + \theta_2 x \quad (9.21)$$

This way of writing $\underline{\underline{X}}$ and $\underline{\theta}$ provides a convenient way of writing models with two or more independent variables $(t$ and t^2, e.g.$)$ such as

$$y = \theta_1 + \theta_2 x + \theta_3 x^2 \quad (9.22)$$

In this case, the independent variable matrix for equation (9.21) is:

$$\underline{\underline{X}} = \begin{pmatrix} 1 & x_1 & x_1^2 \\ 1 & x_2 & x_2^2 \\ \vdots & \vdots & \vdots \\ 1 & x_n & x_n^2 \end{pmatrix} \quad (9.23)$$

An equation for $\underline{\theta}$ based on the experimental data and model predictions can be obtained from the objective function $\Phi(\underline{\theta})$. Equation (9.14) becomes by using equation (9.15)

$$\Phi(\underline{\theta}) = \left(\underline{Y}^* - \underline{\underline{X}}\,\underline{\theta}\right)^T \left(\underline{Y}^* - \underline{\underline{X}}\,\underline{\theta}\right) \quad (9.24)$$

Equation (9.24) can be expanded by recalling that the matrix transpose process is additive [2]

$$\left(\underline{Y}^* - \underline{\underline{X}}\,\underline{\theta}\right)^T = \left(\underline{Y}^*\right)^T - \left(\underline{\underline{X}}\,\underline{\theta}\right)^T \quad (9.25)$$

Substitution of equation (9.25) into equation (9.24) yields

9.2 Least Squares Method

$$\Phi(\underline{\theta}) = \left(\underline{Y}^*\right)^T \left(\underline{Y}^* - \underline{\underline{X}}\underline{\theta}\right) - \left(\underline{\underline{X}}\underline{\theta}\right)^T \left(\underline{Y}^* - \underline{\underline{X}}\underline{\theta}\right) \tag{9.26}$$

Equation (9.26) can be expanded to obtain

$$\Phi(\underline{\theta}) = \left(\underline{Y}^*\right)^T \underline{Y}^* - \left(\underline{Y}^*\right)^T \underline{\underline{X}}\underline{\theta} - \left(\underline{\underline{X}}\underline{\theta}\right)^T \underline{Y}^* + \left(\underline{\underline{X}}\underline{\theta}\right)^T \underline{\underline{X}}\underline{\theta} \tag{9.27}$$

Equation (9.27) can be simplified by recognizing that [2]

$$\left(\underline{\underline{X}}\underline{\theta}\right)^T \underline{Y}^* = \left(\underline{Y}^*\right)^T \left(\underline{\underline{X}}\underline{\theta}\right) \tag{9.28}$$

and that

$$\left(\underline{\underline{X}}\underline{\theta}\right)^T = \underline{\theta}^T \underline{\underline{X}}^T \tag{9.29}$$

Thus, by substituting equations (9.28) and (9.29) into equation (9.27) we obtain

$$\Phi(\underline{\theta}) = \left(\underline{Y}^*\right)^T \underline{Y}^* - 2\left(\underline{Y}^*\right)^T \underline{\underline{X}}\underline{\theta} + \underline{\theta}^T \underline{\underline{X}}^T \underline{\underline{X}}\underline{\theta} \tag{9.30}$$

The next step is to take the derivative of equation (9.30) with respect to $\underline{\theta}$ and set the resulting vector equal to the zero vector $\underline{0}$:

$$\frac{\partial \Phi}{\partial \underline{\theta}} = \underline{0} - \frac{\partial}{\partial \underline{\theta}}\left[2\left(\underline{Y}^*\right)^T \underline{\underline{X}}\underline{\theta}\right] + \frac{\partial}{\partial \underline{\theta}}\left(\underline{\theta}^T \underline{\underline{X}}^T \underline{\underline{X}}\underline{\theta}\right) = \underline{0} \tag{9.31}$$

The first non zero term on the right hand side of equation (9.31) can be written as

$$\frac{\partial}{\partial \underline{\theta}}\left[\left(\underline{Y}^*\right)^T \underline{\underline{X}}\underline{\theta}\right] = \underline{\underline{X}}^T \underline{Y}^* \tag{9.32}$$

where care has been taken to insure that the derivative has been taken properly. That is, according to equation A.75a in Crassidis and Junkins, [2]

$$\frac{\partial}{\partial \underline{x}}\left[\left(\underline{\underline{A}}\underline{x}+\underline{b}\right)^T \underline{\underline{C}}\left(\underline{\underline{D}}\underline{x}+\underline{e}\right)\right] = \underline{\underline{A}}^T \underline{\underline{C}}\left(\underline{\underline{D}}\underline{x}+\underline{e}\right) + \underline{\underline{D}}^T \underline{\underline{C}}^T \left(\underline{\underline{A}}\underline{x}+\underline{b}\right) \tag{9.33}$$

Comparison of equation (9.33) to equation (9.32) shows that in our case, the symbols in equation (9.33) are defined as follows: $\underline{x} = \underline{\theta}$, $\underline{\underline{A}} = \underline{0}$, $\underline{b} = \underline{Y}^*$, $\underline{\underline{C}} = \underline{\underline{I}}$, $\underline{\underline{D}} = \underline{\underline{X}}$, and $\underline{e} = \underline{0}$. Substitution of these symbols into equation (9.33) yields the results given in equation (9.32). Note that the quantity in the square bracket on the left side of equation (9.32) is a scalar. Equation A.71 of Crassidis and Junkins[2] shows that the derivative of a scalar with respect to a vector yields a vector. That is, if f is a scalar and $\underline{x}$ is a $n \times 1$ vector

$$\frac{\partial f}{\partial \underline{x}} = \left[\frac{\partial f}{\partial x_1} \frac{\partial f}{\partial x_2} \cdots \frac{\partial f}{\partial x_n} \right]^T \tag{9.34}$$

In our case,

$$f = \left(\underline{Y}^* \right)^T \underline{\underline{X}} \, \underline{\theta} \tag{9.35}$$

Consider the case where we have two parameters and three data points

$$\underline{\theta} = [\theta_1 \quad \theta_2]^T \tag{9.36}$$

$$\underline{Y}^* = [Y_1^* \quad Y_2^* \quad Y_3^*]^T \tag{9.37}$$

and

$$\underline{\underline{X}} = \begin{bmatrix} 1 & x_1 \\ 1 & x_2 \\ 1 & x_3 \end{bmatrix} \tag{9.38}$$

Substitution of equations (9.36) - (9.38) into equation (9.35) yields

$$f = Y_1^* \left(\theta_1 + \theta_2 x_1 \right) + Y_2^* \left(\theta_1 + \theta_2 x_2 \right) + Y_3^* \left(\theta_1 + \theta_3 x_3 \right) \tag{9.39}$$

Application of equation (9.34) to equation (9.39) yields

$$\frac{\partial f}{\partial \underline{\theta}} = \begin{bmatrix} Y_1^* + Y_2^* + Y_3^* \\ Y_1^* x_1 + Y_2^* x_2 + Y_3^* x_3 \end{bmatrix} \tag{9.40}$$

The right hand side of equation (9.32) becomes

$$\underline{\underline{X}}^T \underline{Y}^* = \begin{bmatrix} 1 & 1 & 1 \\ x_1 & x_2 & x_3 \end{bmatrix} \begin{bmatrix} Y_1^* \\ Y_2^* \\ Y_3^* \end{bmatrix} \tag{9.41}$$

$$\underline{\underline{X}}^T \underline{Y}^* = \begin{bmatrix} Y_1^* + Y_2^* + Y_3^* \\ Y_1^* x_1 + Y_2^* x_2 + Y_3^* x_3 \end{bmatrix} \tag{9.42}$$

Comparison of equations (9.40) and (9.42) reveals that equation (9.32) is correct.

9.2 Least Squares Method

The last term in equation (9.31) can be evaluated by letting

$$\underline{\underline{C}} = \underline{\underline{X}}^T \underline{\underline{X}} \tag{9.43}$$

and recalling that (see equation A.75b in Crassidis and Junkins [2])

$$\frac{\partial}{\partial \underline{\theta}} \left(\underline{\theta}^T \underline{\underline{C}} \, \underline{\theta} \right) = \left(\underline{\underline{C}} + \underline{\underline{C}}^T \right) \underline{\theta} \tag{9.44}$$

Since the matrix $\underline{\underline{X}}^T \underline{\underline{X}}$ is symmetrical, the matrix $\underline{\underline{X}}^T \underline{\underline{X}}$ is equal to its transpose:

$$\underline{\underline{X}}^T \underline{\underline{X}} = \left(\underline{\underline{X}}^T \underline{\underline{X}} \right)^T \tag{9.45}$$

Thus, equation (9.44) and (9.45) yield

$$\frac{\partial}{\partial \underline{\theta}} \left(\underline{\theta}^T \underline{\underline{X}}^T \underline{\underline{X}} \, \underline{\theta} \right) = \left(\underline{\underline{X}}^T \underline{\underline{X}} + \left(\underline{\underline{X}}^T \underline{\underline{X}} \right)^T \right) \underline{\theta} = 2 \underline{\underline{X}}^T \underline{\underline{X}} \, \underline{\theta} \tag{9.46}$$

Note equation (9.46) also requires taking the derivative of a scalar with respect to a vector which yields a vector. That is, in this case,

$$f = \underline{\theta}^T \underline{\underline{X}}^T \underline{\underline{X}} \, \underline{\theta} \tag{9.47}$$

which for our example case becomes

$$f = \begin{bmatrix} \theta_1 & \theta_2 \end{bmatrix} \begin{bmatrix} 1 & 1 & 1 \\ x_1 & x_2 & x_3 \end{bmatrix} \begin{bmatrix} 1 & x_1 \\ 1 & x_2 \\ 1 & x_3 \end{bmatrix} \begin{bmatrix} \theta_1 \\ \theta_2 \end{bmatrix} \tag{9.48}$$

or

$$f = \begin{bmatrix} \theta_1 & \theta_2 \end{bmatrix} \begin{bmatrix} 3 & \sum_{i=1}^{3} x_i \\ \sum_{i=1}^{3} x_i & \sum_{i=1}^{3} x_i^2 \end{bmatrix} \begin{bmatrix} \theta_1 \\ \theta_2 \end{bmatrix} \tag{9.49}$$

or

$$f = \begin{bmatrix} \theta_1 & \theta_2 \end{bmatrix} \begin{bmatrix} 3\theta_1 + \theta_2 \sum_{i=1}^{3} x_i \\ \theta_1 \sum_{i=1}^{3} x_i + \theta_2 \sum_{i=1}^{3} x_i^2 \end{bmatrix} \tag{9.50}$$

or

$$f = \left[3\theta_1^2 + \theta_1\theta_2 \sum_{i=1}^{3} x_i + \theta_1\theta_2 \sum_{i=1}^{3} x_i + \theta_2^2 \sum_{i=1}^{3} x_i^2 \right] \quad (9.51)$$

Applying equation (9.34) to equation (9.51) yields

$$\frac{\partial f}{\partial \begin{bmatrix} \theta_1 \\ \theta_2 \end{bmatrix}} = \begin{bmatrix} \frac{\partial f}{\partial \theta_1} \\ \frac{\partial f}{\partial \theta_2} \end{bmatrix} = 2 \begin{bmatrix} 3\theta_1 + \theta_2 \sum_{i=1}^{3} x_i \\ \theta_1 \sum_{i=1}^{3} x_i + \theta_2 \sum_{i=1}^{3} x_i^2 \end{bmatrix} \quad (9.52)$$

The right hand side of equation (9.46) becomes for our example case

$$2\underline{\underline{X}}^T \underline{\underline{X}} \underline{\theta} = 2 \begin{bmatrix} 3 & \sum_{i=1}^{3} x_i \\ \sum_{i=1}^{3} x_i & \sum_{i=1}^{3} x_i^2 \end{bmatrix} \begin{bmatrix} \theta_1 \\ \theta_2 \end{bmatrix} \quad (9.53)$$

$$2\underline{\underline{X}}^T \underline{\underline{X}} \underline{\theta} = 2 \begin{bmatrix} 3\theta_1 + \theta_2 \sum_{i=1}^{3} x_i \\ \theta_1 \sum_{i=1}^{3} x_i + \theta_2 \sum_{i=1}^{3} x_i^2 \end{bmatrix} \quad (9.54)$$

Comparison of equation (9.52) to equation (9.54) reveals that equation (9.46) is correct. Substitution of equations (9.32) and (9.46) into equation (9.31) yields

$$\frac{\partial \Phi}{\partial \underline{\theta}} = -2\underline{\underline{X}}^T \underline{Y}^* + 2\underline{\underline{X}}^T \underline{\underline{X}} \underline{\theta} = \underline{0} \quad (9.55)$$

Equation (9.55) can be solved for $\underline{\theta}$:

$$\underline{\theta} = \left(\underline{\underline{X}}^T \underline{\underline{X}} \right)^{-1} \underline{\underline{X}}^T \underline{Y}^* \quad (9.56)$$

which is the least squares equation for models that are linear in the parameters $\underline{\theta}$.

Equation (9.56) can also be derived by considering the least squares problem as an over-determined system of equations. That is, we would like to find $\underline{\theta}$ such that

9.2 Least Squares Method

$$\underline{\underline{X}}\,\underline{\theta} = \underline{Y}^* \tag{9.57}$$

but we have an over-determined system since $\underline{\theta}$ is typically a much smaller vector than $\underline{Y}^*$. Consequently, if we multiply equation (9.57) from the left by the transpose of $\underline{\underline{X}}\left(\underline{\underline{X}}^T\right)$ we obtain

$$\underline{\underline{X}}^T \underline{\underline{X}}\,\underline{\theta} = \underline{\underline{X}}^T \underline{Y}^* \tag{9.58}$$

and we realize that $\underline{\underline{X}}^T \underline{\underline{X}}$ is a $n \times n$ square matrix that is the same size $(n \times 1)$ as the vector $\underline{\theta}$ and the vector $\underline{\underline{X}}^T \underline{Y}^*$. Usually, $\left(\underline{\underline{X}}^T \underline{\underline{X}}\right)^{-1}$ exists; consequently, the least squares or normal equation for $\underline{\theta}$ can be obtained:

$$\underline{\theta} = \left(\underline{\underline{X}}^T \underline{\underline{X}}\right)^{-1} \underline{\underline{X}}^T \underline{Y}^* \tag{9.59}$$

Additional discussion about this approach can be found in Lopez (page 772) [3] and Ogunnaike and Ray (Example D.8, page 1230). [4]

9.2.1 Summation Form or Classical Form

The normal least squares equation (9.59) can be determined in summation or classical form [5, 6] by writing the objective function as

$$\Phi(\theta_1, \theta_2) = \sum_{i=1}^{n} (y_i - \theta_1 - \theta_2 x_i)^2 \tag{9.60}$$

where y_i represents the measured experimental value at the independent variable x_i. The derivative of Φ in equation (9.60) with respect to θ_1 is

$$\frac{\partial \Phi}{\partial \theta_1} = -2 \sum_{i=1}^{n} (y_i - \theta_1 - \theta_1 x_i) = 0 \tag{9.61}$$

Equation (9.61) can be written out for three data points $(n = 3)$ for illustration purposes:

$$(y_1 - \theta_1 - \theta_2 x_1) + (y_2 - \theta_1 - \theta_1 x_2) + (y_3 - \theta_1 - \theta_1 x_3) = 0 \tag{9.62}$$

Inspection reveals that equation (9.62) can be simplified to

$$\sum_{i=1}^{n} y_i = n\theta_1 + \theta_2 \sum_{i=1}^{n} x_i \qquad (9.63)$$

The derivative of Φ with respect to θ_2 is

$$\frac{\partial \Phi}{\partial \theta_2} = -2\sum_{i=1}^{n}(y_i - \theta_1 - \theta_2 x_i)x_i = 0 \qquad (9.64)$$

which can be rewritten as

$$\theta_1 \sum_{i=1}^{n} x_i + \theta_2 \sum_{i=1}^{n} x_i^2 = \sum_{i=1}^{n} y_i x_i \qquad (9.65)$$

Next, let $\bar{y}$ be the average or mean value of the experimentally measured values of the dependent variable be defined as follows

$$\bar{y} = \frac{1}{n}\sum_{i=1}^{n} y_i \qquad (9.66)$$

and the average value of the independent variable values be defined in a similar manner

$$\bar{x} = \frac{1}{n}\sum_{i=1}^{n} x_i \qquad (9.67)$$

Equation (9.63) can be rewritten by using $\bar{y}$ and $\bar{x}$ (equations (9.66) and (9.67), respectively) to yield an expression for θ_1 in terms of $\bar{y}$ and $\bar{x}$ for convenience. If we can find an equation for θ_2 in terms of the measured values of y_i and x_i, we can substitute it into equation (9.65) and have expressions for the parameters θ_1 and θ_2 in terms of the measured values. Equation (9.63) can be rewritten as

$$\theta_1 = \bar{y} - \theta_2 \bar{x} \qquad (9.68)$$

Substitution of equation (9.68) into (9.65) yields

$$(\bar{y} - \theta_2 \bar{x})\left(\sum_{i=1}^{n} x_i\right) + \theta_2 \sum_{i=1}^{n} x_i^2 = \sum_{i=1}^{n} y_i x_i \qquad (9.69)$$

or

$$\left(\frac{1}{n}\sum_{i=1}^{n} y_i - \frac{\theta_2}{n}\sum_{i=1}^{n} x_i\right)\left(\sum_{i=1}^{n} x_i\right) + \theta_2 \sum_{i=1}^{n} x_i^2 = \sum_{i=1}^{n} y_i x_i \qquad (9.70)$$

Equation (9.70) can be rewritten as

9.2 Least Squares Method

$$-\frac{\theta_2}{n}\left(\left(\sum_{i=1}^{n}x_i\right)\left(\sum_{i=1}^{n}x_i\right)-n\sum_{i=1}^{n}x_i^2\right)=\sum_{i=1}^{n}y_i x_i -\frac{1}{n}\sum_{i=1}^{n}y_i\sum_{i=1}^{n}x_i \quad (9.71)$$

or

$$\theta_2\left(-\frac{\left(\sum_{i=1}^{n}x_i\right)^2}{n}+\sum_{i=1}^{n}x_i^2\right)=\sum_{i=1}^{n}y_i x_i -\frac{1}{n}\sum_{i=1}^{n}y_i\sum_{i=1}^{n}x_i \quad (9.72)$$

Thus, equation for θ_2 becomes

$$\theta_2=\frac{\sum_{i=1}^{n}y_i x_i -\frac{1}{n}\left(\sum_{i=1}^{n}y_i\sum_{i=1}^{n}x_i\right)}{\sum_{i=1}^{n}x_i^2-\frac{\left(\sum_{i=1}^{n}x_i\right)^2}{n}} \quad (9.73)$$

Next, let

$$S_{xx}=\sum_{i=1}^{n}x_i^2-\frac{\left(\sum_{i=1}^{n}x_i\right)^2}{n}=\sum_{i=1}^{n}(x_i-\bar{x})^2 \quad (9.74)$$

and

$$S_{xy}=\sum_{i=1}^{n}y_i x_i -\frac{\left(\sum_{i=1}^{n}y_i\right)\left(\sum_{i=1}^{n}x_i\right)}{n} \quad (9.75)$$

Thus, equation (9.73) becomes

$$\theta_2=\frac{S_{xy}}{S_{xx}} \quad (9.76)$$

Next, write equation (9.60), the objective function, as a sum of the residuals

$$SS_E=\sum_{i=1}^{n}\left(y_i-(\theta_1+\theta_2 x_i)\right)^2 \quad (9.77)$$

which can be expanded and rewritten as

$$SS_E = \sum_{i=1}^{n} y_i^2 - n\bar{y}^2 - \theta_2 S_{xy} \tag{9.78}$$

Next, let's define

$$S_{yy} = \sum_{i=1}^{n} y_i^2 - n\bar{y}^2 \tag{9.79}$$

so that equation (9.78) becomes

$$SS_E = S_{yy} - \theta_2 S_{xy} \tag{9.80}$$

Note that S_{yy} can also be written as

$$S_{yy} = \sum_{i=1}^{n} (y_i - \bar{y})^2 \tag{9.81}$$

which can be seen by letting $n = 2$ and expanding equation (9.81) to

$$S_{yy} = (y_1 - \bar{y})^2 + (y_2 - \bar{y})^2 \tag{9.82}$$

which becomes

$$S_{yy} = y_1^2 - 2y_1\bar{y} + \bar{y}^2 + y_2^2 - 2y_2\bar{y} + \bar{y}^2 \tag{9.83}$$

or

$$S_{yy} = y_1^2 + y_2^2 - 2y_1\bar{y} - 2y_2\bar{y} + 2\bar{y}^2 \tag{9.84}$$

or

$$S_{yy} = \sum_{i=1}^{n} y_i^2 - 2\sum_{i=1}^{n} y_i \bar{y} + n\bar{y}^2 \tag{9.85}$$

or

$$S_{yy} = \sum_{i=1}^{n} y_i^2 - 2n\bar{y}\,\bar{y} + n\bar{y}^2 \tag{9.86}$$

Thus,

$$S_{yy} = \sum_{i=1}^{n} y_i^2 - n\bar{y}^2 \tag{9.87}$$

9.2 Least Squares Method

Note that SS_E has $n-2$ degrees of freedom since two degrees of freedom were used to determine θ_1 and θ_2. Thus, let the mean square (MS_E) be defined as s^2

$$MS_E = s^2 = \frac{SS_E}{n-2} \tag{9.88}$$

which can be used to estimate the variance (σ^2). We will call s the standard deviation. Next, let the predicted value of the dependent variable be written with a caret, $\hat{y}_i$:

$$\hat{y}_i = \theta_1 + \theta_2 x_i \tag{9.89}$$

Now, let's develop an equation for the sum of the difference between the measured value of the dependent variable y_i and the average value of the observed values $\bar{y}$ in terms of $\hat{y}_i$. Start by writing the identity

$$y_i - \bar{y} = (\hat{y}_i - \bar{y}) + (y_i - \hat{y}_i) \tag{9.90}$$

Next, square both sides of equation (9.90) and sum over all n observations

$$\sum_{i=1}^n (y_i - \bar{y})^2 = \sum_{i=1}^n (\hat{y}_i - \bar{y})^2 + \sum_{i=1}^n (y_i - \hat{y}_i)^2 + 2\sum_{i=1}^n (\hat{y}_i - \bar{y})(y_i - \hat{y}_i) \tag{9.91}$$

The third term on the right hand side of equation (9.91) can be expanded to obtain

$$2\sum_{i=1}^n (\hat{y}_i - \bar{y})(y_i - \hat{y}_i) = 2\sum_{i=1}^n \hat{y}_i (y_i - \hat{y}_i) - 2\bar{y}\sum_{i=1}^n (y_i - \hat{y}_i) \tag{9.92}$$

Now let r_i be the residual:

$$r_i = y_i - \hat{y}_i \tag{9.93}$$

so that equation (9.92) becomes

$$2\sum_{i=1}^n (\hat{y}_i - \bar{y})(y_i - \hat{y}_i) = 2\sum_{i=1}^n \hat{y}_i r_i - 2\bar{y}\sum_{i=1}^n r_i \tag{9.94}$$

Note that both terms on the right hand side of equation (9.94) are zero since the sum of the residuals is always equal to zero,

$$\sum_{i=1}^n r_i = 0 \tag{9.95}$$

and the sum of the residuals weighted by the predicted values $(\hat{y}_i)$ is also zero:

$$\sum_{i=1}^{n} \hat{y}_i r_i = 0 \qquad (9.96)$$

Thus, equation (9.91) simplifies to

$$\sum_{i=1}^{n}(y_i - \bar{y})^2 = \sum_{i=1}^{n}(\hat{y}_i - \bar{y})^2 + \sum_{i=1}^{n}(y_i - \hat{y}_i)^2 \qquad (9.97)$$

Equation (9.97) can be rewritten as

$$\sum_{i=1}^{n}(y_i - \bar{y})^2 = SS_R + SS_E \qquad (9.98)$$

where SS_E is given by equation (9.80) and SS_R is called the regression sum of squares and is defined as

$$SS_R = \sum_{i=1}^{n}(\hat{y}_i - \bar{y})^2 \qquad (9.99)$$

The left hand side of equation (9.97) can be written as (see equation (9.81))

$$S_{yy} = \sum_{i=1}^{n}(y_i - \bar{y})^2 \qquad (9.100)$$

so that equation (9.98) becomes

$$S_{yy} = SS_R + SS_E \qquad (9.101)$$

Substitution of equation (9.80) into equation (9.101) yields an equation for SS_R in terms of θ_2:

$$SS_R = \theta_2 S_{xy} \qquad (9.102)$$

Finally, the coefficient of determination is called R^2 and can be written in terms of SS_E and S_{yy} by using equation (9.102) as follows:

$$R^2 = \frac{SS_R}{S_{yy}} = 1 - \frac{SS_E}{S_{yy}} \qquad (9.103)$$

or in terms of θ_2:

$$R^2 = \theta_2^2 \frac{S_{xy}}{S_{yy}} \qquad (9.104)$$

9.2 Least Squares Method

Values of R^2 close to 1 imply that the variability in y as defined by S_{yy} can be explained by the regression model; however, it is important to note that R^2 has no meaning for cases when x is a controlled variable because the magnitude of R^2 depends on the spacing of x through S_{xy} (see equation(9.104)).

The least squares estimates $\hat{\theta}_1$ and $\hat{\theta}_2$ have several important statistical properties. For example,

$$\hat{\theta}_2 = \frac{S_{xy}}{S_{xx}} = \sum_{i=1}^{n} c_i y_i \qquad (9.105)$$

where

$$c_i = \frac{x_i - \bar{x}}{S_{xx}} \qquad i = 1,..,n \qquad (9.106)$$

This means that $\hat{\theta}_2$ is a linear combination of the observations (y_i). Now since the observations (y_i) are assumed to be random variables that are normally and independently distributed, $\hat{\theta}_2$ is also a normally and independently distributed random variable.

9.2.2 Confidence Intervals: Classical Approach

The confidence intervals for θ_1, θ_2, and the variance σ^2 can be determined by assuming that the errors in measuring y_i are normally and independently distributed so that the distribution of the variable v_1 which is defined as

$$v_1 = \frac{\hat{\theta}_1 - \theta_1}{\sqrt{MS_E \left(\frac{1}{n} + \frac{\bar{x}^2}{S_{xx}} \right)}} \qquad (9.107)$$

follows a student's t distribution with $n-2$ degrees of freedom. Therefore, a $100(1-\alpha)$ percent confidence interval on the intercept θ_1 is given by

$$\hat{\theta}_1 - t_{\alpha/2,\, n-2} \sqrt{MS_E \left(\frac{1}{n} + \frac{\bar{x}^2}{S_{xx}} \right)} \leq \theta_1 \leq \hat{\theta}_1 + t_{\alpha/2,\, n-2} \sqrt{MS_E \left(\frac{1}{n} + \frac{\bar{x}^2}{S_{xx}} \right)} \qquad (9.108)$$

In a similar manner the distribution of the variable v_2

$$v_2 = \frac{\hat{\theta}_2 - \theta_2}{\sqrt{\frac{MS_E}{S_{xx}}}} \quad (9.109)$$

follows a student's t distribution with $n-2$ degrees of freedom, and

$$\hat{\theta}_2 - t_{\alpha/2, n-2}\sqrt{\frac{MS_E}{S_{xx}}} \le \theta_2 \le \hat{\theta}_2 + t_{\alpha/2, n-2}\sqrt{\frac{MS_E}{S_{xx}}} \quad (9.110)$$

Also, if the errors are normally and independently distributed, the sampling distribution of the variable v_3

$$v_3 = \frac{(n-2) MS_E}{\sigma^2} \quad (9.111)$$

is chi-square with $n-2$ degrees of freedom. Thus, for $100(1-\alpha)$ confidence interval in percent

$$\frac{(n-2) MS_E}{\chi^2_{\alpha/2}} \le \sigma^2 \le \frac{(n-2) MS_E}{\chi^2_{1-\alpha/2, n-2}} \quad (9.112)$$

The standard error (se) of the slope $\hat{\theta}_2$ is defined as

$$se(\hat{\theta}_2) = \sqrt{\frac{MS_E}{S_{xx}}} \quad (9.113)$$

and the standard error of the intercept $\hat{\theta}_1$ is

$$se(\hat{\theta}_1) = \sqrt{MS_E \left(\frac{1}{n} + \frac{\bar{x}^2}{S_{xx}} \right)} \quad (9.114)$$

9.2.3 Prediction of New Observations

A new measurement x_0 can be used to predict a new observation, $\hat{y}_0$:

$$\hat{y}_0 = \hat{\theta}_1 + \hat{\theta}_2 x_0 \quad (9.115)$$

where $\hat{y}_0$ is the point estimate of the new observation. Let the difference between value y_0 and the predicted value $\hat{y}_0$ be the random variable, φ, where

$$\varphi = y_0 - \hat{y}_0 \quad (9.116)$$

9.2 Least Squares Method

and φ is normally distributed with mean zero and variance. That is,

$$V(\varphi) = V(y_0 - \hat{y}_0) \qquad (9.117)$$

The variance $V(\varphi)$ in this case includes the variance of the fitted regression line and the variance of the error term:

$$V(\varphi) = \sigma^2 \left[1 + \frac{1}{n} + \frac{(x_0 - \bar{x})^2}{S_{xx}} \right] \qquad (9.118)$$

because y_0 is independent of $\hat{y}_0$. Thus, the $100(1-\alpha)$ percent prediction interval on a future observation at x_0 is

$$\hat{y}_0 - t_{\alpha/2,n-2}\sqrt{MS_E \left(1 + \frac{1}{n} + \frac{(x_0 - \bar{x})^2}{S_{xx}}\right)} \leq y_0 \leq \hat{y}_0 + t_{\alpha/2,n-2}\sqrt{MS_E \left(1 + \frac{1}{n} + \frac{(x_0 - \bar{x})^2}{S_{xx}}\right)} \qquad (9.119)$$

9.2.4 A One Parameter through the Origin Model

A single parameter model that forces the data to go through the origin can be written as

$$y = \beta_1 x + \varepsilon \qquad (9.120)$$

In this case, the objective function can be written as

$$S(\beta_1) = \sum_{i=1}^{n} (y_i - \beta_1 x_i)^2 \qquad (9.121)$$

The normal equation in this case is

$$\hat{\beta}_1 \sum_{i=1}^{n} x_i^2 = \sum_{i=1}^{n} y_i x_i \qquad (9.122)$$

so that

$$\hat{\beta}_1 = \frac{\sum_{i=1}^{n} y_i x_i}{\sum_{i=1}^{n} x_i^2} \qquad (9.123)$$

The mean square error in this case is

$$MS_E = \frac{\sum_{i=1}^{n} (y_i - \hat{y}_i)^2}{n-1} \qquad (9.124)$$

or

$$MS_E = \frac{\sum_{i=1}^{n} y_i^2 - \hat{\beta}_1 \sum_{i=1}^{n} y_i x_i}{n-1} \tag{9.125}$$

For this one parameter case, the $100(1-\alpha)$ percent confidence interval becomes

$$\hat{\beta}_1 - t_{\alpha/2, n-1} \sqrt{\frac{MS_E}{\sum_{i=1}^{n} x_i^2}} \leq \beta_1 \leq \hat{\beta}_1 + t_{\alpha/2, n-1} \sqrt{\frac{MS_E}{\sum_{i=1}^{n} x_i^2}} \tag{9.126}$$

A $100(1-\alpha)$ percent confidence interval for the expected value of the mean response at $x = x_0$ $\left(E(y_1|x_0)\right)$ is

$$\hat{y}_0 - t_{\alpha/2, n-1} \sqrt{\frac{x_0^2 MS_E}{\sum_{i=1}^{n} x_i^2}} \leq E(y_1|x_0)_1 \leq \hat{y}_0 + t_{\alpha/2, n-1} \sqrt{\frac{x_0^2 MS_E}{\sum_{i=1}^{n} x_i^2}} \tag{9.127}$$

The $100(1-\alpha)$ percent prediction interval on a future observation a $x = x_0$ for y_0 is

$$\hat{y}_0 - t_{\alpha/2, n-1} \sqrt{MS_E \left(1 + \frac{x_0^2}{\sum_{i=1}^{n} x_i^2}\right)} \leq y_0 \leq \hat{y}_0 + t_{\alpha/2, n-1} \sqrt{MS_E \left(1 + \frac{x_0^2}{\sum_{i=1}^{n} x_i^2}\right)} \tag{9.128}$$

9.3 Nonlinear Least Squares

The nonlinear least squares method can be used to determine parameter values that appear in model equations such as those given by equations (9.11) – (9.13). However, before doing that let's consider a simpler problem. Bard [7] shows how to determine the pre-exponential factor or Arrhenius constant and the activation energy for reaction (9.1) at different temperatures. In this case, let's assume that we know $C_A(0)$ and form the fraction of reactant remaining, y:

$$y = C_A/C_A(0) \tag{9.129}$$

And set the initial condition to

$$y = 1 \text{ at } t = 0 \tag{9.130}$$

In this case

9.3 Nonlinear Least Squares

$$y = \exp(-k_1 x_1) \tag{9.131}$$

where

$$x_1 = \text{time in hours} \tag{9.132}$$

and

$$k_1 = \theta_1 \exp(-\theta_2/x_2) \tag{9.133}$$

where θ_1 is the Arrhenius constant in hr^{-1} and θ_2 is the activation energy divided by the gas constant R so that the unit of θ_2 is K. Here x_2 is the second independent variable T in units of K.

In this case, we have one dependent variable y, two independent variables x_1 and x_2, and two parameter values $(\theta_1 \text{ and } \theta_2)$. The model equation is

$$y = \exp\left[-\left(\theta_1 \exp(-\theta_2/x_2)\right) x_1\right] \tag{9.134}$$

The experimental data from Bard [7] are presented in Table 9.1.

Table 9.1

Data for Batch Reactor[3]

Experiment number, μ	Time, $x_{\mu1}$ (hr)	Temperature, $x_{\mu2}$ (K)	Fraction A Remaining, y_μ
1	0.1	100	0.980
2	0.2	100	0.983
3	0.3	100	0.955
4	0.4	100	0.979
5	0.5	100	0.993
6	0.05	200	0.626
7	0.1	200	0.544
8	0.15	200	0.455
9	0.2	200	0.225
10	0.25	200	0.167
11	0.02	300	0.566
12	0.04	300	0.317
13	0.06	300	0.034
14	0.08	300	0.016
15	0.1	300	0.066

The next step in the process is to define the scalar objective function Φ that depends on the vector of parameters $\underline{\theta}$

$$\Phi(\underline{\theta}) = \left(\underline{Y}^* - \underline{Y}\right)^T \left(\underline{Y}^* - \underline{Y}\right) \tag{9.135}$$

where

$\underline{Y}^*$ = vector of experimental values for the depdendent variable y (9.136)

$\underline{Y}$ = vector of predicted values of y (from equation 9.19) (9.137)

The minimum of Φ with respect to the parameters $\underline{\theta}$ can be found by setting the derivative of Φ with respect to the $\underline{\theta}$ equal to zero:

$$\frac{\partial \Phi}{\partial \underline{\theta}} = \underline{0} \tag{9.138}$$

Equation (9.134) is nonlinear in the parameter values $\underline{\theta}$. Consequently, let's expand $\underline{Y}$ in a Taylor series about guessed values for the parameter $\underline{\theta}^{(k)}$ where the superscript k indicates iteration number k. Thus,

$$\underline{Y}(\underline{x},\,\underline{\theta}) \approx \underline{Y}\left(\underline{x},\,\underline{\theta}^{(k)}\right) + \left.\frac{\partial \underline{Y}}{\partial \underline{\theta}}\right|_{\underline{\theta}^{(k)}} \Delta\underline{\theta} \tag{9.139}$$

or

$$\underline{Y} = \underline{Y}\left(\underline{x},\underline{\theta}^{(k)}\right) + \underline{\underline{J}}\Big|_{\underline{\theta}=\underline{\theta}^{(k)}} \Delta\underline{\theta} \tag{9.140}$$

where

$$\Delta\underline{\theta} = \underline{\theta}^{(k+1)} - \underline{\theta}^{(k)} \tag{9.141}$$

and for two parameters $(\theta_1 \text{ and } \theta_2)$

$$\underline{\underline{J}} = \begin{bmatrix} \dfrac{\partial Y_1}{\partial \theta_1} & \dfrac{\partial Y_1}{\partial \theta_2} \\ \vdots & \\ \dfrac{\partial Y_n}{\partial \theta_1} & \dfrac{\partial Y_n}{\partial \theta_2} \end{bmatrix} \tag{9.142}$$

where $\partial Y_1/\partial \theta_1$ is referred to as the sensitivity coefficient of the dependent variable Y with respect to parameter number one θ_1 evaluated at the first experimental condition as indicated by the subscript 1 on Y (i.e., Y_1). Again, the subscript n is the total number of data points which can include different independent

9.3 Nonlinear Least Squares

variable values. For example, $n = 15$ for the data in Table 9.1 where y was measured at three different temperatures for five different times. Substitution of equation (9.140) into equation (9.135) yields

$$\Phi(\underline{\Delta\theta}) = \left(\underline{Y}^* - \underline{Y} - \underline{\underline{J}}\underline{\Delta\theta}\right)^T \left(\underline{Y}^* - \underline{Y} - \underline{\underline{J}}\underline{\Delta\theta}\right) \tag{9.143}$$

The minimum of $\Phi(\underline{\Delta\theta})$ given by equation (9.143) can be found by taking its derivative with respect to $\underline{\Delta\theta}$ and setting that vector equal to the zero vector. This can be done by using equation (9.33) above where in this case $\underline{x} = \underline{\Delta\theta}$, $\underline{\underline{A}} = -\underline{\underline{J}}$, $\underline{b} = \underline{Y}^* - \underline{Y}$, $\underline{e} = \underline{Y}^* - \underline{Y}$, $\underline{\underline{C}} = \underline{\underline{I}}$, and $\underline{\underline{D}} = -\underline{\underline{J}}$. Thus,

$$\frac{\partial \Phi}{\partial \underline{\Delta\theta}} = -\underline{\underline{J}}^T \left(-\underline{\underline{J}}\underline{\Delta\theta} + \underline{Y}^* - \underline{Y}\right) - \underline{\underline{J}}^T \left(-\underline{\underline{J}}\underline{\Delta\theta} + \underline{Y}^* - \underline{Y}\right) = 0 \tag{9.144}$$

or

$$\frac{\partial \Phi}{\partial \underline{\Delta\theta}} = 2\underline{\underline{J}}^T \left(\underline{\underline{J}}\underline{\Delta\theta} - \left(\underline{Y}^* - \underline{Y}\right)\right) = 0 \tag{9.145}$$

or

$$\frac{\partial \Phi}{\partial \underline{\Delta\theta}} = -2\underline{\underline{J}}^T \left(\underline{Y}^* - \underline{Y} - \underline{\underline{J}}\underline{\Delta\theta}\right) = \underline{0} \tag{9.146}$$

Equation (9.146) can be divided by 2 and then expanded to yield

$$\underline{\underline{J}}^T \underline{\underline{J}} \underline{\Delta\theta} = \underline{\underline{J}}^T \left(\underline{Y}^* - \underline{Y}\right) + \underline{0} \tag{9.147}$$

which can be solved to obtain an expression for the change in the parameter vector $\underline{\Delta\theta}$:

$$\underline{\Delta\theta} = \left(\underline{\underline{J}}^T \underline{\underline{J}}\right)^{-1} \underline{\underline{J}}^T \left(\underline{Y}^* - \underline{Y}\right) \tag{9.148}$$

This is an equation that can be used in an iterative process to find $\underline{\theta}$ when the model is nonlinear in one or more of the parameters $\underline{\theta}$.

The procedure to determine the values for the elements of $\underline{\theta}$ (θ_1 and θ_2 in this case) consists of the following steps when it is possible to obtain the elements of $\underline{\underline{J}}$ analytically.

1. Guess values for $\underline{\theta}^{(k)}$.
2. Use $\underline{\theta}^{(k)}$ to determine $\underline{\underline{J}}$ according to equation (9.142).
3. Use $\underline{\theta}^{(k)}$ to determine $\underline{Y}$ according to equation (9.134). ($\underline{Y}$ is the vector of values of y obtained from the model equation (equation (9.134) in this case) evaluated at the independent variables 1 through 15 with $\underline{\theta}^{(k)}$ and the experimental data point values for $x_{1,j}$ and $x_{2,j}$ with $j = 1, 15$.)

4. Solve equation (9.148) for $\Delta\underline{\theta}$.
5. Solve equation (9.141) for $\underline{\theta}^{(k+1)}$.
6. Replace $\underline{\theta}^{(k)}$ with $\underline{\theta}^{(k+1)}$.
7. Repeat steps 2 through 6 until each $\left|\dfrac{\Delta\theta_i}{\theta_i}\right|$ becomes small (e.g.,

$\left|\dfrac{\Delta\theta_1}{\theta_1}\right|\Delta\theta_1 \le 10^{-4}$, and $\left|\dfrac{\Delta\theta_2}{\theta_2}\right|\Delta\theta_2 \le 10^{-4}$).

Figure 9.1 shows a flow diagram for this procedure.

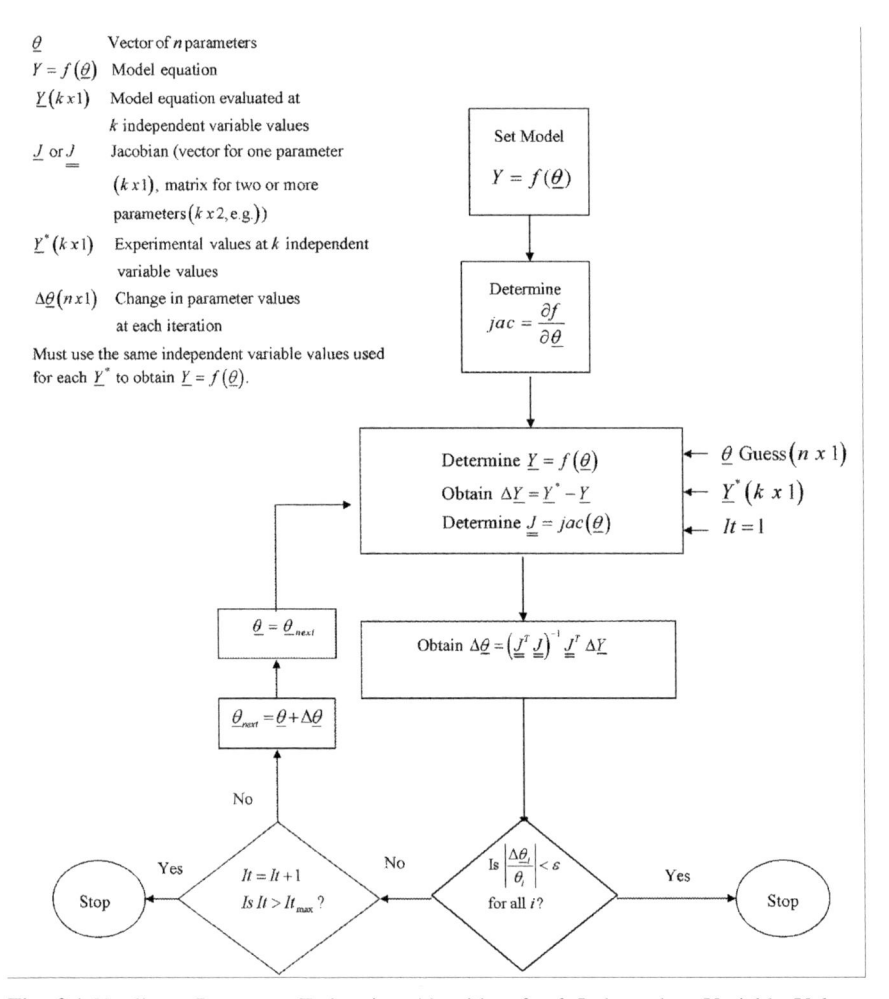

Fig. 9.1 Nonlinear Parameter Estimation Algorithm for k Independent Variable Values, One Dependent Variable (Y) and n Parameters ($\underline{\theta}$)

9.3 Nonlinear Least Squares

A Maple worksheet for this procedure is presented below.

Example 9.1 Parameter Estimation

> restart:

> with (linalg):

Enter experimental data for fraction of reactant remaining for all fifteen data points(yexp, j=1,15).

> yexp:=matrix(15,1,[0.980,0.983,0.955,0.979,0.993,0.626,0.544,0.455, 0.225,0.167,0.566,0.317,0.034,0.016,0.066]):

Enter values for the independent variables (time in hours and temperature in K) for all 15 data points in pairs (tj,tempj, j=1,15).

> x:=matrix(15,2,[0.1,100,0.2,100,0.3,100,0.4,100,0.5,100,0.05,200,0.1,200, 0.15,200,.2,200,0.25,200,0.02,300,0.04,300,0.06,300,0.08,300,0.1,300]):

The first row in the x matrix is:

> x [1,1],x[1,2];

$$0.1, 100$$

Enter the model equation (ymodel) for the fraction of reactant remaining as a function of time, temperature and parameter values theta1(Arrhenius constant, in hr-1) and theta2 (Activation energy divided by R, in K).

> ymodel:=exp(-theta1*t*exp(-theta2/T));

$$ymodel := e^{\left(-\theta 1\, t\, e^{\left(-\frac{\theta 2}{T}\right)}\right)}$$

Use the ymodel equation to predict the values of the dependent variable in ypred.

> ypred:=matrix(15,1,[seq(eval(ymodel,{t=x[j,1],T=x[j,2]}),j=1..15)]):

Note that ypred is a 15 x 1 matrix or vector, the first element of which is:

> ypred[1,1];

$$e^{-0.1\,\theta 1\, e^{-\frac{1}{100}\theta 2}}$$

Use Maple's Jacobian command in 'linalg' to find the elements of J (the sensitivity coefficients of the dependent variable to the parameters, theta1 and theta2). These are the derivates of the dependent variable y with respect to the parameters evaluated at the experimental conditions.

```
> J:=jacobian([seq(ypred[j,1],j=1..15)],[theta1,theta2]):
```

$$J := \begin{vmatrix} -0.1\,e^{-\frac{1}{100}\theta 2}\,e^{-0.1\,\theta 1\,e^{-\frac{1}{100}\theta 2}} & 0.001000000000\,\theta 1\,e^{-\frac{1}{100}\theta 2}\,e^{-0.1\,\theta 1\,e^{-\frac{1}{100}\theta 2}} \\ -0.2\,e^{-\frac{1}{100}\theta 2}\,e^{-0.2\,\theta 1\,e^{-\frac{1}{100}\theta 2}} & 0.002000000000\,\theta 1\,e^{-\frac{1}{100}\theta 2}\,e^{-0.2\,\theta 1\,e^{-\frac{1}{100}\theta 2}} \\ -0.3\,e^{-\frac{1}{100}\theta 2}\,e^{-0.3\,\theta 1\,e^{-\frac{1}{100}\theta 2}} & 0.003000000000\,\theta 1\,e^{-\frac{1}{100}\theta 2}\,e^{-0.3\,\theta 1\,e^{-\frac{1}{100}\theta 2}} \\ -0.4\,e^{-\frac{1}{100}\theta 2}\,e^{-0.4\,\theta 1\,e^{-\frac{1}{100}\theta 2}} & 0.004000000000\,\theta 1\,e^{-\frac{1}{100}\theta 2}\,e^{-0.4\,\theta 1\,e^{-\frac{1}{100}\theta 2}} \\ -0.5\,e^{-\frac{1}{100}\theta 2}\,e^{-0.5\,\theta 1\,e^{-\frac{1}{100}\theta 2}} & 0.005000000000\,\theta 1\,e^{-\frac{1}{100}\theta 2}\,e^{-0.5\,\theta 1\,e^{-\frac{1}{100}\theta 2}} \\ -0.05\,e^{-\frac{1}{200}\theta 2}\,e^{-0.05\,\theta 1\,e^{-\frac{1}{200}\theta 2}} & 0.0002500000000\,\theta 1\,e^{-\frac{1}{200}\theta 2}\,e^{-0.05\,\theta 1\,e^{-\frac{1}{200}\theta 2}} \\ -0.1\,e^{-\frac{1}{200}\theta 2}\,e^{-0.1\,\theta 1\,e^{-\frac{1}{200}\theta 2}} & 0.0005000000000\,\theta 1\,e^{-\frac{1}{200}\theta 2}\,e^{-0.1\,\theta 1\,e^{-\frac{1}{200}\theta 2}} \\ -0.15\,e^{-\frac{1}{200}\theta 2}\,e^{-0.15\,\theta 1\,e^{-\frac{1}{200}\theta 2}} & 0.0007500000000\,\theta 1\,e^{-\frac{1}{200}\theta 2}\,e^{-0.15\,\theta 1\,e^{-\frac{1}{200}\theta 2}} \\ -0.2\,e^{-\frac{1}{200}\theta 2}\,e^{-0.2\,\theta 1\,e^{-\frac{1}{200}\theta 2}} & 0.001000000000\,\theta 1\,e^{-\frac{1}{200}\theta 2}\,e^{-0.2\,\theta 1\,e^{-\frac{1}{200}\theta 2}} \\ -0.25\,e^{-\frac{1}{200}\theta 2}\,e^{-0.25\,\theta 1\,e^{-\frac{1}{200}\theta 2}} & 0.001250000000\,\theta 1\,e^{-\frac{1}{200}\theta 2}\,e^{-0.25\,\theta 1\,e^{-\frac{1}{200}\theta 2}} \\ -0.02\,e^{-\frac{1}{300}\theta 2}\,e^{-0.02\,\theta 1\,e^{-\frac{1}{300}\theta 2}} & 0.00006666666667\,\theta 1\,e^{-\frac{1}{300}\theta 2}\,e^{-0.02\,\theta 1\,e^{-\frac{1}{300}\theta 2}} \\ -0.04\,e^{-\frac{1}{300}\theta 2}\,e^{-0.04\,\theta 1\,e^{-\frac{1}{300}\theta 2}} & 0.0001333333333\,\theta 1\,e^{-\frac{1}{300}\theta 2}\,e^{-0.04\,\theta 1\,e^{-\frac{1}{300}\theta 2}} \\ -0.06\,e^{-\frac{1}{300}\theta 2}\,e^{-0.06\,\theta 1\,e^{-\frac{1}{300}\theta 2}} & 0.0002000000000\,\theta 1\,e^{-\frac{1}{300}\theta 2}\,e^{-0.06\,\theta 1\,e^{-\frac{1}{300}\theta 2}} \\ -0.08\,e^{-\frac{1}{300}\theta 2}\,e^{-0.08\,\theta 1\,e^{-\frac{1}{300}\theta 2}} & 0.0002666666667\,\theta 1\,e^{-\frac{1}{300}\theta 2}\,e^{-0.08\,\theta 1\,e^{-\frac{1}{300}\theta 2}} \\ -0.1\,e^{-\frac{1}{300}\theta 2}\,e^{-0.1\,\theta 1\,e^{-\frac{1}{300}\theta 2}} & 0.0003333333333\,\theta 1\,e^{-\frac{1}{300}\theta 2}\,e^{-0.1\,\theta 1\,e^{-\frac{1}{300}\theta 2}} \end{vmatrix}$$

Note that J is a 15 x 2 matrix. The rows of J consist of two elements. The first element in the first row is the derivative of the model equation (ymodel) with respect to theta1 evaluated at the experimental conditions associated with the first data point (x[1,1] and x[1,2]). We use ypred to obtain and evaluate this element. The second element in the first row of J is the derivate of ymodel with respect to theta2 evaluated at the first data point. The elements in the first row of J are shown next.

9.3 Nonlinear Least Squares

> J[15,1];J[15,2];

$$\left[-0.1\, e^{-\frac{1}{300}\theta 2} \quad e^{-0.1\,\theta 1\, e^{-\frac{1}{300}\theta 2}} \quad 0.00033333333333\, \theta 1\, e^{-\frac{1}{300}\theta 2}\, e^{-0.1\,\theta 1\, e^{-\frac{1}{300}\theta 2}} \right]$$

Next, determine the error between the measured value of y (yexp) and the predicted value of y (ypred). This is needed in the iteration process programmed below for determining the parameter values, theta1 and theta2, via the least squares equation.

> err:=evalm(yexp-ypred):

The variable named err is a 15 x 1 matrix the first element of which is:

> err[1,1];

$$0.980 - e^{-0.1\,\theta 1\, e^{-\frac{1}{100}\theta 2}}$$

Initialize an iteration counter, n_iters.

> n_iters:=0;

$$n\_iters := 0$$

Initialize the parameter vector theta. Initial guesses can often be determined by considering only one of the data points. Another method is to try different initial guesses until the method yields the same converged values for theta1 and theta2.

> theta:=matrix(2,1,[500,1e3]);

$$\theta := \begin{bmatrix} 500 \\ 1000. \end{bmatrix}$$

Initialize the delta theta vector.

> dtheta:=evalm(theta);

$$dtheta := \begin{bmatrix} 500 \\ 1000. \end{bmatrix}$$

Write a 'do' loop that solves for delta theta, updates the parameter values, and uses a while statement to check for convergence or exceeding 100 iterations.

> while max(abs(dtheta[1,1]/theta[1,1]),abs(dtheta[2,1]/theta[2,1]))> 1.0e-7 and
    n_iters <100 do
    Jeval:=eval(J,{theta1=theta[1,1],theta2=theta[2,1]}):

```
JevalT:=transpose(Jeval):
dthea:=evalm(inverse(JevalT&*Jeval)&*JevalT&*eval(err,{theta1=theta[
1,1],theta2=theta[2,1]})):
 theta:=evalm(theta+dthea);
 n_iters:=n_iters+1;
end do:
```

Enter the number of iterations needed to obtain convergence.

```
> n_iters;
```

$$9$$

Print the converged values for theta1 and theta2.

```
> evalm(theta);
```

$$\begin{bmatrix} 813.8721519 \\ 961.0025780 \end{bmatrix}$$

Use the converged values of theta1 and theta2 to determine predicted values of y for plotting, call this plotypred:

```
> plotypred:=eval(ymodel,{theta1=theta[1,1],theta2=theta[2,1]});
```

$$plotypred := e^{-813.8721519\, t\, e^{-\frac{961.0025780}{T}}}$$

Prepare the experimental data points for plotting.

```
> plotdata:=[seq([x[j,1],x[j,2],yexp[j,1]],j=1..15)];
```

$plotdata := [[0.1, 100, 0.980], [0.2, 100, 0.983], [0.3, 100, 0.955], [0.4, 100, 0.979], [0.5, 100, 0.993], [0.05, 200, 0.626], [0.1, 200, 0.544], [0.15, 200, 0.455], [0.2, 200, 0.225], [0.25, 200, 0.167], [0.02, 300, 0.566], [0.04, 300, 0.317], [0.06, 300, 0.034], [0.08, 300, 0.016], [0.1, 300, 0.066]]$

Specify the data (yexp) plot, P1, and the predicted values (ypred), P2.

```
> P1:=plot3d(plotdata,t=0..0.5,T=100..300,axes=boxed,style=point,
symbol=box,symbolize=14,color=black):
```

```
> P2:=plot3d(plotypred,t=0..0.5,T=100..300,axes=boxed,style=wireframe):
```

Display the plots for the data and the predicted values of y on the same plot as functions of time, t, and temperature, T.

```
> display({P1,P2},title="Example 9.1",labels=["t","T","y"],
labeldirections=[horizontal,horizontal,vertical]);
```

9.3 Nonlinear Least Squares

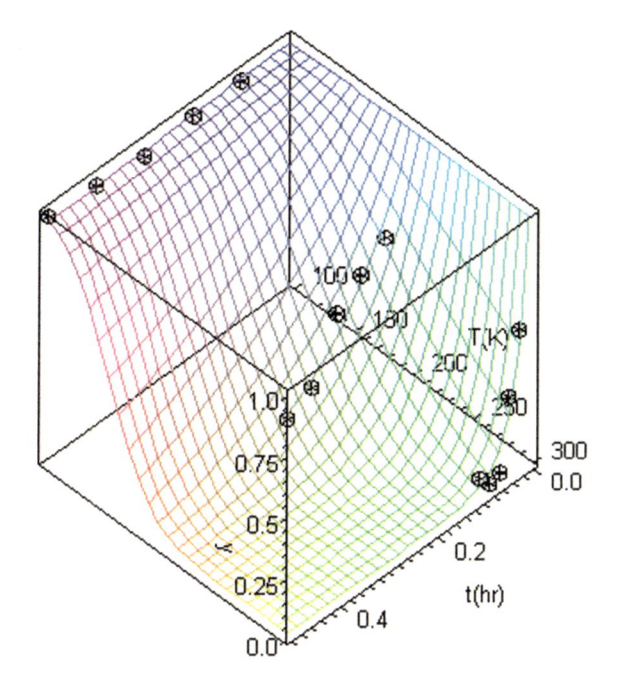

Example 9.1

Inspection of this figure reveals that more experimental data would be helpful particularly for shorter times at higher temperatures. The low temperature (100 K) data are not very useful since y changes so little over the 0.5 hour experiment.

Extension of Example 9.1

Example 9.1 above presents the methodology needed to obtain values for the parameters in an algebraic model that is nonlinear in the parameters. Usually, we will not be able to find an analytical solution to the differential equation model, and we will be forced to solve a differential equation model by numerical methods. This extension is intended to help you learn how to solve a model in differential equation form that is nonlinear in the parameters. That is, we will solve numerically the differential equation model (equation(9.2)) written in dimensionless form where k_1 is given by equation (9.133) to obtain the parameter values by using the data in Example 9.1 above. Again, solve numerically the following differential equation model (where we have used b_1 and b_2 instead of θ_1 and θ_2 as the parameters):

$$\frac{dy}{dt} = -b_1 \exp(-b_2/T) y \qquad (9.149)$$

and

$$y(0) = 1.0 \tag{9.150}$$

Determine the parameter values b_1 and b_2 by using the data given in Example 9.1 and the nonlinear least squares method. Recall that in Example 9.1 we needed the elements of the Jacobian matrix $\underline{\underline{J}}$ (see equation (9.142)). In this case, integrate simultaneously the time dependent sensitivity coefficients (i.e., the Jacobi matrix elements $\partial y/\partial b_1$ and $\partial y/\partial b_2$) and the differential equations. The needed three differential equations can be developed by taking the total derivative (as shown below) of the right hand side of equation (9.149) which we call h:

$$h = -b_1 \exp(-b_2/T) y \tag{9.151}$$

Inspection of equation (9.151) shows that h depends on b_1, b_2 and y at a fixed T. However, we know that the dependent variable y also depends on b_1 and b_2. Thus, sensitivity equations (Jacobian matrix elements $\partial y/\partial b_1$ and $\partial y/\partial b_2$) can be obtained by differentiating both sides of equation (9.151) with respect to $\underline{b}$ by using the chain rule:

$$\frac{\partial}{\partial \underline{b}}\left(\frac{dy}{dt}\right) = \frac{Dh}{D\underline{b}} = \frac{\partial h}{\partial \underline{b}} + \frac{\partial h}{\partial y}\frac{\partial y}{\partial \underline{b}} \tag{9.152}$$

Equation (9.152) becomes (after we exchange the order of differentiation on the left hand side of equation (9.152))

$$\frac{d}{dt}\left(\frac{\partial y}{\partial \underline{b}}\right) = \frac{\partial h}{\partial \underline{b}} + \frac{\partial h}{\partial y}\frac{\partial y}{\partial \underline{b}} \tag{9.153}$$

Next, write equation (9.153) for our case where $\underline{b} = [b_1 \ b_2]^T$ as follows:

$$\frac{d}{dt}\begin{pmatrix} \dfrac{\partial y}{\partial b_1} \\ \dfrac{\partial y}{\partial b_2} \end{pmatrix} = \begin{pmatrix} \dfrac{\partial h}{\partial b_1} \\ \dfrac{\partial h}{\partial b_2} \end{pmatrix} + \begin{pmatrix} \dfrac{\partial h}{\partial y}\dfrac{\partial y}{\partial b_1} \\ \dfrac{\partial h}{\partial y}\dfrac{\partial y}{\partial b_2} \end{pmatrix} \tag{9.154}$$

Next, extract from equation (9.154) the differential equations for $\partial y/\partial b_1$ and $\partial y/\partial b_2$:

$$\frac{d}{dt}\left(\frac{\partial y}{\partial b_1}\right) = \frac{\partial h}{\partial b_1} + \frac{\partial h}{\partial y}\frac{\partial y}{\partial b_1} \tag{9.155}$$

9.4 Hessian Matrix Approach

$$\frac{d}{dt}\left(\frac{\partial y}{\partial b_2}\right) = \frac{\partial h}{\partial b_2} + \frac{\partial h}{\partial y}\frac{\partial y}{\partial b_2} \quad (9.156)$$

By taking the derivatives, equations (9.155) and (9.156) yield

$$\frac{d}{dt}\left(\frac{\partial y}{\partial b_1}\right) = -\exp(-b_2/T)y + \left[-b_1\exp(-b_2/T)\right]\left(\frac{\partial y}{\partial b_1}\right) \quad (9.157)$$

and

$$\frac{d}{dt}\left(\frac{\partial y}{\partial b_2}\right) = (-b_1)\left(\frac{-1}{T}\right)\exp\left(\frac{-b_2}{T}\right)y + \left[-b_1\exp\left(\frac{-b_2}{T}\right)\right]\left(\frac{\partial y}{\partial b_2}\right) \quad (9.158)$$

Thus, we now have three dependent variables, $(y, \partial y/\partial b_1, \text{ and } \partial y/\partial b_2)$ and two independent variables $(t \text{ and } T)$. These three dependent variables have the following initial conditions:

$$y(0) = 1.0 \quad (9.159)$$

$$\left.\frac{\partial y}{\partial b_1}\right|_{t=0} = 0.0 \quad (9.160)$$

$$\left.\frac{\partial y}{\partial b_2}\right|_{t=0} = 0.0 \quad (9.161)$$

Fix T and integrate numerically in time the equations for the dependent variables $(y, \partial y/\partial b_1, \text{ and } \partial y/\partial b_2)$ to find their dependence on time. We need to do this for the three different temperatures. We need to use the values obtained numerically for the model predictions and the elements of the Jacobian. Also, note that you will need to use guessed values for b_1 and b_2 to obtain the needed values for the dependent variable y and the Jacobian elements $\partial y/\partial b_1$ and $\partial y/\partial b_2$. Compare the values of b_1 and b_2 obtained in this manner to the ones obtained in Example 9.1.

9.4 Hessian Matrix Approach

An equation which is essentially the same as equation (9.148) for $\Delta\underline{\theta}$ can be obtained by using a different approach from that used above where we expanded the dependent variable vector $\underline{Y}$ in a Taylor series about an initial vector of the parameter values $\underline{\theta}^{(k)}$ (see equation (9.139)). In this case, instead of expanding the model equations $\underline{Y}$ as shown in equation (9.139), expand the objective function $\Phi(\underline{\theta})$ in a Taylor series about the *k*th guess of the parameter vector of $\underline{\theta}^{(k)}$:

$$\Phi\left(\underline{\theta}^{(k+1)}\right) = \Phi\left(\underline{\theta}^{(k)}\right) + \left(\frac{\partial \Phi}{\partial \underline{\theta}}\bigg|_{\underline{\theta}=\underline{\theta}^{(k)}}\right)^T \Delta\underline{\theta} + \frac{1}{2}(\Delta\underline{\theta})^T \underline{\underline{H}}\bigg|_{\left(\underline{\theta}=\underline{\theta}^{(k)}\right)} \Delta\underline{\theta} \quad (9.162)$$

where the parameter vector $\underline{\theta}^{(k+1)}$ is unknown and

$$\Delta\underline{\theta} = \underline{\theta}^{(k+1)} - \underline{\theta}^{(k)} \quad (9.163)$$

as before. The Hessian matrix $\underline{\underline{H}}$ is defined as

$$\underline{\underline{H}} = \frac{\partial^2 \Phi}{\partial \underline{\theta}\, \partial \underline{\theta}^T} \quad (9.164)$$

We want to find the parameters $\underline{\theta}$ that make the derivative of $\Phi(\underline{\theta})$ with respect to $\underline{\theta}$ equal to the zero vector. Let's do this by taking the derivative of $\Phi\left(\underline{\theta}^{(k+1)}\right)$ given by equation (9.162) with respect to $\Delta\underline{\theta}$ by replacing $\underline{\theta}^{(k+1)}$ with $\Delta\underline{\theta}$ as the unknown vector and setting the resulting equation equal to the zero vector to obtain

$$\frac{\partial \Phi}{\partial \Delta\underline{\theta}} = \frac{\partial}{\partial \Delta\underline{\theta}} \Phi\left(\underline{\theta}^{(k)}\right) + \frac{\partial}{\partial \Delta\underline{\theta}}\left[\left(\frac{\partial \Phi}{\partial \underline{\theta}}\bigg|_{\underline{\theta}=\underline{\theta}^{(k)}}\right)^T \Delta\underline{\theta}\right]$$

$$+ \frac{\partial}{\partial \Delta\underline{\theta}}\left[\frac{1}{2}(\Delta\underline{\theta})^T \underline{\underline{H}}\bigg|_{\underline{\theta}=\underline{\theta}^{k}} \Delta\underline{\theta}\right] = \underline{0} \quad (9.165)$$

Since $\Phi\left(\underline{\theta}^{(k)}\right)$ is a constant and $\underline{\theta}^{(k)}$ is a vector of known constants, the first term in equation (9.165) is the zero vector and the second term in equation (9.165) becomes

9.4 Hessian Matrix Approach

$$\frac{\partial \Phi}{\partial \Delta \underline{\theta}} = \left[\left(\frac{\partial \Phi}{\partial \underline{\theta}} \bigg|_{\underline{\theta}=\underline{\theta}^{(k)}} \right)^T \Delta \underline{\theta} \right] = \frac{\partial \Phi}{\partial \underline{\theta}} \bigg|_{\underline{\theta}=\underline{\theta}^{(k)}} \tag{9.166}$$

where care has been taken to evaluate the derivative carefully (note that the resulting derivative is a column vector, as required). The third term in equation (9.165) becomes by using equation (9.44)

$$\frac{\partial \Phi}{\partial \Delta \underline{\theta}} \left(\frac{1}{2} \Delta \underline{\theta}^T \underline{\underline{H}} \bigg|_{\underline{\theta}=\underline{\theta}^k} \Delta \underline{\theta} \right) = \frac{1}{2} \left(\underline{\underline{H}} \bigg|_{\underline{\theta}=\underline{\theta}^k} + \underline{\underline{H}}^T \bigg|_{\underline{\theta}=\underline{\theta}^{(k)}} \right) \Delta \underline{\theta} \tag{9.167}$$

Equation (9.165) becomes by using equations (9.166) and (9.167)

$$\frac{\partial \Phi}{\partial \Delta \underline{\theta}} = \underline{0} + \frac{\partial \Phi}{\partial \underline{\theta}} \bigg|_{\underline{\theta}=\underline{\theta}^{(k)}} + \frac{1}{2} \left(\underline{\underline{H}} \bigg|_{\underline{\theta}=\underline{\theta}^k} + \underline{\underline{H}}^T \bigg|_{\underline{\theta}=\underline{\theta}^{(k)}} \right) \Delta \underline{\theta} = \underline{0} \tag{9.168}$$

Since $\underline{\underline{H}}$ is a symmetric matrix (see equation (9.164)) and therefore $\underline{\underline{H}} = \underline{\underline{H}}^T$, equation (9.168) becomes

$$\frac{\partial \Phi}{\partial \Delta \underline{\theta}} = \frac{\partial \Phi}{\partial \underline{\theta}} \bigg|_{\underline{\theta}=\underline{\theta}^{(k)}} + \frac{1}{2} \left(2\underline{\underline{H}} \bigg|_{\underline{\theta}=\underline{\theta}^k} \right) \Delta \underline{\theta} = \underline{0} \tag{9.169}$$

Equation (9.169) can be solved for $\Delta \underline{\theta}$:

$$\Delta \underline{\theta} = -\underline{\underline{H}}^{-1} \bigg|_{\underline{\theta}=\underline{\theta}^{(k)}} \frac{\partial \Phi}{\partial \underline{\theta}} \bigg|_{\underline{\theta}=\underline{\theta}^{(k)}} \tag{9.170}$$

or by using equation (9.163) we have

$$\underline{\theta}^{(k+1)} = \underline{\theta}^{(k)} - \underline{\underline{H}}^{-1} \bigg|_{\underline{\theta}=\underline{\theta}^{(k)}} \frac{\partial \Phi}{\partial \underline{\theta}} \bigg|_{\underline{\theta}=\underline{\theta}^{(k)}} \tag{9.171}$$

The next step is to determine expressions for $\frac{\partial \Phi}{\partial \underline{\theta}} \bigg|_{\underline{\theta}=\underline{\theta}^{(k)}}$ and $\underline{\underline{H}}^{-1} \bigg|_{\underline{\theta}=\underline{\theta}^{(k)}}$ in terms of the sensitivity coefficients and the experimental data that can be used in equation (9.171). First, expand equation (9.114) to

$$\Phi(\underline{\theta}) = \left(\underline{Y}^*\right)^T \underline{Y}^* - \left(\underline{Y}^*\right)^T \underline{Y} - \underline{Y}^T \underline{Y}^* + \underline{Y}^T \underline{Y} \qquad (9.172)$$

Next, recall that [2]

$$\left(\underline{Y}^*\right)^T \underline{Y} = \underline{Y}^T \underline{Y}^* \qquad (9.173)$$

so that equation (9.172) becomes

$$\Phi(\underline{\theta}) = \left(\underline{Y}^*\right)^T \underline{Y}^* - 2\left(\underline{Y}^*\right)^T \underline{Y} - \underline{Y}^T \underline{Y} \qquad (9.174)$$

Next, take the derivative of $\Phi(\underline{\theta})$ in equation (9.174) with respect to $\underline{\theta}$

$$\frac{\partial \Phi}{\partial \underline{\theta}} = 0 - 2 \frac{\partial}{\partial \underline{\theta}}\left(\left(\underline{Y}^*\right)^T \underline{Y}\right) + \frac{\partial}{\partial \underline{\theta}} \underline{Y}^T \underline{Y} \qquad (9.175)$$

Next, take the derivative of the second term in equation (9.175) as indicated to obtain

$$-2 \frac{\partial}{\partial \underline{\theta}}\left(\left(\underline{Y}^*\right)^T \underline{Y}\right) = -2 \left(\frac{\partial \underline{Y}}{\partial \underline{\theta}}\right)^T \underline{Y}^* \qquad (9.176)$$

The third term in equation (9.175) becomes

$$\frac{\partial}{\partial \underline{\theta}}\left(\underline{Y}^T \underline{Y}\right) = \left(\frac{\partial \underline{Y}}{\partial \underline{\theta}}\right)^T \underline{Y} + \underline{Y}^T \frac{\partial \underline{Y}}{\partial \underline{\theta}} \qquad (9.177)$$

Equation (9.177) can be simplified by recognizing that

$$\underline{Y}^T \frac{\partial \underline{Y}}{\partial \underline{\theta}} = \left(\frac{\partial \underline{Y}}{\partial \underline{\theta}}\right)^T \underline{Y} \qquad (9.178)$$

Thus, by using equation (9.178), equation (9.177) becomes

$$\frac{\partial}{\partial \underline{\theta}}\left(\underline{Y}^T \underline{Y}\right) = 2 \left(\frac{\partial \underline{Y}}{\partial \underline{\theta}}\right)^T \underline{Y} \qquad (9.179)$$

Substitution of equation (9.176) and (9.179) into equation (9.175) yields

$$\frac{\partial \Phi}{\partial \underline{\theta}} = -2 \left(\frac{\partial \underline{Y}}{\partial \underline{\theta}}\right)^T \underline{Y}^* + 2 \left(\frac{\partial \underline{Y}}{\partial \underline{\theta}}\right)^T \underline{Y} \qquad (9.180)$$

or

9.4 Hessian Matrix Approach

$$\frac{\partial \Phi}{\partial \underline{\theta}} = -2\left(\frac{\partial \underline{Y}}{\partial \underline{\theta}}\right)^T \left(\underline{Y}^* - \underline{Y}\right) \qquad (9.181)$$

Equation (9.181) can be written as

$$\frac{\partial \Phi}{\partial \underline{\theta}} = -2\left(\underline{\underline{J}}\right)^T \left(\underline{Y}^* - \underline{Y}\right) \qquad (9.182)$$

Substitution of equation (9.182) into equation (9.171) yields

$$\underline{\theta}^{(k+1)} = \underline{\theta}^{(k)} + 2\underline{\underline{H}}^{-1}\Big|_{\underline{\theta}=\underline{\theta}^{(k)}} \underline{\underline{J}}\Big|_{\underline{\theta}=\underline{\theta}^{(k)}} \left(\underline{Y}^* - \underline{Y}\Big|_{\underline{\theta}=\underline{\theta}^{(k)}}\right) \qquad (9.183)$$

To obtain $\underline{\underline{H}}$ first write the objective function in summation form:

$$\Phi(\underline{\theta}) = \sum_{i=1}^{n} \left(Y_i^* - Y_i\right)^2 \qquad (9.184)$$

where Y_i^* is element i of $\underline{Y}^*$ and Y_i is element i of $\underline{Y}$ and n is the number of data points. The derivative of the objective function Φ in equation (9.184) with respect to parameter $j(\theta_j)$ yields

$$\frac{\partial \Phi}{\partial \theta_j} = -2\sum_{i=1}^{n}\left(Y_i^* - Y_i\right)\frac{\partial Y_i}{\partial \theta_j} \qquad (9.185)$$

Equation (9.185) can be written for each parameter, so for two parameters $(j = 1, 2)$ equation (9.185) yields

$$\begin{pmatrix} \dfrac{\partial \Phi}{\partial \theta_1} \\ \dfrac{\partial \Phi}{\partial \theta_2} \end{pmatrix} = \begin{pmatrix} -2\sum_{i=1}^{n}\left(Y_i^* - Y_i\right)\dfrac{\partial Y_i}{\partial \theta_1} \\ -2\sum_{i=1}^{n}\left(Y_i^* - Y_i\right)\dfrac{\partial Y_i}{\partial \theta_2} \end{pmatrix} \qquad (9.186)$$

Equation (9.186) can be written in matrix notation as

$$\frac{\partial \Phi}{\partial \underline{\theta}} = -2\left(\underline{\underline{J}}\right)^T \left(\underline{Y}^* - \underline{Y}\right) \qquad (9.187)$$

where $\underline{\underline{J}}$ is given by equation (9.142) for two parameters. Note that equation (9.187) is the same as equation (9.182), as expected.

Next, the Hessian matrix $\underline{\underline{H}}$ (see equation (9.164)) can be written in terms of the sensitivity coefficients. First, in our case we have

$$\underline{\underline{H}} = \frac{\partial \Phi}{\partial \underline{\theta}\, \partial \underline{\theta}^T} \tag{9.188}$$

Equation (9.188) can be written for two parameters:

$$\underline{\underline{H}} = \begin{pmatrix} \dfrac{\partial^2 \Phi}{\partial \theta_1\, \partial \theta_1} & \dfrac{\partial^2 \Phi}{\partial \theta_1\, \partial \theta_2} \\ \dfrac{\partial^2 \Phi}{\partial \theta_2\, \partial \theta_1} & \dfrac{\partial^2 \Phi}{\partial \theta_2\, \partial \theta_2} \end{pmatrix} \tag{9.189}$$

Next, find the elements of $H_{1,1}$ in equation (9.189) by writing equation (9.185) for $j = 1$ then taking the derivative of the resulting equation with respect to θ_1:

$$H_{1,1} = \frac{\partial^2 \Phi}{\partial \theta_1\, \partial \theta_1} = 2 \sum_{i=1}^n \frac{\partial Y_i}{\partial \theta_1} \frac{\partial Y_i}{\partial \theta_1} - \left(Y_i^* - Y_i\right) \frac{\partial^2 Y_i}{\partial \theta_1\, \partial \theta_1} \tag{9.190}$$

Next, write equation (9.185) with $j = 1$ and then take the derivative with respect to θ_2 to obtain $H_{2,1}$:

$$H_{2,1} = \frac{\partial^2 \Phi}{\partial \theta_2\, \partial \theta_1} = 2 \sum_{i=1}^n \frac{\partial Y_i}{\partial \theta_2} \frac{\partial Y_i}{\partial \theta_1} - \left(Y_i^* - Y_i\right) \frac{\partial^2 Y_i}{\partial \theta_2\, \partial \theta_1} \tag{9.191}$$

For simplicity, workers in the past [7] [8] have neglected the second terms on the right hand sides of equations (9.190) and (9.191), and similar terms for the other elements of $\underline{\underline{H}}$. The resulting approximation to $\underline{\underline{H}}$ is often called $\underline{\underline{N}}$ after Newton. [7] Note that by making this assumption we obtain elements of $\underline{\underline{N}}$ that can be written as products of the sensitivity coefficients. That is, for two parameters and three data points

$$\underline{\underline{H}} \approx \underline{\underline{N}} = 2 \begin{pmatrix} \dfrac{\partial Y_1}{\partial \theta_1}\dfrac{\partial Y_1}{\partial \theta_1} + \dfrac{\partial Y_2}{\partial \theta_1}\dfrac{\partial Y_2}{\partial \theta_1} + \dfrac{\partial Y_3}{\partial \theta_1}\dfrac{\partial Y_3}{\partial \theta_1} & \dfrac{\partial Y_1}{\partial \theta_1}\dfrac{\partial Y_1}{\partial \theta_2} + \dfrac{\partial Y_2}{\partial \theta_1}\dfrac{\partial Y_2}{\partial \theta_2} + \dfrac{\partial Y_3}{\partial \theta_1}\dfrac{\partial Y_3}{\partial \theta_2} \\ \dfrac{\partial Y_1}{\partial \theta_2}\dfrac{\partial Y_1}{\partial \theta_1} + \dfrac{\partial Y_2}{\partial \theta_2}\dfrac{\partial Y_2}{\partial \theta_1} + \dfrac{\partial Y_3}{\partial \theta_2}\dfrac{\partial Y_3}{\partial \theta_1} & \dfrac{\partial Y_1}{\partial \theta_2}\dfrac{\partial Y_1}{\partial \theta_2} + \dfrac{\partial Y_2}{\partial \theta_2}\dfrac{\partial Y_2}{\partial \theta_2} + \dfrac{\partial Y_3}{\partial \theta_2}\dfrac{\partial Y_3}{\partial \theta_2} \end{pmatrix} \tag{9.192}$$

or in general

$$\underline{\underline{H}} \approx \underline{\underline{N}} = 2 \underline{\underline{J}}^T \underline{\underline{J}} \tag{9.193}$$

9.5 Confidence Intervals

Finally, by using equations (9.187) and (9.193) equation (9.171) becomes

$$\underline{\theta}^{(k+1)} = \underline{\theta}^{(k)} + 4\left[\left(\underline{\underline{J}}^T \underline{\underline{J}}\right)\Big|_{\underline{\theta}=\underline{\theta}^{(k)}}\right]^{-1} \underline{\underline{J}}^T\Big|_{\underline{\theta}=\underline{\theta}^{(k)}} \left(\underline{Y}^* - \underline{Y}\Big|_{\underline{\theta}=\underline{\theta}^{(k)}}\right) \quad (9.194)$$

Comparison of equation (9.194) to (9.148) shows that the approximate Hessian matrix approach (since we are using $\underline{\underline{N}}$ and not $\underline{\underline{H}}$) leads to the same equation except for the factor of 4. This factor of 4 accelerates the convergence when the parameter values are near their converged values, but can cause additional iterations to be needed for poor initial guesses. Another iterative method that could be used is the Marquardt method (see reference [9] for a discussion).

9.5 Confidence Intervals

The parameters that we have obtained for models that are linear in the parameters (see equation (9.56)) or by iteration for the models that are nonlinear in the parameters are known as point estimates. We would like to determine confidence intervals for these point estimates of the parameters so that we can use these confidence intervals to help us understand better the models we are developing for the systems we are studying. We want to use our parameter estimates to design systems and the more confidence we have in the models we use and the values of the parameters in those models, the more confidence we will have in the system that we design carrying out the objective of the system. For example, if we can develop a model with small confidence intervals for the point estimates of the parameters we will have less uncertainty and more confidence in the design. Essentially, we would like to continue our model development until we have small confidence intervals for the parameters. This may require changing the model or more often the kinetic reaction mechanism for the process of interest. These steps are illustrated in Example 9.5 of Rawlings and Ekerdt.[8]

The procedure that we use to develop confidence intervals is based on making some assumptions about the variables we measure (concentration of A as a function of time, $C_A(t)$, e.g.) and the parameters in the model equation. One assumption is that the measured independent variable (time, t, e.g.) is known exactly. Another assumption is the measured dependent variable values ($C_A(t)$, e.g.) are normally distributed random variables at each value of time. However, we typically have sufficient resources to carry out only two or three replicate experiments; consequently, we alter our assumption to be that the measured dependent variables are random variables whose values are distributed according to the student's t distribution (see Figure 7.7 of Constantinides and Mostoufi [9]). For a large number of replicate experiments the student's t distribution becomes the normal distribution (see Figure 9.4 of Rawlings and Ekerdt [8] and Figure 7.5(a) of Constantinides and Mostoufi [9]). The term used to determine whether or not the student's t distribution and the normal distribution are the same is "degrees of freedom." This term is often defined as the number of observations made in excess of the minimum number needed to estimate an unknown quantity. For example, the sample mean or arithmetic average of the concentration of C_A measured at time $t = 1$ min (e.g.) is

$$C_A(t=1 \text{ min})_{,avg} = \frac{\sum_{i=1}^{n} C_A(t=1 \text{ min})_{,i}}{n} \tag{9.195}$$

where n is the number of times we made the measurement. If we were only able to make one measurement, n would equal 1; and, consequently, we would have only one degree of freedom, and we would not be able to determine the sample variance (σ^2), which is defined as

$$\sigma^2 = \frac{\sum_{i=1}^{n} C_A(t=1 \text{ min})_{,i} - C_A(t=1 \text{ min})_{,avg}}{n-1} \tag{9.196}$$

because n would be equal 1 for only one measurement. The denominator of equation (9.196) is $(n-1)$ because one degree of freedom (i.e., one observation) was used to calculate the average in equation (9.195). The variance for one measurement, of course, makes no sense. Consequently, we would have to make at least two measurements $(n=2)$ to obtain a value for σ^2.

The measured dependent variable is equal to its true value plus measurement error and is also equal to an estimated value plus a residual error, which is often referred to simply as a residual. That is, the measured value of the dependent variable, Y_i^*, is equal to its true value, Y_i, which we will never know for sure, plus a measurement error, υ_i, which we also do not know:

$$\begin{array}{cccc} Y_i^* = & Y_i & + & \upsilon_i \\ \text{measured value} & \text{true value of} & & \text{measurement error} \\ \text{of dependent variable} & \text{dependent variable} & & \text{at } i\text{th condition} \\ \text{at } i\text{th condition} & \text{at } i\text{th condition} & & \end{array} \tag{9.197}$$

Also, the measured dependent variable Y_i^* is equal to its estimated value, $\hat{Y}_i$, plus a residual, ε_i, at condition, i:

$$Y_i^* = \hat{Y}_i + \varepsilon_i \tag{9.198}$$

9.6 Sensitivity Coefficient Equations

The measurement error, v_i, is often assumed to be a normalized, normally distributed random variable with a zero mean and a known variance. The normal probability density function for the measurement error at condition i is

$$p(v_i) = \frac{1}{\sigma\sqrt{2\pi}} \exp\left[-\frac{(v_i - \mu)^2}{2\sigma^2}\right] \tag{9.199}$$

where the mean (μ) of the n measurement errors at the ith experimental condition is

$$\mu = \sum_{j=1}^{n} \frac{(v_i)_j}{n} \tag{9.200}$$

and the variance is

$$\sigma^2 = \sum_{j=1}^{n} \frac{((v_i)_j - \mu)^2}{n-1} \tag{9.201}$$

We can write equation (9.199) in compact form as

$$v_i \sim N(\mu, \sigma^2) \tag{9.202}$$

If we set $\mu = 0$ and $\sigma^2 = 1.0$, for example, equation (9.202) becomes

$$v_i \sim N(0, 1.0) \tag{9.203}$$

Which is plotted in Figure 9.4 of Rawlings and Ekerdt [8] for v_i instead of x.

9.6 Sensitivity Coefficient Equations

Maple can be used to determine the sensitivity equations (see Example A.2 of Rawlings and Ekerdt) for a first order, isothermal constant volume reactor model where

$$\frac{dC_A}{dt} = -kC_A \tag{9.204}$$

and

$$C_A(t=0) = C_{A0} \tag{9.205}$$

We have two parameters C_{A0} and k ($\theta_1 = C_{A0}$ and $\theta_2 = k$) so we will have two sensitivity coefficients and two sensitivity coefficient equations. The first sensitivity coefficient is defined as the rate of change of $C_A(t)$ with respect to the parameter C_{A0}. That is, define

$$S_1(t) = \frac{\partial C_A(t)}{\partial C_{A0}} \tag{9.206}$$

or since $\theta_1 = C_{A0}$

$$S_1 = \frac{\partial C_A}{\partial \theta_1} \tag{9.207}$$

The second sensitivity coefficient is the rate of change of $C_A(t)$ with respect to the parameter k:

$$S_2(t) = \frac{\partial C_A(t)}{\partial k} \tag{9.208}$$

or since $\theta_2 = k$

$$S_2 = \frac{\partial C_A}{\partial \theta_2} \tag{9.209}$$

These sensitivity coefficients were determined by the Maple Jacobian command for this model and evaluated at the experimental times in the Maple worksheet for Example 9.1 (see the $\underline{\underline{J}}$ matrix). The sensitivity coefficient equations for $S_1(t)$ and $S_2(t)$ can be obtained by taking the derivative of the left hand side of equation (9.204) with respect to the parameters and the total derivative of the right hand side of equation (9.204). That is, let the right hand side of equation (9.204) be the function h

$$h = -kC_A \tag{9.210}$$

and take the derivative of equation (9.210) with respect to the parameters which requires that we take the total derivative of h:

$$\frac{\partial}{\partial \underline{\theta}}\left(\frac{dC_A}{dt}\right) = \frac{Dh}{D\underline{\theta}} = \frac{\partial h}{\partial \underline{\theta}} + \frac{\partial h}{\partial C_A}\frac{\partial C_A}{\partial \underline{\theta}} \tag{9.211}$$

9.6 Sensitivity Coefficient Equations

Equation (9.211) can be rewritten by exchanging the order of differentiation on the left hand side:

$$\frac{d}{dt}\left(\frac{\partial C_A}{\partial \underline{\theta}}\right) = \frac{\partial h}{\partial \underline{\theta}} + \frac{\partial h}{\partial C_A}\frac{\partial C_A}{\partial \underline{\theta}} \tag{9.212}$$

Equation (9.212) becomes

$$\frac{d}{dt}\begin{pmatrix}\dfrac{\partial C_A}{\partial \theta_1}\\[4pt]\dfrac{\partial C_A}{\partial \theta_2}\end{pmatrix} = \begin{pmatrix}\dfrac{\partial h}{\partial \theta_1}\\[4pt]\dfrac{\partial h}{\partial \theta_2}\end{pmatrix} + \begin{pmatrix}\dfrac{\partial h}{\partial C_A}\dfrac{\partial C_A}{\partial \theta_1}\\[4pt]\dfrac{\partial h}{\partial C_A}\dfrac{\partial C_A}{\partial \theta_2}\end{pmatrix} \tag{9.213}$$

We can write the elements of equation (9.213) as individual equations:

$$\frac{d}{dt}\left(\frac{\partial C_A}{\partial \theta_1}\right) = \frac{\partial h}{\partial \theta_1} + \frac{\partial h}{\partial C_A}\frac{\partial C_A}{\partial \theta_1} \tag{9.214}$$

and

$$\frac{d}{dt}\left(\frac{\partial C_A}{\partial \theta_2}\right) = \frac{\partial h}{\partial \theta_2} + \frac{\partial h}{\partial C_A}\frac{\partial C_A}{\partial \theta_2} \tag{9.215}$$

We can find $\partial h/\partial \theta_1$, etc., by inspection in this case (or by using Maple to get the total derivative). Thus, equation (9.214) becomes

$$\frac{d}{dt}\left(\frac{\partial C_A}{\partial \theta_1}\right) = 0 + (-k)\frac{\partial C_A}{\partial \theta_1} \tag{9.216}$$

or in terms of $S_1(t)$

$$\frac{d}{dt}S_1 = -kS_1 \tag{9.217}$$

Equation (9.215) becomes

$$\frac{d}{dt}\left(\frac{\partial C_A}{\partial \theta_2}\right) = -C_A + (-k)\frac{\partial C_A}{\partial \theta_2} \tag{9.218}$$

or in terms of $S_2(t)$

$$\frac{d}{dt}S_2 = -C_A - kS_2 \tag{9.219}$$

The initial conditions for the sensitivity coefficients are:

$$S_1(t=0) = \left.\frac{\partial C_A}{\partial \theta_1}\right|_{t=0} = \frac{\partial C_{A0}}{\partial C_{A0}} = 1.0 \qquad (9.220)$$

and

$$S_2(t=0) = \left.\frac{\partial C_A}{\partial \theta_2}\right|_{t=0} = \left.\frac{\partial C_{A0}}{\partial k}\right|_{t=0} = 0.0 \qquad (9.221)$$

We now have three dependent variables, $C_A(t)$, $S_1(t)$, and $S_2(t)$, which we can write in matrix form:

$$\begin{pmatrix} \dfrac{dC_A}{dt} \\ \dfrac{dS_1}{dt} \\ \dfrac{dS_2}{dt} \end{pmatrix} = \begin{pmatrix} -k & 0 & 0 \\ 0 & -k & 0 \\ -1 & 0 & -k \end{pmatrix} \begin{pmatrix} C_A \\ S_1 \\ S_2 \end{pmatrix} \qquad (9.222)$$

The initial conditions are:

$$\begin{pmatrix} C_A(0) \\ S_1(0) \\ S_2(0) \end{pmatrix} = \begin{pmatrix} C_{A0} \\ 1 \\ 0 \end{pmatrix} \qquad (9.223)$$

Maple and the matrix exponential can be used to find that

$$\begin{aligned} C_A(t) &= C_{A0}\exp(-kt) \\ S_1(t) &= \exp(-kt) \\ S_2(t) &= -t\, C_{A0}\exp(-kt) \end{aligned} \qquad (9.224)$$

(Note that we have numbered the parameters differently from Rawlings and Ekerdt.[8] They set $\theta_1 = k$ and $\theta_2 = C_{A0}$). The elements in the first row of the Jacobian matrix (see equation (9.142)) in this case are as follows:

$$J_{11} = \frac{\partial Y_1}{\partial \theta_1} = \left.\frac{\partial Y}{\partial \theta_1}\right|_{t=0} = \left.\frac{\partial C_A}{\partial \theta_1}\right|_{t=0} = S_1(t=0) = \exp(0) = 1.0 \qquad (9.225)$$

9.6 Sensitivity Coefficient Equations

$$J_{12} = \frac{\partial Y_1}{\partial \theta_2} = \frac{\partial Y}{\partial \theta_2}\bigg|_{t=0} = \frac{\partial C_A}{\partial \theta_2}\bigg|_{t=0} = S_2(t=0) = 0.0 \qquad (9.226)$$

The elements in the second row depend on the guessed values for k (e.g., $k = 0.52\,\text{min}^{-1}$) and C_{A0} (e.g., $8\,mol/m^3$) as well as the time for the first data point (e.g., $t = 1\,\text{min}$). In this case the elements in the second row become

$$J_{21} = \frac{\partial Y_2}{\partial \theta_1} = \frac{\partial Y}{\partial \theta_1}\bigg|_{\substack{t=0 \\ k=0.52\,\text{min}^{-1}}} = \frac{\partial C_A}{\partial \theta_1}\bigg|_{\substack{t=0 \\ k=0.52\,\text{min}^{-1}}} = S_1\big|_{\substack{t=0 \\ k=0.52\,\text{min}^{-1}}} = 0.5945 \qquad (9.227)$$

$$J_{22} = \frac{\partial Y_2}{\partial \theta_2} = \frac{\partial Y}{\partial \theta_2}\bigg|_{\substack{t=1\,\text{min} \\ k=0.52\,\text{min}^{-1} \\ C_{A0}=8mol/m^3}} = \frac{\partial C_A}{\partial \theta_2}\bigg|_{\substack{t=1\,\text{min} \\ k=0.52\,\text{min}^{-1} \\ C_{A0}=8mol/m^3}} = S_2\big|_{\substack{t=1\,\text{min} \\ k=0.52\,\text{min}^{-1} \\ C_{A0}=8mol/m^3}} = -4.7562\,\frac{\text{min}\,mol}{m^3} \qquad (9.228)$$

These equations show that the sensitivity coefficients are the elements of the Jacobian matrix. For models that are nonlinear in the dependent variables we may not have analytical expressions for the sensitivity coefficients. In this case, the Jacobian elements, J_{ij}, will be determined numerically.

Maple can be used to solve numerically the material balance equation and sensitivity equations for an nth order reaction case. The material balance equation (model equation) is

$$\frac{dC_A}{dt} = -kC_A^n \qquad (9.229)$$

Use Maple and nonlinear parameter estimation to determine the parameters C_{A0}, k, and n. In this case, the vector of parameters, $\underline{\theta}$ is:

$$\underline{\theta} = \begin{pmatrix} C_{A0} \\ k \\ n \end{pmatrix} \qquad (9.230)$$

Let $S_3(t) = \dfrac{\partial C_A}{\partial n}$ or $S_3(t) = \dfrac{\partial C_A}{\partial \theta_3}$. The system of equations is

$$\frac{d}{dt}\left(\frac{\partial C_A}{\partial \underline{\theta}}\right) = \frac{\partial h}{\partial \underline{\theta}} + \frac{\partial h}{\partial C_A}\frac{\partial C_A}{\partial \underline{\theta}} \qquad (9.231)$$

where

$$h = -kC_A^n \tag{9.232}$$

Equation (9.231) becomes

$$\frac{d}{dt}\begin{pmatrix} \frac{\partial C_A}{\partial \theta_1} \\ \frac{\partial C_A}{\partial \theta_2} \\ \frac{\partial C_A}{\partial \theta_3} \end{pmatrix} = \begin{pmatrix} \frac{\partial h}{\partial \theta_1} \\ \frac{\partial h}{\partial \theta_2} \\ \frac{\partial h}{\partial \theta_3} \end{pmatrix} + \begin{pmatrix} \frac{\partial h}{\partial C_A} & \frac{\partial C_A}{\partial \theta_1} \\ \frac{\partial h}{\partial C_A} & \frac{\partial C_A}{\partial \theta_2} \\ \frac{\partial h}{\partial C_A} & \frac{\partial C_A}{\partial \theta_3} \end{pmatrix} \tag{9.233}$$

or

$$\frac{d}{dt}\left(\frac{\partial C_A}{\partial \theta_1}\right) = \frac{\partial h}{\partial \theta_1} + \frac{\partial h}{\partial C_A}\frac{\partial C_A}{\partial \theta_1} \tag{9.234}$$

$$\frac{d}{dt}\left(\frac{\partial C_A}{\partial \theta_2}\right) = \frac{\partial h}{\partial \theta_2} + \frac{\partial h}{\partial C_A}\frac{\partial C_A}{\partial \theta_2} \tag{9.235}$$

$$\frac{d}{dt}\left(\frac{\partial C_A}{\partial \theta_3}\right) = \frac{\partial h}{\partial \theta_3} + \frac{\partial h}{\partial C_A}\frac{\partial C_A}{\partial \theta_3} \tag{9.236}$$

In this case,

$$h = -kC_A^n \tag{9.237}$$

so that

$$\frac{\partial h}{\partial C_A} = -nkC_A^{n-1} \tag{9.238}$$

and

$$\frac{\partial h}{\partial n} = -kC_A^n \ln(C_A) \tag{9.239}$$

Thus, equation (9.234) becomes by using equation (9.238)

$$\frac{d}{dt}\left(\frac{\partial C_A}{\partial \theta_1}\right) = 0 + \left(-nkC_A^{n-1}\right)\left(\frac{\partial C_A}{\partial \theta_1}\right) \tag{9.240}$$

9.6 Sensitivity Coefficient Equations

or

$$\frac{dS_1}{dt} = -nkC_A^{n-1}S_1 \qquad (9.241)$$

Equation (9.235) becomes by using equation (9.237) for $\partial h/\partial \theta_2$ and equation (9.238)

$$\frac{d}{dt}\left(\frac{\partial C_A}{\partial \theta_2}\right) = -C_A^n + \left(-nkC_A^{n-1}\right)\left(\frac{\partial C_A}{\partial \theta_2}\right) \qquad (9.242)$$

or

$$\frac{dS_2}{dt} = -C_A^n - nkC_A^{n-1}S_2 \qquad (9.243)$$

and equation (9.236) becomes by using equation (9.239)

$$\frac{dS_3}{dt} = -kC_A^n \ln(C_A) - nkC_A^{n-1}S_3 \qquad (9.244)$$

The initial conditions are as follows

$$\begin{pmatrix} C_A(0) \\ S_1(0) \\ S_2(0) \\ S_3(0) \end{pmatrix} = \begin{pmatrix} C_{A0} \\ 1 \\ 0 \\ 0 \end{pmatrix} \qquad (9.245)$$

Maple's 'dsolve' (numeric) can be used to solve the following four differential equations (equations (9.246), (9.247), (9.248) and (9.249))

$$\frac{dC_A}{dt} = -kC_A^n \qquad (9.246)$$

$$\frac{dS_1}{dt} = -nkC_A^{n-1}S_1 \qquad (9.247)$$

$$\frac{dS_2}{dt} = -C_A^n - nkC_A^{n-1}S_2 \qquad (9.248)$$

$$\frac{dS_3}{dt} = -kC_A^n \ln(C_A) - nkC_A^{n-1}S_3 \qquad (9.249)$$

subject to the initial condition given by equation (9.245).

The above process can be used to find $y_1(t)$, $y_2(t)$, b_1, b_2, b_3, and b_4 in Example 7.1 of Constantinides and Mostoufi[9] by using Maple and the Gauss-Newton Method with $w_1 = w_2 = 1.0$ in equations 7.177 and 7.178 in Constantinides and Mostoufi and the data from Run 2 only. The objective function Φ to be minimized in this case is:

$$\Phi = \left(\underline{Y}_1^* - \underline{Y}_1 - \underline{\underline{J}}_1 \underline{\Delta b}\right)^T \left(\underline{Y}_1^* - \underline{Y}_1 - \underline{\underline{J}}_1 \underline{\Delta b}\right) + \left(\underline{Y}_2^* - \underline{Y}_2 - \underline{\underline{J}}_2 \underline{\Delta b}\right)^T \left(\underline{Y}_2^* - \underline{Y}_2 - \underline{\underline{J}}_2 \underline{\Delta b}\right) \quad (9.250)$$

where

$$\underline{\underline{J}}_1 = \begin{pmatrix} \dfrac{\partial Y_{1,1}}{\partial b_1} & \dfrac{\partial Y_{1,1}}{\partial b_2} & \dfrac{\partial Y_{1,1}}{\partial b_3} & \dfrac{\partial Y_{1,1}}{\partial b_4} \\ \dfrac{\partial Y_{1,2}}{\partial b_1} & \dfrac{\partial Y_{1,2}}{\partial b_2} & \dfrac{\partial Y_{1,2}}{\partial b_3} & \dfrac{\partial Y_{1,2}}{\partial b_4} \\ \vdots & & & \\ \dfrac{\partial Y_{1,17}}{\partial b_1} & \dfrac{\partial Y_{1,17}}{\partial b_2} & \dfrac{\partial Y_{1,17}}{\partial b_3} & \dfrac{\partial Y_{1,17}}{\partial b_4} \end{pmatrix} \quad (9.251)$$

and $\dfrac{\partial Y_{1,1}}{\partial b_1}$, $\dfrac{\partial Y_{1,1}}{\partial b_2}$, $\dfrac{\partial Y_{1,1}}{\partial b_3}$, and $\dfrac{\partial Y_{1,1}}{\partial b_4}$ are the sensitivity coefficients for dependent variable number 1 (y_1) to the parameter values b_1, b_2, b_3, and b_4 evaluated at experimental data point number 1, etc. The matrix $\underline{\underline{J}}_2$ has elements that are the sensitivity coefficients for dependent variable number 2 (y_2). One can use the values given as results for b_1, b_2, b_3, and b_4 on page 522 of Constantinides and Mostoufi as the initial guesses for a Maple program. In this case, $\underline{\Delta b}$ is obtained from

$$\underline{\Delta b} = \left[\underline{\underline{J}}_1^T \underline{\underline{J}}_1 + \underline{\underline{J}}_2^T \underline{\underline{J}}_2\right]^{-1} \left[\underline{\underline{J}}_1^T \left(\underline{Y}_1^* - \underline{Y}_1\right) + \underline{\underline{J}}_2^T \left(\underline{Y}_2^* - \underline{Y}_2\right)\right] \quad (9.252)$$

In this case, the sensitivity equations can be written as

$$\dfrac{d}{dt}\left(\dfrac{\partial \underline{y}}{\partial \underline{b}}\right) = \dfrac{\partial \underline{h}}{\partial \underline{b}} + \dfrac{\partial \underline{h}}{\partial \underline{y}} \dfrac{\partial \underline{y}}{\partial \underline{b}} \quad (9.253)$$

where

9.6 Sensitivity Coefficient Equations

$$\underline{y} = [y_1, y_2]^T$$
$$\underline{h} = [h_1, h_2]^T \tag{9.254}$$

and

$$\underline{b} = [b_1 \ b_2 \ b_3 \ b_4]^T \tag{9.255}$$

Here

$$h_1 = b_1 \ y_1 \left(1 - \frac{y_1}{b_2}\right) \tag{9.256}$$

and

$$h_2 = b_3 \ y_1 - b_4 \ y_2 \tag{9.257}$$

Equation (9.253) can be written in expanded form:

$$\frac{d}{dt}\begin{pmatrix} \frac{\partial y_1}{\partial b_1} & \frac{\partial y_1}{\partial b_2} & \frac{\partial y_1}{\partial b_3} & \frac{\partial y_1}{\partial b_4} \\ \frac{\partial y_2}{\partial b_1} & \frac{\partial y_2}{\partial b_2} & \frac{\partial y_2}{\partial b_3} & \frac{\partial y_2}{\partial b_4} \end{pmatrix} = \begin{pmatrix} \frac{\partial h_1}{\partial b_1} & \frac{\partial h_1}{\partial b_2} & \frac{\partial h_1}{\partial b_3} & \frac{\partial h_1}{\partial b_4} \\ \frac{\partial h_2}{\partial b_1} & \frac{\partial h_2}{\partial b_2} & \frac{\partial h_2}{\partial b_3} & \frac{\partial h_2}{\partial b_4} \end{pmatrix} + \begin{pmatrix} \frac{\partial h_1}{\partial y_1} & \frac{\partial h_1}{\partial y_2} \\ \frac{\partial h_2}{\partial y_1} & \frac{\partial h_2}{\partial y_2} \end{pmatrix} \begin{pmatrix} \frac{\partial y_1}{\partial b_1} & \frac{\partial y_1}{\partial b_2} & \frac{\partial y_1}{\partial b_3} & \frac{\partial y_1}{\partial b_4} \\ \frac{\partial y_2}{\partial b_1} & \frac{\partial y_2}{\partial b_2} & \frac{\partial y_2}{\partial b_3} & \frac{\partial y_2}{\partial b_4} \end{pmatrix} \tag{9.258}$$

Each of the derivatives $(\partial y_1/\partial b_1, \partial y_2/\partial b_2, \text{etc.})$ is a sensitivity coefficient, each of which depends on time. Equation (9.258) can be used to write the following eight equations for the time dependence of the sensitivity coefficients:

$$\frac{d}{dt}\left(\frac{\partial y_1}{\partial b_1}\right) = \frac{\partial h_1}{\partial b_1} + \frac{\partial h_1}{\partial y_1}\frac{\partial y_1}{\partial b_1} + \frac{\partial h_1}{\partial y_2}\frac{\partial y_2}{\partial b_1} \tag{9.259}$$

$$\frac{d}{dt}\left(\frac{\partial y_1}{\partial b_2}\right) = \frac{\partial h_1}{\partial b_2} + \frac{\partial h_1}{\partial y_1}\frac{\partial y_1}{\partial b_2} + \frac{\partial h_1}{\partial y_2}\frac{\partial y_2}{\partial b_2} \tag{9.260}$$

$$\frac{d}{dt}\left(\frac{\partial y_1}{\partial b_3}\right) = \frac{\partial h_1}{\partial b_3} + \frac{\partial h_1}{\partial y_1}\frac{\partial y_1}{\partial b_3} + \frac{\partial h_1}{\partial y_2}\frac{\partial y_2}{\partial b_3} \tag{9.261}$$

$$\frac{d}{dt}\left(\frac{\partial y_1}{\partial b_4}\right) = \frac{\partial h_1}{\partial b_4} + \frac{\partial h_1}{\partial y_1}\frac{\partial y_1}{\partial b_4} + \frac{\partial h_1}{\partial y_2}\frac{\partial y_2}{\partial b_4} \tag{9.262}$$

$$\frac{d}{dt}\left(\frac{\partial y_2}{\partial b_1}\right) = \frac{\partial h_2}{\partial b_1} + \frac{\partial h_2}{\partial y_1}\frac{\partial y_1}{\partial b_1} + \frac{\partial h_2}{\partial y_2}\frac{\partial y_2}{\partial b_1} \tag{9.263}$$

$$\frac{d}{dt}\left(\frac{\partial y_2}{\partial b_2}\right) = \frac{\partial h_2}{\partial b_2} + \frac{\partial h_2}{\partial y_1}\frac{\partial y_1}{\partial b_2} + \frac{\partial h_2}{\partial y_2}\frac{\partial y_2}{\partial b_2} \tag{9.264}$$

$$\frac{d}{dt}\left(\frac{\partial y_2}{\partial b_3}\right) = \frac{\partial h_2}{\partial b_3} + \frac{\partial h_2}{\partial y_1}\frac{\partial y_1}{\partial b_3} + \frac{\partial h_2}{\partial y_2}\frac{\partial y_2}{\partial b_3} \tag{9.265}$$

and

$$\frac{d}{dt}\left(\frac{\partial y_2}{\partial b_4}\right) = \frac{\partial h_2}{\partial b_4} + \frac{\partial h_2}{\partial y_1}\frac{\partial y_1}{\partial b_4} + \frac{\partial h_2}{\partial y_2}\frac{\partial y_2}{\partial b_4} \tag{9.266}$$

To simplify the notation, let $\underline{s}$ be the vector of dependent variables. In this case the vector $\underline{s}$ will have 10 elements $(s_1 = y_1, s_2 = y_2, s_3 = \partial y_1/\partial b_1, s_4 = \partial y_1/\partial b_2, s_5 = \partial y_1/\partial b_3, s_6 = \partial y_1/\partial b_4,$ $s_7 = \partial y_2/\partial b_1, s_8 = \partial y_2/\partial b_2, s_9 = \partial y_2/\partial b_3,$ and $s_{10} = \partial y_2/\partial b_4)$. Also, in this case since y_2 does not appear in h_1, $\dfrac{\partial h_1}{\partial y_2}$ is equal to zero and the last term in each of the equations (9.259) through (9.262) drops out. Also, since h_2 does not depend explicitly on b_1 or b_2 the terms $\partial h_2/\partial b_1$ and $\partial h_2/\partial b_2$ in equations (9.263) and (9.264) are zero. The final system of equations is

$$\frac{ds_1}{dt} = b_1 s_1 \left(1 - \frac{s_1}{b_2}\right) \tag{9.267}$$

$$\frac{ds_2}{dt} = b_3 s_1 - b_4 s_2 \tag{9.268}$$

$$\frac{ds_3}{dt} = s_1\left(1 - \frac{s_1}{b_2}\right) + \left(b_1 - \frac{2b_1}{b_2}s_1\right)s_3 \tag{9.269}$$

$$\frac{ds_4}{dt} = \frac{b_1 s_1^2}{b_2^2} + \left(b_1 - \frac{2b_1 s_1}{b_2}\right)s_4 \tag{9.270}$$

$$\frac{ds_5}{dt} = \left(b_1 - \frac{2b_1 s_1}{b_2}\right)s_5 \tag{9.271}$$

9.7 One Parameter Model

$$\frac{ds_6}{dt} = \left(b_1 - \frac{2b_1 s_1}{b_2}\right) s_6 \quad (9.272)$$

$$\frac{ds_7}{dt} = b_3 s_3 - b_4 s_7 \quad (9.273)$$

$$\frac{ds_8}{dt} = b_3 s_4 - b_4 s_8 \quad (9.274)$$

$$\frac{ds_9}{dt} = s_1 + b_3 s_5 - b_4 s_9 \quad (9.275)$$

$$\frac{ds_{10}}{dt} = -s_2 + b_3 s_6 - b_4 s_{10} \quad (9.276)$$

The initial conditions for the dependent variables are as follows:

$$s_1 = 0.18,\ s_2 = 0,\ \text{and}\ s_3 = s_4 = s_5 = s_6 = s_7 = s_8 = s_9 = s_{10} = 0 \quad (9.277)$$

Note that the predicted values for y_1 at each time are in the vector $\underline{Y}_1$. That is,

$$\underline{Y}_1 = \left[y_1(t=0),\ y_1(t=10),\ y_1(t=20),\ \ldots,\ y_1(t=190) \right]^T \quad (9.278)$$

and $\underline{Y}_2$ stores the predicted values of y_2. Note that the elements of $\underline{\underline{J}}_1$ and $\underline{\underline{J}}_2$ are evaluated at each time also. In this case, the elements of $\underline{\underline{J}}_1$ are obtained from solutions of the set of equations $\left(\partial y_1/\partial b_1 = s_3,\ \partial y_1/\partial b_2 = s_4,\ \text{etc.}\right)$. In this case, $\underline{\underline{A}}$ is $\underline{\underline{A}} = \left(\underline{\underline{J}}_1^T \underline{\underline{J}}_1 + \underline{\underline{J}}_2^T \underline{\underline{J}}_2\right)^{-1}$ and the correlation coefficient matrix is $\underline{\underline{R}}$ where the elements of $\underline{\underline{R}}$ are obtained from the elements of $\underline{\underline{A}}$:

$$r_{ij} = \frac{a_{ij}}{\sqrt{a_{ii}\, a_{jj}}} \quad \text{for } i=1 \text{ to } 4 \text{ and } j=1 \text{ to } 4 \quad (9.279)$$

9.7 One Parameter Model

To illustrate the derivation of a confidence interval, let's first determine the parameter value for a one parameter model. That is, for an isothermal, constant volume batch reactor in which the reactant A disappears due to a chemical reaction we can write the reaction rate expression as

$$A \to products \tag{9.280}$$

and the material balance as

$$\frac{dC_A}{dt} = -kC_A \tag{9.281}$$

where we have assumed that the reaction is first order in species A. The solution to equation (9.281) is

$$C_A = C_{A0} \exp(-kt) \tag{9.282}$$

Assume we know C_{A0}. Define $f = \dfrac{C_A}{C_{A0}}$ and let

$$f = \exp(-kt) \tag{9.283}$$

Next, take the natural logarithm of each side of equation (9.283) to obtain

$$\ln f = -kt \tag{9.284}$$

Next, let

$$y = mx \tag{9.285}$$

so that

$$m = -k \tag{9.286}$$

or

$$\theta = -k \tag{9.287}$$

Next, use the following measured values of C_A to find f, then find k using least squares.

t(min)	$C_A(mol/m^3)$	f	$Y^* = \ln f$
0	8.47	1.0	0
1	5.0	0.5903	−0.527
2	2.95	0.3483	−1.055
3	1.82	0.2149	−1.538
4	1.05	0.1240	−2.088
5	0.71	0.08383	−2.479

9.7 One Parameter Model

For a one parameter model the independent variable matrix becomes a vector. That is, for the five data points we have

$$\underline{X} = \begin{pmatrix} x_1 \\ x_2 \\ x_3 \\ x_4 \\ x_5 \end{pmatrix} \tag{9.288}$$

and for this one parameter model $x_1 = t_1$, $x_2 = t_2$, etc. so equation (9.288) is the following vector

$$\underline{X} = \begin{pmatrix} 1 \\ 2 \\ 3 \\ 4 \\ 5 \end{pmatrix} \tag{9.289}$$

For the one parameter model, equation (9.56) simplifies to a scalar equation

$$\hat{\theta} = \left(\underline{X}^T \underline{X}\right)^{-1} \underline{X}^T \underline{Y}^* \tag{9.290}$$

where the carte on theta specifies that $\hat{\theta}$ is a point estimate of θ. In this case

$$\underline{X}^T \underline{X} = \begin{bmatrix} 1 & 2 & 3 & 4 & 5 \end{bmatrix} \begin{bmatrix} 1 \\ 2 \\ 3 \\ 4 \\ 5 \end{bmatrix} \tag{9.291}$$

or

$$\underline{X}^T \underline{X} = 1 + 4 + 9 + 16 + 25 = 55.0 \tag{9.292}$$

Thus, equation (9.290) becomes

$$\hat{\theta} = \left(\frac{1}{55}\right)(1 \quad 2 \quad 3 \quad 4 \quad 5)\begin{bmatrix} -0.527 \\ -1.055 \\ -1.538 \\ -2.088 \\ -2.479 \end{bmatrix} \quad (9.293)$$

Equation (9.293) yields

$$\hat{\theta} = \left(\frac{1}{55}\right)(-0.524 + (-2.110) + (-4.614) + (-8.352) + (-12.393)) \quad (9.294)$$

or

$$\hat{\theta} = -0.509 \quad (9.295)$$

Thus,

$$k = -(-0.509) \text{ min}^{-1} = 0.509 \text{ min}^{-1} \quad (9.296)$$

A confidence interval for θ can be found by using equation (9.126). In this case, equation (9.126) becomes for $\alpha = 0.05$ and $n = 5$

$$\hat{\theta} - t_{0.025,4}\sqrt{\frac{MS_E}{\sum_{i=1}^{n} x_i^2}} \leq \theta \leq \hat{\theta} + t_{0.025,4}\sqrt{\frac{MS_E}{\sum_{i=1}^{n} x_i^2}} \quad (9.297)$$

Equation (9.124) can be used to determine MS_E

$$MS_E = \frac{\sum_{i=1}^{n}\left(Y_i^* - \hat{y}_i\right)^2}{n-1} \quad (9.298)$$

where

$$\hat{y}_i = \hat{\theta} x_i \quad (9.299)$$

Thus,

9.7 One Parameter Model

$$MS_E = (-0.527-(-0.509)(1))^2$$
$$+(-1.055-(-0.509)(2))^2$$
$$+(-1.538-(-0.509)(3))^2 \qquad (9.300)$$
$$+(-2.088-(-0.509)(4))^2$$
$$\underline{+(-2.479-(-0.509)(5))^2}$$
$$5-1$$

$$MS_E = 0.002218 \qquad (9.301)$$

The student's t value for $\alpha/2 = 0.025$ and degrees of freedom $= 4$ can be obtained from a table in many texts (e.g., see Table A.3 of Navidi[10]) or from Maple:

$$t_{0.025,4} = 2.776 \qquad (9.302)$$

Thus, the confidence interval (ci) becomes

$$ci = t_{0.025,4}\sqrt{\frac{MS_E}{\sum_i x_i^2}} \qquad (9.303)$$

$$ci = 2.776\sqrt{\frac{0.002218}{55}} \qquad (9.304)$$

$$ci = 0.0176 \qquad (9.305)$$

Consequently, the parameter value is the point estimate $\hat{\theta} \pm$ the confidence interval ci:

$$\hat{\theta} \pm ci = -0.509 \pm 0.0176 \qquad (9.306)$$

or

$$-0.527 \le \theta \le -0.491 \qquad (9.307)$$

Equation (9.128) can be used to determine the prediction interval for a new observation at $x_0 = 1.5$ min, for example. In this case, we have

$$\hat{y}_0 = \hat{\theta} x_0 \tag{9.308}$$

$$\hat{y}_0 = (-0.509)(1.5) = -0.7636 \tag{9.309}$$

and equation (9.128) becomes

$$-0.7366 - 2.776\sqrt{(0.002218)\left(1 + \frac{(1.5)^2}{55}\right)} \leq y_0 \leq -0.7636 + 2.776\sqrt{(0.002218)\left(1 + \frac{(1.5)^2}{55}\right)} \tag{9.310}$$

$$-0.7636 - 0.1333 \leq y_0 \leq -0.7636 + 0.1333 \tag{9.311}$$

$$-0.897 \leq y_0 \leq -0.630 \tag{9.312}$$

For $x_0 = 4.5$

$$\hat{y}_0 = (-0.509)(4.5) = -2.29 \tag{9.313}$$

and equation (9.128) yields

$$-2.29 - 2.776\sqrt{(0.002218)\left(1 + \frac{(4.5)^2}{55}\right)} \leq y_0 \leq -2.29 + 2.776\sqrt{(0.002218)\left(1 + \frac{(4.5)^2}{55}\right)} \tag{9.314}$$

$$-2.29 - 0.153 \leq y_0 \leq \hat{y} - 2.29 + 0.153 \tag{9.315}$$

9.8 Two Parameter Model

In this case, take the natural logarithm of equation (9.282) to obtain

$$\ln C_A = \ln C_{A0} - kt \tag{9.316}$$

Now let's include C_{A0} (i.e., $\ln C_{A0}$) as an unknown parameter. Let

$$y = \ln C_A \tag{9.317}$$

9.8 Two Parameter Model

so that equation (9.316) becomes

$$y = mx + b \tag{9.318}$$

or

$$y = \theta_1 + \theta_2 x \tag{9.319}$$

where

$$m = -k = \theta_2 \tag{9.320}$$

and

$$x = t \tag{9.321}$$

Also, let

$$\theta_1 = b = \ln C_{A0} \tag{9.322}$$

The data become

$$\begin{array}{ccc} x = t\,(\min) & C_A\,(mol/m^3) & \ln C_A \\ 0 & 8.47 & 2.1365 \\ 1 & 5.0 & 1.6094 \\ 2 & 2.95 & 1.0818 \\ 3 & 1.82 & 0.5988 \\ 4 & 1.05 & 0.0488 \\ 5 & 0.71 & -0.3425 \end{array} \tag{9.323}$$

and the matrix $\underline{\underline{X}}$ becomes

$$\underline{\underline{X}} = \begin{bmatrix} 1 & 0 \\ 1 & 1 \\ 1 & 2 \\ 1 & 3 \\ 1 & 4 \\ 1 & 5 \end{bmatrix} \tag{9.324}$$

and its transpose is

$$\underline{\underline{X}}^T = \begin{bmatrix} 1 & 1 & 1 & 1 & 1 & 1 \\ 0 & 1 & 2 & 3 & 4 & 5 \end{bmatrix} \tag{9.325}$$

Let

$$\underline{\underline{C}} = \underline{\underline{X}}^T \underline{\underline{X}} = \begin{bmatrix} 6 & 15 \\ 15 & 55 \end{bmatrix} \quad (9.326)$$

which yields

$$\underline{\underline{C}}^{-1} = \begin{bmatrix} \dfrac{11}{21} & -\dfrac{1}{7} \\ -\dfrac{1}{7} & \dfrac{2}{35} \end{bmatrix} \quad (9.327)$$

The experimental data are as follows:

$$\underline{Y}^* = \begin{bmatrix} 2.1365 \\ 1.6094 \\ 1.0818 \\ 0.5988 \\ 0.0488 \\ -0.3425 \end{bmatrix} \quad (9.328)$$

Equation (9.56) becomes

$$\begin{bmatrix} \hat{\theta}_1 \\ \hat{\theta}_2 \end{bmatrix} = \underline{\underline{C}}^{-1} \underline{\underline{X}}^T \underline{Y}^* \quad (9.329)$$

or

$$\begin{bmatrix} \hat{\theta}_1 \\ \hat{\theta}_2 \end{bmatrix} = \begin{bmatrix} \dfrac{11}{21} & -\dfrac{1}{7} \\ -\dfrac{1}{7} & \dfrac{2}{35} \end{bmatrix} \begin{bmatrix} 1 & 1 & 1 & 1 & 1 & 1 \\ 0 & 1 & 2 & 3 & 4 & 5 \end{bmatrix} \begin{bmatrix} 2.1365 \\ 1.6094 \\ 1.0818 \\ 0.5988 \\ 0.0488 \\ -0.3425 \end{bmatrix} \quad (9.330)$$

Thus

$$\begin{bmatrix} \hat{\theta}_1 \\ \hat{\theta}_2 \end{bmatrix} = \begin{bmatrix} \dfrac{11}{21} & -\dfrac{1}{7} \\ -\dfrac{1}{7} & \dfrac{2}{35} \end{bmatrix} \begin{bmatrix} 5.1328 \\ 4.0521 \end{bmatrix} \quad (9.331)$$

and

9.8 Two Parameter Model

$$\begin{bmatrix} \hat{\theta}_1 \\ \hat{\theta}_2 \end{bmatrix} = \begin{bmatrix} 2.1097 \\ -0.5017 \end{bmatrix} \tag{9.332}$$

Thus

$$k = -\hat{\theta}_2 \text{ min}^{-1} = 0.5017 \text{ min}^{-1} \tag{9.333}$$

and

$$\ln C_{A0} = \hat{\theta}_1 = 2.1097 \tag{9.334}$$

so that the estimate for the initial concentration is

$$C_{A0} = 8.246 \, mol/m^3 \tag{9.335}$$

Note that the estimated value for C_{A0} is different from the measured value.

The confidence intervals for θ_1 and θ_2 can be determined by using equations (9.108) and (9.110), respectively. In this case the mean squared error (MS_E) is (see equation (9.88))

$$MS_E = s^2 = \frac{SS_E}{n-2} \tag{9.336}$$

which in our case becomes

$$MS_E = \frac{\sum\limits_{i=1}^{n=6}\left(Y_i^* - (\theta_1 + \theta_2 x_i)\right)^2}{6-2} \tag{9.337}$$

or

$$\begin{aligned} MS_E = \Big(& \left[2.1365 - \left(2.1097 - (0.5017)(0)\right)\right]^2 \\ & + \left[1.6094 - \left(2.1097 - 0.5017(1)\right)\right]^2 \\ & + \left[1.0818 - \left(2.1097 - 0.5017(2)\right)\right]^2 \\ & + \left[0.5988 - \left(2.1097 - 0.5017(3)\right)\right]^2 \\ & + \left[0.0488 - \left(2.1097 - 0.5017(4)\right)\right]^2 \\ & + \left[-0.3425 - \left(2.1097 - 0.5017(5)\right)\right]^2 \Big) \Big/ 4 \\ & = 0.00186265 \end{aligned} \tag{9.338}$$

In this case the student's t value for 95% confidence level requires that $\alpha = 0.05$ so that $\alpha/2 = 0.025$ and with 4 degrees of freedom is

$$t_{0.025, 4} = 2.776 \tag{9.339}$$

Recall that equation (9.108) is

$$\hat{\theta}_1 - t_{\alpha/2, n-2}\sqrt{MS_E\left(\frac{1}{n} + \frac{\bar{x}^2}{S_{xx}}\right)} \leq \theta_1 \leq \hat{\theta}_1 + t_{\alpha/2, n-2}\sqrt{MS_E\left(\frac{1}{n} + \frac{\bar{x}^2}{S_{xx}}\right)} \tag{9.340}$$

which in this case becomes

$$2.1097 - 2.776\sqrt{(0.00186265)\left[\frac{1}{6} + \frac{(2.5)^2}{17.5}\right]}$$

$$\leq \theta_1 \leq 2.1097 + 2.776\sqrt{(0.00186265)\left[\frac{1}{6} + \frac{(2.5)^2}{17.5}\right]} \tag{9.341}$$

or

$$2.023 \leq \theta_1 \leq 2.194 \tag{9.342}$$

or according to equation (9.322)

$$7.561 \leq C_{A0} \leq 8.993 \, mol/m^3 \tag{9.343}$$

The confidence interval for θ_2 is (see equation (9.110))

$$\hat{\theta}_2 - t_{\alpha/2, n-2}\sqrt{\frac{MS_E}{S_{xx}}} \leq \theta_2 \leq \hat{\theta}_2 + t_{\alpha/2, n-2}\sqrt{\frac{MS_E}{S_{xx}}} \tag{9.344}$$

or in this case

$$-0.5017 - 2.776\left(\frac{0.00186265}{17.5}\right)^2 \leq \theta_2 \leq -0.5017 + 2.776\left(\frac{0.00186265}{17.5}\right)^2 \tag{9.345}$$

or

$$-0.5303 \leq \theta_2 \leq -0.4731 \tag{9.346}$$

9.8 Two Parameter Model

or according to equation (9.320)

$$0.5303 \le k \le 0.4731 \tag{9.347}$$

Knowing the confidence intervals on θ_1 and θ_2 is useful because when they are small the point estimates of the parameters, $\hat{\theta}_1$ and $\hat{\theta}_2$, are reasonable and the two parameter model may be a good fit to the experimental data. However, we cannot arbitrarily pick values for θ_1 and θ_2 that are within their confidence intervals and be assured that the chosen pair have the probability of a $100(1-\alpha)$ confidence level because we applied a 95% confidence level for each parameter $(\theta_1$ and $\theta_2)$ one at a time when we found their individual confidence intervals. Consequently, we need to construct a joint confidence region [5] for the two parameters that will yield a 95% confidence level for both parameters simultaneously as described next.

The joint confidence region for θ_1 and θ_2 can be constructed (see page 389 of reference 5, page 484 of reference 9, and page 520 reference 8) by using the F factor:

$$\frac{\left(\underline{\theta}-\hat{\underline{\theta}}\right)^T \underline{\underline{X}}^T \underline{\underline{X}} \left(\underline{\theta}-\hat{\underline{\theta}}\right)}{pMS_E} \le F_{\alpha, p, n-p} \tag{9.348}$$

where in our case the number of parameters is p and $p=2$, $\alpha = 0.05$, and $n=6$. From Table A.4 of reference 5 we find that

$$F_{0.05, 2, 4} = 6.94 \tag{9.349}$$

Equation (9.348) becomes

$$\frac{\begin{bmatrix}\theta-2.1097_1 & \theta_2+0.5017\end{bmatrix}\begin{bmatrix}6 & 15 \\ 15 & 55\end{bmatrix}\begin{bmatrix}\theta_1-2.1097 \\ \theta_2+0.5017\end{bmatrix}}{(2)(0.00186265)} = 6.94 \tag{9.350}$$

or

$$6\theta_1^2 - 10.2654\theta_1 + 8.7956 + 30\theta_2\theta_1 - 8.104\theta_2 + 55\theta_2^2 = 0.025854 \tag{9.351}$$

Equation (9.351) is an equation for an ellipse which yields the joint confidence interval for θ_1 and θ_2. That is, the parameter values that exist within this ellipse have a joint probability of being accurate at a 95% confidence level. The Maple worksheet below presents a plot of equation (9.351).

> restart : with(linalg) : with(plots) :
> θ := matrix(2, 1);

$$\theta := \textit{array}(1..2, 1..1, [\,])$$

> th := matrix(2, 1, [2.1097,−0.5017]);

$$th := \begin{bmatrix} 2.1097 \\ -0.5017 \end{bmatrix}$$

> X := matrix(6, 2, [1 , 0 , 1 , 1 , 1, 2 , 1, 3 , 1 , 4 , 1 , 5]);

$$X := \begin{bmatrix} 1 & 0 \\ 1 & 1 \\ 1 & 2 \\ 1 & 3 \\ 1 & 4 \\ 1 & 5 \end{bmatrix}$$

> C := evalm(transpose(X)&*X);

$$C := \begin{bmatrix} 6 & 15 \\ 15 & 55 \end{bmatrix}$$

> x := evalm(θ − th);

$$x := \begin{bmatrix} \theta_{1,1} - 2.1097 \\ \theta_{2,1} + 0.5017 \end{bmatrix}$$

> b := 0.025854;

$$b := 0.025854$$

> joint := evalm(((transpose(x)&*C)&*evalm(x)) − b);

$$\textit{joint} := \big[\big(6\,\theta_{1,1} - 5.1327 + 15\,\theta_{2,1}\big)\big(\theta_{1,1} - 2.1097\big) + \big(15\,\theta_{1,1} - 4.0520 + 55\,\theta_{2,1}\big)\big(\theta_{2,1} + 0.5017\big) - 0.025854\big]$$

> implicitplot(joint[1, 1], θ[2, 1] =−0.55..−0.45, θ[1, 1] = 1.9..2.3);

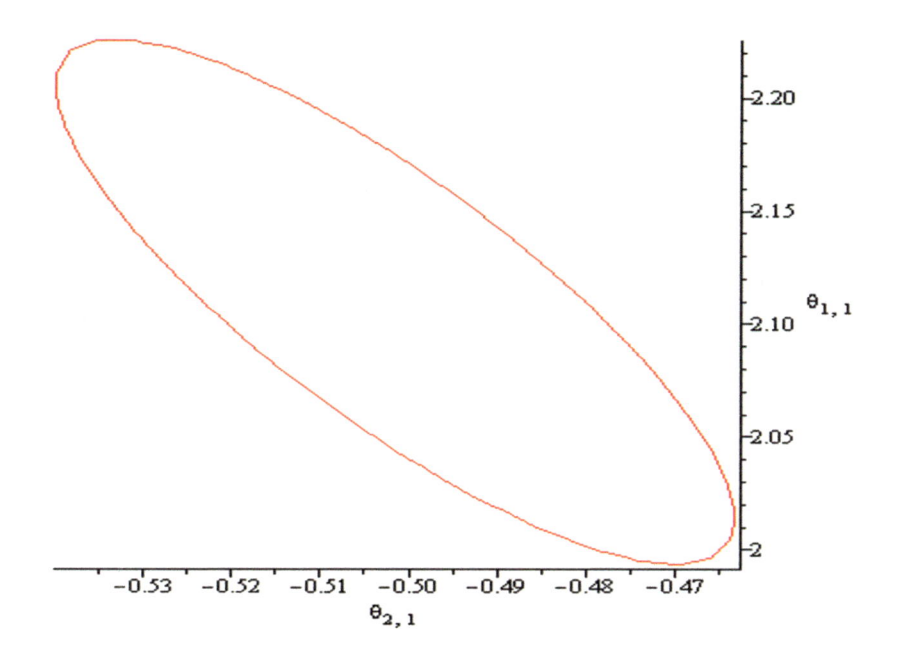

The values inside the ellipse shown in this figure represent the acceptable parameter values for this case.

9.9 Exercise Problems

1. Use the two parameter, three data points case (see equations (9.136), (9.137), and (9.138)) to confirm that equation (9.44) is correct.
2. Confirm by hand that the elements of the Jacobian matrix given in Example 9.1 are correct.
3. Use the matrix exponential method to solve equations (9.149), (9.153), and (9.158) with equations (9.159), (9.160), and (9.161) as the initial conditions. Compare your results to the elements of the Jacobian matrix in Example 9.1.

References

1. Bequette, B.W.: Process Dynamics: Modeling, Analysis, and Simulation. Prentice-Hall PTR, Englewood Cliffs (1998)
2. Crassidis, J.L., Junkins, J.L.: Optimal Estimation of Dynamic Systems. Chapman & Hall/CRC Press, Boca Raton (2004)
3. Lopez, R.J.: Advanced Engineering Mathematics, p. 1158. Addison Wesley, Reading (2001)
4. Ogunnaike, B.A., Harmon Ray, W.: Process Dynamics, Modeling and Control. Oxford University Press, Oxford (1994)

5. Montgomery, D.C., Peck, E.A.: Introduction to Regression Analysis. John Wiley & Sons, Inc., New York (1982)
6. Rafter, J.A., Abell, M.L., Braselton, J.P.: Statistics wtih Maple. Elsevier Science, USA (2003)
7. Bard, Y.: Nonlinear Parameter Estimation. Academic Press, Inc., London (1974)
8. Rawlings, J.B., Ekerdt, J.G.: Chemical Reactor Analysis and Design Fundamentals. Nob Hill Publishing, LLC, Madison (2002)
9. Constantinides, A., Mostoufi, N.: Numerical Methods for Chemical Engineers with MATLAB Applications. Prentice-Hall PTR, Englewood Cliffs (1999)
10. Navidi, W.C.: Statistics for Engineers and Scientists. McGraw-Hill, New York (2006)

Chapter 10
Miscellaneous Topics

10.1 Miscellaneous Topics on Numerical Methods

10.1.1 Introduction

In the previous chapters analytical, symbolic, and semianalytical methods for solving problems in chemical engineering were programmed in Maple. In addition, Maple can be used just as any other programming language like FORTRAN. In this chapter, numerical schemes for solving problems in chemical engineering are programmed in Maple.

10.1.2 Iterative Finite Difference Solution for Boundary Value Problems

Linear Boundary Value Problems were converted to finite difference form and the resulting system of coupled linear algebraic equations was solved symbolically in section 3.1.5. Nonlinear Boundary Value Problems were converted to finite difference form and the resulting system of coupled nonlinear algebraic equations was solved numerically in section 3.2.3 using Maple's fsolve command. Alternatively a recurrence relationship can be obtained for the resulting system of coupled nonlinear algebraic equations. This recurrence relationship can then be used to iterate for finding the solution at the node points. This is best illustrated with the following examples.

Example 10.1. Diffusion with a Second Order Reaction

Examples 3.2.1 and 3.2.3, diffusion with a second order reaction is solved here again. Consider diffusion with a second order reaction in a rectangular pellet (Rice and Do, 1995). The dimensionless concentration is governed by:

$$\frac{d^2u}{dx^2} = \Phi^2 u^2$$

$$\frac{du}{dx}(0) = 0 \qquad (10.1)$$

$$u(1) = 1$$

where Φ is the Thiele modulus. Equation 10.1 can be converted to finite difference form as:

$$\frac{u_{i-1} - 2u_i + u_{i+1}}{h^2} = \Phi^2 u_i, \, i = 1..N$$

$$\frac{-3u_0 + 4u_1 - u_2}{2h} = 0 \qquad (10.2)$$

$$u_{N+1} = 1$$

Equation 10.2 can be used to obtain the recurrence relation as:

$$unew_0 = \frac{4uold_1 - uold_2}{3}$$

$$unew_i = \frac{unew_{i-1} + uold_{i+1} - \Phi^2 h^2 uold_i^2}{2}, \, i = 1..N \qquad (10.3)$$

$$unew_{N+1} = 1$$

Note that in equation 10.3, updated solution at node point i − 1 is used for node point i. An initial guess of $u_i = 0.5$, i = 0..N is taken. u_{N+1} is taken as 1. Equation 10.3 is programmed for $\Phi = 1$ in Maple below:

> restart:

> with(plots):

Warning, the name changecoords has been redefined

The number of node points and length of the domain are entered here:

> N:=10;

$$N := 10$$

> L:=1;

$$L := 1$$

The governing equation is entered here. The nonlinear term in the governing equation is entered as f for convenience:

> ge:=diff(u(x),x$2)-Phi^2*f;

$$ge := \left(\frac{d^2}{dx^2} u(x)\right) - \Phi^2 f$$

10.1 Miscellaneous Topics on Numerical Methods

The boundary conditions are entered here:

> bc1:=diff(u(x),x);

$$bc1 := \frac{d}{dx} u(x)$$

> bc2:=u(x)-1;

$$bc2 := u(x) - 1$$

The governing equation and the boundary conditions are converted to finite difference form here. Note that for finite difference expressions updated values of u, unew are used whenever possible.

> d2ydx2:=(u[m+1]-2*u[m]+unew[m-1])/h^2;

$$d2ydx2 := \frac{u_{m+1} - 2 u_m + unew_{m-1}}{h^2}$$

> dydx:=(u[m+1]-unew[m-1])/2/h;

$$dydx := \frac{1}{2} \frac{u_{m+1} - unew_{m-1}}{h}$$

> dydxf:=(-u[2]+4*u[1]-3*u[0])/(2*h);

$$dydxf := \frac{1}{2} \frac{-u_2 + 4 u_1 - 3 u_0}{h}$$

> dydxb:=(unew[N-1]-4*unew[N]+3*u[N+1])/(2*h);

$$dydxb := \frac{1}{2} \frac{unew_9 - 4 unew_{10} + 3 u_{11}}{h}$$

> eq[0]:=subs(diff(u(x),x)=dydxf,u(x)=u[0],bc1):

> eq[N+1]:=subs(diff(u(x),x)=dydxb,u(x)=u[N+1],bc2):

> for i from 1 to N do
eq[i]:=subs(diff(u(x),x$2)=d2ydx2,diff(u(x),x)=dydx,u(x)=u[i],x=i*h,m=i,ge);od:

Node spacing and the nonlinear function are defined here:

> h:=evalf(L/(N+1));

$$h := 0.09090909091$$

> F(x):=u(x)^2;

$$F(x) := u(x)^2$$

The parameters are entered here.

> pars:={Phi=1};

$$pars := \{ \Phi = 1 \}$$

The recurrence relations are derived here:

> for i from 0 to N+1 do
Unew[i]:=subs(pars,u=uold,f=subs(u(x)=uold[i],pars,F(x))),solve(eq[i],u[i]));od;

$$Unew_0 := -0.3333333333 \; uold_2 + 1.333333333 \; uold_1$$

$$Unew_1 := 0.5000000000 \; uold_2 + 0.5000000000 \; unew_0 - 0.004132231405 \; uold_1^2$$

$$Unew_2 := 0.5000000000 \; uold_3 + 0.5000000000 \; unew_1 - 0.004132231405 \; uold_2^2$$

$$Unew_3 := 0.5000000000 \; uold_4 + 0.5000000000 \; unew_2 - 0.004132231405 \; uold_3^2$$

$$Unew_4 := 0.5000000000 \; uold_5 + 0.5000000000 \; unew_3 - 0.004132231405 \; uold_4^2$$

$$Unew_5 := 0.5000000000 \; uold_6 + 0.5000000000 \; unew_4 - 0.004132231405 \; uold_5^2$$

$$Unew_6 := 0.5000000000 \; uold_7 + 0.5000000000 \; unew_5 - 0.004132231405 \; uold_6^2$$

$$Unew_7 := 0.5000000000 \; uold_8 + 0.5000000000 \; unew_6 - 0.004132231405 \; uold_7^2$$

$$Unew_8 := 0.5000000000 \; uold_9 + 0.5000000000 \; unew_7 - 0.004132231405 \; uold_8^2$$

$$Unew_9 := 0.5000000000 \; uold_{10} + 0.5000000000 \; unew_8 - 0.004132231405 \; uold_9^2$$

$$Unew_{10} := 0.5000000000 \; uold_{11} + 0.5000000000 \; unew_9 - 0.004132231405 \; uold_{10}^2$$

$$Unew_{11} := 1$$

An initial guess of 0.7 is used. Error is initialized to 1.

> for i from 0 to N+1 do uold[i]:=0.7;od:

> iter:=0;err:=1;

$$iter := 0$$

$$err := 1$$

A for loop can be written for the iteration as:

> while err>1e-6 do

> for i from 0 to N+1 do unew[i]:=eval(Unew[i]);od:

> kk:='kk':err:=sqrt(sum((unew[kk]-uold[kk])^2,kk=0..N+1)/(N+2));

> iter:=iter+1:

> for i from 0 to N+1 do uold[i]:=unew[i];od:end:

10.1 Miscellaneous Topics on Numerical Methods

A total of 272 iterations were required and the calculated error is:

> iter;err;

$$272$$

$$0.9953728481 \ 10^{-6}$$

The following plot is obtained:

>plot([seq([i*h,unew[i]],i=0..N+1)],thickness=3,axes=boxed,labels=[x,u]);

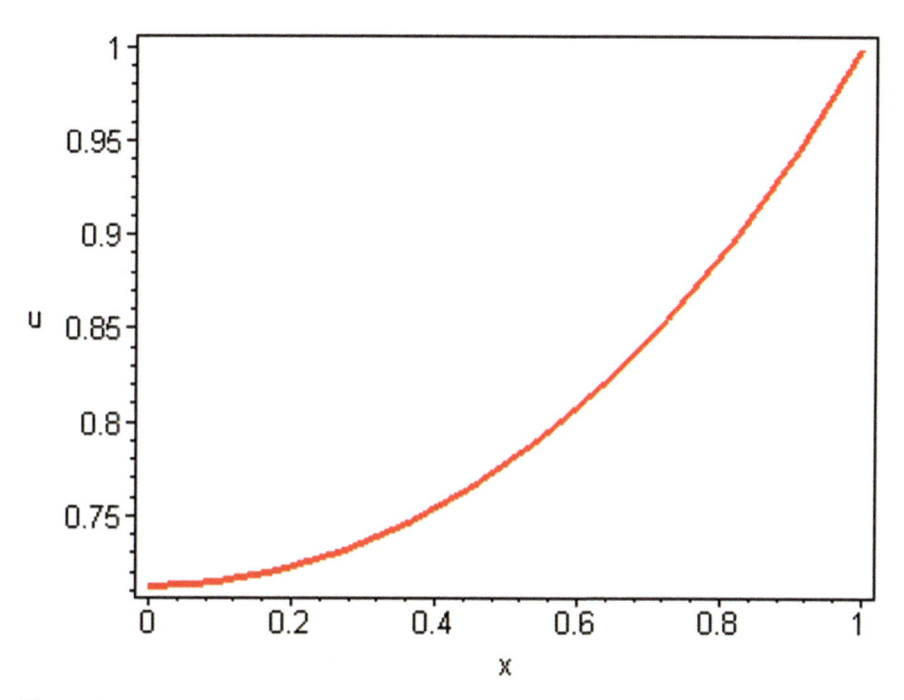

Fig. 10.1

Example 10.2. Nonisothermal Reaction in a Catalyst Pellet – Multiple Steady States

The dimensionless concentration in a non-isothermal catalyst pellet (Villedsen and Michelsen, 1978, example 3.2.2) is governed by:

$$\frac{d^2u}{dx^2} = \Phi^2 u \exp\left(\frac{\gamma\beta(1-u)}{1+\beta(1-u)}\right)$$

$$\frac{du}{dx}(0) = 0 \qquad (10.4)$$

$$u(1) = 1$$

This boundary value problem has multiple solutions for $\Phi = 0.2$, $\beta = 0.8$ and $\gamma = 20$. Equation 10.4 is solved in Maple by modifying the program given for example 10.1 as given below. For this problem, 20 node points are chosen to improve the accuracy.

> N:=20;

$$N := 20$$

> F(x):=u(x)*exp(gamma*beta*(1-u(x))/(1+beta*(1-u(x))));

$$F(x) := u(x) \, e^{\left(\frac{\gamma \beta (1 - u(x))}{1 + \beta (1 - u(x))}\right)}$$

> pars:={Phi=0.2,gamma=20,beta=0.8};

$$pars := \{ \Phi = 0.2, \gamma = 20, \beta = 0.8 \}$$

When an initial guess of 0.95 is used the following plot is obtained:

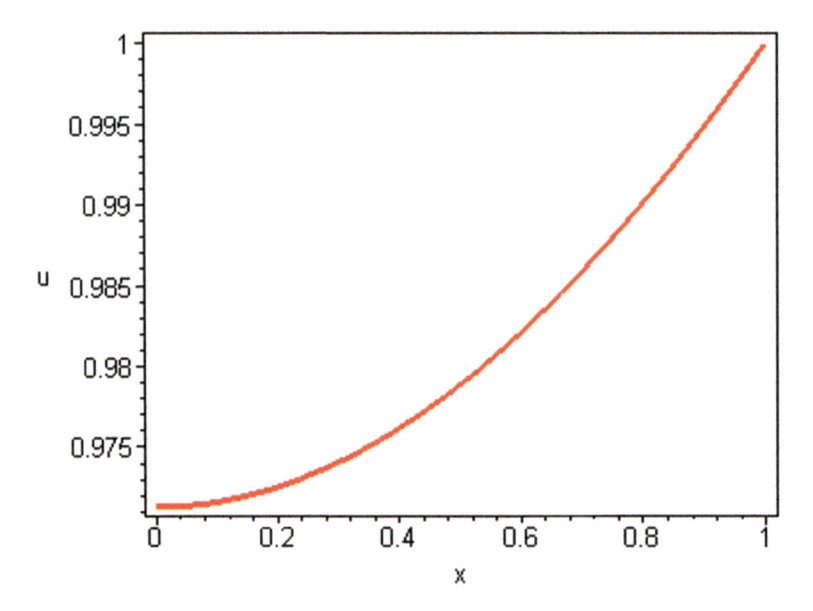

Fig. 10.2

10.1 Miscellaneous Topics on Numerical Methods

When an initial guess of 0.01 is used, the lower steady state is predicted:

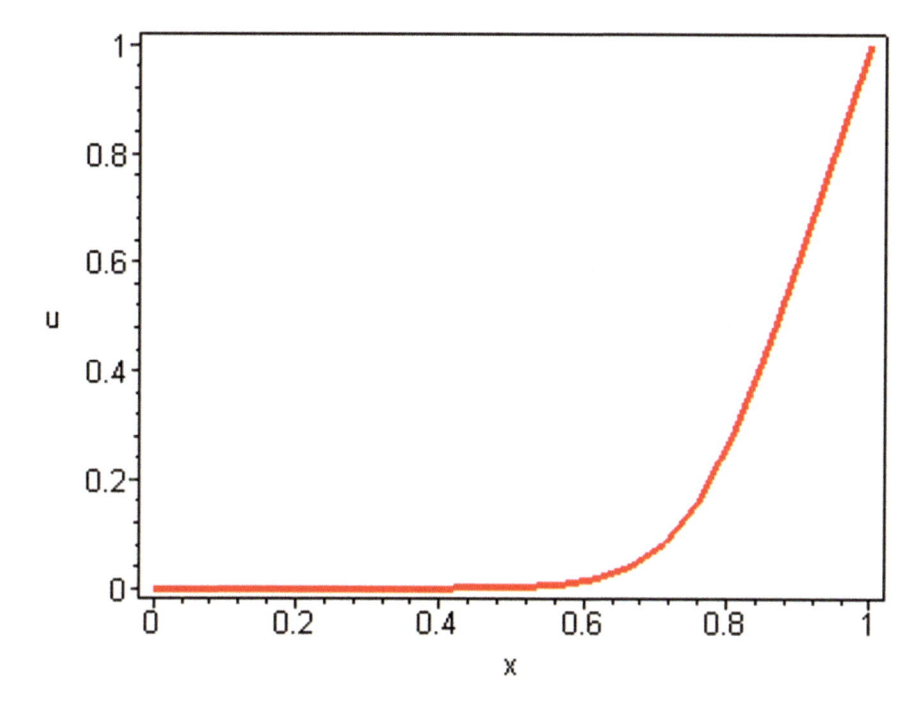

Fig. 10.3

Since this problem is stiff, an error tolerance of 10^{-6} was used. It has to be noted that iterative finite difference method does not predict the middle steady state predicted in example 3.2.2.

10.1.3 Finite Difference Solution for Elliptic PDEs

Steady state linear elliptic PDEs in finite domains are solved by applying finite difference technique in both x and y coordinates in this section. When finite differences are applied, a linear elliptic PDE is converted to a system of linear algebraic equations. This resulting system of linear equations can be directly solved using Maple's solve or fsolve command. This is best illustrated with the following examples.

Example 10.3. Heat Transfer in a Rectangle

Consider the steady state heat transfer problem (Carslaw and Jaeger, 1973) solved in example 7.11. The governing equation for temperature is:

$$\frac{\partial^2 u}{\partial x^2} + \frac{\partial^2 u}{\partial y^2} = 0$$

$$u(0,y) = 0 \text{ and } u(L,y) = 0 \tag{10.5}$$

$$u(x,0) = 0 \text{ and } u(x,H) = 1$$

where L is the length and H is the height of the rectangle. If N interior node points in the x-axis and M interior node points are used in the y-axis equation 10.5 can be converted to finite difference form as:

$$\frac{u_{i-1,j} - 2u_{i,j} + u_{i+1,j}}{h^2} + \frac{u_{i,j-1} - 2u_{i,j} + u_{i,j+1}}{k^2} = 0, \; i = 1..N, \; j = 1..M$$

$$u_{0,j} = 0, \; j = 1..M$$

$$u_{N+1,j} = 0, \; j = 1..M$$

$$u_{i,0} = 0, \; i = 0..N+1$$

$$u_{i,M+1} = 1, \; i = 0..N+1$$

$$\tag{10.6}$$

Equation 10.6 is a system of (N+2)x(M+2) linear equations and solved using Maple's fsolve command below.

> restart;with(plots):

Warning, the name changecoords has been redefined

The governing equation is entered here:

> ge:=diff(u(x,y),x$2)+diff(u(x,y),y$2);

$$ge := \left(\frac{\partial^2}{\partial x^2} u(x,y)\right) + \left(\frac{\partial^2}{\partial y^2} u(x,y)\right)$$

Length and height of the rectangle are entered here:

> L:=1;H:=1;

$$L := 1$$
$$H := 1$$

10.1 Miscellaneous Topics on Numerical Methods

Number of node points in the x-axis (N) and y-axis (M) are entered now:

> N:=10;M:=10;

$$N := 10$$
$$M := 10$$

The boundary conditions at x = 0 and x = L are entered as bc1 and bc2 respectively:

> bc1:=u(x,y)-0;

$$bc1 := u(x, y)$$

> bc2:=u(x,y)-0;

$$bc2 := u(x, y)$$

The boundary conditions at y = 0 and y = H are entered as bc3 and bc4 respectively:

> bc3:=u(x,y)-0;

$$bc3 := u(x, y)$$

> bc4:=u(x,y)-1;

$$bc4 := u(x, y) - 1$$

Next, finite difference expressions are entered for the spatial derivatives in x and y coordiantes.

> dudxf:=1/2*(-u[2,m]-3*u[0,m]+4*u[1,m])/h;

$$dudxf := \frac{1}{2} \frac{-u_{2,m} - 3 u_{0,m} + 4 u_{1,m}}{h}$$

> dudxb:=1/2*(u[N-1,m]+3*u[N+1,m]-4*u[N,m])/h;

$$dudxb := \frac{1}{2} \frac{u_{9,m} + 3 u_{11,m} - 4 u_{10,m}}{h}$$

> dudx:=1/2/h*(u[n+1,m]-u[n-1,m]);

$$dudx := \frac{1}{2} \frac{u_{n+1,m} - u_{n-1,m}}{h}$$

> d2udx2:=1/h^2*(u[n-1,m]-2*u[n,m]+u[n+1,m]);

$$d2udx2 := \frac{u_{n-1,m} - 2 u_{n,m} + u_{n+1,m}}{h^2}$$

> dudyf:=1/2*(-u[n,2]-3*u[n,0]+4*u[n,1])/k;

$$dudyf := \frac{1}{2}\frac{-u_{n,2} - 3u_{n,0} + 4u_{n,1}}{k}$$

> dudyb:=1/2*(u[n,M-1]+3*u[n,M+1]-4*u[n,M])/k;

$$dudyb := \frac{1}{2}\frac{u_{n,9} + 3u_{n,11} - 4u_{n,10}}{k}$$

> dudy:=1/2/k*(u[n,m+1]-u[n,m-1]);

$$dudy := \frac{1}{2}\frac{u_{n,m+1} - u_{n,m-1}}{k}$$

> d2udy2:=1/k^2*(u[n,m-1]-2*u[n,m]+u[n,m+1]);

$$d2udy2 := \frac{u_{n,m-1} - 2u_{n,m} + u_{n,m+1}}{k^2}$$

Boundary conditions are converted to finite difference form below:

> bc1:=subs(diff(u(x,y),x)=dudxf,u(x,y)=u[0,m],x=0,bc1);

$$bc1 := u_{0,m}$$

> bc2:=subs(diff(u(x,y),x)=dudxb,u(x,y)=u[N+1,m],x=L,bc2);

$$bc2 := u_{11,m}$$

> bc3:=subs(diff(u(x,y),y)=dudyf,u(x,y)=u[n,0],y=0,bc3);

$$bc3 := u_{n,0}$$

> bc4:=subs(diff(u(x,y),y)=dudyb,u(x,y)=u[n,M+1],y=H,bc4);

$$bc4 := u_{n,11} - 1$$

Boundary conditions are stored as equations below. A total of N+M+2 equations arise from the boundary conditions.

> for i from 1 to M do eq[0,i]:=subs(m=i,y=i*k,bc1);od:

> for i from 1 to M do eq[N+1,i]:=subs(m=i,y=i*k,bc2);od:

> for i from 0 to N+1 do eq[i,0]:=subs(n=i,x=i*h,bc3);od:

> for i from 0 to N+1 do eq[i,M+1]:=subs(n=i,x=i*h,bc4);od:

Next, the governing equation is converted to finite difference form and stored as algebraic equations. Even though the example chosen does not have first

10.1 Miscellaneous Topics on Numerical Methods

derivatives in x and y, this program is written to accommodate first derivatives in both the governing equation and the boundary conditions.

```
> for i from 1 to N do for j from 1 to M do eq[i,j]:=subs(diff(u(x,y),x$2) = subs(n=i,m=j,d2udx2),diff(u(x,y),y$2) = subs(n=i,m=j,d2udy2), diff(u(x,y),x) = subs(n=i,m=j,dudx),diff(u(x,y),y) = subs(n=i,m=j,dudy),u(x,y)=u[i,j],x=i*h,y=j*k,ge);od;od;
```

```
> h:=evalf(L/(N+1));k:=evalf(H/(M+1));
```

$$h := 0.09090909091$$

$$k := 0.09090909091$$

```
> eqs:=seq(seq(eq[i,j],i=0..N+1),j=0..M+1):
```

An initial guess of 0.5 is used.

```
> vars:=seq(seq(u[i,j]=0.5,i=0..N+1),j=0..M+1):
```

The system of algebraic equations are solved using Maple's fsolve command and the solution obtained is plotted below:

```
> soln:=fsolve({eqs},{vars}):
> assign(soln):
> plotdata := [seq([ seq([i*h,j*k,u[i,j]], i=0..N+1)], j=0..M+1)]:
> surfdata(plotdata,axes=boxed, labels=[x,y,u],orientation=[-120,60] );
```

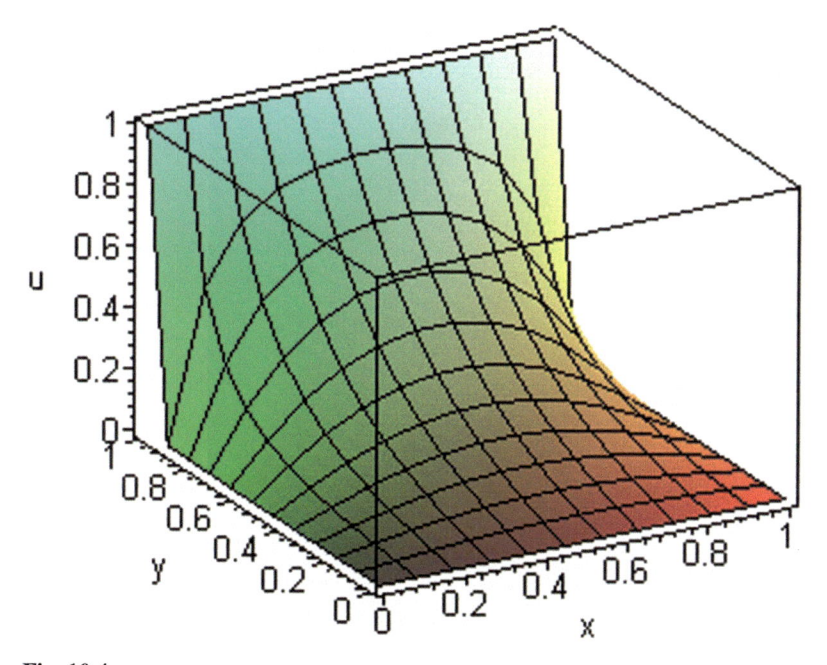

Fig. 10.4

Example 10.4. Heat Transfer in a Cylinder

Consider the steady state heat transfer problem in a cylinder (Carslaw and Jaeger, 1973). The governing equation for temperature is:

$$\frac{\partial^2 u}{\partial x^2} + \frac{1}{x}\frac{\partial u}{\partial x} + \frac{\partial^2 u}{\partial y^2} = 0$$

$$\frac{\partial u}{\partial x}(0,y) = 0 \text{ and } u(L,y) = 0 \tag{10.7}$$

$$u(x,0) = 0 \text{ and } \frac{\partial u}{\partial y}(x,H) = 1$$

Equation 10.7 is solved using the Maple program developed for example 10.3 for L = 1 and H = 1. The results obtained are given below:

> ge:=diff(u(x,y),x$2)+1/x*diff(u(x,y),x)+diff(u(x,y),y$2);

$$ge := \left(\frac{\partial^2}{\partial x^2} u(x, y)\right) + \frac{\frac{\partial}{\partial x} u(x, y)}{x} + \left(\frac{\partial^2}{\partial y^2} u(x, y)\right)$$

> L:=1;H:=1;

$$L := 1$$
$$H := 1$$

> N:=10;M:=10;

$$N := 10$$
$$M := 10$$

> bc1:=diff(u(x,y),x);

$$bc1 := \frac{\partial}{\partial x} u(x, y)$$

> bc2:=u(x,y)-1;

$$bc2 := u(x, y) - 1$$

> bc3:=u(x,y);

$$bc3 := u(x, y)$$

10.1 Miscellaneous Topics on Numerical Methods

> bc4:=diff(u(x,y),y);

$$bc4 := \frac{\partial}{\partial y} \mathrm{u}(x, y)$$

> surfdata(plotdata,axes=boxed, labels=[x,y,u],orientation=[-120,60]);

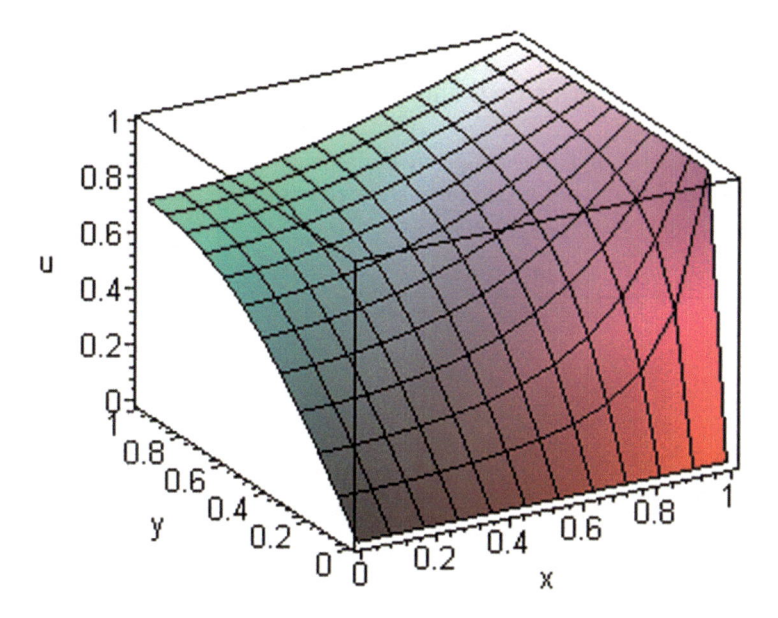

Fig. 10.5

10.1.4 Iterative Finite Difference Solution for Elliptic PDEs

In section 10.1.4 linear elliptic PDEs were solved by solving finite difference equations numerically using Maple's fsolve command. For nonlinear elliptic PDEs, the resulting finite difference expressions are nonlinear. For solving a large number of nonlinear equations fsolve command may not be ideal because fsolve requires a good initial guess and might require a long time. Oftentimes, fsolve command may not yield a result. For handling nonlinear problems, Maple can be used as any other programming language like FORTRAN etc. A recursion relation can be derived for the elliptic PDE using Maple. The resulting recursion can be programmed in Maple to achieve a desired accuracy. In this section Gauss-Jordan iteration (Stress et al) is used for nonlinear elliptic PDEs. This is best illustrated with the following examples.

Example 10.5. Heat Transfer in a Rectangle – Nonlinear Elliptic PDE

Consider a nonlinear steady state heat transfer problem governed by the following elliptic PDE:

$$\frac{\partial^2 u}{\partial x^2} + \frac{\partial^2 u}{\partial y^2} + e^u = 0$$

$$u(0,y) = 0 \text{ and } u(L,y) = 0 \tag{10.8}$$

$$u(x,0) = 0 \text{ and } u(x,H) = 1$$

where L is the length and H is the height of the rectangle. If N interior node points in the x-axis and M interior node points are used in the y-axis equation 10.5 can be converted to finite difference form as:

$$\frac{u_{i-1,j} - 2u_{i,j} + u_{i+1,j}}{h^2} + \frac{u_{i,j-1} - 2u_{i,j} + u_{i,j+1}}{k^2} + \exp(u_{i,j}) = 0, \, i = 1..N, \, j = 1..M$$

$u_{0,j} = 0, \, j = 1..M$

$u_{N+1,j} = 0, \, j = 1..M$

$u_{i,0} = 0, \, i = 0..N+1$

$u_{i,M+1} = 1, \, i = 0..N+1$

$$\tag{10.9}$$

Equation 10.9 is a system of (N+2)x(M+2) nonlinear equations. Equation 10.9 can be used to obtain the recurrence relation:

$$\text{unew}_{i,j} = \frac{\frac{\text{unew}_{i-1,j} + \text{uold}_{i+1,j}}{h^2} + \frac{\text{unew}_{i,j-1} + \text{uold}_{i,j+1}}{k^2} + \exp(\text{uold}_{i,j})}{2\left(\frac{1}{h^2} + \frac{1}{k^2}\right)}, \, i = 1..N, \, j = 1..M$$

$\text{unew}_{0,j} = 0, \, j = 1..M$

$\text{unew}_{N+1,j} = 0, \, j = 1..M$

$\text{unew}_{i,0} = 0, \, i = 0..N+1$

$\text{unew}_{i,M+1} = 1, \, i = 0..N+1$

$$\tag{10.10}$$

Equation 10.10 is programmed using the Maple program given below. N = 10 and M = 10 node points are used for this program. Error is calculated based on the difference of the dependent variables between two successive iterations. Mean of the squares of errors at all the node points is found and a set tolerance of 10^{-12} is used for verifying the convergence.

10.1 Miscellaneous Topics on Numerical Methods

> restart;with(plots):

Warning, the name changecoords has been redefined

The governing equation is entered below. The nonlinear function is stored as f.

> ge:=diff(u(x,y),x$2)+diff(u(x,y),y$2)+f;

$$ge := \left(\frac{\partial^2}{\partial x^2} u(x, y)\right) + \left(\frac{\partial^2}{\partial y^2} u(x, y)\right) + f$$

Length and height of the rectangle are entered here:

> L:=1;H:=1;

$$L := 1$$
$$H := 1$$

Number of node points used in x and y coordinates are entered.

> N:=10;M:=10;

$$N := 10$$
$$M := 10$$

Next, the boundary conditions are entered:

> bc1:=u(x,y);

$$bc1 := u(x, y)$$

> bc2:=u(x,y);

$$bc2 := u(x, y)$$

> bc3:=u(x,y);

$$bc3 := u(x, y)$$

> bc4:=u(x,y);

$$bc4 := u(x, y)$$

Next, finite difference expressions for the governing equation and boundary conditions are entered:

> dudxf:=1/2*(-u[2,m]-3*u[0,m]+4*u[1,m])/h;

$$dudxf := \frac{1}{2} \frac{-u_{2,m} - 3\,u_{0,m} + 4\,u_{1,m}}{h}$$

> dudxb:=1/2*(unew[N-1,m]+3*u[N+1,m]-4*unew[N,m])/h;

$$dudxb := \frac{1}{2} \frac{unew_{9,m} + 3\,u_{11,m} - 4\,unew_{10,m}}{h}$$

> dudx:=1/2/h*(u[n+1,m]-unew[n-1,m]);

$$dudx := \frac{1}{2} \frac{u_{n+1,m} - unew_{n-1,m}}{h}$$

> d2udx2:=1/h^2*(unew[n-1,m]-2*u[n,m]+u[n+1,m]);

$$d2udx2 := \frac{unew_{n-1,m} - 2\,u_{n,m} + u_{n+1,m}}{h^2}$$

> dudyf:=1/2*(-u[n,2]-3*u[n,0]+4*u[n,1])/k;

$$dudyf := \frac{1}{2} \frac{-u_{n,2} - 3\,u_{n,0} + 4\,u_{n,1}}{k}$$

> dudyb:=1/2*(unew[n,M-1]+3*u[n,M+1]-4*unew[n,M])/k;

$$dudyb := \frac{1}{2} \frac{unew_{n,9} + 3\,u_{n,11} - 4\,unew_{n,10}}{k}$$

> dudy:=1/2/k*(u[n,m+1]-unew[n,m-1]);

$$dudy := \frac{1}{2} \frac{u_{n,m+1} - unew_{n,m-1}}{k}$$

> d2udy2:=1/k^2*(unew[n,m-1]-2*u[n,m]+u[n,m+1]);

$$d2udy2 := \frac{unew_{n,m-1} - 2\,u_{n,m} + u_{n,m+1}}{k^2}$$

Next, boundary conditions and governing equation are converted to finite difference form:

> bc1:=subs(diff(u(x,y),x)=dudxf,u(x,y)=u[0,m],x=0,bc1);

$$bc1 := u_{0,m}$$

> bc2:=subs(diff(u(x,y),x)=dudxb,u(x,y)=u[N+1,m],x=L,bc2);

$$bc2 := u_{11,m}$$

> bc3:=subs(diff(u(x,y),y)=dudyf,u(x,y)=u[n,0],y=0,bc3);

$$bc3 := u_{n,0}$$

10.1 Miscellaneous Topics on Numerical Methods

> bc4:=subs(diff(u(x,y),y)=dudyb,u(x,y)=u[n,M+1],y=H,bc4);

$$bc4 := u_{n,\,11}$$

> for i from 1 to M do eq[0,i]:=subs(m=i,y=i*k,bc1);od:

> for i from 1 to M do eq[N+1,i]:=subs(m=i,y=i*k,bc2);od:

> for i from 0 to N+1 do eq[i,0]:=subs(n=i,x=i*h,bc3);od:

> for i from 0 to N+1 do eq[i,M+1]:=subs(n=i,x=i*h,bc4);od:

> for i from 1 to N do for j from 1 to M do eq[i,j]:=subs(diff(u(x,y),x$2) = subs(n=i,m=j,d2udx2),diff(u(x,y),y$2) = subs(n=i,m=j,d2udy2), diff(u(x,y),x) = subs(n=i,m=j,dudx),diff(u(x,y),y) = subs(n=i,m=j,dudy),u(x,y)=u[i,j],x=i*h,y=j*k,ge);od;od;

> h:=evalf(L/(N+1));k:=evalf(H/(M+1));

$$h := 0.09090909091$$

$$k := 0.09090909091$$

Next, the nonlinear function is entered and the recurrence relation is derived for the node points:

> F:=exp(u(x,y));

$$F := e^{u(x,\,y)}$$

> for i from 0 to N+1 do for j from 0 to M+1 do unew[i,j]:=subs(u=uold,f=subs(u(x,y)=uold[i,j],F),solve(eq[i,j],u[i,j]));od;od;

> for i from 0 to N+1 do for j from 0 to M+1 do unew[i,j]:=subs(u[i,j-1]=unew[i,j-1],unew[i,j]);od;od;

> for i from 0 to N+1 do for j from 0 to M+1 do uold[i,j]:=0.5;od;od;

Initially the error is set to 1 and iteration counter is set to zero.

> err:=1;

$$err := 1$$

> iter:=0;

$$iter := 0$$

> while err>1e-12 do

> for i from 0 to N+1 do for j from 0 to M+1 do Unew[i,j]:=eval(unew[i,j]);od;od;

> kk:='kk':jj:='jj':err:=sum(sum((Unew[kk,kk]-uold[kk,kk])^2, kk=0..N+1),jj=0..M+1)/(N+2)/(M+2);

838   10 Miscellaneous Topics

> iter:=iter+1:

> for i from 0 to N+1 do for j from 0 to M+1 do
uold[i,j]:=Unew[i,j];od:od:end:

The program has converged after 136 iterations. The error associated with the problem is:

> iter;err;

$$136$$

$$0.9202589808 \ 10^{-12}$$

The result obtained can be plotted as:

> plotdata := [seq([seq([i*h,j*k,unew[i,j]], i=0..N+1)], j=0..M+1)]:

> surfdata(plotdata,axes=boxed, labels=[x,y,u],orientation=[-120,60]);

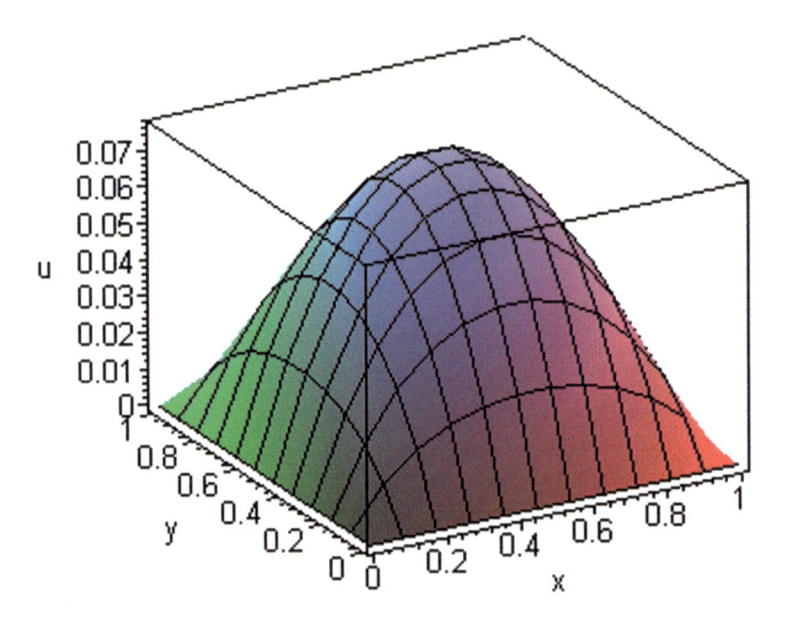

Fig. 10.6

10.1.5 Numerical Method of Lines for First Order Hyperbolic PDEs

Linear first order hyperbolic PDEs were solved analytically in chapter 8 (section 8.1.2). Linear and nonlinear first order hyperbolic PDEs can be solved numerically using numerical method of lines illustrated in chapter 5.2. First order hyperbolic PDEs are usually specified with a boundary condition at x = 0 and an initial condition. In this chapter first order hyperbolic PDEs are solved in the domain

10.1 Miscellaneous Topics on Numerical Methods

x = 0..1. First order hyperbolic PDEs are relatively difficult to solve because of the development of steep gradients at the boundaries and wave propagation. First, spatial derivatives are converted to finite difference form. The resulting system of ODEs is then solved numerically in time. The methodology is illustrated with the following examples.

Example 10.6. Wave Propagation in a Rectangle with Consistent Initial/Boundary Conditions.

Consider wave propagation in a rectangle (Schiesser, 1991). For simplicity and illustration purpose, consistent initial/boundary conditions are taken:

$$\frac{\partial u}{\partial t} + \frac{\partial u}{\partial x} = 0$$
$$u(x,0) = 0 \tag{10.11}$$
$$u(0,t) = t\exp(-t)$$

If N node points in the x-axis (excluding x = 0) equation 10.11 can be converted to finite difference form as:

$$u_0 = t\exp(-t)$$
$$\frac{du_i}{dt} = \frac{u_{i+1} - u_{i-1}}{2h}, \quad i = 1..\,N-1$$
$$\frac{du_N}{dt} = \frac{u_N - u_{N-1}}{h} \tag{10.12}$$
$$h = \frac{1}{N}$$

Note that boundary condition at x = 0 is used at node point i = 0 and governing equation is used for the node points i = 1..N. Central difference is used for the first derivative for node points i = 1..N-1 and backward difference is used for the first derivative for the node point i = N. Equation 10.11 is solved in Maple below:

> restart;

> with(plots):

Warning, the name changecoords has been redefined

The governing equation is entered here:

> ge:=diff(u(x,t),t)=-diff(u(x,t),x);

$$ge := \frac{\partial}{\partial t} u(x,t) = -\left(\frac{\partial}{\partial x} u(x,t)\right)$$

The boundary condition at x = 0 and the initial condition are entered here:

> bc1:=u(x,t)-t*exp(-t);

$$bc1 := u(x, t) - t\,e^{(-t)}$$

> IC:=u(x,0)=0;

$$IC := u(x, 0) = 0$$

Number of node points and length of the domain are entered here:

> N:=10;

$$N := 10$$

> L:=1;

$$L := 1$$

Next, boundary condition and governing equation are converted to finite difference form:

> dydxf:=1/2*(-u[2](t)-3*u[0](t)+4*u[1](t))/h;

$$dydxf := \frac{1}{2}\frac{-u_2(t) - 3\,u_0(t) + 4\,u_1(t)}{h}$$

> dydxb:=1/2*(u[N-2](t)+3*u[N](t)-4*u[N-1](t))/h;

$$dydxb := \frac{1}{2}\frac{u_8(t) + 3\,u_{10}(t) - 4\,u_9(t)}{h}$$

> dydx:=1/2/h*(u[m+1](t)-u[m-1](t));

$$dydx := \frac{1}{2}\frac{u_{m+1}(t) - u_{m-1}(t)}{h}$$

> bc1:=subs(diff(u(x,t),x)=dydxf,u(x,t)=u[0](t),x=0,bc1);

$$bc1 := u_0(t) - t\,e^{(-t)}$$

> eq[0]:=bc1;

$$eq_0 := u_0(t) - t\,e^{(-t)}$$

> for i from 1 to N-1 do eq[i]:=diff(u[i](t),t)= subs(diff(u(x,t),x) = subs(m=i,dydx),u(x,t)=u[i](t),x=i*h,rhs(ge));od:

>eq[N]:=diff(u[N](t),t)=subs(diff(u(x,t),x)=dydxb,u(x,t)=u[N](t),x=L,rhs(ge)):

> u[0](t):=solve(eq[0],u[0](t));

10.1 Miscellaneous Topics on Numerical Methods

$$u_0(t) := \frac{t}{e^t}$$

> h:=L/N;

$$h := \frac{1}{10}$$

> for i from 1 to N do eq[i]:=eval(eq[i]);od:
> eqs:=seq((eq[j]),j=1..N):
> Y:=seq(u[i](t),i=1..N);

$$Y := u_1(t), u_2(t), u_3(t), u_4(t), u_5(t), u_6(t), u_7(t), u_8(t), u_9(t), u_{10}(t)$$

> ICs:=seq(u[i](0)=rhs(IC),i=1..N);

$$ICs := u_1(0) = 0, u_2(0) = 0, u_3(0) = 0, u_4(0) = 0, u_5(0) = 0, u_6(0) = 0, u_7(0) = 0,$$
$$u_8(0) = 0, u_9(0) = 0, u_{10}(0) = 0$$

>sol:=dsolve({eqs,ICs},{Y},type=numeric,output=listprocedure);

$$sol := [\,t = (\,\mathbf{proc}\,(t)\; ...\; \mathbf{end\ proc}\,), u_1(t) = (\,\mathbf{proc}\,(t)\; ...\; \mathbf{end\ proc}\,),$$
$$u_2(t) = (\,\mathbf{proc}\,(t)\; ...\; \mathbf{end\ proc}\,), u_3(t) = (\,\mathbf{proc}\,(t)\; ...\; \mathbf{end\ proc}\,),$$
$$u_4(t) = (\,\mathbf{proc}\,(t)\; ...\; \mathbf{end\ proc}\,), u_5(t) = (\,\mathbf{proc}\,(t)\; ...\; \mathbf{end\ proc}\,),$$
$$u_6(t) = (\,\mathbf{proc}\,(t)\; ...\; \mathbf{end\ proc}\,), u_7(t) = (\,\mathbf{proc}\,(t)\; ...\; \mathbf{end\ proc}\,),$$
$$u_8(t) = (\,\mathbf{proc}\,(t)\; ...\; \mathbf{end\ proc}\,), u_9(t) = (\,\mathbf{proc}\,(t)\; ...\; \mathbf{end\ proc}\,),$$
$$u_{10}(t) = (\,\mathbf{proc}\,(t)\; ...\; \mathbf{end\ proc}\,)\,]$$

> for i to N do U[i]:=subs(sol,u[i](t));od:
> U[0]:=subs(u[1](t)=U[1],u[2](t)=U[2],u[0](t));

$$U_0 := \frac{t}{e^t}$$

The following plots are obtained:

> for i from 0 to N do p[i]:=plot(U[i](t),t=0..2,thickness=3);od:
> display({seq(p[i],i=0..N)},axes=boxed,labels=[t,"u"]);

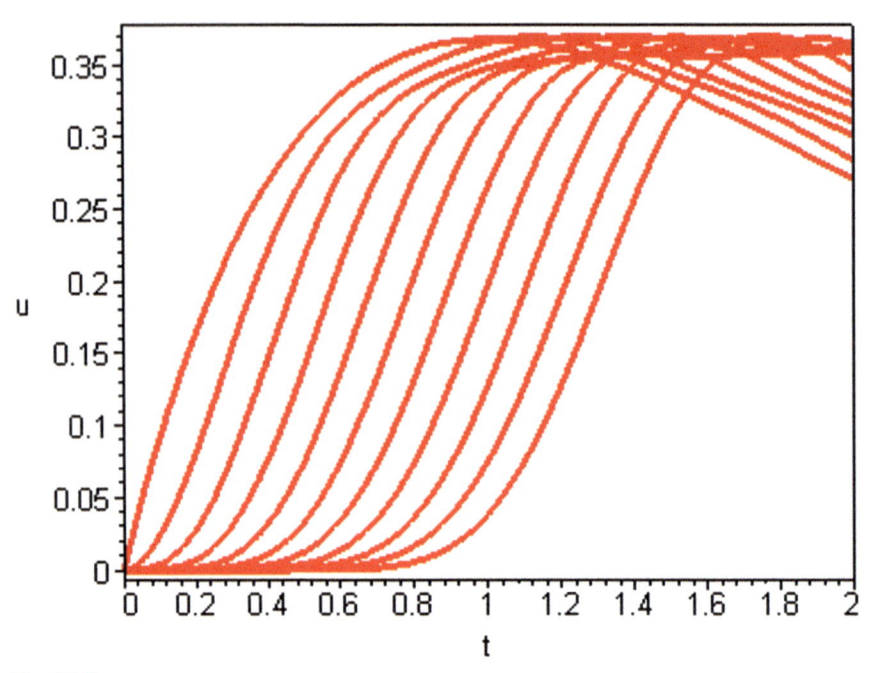

[Figure: plot of u vs t showing family of curves rising from 0 to ~0.35 over t ∈ [0, 2]]

Fig. 10.7

> tf:=2.;

$$tf := 2.$$

> M:=30;

$$M := 30$$

> T1:=[seq(tf*i/M,i=0..M)]:

> PP:=matrix(N+1,M+1);

$$PP := \text{array}(1 .. 11, 1 .. 31, [\ ])$$

> for i from 1 to N+1 do PP[i,1]:=evalf(subs(x=(i-1)*h,rhs(IC)));od:

> for i from 1 to N+1 do for j from 2 to M+1 do PP[i,j]:=evalf(subs(t=T1[j],U[i-1](t)));od;od:

> plotdata := [seq([seq([(i-1)*h,T1[j],PP[i,j]], i=1..N+1)], j=1..M+1)]:

> surfdata(plotdata, axes=boxed, labels=[x,t,u],orientation=[-45,60]);

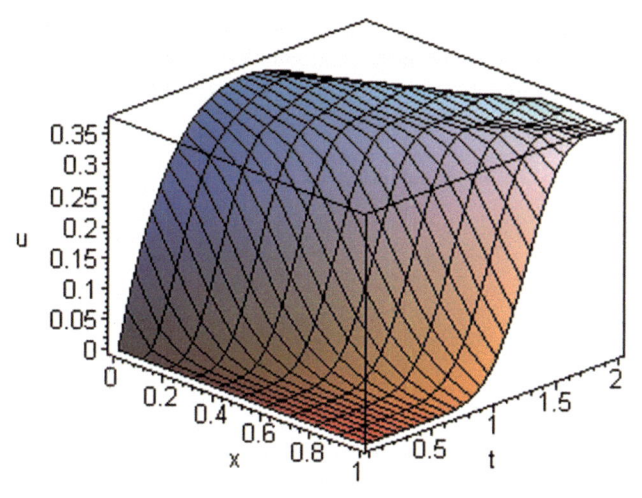

10.1 Miscellaneous Topics on Numerical Methods 843

Fig. 10.8

An analytical solution at x = 1 can be obtained as in chapter 8, section 8.1.2 as:

> ua:=Heaviside(t-1)*(t-1)*exp(1-t);

$$ua := \text{Heaviside}(t-1)(t-1)e^{(1-t)}$$

Analytical solution is compared to the numerical solution as:

> plot([ua,U[N](t)],t=0..5,thickness=3,axes=boxed);

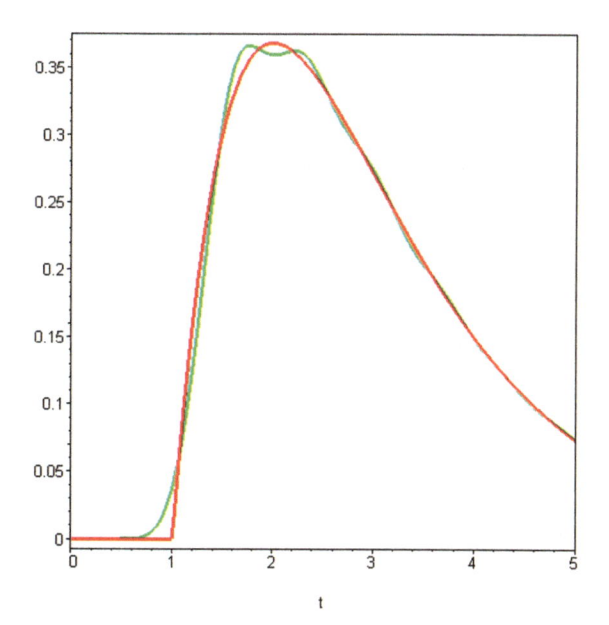

Fig. 10.9

We observe that the numerical solution traces the analytical solution, but steep gradients are smoothened. In addition, the numerical solution slightly oscillates. When node points are increased, the accuracy increases. For N = 20 node points the following plot is obtained:

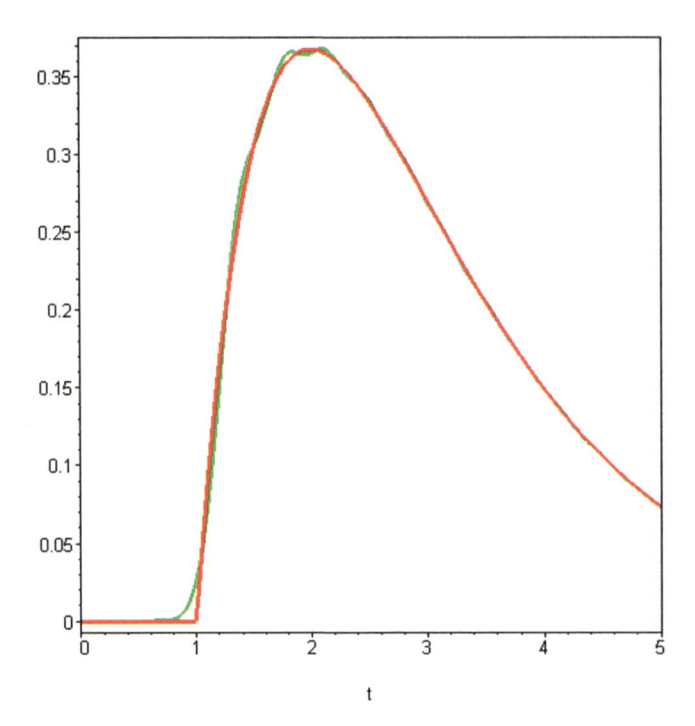

Fig. 10.10

Example 10.7. Wave Propagation in a Rectangle with inconsistent Initial/Boundary Conditions

Consider wave propagation in a rectangle (Schiesser, 1991) with inconsistent initial/boundary conditions:

$$\frac{\partial u}{\partial t} + \frac{\partial u}{\partial x} = 0$$
$$u(x,0) = 0 \tag{10.13}$$
$$u(0,t) = 1$$

Equation 10.13 is solved using the Maple program developed for example 10.6 and the following plots are obtained:

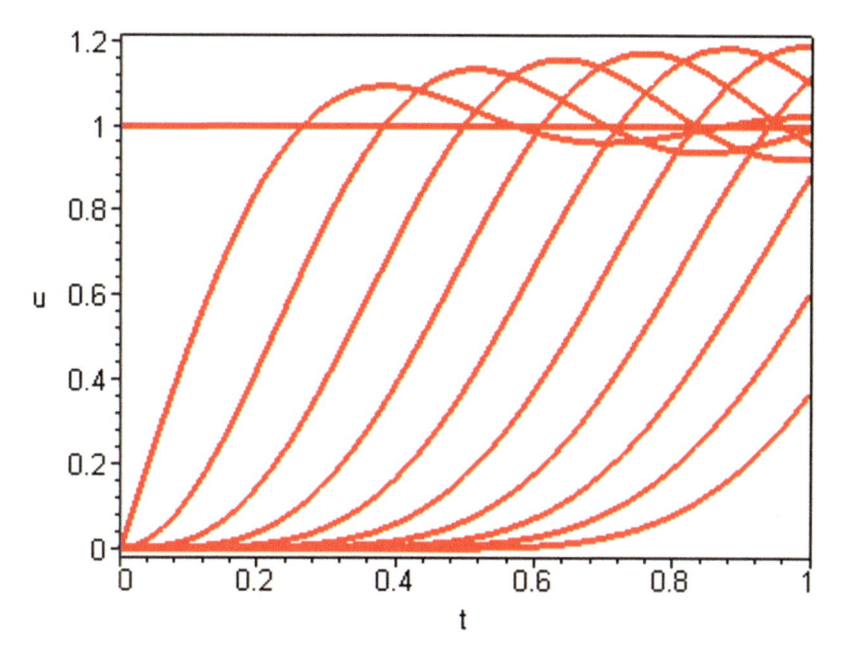

10.1 Miscellaneous Topics on Numerical Methods 845

Fig. 10.11

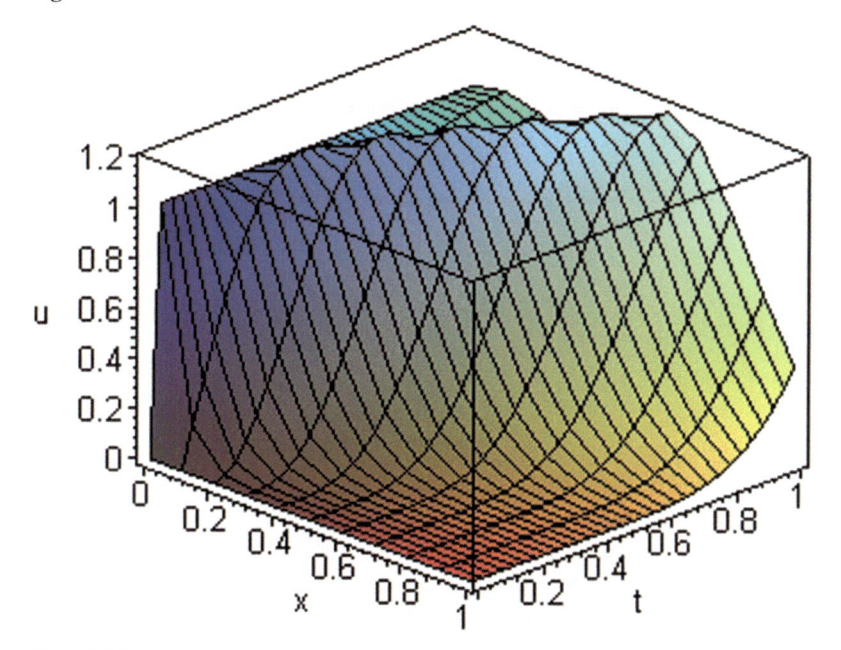

Fig. 10.12

> ua:=Heaviside(t-1);

$$ua := \text{Heaviside}(t-1)$$

> plot([ua,U[N](t)],t=0..4,thickness=3,axes=boxed);

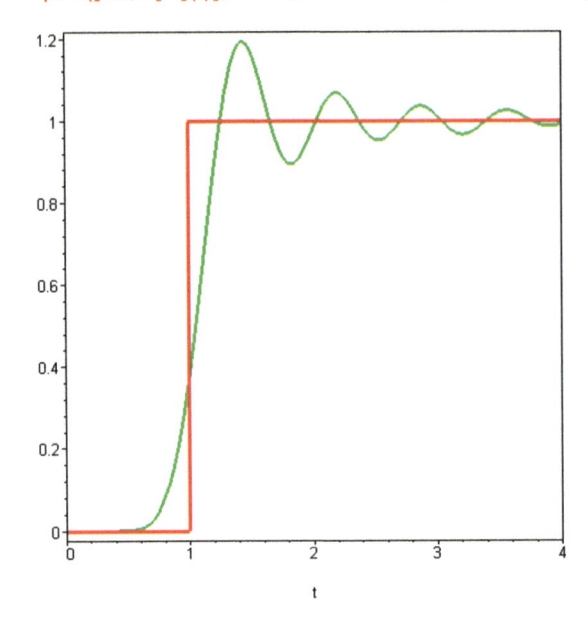

Fig. 10.13

When backward finite difference accurate to the order h is used for the first derivative in the governing equation, the solution does not oscillate, and the following plots are obtained for N = 10 node points.

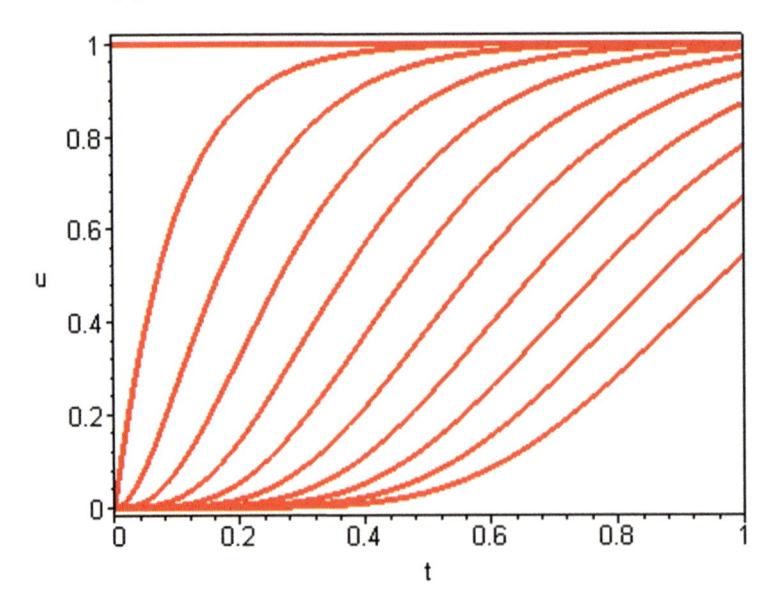

Fig. 10.14

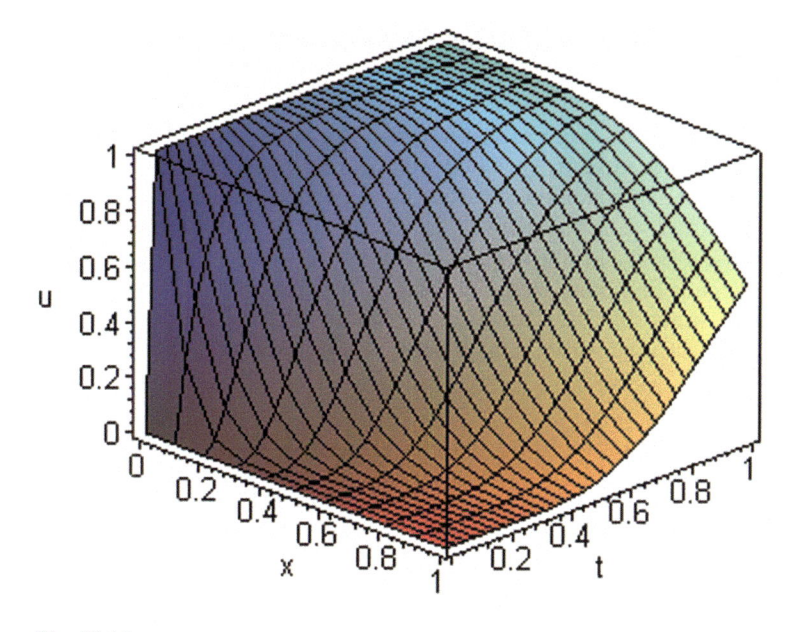

10.1 Miscellaneous Topics on Numerical Methods

Fig. 10.15

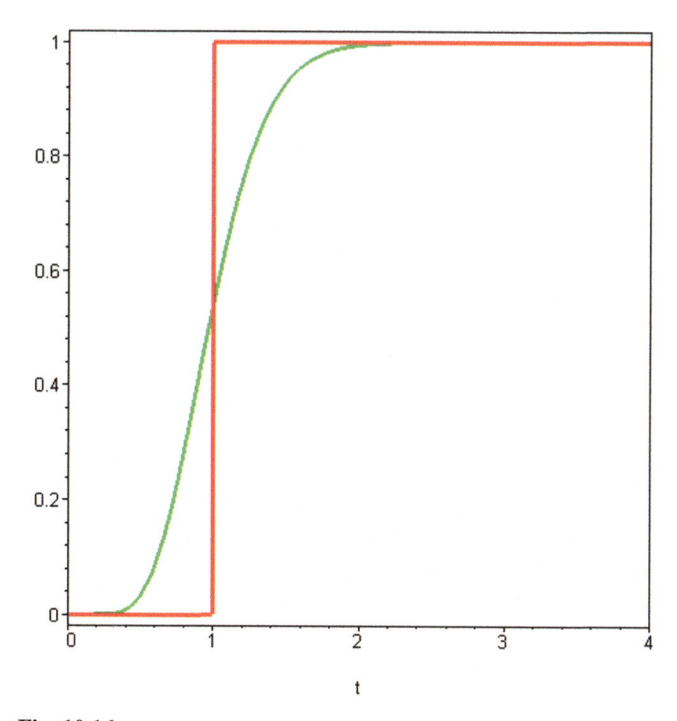

Fig. 10.16

10.1.6 Numerical Method of Lines for Second Order Hyperbolic PDEs

First order hyperbolic PDEs were solved numerically in section 10.1.5. Second order hyperbolic PDEs are usually specified with boundary conditions at x = 0 and x = 1. In addition, initial conditions for both the dependent variable and its time derivative are specified. The methodology is very similar to numerical method of lines for parabolic PDEs described in chapter 5.2. The only difference is that instead of a system of first order ODEs, second order hyperbolic PDEs result in a system of second order ODEs. The resulting system of second order ODEs is solved numerically in time. The methodology is illustrated with the following examples.

Example 10.8. Wave Equation with Consistent Initial/Boundary Conditions

Consider wave equation in a rectangle (Schiesser, 1991). For simplicity and illustration purpose, consistent initial/boundary conditions are taken:

$$\frac{\partial^2 u}{\partial t^2} = \frac{\partial^2 u}{\partial x^2}$$

$$u(x,0) = \sin(\pi x) \; ; \; \frac{\partial u}{\partial t}(x,0) = 0 \qquad (10.14)$$

$$u(0,t) = 0; \; u(1,t) = 0$$

If N interior node points in the x-axis (excluding x = 0 and x = 1 as in chapter 5.2) equation 10.14 can be converted to finite difference form as:

$$\begin{aligned} u_0 &= 0 \\ \frac{d^2 u_i}{dt^2} &= \frac{u_{i+1} - 2u_i + u_{i-1}}{h^2}, \; i = 1..\, N \\ u_{N+1} &= 0 \\ h &= \frac{1}{N+1} \end{aligned} \qquad (10.15)$$

Note that boundary condition at x = 0 is used at node point i = 0, boundary condition at x = 1 is used at node point i = N+1 and governing equation is used for the node points i = 1..N. Equation 10.15 is solved in Maple below:

> restart;

> with(plots):

Warning, the name changecoords has been redefined

10.1 Miscellaneous Topics on Numerical Methods

Governing equation, boundary and initial conditions are entered here:

```
> ge:=diff(u(x,t),t$2)=diff(u(x,t),x$2);
```

$$ge := \frac{\partial^2}{\partial t^2} u(x, t) = \frac{\partial^2}{\partial x^2} u(x, t)$$

```
> bc1:=u(x,t);
```

$$bc1 := u(x, t)$$

```
> bc2:=u(x,t);
```

$$bc2 := u(x, t)$$

```
> IC1:=u(x,0)=sin(Pi*x);
```

$$IC1 := u(x, 0) = \sin(\pi x)$$

```
> IC2:=diff(u(x,t),t)=0;
```

$$IC2 := \frac{\partial}{\partial t} u(x, t) = 0$$

```
> N:=10;
```

$$N := 10$$

```
> L:=1;
```

$$L := 1$$

```
> dydxf:=1/2*(-u[2](t)-3*u[0](t)+4*u[1](t))/h:
> dydxb:=1/2*(u[N-1](t)+3*u[N+1](t)-4*u[N](t))/h:
> dydx:=1/2/h*(u[m+1](t)-u[m-1](t)):
> d2ydx2:=1/h^2*(u[m-1](t)-2*u[m](t)+u[m+1](t)):
> bc1:=subs(diff(u(x,t),x)=dydxf,u(x,t)=u[0](t),x=0,bc1);
```

$$bc1 := u_0(t)$$

```
> bc2:=subs(diff(u(x,t),x)=dydxb,u(x,t)=u[N+1](t),x=1,bc2);
```

$$bc2 := u_{11}(t)$$

```
> eq[0]:=bc1:
> eq[N+1]:=bc2:
> for i from 1 to N do eq[i]:=subs(diff(u(x,t),x$2) =
subs(m=i,d2ydx2),diff(u(x,t),x) = subs(m=i,dydx),u(x,t)=u[i](t),x=i*h,ge);od:
```

850   10 Miscellaneous Topics

Values at the exterior node points (x = 0 and x = 1) are solved as:

> u[0](t):=(solve(eq[0],u[0](t)));
$$u_0(t) := 0$$

> u[N+1](t):=solve(eq[N+1],u[N+1](t));
$$u_{11}(t) := 0$$

> h:=L/(N+1);
$$h := \frac{1}{11}$$

> for i from 1 to N do eq[i]:=eval(eq[i]);od:
> eqs:=seq((eq[j]),j=1..N):
> Y:=seq(u[i](t),i=1..N);

$$Y := u_1(t), u_2(t), u_3(t), u_4(t), u_5(t), u_6(t), u_7(t), u_8(t), u_9(t), u_{10}(t)$$

Initial conditions are entered here:

>ICs:=seq(u[i](0)=rhs(subs(x=i*h,IC1)),i=1..N),
seq(D(u[i])(0)=rhs(subs(x=i*h,IC2)),i=1..N);

$$ICs := u_1(0) = \sin\left(\frac{\pi}{11}\right), u_2(0) = \sin\left(\frac{2\pi}{11}\right), u_3(0) = \sin\left(\frac{3\pi}{11}\right), u_4(0) = \sin\left(\frac{4\pi}{11}\right),$$
$$u_5(0) = \sin\left(\frac{5\pi}{11}\right), u_6(0) = \sin\left(\frac{6\pi}{11}\right), u_7(0) = \sin\left(\frac{7\pi}{11}\right), u_8(0) = \sin\left(\frac{8\pi}{11}\right),$$
$$u_9(0) = \sin\left(\frac{9\pi}{11}\right), u_{10}(0) = \sin\left(\frac{10\pi}{11}\right), D(u_1)(0) = 0, D(u_2)(0) = 0, D(u_3)(0) = 0,$$
$$D(u_4)(0) = 0, D(u_5)(0) = 0, D(u_6)(0) = 0, D(u_7)(0) = 0, D(u_8)(0) = 0,$$
$$D(u_9)(0) = 0, D(u_{10})(0) = 0$$

>sol:=dsolve({eqs,ICs},{Y},type=numeric,output=listprocedure):
> for i to N do U[i]:=subs(sol,u[i](t));od:
> U[0]:=subs(u[1](t)=U[1],u[2](t)=U[2],u[0](t));
$$U_0 := 0$$

> U[N+1]:=subs(u[N](t)=U[N],u[N-1](t)=U[N-1],u[N+1](t));
$$U_{11} := 0$$

> for i from 0 to N+1 do p[i]:=plot(U[i](t),t=0..1,thickness=3);od:

10.1 Miscellaneous Topics on Numerical Methods 851

The solution obtained is plotted below:

> display({seq(p[i],i=0..N+1)},axes=boxed,labels=[t,"u"]);

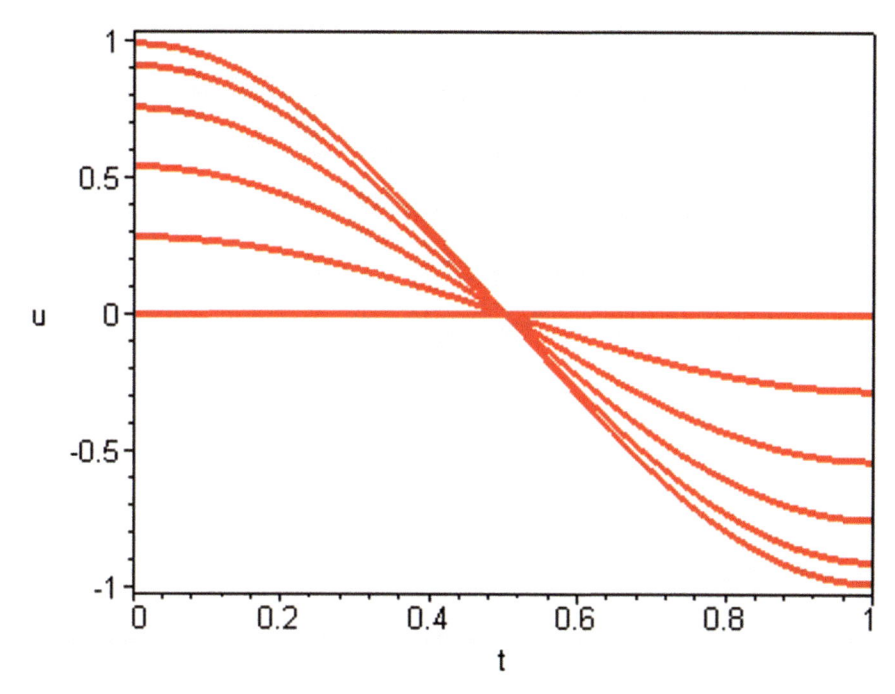

Fig. 10.17

> tf:=2.;

$$tf := 2.$$

> M:=30:

> T1:=[seq(tf*i/M,i=0..M)]:

> PP:=matrix(N+2,M+1):

> for i from 1 to N+2 do PP[i,1]:=evalf(subs(x=(i-1)*h,rhs(IC1)));od:

> for i from 1 to N+2 do for j from 2 to M+1 do
PP[i,j]:=evalf(subs(t=T1[j],U[i-1](t)));od;od:

> plotdata := [seq([seq([(i-1)*h,T1[j],PP[i,j]], i=1..N+2)], j=1..M+1)]:

> surfdata(plotdata, axes=boxed, labels=[x,t,u],orientation=[15,60]);

852   10 Miscellaneous Topics

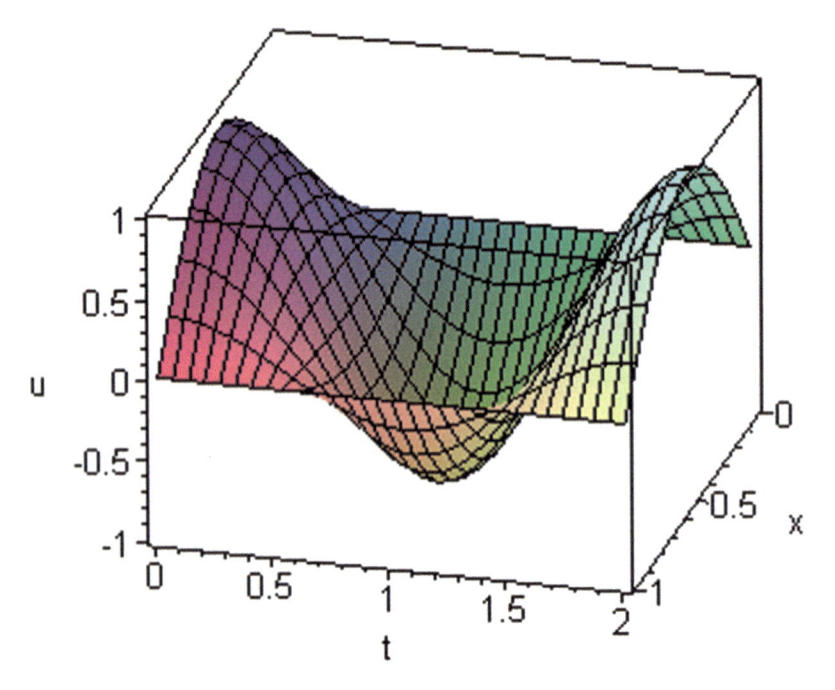

Fig. 10.18

Example 10.9. Wave Equation with Inconsistent Initial/Boundary Conditions

Consider wave equation in a rectangle with inconsistent initial/boundary conditions:

$$\frac{\partial^2 u}{\partial t^2} = \frac{\partial^2 u}{\partial x^2}$$
$$u(x,0) = 0 \; ; \; \frac{\partial u}{\partial t}(x,0) = 0 \qquad (10.16)$$
$$u(0,t) = 1; \; \frac{\partial u}{\partial x}(1,t) = 0$$

Equation 10.16 is solved in Maple and the following plots are obtained:

```
> ge:=diff(u(x,t),t$2)=diff(u(x,t),x$2);
```

$$ge := \frac{\partial^2}{\partial t^2} u(x, t) = \frac{\partial^2}{\partial x^2} u(x, t)$$

10.1 Miscellaneous Topics on Numerical Methods

> bc1:=u(x,t)-1;

$$bc1 := \mathrm{u}(x,t) - 1$$

> bc2:=diff(u(x,t),x);

$$bc2 := \frac{\partial}{\partial x} \mathrm{u}(x,t)$$

> IC1:=u(x,0)=0;

$$IC1 := \mathrm{u}(x,0) = 0$$

> IC2:=diff(u(x,t),t)=0;

$$IC2 := \frac{\partial}{\partial t} \mathrm{u}(x,t) = 0$$

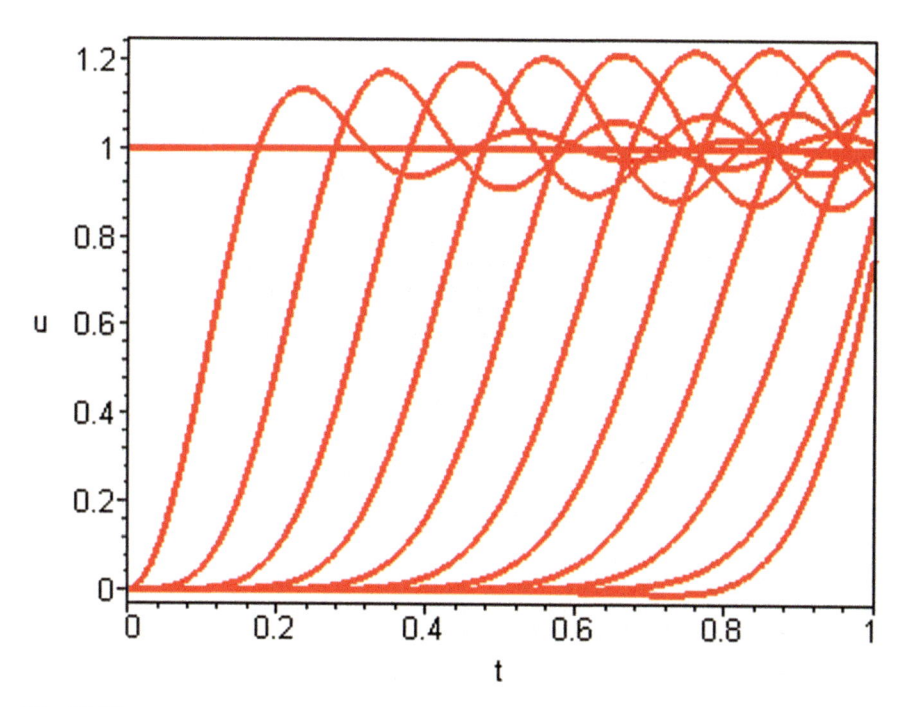

Fig. 10.19

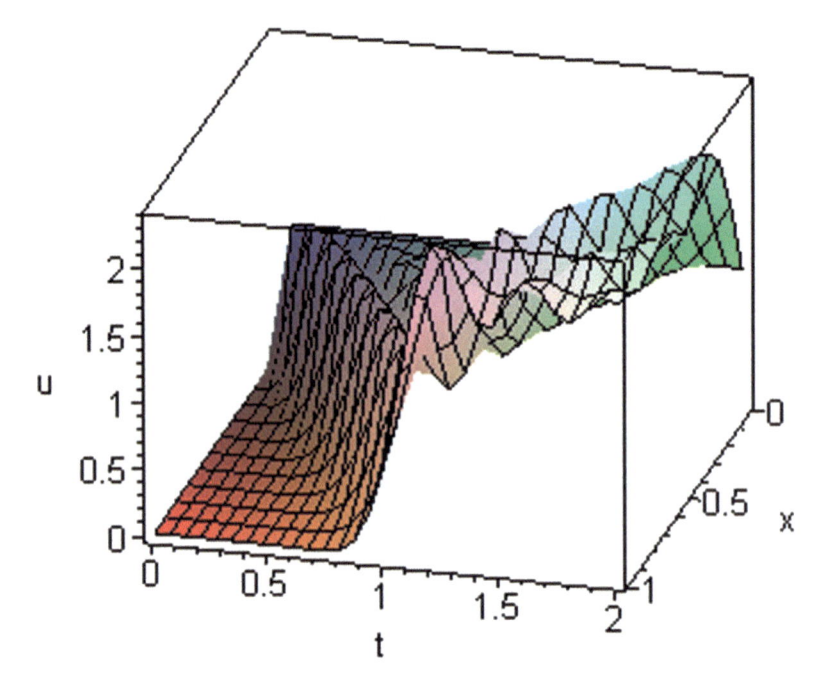

Fig. 10.20

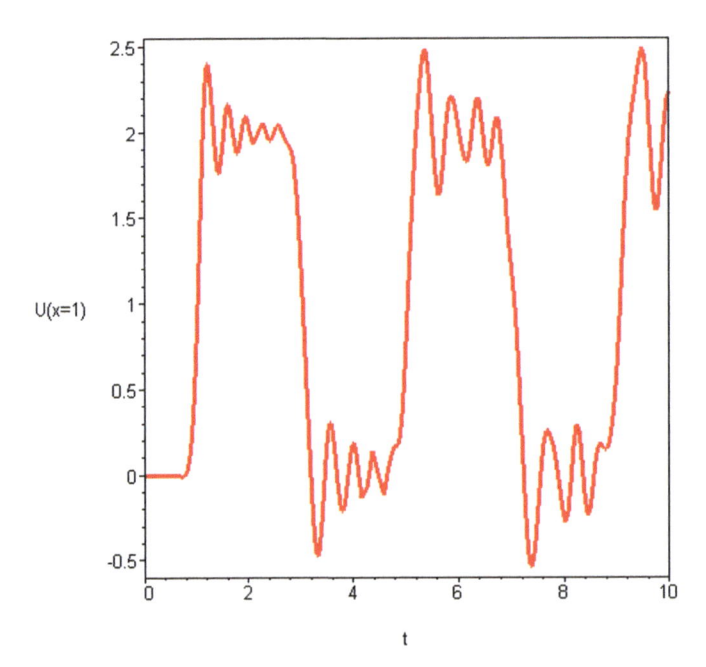

Fig. 10.21

10.1 Miscellaneous Topics on Numerical Methods 855

Compared to example 10.7, we observe oscillations in example 10.8. For hyperbolic PDEs with inconsistent initial/boundary conditions, method of lines is not a good choice. Special numerical methods that involve discretization in both x and t are required for this purpose (Schiesser 1991).

10.1.7 Summary

In this chapter, classical numerical methods for solving BVPs and PDEs were programmed in Maple. In section 10.1.2, nonlinear Boundary Value Problems (ODEs) were solved numerically. First, the given nonlinear BVP in x was converted to a system of nonlinear coupled algebraic equations by applying finite differences for the spatial derivatives in the governing equation and boundary conditions. The resulting system of nonlinear algebraic equations was the solved using iteration by developing recurrence relations. The same methodology was then extended to elliptic PDEs in sections 10.1.3 and 10.1.4 by applying finite difference approximations in both x and y coordinates. In section 10.1.3 linear elliptic PDEs were solved using fsolve command and in section 10.1.4 nonlinear elliptic PDEs were solved using iteration.

First order hyperbolic PDEs were solved using numerical method of lines in section 10.1.5. The methodology involves applying finite differences in x and integrating numerically in time. The same methodology was then extended to second order hyperbolic PDEs in section 10.1.6. A total of nine examples were presented in this chapter.

10.1.8 Exercise Problems

1. Consider diffusion with a second-order reaction in a cylindrical catalyst pellet (exercise problem 2 chapter 3). Solve this problem using recursion technique described in section 10.1.2.
2. Redo exercise problem 3 of chapter 3 using recursion technique described in section 10.1.2.
3. Consider the Graetz problem discussed in example 6.3. Solve this problem by applying finite differences in both directions as described in section 10.1.3.
4. Consider the elliptic PDE with nonlinear boundary condition discussed in example 6.4. Solve this problem by applying finite differences in both directions as described in section 10.1.3.
5. Consider current-distribution problem discussed in example 6.5. Redo this problem using the methodology described in section 10.1.3. Hint: the node spacing in y changes as a function.
6. Consider diffusion with second-order reaction discussed in example 6.7. Redo this problem using the methodology described in section 10.1.3.
7. Redo problem 3 using iterative finite difference technique described in section 10.1.4.
8. Redo problem 4 using iterative finite difference technique described in section 10.1.4.

9. Redo problem 5 using iterative finite difference technique described in section 10.1.4.
10. Redo problem 6 using iterative finite difference technique described in section 10.1.4.
11. Consider diffusion with reaction in a non-isothermal cylindrical pellet (Finlayson, 1980, exercise problem 7, chapter 6). The governing equations and boundary conditions are:

$$\frac{\partial^2 u}{\partial y^2} + \frac{\partial^2 u}{\partial x^2} + \frac{1}{x}\frac{\partial u}{\partial x} = \Phi^2 u \exp\left(\frac{\gamma\beta(1-u)}{1+\beta(1-u)}\right)$$

$$\frac{\partial u}{\partial x}(0,y) = 0 \text{ and } u(1,y) = 1$$

$$\frac{\partial u}{\partial y}(x,0) = 0 \text{ and } u(x,1) = 1$$

Solve this problem using finite difference technique (either using fsolve command [section 10.1.3] or iteration [section 10.1.4]) for $\Phi = 2$, $\gamma = 30$ and $\beta = 0.1$. Which method is more efficient?

12. Complete the details missing in example 10.7.
13. Complete the details missing in example 10.9.
14. Redo example 10.6 if the boundary condition at $x = 0$ is replaced by $u(0,t) = t^2 \exp(-t)$.
15. Redo example 10.8 if the initial condition is replaced by $u(x,0) = x\sin(\pi x)$.
16. In problem 16, how can one change the boundary conditions at $x = 0$ without changing the consistency of boundary/initial conditions? Solve this new problem.

References

1. Carslaw, H.S., Jaeger, J.C.: Conduction of Heat in Solids. Oxford University Press, London (1973)
2. Crank, J.: Mathematics of Diffusion. Oxford University Press, New York (1975)
3. Rice, R.G., Do, D.D.: Applied Mathematics and Modeling for Chemical Engineers. John Wiley & Sons, Inc., New York (1995)
4. Schiesser, W.E.: The Numerical Method of Lines. Academic Press Inc., New York (1991)
5. Varma, A., Morbidelli, M.: Mathematical Methods in Chemical Engineering. Oxford University Press, New York (1997)

Subject Index

analytical method of lines 456
axial conduction and diffusion in a tubular reactor 259
axial conduction 537
axial diffusion 175, 259, 262

Blassius equation, infinite domains 256, 342, 345
boundary value problems 169

classic diffusion problem in a cylindrical catalyst pellet 189
composite domains 26, 425, 437, 452
conduction of heat in a rectangular cooling fan 171
confidence intervals 775, 776, 778, 795, 807, 810, 811, 815, 816, 817
convective diffusion equations 212
convective diffusion problem 175, 196
convective term 201
current density 374, 556, 729
current distribution in an electrochemical cell 336, 339, 507
cylindrical catalyst pellet 203, 209

differential algebraic equations 112
diffusion in a tubular reactor 259
diffusion of a substrate in an enzyme catalyzed reaction with removable singularity 250
diffusion with a convection 175, 196
diffusion with a first order reaction in a semi - infinite plane 181
diffusion with a second order reaction 218, 229, 245, 262
'do loop' 18, 24, 173, 186, 785
'dsolve' 11, 12, 13, 14, 80-84, 89, 94, 95, 98, 99, 101

eigenvalues 8, 9. 24, 43, 45, 166-167
eigenvalue problems 272
eigenvectors 8-9, 38, 167, 437, 438, 515
elliptic partial differential equations 295, 333, 339, 348, 507, 508, 510, 536, 547, 556, 564, 565, 573, 581, 587, 649, 672, 827, 833, 855
exothermal reaction in a sphere 474, 480
exponential matrix 10, 30, 38, 39, 48, 51, 63, 65, 80, 83, 84, 169, 180, 185, 187, 196, 212, 213, 355, 359, 366, 374, 437, 438, 440, 451, 452, 507, 508, 510, 511, 515, 520, 537
exponential matrix method, linear BVP 170

fermentation kinetics 99
finite difference method 827
finite difference solution for boundary value problems 821, 827
finite difference solution for elliptic partial differential equations 827
finite difference solution for nonlinear BVPs 229
first order differential equations 510, 848
first order hyperbolic partial differential equations 678, 838, 839, 848, 855, 858
first order isothermal constant volume ractor model 797
first order linear ordinary differential equations 511
first order ordinary differential equations 355, 457, 507
first order parabolic partial differential equation 685

first order reaction 318, 390
first order irreversible series reactions 31
first order reversible series reactions 37
flux boundary conditions 305
'fsolve' 5, 6, 25, 54, 113, 119, 126, 220, 222, 225, 228, 229, 232, 275, 276, 277, 373, 440, 542, 552, 553, 620, 621, 658, 738, 739, 745, 821, 827, 828, 831, 833, 855, 856

Gear's method 101, 149
genmatrix' 31, 32, 39, 48, 50, 57, 74, 77, 196, 198, 205, 355, 358, 359, 366, 369, 377, 384, 393, 404, 418, 427, 431, 439, 444, 449, 511, 514, 525, 541, 549, 559
Graetz problem 272, 278, 287, 401, 452, 536

heat conduction in a rectangular slab 356
heat conduction mass transfer problem 366
heat conduction with radiation at the surface 214
heat conduction with transient boundary conditions 301
heat transfer 49, 81, 169, 196, 208, 233, 247, 295, 353, 414, 425, 426, 470, 474, 507, 565, 573, 587, 643, 649, 660, 667, 685, 688, 703, 720, 827, 832, 833
heat transfer in a fin 208
heat transfer in a rectangle 508
heat transfer with nonlinear radiation boundary conditions 247
heated tanks in a series 112
heating of fluids in a series of tanks 48
higher order linear ordinary differential equations 63
homogeneous linear ODEs 29
Hull cell, potential distribution 556, 581

in a rectangular slab 296, 325
infinite domains 256
initial value problems 29
inverse Laplace 163, 164, 165
inverse Laplace transforms 17, 18, 24

kinetics 175, 452

Laplace equation 556, 660
Laplace solution for second order system with Dirac forcing function 75
Laplace solution for irreversible series reactions 73

Laplace transform technique for partial differential equations 679
Laplace transform techniques 72, 73, 75, 79, 84, 161, 167, 295, 314, 318, 325, 348, 679, 690, 691, 701, 709, 714, 719, 720, 725, 736, 742, 747, 755
Laplace transformations 16, 17, 24, 26, 72, 73, 155
least squares method 761, 762, 778, 788
linear boundary value problems 170
linear ordinary differential equations 29

matrix exponential by the Laplace transform method 161
matrix exponential method 155
matrix operations 6
matrizant example 184
matrizant method 184
method of lines technique 287, 355, 457
method of lines for parabolic partial differential equations 353, 456
multiple steady states 116, 266
multiple steady states for initial value problems 117
multiple steady states in a catalyst pellet 253, 266
multiple steady states in a catalyst pellet solving BVPs and IVPs 238
multiple steady states in a rectangular catalyst pellet 266

nonhomogeneous ordinary differential equations 457
non - isothermal catalyst pellet, multiple steady states 223
nonlinear boundary value problems 218
nonlinear heat transfer 233
nonlinear least squares method 778, 788
nonlinear ordinary differential equations 87
nonlinear radiation boundary condition 470, 547
numerical nethod of lines 480, 491, 501, 502, 536
numerical method of lines for elliptic PDEs 507, 564
numerical method of lines for nonlinear coupled PDEs 480
numericak nethod of lines for moving boundary problems 491

Subject Index

numerical method of lines for parabolic PDEs 353, 456, 469
numerical method of ines for stiff nonlinear PDEs 474
numerical solution coupled BVPs using Maple's 'dsolve' command 259
numerical solution for heat transfer for nonlinear elliptical PDEs 573

one parameter model 807, 809

parabolic partial differential equations 297, 324, 348, 353, 355, 365, 456, 457, 507, 587, 672, 679, 685, 689, 701, 719, 742, 747, 755
parabolic velocity profile 272
parameter estimation 761, 783, 801
phase plane analysis 149
plane flow past a flat plate 342
plotting with Maple 3

radiation boundary condition at the surface 314, 474, 618, 736
read data into Maple from Text File 23
removable singularity 250, 281, 287
Rosenbrock 101, 456, 457, 477
Runge - Kutta methods 107, 110, 149, 456, 457, 458, 474

second order boundary value problems 572
second order differential equations 195
second order equations 11, 272
second order hyperbolic partial differential equations 848, 855
second order ordinary differential equations 19, 65, 75, 508, 509, 511, 565, 848
second order reaction 88, 97, 218, 229, 245, 253, 262, 458, 573, 821
semianalytical method for elliptic PDEs 507, 536, 547, 556,
semianalytical method for the Graetz problem 401
semianalytical method for homogeneous PDEs 353
semianalytical method for nonhomogeneous PDEs 365

semianalytical method for parabolic partial differential equations 353, 456
semianalytcal method for PDEs in composite domains 425
semianalytical method for PDEs with known initial profiles 414
semi - infinite domains 180, 181, 295
semi - infinite plane 250
separation of variables 272, 588, 597, 609, 618, 623, 624, 625, 635, 643, 649, 672, 690, 714, 755
series solutions for nonlinear BVPs 218
series solutions for nonlinear ODEs 98
shooting technique for boundary value problems 233
simultaneous first order reaction 175
simultaneous series reactions 88
singularity 497, 696
solving linear BVP's using Maple's 'dsolve' command 208
solving linear ODEs using Maple 81
solving linear ODEs Using Maple's 'dsolve' command 80
solving nonlinear BPV's using Maple's 'dsolve' command 244
solving nonlinear ODEs using Maple's 'dsolve' command 94
solving systems of ODEs using the Laplace transform method 72
source term 311
spherical catalyst pellet 210
steady state solutions 124
steady states 266
stiff nonlinear partial differential equations 107, 457
stop conditions 103
symbolic finite difference solutions for linear BVP 195
systems of ODEs 72

text files in Maple 19
thermal conductivity 297, 353, 456, 507, 587, 679
thermal diffusivity 295, 296, 314, 353, 426, 456, 507, 587, 679
time varying input to a CSTR with a series reaction 55
transient boundary conditions 301

transient heat conduction or mass
    transfer 456, 580, 679
two parameter model 812, 817
tubular reactor 259

unsteady state diffusion 318

variable diffusivity 217, 340, 464

wave propagation 679, 682,
    848, 852
'while loop' 19, 24